AF356595

Plasma Biology

Plasma Biology

Editors

Akikazu Sakudo
Yoshihito Yagyu

MDPI • Basel • Beijing • Wuhan • Barcelona • Belgrade • Manchester • Tokyo • Cluj • Tianjin

Editors
Akikazu Sakudo
School of Veterinary Medicine
Okayama University of Science
Imabari
Japan

Yoshihito Yagyu
Department of Electrical and
Electric Engineering
Sasebo National College of
Technology
Sasebo
Japan

Editorial Office
MDPI
St. Alban-Anlage 66
4052 Basel, Switzerland

This is a reprint of articles from the Special Issue published online in the open access journal *International Journal of Molecular Sciences* (ISSN 1422-0067) (available at: www.mdpi.com/journal/ijms/special_issues/Plasma_Biology).

For citation purposes, cite each article independently as indicated on the article page online and as indicated below:

LastName, A.A.; LastName, B.B.; LastName, C.C. Article Title. *Journal Name* **Year**, *Volume Number*, Page Range.

ISBN 978-3-0365-1568-7 (Hbk)
ISBN 978-3-0365-1567-0 (PDF)

Contents

Akikazu Sakudo and Yoshihito Yagyu
Plasma Biology
Reprinted from: *Int. J. Mol. Sci.* **2021**, *22*, 5441, doi:10.3390/ijms22115441 . . . . **1**

Soon Young Hwang, Ngoc Hoan Nguyen, Tae Jung Kim, Youngsoo Lee, Mi Ae Kang and Jong-Soo Lee
Non-Thermal Plasma Couples Oxidative Stress to TRAIL Sensitization through DR5 Upregulation
Reprinted from: *Int. J. Mol. Sci.* **2020**, *21*, 5302, doi:10.3390/ijms21155302 . . . . **5**

Dominik Terefinko, Anna Dzimitrowicz, Aleksandra Bielawska-Pohl, Aleksandra Klimczak, Pawel Pohl and Piotr Jamroz
The Influence of Cold Atmospheric Pressure Plasma-Treated Media on the Cell Viability, Motility, and Induction of Apoptosis in Human Non-Metastatic (MCF7) and Metastatic (MDA-MB-231) Breast Cancer Cell Lines
Reprinted from: *Int. J. Mol. Sci.* **2021**, *22*, 3855, doi:10.3390/ijms22083855 . . . . **25**

Anna Dzimitrowicz, Piotr Jamroz, Pawel Pohl, Weronika Babinska, Dominik Terefinko, Wojciech Sledz and Agata Motyka-Pomagruk
Multivariate Optimization of the FLC-dc-APGD-Based Reaction-Discharge System for Continuous Production of a Plasma-Activated Liquid of Defined Physicochemical and Anti-Phytopathogenic Properties
Reprinted from: *Int. J. Mol. Sci.* **2021**, *22*, 4813, doi:10.3390/ijms22094813 . . . . **49**

Charlotta Bengtson and Annemie Bogaerts
The Quest to Quantify Selective and Synergistic Effects of Plasma for Cancer Treatment: Insights from Mathematical Modeling
Reprinted from: *Int. J. Mol. Sci.* **2021**, *22*, 5033, doi:10.3390/ijms22095033 . . . . **71**

Sander Bekeschus, Dorothee Meyer, Kevin Arlt, Thomas von Woedtke, Lea Miebach, Eric Freund and Ramona Clemen
Argon Plasma Exposure Augments Costimulatory Ligands and Cytokine Release in Human Monocyte-Derived Dendritic Cells
Reprinted from: *Int. J. Mol. Sci.* **2021**, *22*, 3790, doi:10.3390/ijms22073790 . . . . **107**

Tom Zimmermann, Lisa A. Gebhardt, Lucas Kreiss, Christin Schneider, Stephanie Arndt, Sigrid Karrer, Oliver Friedrich, Michael J. M. Fischer and Anja-Katrin Bosserhoff
Acidified Nitrite Contributes to the Antitumor Effect of Cold Atmospheric Plasma on Melanoma Cells
Reprinted from: *Int. J. Mol. Sci.* **2021**, *22*, 3757, doi:10.3390/ijms22073757 . . . . **121**

Ewa Tyczkowska-Sieroń, Tadeusz Kałużewski, Magdalena Grabiec, Bogdan Kałużewski and Jacek Tyczkowski
Genotypic and Phenotypic Changes in *Candida albicans* as a Result of Cold Plasma Treatment
Reprinted from: *Int. J. Mol. Sci.* **2020**, *21*, 8100, doi:10.3390/ijms21218100 . . . . **135**

Lyubomir Haralambiev, Ole Neuffer, Andreas Nitsch, Nele C. Kross, Sander Bekeschus, Peter Hinz, Alexander Mustea, Axel Ekkernkamp, Denis Gümbel and Matthias B. Stope
Inhibition of Angiogenesis by Treatment with Cold Atmospheric Plasma as a Promising Therapeutic Approach in Oncology
Reprinted from: *Int. J. Mol. Sci.* **2020**, *21*, 7098, doi:10.3390/ijms21197098 . . . . **151**

Akikazu Sakudo and Tatsuya Misawa
Antibiotic-Resistant and Non-Resistant Bacteria Display Similar Susceptibility to Dielectric
Barrier Discharge Plasma
Reprinted from: *Int. J. Mol. Sci.* **2020**, *21*, 6326, doi:10.3390/ijms21176326 . . . . **167**

Jae-Sung Kwon, Sung-Hwan Choi, Eun Ha Choi, Kwang-Mahn Kim and Paul K. Chu
Enhanced Osteogenic Differentiation of Human Mesenchymal Stem Cells on
Amine-Functionalized Titanium Using Humidified Ammonia Supplied Nonthermal
Atmospheric Pressure Plasma
Reprinted from: *Int. J. Mol. Sci.* **2020**, *21*, 6085, doi:10.3390/ijms21176085 . . . . **179**

**Agata Przekora, Maïté Audemar, Joanna Pawlat, Cristina Canal, Jean-Sébastien Thomann,
Cédric Labay, Michal Wojcik, Michal Kwiatkowski, Piotr Terebun, Grazyna Ginalska,
Sophie Hermans and David Duday**
Positive Effect of Cold Atmospheric Nitrogen Plasma on the Behavior of Mesenchymal Stem
Cells Cultured on a Bone Scaffold Containing Iron Oxide-Loaded Silica Nanoparticles Catalyst
Reprinted from: *Int. J. Mol. Sci.* **2020**, *21*, 4738, doi:10.3390/ijms21134738 . . . . **197**

Xu Han, James Kapaldo, Yueying Liu, M. Sharon Stack, Elahe Alizadeh and Sylwia Ptasinska
Large-Scale Image Analysis for Investigating Spatio-Temporal Changes in Nuclear DNA
Damage Caused by Nitrogen Atmospheric Pressure Plasma Jets
Reprinted from: *Int. J. Mol. Sci.* **2020**, *21*, 4127, doi:10.3390/ijms21114127 . . . . **219**

Manish Adhikari, Bhawana Adhikari, Bhagirath Ghimire, Sanjula Baboota and Eun Ha Choi
Cold Atmospheric Plasma and Silymarin Nanoemulsion Activate Autophagy in Human
Melanoma Cells
Reprinted from: *Int. J. Mol. Sci.* **2020**, *21*, 1939, doi:10.3390/ijms21061939 . . . . **237**

**Fusheng Sun, Xiaoxue Xie, Yufan Zhang, Jiangwei Duan, Mingyu Ma, Yaqiong Wang, Ding
Qiu, Xinpei Lu, Guangxiao Yang and Guangyuan He**
Effects of Cold Jet Atmospheric Pressure Plasma on the Structural Characteristics and
Immunoreactivity of Celiac-Toxic Peptides and Wheat Storage Proteins
Reprinted from: *Int. J. Mol. Sci.* **2020**, *21*, 1012, doi:10.3390/ijms21031012 . . . . **255**

**Pavol Zubor, Yun Wang, Alena Liskova, Marek Samec, Lenka Koklesova, Zuzana Dankova,
Anne Dørum, Karol Kajo, Dana Dvorska, Vincent Lucansky, Bibiana Malicherova, Ivana
Kasubova, Jan Bujnak, Milos Mlyncek, Carlos Alberto Dussan, Peter Kubatka, Dietrich
Büsselberg and Olga Golubnitschaja**
Cold Atmospheric Pressure Plasma (CAP) as a New Tool for the Management of Vulva Cancer
and Vulvar Premalignant Lesions in Gynaecological Oncology
Reprinted from: *Int. J. Mol. Sci.* **2020**, *21*, 7988, doi:10.3390/ijms21217988 . . . . **277**

Chayanaphat Chokradjaroen, Jiangqi Niu, Gasidit Panomsuwan and Nagahiro Saito
Insight on Solution Plasma in Aqueous Solution and Their Application in Modification of Chitin
and Chitosan
Reprinted from: *Int. J. Mol. Sci.* **2021**, *22*, 4308, doi:10.3390/ijms22094308 . . . . **321**

Dušan Braný, Dana Dvorská, Erika Halašová and Henrieta Škovierová
Cold Atmospheric Plasma: A Powerful Tool for Modern Medicine
Reprinted from: *Int. J. Mol. Sci.* **2020**, *21*, 2932, doi:10.3390/ijms21082932 . . . . **345**

International Journal of
Molecular Sciences

Editorial

Plasma Biology

Akikazu Sakudo [1,*] **and Yoshihito Yagyu** [2]

[1] School of Veterinary Medicine, Okayama University of Science, Imabari, Ehime 794-8555, Japan
[2] Department of Electrical and Electric Engineering, Sasebo National College of Technology, Sasebo, Nagasaki 857-1193, Japan; yyagyu@sasebo.ac.jp
* Correspondence: akikazusakudo@gmail.com

check for **updates**

Citation: Sakudo, A.; Yagyu, Y. Plasma Biology. *Int. J. Mol. Sci.* **2021**, 22, 5441. https://doi.org/10.3390/ijms22115441

Received: 13 May 2021
Accepted: 17 May 2021
Published: 21 May 2021

Publisher's Note: MDPI stays neutral with regard to jurisdictional claims in published maps and institutional affiliations.

It is now more than 90 years since Irving Langmuir used the technical term "plasma" to describe an ionized gas [1]. Plasma technology has recently expanded to encompass more and more of our daily lives [2]. For example, plasma disinfection/sterilization contributes to public health in the field of medicine and dentistry. Moreover, advances in plasma technology are being exploited to accelerate wound healing, as well as for the development of novel forms of tumor treatment. In the agricultural sector, plasma technology could contribute to higher crop yields by enhancing seed germination and the growth of plants, as well as food preservation and disinfection. Plasma technology could also be utilized in environmental applications, including water treatment/remediation and the treatment of exhaust gases. As such, plasma will be a supportive technology in the provision of clean energy to help achieve sustainable development goals (SDGs). Indeed, the broad potential of plasma technology is only just being realized. However, a large potion of plasma's the mechanisms of action in biological applications remains unclear.

On the basis of the background outlined above, this Special Issue entitled, "Plasma Biology", of *International Journal of Molecular Sciences*, includes 14 original articles and 3 reviews providing new insights into the application and molecular mechanisms of plasma biology.

The solutions treated with plasma are referred to as "plasma-activated medium (PAM)" [3], plasma-treated medium (PTM)" [4], "plasma-activated liquid (PAL)" [5], or "plasma-treated liquid (PTL)" [6]. These solutions contain reactive chemical species with a short half-life (e.g., reactive oxygen species (ROS) and reactive nitrogen species (RNS)) that can act as disinfectants [2] or anticancer agents [2].

Hwang et al. [3] examined the efficacy of PAM and tumor necrosis factor-related apoptosis-inducing ligand (TRAIL) in combination (PAM/TRAIL) as an anticancer therapy. PAM/TRAIL showed synergistic effects on growth inhibition in TRAIL-resistant cancer cells via enhanced apoptosis. The antioxidant N-acetylcysteine was found to prevent PAM/TRAIL-induced cancer cell apoptosis, suggesting that ROS is related to the induction of PAM/TRAIL-mediated apoptosis.

Terefinko et al. [4] demonstrated PTM, produced by a cold atmospheric pressure plasma (CAP)-based reaction-discharge system, induces apoptosis in cancer cells, especially metastatic cells.

Dzimitrowicz et al. [5] reported the use of PAL, produced by a direct current atmospheric pressure glow discharge that is generated in contact with a flowing liquid cathode (FLC-dc-APGD). PAL was shown to display antibacterial action against *Dickeya solani* and *Pectobacterium atrosepticum*, which are important plant pathogens. Moreover, the mechanism of action of PAL was possibly due to the presence of ROS and RNS.

Bengtson and Bogaerts [6] developed a mathematical model to investigate the key chemical species involved in the cellular response to PTL, especially hydrogen peroxide (H_2O_2). The model can be used to quantify the selective and synergistic anticancer effect of PTL in both susceptible and resistant cells.

In a report by Bekeschus et al. [7], human monocyte-derived dendritic cells (moDCs) were subjected to plasma treatment. Markers, such as CD25, CD40, and CD83, were shown to be activated by this treatment, which is crucial for T cell co-stimulation. Moreover, the plasma treatment increased cytokine levels such as interleukin (IL)-1α, IL-6, and IL-23. These findings suggest that plasma treatment augments costimulatory ligand and cytokine expression in human moDCs.

Zimmermann et al. [8] demonstrated that treatment with a MiniFlatPlaster CAP device induced elevated levels of nitrite and nitrate as well as enhanced acidification. These results highlight the impact of acidified nitrite on melanoma cells and confirm the importance of RNS during CAP treatment.

Tyczkowska-Sieroń et al. [9] treated *Candida albicans* with a sublethal dose of CAP. Subsequent analysis of the CAP treated yeast identified six single-nucleotide variants, six insertions, and five deletions, as well as the decreased or increased activity of the corresponding enzymes.

Haralambiev et al. [10] examined the use of two different CAP devices, kINPen and MiniJet. A human endothelial cell line (HDMEC) was treated directly and indirectly with CAP. Direct CAP treatment of HDMEC resulted in robust growth-inhibition, whereas indirect CAP treatment did not. Both the migration and tube formation of HDMEC were significantly inhibited after CAP-treatment. In addition, both CAP devices induced HDMEC apoptosis.

Sakudo et al. [11] investigated whether antibiotic-resistant and non-resistant bacteria display differential susceptibility to treatment with plasma [12,13]. *Escherichia coli*, with or without a plasmid that includes an ampicillin resistance gene and chloramphenicol acetyltransferase (CAT) gene, were treated with a dielectric barrier discharge (DBD) plasma torch. The plasma treatment was found to degrade the lipopolysaccharide (LPS) and DNA of the bacteria as well as CAT. Furthermore, the plasma treatment was equally effective against antibiotic-resistant and non-resistant bacteria. This finding suggests that plasma treatment is effective against bacterial strains resistant to conventional antibiotic therapy.

Kwon et al. studied the attachment of human mesenchymal stem cells (hMSCs) to the surface of titanium modified by plasma treatment generated using either nitrogen (N-P), air (A-P), or humidified ammonia (NA-P) [14]. N-P, A-P, and NA-P plasma treatment resulted in a surface with increased hydrophilicity, which promoted enhanced cell attachment compared with the untreated control (C-P). Furthermore, greater cell proliferation was observed on surfaces treated with A-P or NA-P than with C-P or N-P. In addition, the NA-P treated surface resulted in a higher level of alkaline phosphatase activity and osteocalcin expression than the other samples.

Przekora et al. [15] produced silica nanoparticles (NPs) and FexOy/NPs, which are formed of silica NPs decorated with iron oxide (FexOy to denote magnetite + maghemite phase) and embedded them in the polysaccharide matrix of chitosan/curdlan/hydroxyapatite biomaterial. The combined action of CAP and the materials produced on the proliferation and osteogenic differentiation of human adipose tissue-derived mesenchymal stem cells (ADSCs) was then studied. Plasma activation of FexOy/NPs-loaded biomaterial promoted the formation of ROS, resulting in an enhancement of stem cell proliferation without inhibition of osteogenic differentiation.

Han et al. [16] performed large-scale image analysis to investigate the effect of treatment with a plasma jet generating CAP on the cell-cycle stage and quantify damage to nuclear DNA in single cells. S phase cells were found to be more susceptible to DNA damage by the plasma jet treatment than either G1 or G2 phase cells.

Adhikari et al. [17] examined the co-effect of CAP and silymarin nanoemulsion (SN) treatment on the autophagy pathway in a human melanoma cell line. The results showed that CAP and SN together induced autophagy via the phosphoinositide 3-kinase/mechanistic target of rapamycin (PI3K/mTOR) and epidermal growth factor receptor (EGFR) pathways. Moreover, plasma treatment modulated the expression of transcription factors (ZKSCAN3, TFEB, FOXO1, CRTC2, and CREBBP) and specific genes (BECN-1, AMBRA-1, MAP1LC3A, and SQSTM) related to the induction of autophagy.

Sun et al. [18] used high-resolution liquid chromatography mass spectrometry/mass spectrometry (HR-LC-MS/MS) to investigate the structural properties and immunoreactivity of celiac-toxic peptides and wheat storage proteins modified by a plasma jet. The results indicated that plasma jet treatment reduced and modified celiac-toxic peptides by backbone cleavage of QQPFP and PQPQLPY at specific proline and glutamine residues, followed by hydroxylation at the aromatic ring of phenylalanine and tyrosine residues. The immunoreactivity of gliadin extract was reduced by the plasma jet treatment. These observations suggest that the plasma jet could initiate the depolymerization of gluten polymer.

A review by Zubor et al. [19] focuses on the potential of CAP for the management of vulvar cancer. Although no reports have been published concerning the effect of CAP on vulvar cancer cells, progress has been made in gynaecological oncology and in other types of cancer. The review highlights an understudied area that should be a focus for future research.

A review by Chokradjaroen et al. [20] provides insight into solution plasma in aqueous solutions and the potential application of this technology in modifying chitin and chitosan, including degradation and deacetylation.

A review by Braný et al. [21] gives an overview of the current status and future perspectives of CAP in the field of modern medicine. This review article is highly cited and received 22 citations as of 17 May 2021 according to Crossref.

Finally, the Editors are delighted to have had the honor of organizing this Special Issue for *International Journal of Molecular Sciences*, which highlights the research of eminent scientists in the field of plasma biology. The Editors would like to thank all the contributors to this Special Issue for their commitment and enthusiasm during the compilation of the respective articles. The Editors also wish to thank Kaitlyn Wu and other members of the editorial staff at Multidisciplinary Digital Publishing Institute (MDPI) for their professionalism and dedication. Hopefully, readers will enjoy this Special Issue and be inspired with new ideas for future research.

Author Contributions: Conceptualization, A.S. and Y.Y.; writing—original draft preparation, A.S.; writing—review and editing, A.S. and Y.Y. All authors have read and agreed to the published version of the manuscript.

Funding: This research received no external funding.

Institutional Review Board Statement: Not applicable.

Informed Consent Statement: Not applicable.

Data Availability Statement: Not applicable.

Conflicts of Interest: The authors declare no conflict of interest.

References

1. Langmuir, I. Oscillations in ionized gases. *Proc. Natl. Acad. Sci. USA* **1928**, *14*, 627–637. [CrossRef] [PubMed]
2. Sakudo, A.; Yagyu, Y.; Onodera, T. Disinfection and Sterilization Using Plasma Technology: Fundamentals and Future Perspectives for Biological Applications. *Int. J. Mol. Sci.* **2019**, *20*, 5216. [CrossRef]
3. Hwang, S.Y.; Nguyen, N.H.; Kim, T.J.; Lee, Y.; Kang, M.A.; Lee, J.-S. Non-Thermal Plasma Couples Oxidative Stress to TRAIL Sensitization through DR5 Upregulation. *Int. J. Mol. Sci.* **2020**, *21*, 5302. [CrossRef] [PubMed]
4. Terefinko, D.; Dzimitrowicz, A.; Bielawska-Pohl, A.; Klimczak, A.; Pohl, P.; Jamroz, P. The Influence of Cold Atmospheric Pressure Plasma-Treated Media on the Cell Viability, Motility, and Induction of Apoptosis in Human Non-Metastatic (MCF7) and Metastatic (MDA-MB-231) Breast Cancer Cell Lines. *Int. J. Mol. Sci.* **2021**, *22*, 3855. [CrossRef] [PubMed]
5. Dzimitrowicz, A.; Jamroz, P.; Pohl, P.; Babinska, W.; Terefinko, D.; Sledz, W.; Motyka-Pomagruk, A. Multivariate Optimization of the FLC-dc-APGD-Based Reaction-Discharge System for Continuous Production of a Plasma-Activated Liquid of Defined Physicochemical and Anti-Phytopathogenic Properties. *Int. J. Mol. Sci.* **2021**, *22*, 4813. [CrossRef]
6. Bengtson, C.; Bogaerts, A. The Quest to Quantify Selective and Synergistic Effects of Plasma for Cancer Treatment: Insights from Mathematical Modeling. *Int. J. Mol. Sci.* **2021**, *22*, 5033. [CrossRef]

7. Bekeschus, S.; Meyer, D.; Arlt, K.; von Woedtke, T.; Miebach, L.; Freund, E.; Clemen, R. Argon Plasma Exposure Augments Costimulatory Ligands and Cytokine Release in Human Monocyte-Derived Dendritic Cells. *Int. J. Mol. Sci.* **2021**, *22*, 3790. [CrossRef] [PubMed]
8. Zimmermann, T.; Gebhardt, L.A.; Kreiss, L.; Schneider, C.; Arndt, S.; Karrer, S.; Friedrich, O.; Fischer, M.J.M.; Bosserhoff, A.-K. Acidified Nitrite Contributes to the Antitumor Effect of Cold Atmospheric Plasma on Melanoma Cells. *Int. J. Mol. Sci.* **2021**, *22*, 3757. [CrossRef]
9. Tyczkowska-Sieroń, E.; Kałużewski, T.; Grabiec, M.; Kałużewski, B.; Tyczkowski, J. Genotypic and Phenotypic Changes in *Candida albicans* as a Result of Cold Plasma Treatment. *Int. J. Mol. Sci.* **2020**, *21*, 8100. [CrossRef]
10. Haralambiev, L.; Neuffer, O.; Nitsch, A.; Kross, N.C.; Bekeschus, S.; Hinz, P.; Mustea, A.; Ekkernkamp, A.; Gümbel, D.; Stope, M.B. Inhibition of Angiogenesis by Treatment with Cold Atmospheric Plasma as a Promising Therapeutic Approach in Oncology. *Int. J. Mol. Sci.* **2020**, *21*, 7098. [CrossRef]
11. Sakudo, A.; Misawa, T. Antibiotic-Resistant and Non-Resistant Bacteria Display Similar Susceptibility to Dielectric Barrier Discharge Plasma. *Int. J. Mol. Sci.* **2020**, *21*, 6326. [CrossRef]
12. Yamashiro, R.; Misawa, T.; Sakudo, A. Key Role of Singlet Oxygen and Peroxynitrite in Viral RNA Damage during Virucidal Effect of Plasma Torch on Feline Calicivirus. *Sci. Rep.* **2018**, *8*, 17947. [CrossRef]
13. Sakudo, A.; Miyagi, H.; Horikawa, T.; Yamashiro, R.; Misawa, T. Treatment of *Helicobacter pylori* with Dielectric Barrier Discharge Plasma Causes UV Induced Damage to Genomic DNA Leading to Cell Death. *Chemosphere* **2018**, *200*, 366–372. [CrossRef]
14. Kwon, J.-S.; Choi, S.-H.; Choi, E.H.; Kim, K.-M.; Chu, P.K. Enhanced Osteogenic Differentiation of Human Mesenchymal Stem Cells on Amine-Functionalized Titanium Using Humidified Ammonia Supplied Nonthermal Atmospheric Pressure Plasma. *Int. J. Mol. Sci.* **2020**, *21*, 6085. [CrossRef]
15. Przekora, A.; Audemar, M.; Pawlat, J.; Canal, C.; Thomann, J.-S.; Labay, C.; Wojcik, M.; Kwiatkowski, M.; Terebun, P.; Ginalska, G.; et al. Positive Effect of Cold Atmospheric Nitrogen Plasma on the Behavior of Mesenchymal Stem Cells Cultured on a Bone Scaffold Containing Iron Oxide-Loaded Silica Nanoparticles Catalyst. *Int. J. Mol. Sci.* **2020**, *21*, 4738. [CrossRef]
16. Han, X.; Kapaldo, J.; Liu, Y.; Stack, M.S.; Alizadeh, E.; Ptasinska, S. Large-Scale Image Analysis for Investigating Spatio-Temporal Changes in Nuclear DNA Damage Caused by Nitrogen Atmospheric Pressure Plasma Jets. *Int. J. Mol. Sci.* **2020**, *21*, 4127. [CrossRef]
17. Adhikari, M.; Adhikari, B.; Ghimire, B.; Baboota, S.; Choi, E.H. Cold Atmospheric Plasma and Silymarin Nanoemulsion Activate Autophagy in Human Melanoma Cells. *Int. J. Mol. Sci.* **2020**, *21*, 1939. [CrossRef]
18. Sun, F.; Xie, X.; Zhang, Y.; Duan, J.; Ma, M.; Wang, Y.; Qiu, D.; Lu, X.; Yang, G.; He, G. Effects of Cold Jet Atmospheric Pressure Plasma on the Structural Characteristics and Immunoreactivity of Celiac-Toxic Peptides and Wheat Storage Proteins. *Int. J. Mol. Sci.* **2020**, *21*, 1012. [CrossRef]
19. Zubor, P.; Wang, Y.; Liskova, A.; Samec, M.; Koklesova, L.; Dankova, Z.; Dørum, A.; Kajo, K.; Dvorska, D.; Lucansky, V.; et al. Cold Atmospheric Pressure Plasma (CAP) as a New Tool for the Management of Vulva Cancer and Vulvar Premalignant Lesions in Gynaecological Oncology. *Int. J. Mol. Sci.* **2020**, *21*, 7988. [CrossRef]
20. Chokradjaroen, C.; Niu, J.; Panomsuwan, G.; Saito, N. Insight on Solution Plasma in Aqueous Solution and Their Application in Modification of Chitin and Chitosan. *Int. J. Mol. Sci.* **2021**, *22*, 4308. [CrossRef]
21. Braný, D.; Dvorská, D.; Halašová, E.; Škovierová, H. Cold Atmospheric Plasma: A Powerful Tool for Modern Medicine. *Int. J. Mol. Sci.* **2020**, *21*, 2932. [CrossRef] [PubMed]

International Journal of
Molecular Sciences

Article

Non-Thermal Plasma Couples Oxidative Stress to TRAIL Sensitization through DR5 Upregulation

Soon Young Hwang [1,†], Ngoc Hoan Nguyen [1,†], Tae Jung Kim [2], Youngsoo Lee [3], Mi Ae Kang [1,*] and Jong-Soo Lee [1,*]

[1] Department of Life Sciences, College of Natural Sciences, Ajou University, Suwon 16499, Korea; hwang630@ajou.ac.kr (S.Y.H.); hoanbiology@gmail.com (N.H.N.)
[2] Department of Electrical and Computer Engineering, College of Information and Technology, Ajou University, Suwon 16499, Korea; prosecutor33@ajou.ac.kr
[3] Department of Biomedical Sciences, Ajou University Graduate School of Medicine, Suwon 16499, Korea; ysoolee@ajou.ac.kr
* Correspondence: makang@ajou.ac.kr (M.A.K.); jsjlee@ajou.ac.kr (J.-S.L.); Tel.: +82-31-219-2629 (M.A.K.); +82-31-219-1886 (J.-S.L.); Fax: +82-31-219-1615 (M.A.K. & J.-S.L.)
† These authors contributed equally.

Received: 8 July 2020; Accepted: 23 July 2020; Published: 26 July 2020

Abstract: Tumor necrosis factor-related apoptosis-inducing ligand (TRAIL) induces apoptosis in various tumor cells without affecting most normal cells. Despite being in clinical testing, novel strategies to induce TRAIL-mediated apoptosis are in need to overcome cancer cell unresponsiveness and resistance. Plasma-activated medium (PAM) markedly stimulates reactive oxygen/nitrogen species (ROS/RNS)-dependent apoptosis in cancer cells. We investigate the capability of PAM and TRAIL (PAM/TRAIL) combination therapy to overcome TRAIL resistance and improve the anticancer efficacy of TRAIL. The combinatorial treatment of PAM and TRAIL shows synergistic effects on growth inhibition in TRAIL-resistant cancer cells via augmented apoptosis by two attributes. DR5 (TRAIL-R2) transcription by CHOP is upregulated in a PAM-generated ROS/RNS-dependent manner, and PAM itself upregulates PTEN expression mediated by suppression of miR-425 which is involved in Akt inactivation, leading to increased apoptosis induction. Treatment of cancer cell lines with the antioxidant N-acetylcysteine reduces the extent of membrane dysfunction and the expression of both CHOP-DR5 and miR-425-PTEN axes, attenuating PAM/TRAIL-induced cancer cell apoptosis. These data suggest that PAM/TRAIL treatment is a novel approach to sensitizing cancer cells to TRAIL-induced apoptosis and overcoming TRAIL resistance. PAM is a promising candidate for further investigations as a chemotherapeutic sensitizer in the treatment of cancer.

Keywords: plasma-activated medium; TRAIL; DR5; apoptosis; ROS/RNS

---

1. Introduction

Apoptosis can be triggered through both intrinsically and extrinsically initiated pathways. Most current chemotherapeutic strategies target the dysregulation of apoptotic pathways in cancer cells. Intrinsic pathways are initiated at the mitochondria level in a p53-dependent manner [1]. Conventional radio- and chemotherapies aim mainly at the p53-dependent intrinsic apoptotic pathway. However, more than half of human cancers carry the loss of p53 function, and are either initially resistant or eventually acquire resistance to these treatments. Thus many cancers continue to survive and thrive because of the lack of p53 function in intrinsic cellular apoptotic mechanisms. In contrast, p53 appears to be dispensable for extrinsic apoptotic pathways in most cancers. Extrinsic pathways are triggered by the binding of death ligands to death receptors (DRs) found in the cellular membrane.

DRs contain death domains endowing these receptors a role in apoptosis, among other non-apoptotic roles. Hence, DR-mediated apoptosis may represent a better target for the treatment of cancers that harbor p53 mutations.

TNF-related apoptosis-inducing ligand (TRAIL, also known as Apo2 ligand) is a member of the tumor necrosis factor (TNF) family of cytokines that binds to DRs to induce apoptosis [1]. TRAIL could be used as a chemotherapeutic agent because it induces apoptosis in cancer cells but not in normal cells [2]. However, the promising preclinical results have not successfully translated into clinical trials [3–5], since most primary cancers and multiple cancer cell lines are TRAIL-resistant [5]. However, TRAIL can trigger non-apoptotic signaling pathways in certain TRAIL-resistant cancer cells [5]. TRAIL-sensitizing strategies targeting TRAIL-activated non-apoptotic pathways could be effective in overcoming TRAIL resistance in cancer cells.

TRAIL triggers extrinsic apoptosis by binding to the death receptors TRAIL-R1 (DR4) and TRAIL-R2 (DR5, also called Apo2, KILLER, or TRICK2). However, several factors account for cancer resistance to apoptotic and non-apoptotic TRAIL signalings, which may provide opportunities to overcome TRAIL resistance. TRAIL-related receptors TRAIL-R3, TRAIL-R4, and osteoprotegerin (TRAIL-R5 or OPG) lack death domains and serve as decoy receptors [6,7]. These decoy receptors compete with DR4 and DR5 for TRAIL-binding, and reducing opportunities for TRAIL induction of apoptotic pathways [6]. Activation of the PI3/Akt and Erk-mediated pathways induces the survival and proliferation of cancer cells, eventually leading to TRAIL resistance [8]. Upregulation of cFLIP, a caspase-8 inhibitor [9] and mutations in the *Bax* and *Bak* genes [10] have been associated with decreased TRAIL-induced apoptosis in cancers. These findings have resulted in the development of promising TRAIL-sensitizing treatment strategies, including DR4/5 induction, Akt and Erk pathway inhibition, and the repression of cFLIP expression. Various anti-cancer drugs used in traditional chemotherapy, such as bortezomib, doxorubicin, valproic acids, or decitabin enhance the TRAIL sensitivity in the cancer cells, but these chemicals also exhibit cytotoxic effects in normal cells [11,12]. Therefore, there are continuing urgent needs to identify novel agents that can be used in combination with TRAIL to improve apoptotic efficacy and to overcome TRAIL resistance in cancer cells.

A number of studies have focused on oxidative agents which potentiate TRAIL-mediated apoptosis in a reactive oxygen species (ROS)-dependent manner [2,8,9,13–16]. These oxidative agents can promote diverse effects, such as inducing the upregulation of DR5, and promoting TRAIL-induced apoptosis [9,14–16]; also inhibiting oncogenic pathways and/or activating apoptotic pathways such as the NF-κB-mediated oncogenic signaling pathway and the ROS-mediated JNK-CHOP pathway [9,14–16]; furthermore, inducing ROS-dependent apoptosis via PTEN-mediated Akt inactivation and p53 activation [8]; and inducing cell membrane depolarization and disruption of intracellular ion homeostasis, possibly via impairment of ion channels or transporters for Na^+, K^+, Cl^-, and Ca^{2+} [2].

Nonthermal (room temperature) plasma generated from microplasma jet devices is comprised of charged particles, some of which are reactive. Nonthermal plasma has recently emerged as a therapeutic agent for clinical applications such as in vivo antiseptics, wound healing, dermatology, dentistry, and cancer treatment. Such therapeutic applications have shaped the concept of plasma medicine. In previous studies, plasma was shown to efficiently induce apoptosis in cancer cells by disrupting mitochondrial membrane potentials and promoting mitochondrial ROS accumulation, consequently leading to ROS-dependent apoptotic cell death [17–21]. Moreover, plasma treatment does not significantly affect healthy cells [15–19]. So, it has been proposed that the level of plasma-generated ROS/RNS is high enough to induce cell death in cancer cells, but not in normal cells upon the same plasma-activated medium (PAM) treatment [17].

We investigated if PAM in combination with TRAIL (PAM/TRAIL sensitization) can induce apoptosis in TRAIL-resistant cancer cells. PAM/TRAIL sensitization has upregulated DR5 expression and membrane dysfunction, inducing ROS-dependent apoptosis of cancer cells. PAM/TRAIL sensitization could serve as a novel strategy to overcome TRAIL resistance in cancer cells. PAM is a promising candidate for further investigations as a chemotherapeutic sensitizer in the treatment of cancer.

2. Results

2.1. PAM Synergistically Enhances the Anticancer Efficacy of TRAIL

Previous reports have demonstrated that oxidative agents induce TRAIL sensitization [2] and that plasma mediates ROS-induced apoptosis of cancer cells [17–21]. Thus, to explore a new method for TRAIL sensitization, we generated PAM by spraying air plasma at atmospheric pressure onto the surface of DMEM media for 10 min (Figure 1a) [17,22]. We first determined levels of ROS (H_2O_2, hydrogen peroxide) and RNS (NO, nitrogen oxide) to be approximately 10 and 160 μM, respectively, in PAM (Supplementary Figure S1a,b). Next, we examined the TRAIL sensitizing effects of PAM in cervical cancer HeLa cells (Figure 1b–e). HeLa cells treated with either TRAIL alone (10–100 ng/mL) or with a 50-fold dilution of PAM alone did not affect cell viability (Figure 1b). The results of these experiments demonstrate that subtoxic doses of TRAIL and PAM are 10–100 ng/mL and 5- to 50-fold dilutions, respectively, when applied separately. However, HeLa cell viability was significantly reduced by treatment of PAM at various concentrations with a fixed TRAIL concentration or vice versa (Figure 1b). Flow cytometric analysis of annexin V/propidium iodide stained cells revealed that apoptosis was significantly induced in HeLa cells by co-treatment with a five-fold dilution of PAM and 20 ng/mL TRAIL for 24 h, but not with single treatment (Figure 1c). Furthermore, PAM/TRAIL treatment of HeLa cells significantly induced cleavage of the apoptotic markers such as caspase-3 and PARP (Figure 1d). DNA fragmentation was induced only by PAM/TRAIL treatment, but not by treatment with PAM or TRAIL separately (Figure 1e). These results demonstrate that PAM/TRAIL co-treatment of HeLa cells increases apoptosis.

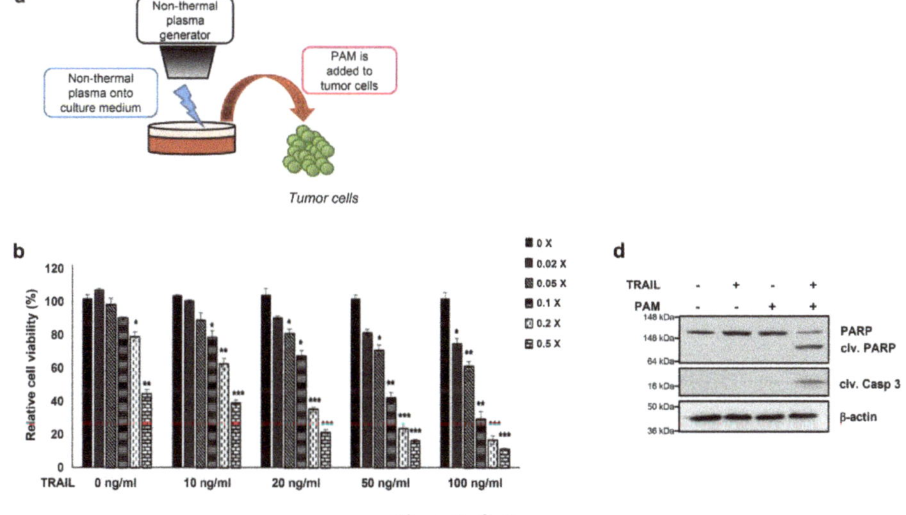

Figure 1. *Cont.*

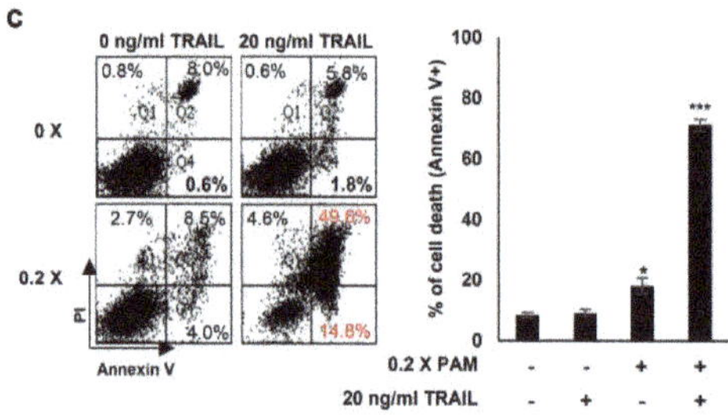
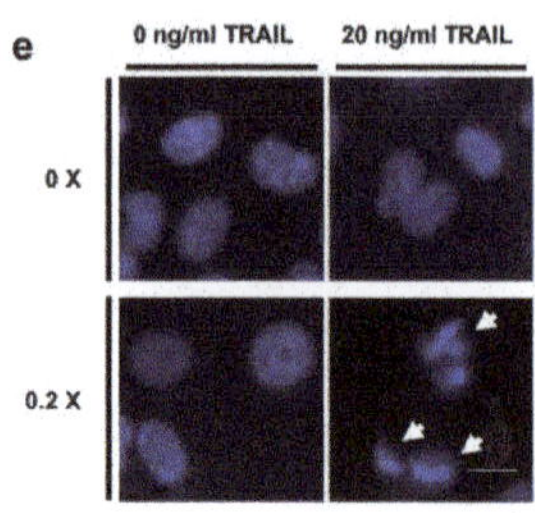

Figure 1. PAM and tumor necrosis factor-related apoptosis-inducing ligand (TRAIL) synergistically induce cancer cell death. (**a**) Preparation of plasma activated medium (PAM) using a non-thermal plasma generator system. PAM was generated by exposing DMEM or RPMI-1640 medium to non-thermal plasma jet at a distance of approximately 2 cm. (**b**) HeLa cells were treated with TRAIL at various concentrations (10 to 100 ng/mL) in the absence or presence of serially diluted PAM as indicated. Cell viability was assessed after 24 h of treatment using the MTT assay. * $p < 0.05$, ** $p < 0.01$, *** $p < 0.001$. (**c**) Cell death was analyzed using fluorescence-activated cell sorting (FACS) following Annexin V and propidium iodide (PI) staining. Left panel shows a representative image of necrosis and early and late apoptosis. Right panel shows statistical analysis of cell death. * $p < 0.05$, *** $p < 0.001$). (**d**) PAM promoted TRAIL-mediated apoptosis. Immunoblotting was performed using antibodies directed against cleaved caspase 3, PARP, and actin. (**e**) Condensation and fragmentation of the nuclei were detected via 4′,6-diamidino-2-phenylindole (DAPI) staining. Arrows: fragmented nuclei. Scale bar: 10 µm.

2.2. PAM/TRAIL Treatment Induces Apoptosis in TRAIL-Resistant Cancer Cells but Not in Normal Cells

We investigated the cytotoxicity of PAM and/or TRAIL on multiple cell lines to determine if PAM can stimulate TRAIL-mediated apoptosis in TRAIL-resistant cancer cell lines. PAM/TRAIL treatment of HeLa, A549, or HepG2 cell lines resulted in cell growth inhibition and induction of cell death (Tables 1 and 2, Supplementary Figure S2a,b). PAM/TRAIL treatment also reduced cell viability in the TRAIL-resistant DU145 cancer cell line (Supplementary Figure S3a). In contrast, the combined treatment did not significantly affect cell viability in the non-cancer human dermal fibroblast (HDF) cell line (Supplementary Figure S3a). The broad-spectrum caspase inhibitor zVAD abrogated apoptosis in cancer cells co-treated with PAM and TRAIL (Tables 1 and 2, Supplementary Figure S2a,b), confirming that PAM/TRAIL treatment induces apoptosis in various TRAIL-resistant cancer cells but not in normal cells.

Table 1. Statistical analysis of MTT cell viability assay results (mean ± SD (%)).

Cell Line	Vehicle	TRAIL	zVAD	DR5/Fc	NAC	PAM	PAM/TRAIL Vehicle	zVAD	DR5/Fc	NAC
HeLa	100 ± 2.3	99.1 ± 1.6	96.4 ± 4.3	92.7 ± 0.5	93.6 ± 3.3	69.0 ± 1.6 **	25.6 ± 0.5 ***	53.3 ± 1.8 ***	55.3 ± 3.9 *	40.2 ± 3.3 ***
A549	100 ± 1.3	99.0 ± 0.6	103.8 ± 6.1	101.0 ± 4.8	106.1 ± 2.7	78.2 ± 0.8 *	39.9 ± 0.2 ***	58.8 ± 1.0 **	61.7 ± 3.2 ***	56.6 ± 2.2 ***
HepG2	100 ± 2.7	96.7 ± 2.8	108.7 ± 5.0	95.8 ± 2.9	108.7 ± 7.6	76.5 ± 3.3 *	32.8 ± 0.6 ***	44.1 ± 0.2 **	45.9 ± 2.3 ***	48.9 ± 2.3 ***

HeLa, A549 or HepG2 cells were pretreated with 25 µM of zVAD, 20 ng/mL DR/Fc or 5 mM NAC for 1 h before TRAIL treatment in the absence or presence of PAM. After 24 h, growth inhibition was monitored via the MTT assay. DMSO was used as a vehicle. Statistical analysis was performed using a one-way ANOVA test followed by Dunnett's test for comparisons. *, **, and *** indicate significant differences from the control group (* $p < 0.05$, ** $p < 0.01$, *** $p < 0.001$).

Table 2. Statistical analysis of apoptosis assay results (mean ± SD (%)).

Cell Line							PAM/TRAIL			
	Vehicle	TRAIL	zVAD	DR5/Fc	NAC	PAM	Vehicle	zVAD	DR5/Fc	NAC
HeLa	7.0 ± 0.5	7.7 ± 0.3	8.2 ± 1.3	8.3 ± 1.3	8.2 ± 0.9	21.1 ± 1.4 *	64.3 ± 1.6 ***	27.4 ± 1.2 *	19.2 ± 1.1	27.2 ± 2.0 *
A549	9.6 ± 0.2	14.9 ± 0.5	11.6 ± 0.6	12.0 ± 0.4	11.6 ± 1.1	24.6 ± 0.4 *	61.3 ± 2.1 ***	31.7 ± 1.4 *	15.6 ± 0.6	29.4 ± 1.4 *

HeLa and A549 cells were pretreated with 25 µM of zVAD, 20 ng/mL DR/Fc or 5 mM NAC for 1 h before TRAIL treatment in the absence or presence of PAM. After 24 h incubation, cell death was determined by FACS analysis following Annexin V and propidium iodide staining. DMSO was used as a vehicle. Statistical analysis was performed using a one-way ANOVA test followed by Dunnett's test for comparisons. * and *** indicate significant differences from the control group (* $p < 0.05$, *** $p < 0.001$).

2.3. PAM/TRAIL Treatment Induces Apoptosis via DR5 Upregulation

We investigated the expression of apoptosis-related proteins, Bcl-2 and c-FLIP, and the TRAIL receptors DR4 and DR5 to elucidate the molecular mechanisms underlying PAM-mediated TRAIL sensitization. PAM/TRAIL treatment increased DR4 levels and significantly increased DR5 levels but did not affect the expression levels of anti-apoptotic Bcl-2 proteins in HeLa cells (Figure 2a). Anti-apoptotic cFLIP protein levels were slightly decreased in PAM/TRAIL-treated cells (Figure 2a). We investigated the effects of PAM and PAM/TRAIL treatments on the transcriptional expressions of DR5, DR4, cFLIP, and Bcl-2. PAM/TRAIL treatment significantly increased DR5 mRNA compared to that of PAM alone (Figure 2b). However, contrast to DR5 (approximately 4–12 fold increase), mRNA levels of DR4, cFLIP, and Bcl-2 were not significantly different in treatments of PAM, TRAIL, or PAM/TRAIL combination in HeLa cells (Supplementary Figure S4). Also the DR5 transcription induced by PAM/TRAIL treatment was detected in A549 cells (Supplementary Figure S5), further validating that the combinational treatment of PAM and TRAIL upregulates DR5 transcription (Figure 2a,b).

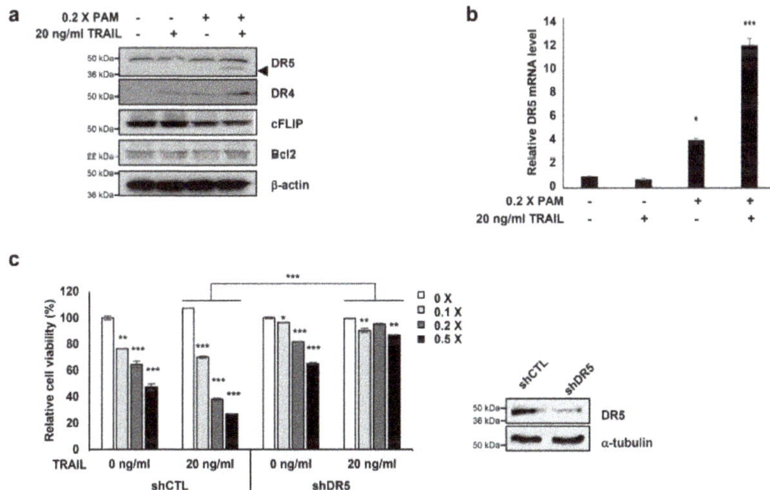

Figure 2. Co-treatment of PAM with TRAIL induces apoptosis via DR5 upregulation. (**a**) Combined treatment of PAM with TRAIL upregulates DR5 protein levels and downregulates c-FLIP protein levels in HeLa cells. Cells were treated with 0.2× PAM, TRAIL (20 ng/mL), or PAM/TRAIL (0.2× PAM with 20 ng/mL TRAIL). Cell extracts were prepared for immunoblotting of DR5, DR4, c-FLIP, Bcl-2, and β-actin. For immunoblotting, β-actin used as loading control. (**b**) DR5 mRNA levels were determined by qRT-PCR. GAPDH used as an internal control. (**c**) Silencing of *DR5* gene was abrogated synergistic effects of TRAIL to PAM-mediated cell survival inhibition. shCTL or shDR5-transfected HeLa cells were treated 0 or 20 ng/mL TRAIL with serial diluted PAM for 24 h. Cellular growth inhibition was determined via the MTT assay. * $p < 0.05$, ** $p < 0.01$, *** $p < 0.001$.

To examine the functional role of DR5 in PAM/TRAIL-induced cytotoxicity, we tested the effects of the DR5-specific blocking chimeric antibody (DR5/Fc) and DR5 knockdown. Pretreatment with DR5/Fc (20 ng/mL) significantly reduced growth inhibition in PAM/TRAIL-treated A549, HeLa, HepG2, and DU145 cells by at least 45% (Table 1 and Supplementary Figure S3a). Also DR5/Fc pretreatment ameliorated the PAM/TRAIL-induced apoptosis in HeLa and A549 cells (Table 2 and Supplementary Figure S3b). Knockdown of DR5 protected HeLa cells from cell death induced by PAM/TRAIL treatment (Figure 2c). These results demonstrate that DR5 upregulation is required for PAM/TRAIL-induced apoptosis.

2.4. CHOP Mediates DR5 Upregulation Induced by PAM/TRAIL Treatment

The transcription factor CCAAT/enhancer binding protein (C/EBP) homologous protein (CHOP) is known to activate DR5 transcription [23] resulting in TRAIL sensitization in various cancer cells [16,24]. Our investigations showed that PAM or PAM/TRAIL treatment induced CHOP expression at both protein and mRNA levels (Figure 3a,b), concomitant with the increase in transcriptional and protein expressions of DR5 (Figure 2a,b). We investigated if treatment with PAM/TRAIL upregulates CHOP expression, which in turn increases DR5 expression. Actinomycin D abrogated PAM/TRAIL-induced up-regulation of both CHOP and DR5 transcriptions (Figure 3c). We used CHOP knockdown to investigate the relationship between PAM/TRAIL treatment and DR5 mRNA upregulation. In the absence of CHOP, PAM/TRAIL treatment did not increase expression of DR5 mRNA or protein (Figure 3d,e). To see whether CHOP is implicated in the PAM/TRAIL-induced DR5 transcription, we performed ChIP-qPCR assay. CHOP was recruited at the DR5 promoter (−276 to −264) [25] in the presence of PAM/TRAIL (Figure 3f). These results demonstrate that PAM/TRAIL treatment induces CHOP-mediated upregulation of DR5 expression.

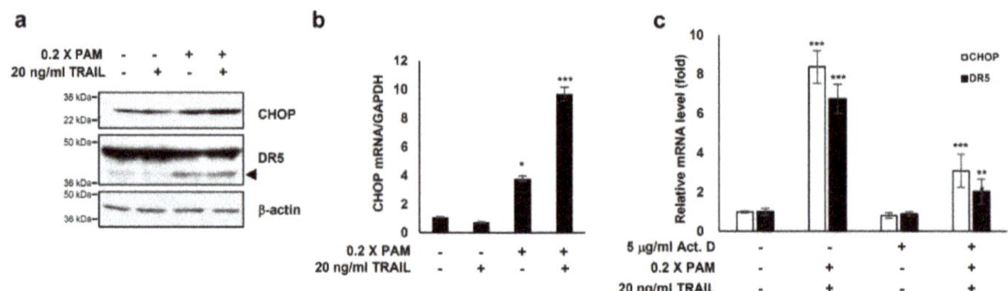

Figure 3. *Cont.*

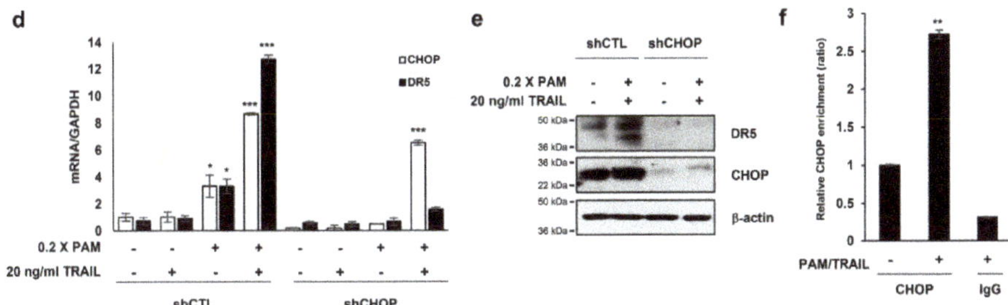

Figure 3. PAM/TRAIL induces DR5 upregulation through CHOP-mediated transcription. (**a**) HeLa cells were treated with 0.2× PAM alone or PAM/TRAIL. After 24 h, proteins in cell lysates were separated by sodium dodecyl sulfate polyacrylamide gel electrophoresis and immunoblotting with antibodies against CHOP, DR5, and β-actin. (**b**) HeLa cells were treated as described in (**a**). After 24 h, CHOP mRNA levels were determined via qRT-PCR. GAPDH was used as an internal control. (**c**) HeLa cells were co-treated with PAM/TRAIL for 12 h. Afterwards, the culture was replaced with fresh medium, pretreated with or without 5 μg/mL actinomycin D (Act D) for 30 min, and co-treated with or without PAM/TRAIL for 10 h. mRNA levels of CHOP and DR5 were determined via qPCR. (**d,e**) HeLa cells were transfected with a plasmid expressing either CHOP shRNA (shCHOP) or control shRNA (shCTL). At 24 h after transfection, cells were treated with PAM, TRAIL, or PAM/TRAIL for an additional 24 h. (**d**) Levels of CHOP and DR5 mRNA were determined via qRT-PCR. (**e**) Protein levels of CHOP and DR5 were determined by immunoblot analysis using antibodies against DR5 and CHOP. (**f**) Chromatin immunoprecipitation (ChIP) was performed using anti-CHOP antibody in HeLa cells following PAM/TRAIL treatment for 8 h. The IgG was used to control for antibody specificity. qRT-PCR was carried out using primers surrounding the CHOP binding sites in the DR5 promoter. * $p < 0.05$, ** $p < 0.01$, *** $p < 0.001$.

2.5. Plasma-Activated Medium (PAM) Promotes Membrane-Bound DR5 Redistribution

Next we investigated the effect of PAM treatment on the expression and clustering of DR5 on the membrane, similarly to the case of the membrane-bound death receptor Fas whose forced redistribution and aggregation into ceramide-rich lipid platforms enhance FasL-mediated apoptosis [26]. Undiluted and 0.5× PAM enhanced the membrane clustering of DR5 in HeLa and HT-29 cells, inducing its redistribution in a dose- and time-dependent manner (Figure 4a,b).

Because the ligand TRAIL is required for DR5 membrane clustering [27] and we showed that the lethal-dose (undiluted and 0.5×) of PAM promoted DR5 clustering on the membrane (Figure 4a,b), we examined whether PAM could induce TRAIL expression. We evaluated the TRAIL mRNA levels with and without PAM treatment. The TRAIL mRNA levels were barely affected by the sublethal- (0.2×) and lethal (0.5×) doses of PAM (Supplementary Figure S6), indicating that the PAM-mediated DR5 redistribution on the membrane may occur in a TRAIL-independent manner. Since a previous study showed that up-regulated DR5 induces TRAIL-independent apoptosis via caspase 8 [28], we examined whether the overexpressed DR5 by PAM (Figure 4a,b) induces caspase 8-dependent apoptosis. In the 1 × PAM-treated HeLa cells, cleaved caspase 8 was increased (Figure 4c), suggesting that the PAM-DR5-caspase 8 axis is distinct from TRAIL-induced apoptosis.

The observation of PAM-induced DR5 redistribution on the membrane (Figure 4a,b), led us to question whether PAM may induce membrane alternations, also. Previously we reported that plasma-induced ROS/RNS disturbs mitochondrial membrane potential [19] and impairs cellular membrane through coincident lipid oxidation, altered electrical conductivity, and membrane roughening [20]. Impairment of cellular membranes [20] could compromise the maintenance of transmembrane ion gradients, rendering cells vulnerable to extracellular ion stress. We investigated cell viability with and without PAM under conditions of high [K$^+$], a membrane depolarizing agent. PAM increased cell death as [K$^+$] increased (Figure 4d). The addition of the antioxidants glutathione or N-acetyl-cysteine (NAC) significantly reduced PAM-induced cell death under conditions of high [K$^+$] (Figure 4e), indicating that PAM-induced oxidative stress contributes to cell death. We assessed intracellular Ca^{2+} concentrations in the presence or absence of PAM to investigate if PAM-induced cellular membrane damage might disrupt ion homeostasis. Exposure of A549 cells to PAM resulted in an elevation of Ca^{2+} by approximately 1.5-fold compared to untreated cells (Figure 4f). PAM-induced accumulation of intracellular Ca^{2+} was synergistically enhanced in U2OS cells in the presence of 50 mM KCl (Figure 4g). These results indicate that PAM compromises cellular membranes, disrupting intracellular Ca^{2+} ion homeostasis.

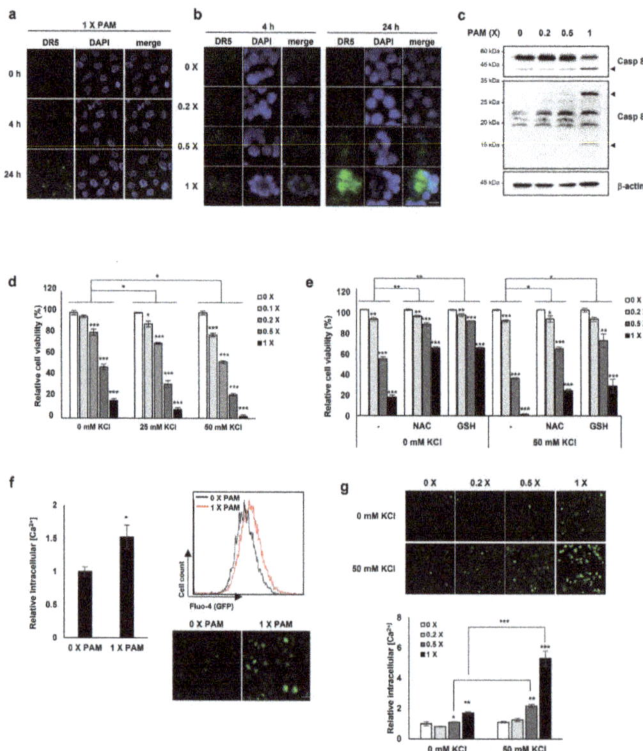

Figure 4. PAM promotes membrane-bound DR5 redistribution. HeLa cervical cancer cells (**a**) and HT-29 colorectal cancer cells (**b**) were treated with serially diluted PAM for 4 or 24 h. Original magnification: × 630. Scale bar: 10 µm. After PAM treatment, cells were fixed with 4% paraformaldehyde for 15 min followed by immunostaining with anti-DR5 antibody. DR5 clustering was detected by fluorescence microscopy. (**c**) HeLa cells were treated with PAM for 24 h, as indicated. Caspase 8 protein levels were determined by immunoblotting. β-actin was used as a loading control. Arrow head: cleaved form of caspase 8. (**d**) The effects of PAM on A549 cancer cells. A549 lung cancer cells were treated with serially diluted PAM supplemented with 0, 25, or 50 mM KCl for 24 h. Cell viability was measured using the MTT assay. (**e**) The antioxidants, N-acetyl-cysteine (NAC), and reduced glutathione (GSH) abrogate the impaired ion homeostasis induced by PAM. A549 cells were treated with serially diluted PAM in the absence or presence of 50 mM KCl supplemented with 2 mM NAC or 1 mM GSH for 24 h. (**f**) Intracellular concentration of Ca^{2+} was measured following PAM treatment of A549 cells. After incubation with 2 µM Fluo-4, AM for 45 min, intracellular Ca^{2+} concentration was analyzed by FACS analysis or by fluorescence microscopy. Scale bar: 50 µm. (**g**) U2OS cells were treated with serially diluted PAM in the absence or presence of 50 mM KCl for 24 h, and then the level of intracellular Ca^{2+} was evaluated. * $p < 0.05$, ** $p < 0.01$, *** $p < 0.001$. Scale bar: 50 µm.

2.6. ROS is Implicated in PAM/TRAIL Sensitization

We investigated the involvement of PAM/TRAIL-sensitization in the course of intracellular ROS production. PAM/TRAIL treatment revealed a significant increase in intracellular ROS levels in HeLa cells as determined by MitoSOX-based fluorescent microscopy (Figure 5a) and H_2DCFDA-based FACS analysis (Figure 5b). We next investigated the role of ROS in PAM-induced DR5 mRNA expression, as PAM induced the DR5 expression levels (Figure 2a,b). Pretreatment of HeLa cells with the antioxidant NAC abrogated PAM-induced DR5 transcription (Figure 5c), indicating that PAM-induced ROS is involved in DR5 mRNA upregulation. Furthermore, NAC pretreatment attenuated PAM/TRAIL-induced growth inhibition and cell death in cancer cells, including HeLa, A549, and HepG2 (Tables 1 and 2, Supplementary Figure S7). Because PAM/TRAIL treatment induced generation of ROS that could be originated from mitochondria (Figure 5a), we investigated the consequence of mitochondrial ROS inhibition in the presence of PAM/TRAIL treatment. MnTBAP and Ebselen, mimetic agents for superoxide dismutase (SOD) and glutathione peroxidase, respectively, abrogated PAM/TRAIL-induced cytotoxic effects in HeLa cells (Supplementary Figure S8). These results indicate that DR5 upregulation and TRAIL sensitization are dependent on PAM-induced ROS generation.

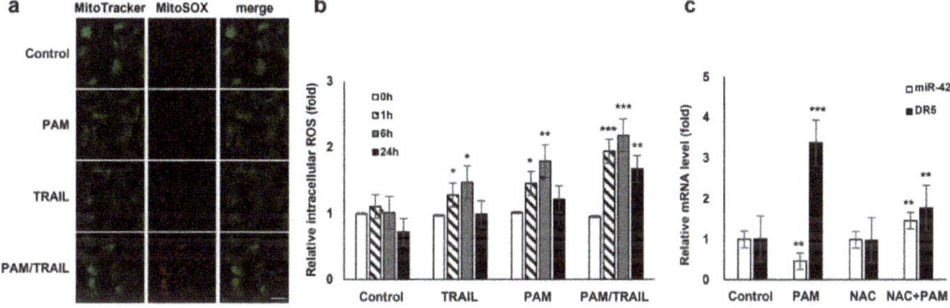

Figure 5. ROS generation is involved in PAM-mediated TRAIL sensitization. (**a**) Reactive oxygen species (ROS) production in mitochondria was induced by treatment with PAM/TRAIL. HeLa cells were treated with PAM, TRAIL, or PAM/TRAIL, after which mitochondrial ROS accumulation was assessed after 24 h by staining the cells with MitoTracker Green and MitoSox Red under a fluorescent microscope. Scale bar: 10 μm. (**b**) Intracellular ROS generation was induced by treatment with PAM/TRAIL. Cells were treated as in (**a**) for the indicated times. Intracellular ROS were quantified via the H_2DCF-DA assay. * $p < 0.05$, ** $p < 0.01$, *** $p < 0.001$. (**c**) PAM-induced ROS generation is required for DR5 upregulation and miR-425 downregulation. Pretreatment with a surrogate antioxidant, NAC, reversed PAM-induced DR5, and miR-425 transcriptional modulation. ** $p < 0.01$, *** $p < 0.001$.

2.7. PAM Sensitizes Cancer Cells to TRAIL-Induced Apoptosis via Modulation of miR-425-PTEN-Akt Axis

Several studies have reported that Akt inactivation induces TRAIL sensitization in TRAIL-resistant cancers [13]. We investigated the involvement of PTEN-Akt signaling in PAM/TRAIL-induced apoptosis. Following PAM or PAM/TRAIL treatment, levels of PTEN protein, an upstream phosphatase of Akt signal pathway, were markedly increased, and accordingly phosphorylated Akt levels were decreased in HeLa cells (Figure 6a). These results demonstrate that PAM-induced PTEN upregulation, leading to Akt inactivation, contributes to TRAIL-mediated apoptosis by PAM.

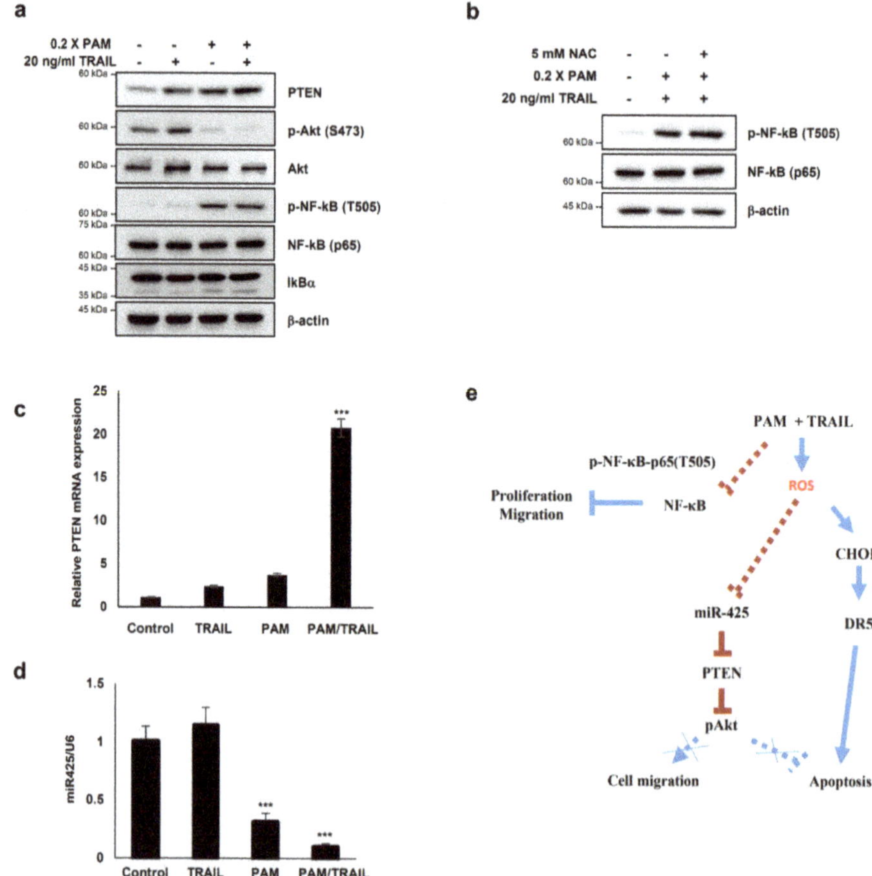

Figure 6. PAM sensitizes cancer cells to TRAIL-induced apoptosis via modulation of the miR-425-PTEN-Akt axis. (**a–d**) HeLa cells were treated with TRAIL with or without PAM for 24 h. (**a**) PTEN, phospho-Akt1 (S473), Akt1, NF-κB, phospho-NF-κB (T505) and IκBα protein levels were determined by immunoblotting. β-actin was used as a loading control. (**b**) HeLa cells were pretreated with 5 mM NAC for 1 h before TRAIL treatment in the presence of PAM. After 24 h, phosphor-NF-κB (T505) and NF-κB protein levels were observed by immunoblot analysis. (**c**) PTEN mRNA levels were determined by qRT-PCR. *** $p < 0.001$. (**d**) miR-425 levels were determined via qRT-PCR. *** $p < 0.001$. (**e**) Proposed model for PAM-mediated TRAIL sensitization. Our work has revealed that combinational treatment with PAM and TRAIL stimulates ROS-dependent apoptosis by two attributes; (1) via CHOP-mediated upregulation of DR5 transcription (solid blue line), and (2) via miR-425 suppression-mediated PTEN upregulation, resulting in Akt inactivation (solid red line). In this study, PAM and TRAIL cotreatment induces inhibitory phosphorylation of NF-κB (solid red line). Previous studies have provided evidences that active NF-κB and phospho-Akt promote proliferation and migration (dotted blue lines).

In order to determine whether PAM- and PAM/TRAIL-induced PTEN expression (Figure 6a) occurs at the transcriptional level, we evaluated the mRNA levels of PTEN in the PAM and TRAIL treated HeLa cells. PAM- and PAM/TRAIL-treatment increased PTEN mRNA levels, to approximately 4- and 21-fold, respectively (Figure 6c). Based on the previous studies demonstrating that miR-425-mediated suppression of PTEN promotes cancer aggressiveness and is associated with tumorigenesis and malignancy in many cancers [29–31], we tested whether PAM- and PAM/TRAIL-treatment can downregulate the miR-425 transcription. The levels of miR-425 significantly decreased in PAM- and PAM/TRAIL-treated HeLa cells (Figure 6d), concomitantly with the increased PTEN mRNA (Figure 6c). PAM-mediated suppression of miR-425 was abrogated by NAC (Figure 5c) indicating that downregulation of miR-425 is possibly mediated through PAM-induced ROS generation, similar to NAC ablation of PAM-induced DR5 upregulation. In order to determine whether suppression of miR-425 mediates PTEN transcription in response to PAM or PAM/TRAIL treatment, we investigated levels of PTEN protein in miR-425-transfeccted HeLa cells to confirm the transcription response of PTEN to miR-425. The increase in PTEN protein by PAM or PAM/TRAIL treatment was reduced in miR-425-transfected HeLa cells (Supplementary Figure S9a). PAM treatment induced the downregulation of miR-425 and the increase in TRAIL sensitization in different TRAIL-resistant cell lines, including DU145, U87-MG, MCF7, and MDA-MB-435 (Supplementary Figure S9b). These results show that PAM/TRAIL treatment can enhance PTEN transcription through suppression of miR-425, increasing PTEN protein levels, leading to Akt inactivation and apoptosis.

Furthermore, one of the well-known regulatory mechanisms for tumor cells is phosphorylation of NF-κB RelA (p65) at T505, which mediates cellular apoptosis and inhibits proliferation and migration through differential gene regulation [31]. Hence, we tested whether PAM- and PAM/TRAIL-treatment regulates phosphorylation of p65 at T505. The phospho-p65 at T505 was induced by PAM- and PAM/TRAIL treatment (Figure 6a). Next, we examined whether the phosphorylation of p65 at T505 is regulated by activation or translocation of NF-κB. The protein levels of IκBα, an inhibitory regulator of NF-κB was not altered by treatment of PAM/TRAIL (Figure 6a), suggesting that the phosphorylation of p65 at T505 is not relevant to the canonical NF-κB activation. Together, these results suggest that PAM/TRAIL-induced p65 phosphorylation may contribute to cancer cell apoptosis.

To investigate whether the PAM-induced p65 phosphorylation associated with NF-κB inhibition depends on ROS generated in PAM, we evaluated the level of p65 phosphorylation at T505 in the NAC-pretreated HeLa cells prior to PAM/TRAIL-treatment. The level of phospho-p65 (T505) was not affected by NAC (Figure 6b), suggesting that the inhibitory phosphorylation of p65 may not depend on the ROS generated in PAM.

3. Discussion

The selective induction of apoptosis by TRAIL and agonistic TRAIL receptor antibodies in cancer cells has led to their clinical development as promising anticancer therapeutics. However, in many cancers, TRAIL resistance is intrinsic or acquired during the course of TRAIL treatment [8,9]. To overcome this resistance, it is important to identify effective TRAIL sensitizers that target tumor heterogeneity in the TRAIL pathway for patient-tailored therapy [32]. In this study, we explored the ability of PAM to potentiate TRAIL-induced ROS-mediated apoptosis of cancer cells. We investigated the effects of PAM on TRAIL sensitization and the molecular mechanisms behind TRAIL-resistant A549 and HeLa cancer cells. PAM triggers ROS/RNS generation, leading to transcriptional upregulation of DR5 via CHOP-mediated transcriptional activation, disruption of intracellular Ca^{2+} homeostasis, and inactivation of Akt via PTEN upregulation. The role of PAM/TRAIL sensitization, and its underlying mechanisms involving the generation of ROS/RNS [21] is further supported by results showing that antioxidants, such as NAC and glutathione which can prevent cellular accumulation of ROS/RNS [21], reduce TRAIL sensitization, mitigate DR5 upregulation, and reduce intracellular Ca^{2+} aberration.

PAM has potent anti-cancer activity because of its reactive oxygen and nitrogen species and is effective regardless of tumor genetic variations [21]. In recent years, plasma has been used in fields such as medical disinfection and antiseptics, blood clotting, dentistry, and skin care. Its use is steadily increased in immunotherapy, wound regeneration, and cancer treatment. Accumulating experimental and animal evidence demonstrates plasma is a safe medicinal agent that does not have adverse effects on healthy normal cells [17,18,21]. Recent studies have shown that plasma exhibits anticancer efficacy against a broad range of cancers and that PAM addresses the problem of limited penetrance in cell culture studies and rodent models [17,18,21]. Here, we show that PAM/TRAIL treatment induces apoptosis in TRAIL-resistant cancer cells without side effects on normal cells.

Our current findings demonstrate that CHOP induction by PAM induces transcriptional activation of DR5, leading to DR5 upregulation and further induction of TRAIL-mediated apoptosis. Incubation of cells with antibodies against DR5 attenuated PAM/TRAIL-induced reduction in cell viability, indicating that PAM is capable of provoking apoptosis in TRAIL-resistant cancer cells. Our results show that PAM induced the expression of the DR5 transcriptional activator, CHOP, resulting in increasing DR5 expression. In cells lacking CHOP, PAM-induced DR5 upregulation and TRAIL-induced apoptosis were reduced, confirming that DR5 upregulation by PAM is mediated through CHOP induction. Our results are consistent with previous findings indicating that CHOP is required for PAM-induced DR5 transcriptional upregulation [16,23,33].

The oxidative stress-dependent effects of PAM on DR5 clustering (Figure 4a,b) and membrane dysfunction (Figure 4d) [19–21] are similar to interaction mechanisms between oxidative stress and TRAIL-induced apoptosis [2,8,9,15]. PAM elevates intracellular Ca^{2+} levels possibly through impairment of ion channel or transporters and permeability transitions. PAM-induced membrane dysfunction was enhanced under high-salt (KCl) stress, indicating a lack of membrane integrity (Figure 4d). K^+ potentiates TRAIL-induced apoptosis in human tumor cells, including leukemia, melanoma, and lung cancer cells via mitochondria-derived ROS (mROS) accumulation [34]. Inactivation of Na^+-K^+-ATPase and sustained membrane depolarization were observed during Fas-induced apoptosis [35,36], initiated by intracellular glutathione depletion and H_2O_2 generation [37]. ROS have been shown to stimulate external Ca^{2+} entry into the cytoplasm, leading to the generation of intracellular mROS and mitochondrial damage [38,39]. Mitochondrial damage induces a release of cytochrome c and other apoptotic protein factors that enhance necrosis and apoptotic cell death [36,40]. Depolarization can act in both pro- and anti-apoptotic processes, depending on the cell types and apoptotic stimuli involved. Suzuki Y et al. reported that the depolarization appears to be a prerequisite event for TRAIL-induced apoptosis, because TRAIL induces minimal cytotoxicity despite the substantial expression of DR4 and DR5 in the cellular membrane [41]. Our results indicate that PAM first induces ROS-mediated membrane depolarization followed by DR5 clustering and TRAIL-induced apoptosis.

TRAIL-induced DR4 or DR5 trimerization activates the extrinsic apoptotic pathway [42–44]. Once trimerized, DR4 or DR5 serves as the aggregation point for a multimeric protein structure called "death-induced signaling complex (DISC)" that is comprised of the ligated DR4/5, Fas-associated death-domain protein (FADD), and procaspases 8 and 10. Caspases play essential roles in programmed cell death including apoptosis and necroptosis. When the caspase 8 activity is inhibited by zVAD, necroptosis is initiated [45]. Also, numerous necroptosis-insensitive cancer cell lines, such as HeLa, HCT116, and OVCAR4 (human ovarian cancer) cells do not have an effective necroptotic machinery. The necroptosis-sensitive cells, such as A549 and HepG2, are responsive to a few specific necroptosis inducers, but not to all necroptosis inducers, revealing that necroptosis can occur by specific stress signals in a controlled manner [45]. Therefore, we tested whether the growth inhibition by PAM/TRAIL treatment was caused by necroptosis as well as apoptosis. As shown in Supplementary Figure S2c, necrostatin-1(Nec-1, an inhibitor of necroptosis) had little effects on the PAM/TRAIL-induced growth inhibition in the both TRAIL-unresponsive HeLa (necroptotic-insensitive) and A549 (necroptotic-sensitive) cells. Also phosphorylation of RIP3 and MLKL (necroptosis markers) were not changed by treatment of zVAD or Nec-1 in the PAM/TRAIL-treated HeLa cells (Supplementary Figure S2d). In the Table 1 and Supplementary Figure S2, the inhibitory effect of zVAD on the PAM/TRAIL-mediated cell death in HepG2 cells was slightly lower than that in HeLa and A549 cells, suggesting that PAM/TRAIL may induce necroptosis in HepG2 cells. Additionally, DR5/Fc inhibited more effectively the PAM/TRAIL-mediated cell death in HeLa and A549 cells, compared with HepG2 cells. Taken together, these data indicate that these inhibitors (zVAD, Nec-1, and DR5/Fc) have differential effects depending on the type of cell lines. However, the inhibitory effects of NAC on the

PAM/TRAIL-mediated cell death were comparable in the three tested cell lines, supporting that the PAM/TRAIL-mediated cell death occur in an ROS-dependent manner, in the tested cell lines.

The microRNAs miR-221, -222, and -425 are known to target PTEN and induce TRAIL resistance in cancer cells [30,33]. Here, we found that PAM treatment leads to upregulation of PTEN and reduction of phosphorylated Akt levels, eventually sensitizing cells to TRAIL-induced apoptosis through inactivation of the Akt pathway. Consistent with earlier studies [30,33], PAM treatment suppressed miR-425 expression, resulting in increased PTEN expression and leading to TRAIL-inducing apoptosis mediated by Akt inactivation. In addition, PAM treatment decreased the expression of the cell survival protein c-FLIP, which is linked to TRAIL resistance.

The NF-κB pathway contributes to the growth, survival, and malignancy of numerous cancer cell types while also affecting the response of tumors to chemotherapy and radiotherapy [46]. Post-translational modifications are crucial for the functions of NF-κB subunits [47], such as phosphorylations of p65 at T254, S276, S281, S316, S468 lead to cell proliferation, migration, and ubiquitination [48–52]. In contrast, threonine505 (T505) residue in p65 with an important inhibitory regulation role is phosphorylated by Chk1 in response to cisplatin. The inhibitory p65 phosphorylation at T505 leads to induction of the tumor suppressor p14ARF, which contributes to tumor suppression [53]. We found that phosphorylation of NF-κB at T505 was increased by PAM or PAM/TRAIL treatment (Figure 6a), suggesting the possibility that PAM or PAM/TRAIL may suppress migration and proliferation of cancer cells via inhibition of NF-κB signal pathway.

Consistent with prior studies, we found that PAM treatment induced the generation of intracellular and mitochondrial ROS, which likely mediates the upregulation of DR5 and sensitizes cells to TRAIL-induced apoptosis [9,54–56]. We found that the antioxidant NAC significantly ameliorated PAM-induced TRAIL-mediated apoptosis with concomitant recovery of Ca^{2+} homeostasis and the abolishment of DR5 upregulation, confirming that ROS/RNS plays a crucial role in PAM-induced TRAIL sensitization. Our results demonstrate that PAM induces ROS/RNS production, which in turn leads to CHOP-induced DR5 upregulation, and consequently gives rise to the TRAIL-induced apoptosis of cancer cells.

In summary, PAM may be beneficial for sensitization of TRAIL-resistant cancer cells leading to apoptosis. Clinically, resistance to anticancer drugs is one of major causes in treatment failure due to primary or acquired non-responsiveness of cancer cells. We demonstrated that various cancer cells including TRAIL resistant cells can be sensitized by PAM, while PAM has no cytotoxic effects on non-cancerous cells. Moreover, in xenograft formation using A549 adenocarcinoma in the mouse model, the anticancer effect of PAM was detected by assessing tumor volume. Thus, further studies to verify anticancer effects of PAM/TRAIL using in vivo tumor model have important implications for the development of novel strategies in cancer therapy.

4. Materials and Methods

4.1. Reagents and Antibodies

The following reagents were used in the experiments: recombinant protein human TRAIL/Apo2L (KOMA Biotech, Seoul, Korea); FITC Annexin V Apoptosis Kit I (BD Pharmingen, Franklin Lakes, NJ, USA); z-VAD-fmk (Promega, Madison, WI, USA); N-acetyl-cysteine, reduced glutathione, Mn(III) tetrakis (4-benzoic acid) porphyrin chloride (MnTBAP), Ebselen, necrostatin-1 (Sigma Aldrich, St. Louis, MO, USA); MitoTracker™ Green FM, MitoSOX™ Red, Fluo-4 AM, 2′,7′-dichlorodihydrofluorescein diacetate (H₂DCFDA; Invitrogen, Carlsbad, CA, USA). We used antibodies against β-actin, IκBα (Santa Cruz Biotechnology, Inc., Dallas, TX, USA), DR4, phospho-RIP3 (T231/S232), phospho-MLKL (Abcam, Cambridge, UK), CHOP, caspase-3, caspase-8, p65 (Cell Signaling Technology, Danvers, MA, USA), poly(ADP-ribose) polymerase (BD Pharmingen), DR5 (KOMA Biotech and Abcam), α-tubulin (Millipore, Burlington, MA, USA), c-FLIP (Alexis, Farmingdale, NY, USA), phospho-p65

(T505) (Bioss antibodies, Woburn, MA, USA), and horseradish peroxidase (HRP)-conjugated anti-rabbit or anti-mouse IgG antibodies (Enzo Life Sciences, Farmingdale, NY, USA).

4.2. Immunostaining of DR5

HeLa and HT-29 cells were treated with serially diluted PAM for 4 or 24 h. After treatment, cells were fixed for 15 min with 4% paraformaldehyde and blocked in PBS containing 10% FBS for 30 min. Cells were incubated with antibody against DR5 (Abcam) for 16 h at 4 °C and then with FITC-conjugated secondary antibodies (Invitrogen) for 1 h at room temperature. After counterstaining with 4′6-diamidino-2-phenylindole (DAPI), immunofluorescence images were acquired using a fluorescence microscope (Eclipse Ti-S, Nikon Instruments Inc., Walt Whitman Road Melville, NY, USA).

4.3. Cell Culture and Treatment with Plasma-Activated Medium (PAM) and TRAIL

HeLa, A549, HepG2, U2OS, and HDF cells were obtained from American Type Culture Collection (ATCC) and cultured in DMEM (WELGENE, Kyung-san, Korea) supplemented with 10% fetal bovine serum (FBS) and antibiotics (Life Technologies, Carlsbad, CA, USA). HCT116 and HT-29 cells were cultured in RPMI-1640 containing 10% FBS and antibiotics. Plasma-activated medium (PAM) was produced using a microplasma jet device at atmospheric pressure [17,18,43] and discharging the plasma jet onto a liquid such as the mammalian cell culture medium DMEM (Figure 1a) [21]. Using the air plasma-jet system (AMED, Seoul, Korea), we jetted non-thermal air plasma 2 cm above the surface of growth medium, in a chamber of a 12-well-plate for 10 min at atmospheric pressure and room temperature to generate PAM [22]. Cells were washed with DPBS (Life Technologies) and covered with fresh culture medium that was either untreated or treated with undiluted or diluted plasma-activated medium (PAM) as indicated [21,43]. TRAIL was added to the culture medium at the indicated concentrations, either alone or in combination with PAM. Detection of apoptosis and the MTT colorimetric assay were performed as previously described [17,18,21,22].

4.4. Detection of Nuclei Condensation and Fragmentation

Cells were fixed with 1% paraformaldehyde, followed by staining with 300 nM DAPI for 5 min. Morphology of nuclei were observed under a fluorescence microscope (Eclipse Ti-S, Nikon Instruments Inc., Walt Whitman Road Melville, NY, USA).

4.5. Quantification of Mitochondrial and Intracellular ROS

Mitochondrial and intracellular ROS measurements were performed as previously described [17] according to the manufacturer's protocol. Relative ROS levels were expressed as arbitrary fluorescence units.

4.6. Quantification of ROS and RNS of PAM

Concentration of H_2O_2 in PAM was measured using the Amplex Red Hydrogen Peroxide/Peroxidase Assay Kit (Invitrogen) and colorimetric intensity was measured on a microplate reader (Bio-Rad, Hercules, CA, USA) at 540/595 nm, according to the manufacturer's protocol. Production of RNS in PAM was measured by the Griess assay [57].

4.7. Measurement of Intracellular Ca^{2+}

A549 and U2OS cells were treated with PAM for 24 h, followed by staining with 2 μM Fluo-4 AM for 45 min. Cells were harvested by trypsinization and suspension in PBS. Intracellular Ca^{2+} was analyzed by flow cytometry and fluorescence microscopy.

4.8. RNA Isolation and Quantitative Real-Time PCR (qRT-PCR)

Total RNA was isolated using the RNeasy Min Kit (Qiagen, Carlsbad, CA, USA), and cDNA was synthesized using M-MLV reverse transcriptase (Promega) according to the manufacturer's instructions. The following primers were used for amplification of human DR5, PTEN, DR4, cFLIP, Bcl-2, TRAIL, and GAPDH: DR5 (sense): 5′-GTC ACA GTT GCA GCC GTA GT-3′ and (antisense) 5′-TGC CTT TCA GGT AAG GAA GG-3′; PTEN (sense): 5′-GAT GTG GCG GGA CTC TTT AT-3′ and (antisense): 5′-AGC GGC TCA ACT CTC AAA CT-3′; DR4 (sense): 5′-AGA GAG AAG TCC CTG CAC CA-3′ and (antisense): 5′-GTC ACT CCA GGG CGT ACA AT-3′; cFLIP (sense): 5′-GCA AGA CCC TTG TGA GCT TC-3′ and (antisense): 5′-TCG CCT CAC TCT GTA GAG CA-3′; Bcl-2 (sense): 5′-GAG GAT TGT GGC CTT CTT TG-3′ and (antisense): 5′-ACA GTT CCA CAA AGG CAT CC-3′; TRAIL (sense): 5′-GGA ACC CAA GGT GGG TAG AT-3′ and (antisense): 5′-TCT CAC CAC ACT GCA ACC TC-3′; GAPDH (sense): 5′-GTC AAC GGA TTT GGT CTG TAT T-3′ and (antisense) 5′-AGT CTT CTG GGT GGC AGT GAT-3′. The following primers were used for amplification of miR-425: Specific primer: 5′-TGG ACC AGA ATG ACA CGA TCA CTC C-3′ and Universal reverse primer: 5′-GTG CAG GGT CCG AGG T-3′. Real-time PCR amplification was carried out using Qiagen Rotor-Gene Q system with the following cycling conditions: 95 °C for 10 min; 40 cycles of 95 °C for 15 s, 60 °C for 30 s, and 72 °C for 30 s; and a final extension step at 72 °C for 10 min.

4.9. Immunoblotting

Immunoblotting experiments were performed as previously described [19]. Representative results from at least three independent experiments are shown in the figures.

4.10. Plasmid Construction and Transfection

Double-stranded small interfering RNA (siRNA) against CHOP was generated using a pSUPER.retro.puro, an H1 promoter-driven RNA interference retroviral vector (Oligoengine, Seattle, WA, USA). The siRNA was designed to target CHOP (5′-GAT CGA CGT GTA GTG AAT G-3′) and DR5 (5′-GAC CCT TGT GCT CGT TGT C-3′). The miR-425 (5′-AAU GAC ACG AUC ACU CCC GUU GA-3′) was synthesized by Geneolution, Inc. (Seoul, Korea). Transfections were performed using the Effectene Kit (Qiagen) and Lipofectamine 2000 (Invitrogen) using manufacture's manual.

4.11. Chromatin Immunoprecipitation (ChIP) Assay

ChIP assay with CHOP antibody (Cell signaling) was performed using the Chromatin Immunoprecipitation Assay Kit (Millipore) according to the manufacturer's manual. The precipitates were analyzed by qRT-PCR using the primers 5′-AGG TTA GTT CCG GTC CCT TC-3′ and 5′-CAA CTG CAA ATT CCA CCA CA-3′ to amplify a DR5 promoter fragment containing CHOP binding site (-276 to -264).

4.12. Statistical Analysis

All data were expressed as mean ± standard deviation (SD) of at least three replicates. The Student's *t*-test was used for comparisons between two groups. One-way or two-way ANOVA analysis followed by post-hoc test was used for comparisons between multiple groups. Differences were considered statistically significant at $p < 0.05$ (in figures: * $p < 0.05$, ** $p < 0.01$, *** $p < 0.001$).

5. Conclusions

Our results provide evidence that PAM synergistically enhances the efficacy of TRAIL-induced apoptosis in TRAIL-resistant cells by triggering ROS/RNS generation, which in turn upregulates DR5 expression via CHOP-mediated transcription, disturbs Ca^{2+} homeostasis, and inactivates Akt-mediated TRAIL resistance via suppression of miRNAs targeting PTEN. The findings that the oxidative stress-dependent PAM effects recapitulate the physiological condition in TRAIL-sensitive cells importantly provide a molecular basis for TRAIL sensitization. Taken together, these data highlight an oxidative stress-mediated mechanism through which CHOP-mediated DR5 upregulation, disturbance of ion homeostasis, and miRNA-mediated PTEN upregulation could promote TRAIL-induced apoptosis (Figure 6e). Thus, co-treatment with PAM and TRAIL serves as a novel combinational therapeutic approach to overcome TRAIL-resistant cancers.

Supplementary Materials: Supplementary materials can be found at http://www.mdpi.com/1422-0067/21/15/5302/s1.

Author Contributions: Investigation—S.Y.H., N.H.N., T.J.K., Y.L., and M.A.K.; writing—original draft, S.Y.H. and N.H.N.; writing—review and editing, Y.L., M.A.K., and J.-S.L.; supervision, M.A.K. and J.-S.L.; funding acquisition, M.A.K. and J.-S.L. All authors have read and agreed to the published version of the manuscript.

Funding: This work was supported by the National Research Foundation of Korea (NRF), grant number NRF-2017R1D1A1B03031171, 2017R1A2B4010146 and 2020R1A2C2011302.

Acknowledgments: We thank all members of the Lee lab for helpful advice. Sang Sik Yang (Ajou University) generously supplied a microplasma jet system.

Conflicts of Interest: There are no conflicts of interest and no competing financial interests.

Abbreviations

TRAIL	Tumor necrosis factor-related apoptosis-inducing ligand
PAM	plasma-activated medium
ROS	reactive oxygen species
RNS	reactive nitrogen species
NAC	N-acetyl-cysteine
GSH	glutathione
DR	death receptor
TNF	tumor necrosis factor
HDF	human dermal fibroblast
c-FLIP	cellular FLICE (FADD-like IL-1β-converting enzyme)-inhibitory protein
CHOP	CCAAT/enhancer binding protein (C/EBP) homologous protein
SOD	superoxide dismutase
PTEN	phosphatase and Tensin homolog deleted on Chromosome 10
mROS	mitochondrial reactive oxygen species

References

1. Fridman, J.S.; Lowe, S.W. Control of apoptosis by p53. *Oncogene* **2003**, *22*, 9030–9040. [CrossRef] [PubMed]
2. Voltan, R.; Secchiero, P.; Casciano, F.; Milani, D.; Zauli, G.; Tisato, V. Redox signaling and oxidative stress: Cross talk with TNF-related apoptosis inducing ligand activity. *J. Biochem. Cell Biol.* **2016**, *81*, 364–374. [CrossRef] [PubMed]
3. Dimberg, L.Y.; Anderson, C.K.; Camidge, R.; Behbakhr, K.; Thorburn, A.; Ford, H.L. On the TRAIL to successful cancer therapy? Predicting and couteracting resistance against TRAIL-based tehrapeutic. *Oncogene* **2013**, *32*, 1341–1350. [CrossRef] [PubMed]
4. Allen, J.E.; Krigsfeld, G.; Mayes, P.A.; Patel, L.; Dicker, D.T.; Patel, A.S.; Dolloff, N.G.; Messaris, E.; Scata, K.A.; Wang, W.; et al. Dual inactivation of Akt and Erk by TIC10 signals Foxo3a nuclear translocation, TRAIL gene induction, and potent antitumor effects. *Sci. Transl. Med.* **2013**, *5*. [CrossRef] [PubMed]
5. Von Karstedt, S.; Montinaro, A.; Walczak, H. Exploring the TRAILs less travelled: TRAIL in cancer biology and therapy. *Nat. Rev. Cancer* **2017**, *17*, 352–366. [CrossRef] [PubMed]
6. LeBlanc, H.N.; Ashkenazi, A. Apo2L/TRAIL and its death and decoy receptors. *Cell Death Differ.* **2003**, *10*, 66–75. [CrossRef] [PubMed]
7. Sessler, T.; Healy, S.; Samali, A.; Szegezdi, E. Structural determinants of DISC function: New insights into death receptor-mediated apoptosis signalling. *Pharmacol. Ther.* **2013**, *140*, 186–199. [CrossRef]
8. Xu, J.; Zhou, J.Y.; Wei, W.Z.; Wu, G.S. Activation of the Akt survival pathway contributes to TRAIL resistance in cancer cells. *PLoS ONE* **2010**, *5*, e10226. [CrossRef]

9.   Trivedi, R.; Maurya, R.; Mishra, D.P. Medicarpin, a legume phytoalexin sensitizes myeloid leukemia cells to TRAIL-induced apoptosis through the induction of DR5 and activation of the ROS-JNK-CHOP pathway. *Cell Death Dis.* **2014**, *5*, e1465. [CrossRef]

10.   Kholoussi, N.M.; El-Nabi, S.E.; Esmaiel, N.N.; Abd El-Bary, N.M.; El-Kased, A.F. Evaluation of Bax and Bak gene mutations and expression in breast cancer. *Biomed. Res. Int.* **2014**, *2014*, 249372. [CrossRef]

11.   Amelio, I.; Gostev, M.; Knight, R.A.; Willis, A.E.; Melino, G.; Antonov, A.V. DRUG-SURV: A resource for repositioning of approved and experimental drugs in oncology based on patient survival information. *Cell Death Dis.* **2014**, *5*, e1051. [CrossRef] [PubMed]

12.   Trivedi, R.; Mishra, D.P. Trailing TRAIL resistance: Novel targets for TRAIL sensitization in cancer cells. *Front. Oncol.* **2015**, *5*, 69. [CrossRef] [PubMed]

13.   Cao, W.; Li, X.; Zheng, S.; Wong, Y.S.; Chen, T. Selenocysteins derivative overcomes TRAIL resistance in melanoma cells: Evidence for ROS-depedent synergism and sinaling crosstalk. *Oncotarget* **2014**, *5*, 7431–7445. [CrossRef] [PubMed]

14.   Yang, E.S.; Woo, S.M.; Choi, K.S.; Kwon, T. Acrolein sensitizes human renal cancer Caki cells to TRAIL-induced apoptosis via ROS-mediated up-regulation of death receptor-5 (DR5) and down-regulation of Bcl-2. *Exp. Cell Res.* **2011**, *317*, 2592–2601. [CrossRef] [PubMed]

15.   Yoon, M.J.; Kang, Y.J.; Kim, E.H.; Lee, J.A.; Lim, J.H.; Kwon, T.K.; Choi, K.S. Monensin, a polyeher ionophore antibiotic, overcomes TRAIL resistance in glioma cells via endoplasmic reticulum stress, DR5 upregulation and c-FLIP downregulation. *Carcinogenesis* **2013**, *34*, 1918–1928. [CrossRef] [PubMed]

16.   Yi, L.; Zongyuan, Y.; Cheng, G.; Lingyun, Z.; Guilian, Y.; Wei, G. Quercetin enhances apoptotic effect of tumor necrosis factor-related apoptosis-inducing ligand (TRAIL) in ovarian cancer cells through reactive oxygen species (ROS) mediated CCAAT enhancerbinding protein homologous protein (CHOP)-death receptor 5 pathway. *Cancer Sci.* **2014**, *105*, 520–527. [CrossRef]

17.   Ahn, H.J.; Kim, K.I.; Hoan, N.N.; Kim, C.H.; Moon, E.; Choi, K.S.; Yang, S.S.; Lee, J.S. Targeting cancer cells with reactive oxygen and nitrogen species generated by atmospheric-pressure air plasma. *PLoS ONE* **2014**, *9*, e86173. [CrossRef]

18.   Kang, S.U.; Cho, J.H.; Chang, J.W.; Shin, Y.S.; Kim, K.I.; Park, J.K.; Yang, S.S.; Lee, J.S.; Moon, E.; Lee, K. Nonthermal plasma induces head and neck cancer cell death: The potential involvement of mitogen-activated protein kinase-dependent mitochondrial reactive oxygen species. *Cell Death Dis.* **2014**, *5*, e1056. [CrossRef]

19.   Ahn, H.J.; Kim, K.I.; Kim, G.; Moon, E.; Yang, S.S.; Lee, J.S. Atmospheric-pressure plasma jet induces apoptosis involving mitochondria via generation of free radicals. *PLoS ONE* **2011**, *6*, e28154. [CrossRef]

20.   Kim, K.I.; Ahn, H.J.; Lee, J.H.; Yang, S.S.; Lee, J.S. Cellular membrane collapse by atmospheric-pressure plasma jet. *Appl. Phys. Lett.* **2014**, *104*, 13701. [CrossRef]

21.   Nguyen, N.H.; Park, H.J.; Yang, S.S.; Choi, K.S.; Lee, J.S. Anti-cancer efficacy of nonthermal plasma dissolved in a liquid, lquid plasma in heterogeneous cancer cells. *Sci. Rep.* **2016**, *6*, 29020. [CrossRef] [PubMed]

22.   Nguyen, N.H.; Park, H.J.; Hwang, S.Y.; Lee, J.S.; Yang, S.S. Anticancer efficacy of long-term stored plasma-activated medium. *Appl. Sci.* **2019**, *9*, 801. [CrossRef]

23.   Kim, E.H.; Yoon, M.J.; Kim, S.U.; Kwon, T.K.; Shon, S.; Choi, K.S. Arsenic trioxide sensitizes human glioma cells, but not normal astrocytes, to TRAIL-induced apoptosis via CCAAT/enhancer-binding protein homologous protein-dependent DR5 up-regulation. *Cancer Res.* **2008**, *68*, 266–275. [CrossRef] [PubMed]

24.   Su, R.Y.; Chi, K.H.; Huang, D.Y.; Tai, M.H.; Lin, W.W. 15-deoxy-Delta12,14-prostaglandin J2 up-regulates death receptor 5 gene expression in HCT116 cells: Involvement of reactive oxygen species and C/EBP homologous transcription factor gene transcription. *Mol. Cancer Ther.* **2008**, *7*, 3429–3440. [CrossRef]

25.   Yamaghci, H.; Wang, H.G. CHOP is involved in endoplasmic reticulum stress-induced apoptosis by enhancing DR5 expression in human carcinoma cells. *J. Biol. Chem.* **2004**, *279*, 45495–45502. [CrossRef]

26.   Grassme, H.; Schwarz, H.; Gulbins, E. Molecular mechanisms of ceramide-mediated CD95 clustering. *Biochem. Biophys. Res. Commun.* **2001**, *284*, 1016–1030. [CrossRef]

27.   Pang, L.; Fu, T.M.; Zhao, W.; Zhao, L.; Chen, W.; Qiu, C.; Liu, W.; Liu, Z.; Piai, A.; Fu, Q.; et al. Higher-order clustering of the transmembrane anchor of DR5 drives signaling. *Cell* **2019**, *176*, 1477–1489. [CrossRef]

28.   Lu, M.; Lawrence, D.A.; Marsters, S.; Acosta-Alvear, D.; Kimmig, P.; Mendez, A.S.; Paton, A.W.; Paton, J.C.; Walter, P.; Ashkenazi, A. Opposing unfolded-protein-response signals converge on death receptor 5 to control apoptosis. *Science* **2014**, *345*, 98–101. [CrossRef]

29.	Zhang, J.G.; Wang, J.J.; Zhao, F.; Liu, Q.; Jiang, K.; Yang, G.H. MicroRNA-21 (miR-21) represses tumor suppressor PTEN and promotes growth and invasion in non-small cell lung cancer (NSCLC). *Clin. Chim. Acta* **2010**, *411*, 846–852. [CrossRef]

30.	Ma, J.; Liu, J.; Wang, Z.; Gu, X.; Fan, Y.; Zhang, W.; Xu, L.; Zhang, J.; Cai, D. NF-κB-depedent microRNA-425 upregulation promotes gastric cancer cell growth by targeting PTEN upon IL-1β induction. *Mol. Cancer* **2014**, *13*, 40. [CrossRef]

31.	Li, C.; Song, L.; Zhang, Z.; Bai, X.X.; Cui, M.F.; Ma, L.J. MicroRNA-21 promotes TGF-β1 induced epithelial-mesenchymal transition in gastric cancer through upregulation PTEN expression. *Oncotarget* **2016**, *7*, 66989–67003. [CrossRef] [PubMed]

32.	Spencer, S.L.; Gaudet, S.; Albeck, J.G.; Burke, J.M.; Sorger, P.K. Non-genetic origins of cell-to-cell variability in TRAIL-induced apoptosis. *Nature* **2009**, *459*, 428–432. [CrossRef] [PubMed]

33.	Garofalo, M.; Leva, G.D.; Romano, G.; Nuovo, G.; Shu, S.S.; Ngankeu, A.; Taccioli, C.; Pichiorri, F.; Alder, H.; Secchiero, P.; et al. miR-221&222 regulate TRAIL resistance and enhance tumorigenicity through PTEN and TIMP3 downregulation. *Cancer Cell* **2009**, *16*, 498–509. [CrossRef] [PubMed]

34.	Suzuki-Karasaki, M.; Ochiai, T.; Suzuki-Karasaki, Y. Crosstalk between mitochondrial ROS and depolarization in the potentiation of TRAIL-induced apoptosis in human tumor cells. *Int. J. Oncol.* **2014**, *44*, 616–628. [CrossRef]

35.	Bortner, C.D.; Gomez-Angelats, M.; Cidlowski, J.A. Plasma membrane depolarization without repolarization is an early molecular event in anti-Fas-induced apoptosis. *J. Biol. Chem.* **2001**, *276*, 4304–4314. [CrossRef] [PubMed]

36.	Yin, W.; Li, X.; Feng, S.; Cheng, W.; Tang, B.; Shi, Y.L.; Hua, Z.C. Plasma membrane depolarization and Na,K-ATPase impairment induced by mitochondrial toxins augment leukemia cell apoptosis via a novel mitochondrial amplification mechanism. *Biochem. Pharmacol.* **2009**, *78*, 191–202. [CrossRef] [PubMed]

37.	Yin, W.; Cheng, W.; Shen, W.; Shu, L.; Zhao, J.; Zhang, J.; Hua, Z.C. Impairment of Na$^+$,K$^+$-ATPase in CD95(APO-1)-induced human T-cell leukemia cell apoptosis mediated by glutathione depletion and generation of hydrogen peroxide. *Leukemia* **2007**, *21*, 1669–1678. [CrossRef]

38.	La Rovere, R.M.; Roest, G.; Bultynchk, G.; Parys, J.B. Intracellular Ca^{2+} signaling and Ca^{2+} microdomains in the control of cell survival, apoptosis and autophagy. *Cell Calcium.* **2016**, *60*, 74–87. [CrossRef]

39.	Gorlach, A.; Bertram, K.; Hudecova, S.; Krizanova, O. Calcium and ROS: A mutual interplay. *Redox Biol.* **2015**, *6*, 260–271. [CrossRef]

40.	Kroemer, G.; Dallaporta, B.; Resche-Rigon, M. The mitochondrial death/life regulator in apoptosis and necrosis. *Annu. Rev. Physiol.* **1998**, *60*, 619–642. [CrossRef]

41.	Suzuki, Y.; Inoue, T.; Murai, M.; Suzuki-Karasaki, M.; Ochiai, T.; Ra, C. Depolarization potentiates TRAIL-induced apoptosis in human melanoma cells: Role for ATP-sensitive K$^+$ channels and endoplasmic reticulum stress. *Int. J. Oncol.* **2012**, *41*, 465–475. [CrossRef]

42.	Testa, U. Apoptotic mechanisms in the control of erythropoiesis. *Leukemia* **2004**, *18*, 1176–1199. [CrossRef] [PubMed]

43.	Kischkel, F.C.; Lawrence, D.A.; Chuntharapai, A.; Schow, P.; Kim, K.J.; Ashkenazi, A. Apo2L/TRAIL-dependent recruitment of endogenous FADD and caspase-8 to death receptors 4 and 5. *Immunity* **2000**, *12*, 611–620. [CrossRef]

44.	Sprick, M.R.; Weigand, M.A.; Rieser, E.; Rauch, C.T.; Juo, P.; Blenis, J.; Krammer, P.H.; Walczak, H. FADD/MORT1 and caspase-8 are recruited to TRAIL receptors 1 and 2 and are essential for apoptosis mediated by TRAIL receptor 2. *Immunity* **2000**, *12*, 599–609. [CrossRef]

45.	Su, Z.; Yang, Z.; Xie, L.; DeWitt, J.P.; Chen, Y. Cancer therapy in the necroptosis era. *Cell Death Differ.* **2016**, *23*, 748–756. [CrossRef] [PubMed]

46.	Kim, H.J.; Hawke, N.; Baldwin, A.S. NF-kappaB and IKK as therapeutic targets in cancer. *Cell Death Differ.* **2006**, *13*, 738–747. [CrossRef] [PubMed]

47.	Perkins, N.D. Post-translational modifications regulating the activity and function of the nuclear factor kappa B pathway. *Oncogene* **2006**, *25*, 6717–6730. [CrossRef]

48.	Ryo, A.; Suizu, F.; Yoshida, Y.; Perrem, K.; Liou, Y.C.; Wulf, G.; Rottapel, R.; Yamaoka, S.; Lu, K.P. Regulation of NF-kappaB signaling by Pin1-dependent prolyl isomerization and ubiquitin-mediated proteolysis of p65/RelA. *Mol. Cell* **2003**, *12*, 1413–1426. [CrossRef]

49. Joo, J.H.; Jetten, A.M. NF-kappaB-dependent transcriptional activation in lung carcinoma cells by farnesol involves p65/RelA(Ser276) phosphorylation via the MEK-MSK1 signaling pathway. *J. Biol. Chem.* **2008**, *283*, 16391–16399. [CrossRef]

50. Nowak, D.E.; Tian, B.; Jamaluddin, M.; Boldogh, I.; Vergara, L.A.; Choudhary, S.; Brasier, A.R. RelA Ser276 phosphorylation is required for activation of a subset of NF-kappaB-dependent genes by recruiting cyclin-dependent kinase 9/cyclin T1 complexes. *Mol. Cell Biol.* **2008**, *28*, 3623–3638. [CrossRef]

51. Hochrainer, K.; Racchumi, G.; Anrather, J. Site-specific phosphorylation of the p65 protein subunit mediates selective gene expression by differential NF-κB and RNA polymerase II promoter recruitment. *J. Biol. Chem.* **2013**, *288*, 285–293. [CrossRef]

52. Huang, C.Y.; Fong, Y.C.; Lee, C.Y.; Chen, M.Y.; Tsai, H.C.; Hsu, H.C.; Tang, C.H. CCL5 increases lung cancer migration via PI3K, Akt and NF-kappaB pathways. *Biochem. Pharmacol.* **2009**, *77*, 794–803. [CrossRef] [PubMed]

53. Msaki, A.; Sanchez, A.M.; Koh, L.F.; Barre, B.; Rocha, S.; Perkins, N.D.; Johnson, R.F. The role of RelA (p65) threonine 505 phosphorylation in the regulation of cell growth, survival and migration. *Mol. Biol. Cell* **2011**, *22*, 3032–3040. [CrossRef] [PubMed]

54. Ahir, M.; Bhattacharya, S.; Kamakar, S.; Mukhopadhyay, A.; Mukherjee, S.; Ghosh, S.; Chattopadhyay, S.; Patra, P.; Adhikary, A. Tailoerd-CuO-nanowire decorated with folic acid mediated coupling of the mitochondrial-ROS generation and miR425-PTEN axis in furnishing potent anti-cancer activity in human tripe negative breast carcinoma cells. *Biomaterials* **2016**, *76*, 115–132. [CrossRef] [PubMed]

55. Prasad, S.; Yadav, V.R.; Ravindran, J.; Aggarwal, B.B. ROS and CHOP are critical for dibenzylideneacetone to sensitize tumor cells to TRAIL through induction of death receptors and downregulation of cell survival proteins. *Cancer Res.* **2011**, *71*, 538–549. [CrossRef] [PubMed]

56. Sung, B.; Prasad, S.; Ravindran, J.; Yadav, V.R.; Aggarwal, B.B. Capsazepine, a TRPV1 antagonist, sensitizes colorectal cancer cells to apoptosis by TRAIL through ROS-JNK-CHOP-mediated upregulation of death receptors. *Free Radic. Biol. Med.* **2012**, *53*, 1977–1987. [CrossRef] [PubMed]

57. Sun, J.; Zhang, X.; Broderick, M.; Fein, H. Measurement of nitric oxide production in biological systems by using Griess Reaction Assay. *Sensors* **2003**, *3*, 276–284. [CrossRef]

International Journal of
Molecular Sciences

Article

The Influence of Cold Atmospheric Pressure Plasma-Treated Media on the Cell Viability, Motility, and Induction of Apoptosis in Human Non-Metastatic (MCF7) and Metastatic (MDA-MB-231) Breast Cancer Cell Lines

Dominik Terefinko [1,2,*], Anna Dzimitrowicz [1,*], Aleksandra Bielawska-Pohl [2], Aleksandra Klimczak [2], Pawel Pohl [1] and Piotr Jamroz [1]

[1] Department of Analytical Chemistry and Chemical Metallurgy, Faculty of Chemistry, Wroclaw University of Science and Technology, Wybrzeze St. Wyspianskiego 27, 50-370 Wroclaw, Poland; pawel.pohl@pwr.edu.pl (P.P.); piotr.jamroz@pwr.edu.pl (P.J.)
[2] Laboratory of Biology of Stem and Neoplastic Cells, Hirszfeld Institute of Immunology and Experimental Therapy, Polish Academy of Sciences, Weigla 12, 53-114 Wroclaw, Poland; aleksandra.bielawska-pohl@hirszfeld.pl (A.B.-P.); aleksandra.klimczak@hirszfeld.pl (A.K.)
* Correspondence: dominik.terefinko@pwr.edu.pl (D.T.); anna.dzimitrowicz@pwr.edu.pl (A.D.)

check for updates

Citation: Terefinko, D.; Dzimitrowicz, A.; Bielawska-Pohl, A.; Klimczak, A.; Pohl, P.; Jamroz, P. The Influence of Cold Atmospheric Pressure Plasma-Treated Media on the Cell Viability, Motility, and Induction of Apoptosis in Human Non-Metastatic (MCF7) and Metastatic (MDA-MB-231) Breast Cancer Cell Lines. *Int. J. Mol. Sci.* **2021**, *22*, 3855. https://doi.org/10.3390/ijms22083855

Academic Editors: Akikazu Sakudo and Yoshihito Yagyu

Received: 3 March 2021
Accepted: 5 April 2021
Published: 8 April 2021

Publisher's Note: MDPI stays neutral with regard to jurisdictional claims in published maps and institutional affiliations.

Abstract: Breast cancer remains the most common type of cancer, occurring in middle-aged women, and often leads to patients' death. In this work, we applied a cold atmospheric pressure plasma (CAPP)-based reaction-discharge system, one that is unique in its class, for the production of CAPP-activated media (DMEM and Opti-MEM); it is intended for further uses in breast cancer treatment. To reach this aim, different volumes of DMEM or Opti-MEM were treated by CAPP. Prepared media were exposed to the CAPP treatment at seven different time intervals and examined in respect of their impact on cell viability and motility, and the induction of the apoptosis in human non-metastatic (MCF7) and metastatic (MDA-MB-231) breast cancer cell lines. As a control, the influence of CAPP-activated media on the viability and motility, and the type of the cell death of the non-cancerous human normal MCF10A cell line, was estimated. Additionally, qualitative and quantitative analyses of the reactive oxygen and nitrogen species (RONS), generated during the CAPP operation in contact with analyzed media, were performed. Based on the conducted research, it was found that 180 s (media activation time by CAPP) should be considered as the minimal toxic dose, which significantly decreases the cell viability and the migration of MDA-MB-231 cells, and also disturbs life processes of MCF7 cells. Finally, CAPP-activated media led to the apoptosis of analyzed cell lines, especially of the metastatic MDA-MB-231 cell line. Therefore, the application of the CAPP system may be potentially applied as a therapeutic strategy for the management of highly metastatic human breast cancer.

Keywords: non-thermal plasma; reactive oxygen and nitrogen species; biological activity; breast cancer

1. Introduction

The incidence of cancer in the world population is growing. In the group of females, breast cancer is one of the most common types of cancers, following lung cancer [1,2]. According to the National Breast Cancer Foundation, 276,480 new cases of invasive breast cancer are predicted to be diagnosed in 2020 [2]. Additionally, a highly metastatic character of breast cancer decreases the survival rate and the convalescence perspective for diagnosed patients, and is often associated with a lack of effective treatment modalities. Nowadays, several therapies, including modern chemotherapy, radiotherapy, or immunotherapy, are used to treat breast cancer, but they are often associated with pain, infections, amputations, deformations, and other side effects [3]. For that reason, there is a high need for new, fast, and non-invasive therapies used for breast cancer treatment.

25

An encouraging and quite new approach to breast cancer treatment is a targeted drug-based therapy [4]. This therapy can be supported by gold nanoparticles (AuNPs), used as carriers for specific drugs [5,6]. Unique optical and structural properties of AuNPs in addition to their high surface-to-volume ratio enable to attach various inhibitors or proteins to their surface and carry them to a defined target in the body [6]. As was found by Devi et al. [6], a target-based therapy is characterized by very promising rates of survival and low side effects after its application [6]. Another method used for reducing breast cancer volume is hyperthermia [6]. In this process, the temperature of tissues is locally raised by an artificial heat source, leading to damage of cancerous cells [7,8]. A very promising strategy for making the hyperthermia process more effective is the application of magnetic nanoparticles [7,8]. In magnetic hyperthermia, a magnetic nanofluid is transported via blood to a targeted tissue, and then the local heat is induced in a radiofrequency magnetic field [8]. From a long-term perspective, such application of magnetic nanoparticles inside tissues can, however, be hazardous because of the absorption of nanoparticles into the intracellular environment, especially in the case of lymphatic tissues of the intestine, which lead to the generation of the oxidative stress, cell destruction, and genotoxicity [9]. A remedy to this drawback could be the utilization of cold atmospheric pressure plasmas (CAPPs), those being a new, very promising and effective treatment modality in breast cancer therapy [10–22].

Significant biological effects of CAPP were identified at the end of the last decade as a result of reactive oxygen and nitrogen species (RONS) production. Because hydrogen peroxide (H_2O_2), nitric oxide (NO), nitrate (NO_3^-), nitrite (NO_2^-), ammonia (NH_4^+) ions, hydroxyl ($OH^\bullet$), and hydrogen ($H^\bullet$) radicals are produced during the CAPP operation with water or cell culture media [10,23], the CAPP-based approach is used in a wide range of applications such as wound healing [24], inactivation of parasites and foreign organisms [25], dentistry [26], seed germination [27], nanoparticles synthesis [28], blood coagulation [29], and so on. Concerning breast cancer therapy, several reaction-discharge systems have been developed and studied, including dielectric barrier discharges (DBDs) [12–15,17,19–21], atmospheric pressure plasma jets (APPJs) [10,16,18], and radio-frequency discharge [22]. CAPP sources applied in these systems have been used for the direct [10,11,13,14,16–22] or indirect [12,14,15,20] treatment of cell lines. In the first case, selected cell lines, i.e., MDA-MB-231 [10–12,14,16,20,22], MCF-7 [10,13,15,17,18,21], SKBR3 [15], AMN3 [21], AMJ13 [19,21], and MCF10A [11,18], were directly treated by proper CAPP sources for specific times. Next, the effect of such direct CAPP treatment on the cell lines' viability was assessed; however, it exhibited some limitations, mostly related to its insufficient efficiency [10,11,13,14,16–22]. Additionally, in some cases, the application of the direct CAPP treatment on cell lines was restricted due to problems with the transportation of the whole reaction-discharge system to a place of cell line irradiation. In most cases, the CAPP-based reaction-discharge systems were expensive and not very mobile.

A remedy to these drawbacks was the indirect CAPP-based method used for water [12] or cell culture media [14,15,20] activation. The resultant CAPP-activated liquids (DMEM, L-15/DMEM, DMEM/F-12, RPMI 1640, serum free medium) were used then in the MDA-MB-231 [10–12,14,16,20,22], MCF-7 [10,13,15,17,18,20,21], MCF10A [11,15,18,20], and AMJ13 [19] and SKBR3 [15] cell lines treatment. Indeed, they could be easily stored and transported for further application to cancerous cell lines, causing their death due to an efficient transfer of RONS [14]. In addition, the selectivity of the CAPP-activated media was that the cancerous cell lines were more vulnerable to the oxidative stress caused by RONS than the normal cell lines, which were not so sensitive to this oxidative stress [30]. What is more, in the case of the application of the CAPP-activated media towards the cancerous cell lines, there was no physical effect on the cell lines associated with the damages caused by a gas flux (CAPP discharge gas) and the electromagnetic field (from a CAPP source) [31,32].

In this study, we examined the influence of two CAPP-treated media, including DMEM and Opti-MEM, on the biological activity of three cell lines: The highly metastatic human breast cancer MDA-MB-231 cell line, the non-invasive human breast cancer MCF7 cell line, and the non-cancerous human normal MCF10A cell line. To reach this aim, we developed and used a special portable CAPP-based reaction-discharge system, which had already been successfully applied for the activation of normal human skin cell lines [33]. To the best of our knowledge, this was the first mobile and inexpensive in-operation CAPP-based reaction-discharge system in which a DBD plasma jet was applied for human cell line treatment. A broad spectrum of confirmed utilizations of the studied CAPP-based reaction-discharge system deserves special attention. In the previous study, it was described that the direct CAPP-treatment of human normal cell lines results in the enhancement of their proliferation ability, supporting wound healing processes. From this perspective, we suggest that the modulation of the CAPP-treatment regimen, as well as the treatment time and the targeted liquid environment, may influence the cancer biological response. Furthermore, we have studied the scavenger effect of the fetal bovine serum (FBS) addition to the CAPP-treated media in order to assess its anticancer activity towards the analyzed human cell lines. We experimentally inspected how the quantitative and qualitative composition of the produced reactive oxygen species can provoke a different biological response. Finally, we performed a detailed evaluation of processes and interactions occurring in the CAPP-treated media interfaces to better assess the composition of major active components and correlate them with biological studies.

2. Results and Discussion

2.1. Effect of the CAPP-Activated Media on the Defined Biological Activities

The CAPP-activated media, including DMEM and Opti-MEM, were used for the assessment of the selected biological functions of the target breast cancer cell lines such as cell viability, cell migration, and cell death type (apoptosis/necrosis).

2.1.1. Effect of the CAPP-Activated Media on Cell Viability

Cell viability after the application of the CAPP-activated medium was assessed using an MTT assay. The minimal toxic dose for the normal cell line (MCF10A) was estimated as the medium treatment time by CAPP, which resulted in decreasing the cell line viability. The disorders in cell viability of the MCF10A cells were noted when the FBS was present in both culture media during the CAPP-activation process (Figure 1a,b). On the other hand, it was found that the MCF10A cells showed high resistance to the CAPP-activated media, where the FBS was introduced after the preparation step (Figure 2a,b). Similar observations were found by other scientists, but in these cases the normal human cell lines (MCF10A) were directly treated by CAPP [15,18,20]. A preliminary screening of the time of the CAPP exposition on the cell culture media showed that the application of these CAPP-activated media (CAPP treatment time within 45–150 s) on the human breast cancer cell lines such as MDA-MB-231 and MCF7 did not affect their mitochondrial activity and proliferation rate (data not shown). Based on these observations, the analyzed medium was exposed to CAPP for 150 s, 180 s, 210 s, or 240 s (Figure 1). The selected CAPP-activated media (groups V and VI) exhibited a significant reduction in the mitochondrial activity of the MDA-MB-231 and MCF7 cell lines (* $p < 0.01$; ** $p < 0.001$; *** $p < 0.0004$, Figure 1). In more detail, the CAPP-activated Opti-MEM significantly affected the cell viability of the MDA-MB-231 cell line incubated for one day, when the longest CAPP-treatment time (240 s—* $p < 0.01$) of the analyzed medium was used. During two days of incubation of the MDA-MB-231 cell line in the CAPP-activated Opti-MEM, it exhibited a great impact on a decrease of the cell viability, especially in the case of the CAPP treatment times of 180 s and 240 s (** $p < 0.001$, *** $p < 0.0004$, respectively). Nevertheless, for the MCF7 cell line, a significant reduction in the cell viability following the one-day incubation was observed for the CAPP treatment times of 180 s and 240 s (* $p < 0.01$, ** $p < 0.01$, respectively). Similar effects were observed following the two-day incubation, leading to a more prominent cell viability reduction for the CAPP treatment times of 150 s, 180 s, 210 s, and 240 s (** $p < 0.001$, * $p < 0.01$, * $p < 0.01$, **** $p < 0.0004$, respectively). Based on these results, it was supposed that the reduction of the applied medium volume due to its exposure to CAPP might increase the concentration of RONS [34] and induce a biological impact on the analyzed cell lines (Groups I, II, V, and VI). Additionally, by increasing the organic content in the cell culture medium through the use of the Opti-MEM CAPP-activated medium, its toxic effect towards both human breast cancer cell lines was noticed. Accordingly, the exposure of a proper medium to the CAPP source for 180 s resulted in producing the CAPP-activated medium recognized as the one with a minimal toxic dose.

The anti-proliferative properties of the CAPP-activated media are currently also the focus of other research groups, and conclusions similar to our results were found in references [10,15,18–20]. Bekeschus et al. [10] cultured the MDA-MB-231 and SW 480 cancer cell lines in the DMEM (supplemented with 10% FBS) and directly them exposed to CAPP, which significantly reduced the viability of those cancer cells [10]. Xiang et al. showed that the media irradiated by CAPP (DMEM and DMEM/F-12) exhibited a selective character towards the highly metastatic human breast cancer cell lines [20]. Moreover, it was established that the CAPP-activated media slightly increase the cell viability of the non-cancerous MCF10A and non-metastatic MCF7 cell lines [20]. Results opposite to those reported in [20] were described by Mokhtari et al. [15]. In this case, the biological activity of the CAPP-activated DMEM was demonstrated towards different human cancer cell lines, including the human breast cancer cell lines MCF7 and SKBR3, the human lung adenocarcinoma cell line A-549, the human colon carcinoma cell line SW742, the human pancreatic cancer cell line ASPC-1, and the human primary osteogenic sarcoma cell line G-292. The biological models of the non-cancerous cell lines were also prepared using the human gland cells MCF10A and the skin fibroblast cells FMGB-1 [15]. It was concluded that the obtained CAPP-treated medium showed a significant reduction in the MCF7 cell lines cell viability, while the normal MCF10A cell lines remained untouched by CAPP for a shorter irradiation time [15]. Comparable results for these cell lines (MCF7 and MCF10A) were also presented in the research on the reduction of the cell viability following the direct CAPP treatment. In more detail, the treatment time of 60 s led to a significant decrease in the proliferation rate of the MCF7 cell line, while the MCF10A cells remained uninjured [18].

Because the FBS acts as an $OH^\bullet$ and H_2O_2 scavenger [35], another set of experiments was performed, where the FBS was immediately added to the proper CAPP-activated medium. The proliferation activity of the analyzed cancer cell lines (MDA-MB-231 and MCF7), incubated with the CAPP-activated medium (CAPP treatment times of 150 s, 180 s, 210 s, or 240 s), to which the FBS was added after the CAPP treatment (Groups III, IV, VII, VIII), is given in Figure 2. In the graph "c" (Figure 2), a comparison for the MDA-MB-231 cell line incubated in the CAPP-activated DMEM (3.0 mL, without the 3% FBS) is shown. A significant decrease in cell viability for the treatment times of 180 s and 240 s (day 1) (** $p < 0.0014$) and for the treatment times of 180 s, 210 s, and 240 s (day 2) was observed (* $p < 0.013$; * $p < 0.013$; ** $p < 0.0014$). A comparable observation was made for the CAPP-activated DMEM (1.5 mL, without the 3% FBS), as is shown in Figure 2d. Moreover, the most prominent decrease in the proliferation rate was for the 180 s and 240 s CAPP-activation times (* $p < 0.013$; ** $p < 0.0014$, respectively) of the analyzed medium. The presence of the FBS in the DMEM during the CAPP activation did not affect the cell viability of the MDA-MB-231 cell line. As can be seen in Figure 2e,f, the CAPP-activated Opti-MEM (3.0 mL or 1.5 mL, with the absence of 3% FBS) significantly affected the MDA-MB-231 cell line viability in the one-day incubation for the CAPP-activation times of 180 s and 240 s (** $p < 0.0014$; *** $p < 0.0002$). For the two-day incubation of the MDA-MB-231 cell line with the CAPP-activated Opti-MEM (3.0 mL or 1.5 mL), the respective cell viability was reduced for the medium for all the CAPP-activation times: 150 s, 180 s, 210 s, and 240 s (* $p < 0.013$; ** $p < 0.0014$; *** $p < 0.0002$). Finally, the use of the CAPP-activated DMEM (see Figure 2g,h) resulted in no difference in the case of the reduction in the cell viability of the MCF7 cell line. However, the preparation of the CAPP-activated Opti-MEM (3.0 mL) led to a decrease in the proliferation rate when the 3% FBS was not added during the CAPP-activated Opti-MEM preparation.

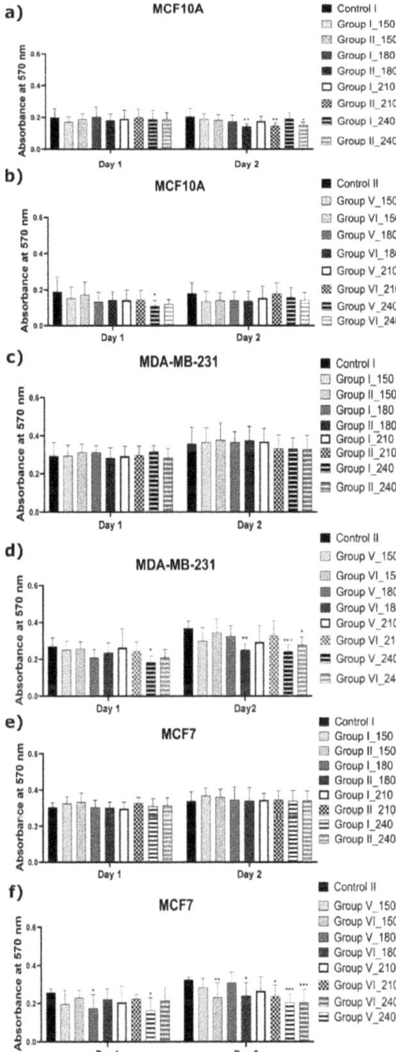

Figure 1. The MTT assay for the proliferation of human metastatic breast cancer (MDA-MB-231), human non-metastatic breast cancer (MCF7), and non-cancerous human normal (MCF10A) cell lines, incubated from 0 to 2 days in the cold atmospheric pressure plasma (CAPP)-activated media (exposure times of 150 s, 180 s, 210 s, or 240 s). (**a**,**c**,**e**) The MCF10A, MDA-MB-231, and MCF7 cell lines incubated in 3.0 mL (Group I) or in 1.5 mL (Group II) of the CAPP-activated DMEM. (**b**,**d**,**f**) The MCF10A, MDA-MB-231, and MCF7 cell lines incubated in 3.0 mL (Group V) or 1.5 mL (Group VI) of the CAPP-activated Opti-MEM. In both cases, fetal bovine serum (FBS) was added to the analyzed medium before the CAPP treatment. As a control, cells untreated by a CAPP-activated medium were used (Control I as the untreated DMEM and Control II as the untreated Opti-MEM). Data are means ± SD values for three independent experiments conducted in triplicates. The statistical calculation was performed by a comparison of all investigated groups versus the control using one-way analysis of variance (ANOVA) with the Dunnett's post hoc test. (* $p < 0.01$; ** $p < 0.001$; *** $p < 0.0004$).

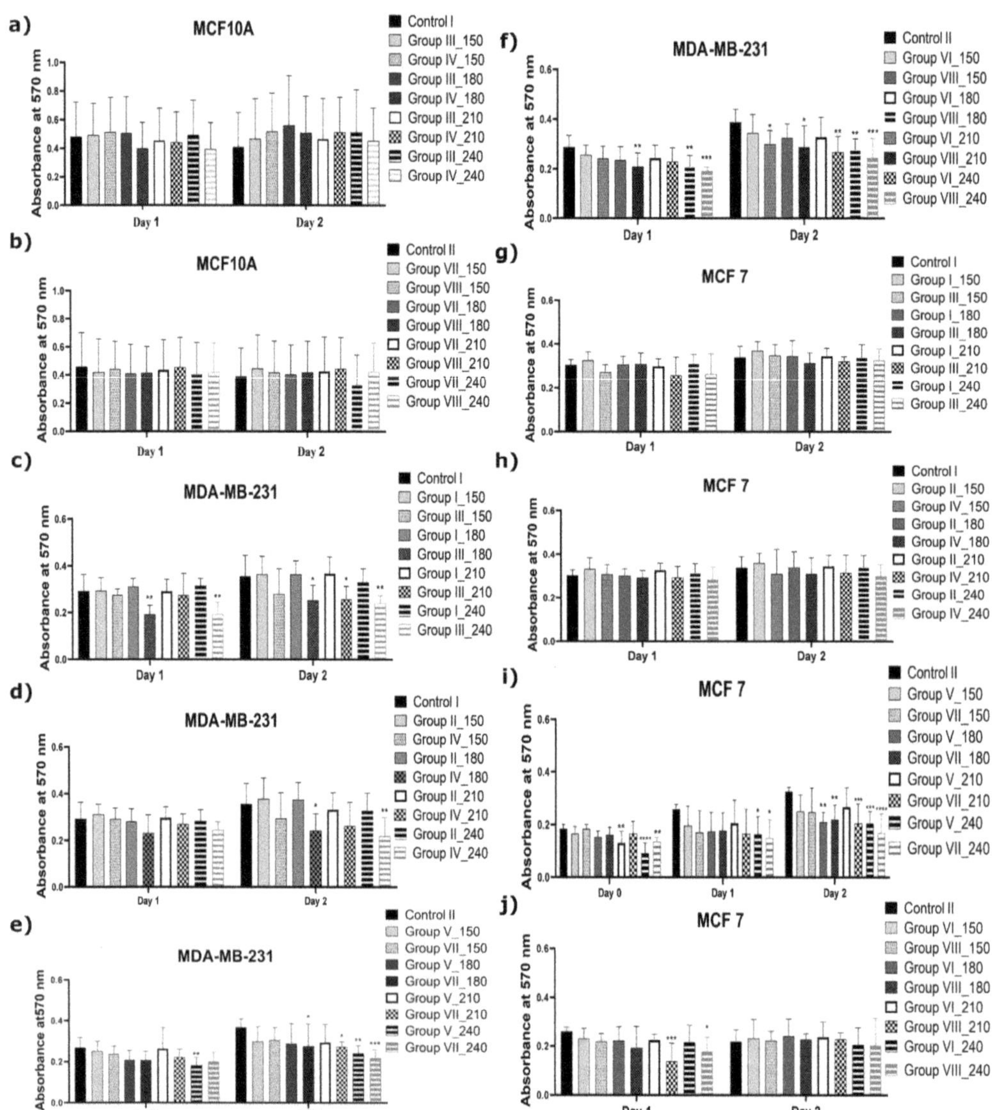

Figure 2. The comparison of the impact of the addition of FBS after the production of the CAPP-activated medium on the proliferative ability of human metastatic breast cancer (MDA-MB-231), human non-metastatic breast cancer (MCF7), and non-cancerous human normal (MCF10A) cell lines. The cell lines were incubated in the CAPP-activated medium obtained by CAPP treatment for 150 s, 180 s, 210 s, or 240 s. (**a,c,g**) The MCF10A, MDA-MB-231, and MCF7 cell lines treated with 3.0 mL of the CAPP-activated DMEM with the FBS (Group I) or without the FBS (Group III); (**a,d,h**) the MCF10A, MDA-MB-231, and MCF7 cell lines treated with 1.5 mL of the CAPP-activated DMEM with the FBS (Group II) or without the FBS (Group IV); (**b,e,i**) the MDA-MB-231 and MCF7 cell lines treated with 3.0 mL of the CAPP-activated Opti-MEM with the FBS (Group V) or without the FBS (Group VII); (**b,f,j**) the MDA-MB-231 and MCF7 cell lines treated with 1.5 mL of the CAPP-activated Opti-MEM with the FBS (Group VI) or without the FBS (Group VIII). A culture medium that was not treated with CAPP was used as control (Controls I and II). Data are means ± SD values for three independent experiments conducted in triplicates. The statistical calculation was performed by the comparison of all the investigated groups versus the control using one-way ANOVA with the Dunnett's post hoc test (* $p < 0.013$; ** $p < 0.0014$; *** $p < 0.0002$; **** $p < 0.0001$).

Concerning the role of the addition of FBS to the CAPP-activated medium, it was established that the produced reaction mixtures significantly enhanced the toxic effect against the MDA-MB-231 as well as MCF7 cell lines. In this case, the most prominent decrease in the proliferation rate of the metastatic cancer cell line MDA-MB-231 was observed. Comparable results were described by Rodder et al. [35], who found that an increase of FBS concentration in the analyzed solutions during the CAPP treatment led to a decrease in H_2O_2 concentration, minimizing the death cell population [35].

2.1.2. Study of the Inhibition Effect of Two Different CAPP-Activated Media on the Migration Ability

The estimation of cell migration after the application of a proper CAPP-activated medium was assessed using the scratch test. For this test, 1.5 mL of the DMEM or the Opti-MEM, activated by CAPP for 180 s or 210 s, were selected. The times of 180 s and 210 s were chosen because they provided a minimum toxic dose. After the CAPP-activated medium production, the FBS was added to reach its final concentration of 3%. The images of the prepared scratches were acquired within 30 h of cell line treatment by the proper CAPP-activated medium.

As can be seen from Figure 3, the scratch closure calculations of the normal human breast cell line (MCF10A) showed that the incubation in both types of CAPP-activated medium had no significant impact on cell migration, which was in line with previously presented results for the proliferation assay (see Section 2.1.1 for more details). In the case of the non-metastatic breast cancer cell line (MCF7), some significant decreases in the scratch closure area were observed for the CAPP-activated medium DMEM, following 8 h of experiments, for both selected CAPP treatment times (180 s or 210 s in case of DMEM, Figure 3c) (** $p < 0.002$, ** $p < 0.002$, respectively). For a longer treatment time (210 s) of the DMEM, a similar inhibition of the MCF7 cell line motility was found (Figure 3) (** $p < 0.002$). The experiments were conducted for 24 h, because after these CAPP treatment times, some insignificant changes in the relative wound closure area (RWC) were detected in two kinds of prepared CAPP-activated media (Figure 3). Concerning the MDA-MB-231 cell line, the most prominent reduction in cell motility was observed when the CAPP-activated DMEM was used for a shorter CAPP-activation time. In this case, the DMEM was activated by CAPP for 180 s or 210 s, and the scratch closure area was observed for 30 h (after 24 h, strong inhibition of the relative scratch closure area was detected, see Figure 3b) (**** $p < 0.0001$). Similar results were established for the Opti-MEM activated by CAPP for 180 s or 210 s (Figure 3b) (** $p < 0.002$; *** $p < 0.001$). Intriguingly, for the DMEM activated by CAPP for 180 s, the cell migration was inhibited after 30 h (* $p < 0.016$), while for the Opti-MEM activated via CAPP for 210 s; the cell migration was significantly inhibited (* $p < 0.016$). Similar results were reported by Xiang et al. [20], who studied the effect of the CAPP-activated medium on the migration ability of the MDA-MB-231 cells. Our results were also in good correlation with those reported by other research groups [22,36].

2.1.3. Induction of the Programmed Cell Death

In order to assess the cell death type after the application of the CAPP-activated proper medium, the Annexin V/PI test was used for the identification of the population of the apoptotic and necrotic cells (Figure 4).

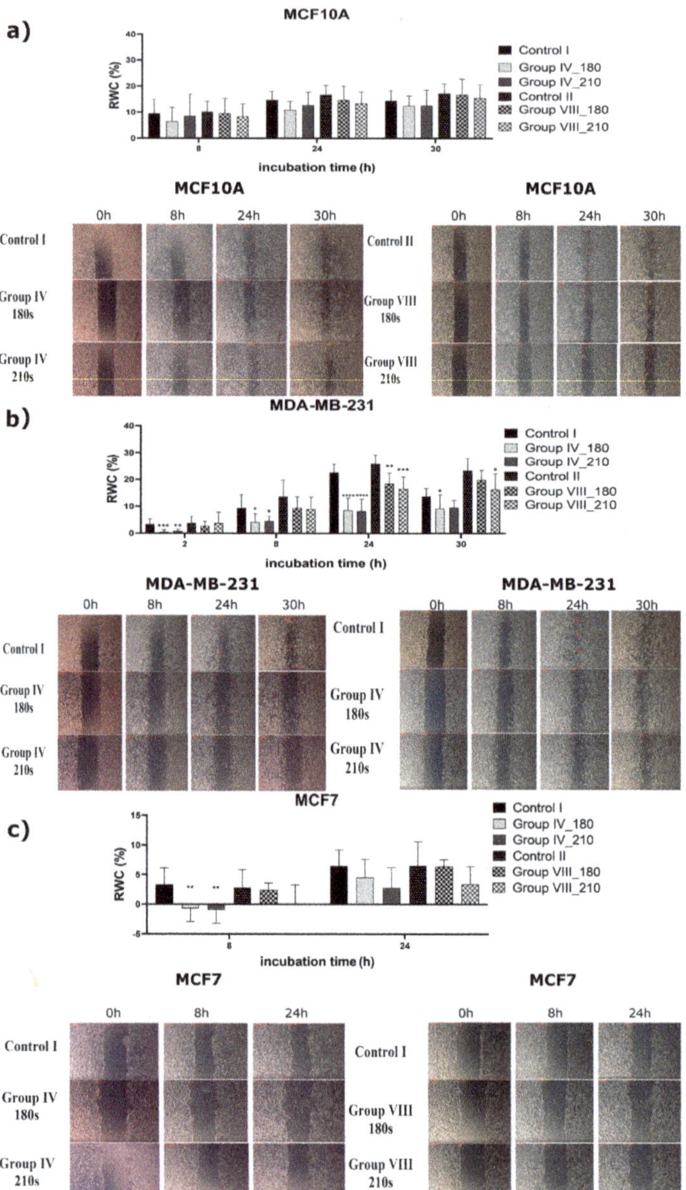

Figure 3. The results of the scratch test assay employed to assess the cells migration for the human breast cell line MCF10A (**a**) and the human cancer cell lines, i.e., MDA-MB-231 (**b**) and MCF7 (**c**). The grown confluent layers were scratched, and old culture media were substituted by: (i) 1.5 mL of the DMEM activated by CAPP for 180 s or 210 s (Group IV) or (ii) 1.5 mL of the Opti-MEM activated by CAPP for 180 s or 210 s (Group VIII). The relative wound closure (RWC) area was calculated following 30 h of the experiment as described in the "Materials and Methods" section. The calculated results are presented as means ± SD values for three independent experiments performed in duplicates. The statistical calculation was performed by the comparison of all the investigated groups versus the control using one-way ANOVA with the Dunnett's post hoc test (* $p < 0.016$, ** $p < 0.002$, *** $p < 0.001$, **** $p < 0.0001$).

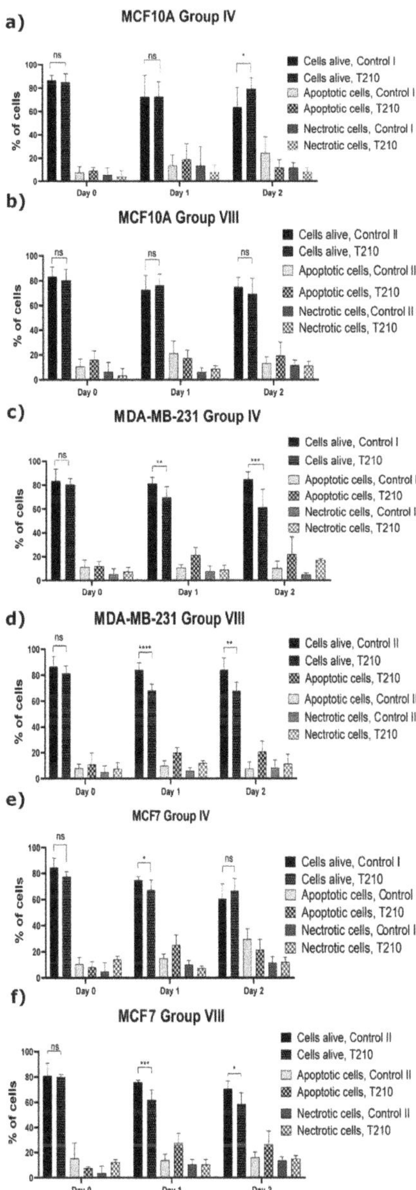

Figure 4. The results on the population of the alive, apoptotic, and necrotic cells of the human breast cell line MCF10A (**a**,**b**) and the human cancer cell lines, i.e., MDA-MB-231 (**c**,**d**), and MCF7 (**e**,**f**). The MCF10A (**a**), MDA-MB-231 (**c**), and MCF7 (**e**) cell lines cultured for 2 days in 1.5 mL of the CAPP-activated DMEM for 210 s (Group IV), as well as 1.5 mL of the CAPP-activated Opti-MEM for 210 s (Group VIII): MCF10A (**b**), MDA-MB-231 (**d**), MCF7 (**f**). The calculated results presented as means $\pm$ SD values for three independent experiments performed in duplicates. The statistical calculation was performed by comparison of all investigated groups versus the control using the one-way ANOVA with the Dunnett's post hoc test (* $p < 0.015$, ** $p < 0.0015$, *** $p < 0.0004$, **** $p < 0.0001$, n.s. not significant).

The population of the alive MCF10A cells remained unchanged after culturing them in the CAPP-activated medium (Figure 4b). Additionally, the population of the alive cells incubated in the CAPP-activated DMEM for two days significantly increased as compared to the control group (CAPP treatment time of 210 s, Figure 4a) (* $p < 0.015$). Based on a culturing guide [37] for the MCF10A cell line, the DMEM culture medium is recommended, suggesting a positive effect on the normal cells. In the case of the non-metastatic breast cancer cell line (MCF7), the cells remained significantly changed in the population of the alive cells incubated in the CAPP-activated medium with a lower organic contribution in the first day of the analysis (DMEM, Figure 4e) (* $p < 0.015$). On the other hand, the application of the CAPP-activated Opti-MEM (Figure 4f) significantly decreased the percentage of the alive cells as a result of their apoptosis after one and two days of the above-described experiment (day 1—from 76.75% to 61.83%, *** $p < 0.0004$; day 2—from 74.00% to 58.50%, * $p < 0.015$). Considering the metastatic breast cancer cell line (MDA-MB-231), the presence of a lower organic matter in the DMEM (Figure 4c) resulted in a prominent induction of the apoptosis, especially after one day and two days of the culturing in the CAPP-activated medium (day 1—from 81.17% to 69.75%, ** $p < 0.0015$; day 2—from 85.00% to 60.88%, *** $p < 0.0004$). The metastatic breast cancer cell line (MDA-MB0231) treated with the CAPP-activated Opti-MEM (Figure 4d) exhibited the most prominent reduction of the alive cell population after the one-day experiment (day 1—from 84.00% to 68.12%, **** $p < 0.0001$; day 2—from 84.00% to 67.86%, ** $p < 0.0015$). It is worth mentioning that a greater reduction of the alive cells was observed in all cases for the MDA-MB-231 cell line; which was incubated in the CAPP-activated medium, following the one- and two-day experiment (** $p < 0.002$, *** $p < 0.001$, **** $p < 0.0001$, ** $p < 0.002$ respectively). The obtained results are in line with the previously presented results related to migration ability measurement, providing a good explanation of the observed phenomena (see Section 2.1.2).

The selective induction of the apoptosis was also confirmed by others, who described the application of the CAPP-activated DMEM towards the human metastatic breast cancer MDA-MB-231 cell line [20]. Accordingly, a significant drop of the apoptosis rate was only observed in the case of the metastatic MDA-MB-231 cell line, while the MCF10A and MCF7 cells showed no differences in their apoptosis rate. For the direct CAPP treatment, it was confirmed that the irradiation induced the apoptosis in the MCF7 cell line [13,17], not damaging the MCF10A cell line [18].

Figure 5 shows a summary of the results obtained during all the biological experiments. As can be seen from the figure, in a great majority of cases, the biological activity of the MCF10A cell lines was not disturbed, showing that the CAPP-activated media has no harmful effects on the human normal cell line. Concerning the next cell line, i.e., MCF7, it can be clearly seen that any perturbations in their viability occur only for the CAPP-activated Opti-MEM, with a minimal toxic dose established to be 180 s of CAPP activation. Moreover, the generation of the CAPP-activated media in a smaller medium volume greatly impacts the activity. Additionally, the absence of the FBS during CAPP activation also improves the biological activity of the culture media. In general, interference in the cell viability increases with the time of the experiment. These observations were confirmed by the scratch test, where a significant inhibition in the cell viability was found only at the beginning of the observation and, what is quite interesting, only in the case of the CAPP-activated DMEM. However, this pattern was observed in the assessment of the cell death type, where a more significant decrease in the population of the alive cells was noted, following the first day of the experiment. The influence of the CAPP-activated media composition on MDA-MB-231 viability was as follows: The DMEM without FBS contribution during CAPP activation harms the cells with a minimal toxic dose of 180 s, indicating a major impact on the second day of the experiment. The CAPP-activated Opti-MEM leads to the disruption of cell viability in most studied groups, especially during the second day of the test, with the same minimal toxic dose. Interestingly, the cell motility was strongly inhibited throughout the experiment for the DMEM CAPP activation, while in the case of the Opti-MEM, it occurs only for a longer observation time. Finally, in the case of a decrease in the alive cell population, the results are comparable.

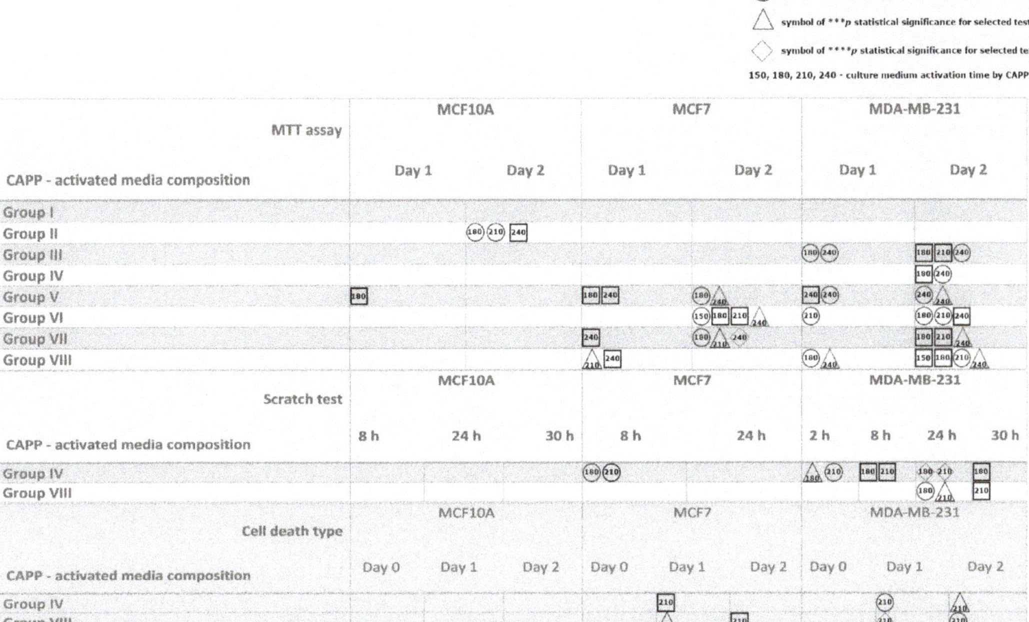

Figure 5. A graphical summary of the biological results successively obtained from the MTT assay (Figures 1 and 2), the scratch test (Figure 3), and the cell death type appearance (Figure 4). □—the * p statistical significance for the selected test; ○—the ** p statistical significance for the selected test; △—the *** p statistical significance for the selected test; ◇—the **** p statistical significance for the selected test in all the investigated CAPP-activated media compositions. The numbers given inside the symbols (150, 180, 210, and 240 s) describe the CAPP-activation times of the culture media.

2.2. Processes and Reactions Leading to the Production of the CAPP-Activated Media with Different Biological Activities

Interactions between CAPP and liquids result in cascading reactions, leading to the production of a cocktail of various RONS. A special attention should be paid to the long-lived RONS such as NO_2^-, NH_4^+, NO_3^-, and H_2O_2. The qualitative and quantitative determination of these RONS is necessary to reveal the CAPP reactions and processes, responsible for the anticancer activity of the cell culture media activated by CAPP.

2.2.1. Identification of the RONS in the Gaseous Phase of CAPP during the Production of the CAPP-Activated Media

To identify the RONS produced in the gaseous phase of CAPP, generated during DMEM and Opti-MEM activation, optical emission spectrometry (OES) was used. As can been seen from Figure 6, quite similar RONS were identified in the case of CAPP use for the activation of both media. Accordingly, the following reactive species, i.e., NO, N_2, N_2^+, NH, O, H, He, and OH, were observed in the OES spectra of CAPP (Figure 6). In the 200–260 nm region the γ-system of NO ($A^2\Sigma^+ - X^2\Pi$), with band heads at 226.9 nm (0−0), 237.0 nm (0−1), and 247.9 nm (0−3), was identified. Numerous rotational–vibrational bands of the N_2 molecule, belonging to the $C^3\Pi_u - B^3\Pi_g$ system, with the most intense band heads at 315.9 nm (1−0), 337.1 nm (0−0), 357.7 nm (0−1), and 380.4 nm (0−2), were also clearly observed. The bands of the OH radical, belonging to the $A^2\Sigma-X^2\Pi$ system, with the intense band heads at 309,4 nm (0−0) and 286.1 nm (0−1), were also identified. Finally, the bands of the N_2^+ molecule, belonging to the $B^2\Sigma^+_u - X^2\Sigma^+_g$ system, were identified with the band heads at 391.4 nm (0−0) and 427.8 nm (0−1). Additionally, numerous lines of He I at 388.8 nm, 587.5 nm, 667.8 nm, 706.5 nm, and 728.1 nm were noted. There were also found, in the emission spectra of CAPP, during the media activation, H I lines at 486.1 nm and 656.2 nm as well as O I lines at 777.2 nm, 777.4 nm, and 844.6 nm. Based on this OES qualitative characterization, production of the selected RONS during DMEM and Opti-MEM activation was confirmed.

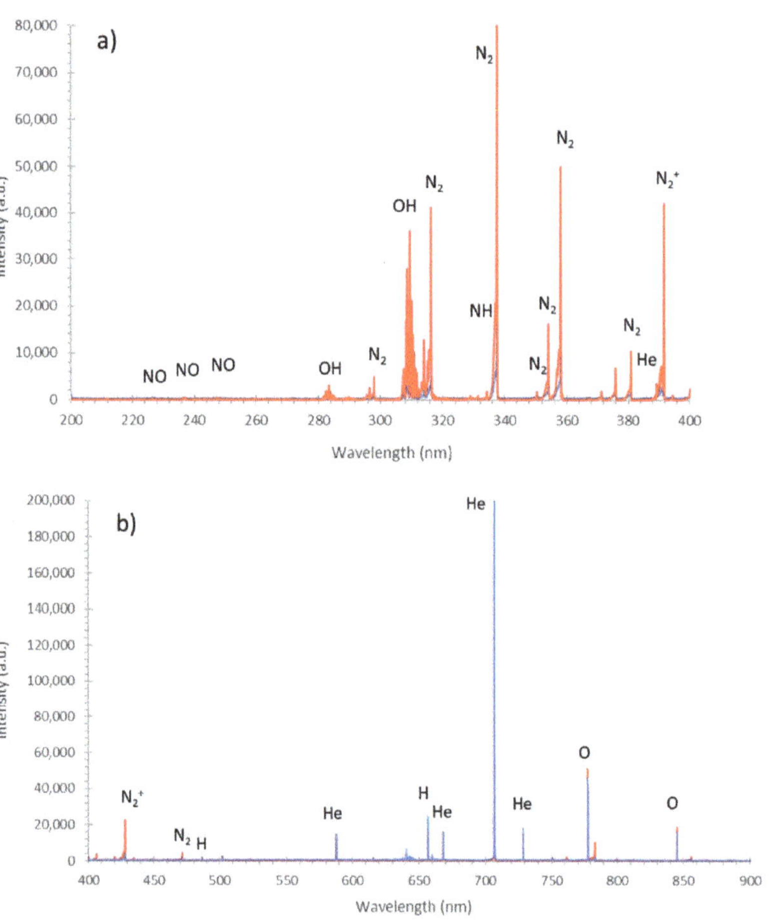

Figure 6. The emission spectra of CAPP generated in contact with the DMEM (red line) and the Opti-MEM (blue line) in the (**a**) 200–400 nm and (**b**) 400–900 nm spectral regions.

2.2.2. Determination of the RONS Concentration in the Liquid Phase of the CAPP-Activated Medium

To quantitatively determine the concentration of the selected long-term RONS, including the NO_2^-, NH_4^+, and NO_3^- ions (Figure 7a) as well as H_2O_2 (Figure 7b), in the analyzed cell culture media, colorimetric methods were used. It was established that the concentration of all measured RONS changed after CAPP activation of the DMEM (without phenyl red). Moreover, the concentration of H_2O_2 varied in different CAPP-activated media. A minor change was observed for the NO_2^- ions. In this case, their concentration for the CAPP-activated medium (time of the medium activation: 180 s) and the untreated one was comparable, i.e., 10.0 ± 0.1 mg L^{-1} and 9.50 ± 0.05 mg L^{-1}, respectively. A much bigger change was noted for the NH$_4^+$ ions. In this case, the obtained results were as follows: 19.5 ± 0.4 mg L^{-1} and 10.0 ± 0.6 mg L^{-1} for the CAPP-activated medium (time of the medium activation: 180 s) and the untreated one, respectively (**** $p < 0.0001$). A major change was noted for the NO$_3^-$ ions. Here, the concentrations of these ions for the CAPP-activated medium (time of the medium activation: 180 s) and the untreated one were estimated to be 1.20 ± 0.08 mg L^{-1} and 0.15 ± 0.04 mg L^{-1}, respectively (*** $p < 0.006$). This eight-times-higher concentration of the NO$_3^-$ ions might be partly responsible for the anticancer activity of the produced CAPP-activated medium. The CAPP activation of the culture media without the addition of the FBS following the preparation results in increased H$_2$O$_2$ production, while the presence of FBS during the CAPP treatment decreases the H$_2$O$_2$ content in all the analyzed media (1.14 ± 0.01 mg L^{-1} versus 0.99 ± 0.03 mg L^{-1} in the DMEM (* $p < 0.026$), 1.41 ± 0.10 mg L^{-1} versus 1.21 ± 0.05 mg L^{-1} in the DPBS, and 3.31 ± 0.03 mg L^{-1} versus 2.66 ± 0.04 mg L^{-1} in the Opti-MEM (** $p < 0.009$), respectively). Comparing the CAPP activation of the DMEM (Groups II and IV) to the DBPS with the lowest organic content, it appears that a higher H$_2$O$_2$ concentration is obtained when the organic content decreases (1.14 ± 0.01 mg L^{-1} versus 1.41 ± 0.03 mg L^{-1}). However, this regularity was not observed in the case of the Opti-MEM culture media with the highest organic content (1.41 ± 0.03 mg L^{-1} versus 3.31 ± 0.03 mg L^{-1}). A possible explanation for this abnormality can be associated with the color of this media, which can interfere at a maximum absorbance at 450 nm. On the other hand, the addition of the FBS to the Opti-MEM after CAPP activation significantly elevated the H$_2$O$_2$ concentration (2.66 ± 0.04 mg L^{-1} versus 3.31 ± 0.03 mg L^{-1}; ** $p < 0.009$). The quantitative analysis of the RONS during the CAPP activation of 3.0 mL of the DMEM for 5 min was also reported by Trizio et al. [38]. Despite evident differences in the media volume and the exposition time of the DMEM to DBD, comparable to our results, differences in the concentration of the NO$_2^-$ ions (0.40 mg L^{-1} versus 0.50 mg L^{-1}) and the NO$_3^-$ ions (1.90 mg L^{-1} versus 1.05 mg L^{-1}) were established in the cited work in the case of the CAPP-activated medium and the untreated one. As here, the concentration of the NO$_2^-$ ions were at the same level, while in the case of the NO$_3^-$ ions, their concentration in the CAPP-activated medium was higher than that determined in the untreated medium. In addition, the concentration of the sum of the NO$_3^-$ and NO$_2^-$ ions, produced in 2.0 mL of the CAPP-activated DMEM for 120 s and used concerning the MCF7 cell line [12], was 0.52 mM, which is quite similar to our findings. Familiar concentrations of H$_2$O$_2$ following the CAPP treatment of the RPMI with the addition of the FBS were reported by Rodder et al. [34]. In this case, 1 mL of the RPMI with the added FBS was treated by CAPP for 20 s, which results in the production of 1.7 mg L^{-1} of H$_2$O$_2$. Moreover, in the paper by Yadav et al. [35], 300 s treatment of 1 mL of RPMI or MEM with the added FBS results in 1.43 mg L^{-1} and 1.56 mg L^{-1} of H$_2$O$_2$, respectively. These results correlate well with our studies, where higher volumes of the culture media were irradiated by a shorter treatment time.

To reveal the impact of the media activation with the aid of CAPP on the cancer cell lines, the plasma–liquid interactions were discussed. At the beginning of the CAPP irradiation of the liquid medium, the broad spectrum of RONS was likely produced in the gas phase, including NO, N$_2$, N$_2^+$, NH, O, and OH (Figure 6, according to the results collected by us) and superoxide anions (O$_2^-$), atomic oxygen (O), singlet oxygen (^{1}O$_2$), ozone (O$_3$), oxonium ions (H$_2$O$^+$), in addition to the UV radiation (according to the literature reports [39,40]). These high energy molecules and reactive species interacted with different constituents in the solution, generating a broad spectrum of short life-time (peroxynitrite) and a long life-time species (H$_2$O$_2$, NO$_2^-$, NO$_3^-$, and organic peroxides). Because the study conducted by us was based on the indirect CAPP treatment and the preparation of the CAPP-activated media, which were then transferred into cells, the impact of the short life-time reactive species did not matter in the case of the CAPP–cells interactions. The latter reactive species likely took part in further cascade reactions, leading to an increase in the long life-time reactive species production. For this reason, in most of the published papers on the CAPP anticancer treatment, the major RONS such as NO_2^-, NO_3^-, H_2O_2, and organic peroxides were measured [41] as presented in Figure 7.

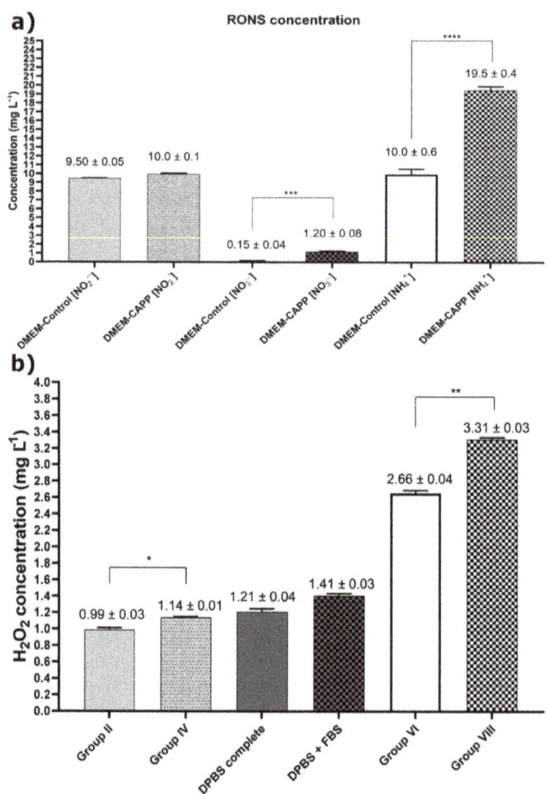

Figure 7. The concentrations of (**a**) the NO_2^-, NO_3^-, and NH_4^+ ions and (**b**) H_2O_2 determined in 1.50 mL of the CAPP-treated DMEM (for 180 s). The analyzed experimental groups were as follows: DMEM Groups. II (with FBS during the activation) and IV (without FBS during the activation); DPBS (with FBS during the activation), DPBS + FBS (without FBS during the activation), Groups VI (with FBS during the activation) and VIII (without FBS during the activation). The calculated results are presented as means ± SD values for three independent experiments. The statistical comparison was carried out using the one-way ANOVA analysis with the Tukey's post hoc test (* $p < 0.026$, ** $p < 0.009$, *** $p < 0.006$, **** $p < 0.0001$).

Considering the different organic content of CAPP-activated culture media, it seemed that the obtained biological response and the concentration of the generated RONS could be tailored. In general, the increase of the organic contribution to a certain level resulted in elevating the production rate of H_2O_2, NO_2^-, NO_3^- in various proportions [42]. On the other hand, the organic admixtures exceeding a certain level were responsible for the scavenger phenomena of certain RONS as well as an inhibited biological response of certain cell lines. A familiar situation was observed in the present study for the measurement of the H_2O_2 concentration (Figure 7b), where for the CAPP-activated DMEM the lowest H_2O_2 content was observed. In detail, for the CAPP-activated DPBS (with the lowest organic content) and the Opti-MEM, the H_2O_2 concentration was significantly elevated. From this perspective, the addition of the FBS during the CAPP-activation step should be considered as one of the most important factors responsible for the increased H_2O_2 production. Finally, the overall broad spectrum of the reactive constituents and a variability of possible by-products obtained during the CAPP activation of the complex culture medium are hard to examine in a chemical way. Therefore, the effect of the generated RONS on cells should be primarily investigated with reference to their biological response.

The presented results clearly showed that the CAPP-activated DMEM and Opti-MEM do not harm the human normal cell line while significantly they affect the cell viability of the cancer cells. This selective CAPP action can be associated with a differentiated content of cholesterol fractions in the cell membrane, playing a crucial role in maintaining membrane integrity and fluidity. Cholesterol fractions are the first barrier against the RONS, which can oxidise them, creating some holes in them for further penetration inside the cells. It was previously shown that the MCF10A cell lines, which pose a significantly lower level of cholesterol than MCF7 and MDA-MB-231 cell lines, were resistant to the MBCD-induced apoptosis, while the breast cancer cell lines were sensitive [43]. The facilitated permeability of the RONS inside the cells due to membrane oxidation what creates some holes, results in the accumulation of, e.g., H_2O_2, leading to oxidative stress, which is greatly harmful for the MCF7 and MDA-MB-231 cell lines but not for the MCF10A cell line [44]. Additionally, cancer cells abundantly express aquaporins (AQP) as compared to normal cells, i.e., AQP1, AQP3, and AQP5. Aquaporins, being membrane proteins that create channels in the membrane, facilitate the transport of water and glycerol, and correlate with the metastatic character of human breast cancer cells and their aggressiveness [45]. They can also promote H_2O_2 permeation [46]. Finally, it was confirmed that the targeted therapy related to a mutant p53 gene is a promising treatment against breast cancer and can be correlated with observed CAPP-activated DMEM and Opti-MEM phenomena [47].

3. Materials and Methods

3.1. CAPP-Based Reaction-Discharge System Used for CAPP-Activated Media Production

To obtain the CAPP-activated media, including the DMEM and the Opti-MEM, a CAPP-based reaction-discharge system, previously developed and optimized in our research group, was used [33]. In this CAPP-based reaction-discharge system, a DBD plasma jet is sustained in He and used as a CAPP source. As is shown in Figure 8, the main corpus of the applied CAPP-based reaction-discharge system consisted of an E-57 epoxy resin with an immersed quartz tube and two ring-shaped tungsten electrodes attached. From the inside, the corpus was covered with a CORIAN insulator packed into a ceramic material. The He-CAPP cone was formed and extended beyond the ceramic cover as much as 38 mm. A HV potential was supplied to the reaction-discharge system using a portable power supply (Dora Electronics Equipment, Wilczyce, Poland). The optimal operating conditions of this reaction-discharge system, leading to the CAPP-activated media production, were as follows: The frequency of the modulation: 2.14 kHz; the duty cycle: 74.29%; and the He flow rate 10.6 L min^{-1}. Under these optimal operating conditions, the gas temperature (based on the OH (0-0) radical emission spectrum) was about 37 °C (310 K) [33]. The additional electrical parameters, i.e., the voltage amplitude, the voltage waveform shape, and the general frequency, were assessed using a digital two-channel storage oscilloscope (Tektronix, TBS 1000, Beaverton, OR, USA) and were as follows: (i) the voltage amplitude: 6 kV; (ii) the voltage waveform: square wave; and (iii) the general frequency: 66.45 kHz. The distance between the activated medium and the tip of CAPP was measured using a digital caliper and was equal to 25.00 mm.

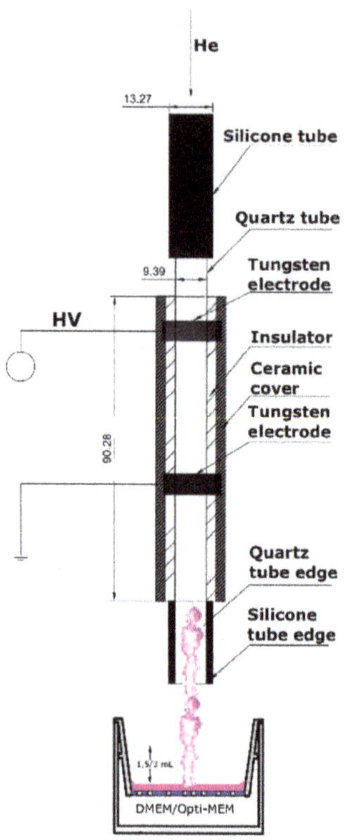

Figure 8. The CAPP-based reaction-discharge system used for the CAPP-activated media production.

3.2. Breast Cancer Cell Lines and Their Culture Conditions

To assess the biological effects of the CAPP-activated media in the in vitro models of breast cancer, two cell lines were selected: The human non-metastatic breast adenocarcinoma MCF7 (ATCC® HTB-22TM) cell line [48] and the human metastatic MDA-MB-231 (ATCC® HTB-22TM) cell line (delivered from a pleural effusion of a middle-aged Caucasian female) [49]. The MCF 7 cell line is non-invasive and exhibits a low proliferation rate in contrast to the MDA-MB-231 cell line, which is considered highly aggressive and invasive, possessing elevated expression of p53 protein. Furthermore, the non-cancerous human normal MCF10A cell line (ATCC® CRL-10317TM), originating from the spontaneously immortalized benign proliferative breast tissue, was chosen as the non-carcinogenic in vitro model [50]. All the analyzed cell lines were cultured in the Opti-MEM with the Gluta-MAX medium (Thermo Fisher Scientific Inc., Grand Island, NE; New York, NY, USA), supplemented with 3% of the FBS (Gibco, Origin: Brazil, Campinas, Brazil), a 100 U mL^{-1} penicillin solution (Sigma-Aldrich, Steinheim, Germany), and a 100 µg mL^{-1} streptomycin solution (Sigma-Aldrich, Steinheim, Germany). Additionally, the normal MCF10A cell line was supplemented with insulin (Sigma-Aldrich, Steinheim, Germany), an epidermal growth factor (EGF, 20 µg mL^{-1}), a CorningTM Endothelial Cell Growth Supplement (ECGS, 50 µg mL^{-1}), and hydrocortisone (0.5 µg mL^{-1}). EGF, ECGS, and hydrocortisone were obtained from Sigma-Aldrich (Sigma-Aldrich, Steinheim, Germany). The analyzed cell lines were incubated under the following conditions: Temperature 37 °C, 5% CO$_2$, and 95% air atmosphere. After the preparation and incubation of the cell lines, they were routinely passaged using a 0.05% Trypsin/0.02% EDTA (*w*/*v*) solution (IITE PAN, Wroclaw, Poland).

To evaluate the effect of the selected CAPP-activated media (see Table 1 for more details) on the biological activity of cells, including their viability, migration rate, and the type of cell death, the following procedure was used. Firstly, the defined volume (3.0 mL or 1.5 mL) of the selected media (DMEM or Opti-MEM with GlutaMAX), supplemented with a 100 U mL^{-1} penicillin solution, and a 100 µg mL^{-1} streptomycin solution, were introduced to the 12-well plates. Considering the hypothesis that the presence of the FBS in the culture medium during the CAPP treatment can act as a scavenger of H_2O_2, in the first step of the experiments (Table 1, experimental Groups I, II, V, and VI), 3% of the FBS was added to the medium before the CAPP treatment, opposite to the second step of the experiments in which 3% of the FBS was introduced to the analyzed media immediately after the CAPP treatment (Table 1, experimental Groups III, IV, VII, and VIII).

Table 1. The summary of the CAPP-activated media composition.

Group	CAPP-Activated Media Preparation Procedure
Control I	Cell incubated in a complete untreated culture medium DMEM
Control II	Cell incubated in a complete untreated culture medium Opti-MEM
Group I	Cell incubated in the DMEM irradiated in volume 3.0 mL supplemented with 3% FBS
Group II	Cell incubated in the DMEM irradiated in volume 1.5 mL supplemented with 3% FBS
Group III	Cell incubated in the DMEM irradiated in volume 3.0 mL without the 3% FBS
Group IV	Cell incubated in the DMEM irradiated in volume 1.5 mL without the 3% FBS
Group V	Cell incubated in the Opti-MEM irradiated in volume 3.0 mL supplemented with 3% FBS
Group VI	Cell incubated in the Opti-MEM irradiated in volume 1.5 mL supplemented with 3% FBS
Group VII	Cell incubated in the Opti-MEM irradiated in volume 3.0 mL without 3% FBS
Group VIII	Cell incubated in the Opti-MEM irradiated in volume 1.5 mL without 3% FBS

In order to produce the particular type of CAPP-activated medium (see Table 1), the CAPP-based reaction-discharge system was used (see Section 3.1, CAPP-based reaction-discharge system used for CAPP-activated media production, for more details). The distance between the tip of the CAPP and the analyzed medium was set to 25.0 mm, remaining in direct contact with the culture media. The medium was activated by CAPP within 45 s, 90 s, 120 s, 150 s, 180 s, 210 s, or 240 s. Finally, the freshly prepared CAPP-activated medium (supplemented before or after the CAPP activation with 3% of the FBS) was collected in separate plastic tubes and then directly transferred into the prepared cell lines for the further biological analysis.

3.3. Biological Activities of the CAPP-Activated Media Concerning the Human Breast Cancer Cell Lines

3.3.1. Determination of the Cell Viability

In order to determinate the influence of the CAPP-activated media on cell viability, an MTT (3-(4.5-dimethylthialzol-2-yl)-2.5-diphenyl tetrazolium bromide) (Thermo Fisher Scientific Inc., Grand Island, NE, New York, NY, USA) assay was carried out [51]. Initially, two cell lines were chosen: MCF7 and MDA-MB-231 (see Section 3.2, breast cell lines and their culture conditions, for more details). As a control, the non-cancerous MCF10A cell line was chosen. The MTT assay was performed as follows: 5×10^3 of given cells (per well) were plated in 100 µL (per well) of the complete medium in the flat-bottomed 96-well plates. Then, the cells were incubated at 37 °C, with 5% CO_2 and 95% air atmosphere, for 24 h. After this time, the completed medium was carefully removed from each well and refilled with a defined volume of the proper CAPP-activated medium (Table 1, experimental groups, Groups I, II, V, and VI), or with the CAPP-activated medium supplemented with 3% of the FBS (Table 1, experimental groups, Groups III, IV, VII, and VIII) for the indicated time period (day 0, day 1, and day 2). Next, 10 µL of the MTT reagent was added to each well to reach a final concentration of 10 mg mL^{-1}. Then, the obtained reaction mixtures were incubated for 3 h at 37 °C, with 5% CO_2 and 95% air atmosphere, maintaining darkness. After 3 h of incubation, the obtained reaction mixture was thoroughly removed from each well and carefully dried. Empty wells were filled with 100 µL of DMSO (POCH SA, Gliwice, Poland) per well. After 45 s of the reaction with DMSO, a purple color appeared in each well. The described change in color of the analyzed reaction mixtures was associated with the formation of formazan crystals. Next, the absorbance of the latter mixtures at 570 nm was measured using a Victor 2 microplate reader (Perkin Elmer, Woodbridge, Vaughan, ON, Canada). The experiments were done in triplicate at least three times.

3.3.2. Determination of the Cell Migration

To assess the influence of the CAPP-activated media on cell migration, a scratch test was conducted [52]. Two breast cancer cell lines were chosen: MCF7 and MDA-MB-231 (see Section 3.2, breast cell lines and their culture conditions, for more details), and the MCF10A cell line served as a control. Firstly, the cell lines, with a concentration of 1.5×10^5 cells per well, were separately seeded into the 24-well plates. Next, the incubated cells were suspended in 350 µL of the Opti-MEM culture medium supplemented with 3% of the FBS, a 100 U mL^{-1} penicillin solution, and a 100 µg mL^{-1} streptomycin solution, and cultured to produce a confluent cell monolayer. Then, the cells were incubated for 24 h at 37 °C, with 5% CO_2 and 95% air atmosphere. After this time, the confluent surfaces of the cells were scratched using a 200 µL sterile pipette tip to prepare the straight dashes for the scratch assay. Then, the cells detached during the scratch preparation were removed with the complete culture medium. Next, the attached cell monolayer was filled using 350 µL of the proper CAPP-activated media (see Table 1 for more details) treated by CAPP either for 180 s or 210 s. Afterwards, the 24-well plates, containing cell monolayers, were incubated for 30 h at 37 °C, under the 5% CO_2 and 95% air atmosphere. The experiments were carried out three times in duplicate. The cell migration was acquired by taking digital pictures at 0, 2, 8, 24, and 30 h after the scratches. The closure area was calculated using ZEN 3.1 blue edition software (Carl Zeiss Microscopy GmbH, Jena, Germany) and presented as the relative wound closure (RWC, %), as previously described in reference [53].

3.3.3. Estimation of the Cell Death Type

To assess the cytotoxic effect of the CAPP-activated media on the analyzed human breast cancer cell lines, the cell death Annexin V/PI test was performed (eBioscienceTM Annexin V Apoptosis Detection kit APC, Invitrogen, Thermo Fisher Scientific Inc., Grand Island, NE; New York, NY, USA) [54,55]. In this experiment, the proper cell lines, including MCF7, MDA-MB-231, and MCF10A, were seeded onto the 24-well plates with a density of the 1.5×10^5 cells per well. Then, the prepared cell lines were suspended in 350 µL of the Opti-MEM medium supplemented with 3% of the FBS, a 100 U mL^{-1} penicillin solution, and a 100 µg mL^{-1} streptomycin solution. Next, the seeded well plates were incubated for 30 h at 37 °C, with 5% CO_2 and 95% air atmosphere. After 30 h of incubation, the complete medium was removed from each well and refilled by the freshly CAPP-activated medium, prepared after 210 s irradiation (Table 1, experimental groups, Groups III, IV, VII, and VIII). Afterwards, all the cells from the 24-well plates were detached by using a 0.05% Trypsin/0.02% EDTA solution and washed, firstly with 1.00 mL of PBS (Gibco, Thermo Fisher Scientific Inc., Grand Island, NE; New York, NY, USA) and secondly with a buffer binding solution (included in kits, eBioscienceTM Annexin V Apoptosis Detection kit APC, Invitrogen, Thermo Fisher Scientific Inc., Grand Island, NE; New York, NY, USA). Next, the cell pellets were vigorously agitated and suspended in an Annexin V dye solution (according to the instructions) and then incubated for 15 min at room temperature. After the incubation, the resultant solutions were filled with 1.00 mL of the buffer binding solution and then centrifuged. In the final step, propidium iodide (PI, according to the instructions) was added to the analyzed pellets. The same procedure was employed in the case of the cells not treated with the CAPP-activated media (controls). However, after the previously described procedure of cell seeding and incubation, the complete medium was replaced by a fresh one. The prepared controls were further used for the Annexin V/PI assay, as was described before. Next, the stained cells were assessed by a FACS-Calibur flow cytometry instrument (Becton-Dickinson, Franklin Lakes, NJ, USA). The apoptotic cell population (Annexin V positive, PI negative), the necrotic cells (Annexin V negative, PI positive), and the alive cells (Annexin V negative, PI negative) were measured using the FL4 (λ_{em} = 660 nm) and FL2 (λ_{em} = 535 nm) modes. The results were analyzed using a Flowing Software 2 program (Flowing Software ver. 2.5.1, Flowing Software, Turku, Finland). The same experimental protocol was conducted for the group of cell lines exposed to the CAPP-activated media for 24 h and 48 h. The results were presented as means ±SD values for three independent experiments, each carried out in duplicates. In order to perform the statistical analyses, one-way ANOVA was applied. Additionally, GraphPad Prism 8 software (GraphPad Prism version 8.0.1 for Windows, GraphPad Software, San Diego, CA, USA) was employed to present the final results.

3.4. Estimation of the CAPP-Derived Active Constituents Leading to the Anticancer Activity of the Analyzed Media

For the qualitative and quantitative determination of the RONS, the 1.50 mL portions of the analyzed media were placed into the wells of the 24-well plates, and supplemented with a 100 U mL^{-1} penicillin solution and a 100 μg mL^{-1} streptomycin solution. Next, the analyzed media were activated by CAPP, operated under the working conditions given in Section 3.1, CAPP-based reaction-discharge system used for CAPP-activated media production. During the CAPP activation, the qualitative determination of the RONS in the gas phase produced during the CAPP operation was conducted using the OES measurements. On the other hand, the quantitative measurement of these species in the liquid phase was performed after 180 s of the proper medium irradiation by CAPP.

3.4.1. Identification of the Reactive Oxygen and Nitrogen Species Using Optical Emission Spectrometry

The OES measurements were made to identify the RONS generated in the plasma-gas phase as a result of the CAPP–liquid (medium) interactions and acquired during the medium activation by CAPP. The interactions between CAPP and DMEM (without 3% FBS) and between CAPP and Opti-MEM (without 3% FBS) were analyzed. The radiation emitted by the CAPP system in the near-liquid surface zone was imaged with the aid of a UV achromatic lens (f = 60) on the entrance slit (10 μm) of a Shamrock SR-500i spectrometer (Andor, Belfast, United Kingdom). An OES spectrometer was equipped with two holographic gratings (1800 and 1200 lines nm^{-1}) and a Newton DU-920P-0E (Andor, Belfast, United Kingdom) CCD camera. The data processing and the OES spectrometer acquisition parameters were maintained with Solis software (Andor, Belfast, United Kingdom). The CCD camera was operated in the full vertical binning (FVB) mode; a 1-s integration time was applied. The OES spectra were acquired in the 200–900 nm spectral range. For the spectral range above 400 nm, a PG-5 filter (Zeiss Jena, Jena, Germany) was additionally applied for eliminating second order radiation.

3.4.2. Determination of the Selected RONS Concentration Produced in the CAPP-Activated Media

The colorimetric methods were used to quantitatively determine the concentration of the NO_2^-, NO_3^-, and NH_4^+ ions in the CAPP-treated (for 180 s) DMEM. In this case, the CAPP activation was performed for the DMEM without the phenol red dye because it is dark-pink and could affect the colorimetric measurements. For that reason, the colorless medium, which has a composition quite similar to the complete DMEM, was used. The following protocols were taken for the determination of the RONS concentration:

(i) NO_2^- ions: To determinate the concentration of the NO_2^- ions in the CAPP-activated medium, a HANNA HI 96708 spectrophotometer (HANNA Instruments, Olsztyn, Poland) was used. The measurements were performed according to the protocol suggested by the manufacturer and applying all the reagents and solutions provided by the manufacturer. As a control, the content of the NO_2^- ions in the medium not activated by CAPP was assessed.

(ii) NO_3^- ions: To assess the concentration of the NO_3^- ions in the CAPP-activated medium, a HANNA HI 96728 spectrophotometer (HANNA Instruments, Olsztyn, Poland) was used. The measurements were conducted according to the protocol suggested by the manufacturer and using all the reagents and solutions provided by the manufacturer. As a control, the content of the NO_3^- ions in the medium not activated by CAPP was assessed.

(iii) NH_4^+ ions: To estimate the content of the NH_4^+ ions in the analyzed medium, Nessler's method was used [56]. In this method, the reaction with Nessler's (K_2HgI_4) reagent (Sigma-Aldrich, Steinheim, Germany) was used for the determination of the NH_4^+ ions. In this case, the absorbance of the obtained product, i.e., $[(Hg-O-Hg)NH_2]$, was measured spectrophotometrically at 420 nm, using an Analytik Jena AG UV/Vis Specord 210 Plus (Jena, Germany). External calibrations with the simple standard solutions were used for the quantification. As a control, the content of the NH_4^+ ions in the medium not activated by CAPP was assessed. All experiments were performed in triplicates.

(iv) H_2O_2: To measure the total concentration of H_2O_2 in the prepared CAPP-activated media, the spectrophotometric method with ammonium metavanadate (NH_4VO_3) was adopted [57]. Following the reaction of H_2O_2 with NH_4VO_3 (Avantor Pefrormance Materials, Gliwice, Poland) in a H_2SO_4 solution (Avantor Pefrormance Materials, Gliwice, Poland), the absorbance of the resultant peroxovanadium cations formed in the solutions was measured at 450 nm with the aid of an Analytik Jena AG UV/Vis Specord 210 Plus (Jena, Germany). The contents of NH_4VO_3 and H_2SO_4 in these solutions were 6.2 mmol L^{-1} and 0.058 mol L^{-1}, respectively. The H_2O_2 concentration was calculated in the culture medium with the organic content and in a DPBS solution. The external calibrations with the simple standard solutions were used for the quantification. As a control, the content of H_2O_2 in the respective procedural blank solution was assessed.

3.5. Statistical Analysis

The graphs (Figures 1–4) and the statistical analysis for all the biological tests were performed using Prism 8.0 (GraphPad Software, San Diego, CA, USA). The comparison of the investigated groups versus the control group was made using one-way ANOVA with Dunnett's post hoc test. Figure 7 was constructed and the relevant statistical analyses for the chemical analysis were performed using Prism 8.0 (GraphPad Software, San Diego, CA, USA) as well. The comparison of the investigated groups versus the controls group was made using one-way ANOVA with Tukey's post hoc test.

4. Conclusions

In the present study, we tried to answer the question whether the chemical composition of the CAPP-treated media is important for their anticancer activity towards the breast cancer cell lines or not. Based on the present research, it was found that the chemical composition of the CAPP-activated medium is indeed important for their anticancer activity. More specifically:

(i)   The resultant CAPP-activated medium did not exhibit the apoptotic effect on the normal MCF10A cell line, developing an opportunity to successfully design a selective approach against the human breast cancer cells.

(ii)   The resultant CAPP-activated medium had a harmful effect on the MCF7 and MDA-MB-231 cancer cell lines.

(iii)   The presence of the FBS during the CAPP-activated media preparation negatively affected the biological response of the MCT7 and MDA-MB-231 cell lines, causing a minor decrease in their viability and disrupting the cell viability of the MCF10A cells.

(iv)   The choice of the proper culture medium for the production of the CAPP-activated media with the highest biological impact is a crucial step. For the selected biological models, the culture medium with a lower content of the organic matter in the CAPP-activated DMEM resulted in a significant drop in the cell viability of the MCF7 and MDA-MB-231 cancer cells as well as in the inhibition of their motility.

(v)   The disturbance in the life processes of the breast cancer cell lines was associated with the induction of the apoptosis by the CAPP-activated media. The largest population of the cells with the apoptotic pathway as well as the strongest inhibition in the cell migration was observed for the MDA-MB-231 cancer cells. This led us to the conclusion that this cell line is more sensitive to the CAPP-activated media. We believe that the studies carried out by us could be the base for alternative therapy, dedicated to the highly aggressive human breast cancer.

5. Patents

The construction of the portable He-DBD-based reaction-discharge system is protected by Polish Patent Application no. P 429275 (UP RP, 14 March 2019).

Author Contributions: D.T., A.D., A.B.-P., A.K. and P.J. conceptualized all work. D.T., A.D. and P.J. prepared CAPP-based reaction-discharge system for further assays. D.T. and A.B.-P. performed the test associated with the biological activity of CAPP-based reaction-discharge systems. D.T., A.D. and P.J. conducted experiments related to the RONS identification and determination of their concentration. D.T., A.D. and P.J. wrote the presented paper. A.B.-P., A.K. and P.P. revised the presented work and took a part in discussion. P.P. and P.J. covered costs associated with the article proceeding charges. All authors have read and agreed to the published version of the manuscript.

Funding: This study was co-financed by a statutory activity subsided from the Polish Ministry of Science and Higher Education for the Faculty of Chemistry of Wroclaw University of Science and Technology. The presented research was also supported by European Social Found for the interdisciplinary doctoral studies and internship, program BioTechNan (15/US/BIO/2018). The scientific activity of Dr. Anna Dzimitrowicz is supported by Ministry of Science and Higher Education (UMO-532/STYP/13/2018), program Outstanding Young Scientist.

Institutional Review Board Statement: Not applicable.

Informed Consent Statement: Not applicable.

Acknowledgments: Authors would like to thank Elzbieta Wojdat and Eng. Jerzy Dora for technical support.

Conflicts of Interest: The Authors declare no conflict of interest.

References

1.   Ghoncheh, M.; Pournamdar, Z.; Salehiniya, H. Incidence and Mortality and Epidemiology of Breast Cancer in the World. *Asian Pac. J. Cancer Prev.* **2016**, *17*, 43–46. [CrossRef]
2.   2020 Breast Cancer Statistics. Available online: https://www.nationalbreastcancer.org/wp-content/uploads/2020-Breast-Cancer-Stats.pdf (accessed on 17 November 2020).
3.   Rostami, R.; Mittal, S.; Rostami, P.; Tavassoli, F.; Jabbari, B. Brain metastasis in breast cancer: A comprehensive literature review. *J. Neurooncol.* **2016**, *127*, 407–414. [CrossRef] [PubMed]
4.   Kamdje, A.H.N.; Etet, P.F.S.; Vecchio, L.; Tagne, R.S.; Amvene, J.M.; Muller, J.M.; Krampera, M.; Lukong, K.E. New targeted therapies for breast cancer: A focus on tumor microenvironmental signals and chemoresistant breast cancers. *World J. Clin. Cases* **2014**, *2*, 769. [CrossRef] [PubMed]
5.   Calavia, P.G.; Chambrier, I.; Cook, M.J.; Haines, A.H.; Field, R.A.; Russell, D.A. Targeted photodynamic therapy of breast cancer cells using lactose-phthalocyanine functionalized gold nanoparticles. *J. Colloid Interface Sci.* **2018**, *512*, 249–259. [CrossRef]
6.   Devi, L.; Gupta, R.; Jain, S.K.; Singh, S.; Kesharwani, P. Synthesis, characterization and in vitro assessment of colloidal gold nanoparticles of Gemcitabine with natural polysaccharides for treatment of breast cancer. *J. Drug Deliv. Sci. Technol.* **2020**, *56*, 101565. [CrossRef]
7.   Maluta, S.; Kolff, M.W. Role of Hyperthermia in Breast Cancer Locorgional Resurrence: A Review. *Breast Care* **2015**, *10*, 408. [CrossRef]
8.   Bañobre-López, M.; Teijeiro, A.; Rivas, J. Magnetic nanoparticle-based hyperthermia for cancer treatment. *Rep. Pract. Oncol. Radiother.* **2013**, *18*, 397–400. [CrossRef]
9.   Biazar, E.; Majdi, A.; Zafari, M.; Avar, M.; Aminifard, S.; Zaeifi, D.; Ai, J.; Jafarpour, M.; Montazeri, M.; Rad, H.G. Nanotoxicology and nanoparticle safety in biomedical designs. *Int. J. Nanomed.* **2011**, *6*, 1117–1127. [CrossRef]

10. Bekeschus, S.; Lippert, M.; Diepold, K.; Chiosis, G.; Seufferlein, T.; Azoitei, N. Physical plasma-triggered ROS induces tumor cell death upon cleavage of HSP90 chaperone. *Sci. Rep.* **2019**, *9*, 4112. [CrossRef]

11. Liu, Y.; Tan, S.; Zhang, H.; Kong, X.; Ding, L.; Shen, J.; Lan, Y.; Cheng, C.; Zhu, T.; Xia, W. Selective effects of non-thermal atmospheric plasma on triple-negative breast normal and carcinoma cells through different cell signaling pathways. *Sci. Rep.* **2017**, *7*, 1–12. [CrossRef]

12. Subramanian, P.S.G.; Jain, A.; Shivapuji, A.M.; Sundaresan, N.R.; Dasappa, S.; Rao, L. Plasma-activated water from a dielectric barrier discharge plasma source for the selective treatment of cancer cells. *Plasma Process. Polym.* **2020**, *17*, 1900260. [CrossRef]

13. Zhang, H.; Zhang, J.; Ma, J.; Shen, J.; Lan, Y.; Liu, D.; Xia, W.-D.; Xu, D.; Cheng, C. Differential sensitivities of HeLa and MCF-7 cells at G1-, S-, G2- and M-phase of the cell cycle to cold atmospheric plasma. *J. Phys. D Appl. Phys.* **2020**, *53*, 125202. [CrossRef]

14. Jezeh, M.A.; Tayebi, T.; Khani, M.R.; Niknejad, H.; Shokri, B. Direct cold atmospheric plasma and plasma-activated medium effects on breast and cervix cancer cells. *Plasma Process. Polym.* **2020**, *17*, 1900241. [CrossRef]

15. Mokhtari, H.; Farahmand, L.; Yaserian, K.; Jalili, N.; Majidzadeh, A.K. The antiproliferative effects of cold atmospheric plasma-activated media on different cancer cell lines, the implication of ozone as a possible underlying mechanism. *J. Cell. Physiol.* **2019**, *234*, 6778–6782. [CrossRef] [PubMed]

16. Gurung, J.P.; Subedi, D.P.; Shrestha, R.; Shrestha, B.G. Application of Atmospheric Pressure Argon Plasma Jet (APAPJ) in Biomedical Science and Engineering. *J. Trop. Life Sci.* **2020**, *10*, 149–154. [CrossRef]

17. Park, S.; Kim, H.; Ji, H.W.; Kim, H.W.; Yun, S.H.; Choi, E.H.; Kim, S.J. Cold Atmospheric Plasma Restores Paclitaxel Sensitivity to Paclitaxel-Resistant Breast Cancer Cells by Reversing Expression of Resistance-Related Genes. *Cancers* **2019**, *11*, 2011. [CrossRef]

18. Mirpour, S.; Ghomi, H.; Piroozmand, S.; Nikkhah, M.; Tavassoli, S.H.; Azad, S.Z. The Selective Characterization of Nonthermal Atmospheric Pressure Plasma Jet on Treatment of Human Breast Cancer and Normal Cells. *IEEE Trans. Plasma Sci.* **2014**, *42*, 315–322. [CrossRef]

19. Adil, B.H.; Al-Shammri, A.M.; Murbat, H.H. Cold Atmospheric Plasma generated by FE-DBD Scheme cytotoxicity against Breast Cancer cells. *Res. J. Biotech.* **2019**, *14*, 192.

20. Xiang, L.; Xu, X.; Zhang, S.; Cai, D.; Dai, X. Cold atmospheric plasma conveys selectivity on triple negative breast cancer cells both in vitro and in vivo. *Free Radic. Biol. Med.* **2018**, *124*, 205–213. [CrossRef]

21. Adil, B.H.; Al-Shammari, A.M.; Murbit, H.H. Breast cancer treatment using cold atmospheric plasma generated by the FE-DBD scheme. *Clin. Plasma Med.* **2020**, *19*, 100103. [CrossRef]

22. Mehrabifard, R.; Mehdian, H.; Hajisharifi, K.; Amini, E. Improving Cold Atmospheric Pressure Plasma Efficacy on Breast Cancer Cells Control-Ability and Mortality Using Vitamin C and Static Magnetic Field. *Plasma Chem. Plasma Process.* **2020**, *40*, 511–526. [CrossRef]

23. Chauvin, J.; Judee, F.; Yousfi, M.; Vicendo, P.; Merbahi, N. Analysis of reactive oxygen and nitrogen species generated in three liquid media by low temperature helium plasma jet. *Sci. Rep.* **2017**, *7*, 4562. [CrossRef]

24. Kleineidam, B.; Nokhbehsaim, M.; Deschner, J.; Wahl, G. Effect of cold plasma on periodontal wound healing—An in vitro study. *Clin. Oral Investig.* **2019**, *23*, 1941–1950. [CrossRef] [PubMed]

25. Fridman, G.; Friedman, G.; Gutsol, A.; Shekhter, A.B.; Vasilets, V.N.; Fridman, A. Applied Plasma Medicine. *Plasma Process. Polym.* **2008**, *5*, 503–533. [CrossRef]

26. Sladek, R.; Stoffels, E.; Walraven, R.; Tielbeek, P.; Koolhoven, R.A. Plasma Treatment of Dental Cavities: A Feasibility Study. *IEEE Trans. Plasma Sci.* **2004**, *32*, 1540–1543. [CrossRef]

27. Ling, L.; Jiafeng, J.; Jiangang, L.; Minchong, S.; Xin, H.; Hanliang, S.; Yuanhua, D. Effects of cold plasma treatment on seed germination and seedling growth of soybean. *Sci. Rep.* **2014**, *4*, 5859. [CrossRef] [PubMed]

28. Dzimitrowicz, A.; Cyganowski, P.; Pohl, P.; Jermakowicz-Bartkowiak, D.; Terefinko, D.; Jamroz, P. Atmospheric Pressure Plasma-Mediated Synthesis of Platinum Nanoparticles Stabilized by Poly(vinylpyrrolidone) with Application in Heat Management Systems for Internal Combustion Chambers. *Nanomaterials* **2018**, *8*, 619. [CrossRef] [PubMed]

29. Kurosawa, M.; Takamatsu, T.; Kawano, H.; Hayashi, Y.; Miyahara, H.; Ota, S.; Okino, A.; Yoshida, M. Endoscopic Hemostasis in Porcine Gastrointestinal Tract Using CO2 Low-Temperature Plasma Jet. *J. Surg. Res.* **2019**, *234*, 334–342. [CrossRef]

30. Trachootham, D.; Alexandre, J.; Huang, P. Targeting cancer cells by ROS-mediated mechanisms: A radical therapeutic approach? *Nat. Rev. Drug Discov.* **2009**, *8*, 579. [CrossRef]

31. Vermeylen, S.; De Waele, J.; Vanuytsel, S.; De Backer, J.; Van Der Paal, J.; Ramakers, M.; Leyssens, K.; Marcq, E.; Van Audenaerde, J.; Smits, E.L.J.; et al. Cold atmospheric plasma treatment of melanoma and glioblastoma cancer cells. *Plasma Process. Polym.* **2016**, *13*, 1195–1205. [CrossRef]

32. Malyavko, A.; Yan, D.; Wang, Q.; Klein, A.L.; Patel, K.C.; Sherman, J.H.; Keidar, M. Cold atmospheric plasma cancer treatment, direct versus indirect approaches. *Mater. Adv.* **2020**, *1*, 1494–1505. [CrossRef]

33. Dzimitrowicz, A.; Bielawska-Pohl, A.; Jamroz, P.; Dora, J.; Krawczenko, A.; Busco, G.; Grillon, C.; Kieda, C.; Klimczak, A.; Terefinko, D.; et al. Activation of the Normal Human Skin Cells by a Portable Dielectric Barrier Discharge-Based Reaction-Discharge System of a Defined Gas Temperature. *Plasma Chem. Plasma Process.* **2020**, *40*, 79. [CrossRef]

34. Nie, L.; Yang, Y.; Duan, J.; Sun, F.; Lu, X.P.; He, G. Effect of tissue thickness and liquid composition on the penetration of long-lifetime reactive oxygen and nitrogen species (RONS) generated by a plasma jet. *J. Phys. D Appl. Phys.* **2018**, *51*, 345204. [CrossRef]

35. Rödder, K.; Moritz, J.; Miller, V.; Weltmann, K.-D.; Metelmann, H.-R.; Gandhirajan, R.; Bekeschus, S. Activation of Murine Immune Cells upon Co-culture with Plasma-treated B16F10 Melanoma Cells. *Appl. Sci.* **2019**, *9*, 660. [CrossRef]

36. Wang, M.; Holmes, B.; Cheng, X.; Zhu, W.; Keidar, M.; Zhang, L.G. Cold Atmospheric Plasma for Selectively Ablating Metastatic Breast Cancer Cells. *PLoS ONE* **2013**, *8*, e73741. [CrossRef]

37. Debath, J.; Muthuswamy, S.K.; Brugge, J.S. Morphogenesis and oncogenesis of MCF-10A mammary epithelial acini grown in three-dimensional basement membrane cultures. *Methods* **2003**, *30*, 256. [CrossRef]

38. Trizio, I.; Sardella, E.; Rizzi, V.; Dilecce, G.; Cosma, P.; Schmidt, M.; Von Woedtke, T.; Gristina, R.; Favia, P. Characterization of Reactive Oxygen/Nitrogen Species Produced in PBS and DMEM by Air DBD Plasma Treatments. *Plasma Med.* **2016**, *6*, 13–19. [CrossRef]

39. Kaushik, N.K.; Ghimire, B.; Li, Y.; Adhikari, M.; Veerana, M.; Kaushik, N.; Jha, N.; Adhikari, B.; Lee, S.-J.; Masur, K.; et al. Biological and medical applications of plasma-activated media, water and solutions. *Biol. Chem.* **2019**, *400*, 39–62. [CrossRef]

40. Bruggeman, P.J.; Kushner, M.J.; Locke, B.R.; Gardeniers, J.G.E.; Graham, W.G.; Graves, D.B.; Hofman-Caris, R.C.H.M.; Maric, D.; Reid, J.P.; Ceriani, E.; et al. Plasma–liquid interactions: A review and roadmap. *Plasma Sources Sci. Technol.* **2016**, *25*, 053002. [CrossRef]

41. Terefinko, D.; Dzimitrowicz, A.; Bielawska-Pohl, A.; Klimczak, A.; Pohl, P.; Jamroz, P. Biological Effects of Cold Atmospheric Pressure Plasma on Skin Cancer. *Plasma Chem. Plasma Process.* **2021**, *41*, 507–529. [CrossRef]

42. Griseti, E.; Merbahi, N.; Golzio, M. Ani-cancer potential of two plasma-activated liquids: Implication of long-lived reactive oxygen and nitrogen species. *Cancers* **2020**, *12*, 721. [CrossRef] [PubMed]

43. Li, Y.C.; Park, M.J.; Ye, S.-K.; Kim, C.-W.; Kim, Y.-N. Elevated Levels of Cholesterol-Rich Lipid Rafts in Cancer Cells Are Correlated with Apoptosis Sensitivity Induced by Cholesterol-Depleting Agents. *Am. J. Pathol.* **2006**, *168*, 1107–1118. [CrossRef]

44. Warleta, F.; Campos, M.; Allouche, Y.; Sánchez-Quesada, C.; Ruiz-Mora, J.; Beltrán, G.; Gaforio, J.J. Squalene protects against oxidative DNA damage in MCF10A human mammary epithelial cells but not in MCF7 and MDA-MB-231 human breast cancer cells. *Food Chem. Toxicol.* **2010**, *48*, 1092–1100. [CrossRef]

45. Shi, Z.; Zhang, T.; Luo, L.; Zhao, H.; Cheng, J.; Xiang, J.; Zhao, C. Aquaporins in human breast cancer: Identification and involvement in carcinogenesis of breast cancer. *J. Surg. Oncol.* **2012**, *106*, 267–272. [CrossRef] [PubMed]

46. Rodrigues, C.; Pimpão, C.; Mósca, A.F.; Coixao, A.S.; Lopes, D.; Da Silva, I.V.; Pedersen, P.A.; Antunes, F.; Soveral, G. Human Aquaporin-5 Facilitates Hydrogen Peroxide Permeation Affecting Adaption to Oxidative Stress and Cancer Cell Migration. *Cancers* **2019**, *11*, 932. [CrossRef] [PubMed]

47. Synnott, N.; Murray, A.; McGowan, P.; Kiely, M.; Kiely, P.; O'Donovan, N.; O'Connor, D.; Gallagher, W.; Crown, J.; Duffy, M. Mutant p53: A novel target for the treatment of patients with triple-negative breast cancer? *Int. J. Cancer* **2017**, *140*, 234–246. [CrossRef] [PubMed]

48. Soule, H.D.; Vazquez, J.; Long, A.; Albert, S.; Brennan, M. A Human Cell Line from a Pleural Effusion Derived from a Breast Carcinoma. *J. Natl. Cancer Inst.* **1973**, *51*, 1409–1416. [CrossRef] [PubMed]

49. Cailleau, R.; Olive, M.; Cruciger, Q.V. Long-term human breast carcinoma cell lines of metastatic origin: Preliminary characterization. *In Vitro* **1978**, *14*, 911. [CrossRef]

50. Soule, H.D.; Maloney, T.M.; Wolman, S.R.; Peterson, W.D., Jr.; Brenz, R.; McGrath, C.M.; Russo, J.; Pauley, R.J.; Jones, R.F.; Brooks, S.C. Isolation and characterization of a spontaneously immortalized human breast epithelial cell line, MCF-10. *Cancer Res.* **1990**, *50*, 6075. [PubMed]

51. Stockert, J.C.; Horobin, R.W.; Colombo, L.L.; Blázquez-Castro, A. Tetrazolium salts and formazan products in Cell Biology: Viability assessment, fluorescence imaging, and labeling perspectives. *Acta Histochem.* **2018**, *120*, 159–167. [CrossRef] [PubMed]

52. Rodriguez, L.G.; Wu, X.; Guan, J.-L. Wound-Healing Assay. *Cell Migr.* **2005**, *294*, 23–30. [CrossRef]

53. Beyeler, J.; Schnyder, I.; Katsaros, C.; Chiquet, M. Accelerated Wound Closure In Vitro by Fibroblasts from a Subgroup of Cleft Lip/Palate Patients: Role of Transforming Growth Factor-α. *PLoS ONE* **2014**, *9*, e111752. [CrossRef]

54. Van Engeland, M.; Nieland, L.J.W.; Ramaekers, F.C.; Schutte, B.; Reutelingsperger, C.P.M. Annexin V-Affinity assay: A review on an apoptosis detection system based on phosphatidylserine exposure. *Cytometry* **1998**, *31*, 1. [CrossRef]

55. Crowley, L.C.; Marfell, B.J.; Scott, A.P.; Waterhouse, N.J. Quantitation of Apoptosis and Necrosis by Annexin V Binding, Propidium Iodide Uptake, and Flow Cytometry. *Cold Spring Harb. Protoc.* **2016**, *2016*, 953. [CrossRef]

56. Morrison, G.R. Microchemical determination of organic nitrogen with nessler reagent. *Anal. Biochem.* **1971**, *43*, 527–532. [CrossRef]

57. Rubio-Clemente, A.; Cardona, A.; Chica, E.; Penuela, G.A. Sensitive spectrophotometric determination of hydrogen peroxide in aqueous samples from advanced oxidation processes: Evaluation of possible interferences. *Afinidad* **2017**, *74*, 578.

International Journal of
Molecular Sciences

Article

Multivariate Optimization of the FLC-dc-APGD-Based Reaction-Discharge System for Continuous Production of a Plasma-Activated Liquid of Defined Physicochemical and Anti-Phytopathogenic Properties

Anna Dzimitrowicz [1,*,†], Piotr Jamroz [1,†], Pawel Pohl [1], Weronika Babinska [2], Dominik Terefinko [1], Wojciech Sledz [2,‡] and Agata Motyka-Pomagruk [2,*,‡]

1 Department of Analytical Chemistry and Chemical Metallurgy, Wroclaw University of Science and Technology, 27 Wybrzeze St. Wyspianskiego, 50-370 Wroclaw, Poland; piotr.jamroz@pwr.edu.pl (P.J.); pawel.pohl@pwr.edu.pl (P.P.); dominik.terefinko@pwr.edu.pl (D.T.)
2 Laboratory of Plant Protection and Biotechnology, Intercollegiate Faculty of Biotechnology University of Gdansk and Medical University of Gdansk, University of Gdansk, 58 Abrahama, 80-307 Gdansk, Poland; weronika.babinska@phdstud.ug.edu.pl (W.B.); wojciech.sledz@biotech.ug.edu.pl (W.S.)
* Correspondence: anna.dzimitrowicz@pwr.edu.pl (A.D.); agata.motyka@biotech.ug.edu.pl (A.M.-P.); Tel.: +48-71-320-2815 (A.D.); +48-58-523-6330 (A.M.-P.)
† These Authors equally contributed to this work.
‡ These Authors also contributed equally to this work.

Citation: Dzimitrowicz, A.; Jamroz, P.; Pohl, P.; Babinska, W.; Terefinko, D.; Sledz, W.; Motyka-Pomagruk, A. Multivariate Optimization of the FLC-dc-APGD-Based Reaction-Discharge System for Continuous Production of a Plasma-Activated Liquid of Defined Physicochemical and Anti-Phytopathogenic Properties. *Int. J. Mol. Sci.* **2021**, *22*, 4813. https://doi.org/10.3390/ijms22094813

Academic Editor: Phillip E. Klebba

Received: 26 March 2021
Accepted: 27 April 2021
Published: 1 May 2021

Publisher's Note: MDPI stays neutral with regard to jurisdictional claims in published maps and institutional affiliations.

Abstract: To the present day, no efficient plant protection method against economically important bacterial phytopathogens from the *Pectobacteriaceae* family has been implemented into agricultural practice. In this view, we have performed a multivariate optimization of the operating parameters of the reaction-discharge system, employing direct current atmospheric pressure glow discharge, generated in contact with a flowing liquid cathode (FLC-dc-APGD), for the production of a plasma-activated liquid (PAL) of defined physicochemical and anti-phytopathogenic properties. As a result, the effect of the operating parameters on the conductivity of PAL acquired under these conditions was assessed. The revealed optimal operating conditions, under which the PAL of the highest conductivity was obtained, were as follows: flow rate of the solution equaled 2.0 mL min^{-1}, the discharge current was 30 mA, and the inorganic salt concentration (ammonium nitrate, NH_4NO_3) in the solution turned out to be 0.50% (m/w). The developed PAL exhibited bacteriostatic and bactericidal properties toward *Dickeya solani* IFB0099 and *Pectobacterium atrosepticum* IFB5103 strains, with minimal inhibitory and minimal bactericidal concentrations equaling 25%. After 24 h exposure to 25% PAL, 100% ($1-2 \times 10^6$) of *D. solani* and *P. atrosepticum* cells lost viability. We attributed the antibacterial properties of PAL to the presence of deeply penetrating, reactive oxygen and nitrogen species (RONS), which were, in this case, OH, O, O_3, H_2O_2, HO_2, NH, N_2, N_2^+, NO_2^-, NO_3^-, and NH_4^+. Putatively, the generated low-cost, eco-friendly, easy-to-store, and transport PAL, exhibiting the required antibacterial and physicochemical properties, may find numerous applications in the plant protection sector.

Keywords: non-thermal atmospheric pressure plasma; reactive oxygen and nitrogen species; *Pectobacteriaceae*; *Dickeya* spp.; *Pectobacterium* spp.; antibacterial; plant protection; agriculture

1. Introduction

Plant diseases remain a constant threat to agriculture, forestry, and food processing [1]. Among bacterial phytopathogens of the highest economic importance, *Pectobacterium* and *Dickeya* spp. from the *Pectobacteriaceae* family [2] are often listed [3–7]. These microorganisms affect various crops, vegetables, and ornamental plants with the symptoms of soft rot and/or blackleg [8,9]. The ubiquitous presence [10,11] of soft rot *Pectobacteriaceae* (SRP) is responsible for significant economic losses, especially in the European potato production sector that suffers each year from a 46 M euro damage [12].

It is worth considering that no efficient control measures against *Pectobacterium* and *Dickeya* spp. have been introduced to the market yet [13] even though a vast number of potential SRP control procedures involving physical (hot water, steam, hot dry air, and UV light), chemical (antibiotics, hydrogen peroxide, essential oils, nanoparticles, antimicrobial peptides, organic, and inorganic compounds) in addition to biological (antagonistic bacteria, fungi, and lytic phages) treatments have been studied [13,14]. Unfortunately, since none of the above-mentioned approaches fulfilled all the requirements of effectiveness, high-throughput, low cost, and no damage to the plant host or environment, therefore, they have not been implemented into practice [13,14].

In search for novel methods to combat SRP, we focused our attention on non-thermal atmospheric pressure plasmas (NTAPs) [15]. NTAPs could be applied against microorganisms either directly or indirectly [16]. Direct action of NTAP requires transportation of the usually non-portable plasma source and is believed to be expensive in exploitation. Besides, such exposure might lower the quality of agricultural goods, i.e., lead to changes in color, surface topography, or diminishment in the contents of bioactive compounds [17]. Hence, we developed an indirect NTAP-based SRP eradication method based on the usage of an easy-to-store and eco-friendly plasma-activated liquid (PAL).

The contact of NTAP with an aqueous solution triggers differences in its pH, redox potential, and conductivity. In addition, it leads to a variation in the profile and the concentration of reactive oxygen and nitrogen species (RONS) [17]. Numerous long-lived RONS, e.g., hydrogen peroxide (H_2O_2), nitrate (NO_3^-), ammonium (NH_4^+), and nitrite (NO_2^-), in addition to short-lived reactive species such as hydroxyl radicals ($\bullet OH$), peroxynitrite anions ($ONOO^-$), atomic oxygen (O), nitrogen oxide radicals ($\bullet NO$), and superoxide anions ($O_2\bullet^-$) of well-documented antimicrobial properties are produced in such a medium [18]. The type and the concentration of the above-listed RONS might be linked with the operating parameters of NTAPs, including plasma source, the discharge atmosphere, the current-voltage characteristics, the NTAP exposure time (in the case of stationary reaction-discharge systems), or the introduced solution flow rate (concerning continuous-flow reaction-discharge systems).

Several previous works aimed at the exploitation of RONS enclosed in PALs [19] for the inactivation of microorganisms. For instance, antibacterial action of PALs has been reported before against the model microorganism *Escherichia coli* [20–22], human and animal pathogens like *Staphylococcus aureus* [21,23], *Bacillus cereus* [24], *Enterococcus faecalis* [25], *Klebsiella pneumoniae* [22], *Acinetobacter baumannii* [22], *Pseudomonas aeruginosa* [22], common food spoilage agents such as *Listeria monocytogenes* [26], *Leuconostoc mesenteroides* [27], *Hafnia alvei* [27], and Salmonella Typhimurium [26] treated as planktonic cells [20,21,24,26,27] in the form of complex biofilm 3D structures [21,22,25–28] or as food products' surface contaminants [23,29–33]. However, little attention has been attributed to evaluating the potency of PALs for the eradication of plant pathogens instead of the food spoilage-agents [34]. So far, plasma activated waters (PAWs) have been tested against *Colletotrichum gloeosporioides* [35] causing bitter rot in numerous crops, *Xanthomonas vesicatoria* responsible for the leaf spot on tomatoes [36], *Colletotrichum alienum* being a post-harvest pathogen of avocado [37], and *Penicillium italicum* triggering blue and green molds on citrus fruits [38]. Concerning SRP, to the best of our knowledge, solely a direct impact of plasma generated either by electric discharges in a gliding arc reactor in a stationary phase [39] or a direct current atmospheric pressure glow discharge obtained in contact with a flowing cathode [40] has been reported as an effective method to combat these pests.

Even though water (tap, deionized, distilled [37]) is most commonly utilized for the generation of PALs of antimicrobial properties, phosphate-buffered saline, saline, and other NTAP-treated media [41] have also been investigated. In this work, according to the suggestions of Graves et al. [42], we aimed at combining the antibacterial action of PALs with nutritious features of a plasma-activated fertilizer. Having in mind that nitrogen fertilizers are most frequently utilized for boosting the yield and quality of the crops [43], we decided to base the formulation of the herein-reported PAL on an ammonium nitrate (NH_4NO_3) aqueous solution.

It is worth considering that various NTAP sources, including atmospheric pressure plasma jet [44], gliding arc electric discharge [27], and transient spark discharge [45,46] have been employed so far to generate PALs of documented antimicrobial properties. Unfortunately, the majority of the already conducted studies described the production of PALs in stationary NTAP-based reaction-discharge systems, which means that solely a predefined volume of a liquid has been exposed to the plasma [27,44]. Regarding a circle working mode, as far as we are aware, only transient spark discharge-based reaction-discharge systems [45,46] or streamer corona discharge-based systems [46] have been implemented in such a way for the acquisition of PALs [45,46]. Notably, no high-throughput reaction-discharge system, employing direct current atmospheric pressure glow discharge in contact with a flowing liquid cathode (FLC-dc-APGD), was utilized before for the production of PAL designated for future agricultural uses.

Thus, we undertook the development of a procedure for the acquisition of PAL of defined physicochemical properties, by implementing the design of experiments (DOE) followed by response surface methodology (RSM) approaches intended for a multivariate optimization of the operating parameters of the FLC-dc-APGD-based reaction-discharge system. The continuous-flow character of the FLC-dc-APGD system developed by our research group may be considered a novelty in contrast of the majority of previous studies on the stationary reaction-discharge systems [45,46]. In addition, the herein applied multivariate optimization approach is innovative in comparison to the one-factor-at-time (OFAT) method frequently used at the optimization stage for evaluating the impact of the discharge gas flow rate [45] or the gas mixture [45] on the physicochemical composition of the analyzed liquids. Here, we established the optimal operating conditions for continuous production of PAL based on the response surface regression models. For the first time, we demonstrated antibacterial properties of PAL against plant pathogens of high economic importance, which, in this case, belong to two diverse SRP species, namely *Dickeya solani* and *Pectobacterium atrosepticum*. The observed antimicrobial action of plasma-treated solution was attributed, according to the detailed qualitative and quantitative analyses, to RONS and solvated electrons (e^-_{aq}) produced in PAL. In summary, this work is a response to a challenge associated with the effective production of PAL of required antibacterial and physicochemical properties, which will be easy to collect and store. A pioneering character of this research results from the usage of the reaction-discharge system working in a flow-through mode for an efficient generation of a fertilizer-based PAL of potent antibacterial properties, which were documented for the first time toward SRP.

2. Results and Discussion

2.1. Response Surface Regression Models Describing the Effect of the Operating Parameters of the FLC-dc-APGD System on Electrical Conductivity of the Resultant NH_4NO_3-Based PAL

All experimental treatments listed in the Box-Behnken Design (BBD) matrix (see Table 1) were carried out in one block, according to the randomized run order. σ_{1h} and σ_{24h} of all the resultant PALs were measured after 1 and 24 h, respectively. Since three independent aliquots were taken for each sample treatment, means of σ_{1h} and σ_{24h} for each PAL were calculated along with variances for these three-point measurement series. Initially, scatter plots of the previously mentioned variances versus the respective means of σ_{1h} and σ_{24h} were used to judge whether variability of the means of σ_{1h} and σ_{24h} between treatments was greater than variability of these means within single treatments. Scatter plots of the means of σ_{1h} and σ_{24h} versus the randomized run order were used to look for any patterns or trends. Since variability of the means of σ_{1h} and σ_{24h} between treatments was higher than variability within these treatments, and no trends or patterns were observed in both datasets, it was concluded that the differences in the measured values of σ_{1h} and σ_{24h} resulted only from changes in the settings of the operating parameters of the FLC-dc-APGD system.

Table 1. Box-Behnken response surface design with actual and (coded) values of operating parameters related to application of a continuous-flow FLC-dc-APGD reaction discharge system for the production of NH_4NO_3-based PALs having certain electrical conductivity measured after 1 (σ_{1h}) and 24 h (σ_{24h}).

Order		A, mL min^{-1}	B, mA	C, %	σ_{1h}, mS cm^{-1}	σ_{24h}, mS cm^{-1}
Standard	Run					
11	1	6.0 (+1)	40 (0)	0.1 (−1)	2.230	2.310
5	2	2.0 (−1)	30 (−1)	0.3 (0)	5.180	5.230
13	3 [a]	4.0 (0)	40 (0)	0.3 (0)	5.000	5.200
7	4	6.0 (+1)	30 (−1)	0.3 (0)	4.810	5.060
4	5	4.0 (0)	50 (+1)	0.5 (+1)	7.960	8.090
2	6	4.0 (0)	50 (+1)	0.1 (−1)	2.240	2.260
15	7 [a]	4.0 (0)	40 (0)	0.3 (0)	5.130	5.270
3	8	4.0 (0)	30 (−1)	0.5 (+1)	7.800	7.880
10	9	2.0 (−1)	40 (0)	0.5 (+1)	8.400	8.820
6	10	2.0 (−1)	50 (+1)	0.3 (0)	5.120	5.170
8	11	6.0 (+1)	50 (+1)	0.3 (0)	5.320	5.340
12	12	6.0 (+1)	40 (0)	0.5 (+1)	7.510	7.880
9	13	2.0 (−1)	40 (0)	0.1 (−1)	2.230	2.270
14	14 [a]	4.0 (0)	40 (0)	0.3 (0)	5.240	5.300
1	15	4.0 (0)	30 (−1)	0.1 (−1)	2.100	2.120

A: The flow rate of the FLC solution (NH_4NO_3) (in mL min^{-1}). B: The discharge current of the FLC-dc-APGD system (in mA). C: The NH_4NO_3 concentration in the FLC solution (in %). [a] Center points at A = 4.0 mL min^{-1} (0), B = 40 mA (0), C = 0.3% (0).

Means of σ_{1h} and σ_{24h}, assessed for the Box-Behnken response surface design were fitted next with full quadratic functions, including linear (A, B, C), square (A^2, B^2, C^2), and combined (A × B, A × C, B × C) terms. A backward-elimination-of-terms algorithm at $\alpha = 0.05$ was used to select only these terms that were statistically significant and contributed to the changes in σ_{1h} and σ_{24h} of PALs produced in the FLC-dc-APGD-based reaction-discharge system. The model hierarchy was kept in this case, leaving all lower-order terms that comprised higher-order terms. Statistically significant terms in the regression models of σ_{1h} and σ_{24h} along with the respective *p*-values confirming this significance and the lack-of-fit test are gathered in Table 2. Detailed analyses of variance (ANOVA) summaries with the statistics for both response surface regression models are given in Table 3. Values of R^2, adjusted R^2, and predicted R^2 are included in Table 2 to point out goodness-of-fit and the forecast performance of both models.

Table 2. *p*-values for response surface regression models as well as linear and square effects of parameters A, B, and C along with their two-way interactions included in these models to describe changes in σ_{1h} and σ_{24h} (in mS cm^{-1}) of PALs produced by using the continuous-flow FLC-dc-APGD reaction-discharge system. Statistically significant terms included in the developed regression models are given in brackets.

	p-Values					R^2, %	R^2 Adjusted, %	R^2 Predicted, %	S
	Model	Linear	Square	Two-Way Interactions	Lack-of-Fit				
σ_{1h}	0.000	0.000 (A, C)	-	0.016 (A × C)	0.070	99.5	99.4	98.7	0.173
σ_{24h}	0.000	0.000 (A, B, C)	-	0.004 (A × B, A × C)	0.626	99.8	99.7	99.3	0.114
Regression equations modelling the effect of examined parameters (A, B, and C) [a]									
σ_{1h}	$0.258 - 0.128 \times A + 17.269 \times C - 0.613 \times A \times C$								
σ_{24h}	$1.159 - 0.184 \times A - 0.019 \times B + 16.519 \times C + 0.007 \times A \times B + 0.556 \times A \times C$								

σ_{1h}—Electrical conductivity measured after 1 h. σ_{24h}—Electrical conductivity measured after 24 h. A: The flow rate of the FLC solution (NH_4NO_3) (in mL min^{-1}). B: The discharge current of the FLC-dc-APGD system (in mA). C: The NH_4NO_3 concentration in the FLC solution (in %). R^2: Coefficient of determination. S: Residual standard deviation. [a] A backward-elimination-of-terms algorithm at $\alpha = 0.05$ was applied to determine statistically significant terms in response to surface regression models.

Table 3. ANOVA statistics and results of the lack-of-fit test for the response surface regression models established using the backward-elimination-of-terms algorithm (at $\alpha = 0.05$) for production of PALs in the continuous-flow FLC-dc-APGD reaction-discharge system.

Source of Data	DF	Adjusted SS	Adjusted MS	F-Value [a]	*p*-Value
Electrical conductivity measured after 1 h (σ_{1h})					
Model	3	70.612	23.537	785.36	$0.000 < 0.05$
Linear	2	65.172	32.586	1087.29	$0.000 < 0.05$
A	1	0.101	0.101	3.28	0.093
C	1	65.071	65.071	2171.20	$0.000 < 0.05$
Two-way interactions	1	0.240	0.240	8.01	$0.016 < 0.05$
A×C	1	0.240	0.240	8.01	$0.016 < 0.05$
Error	11	0.330	0.030		
Lack-of-fit	9	0.324	0.036	13.69	$0.070 > 0.05$
Pure error	2	0.005	0.003		
Total	14	70.942			
Electrical conductivity measured after 24 h (σ_{24h})					
Model	5	65.870	13.174	1006.94	$0.000 < 0.05$
Linear	3	60.683	20.228	1546.08	$0.000 < 0.05$
A	1	0.221	0.221	16.92	$0.003 < 0.05$
B	1	0.024	0.024	1.84	0.208
C	1	60.399	60.399	4616.53	$0.000 < 0.05$
Two-way interactions	2	0.279	0.140	10.67	$0.004 < 0.05$
A×B	1	0.081	0.081	6.21	$0.034 < 0.05$
A×C	1	0.198	0.198	15.14	$0.004 < 0.05$
Error	9	0.118	0.013		
Lack-of-fit	7	0.089	0.013	0.88	$0.625 > 0.05$
Pure error	2	0.029	0.014		
Total	14	65.987			

DF: Degrees of freedom. SS: The sum of squares. MS: The mean of squares. A: The flow rate of the FLC solution (NH_4NO_3). B: The discharge current of the FLC-dc-APGD system. C: The NH_4NO_3 concentration in the FLC solution. [a] The value of the F-test for comparing model variance with residual variance.

Regression models of σ_{1h} and σ_{24h} were statistically significant since *p*-values equaled 0.000 (see Table 2). *p*-values for the lack-of-fit test were higher than $\alpha = 0.05$, particularly in case of the model for σ_{24h}, confirming that there were no reasons to reject both regression models. R^2 values, showing the degree to which the selected operating parameters explained variance in the collected datasets, were higher than 99%. All the above-listed measures pointed out that the developed response surface regression models properly described the relationships between the examined parameters of the FLC-dc-APGD system and the σ_{1h} and σ_{24h} measures for PALs produced with this system. High values of the predicted R^2 indicated that both models showed the ability to reliably predict responses for a new set of the operating parameters. Interestingly, in the case of both models (Table 2), the square terms were insignificant, while the term B, i.e., the discharge current of the FLC-dc-APGD system, did not contribute to the overall values of σ_{1h} and σ_{24h}.

Finally, residual plots, i.e., normal probability plots and scatter plots of regular residuals versus the run order for both regression models confirmed their goodness-of-fit (Figure 1). Fairly normal distribution of the residuals in the case of the normal probability plots in addition to random and uncorrelated patterns of the residuals in the scatter plots versus the run order additionally indicated that the examined operating parameters of the FLC-dc-APGD system used for the production of PALs affected the measured σ_{1h} and σ_{24h} in a systematic way.

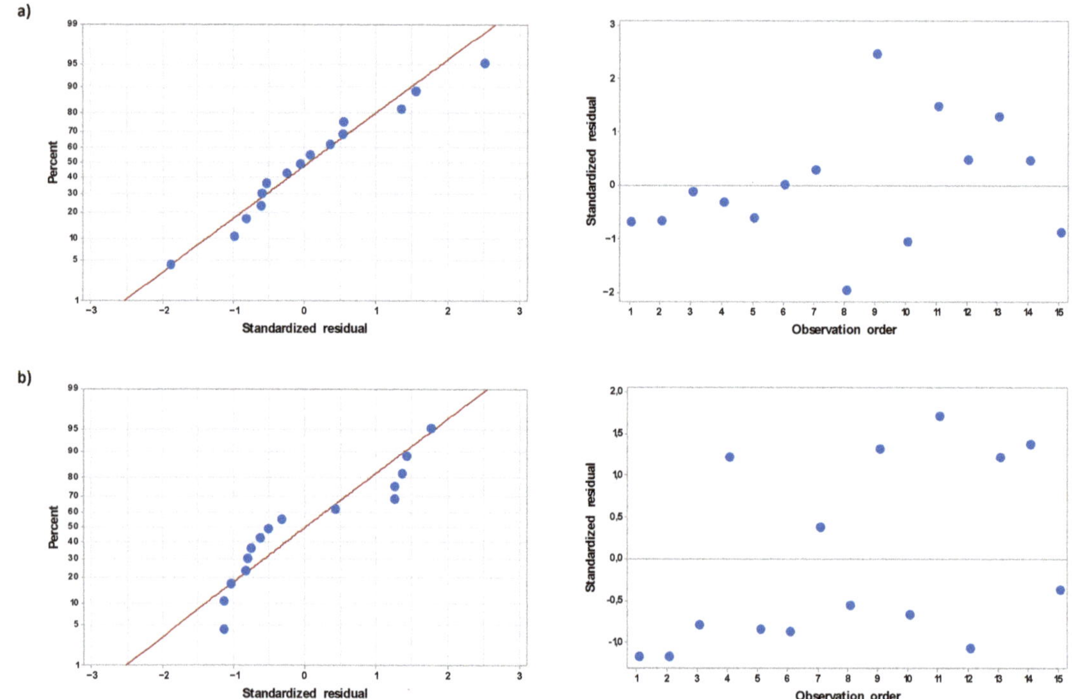

Figure 1. Normal probability plots and scatter plots of residuals versus observation (run) order for surface regression models predicting electrical conductivity of PALs produced using the continuous-flow FLC-dc-APGD system, as measured after (**a**) 1 h (σ_{1h}) and (**b**) 24 h (σ_{24h}).

2.2. Selection of the Optimal Experimental Conditions for Production of the NH_4NO_3-Based PALs of the Highest Electrical Conductivity

The ANOVA statistics of the developed response surface regression models of σ_{1h} and σ_{24h}, as measured in PALs produced in the continuous-flow FLC-dc-APGD reaction-discharge system, fitted the measured data well and described variation in both responses due to changes in the operating parameters of the applied reaction-discharge system. For that reason, both regression models were used for selecting a combination of the settings of the operating parameters of the FLC-dc-APGD system that was enabled to produce PALs of the highest σ_{1h} and σ_{24h}. For this purpose, desirability functions of σ_{1h} and σ_{24h}, i.e., $d(\sigma_{1h})$ and $d(\sigma_{24h})$, were applied and the values of these functions at given settings of the operating parameters of the reaction-discharge system provided the composite desirability (D) value. The latter D value was used to point how well a certain combination of the parameters satisfied the optimization goal, i.e., production of PALs with maximal σ_{1h} and σ_{24h} values. The highest values of $d(\sigma_{1h})$, $d(\sigma_{24h})$, and D, being 0.992, 0.957, and 0.974, respectively, were found for the following combination of the experimental parameters: A = 2.0 mL min^{-1}, B = 30 mA, and C = 0.50%. For these optimal operating parameters, the values of σ_{1h} and σ_{24h} of PALs were predicted by the models to be 8.347 $\pm$ 0.111 mS cm^{-1} and 8.535 $\pm$ 0.130 mS cm^{-1}, respectively. Both values were within the range of the lower and the upper value of σ measured in the experiment, i.e., 2.10–8.40 mS cm^{-1} for σ_{1h} and 2.12–8.82 mS cm^{-1} for σ_{24h}.

The optimization plot, showing how different settings of the operating parameters affect the predicted response of σ_{1h} and σ_{24h}, are given in Figure 2. It appears that an increase of the flow rate of the FLC solution (parameter A) in both models resulted in a gradual decrease of the σ_{1h} and σ_{24h} values, hence, in a lower production rate of the long-lived reactive nitrogen species (RNS) like: NO_2^-, NO_3^-, and NH_4^+. This could be likely the consequence of a relatively shorter time of exposure of the surface of the FLC solution to the discharge when the flow rate of the FLC solution increased. Similar results concerning the effect of the flow rate of the FLC solution on the concentration of the generated NO_2^-, NO_3^-, and NH_4^+ ions in a comparable discharge system were previously reported by Jamroz et al. [47]. In the latter device, in spite of an increase in the energy yield of the discharge, the water evaporation rate was reduced and, hence, the concentration of water vapor and H_2O^+ ions in the discharge phase was also diminished. As a result, the concentration of the products of dissociate electron-recombination of the H_2O^+ ions and electron-impact dissociation of the water molecules themselves, i.e., $\bullet$H and $\bullet$OH radicals as well as other O reactive species had to be lower in these conditions as well [48,49]. The main source of NO_2^-, NO_3^-, and NH_4^+ ions in the liquid treated by APGD systems operated in contact with this solution playing the role of the cathode of the discharge system, are nitric oxide (NO), nitric dioxide (NO_2), and nitric hydride (NH) species, respectively [47,48]. These di-atomic and tri-atomic molecules readily react with $\bullet$OH and $\bullet$H radicals in the discharge phase according to the following reactions: $NO + OH = HNO_2$, $NO_2 + OH = HNO_3$, $NH_x + H = NH_{x+1}$, where x = 0–2 [48,50]. When the concentration of $\bullet$H and $\bullet$OH radicals in the discharge phase is decreased, the concentration of the NO, NO_2, and NH molecules is also lower because they are obtained through reactions between active N_2 molecules with the above-mentioned $\bullet$H radicals and free active O radicals [47,48,50]. Surprisingly, the effect of the discharge current (parameter B) was negligible. Certainly, water evaporation from the surface of the FLC solution as well as acceleration of different particles in the discharge phase is likely higher when the discharge current is increasing [51]. In these conditions, collision processes are more intensive and the population of the excited species rises, as measured by different optical temperatures [51]. Finally, augmentation in the concentration of NH_4NO_3 in the FLC solution led to a linear increase in σ, likely as a result of the elevation in the concentration of electric charge carriers in the liquid.

The appropriateness of the optimal operating parameters of the FLC-dc-APGD system for the production of NH_4NO_3-based PALs with the highest values of σ_{1h} and σ_{24h} was verified in the additional independent experiment. Accordingly, solutions of NH_4NO_3 were treated by FLC-dc-APGD in the applied continuous-flow reaction-discharge system, operated under the above-defined conditions, i.e., A = 2.0 mL min^{-1}, B = 30 mA, and C = 0.50%. The electrical conductivity of the collected aliquots of the resultant PALs measured after 1 and 24 h equaled 8.94 mS cm^{-1} for σ_{1h} and 9.11 mS cm^{-1} for σ_{24h}, respectively. These values turned out to be comparable to those predicted by the developed regression models. Relative errors between them were 7% (σ_{1h}) and 6% (σ_{24h}). Therefore, the reliability of both response surface regression models and the appropriateness of the optimal operating conditions of the FLC-dc-APGD system for the production of the NH_4NO_3-based PALs of pre-defined σ have been confirmed.

Since there are several operating parameters, including the discharge current, the flow rate, and the concentration of the utilized precursor, which might have an effect on the physicochemical properties of the resultant PAL, a multivariate optimization involving DOE along with RSM was applied to find the optimal operating conditions for the production of the PAL of certain desirable properties. In comparison to previous works [45,46], this kind of the multiparameter optimization approach allowed us not only to limit the number of the required experiments, but also to estimate prospective interactions between the studied operating conditions and their effect on physicochemical properties of the PAL. In such a way, efficient production of the PAL having the required physicochemical parameters (e.g., conductivity) could have been achieved.

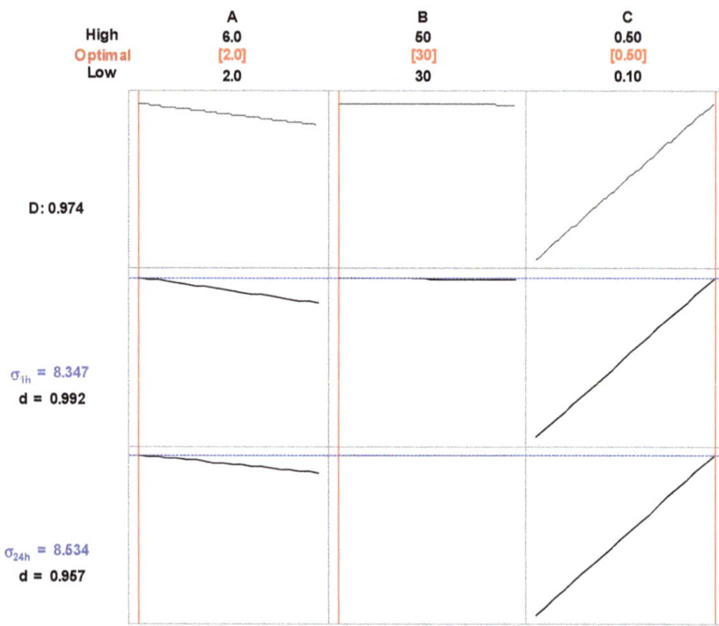

Figure 2. Optimization plots–the effect of the flow rate of the FLC solution (**A**), the discharge current of the FLC-dc-APGD system (**B**), and the NH_4NO_3 concentration (**C**) on electrical conductivity of PALs produced in the continuous-flow FLC-dc-APGD system, as measured after 1 h (σ_{1h}) and 24 h (σ_{24h}).

2.3. Antibacterial Action Against Plant Pathogens of 0.5% NH_4NO_3 Treated in the Optimized FLC-dc-APGD System

It is worth noticing that both species selected for this study, namely *P. atrosepticum* and *D. solani*, belong to the highly homogenous [52–55] representatives of the *Pectobacteriaceae* family. *P. atrosepticum* strains have been commonly isolated from seed potato fields in temperate climate regions for more than a century [56], while *D. solani* is believed to be an emerging pathogen of high economic importance [7].

The 24 h exposure of *D. solani* IFB0099 and *P. atrosepticum* IFB5103 cells to 25% or 50% PAL was potent enough to inhibit the growth of the investigated phytopathogens in TSB (Trypticase Soy Broth) medium (Table 4). The application of 1% and 10% PAL did not result in growth inhibition of the tested plant pathogenic bacteria (Table 4). The included control samples did not only prove a lack of the antibacterial properties of the 0.5% NH_4NO_3 solution not treated by plasma, but also confirmed the viability of the utilized bacterial cells in addition to the sterility of the solutions, the water, and the media used. Therefore, *D. solani* IFB0099 and *P. atrosepticum* IFB5103 turned out to be highly susceptible to the plasma-treated 0.5% NH_4NO_3 solution with minimal inhibitory concentration (MIC) equaling 25% (Table 4).

The incubation (24 h, 28 °C) of *D. solani* IFB0099 or *P. atrosepticum* IFB5103 suspensions on TSA (Trypticase Soy Agar) media after subjection of bacterial cells either to 25% or 50% PAL for 24 h resulted in the observed absence of bacterial colonies (Table 4). However, the utilization of 1% or 10% PAL in an analogous experiment was not effective enough to kill the cells of the investigated phytopathogens (Table 4). The utilized control samples did not only demonstrate a lack of bactericidal action of the 0.5% NH_4NO_3 solution untreated by plasma, but also proved the viability of bacterial cells in addition to the sterility of the solutions, the water, and the media used. Thus, 25% concentrated PAL was potent enough to eradicate the studied phytopathogens and was designated as minimal bactericidal concentration (MBC) (Table 4). In other words, 100% ($1-2 \times 10^6$) cells of *D. solani* IFB0099 and *P. atrosepticum* IFB5103 included in this experiment lost viability due to 24 h of exposure to the developed PAL.

Table 4. Antibacterial properties of the plasma-treated 0.5% NH_4NO_3 solution against phytopathogens.

Bacterial Strain	Assay	Concentration of PAL			
		1%	10%	25%	50%
Dickeya solani	MIC	+	+/−	-	-
IFB0099	MBC	+	+/−	-	-
Pectobacterium atrosepticum	MIC	+	+/−	-	-
IFB5103	MBC	+	+/−	-	-

PAL: Plasma-activated liquid. MIC: Minimal inhibitory concentration. MBC: Minimal bactericidal concentration. +: growth of bacterial cells was observed/lack of antibacterial action. −: no bacterial growth/antibacterial properties were observed. The experiment was repeated three times with two technical repetitions in each. The following control samples were included: MIC and MBC assays performed without the addition of PAL (either 0.5% NH_4NO_3 solution or water utilized for diluting PAL was applied instead). In addition, control samples evaluating the viability of bacterial cells (TSB medium + 0.5 McF bacterial suspension) or sterility of the used components (just TSB medium, TSB + PAL, TSB + 0.5% NH_4NO_3 solution, TSB + water applied for dilutions, TSB + 0.85% NaCl used for the preparation of bacterial suspensions) were utilized in the MIC and MBC procedures.

Among phytopathogens previously subjected to the action of PALs, the causative agents of fungal diseases dominate [35,37,38]. Wu et al. [35] utilized either oxygen or air as working gases to obtain PAWs in a corona plasma jet-based system. The resultant efficacy of *Colletotrichum gloeosporioides* inactivation reached 56% or 96% (post 10 or 30 min of the NTAP treatment, respectively) in terms of the air-PAW, while the action of the oxygen-PAW resulted in a loss of cell viability in 15% or 55% (after 10 or 30 min of exposure, respectively). Regarding the work of Siddique et al. [37], a Dyne-A-Mite HP surface treatment plasma machine was utilized for the generation of PAWs, based on tap, deionized, or distilled water, that were subjected to the action of the discharge either for 30 or 60 min. From the obtained PAWs, the ones based on deionized and distilled water post 60 min exposure to plasma that were diluted to 50% concentration turned out to successfully inhibit germination of the phytopathogenic conidia [37]. Moreover, Guo et al. [38] applied a dielectric barrier discharge (DBD) plasma device for the acquisition of PAW15, PAW30, PAW45, and PAW60, named according to the provided irradiation time. Statistically significant, i.e., 0.75, 1.3, and 3.3 log reductions resulted from 30 min of incubation in PAW30, PAW45, and PAW60 of *Penicillium italicum* conidia that had been attached to the surface of kumquats. No notable variations in the color, thickness, or the total contents of ascorbic acid, flavonoids, and carotenoids in the fruit have been noted [38]. To the best of our knowledge, from the group of bacterial plant pathogens, solely *Xanthomonas vesicatoria*, has been subjected to the action of PAL before [36]. In that study, no direct antibacterial action of water activated by DBD was noted in vitro. However, the enhancement of host plant defense systems leading to the limitation of the disease symptoms development on tomatoes was observed [36]. In view of the former research, we demonstrated, for the first time, antibacterial properties of PAL toward economically important bacterial phytopathogens. The higher susceptibility to PAL of soft rot *Pectobacteriaceae* in contrast to *Xanthomonas vesicatoria* might not only be associated with the plasma source and experimental setup, but also the utilized microbial density and culture state in addition to such apparent factors as the studied species and even the strain [57]. It was previously established that antibacterial properties of PALs have been associated with changes in pH, oxidation-reductive potential, UV radiation, shock waves, photons, electric fields, and, most of all, the generated RONS [17,28,58]. Since extensive characterization and elucidation of the complex plasma chemistry is required not only to reveal the antimicrobial mechanism of action of the obtained PAL, but also to make a step in reaching a generally regarded as safe (GRAS) status [17,26] for the developed formulation, a profound physicochemical study of the produced RONS has been conducted.

2.4. Examination of Interactions and Processes Leading to Acquisition of NH₄NO₃-Based PAL of the Defined Physicochemical and Antibacterial Properties

The following qualitative and quantitative measurements of RONS, present in the FLC-dc-APGD-treated 0.5% NH_4NO_3 solution in contrast to the unexposed control sample, have been performed.

In qualitative evaluation, optical emission spectrometry (OES) (Figure 3) was applied to identify numerous bands of NO and N_2 molecules, belonging to the γ-system ($A^{2\Sigma+}$-$X^2\Pi$) and the second positive system ($C^3\Pi_u$-$B^3\Pi_g$), respectively. Additionally, two strong bands of OH molecules (with heads at 281.1 nm (1-0) and 306.4 nm (0-0), belonging to the $A^2\Sigma$-$X^2\Pi$ system), as well as the band of NH molecules (with the head at 336.0 nm, belonging to the $A^3\Pi_g$-$X^3\Sigma^-$ (0-0) system), were detected. The band of N_2^+ of the first negative system ($B^2\Sigma^+_u$-$X^2\Sigma^+_g$) (with heads at 391.4 nm (0-0) and 427.1 nm (0-1)) was also observed. In the ranges of 400–500 nm and 550–900 nm, several bands of N_2 molecules, belonging to the second positive system ($C^3\Pi_u$-$B^3\Pi_g$) and first positive system ($B^3\Pi_g$–$A^3\Sigma^+_u$), respectively, were clearly visible. Moreover, two lines of hydrogen (H), i.e., the strong one Hα at 656.28 nm and a very weak Hβ at 486.13 nm, in addition to O lines (at 777.2, 772.4, and 844.6 nm) were identified. Based on these OES results, it has been concluded that the following RONS: OH, NH, N_2, NO_x, N_2^+, and O, were excited in the plasma (gas) phase of FLC-dc-APGD.

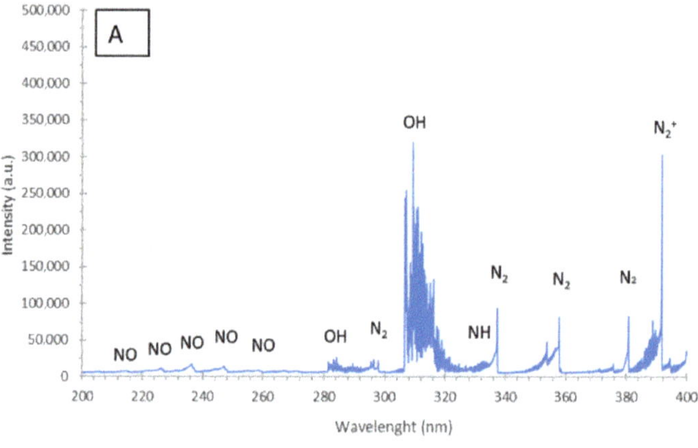
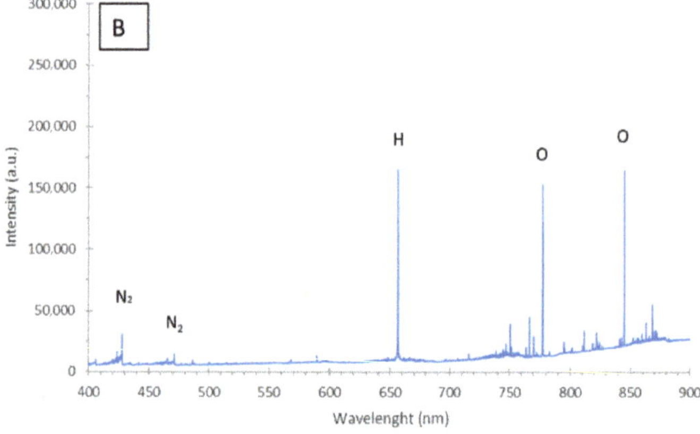

Figure 3. The optical emission spectra acquired in two spectral ranges: (**A**) 200–400 nm and (**B**) 400–900 nm during the FLC-dc-APGD treatment of the 0.5% NH_4NO_3 solution under the optimal operating conditions.

Subsequently, quantitative analyses of the selected RNS were carried out for 0.5% NH_4NO_3 treated solution under optimal operating conditions by FLC-dc-APGD, in contrast to the untreated control sample. In the case of all investigated RNS, including NO_2^-, NO_3^-, and NH_4^+, their concentrations were shown to change after the exposure of the solution to FLC-dc-APGD. The content of NO_2^- increased twice (from 4.5 to 9.3 mg L^{-1}) post the NTAP action. In addition, the concentration of NO_3^- was elevated from 1195 to 1624 mg L^{-1} after the FLC-dc-APGD-treatment. Similarly, the concentration of NH_4^+ was augmented from 710 to 957 mg L^{-1} due to the applied treatment.

Then the total amount of reactive oxygen species (ROS, Figure 4), involving OH, O, O_3, H_2O_2, and HO_2, was determined in the NTAP-treated 0.5% NH_4NO_3 solution with the use of the color-forming reaction with the KI-starch reagent [59]. It was found that the absorbance of the PAL-KI starch system remained stable over time after the NTAP operation (a mean absorbance value was ≈ 0.6), suggesting that the long-lived ROS were present in this solution at the concentration of 28.79; 19.76; 25.63; 25.21; 24.85 mg L^{-1} at subsequent time points. As expected, the measured absorbance for the control sample (untreated 0.5% NH_4NO_3 solution) was close to 0 and unchanged in time (Figure 4).

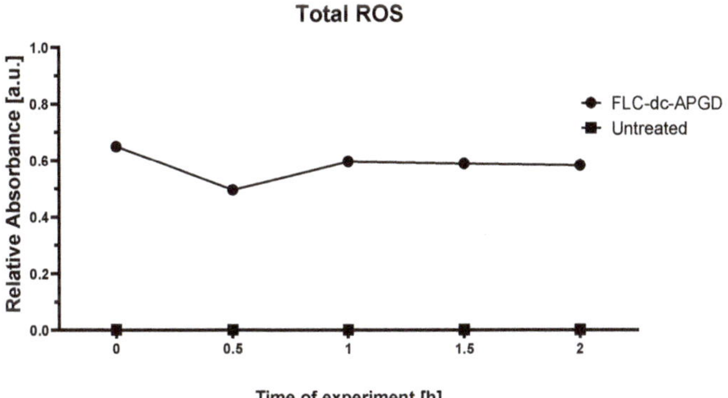

Figure 4. The absorbance measured at 590 nm for the PAL produced in the FLC-dc-APGD-based reaction-discharge system and the untreated 0.5% NH_4NO_3 solution.

To delve further into the topic of ROS generated in the resulting PAL and establish the spatial distribution of these molecules in the volume of the liquid (Figure 5), the interactions between FLC-dc-APGD and a droplet of the 0.5% NH_4NO_3 solution placed on the KI-starch gel were studied. At the beginning of the action of FLC-dc-APGD (stage I, Figure 5) initiated under optimal operating conditions (the flow rate of solution = 2.0 mL min^{-1}, the discharge current = 30 mA, and NH_4NO_3 concentration = 0.50% (m/v)), the oxidation process occurred in a ring-shaped pattern characterized by the intact and colored areas of 5.14 mm and 9.60 mm in diameter, respectively. The penetration depth equaled in this case ~1 mm. Therefore, the oxidation processes occurred in the first layers. The expansion of the previously mentioned ring-shaped region was observed following the prolongation of the contact between the 0.5% NH_4NO_3 solution placed on KI-starch gel and FLC-dc-APGD for the average time in the flowing regime (stage II, Figure 5). In this case, the diameter of the central intact area did not change. However, the colored region enlarged to 11.88 mm. Furthermore, the penetration depth through the gels was higher, reaching 3.72 mm. Finally, the most prominent spatial distribution of the oxidative area was detected after the prolonged NTAP treatment (stage III, Figure 5), leading to generation of a colored area of 15.74 mm without any intact fragment. The produced ROS penetrated the KI-starch gel deeply (5.63 mm). In essence, the NTAP exposure led to the generation of ROS in the entire volume of the irradiated droplets, leading to the acquisition of a stable in time PAL of high oxidative potential. To the best of our knowledge, spatial distribution of total ROS generated during the NTAP operation in the continuous-flow reaction-discharge system has not been examined yet. However, the resultant ring-shaped oxidative area typically occurs in other NTAP sources, e.g., in a DBD system [60] or in an atmospheric pressure plasma jet [61], presenting similar patterns and a comparable penetration depth into the KI-starch gel.

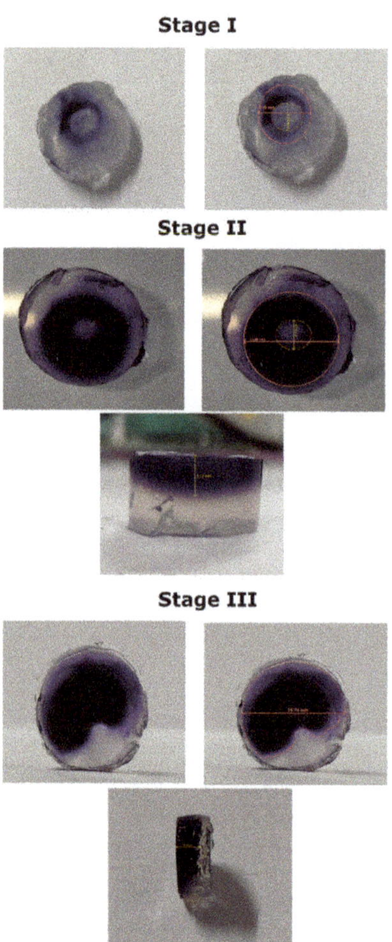

Figure 5. Spatial distribution of ROS in the KI-starch gels. Stage I—photograph from initial generation of FLC-dc-APGD (1 s). Stage II—changes recorded after an average time (2 s) of FLC-dc-APGD generation in a flowing mode. Stage III—variation resulting from a prolonged (4 s) FLC-dc-APGD irradiation. Prior to NTAP exposure, the gels had been covered with a droplet of 0.5% (m/v) NH_4NO_3. Navy blue tint results from effective ROS generation. The gels were photographed from the front, back, and side views.

A synergistic antibacterial action of the detected ROS and RNS, i.e., OH, O, O_3, H_2O_2, HO_2, NH, N_2, N_2^+, NO_2^-, NO_3^-, and NH_4^+ might be postulated. ROS destroy microbial cell membranes, cell wall, DNA, RNA, proteins, carbohydrates, and other components of the intracellular machinery [17,28,62]. Active and structural biomolecules are oxidized and chemical bonds within peptidoglycan of the cell wall break. Cytoplasmic leakage resulting from the loss of integrity of the depolarized outer membrane follows. Regarding RNS, these radicals trigger not only lipid peroxidation associated with the loss of a hydrogen atom from a methylene group, leading to cross-linkage of the fatty acid side chain and creation of pores in the cellular membrane [62], but also oxidize tyrosine residues of proteins, thiols, and unsaturated fatty acids [17]. It is worth considering that NTAP-generated RONS might reach the inner part of the bacterial cell by active transport as well as due to infiltration through pores resulting from the lipid peroxidation process. An intracellular concentration of the active species may also rise post subjection to oxidative and nitrosative stresses [19,62].

Notably, the varying antibacterial impact of the PAL-related RONS in the case of human pathogens was attributed to Gram classification of the treated microbe, physiological condition of the culture, and whether planktonic cells or biofilms were exposed to the plasma-activated media [28]. Different bacterial responses to active species may also be linked with diverse efficacy of microbial oxidative and nitrosative stress response systems [19], which rely on specific enzymes, including catalases, peroxidases, and superoxide dismutases, in addition to the production of smaller molecules like thioredoxin, glutaredoxin, or glutathione [28]. Importantly, the herein reported RONS of a long-lived character were demonstrated to penetrate deeply into the activated liquid and the included *D. solani* and *P. atrosepticum* strains were unable to tackle this threat.

Referring to future applications, a high-voltage NTAP source, i.e., FLC-dc-APGD, operated in the continuous-flow reaction-discharge system has been utilized for the generation of PAL of defined physicochemical properties. In more detail, the type and concentration of the acquired RONS may be controlled by the current-voltage parameters. Due to the operation of the reaction-discharge system under the stated optimal conditions (discharge current: 50 mA, concentration of NH_4NO_3: 0.5% (m/v), flow rate of solution: 2.0 mL min^{-1}), a high-throughput production of PAL (120 mL h^{-1}) has been achieved. It is worth underlining that 1 mL of PAL was subjected to FLC-dc-APGD just for 30 s, which was a much shorter exposure time in comparison to other devices used for the production of the previously reported PALs of antimicrobial properties [37,38]. Furthermore, reaching physicochemical stability of the developed PAL was necessary to aim for the application of the developed active solution not only for decontamination, but also for the extension of storage time and shelf life of agricultural goods and other food products [58].

3. Materials and Methods

3.1. Reagents and Solutions

As a PAL solution precursor, ammonium nitrate NH_4NO_3 (Archem, Kamieniec Wroclawski, Poland) was used. To perform quantitative analyses of RONS, produced under the optimal operating conditions of PAL fabrication, commercially available reagents, i.e., Nitrite HR Reagent HI93708-0 (Hanna Instruments, Salaj, Romania) and Nitrate Reagent HI93728-0 (Hanna Instruments, Salaj, Romania) along with the Nessler's reagent (Pol-Aura, Zabrze, Poland), potassium iodide (Avantor Performance Materials, Gliwice, Poland), starch (Sigma-Aldrich, Steinheim, Germany), and agar (BTL, Lodz, Poland) were used. Re-distilled water was used throughout. All of the reagents were of an analytical grade or better.

3.2. Production of the Plasma-Treated Liquid by FLC-dc-APGD

The PAL was obtained in the continuous flow reaction-discharge system, previously developed by our group [63], employing FLC-dc-APGD as an NTAP source (Figure 6). In more detail, FLC-dc-APGD was generated in a 2.0 mm gap between a pin-type tungsten anode and NH_4NO_3 solution of a defined concentration (0.1–0.5% (m/v)) acting as the FLC. The flow rate of the introduced solution depended on the working conditions of the system, as shown in detail in the BBD matrix (Table 1). To control the flow rate of the NH_4NO_3 solution, a four-channel peristaltic pump (Masterflex L/S, Cole Palmer, Vernon Hills, IL, USA) was used. To link this pump with a discharge chamber, a silicone hose was connected to a quartz capillary (OD = 4.00 mm) onto which a graphite tube (OD = 6.0 mm) was mounted. Additionally, a platinum wire was attached to the quartz-graphite tube and connected to HV dc power supply (Dora Electronics Equipment, Wroclaw, Poland), which enabled us to set a certain discharge current value (30–50 mA, according to the BBD matrix, see Table 1). The discharge current values of 30–50 mA were stabilized by applying a 10 kΩ ballast resistor (Tyco Electronics, USA). The supplied voltage was maintained in the range of 1100–1400 V. After exposure of the NH_4NO_3 solution to the FLC-dc-APGD operation, 10 mL of such NTAP-treated NH_4NO_3 solution was collected to polypropylene tubes (Sarstedt, Numbrecht, Germany) and stored at 4 °C for further analyses.

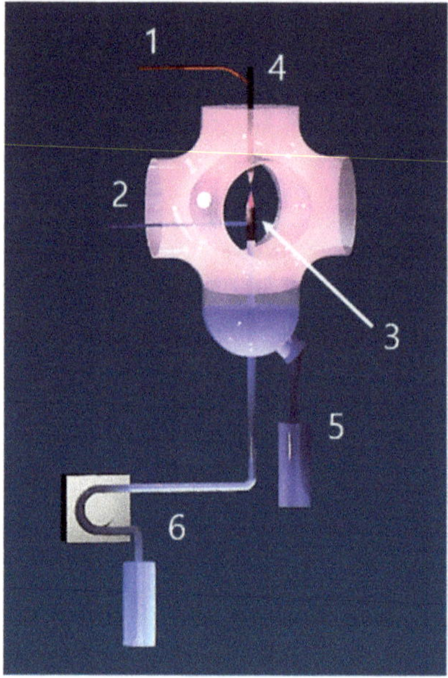

Figure 6. The NTAP-based reaction-discharge system used for PAL production. A layout of the NTAP-based reaction-discharge system applied for continuous production of PAL. (1)—negative potential supplied to a metallic anode. (2)—positive potential supplied to a flowing liquid cathode through a Pt wire. (3)—FLC-dc-APGD being a NTAP source. (4)—a metallic anode. (5)—PAL. (6)—a four channel peristaltic pump used for pumping the NH_4NO_3 solution.

3.3. Multiparameter Optimization of the Operating Parameters of FLC-dc-APGD System

In order to produce PAL of defined electrical conductivity (σ, in mS cm^{-1}) in the proposed continuous-flow reaction-discharge system, the operating parameters of this system were optimized by a multiparameter approach, i.e., the RSM, using a Minitab 17 software. A BBD was used for this purpose and the optimization procedure looked as follows: (a) all experiments according to treatments listed in the BBD matrix (Table 1) were carried out, (b) response surface regression models expressing changes of the response, i.e., σ of the resulting PALs measured after 1 and after 24 h, were developed, (c) abilities to describe the measured values of σ_{1h} and σ_{24h} and predict new values by these models were assessed, and, finally, (d) the optimal settings of the operating parameters of the reaction-discharge system providing the highest values of σ_{1h} and σ_{24h} for the produced PALs were selected.

The operating parameters, initially verified to have an effect on physicochemical properties of the PAL and selected for the optimization experiment by the BBD, were as follows: the flow rate of the NH_4NO_3 solution (A, in mL min^{-1}), the discharge current of FLC-dc-APGD sustained in contact with the NH_4NO_3 solution (B, in mA), and the concentration of NH_4NO_3 in this solution (C, in %). The response surface design included

15 randomized treatments (see Table 1) at three different levels (codded values in brackets) of the previously mentioned parameters, i.e., 2.0 mL min^{-1} (-1), 4.0 mL min^{-1} (0), and 6.0 mL min^{-1} (+1) for A, 30 mA (-1), 40 mA (0), and 50 mA (+1) for B, and 0.1% (-1), 0.3% (0), and 0.5% (+1) for C. Three center points were included in this response surface design. Ranges of the operating parameters were selected to provide a stable and reproducible operation of the FLC-dc-APGD system. All treatments were carried out in one block. For each treatment, three portions (10 mL) of the PAL produced at the given experimental conditions were collected into polypropylene vials (Sarstedt, Numbrecht, Germany) and, then, average values of σ_{1h} and σ_{24h}, measured in PALs with the use of an Elmetron CPC-505 pH/conductivity-meter (Zabrze, Poland), were used to develop appropriate response surface regression models. σ_{1h} and σ_{24h} were fitted with full quadratic functions with linear (A, B, C) and square (A^2, B^2, C^2) terms of the studied parameters, and their two-way interaction terms (AB, AC, BC). A general equation for the response surface regression models were given as: $\sigma = c_0 + c_1 \times A + c_2 \times B + c_3 \times C + c_{11} \times A^2 + c_{22} \times B^2 + c_{33} \times C^2 + c_{12} \times A \times B + c_{13} \times A \times C + c_{23} \times B \times C$, where c_0 is the intercept while c_1–c_{33} are regression coefficients. A backward-elimination-of-terms algorithm at $\alpha = 0.05$ was used to select statistically significant terms in these equations describing σ_{1h} and σ_{24h}.

Reliability of the developed response surface regression models for both responses, i.e., σ_{1h} and σ_{24h}, was checked by using ANOVA. In this case, goodness-of-fit (a degree of explanation of variance in measured responses by input variables) of established modeling functions was assessed through coefficients of determination (R^2). Their general explanatory power was evaluated by adjusted R^2 values, while the prediction power (ability to predict responses for new sets of variables) was given by the predicted R^2 values. In addition, *p*-values were given to point out the statistical significance (at $\alpha = 0.05$) of the regression models, and certain linear, square, and two-way interaction terms included in them. Finally, residuals associated with the developed regression models were visually analyzed to find any clear heteroscedasticities and non-normalities (normal probability plots) or correlations and trends (scatter plots of normalized residuals versus the run order).

The optimal settings of the operating parameters, leading to continuous production of PALs with the highest values of σ_{1h} and σ_{24h} in the applied flow-through reaction-discharge system, were selected on the basis of values of the individual d(σ_{1h}) and d(σ_{24h}) functions. The values of these functions changed from 0 (no satisfaction of the optimization goal) to 1 (full satisfaction of the optimization goal). The optimization goal was established in order to find such a set of parameters at which both σ_{1h} and σ_{24h} would be maximized. Hence, the D, being [d(σ_{1h}) $\times$ d(σ_{24h})]$^{\frac{1}{2}}$, was used to find this set of parameters and operate the FLC-dc-APGD system to produce NH$_4$NO$_3$-based PALs of pre-defined physicochemical properties in a controlled manner.

3.4. Assessment of Antibacterial Properties of 0.5% NH$_4$NO$_3$ Activated in the Optimized FLC-dc-APGD System

Plant pathogenic bacteria (Table 5) utilized for evaluating antimicrobial properties of the 0.5% NH$_4$NO$_3$-based PAL were kept as frozen stocks in 40% glycerol at -80 °C at the Intercollegiate Faculty of Biotechnology University of Gdansk and Medical University of Gdansk in Poland.

Prior to the experiment, the inoculation loop full of the bacterial biomass (Table 5), collected from the frozen stock, was streaked on a Trypticase Soy Agar (TSA, BTL, Poland). This culture was conducted for 24 h at 28 °C. Subsequently, a single bacterial colony was picked from the TSA plate and utilized for inoculation of 5 mL of a Trypticase Soy Broth (TSB, BTL, Poland) liquid medium. TSB containing bacterial cells was incubated at 28 °C with 120 rpm shaking for 24 h in order to obtain the overnight bacterial culture of a given strain. This culture was then centrifuged for 10 min at 6500 rpm and the resultant bacterial pellet was washed twice in sterile 0.85% NaCl solution. Afterward, the optical density of bacterial suspension was adjusted to 0.5 in McFarland (McF, approx. $1-2 \times 10^8$ colony forming units mL^{-1} [66]) scale with a densitometer DEN-1B (Biosan, Latvia).

Table 5. Phytopathogenic bacteria used in this study.

Bacterial Species	Strain Nos [a]	Disease Caused	Host	Year of Isolation	Country of Isolation	Reference
Dickeya solani	IFB0099, IPO2276, LMG28824.	Blackleg and soft rot	*Solanum tuberosum*	2005	Poland	Slawiak et al. [64]
Pectobacterium atrosepticum	IFB5103, SCRI1086.			1985	Canada	SCRI collection [65]

[a] The listed numbers originate from the following international bacterial collections: IFB—collection of Intercollegiate Faculty of Biotechnology University of Gdansk and Medical University of Gdansk (Gdansk, Poland), IPO—collection of the Institute for Phytopathological Research (Wageningen, The Netherlands), LMG—collection of the Laboratory of Microbiology in Gent (Gent, Belgium), SCRI—The James Hutton Institute bacterial collection (Dundee, Scotland).

MIC defined as the concentration of the active substance potent enough to prevent multiplication of microorganisms, in addition to MBC, referring to the formulation that causes loss of bacterial cell viability due to the action of 0.5% NH_4NO_3 solution treated in the optimized operating conditions of the FLC-dc-APGD system, were determined in 96 well-plates as follows. A total of 90 μL of the TSB medium, 10 μL of the 0.5 McF suspension of the tested phytopathogenic cells (Table 5), prepared as given above, and 100 μL of the NTAP-treated 0.5% NH_4NO_3 solution dissolved in sterile distilled water to reach the final concentration of 1%, 10%, 25%, or 50%, were introduced to each well. Post preparation, basal absorbance of the contents of the wells was measured with an EnVision 2105 Multimode Plate Reader (PerkinElmer, Waltham, MA, USA). The 96-well plates were incubated for 24 h at 28 °C and the absorbance measurement was then repeated. The lowest concentration of NTAP-treated 0.5% NH_4NO_3 allowing for inhibition of the bacterial growth was stated as MIC. Afterward, 5 μL of the contents of the wells that showed no visible growth of the studied microorganism was plated on TSA medium. If bacterial colonies did not grow after a 24 h incubation at 28 °C, the lowest concentration allowing for such an observation was described as MBC. This experiment was conducted in three repetitions with each consisting of two technical repeats. Concerning control samples, MIC and MBC assays have also been performed without the addition of PAL (either non-treated by plasma 0.5% NH_4NO_3 solution or water utilized for diluting PAL was applied to the 96-wells plate instead). Moreover, controls evaluating the viability of bacterial cells (TSB medium + 0.5 McF bacterial suspension) or the sterility of the utilized components (just TSB medium, TSB + PAL, TSB + 0.5% NH_4NO_3 solution, TSB + water used for dilutions, TSB + 0.85% NaCl utilized for preparation of bacterial suspensions) were included in the MIC and MBC procedures.

3.5. Plasma Reactive Species Responsible for the Antibacterial Action of the PAL

To decipher the mechanism of action of the NTAP-treated 0.5% NH_4NO_3 against plant pathogenic bacteria, the following qualitative and quantitative analyses were performed.

The RONS, which have been generated in the NH_4NO_3 solution treated by FLC-dc-APGD under the optimal operating conditions, were qualitatively identified by OES. The OES spectrum was acquired in the range from 200–900 nm with a Shamrock SR-500i (Andor, Belfast, United Kingdom) instrument, supported by a Newton DU-920P-OE CCD camera (Andor, Belfast, United Kingdom). The radiation emitted by FLC-dc-APGD during the production of the plasma-treated 0.5% NH_4NO_3 solution was collimated onto a 10 μm slit. The gaining time of the CCD camera was set to 0.50 s. The Solis software (version 2.5) was used for data acquisition and analysis.

In the subsequent quantitative analyses of RNS, the concentrations of NO_2^- and NO_3^- generated in the NTAP-treated 0.5% NH_4NO_3 solution, in contrast to the untreated control, were assessed immediately after exposure of the liquid to dc-APGD. Here, manufacturer's protocols attached to the commercial colorimetric assays, respectively, HANNA HI 96,708 and HANNA HI 96,728 (Hanna Instruments, Salaj, Romania), were used. The

Nessler's method [47] was applied to determine the concentrations of NH_4^+ in the PAL produced under the optimal conditions and the corresponding untreated control sample. The absorbance of the product of a reaction between ammonia and the Nessler's reagent, i.e., $[(Hg\text{-}O\text{-}Hg)NH_2]$, was measured at 420 nm by using an Analytik Jena Specord 210 Plus UV/Vis absorption spectrophotometer (Jena, Germany).

Moreover, the method described by Uchida et al. [59] was utilized for the estimation of the total concentration of ROS. In more detail, 2.9 mL of the tested 0.5% NH_4NO_3 solution (either NTAP-activated under the optimal operating conditions or the untreated control sample) was mixed with 100 μL of a KI-starch reagent solution, consisting of a 0.3% potassium iodide solution and 0.5% starch solution [59] post (t1 = 0 h; t2 = 0.5 h; t3 = 1 h; t4 = 1.5 h; t5 = 2 h) NTAP-treatment of the experimental sample. In these reactions, the KI-starch suspension acts as the source of I^- ions, reacting with all ROS present in the analyzed samples, whose oxidative potential is higher than this for iodide (0.54 V). The presence of starch in the reactant suspension results in a deeper blue tint of the starch-iodide complex exposed to ROS. The total amount of oxidative agents generated during NTAP exposure was evaluated from absorbance values at 600 nm with the use of Analytik Jena Specord 210 Plus UV/Vis absorption spectrophotometer (Jena, Germany). According to the measured absorbance values at 600, the quantitative analysis of the total ROS concentration was based on to a five-point calibration curve. For this purpose, the standard H_2O_2 solution was serially diluted, obtaining: 1.75, 4.3, 8.25, 12.75, and 17 mg L^{-1} H_2O_2 working solutions. Next, the prepared solutions were mixed with the KI-starch reagent in the same manner as was previously described and the absorbance values at 600 nm were acquired.

Finally, the spatial distribution of total ROS concentration was visualized with the use of a KI-starch gel as reported by Uchida et al. [59]. For this purpose, a KI-starch reaction suspension was prepared by dissolution of 0.3% (m/v) potassium iodide, 0.5% (m/v) starch, and 1.2% agar (m/v) (BTL, Lodz, Poland) [59] in the volume of 200 mL of deionized water. Subsequently, the mixture was heated under ambient air conditions without reaching the boiling point. Afterward, the prepared KI-starch reaction suspension was poured on standard Petri dish plates to obtain 8 mm thick solid disks from which 19 mm-diameter smaller disks have been cut. The resultant single disk has been placed in the central part beneath the FLC electrode (Figure 6, (3)-FLC-dc-APGD) and its surface has been wetted with the droplet of non-treated 0.5% (m/v) NH_4NO_3 solution to ignite NTAP. The impact of FLC-dc-APGD on ROS generation was captured as colorimetric changes in three stages: (1) at the moment of glow discharge ignition (1 s), (2) post an average contact (2 s) with droplets of NH_4NO_3 solution in a flowing mode, and (3) after prolonged (4 s) NTAP action. An increase in total ROS concentration was recognized by the occurrence of a dark blue tint appearing on the KI-starch disks.

4. Conclusions

In view of a high need for novel, innovative, and eco-friendly plant protection methods against SRP, a multivariate optimization of FLC-dc-APGD-based reaction-discharge system allowed us to define the optimal operating parameters, i.e., flow rate of the FLC as 2.0 mL min^{-1}, the discharge current as 30 mA, and the inorganic salt concentration in the FLC solution as 0.50, to be utilized for a high-throughput production of NH_4NO_3-based PAL of the highest conductivity. The developed PAL exhibited antibacterial properties toward *D. solani* and *P. atrosepticum* basing on the established MIC and MBC values of 25%. We attributed the antimicrobial action of PAL to the following RONS: OH, O, O_3, H_2O_2, HO_2, NH, N_2, N_2^+, NO_2^-, NO_3^-, and NH_4^+. Since the generated PAL is easy to store and transport, it may find numerous applications in the agricultural sector, such as limiting the occurrence and spread of devastating bacterial phytopathogens.

5. Patents

Patent no. 236665 attributed on 13.10.2020 by the Patent Office of the Republic of Poland resulted from the work reported in this manuscript.

Author Contributions: A.D. and P.J. adopted the reaction-discharge system for high-throughput production of plasma-activated liquids. A.D., P.J., and P.P. performed the multivariate optimization of the employed NTAP-based reaction-discharge system. A.D., P.J., and D.T. conducted analyses associated with identification of the RNS and ROS of interest in the plasma-treated solution under the defined operating conditions. P.P. carried out statistical analyses. W.B. and W.S. performed microbiological experiments, while A.M.-P. analyzed the collected data. A.D., P.J., D.T., P.P., and A.M.-P. wrote the presented manuscript. A.D. and A.M.-P. prepared the final version of the manuscript, which has been accepted by all the other authors. All authors have read and agreed to the published version of the manuscript.

Funding: The presented work was financed by the National Science Centre (NCN), Poland, under the grant agreement no. UMO-2019/33/B/NZ9/00940 awarded to W.S. The research activity of Dr. Anna Dzimitrowicz is co-financed by the Polish Ministry of Science and Higher Education, program Outstanding Young Scientists (532/STYP/13/2018).

Institutional Review Board Statement: Not applicable.

Informed Consent Statement: Not applicable.

Acknowledgments: Authors would like to thank Piotr Cyganowski for assistance in preparation of some graphics and Ewa Łojkowska for taking part in the discussion and comment on the final version of this manuscript.

Conflicts of Interest: The authors declare no conflict of interest. The funders had no role in the design of the study; in the collection, analyses, or interpretation of data; in the writing of the manuscript, or in the decision to publish the results.

Abbreviations

ANOVA	analysis of variance
BBD	Box-Behnken Design
D	composite desirability
DBD	dielectric-barrier discharge
dc-APGD	direct current atmospheric pressure glow discharge
DOE	design of experiments
FLC	flowing liquid cathode
GRAS	generally regarded as safe
MBC	minimal bactericidal concentration
McF	McFarland
MIC	minimal inhibitory concentration
NTAP	non-thermal atmospheric pressure plasma
OES	optical emission spectrometry
OFAT	one-factor-at-time
PAL	plasma-activated liquid
PAW	plasma-activated water
RONS	reactive oxygen and nitrogen species
ROS	reactive oxygen species
RNS	reactive nitrogen species
RSM	response surface methodology
SRP	soft rot *Pectobacteriaceae*
TSA	Trypticase Soy Agar
TSB	Trypticase Soy Broth

References

1. Kannan, V.R.; Bastas, K.K.; Antony, R. Plant pathogenic bacteria. An overview. In *Sustainable Approaches to Controlling Plant Pathogenic Bacteria*; Kannan, V.R., Bastas, K.K., Eds.; CRC Press: Boca Raton, FL, USA, 2015; pp. 1–16. ISBN 978-1-4822-4054-2.
2. Adeolu, M.; Alnajar, S.; Naushad, S.; Gupta, R.S. Genome-based phylogeny and taxonomy of the '*Enterobacteriales*': Proposal for *Enterobacterales* ord. nov. divided into the families *Enterobacteriaceae*, *Erwiniaceae* fam. nov., *Pectobacteriaceae* fam. nov., *Yersiniaceae* fam. nov., *Hafniaceae* fam. nov., *Morganellaceae* fam. nov., and *Budviciaceae* fam. nov. *Int. J. Syst. Evol. Microbiol.* **2016**, *66*, 5575–5599. [CrossRef] [PubMed]

3. Mansfield, J.; Genin, S.; Magori, S.; Citovsky, V.; Sriariyanum, M.; Ronald, P.; Dow, M.; Verdier, V.; Beer, S.V.; Machado, M.A.; et al. Top 10 plant pathogenic bacteria in molecular plant pathology. *Mol. Plant Pathol.* **2012**, *13*, 614–629. [CrossRef]

4. Motyka, A.; Zoledowska, S.; Sledz, W.; Lojkowska, E. Molecular methods as tools to control plant diseases caused by *Dickeya* and *Pectobacterium* spp: A minireview. *N. Biotechnol.* **2017**, *39*, 181–189. [CrossRef] [PubMed]

5. Zoledowska, S.; Motyka, A.; Zukowska, D.; Sledz, W.; Lojkowska, E. Population structure and biodiversity of *Pectobacterium parmentieri* isolated from potato fields in temperate climate. *Plant Dis.* **2018**, *102*, 154–164. [CrossRef] [PubMed]

6. Potrykus, M.; Golanowska, M.; Sledz, W.; Zoledowska, S.; Motyka, A.; Kolodziejska, A.; Butrymowicz, J.; Lojkowska, E. Biodiversity of *Dickeya* spp. isolated from potato plants and water sources in temperate climate. *Plant Dis.* **2016**, *100*, 408–417. [CrossRef]

7. Toth, I.K.; van der Wolf, J.M.; Saddler, G.; Lojkowska, E.; Helias, V.; Pirhonen, M.; Tsror (Lahkim), L.; Elphinstone, J.G. *Dickeya* species: An emerging problem for potato production in Europe. *Plant Pathol.* **2011**, *60*, 385–399. [CrossRef]

8. Perombelon, M.C.M. Potato diseases caused by soft rot erwinias: An overview of pathogenesis. *Plant Pathol.* **2002**, *51*, 1–12. [CrossRef]

9. Perombelon, M.C.M.; Kelman, A. Ecology of the soft rot Erwinias. *Annu. Rev. Phytopathol.* **1980**, *18*, 361–387. [CrossRef]

10. Van Gijsegem, F.; Toth, I.K.; van der Wolf, J.M. Outlook—Challenges and perspectives for management of diseases caused by *Pectobacterium* and *Dickeya* species. In *Plant Diseases Caused by Dickeya and Pectobacterium Species*; Van Gijsegem, F., van der Wolf, J.M., Toth, I.K., Eds.; Springer Nature: Cham, Switzerland, 2021; pp. 283–289. ISBN 978-3-030-61459-1.

11. Van der Wolf, J.M.; Acuña, I.; De Boer, S.H.; Brurberg, M.B.; Cahill, G.; Charkowski, A.O.; Coutinho, T.; Davey, T.; Dees, M.W.; Degefu, Y.; et al. Diseases caused by *Pectobacterium* and *Dickeya* species around the world. In *Plant Diseases Caused by Dickeya and Pectobacterium Species*; Van Gijsegem, F., van der Wolf, J.M., Toth, I.K., Eds.; Springer Nature: Cham, Switzerland, 2021; pp. 215–262. ISBN 978-3-030-61458-4.

12. Dupuis, B.; Nkuriyingoma, P.; Van Gijsegem, F. Economic impact of *Pectobacterium* and *Dickeya* species on potato crops: A review and case study. In *Plant Diseases Caused by Dickeya and Pectobacterium Species*; Van Gijsegem, F., van der Wolf, J.M., Toth, I.K., Eds.; Springer Nature: Cham, Switzerland, 2021; pp. 263–282. ISBN 978-3-030-61458-4.

13. Czajkowski, R.; Pérombelon, M.C.M.; van Veen, J.A.; van der Wolf, J.M. Control of blackleg and tuber soft rot of potato caused by *Pectobacterium* and *Dickeya* species: A review. *Plant Pathol.* **2011**, *60*, 999–1013. [CrossRef]

14. Van der Wolf, J.M.; De Boer, S.H.; Czajkowski, R.; Cahill, G.; Van Gijsegem, F.; Davey, T.; Dupuis, B.; Ellicott, J.; Jafra, S.; Kooman, M.; et al. Management of diseases caused by *Pectobacterium* and *Dickeya* species. In *Plant Diseases Caused by Dickeya and Pectobacterium Species*; Van Gijsegem, F., van der Wolf, J.M., Toth, I.K., Eds.; Springer Nature: Cham, Switzerland, 2021; pp. 175–214. ISBN 978-3-030-61458-4.

15. Brány, D.; Dvorská, D.; Halašová, E.; Škovierová, H. Cold Atmospheric Plasma: A Powerful Tool for Modern Medicine. *Int. J. Mol. Sci.* **2020**, *21*, 2932. [CrossRef]

16. Lackmann, J.W.; Bandow, J.E. Inactivation of microbes and macromolecules by atmospheric-pressure plasma jets. *Appl. Microbiol. Biotechnol.* **2014**, *98*, 6205–6213. [CrossRef] [PubMed]

17. Thirumdas, R.; Kothakota, A.; Annapure, U.; Siliveru, K.; Blundell, R.; Gatt, R.; Valdramidis, V.P. Plasma activated water (PAW): Chemistry, physico-chemical properties, applications in food and agriculture. *Trends Food Sci. Technol.* **2018**, *77*, 21–31. [CrossRef]

18. Gorbanev, Y.; Privat-Maldonado, A.; Bogaerts, A. Analysis of Short-Lived Reactive Species in Plasma-Air-Water Systems: The Dos and the Do Nots. *Anal. Chem.* **2018**, *90*, 13151–13158. [CrossRef]

19. Zhang, Q.; Liang, Y.; Feng, H.; Ma, R.; Tian, Y.; Zhang, J.; Fang, J. A study of oxidative stress induced by non-thermal plasma-activated water for bacterial damage. *Appl. Phys. Lett.* **2013**, *102*, 203701. [CrossRef]

20. Traylor, M.J.; Pavlovich, M.J.; Karim, S.; Hait, P.; Sakiyama, Y.; Clark, D.S.; Graves, D.B. Long-term antibacterial efficacy of air plasma-activated water. *J. Phys. D. Appl. Phys.* **2011**, *44*, 472001. [CrossRef]

21. Chen, T.-P.; Su, T.-L.; Liang, J. Plasma-Activated Solutions for Bacteria and Biofilm Inactivation. *Curr. Bioact. Compd.* **2017**, *13*, 59–65. [CrossRef]

22. Bălan, G.G.; Roşca, I.; Ursu, E.L.; Doroftei, F.; Bostănaru, A.C.; Hnatiuc, E.; Năstasă, V.; Şandru, V.; Ştefănescu, G.; Trifan, A.; et al. Plasma-activated water: A new and effective alternative for duodenoscope reprocessing. *Infect. Drug Resist.* **2018**, *11*, 727–733. [CrossRef]

23. Ma, R.; Wang, G.; Tian, Y.; Wang, K.; Zhang, J.; Fang, J. Non-thermal plasma-activated water inactivation of food-borne pathogen on fresh produce. *J. Hazard. Mater.* **2015**, *300*, 643–651. [CrossRef] [PubMed]

24. Bai, Y.; Idris Muhammad, A.; Hu, Y.; Koseki, S.; Liao, X.; Chen, S.; Ye, X.; Liu, D.; Ding, T. Inactivation kinetics of *Bacillus cereus* spores by Plasma activated water (PAW). *Food Res. Int.* **2020**, *131*, 109041. [CrossRef]

25. Pan, J.; Li, Y.L.; Liu, C.M.; Tian, Y.; Yu, S.; Wang, K.L.; Zhang, J.; Fang, J. Investigation of Cold Atmospheric Plasma-Activated Water for the Dental Unit Waterline System Contamination and Safety Evaluation in Vitro. *Plasma Chem. Plasma Process.* **2017**, *37*, 1091–1103. [CrossRef]

26. Smet, C.; Govaert, M.; Kyrylenko, A.; Easdani, M.; Walsh, J.L.; Van Impe, J.F. Inactivation of Single Strains of *Listeria monocytogenes* and Salmonella Typhimurium Planktonic Cells Biofilms with Plasma Activated Liquids. *Front. Microbiol.* **2019**, *10*, 1539. [CrossRef] [PubMed]

27. Kamgang-Youbi, G.; Herry, J.-M.; Meylheuc, T.; Brisset, J.-L.; Bellon-Fontaine, M.-N.; Doubla, A.; Naïtali, M. Microbial inactivation using plasma-activated water obtained by gliding electric discharges. *Lett. Appl. Microbiol.* **2009**, *48*, 13–18. [CrossRef] [PubMed]

28. Mai-Prochnow, A.; Zhou, R.; Zhang, T.; Ostrikov, K.K.; Mugunthan, S.; Rice, S.A.; Cullen, P.J. Interactions of plasma-activated water with biofilms: Inactivation, dispersal effects and mechanisms of action. *NPJ Biofilms Microbiomes* **2021**, 7, 1–12. [CrossRef] [PubMed]

29. Xu, Y.; Tian, Y.; Ma, R.; Liu, Q.; Zhang, J. Effect of plasma activated water on the postharvest quality of button mushrooms, *Agaricus bisporus*. *Food Chem.* **2016**, 197, 436–444. [CrossRef]

30. Xiang, Q.; Zhang, R.; Fan, L.; Ma, Y.; Wu, D.; Li, K. Microbial inactivation and quality of grapes treated by plasma-activated water combined with mild heat. *LWT* **2020**, 126, 109336. [CrossRef]

31. Chen, C.; Liu, C.; Jiang, A.; Guan, Q.; Sun, X.; Liu, S.; Hao, K.; Hu, W. The Effects of Cold Plasma-Activated Water Treatment on the Microbial Growth and Antioxidant Properties of Fresh-Cut Pears. *Food Bioprocess Technol.* **2019**, 12, 1842–1851. [CrossRef]

32. Xiang, Q.; Liu, X.; Liu, S.; Ma, Y.; Xu, C.; Bai, Y. Effect of plasma-activated water on microbial quality and physicochemical characteristics of mung bean sprouts. *Innov. Food Sci. Emerg. Technol.* **2019**, 52, 49–56. [CrossRef]

33. Ma, R.; Yu, S.; Tian, Y.; Wang, K.; Sun, C.; Li, X.; Zhang, J.; Chen, K.; Fang, J. Effect of Non-Thermal Plasma-Activated Water on Fruit Decay and Quality in Postharvest Chinese Bayberries. *Food Bioprocess Technol.* **2016**, 9, 1825–1834. [CrossRef]

34. Adhikari, B.; Pangomm, K.; Veerana, M.; Mitra, S.; Park, G. Plant Disease Control by Non-Thermal Atmospheric-Pressure Plasma. *Front. Plant Sci.* **2020**, 11, 77. [CrossRef]

35. Wu, M.C.; Liu, C.T.; Chiang, C.Y.; Lin, Y.J.; Lin, Y.H.; Chang, Y.W.; Wu, J.S. Inactivation Effect of *Colletotrichum Gloeosporioides* by Long-Lived Chemical Species Using Atmospheric-Pressure Corona Plasma-Activated Water. *IEEE Trans. Plasma Sci.* **2019**, 47, 1100–1104. [CrossRef]

36. Perez, S.M.; Biondi, E.; Laurita, R.; Proto, M.; Sarti, F.; Gherardi, M.; Bertaccini, A.; Colombo, V. Plasma activated water as resistance inducer against bacterial leaf spot of tomato. *PLoS ONE* **2019**, 14, e0217788. [CrossRef]

37. Siddique, S.S.; Hardy GE, S.J.; Bayliss, K.L. Plasma-activated water inhibits *in vitro* conidial germination of *Colletotrichum alienum*, a postharvest pathogen of avocado. *Plant Pathol.* **2021**, 70, 367–376. [CrossRef]

38. Guo, J.; Qin, D.; Li, W.; Wu, F.; Li, L.; Liu, X. Inactivation of *Penicillium italicum* on kumquat *via* plasma-activated water and its effects on quality attributes. *Int. J. Food Microbiol.* **2021**, 343, 109090. [CrossRef] [PubMed]

39. Moreau, M.; Feuilloley, M.G.J.; Orange, N.; Brisset, J.-L. Lethal effect of the gliding arc discharges on *Erwinia* spp. *J. Appl. Microbiol.* **2005**, 98, 1039–1046. [CrossRef]

40. Motyka, A.; Dzimitrowicz, A.; Jamroz, P.; Lojkowska, E.; Sledz, W.; Pohl, P. Rapid eradication of bacterial phytopathogens by atmospheric pressure glow discharge generated in contact with a flowing liquid cathode. *Biotechnol. Bioeng.* **2018**, 115, 1581–1593. [CrossRef] [PubMed]

41. Xiang, Q.; Fan, L.; Li, Y.; Dong, S.; Li, K.; Bai, Y. A review on recent advances in plasma-activated water for food safety: Current applications and future trends. *Crit. Rev. Food Sci. Nutr.* **2020**, 1–20. [CrossRef]

42. Graves, D.; Bakken, L.; Jensen, M.; Ingels, R. Plasma activated organic fertilizer. *Plasma Chem. Plasma Process.* **2019**, 39, 1–19. [CrossRef]

43. Zalewski, A. Zmiany na rynku nawozów azotowych w Polsce w latach 2000–2010. *J. Agribus. Rural Dev.* **2013**, 4, 257–267.

44. Zhao, Y.M.; Ojha, S.; Burgess, C.M.; Sun, D.W.; Tiwari, B.K. Inactivation efficacy and mechanisms of plasma activated water on bacteria in planktonic state. *J. Appl. Microbiol.* **2020**, 129, 1248–1260. [CrossRef]

45. Kučerová, K.; Machala, Z.; Hensel, K. Transient Spark Discharge Generated in Various N_2/O_2 Gas Mixtures: Reactive Species in the Gas and Water and Their Antibacterial Effects. *Plasma Chem. Plasma Process.* **2020**, 40, 749–773. [CrossRef]

46. Machala, Z.; Tarabová, B.; Sersenová, D.; Janda, M.; Hensel, K. Chemical and antibacterial effects of plasma activated water: Correlation with gaseous and aqueous reactive oxygen and nitrogen species, plasma sources and air flow conditions. *J. Phys. D. Appl. Phys.* **2019**, 52, 034002. [CrossRef]

47. Jamróz, P.; Gręda, K.; Pohl, P.; Żyrnicki, W. Atmospheric pressure glow discharges generated in contact with flowing liquid cathode: Production of active species and application in wastewater purification processes. *Plasma Chem. Plasma Process.* **2014**, 34, 25–37. [CrossRef]

48. Jamróz, P.; Zyrnicki, W.; Pohl, P. The effect of a miniature argon flow rate on the spectral characteristics of a direct current atmospheric pressure glow micro-discharge between an argon microjet and a small sized flowing liquid cathode. *Spectrochim. Acta Part B At. Spectrosc.* **2012**, 73, 26–34. [CrossRef]

49. Jamroz, P.; Greda, K.; Pohl, P. Development of direct-current, atmospheric-pressure, glow discharges generated in contact with flowing electrolyte solutions for elemental analysis by optical emission spectrometry. *TrAC Trends Anal. Chem.* **2012**, 41, 105–121. [CrossRef]

50. Machala, Z.; Tarabova, B.; Hensel, K.; Spetlikova, E.; Sikurova, L.; Lukes, P. Formation of ROS and RNS in Water Electro-Sprayed through Transient Spark Discharge in Air and their Bactericidal Effects. *Plasma Process. Polym.* **2013**, 10, 649–659. [CrossRef]

51. Wang, J.; He, M.; Zheng, P.; Chen, Y.; Mao, X. Comparison of the Plasma Temperature and Electron Number Density of the Pulsed Electrolyte Cathode Atmospheric Pressure Discharge and the Direct Current Solution Cathode Glow Discharge. *Anal. Lett.* **2019**, 52, 697–712. [CrossRef]

52. Śledź, W.; Jafra, S.; Waleron, M.; Lojkowska, E. Genetic diversity of *Erwinia carotovora* strains isolated from infected plants grown in Poland. *EPPO Bull.* **2000**, 30, 403–407. [CrossRef]

53. Waleron, M.; Waleron, K.; Podhajska, A.J.; Łojkowska, E. Genotyping of bacteria belonging to the former *Erwinia* genus by PCR-RFLP analysis of a *recA* gene fragment. *Microbiology* **2002**, 148, 583–595. [CrossRef]

54. Golanowska, M.; Potrykus, M.; Motyka-Pomagruk, A.; Kabza, M.; Bacci, G.; Galardini, M.; Bazzicalupo, M.; Makalowska, I.; Smalla, K.; Mengoni, A.; et al. Comparison of highly and weakly virulent *Dickeya solani* strains, with a view on the pangenome and panregulon of this species. *Front. Microbiol.* **2018**, *9*, 1940. [CrossRef]

55. Motyka-Pomagruk, A.; Zoledowska, S.; Misztak, A.E.; Sledz, W.; Mengoni, A.; Lojkowska, E. Comparative genomics and pangenome-oriented studies reveal high homogeneity of the agronomically relevant enterobacterial plant pathogen *Dickeya solani*. *BMC Genomics* **2020**, *21*, 449. [CrossRef]

56. Van Gijsegem, F.; Toth, I.K.; van der Wolf, J.M. Soft Rot *Pectobacteriaceae*: A brief overview. In *Plant Diseases Caused by Dickeya and Pectobacterium Species*; Van Gijsegem, F., van der Wolf, J.M., Toth, I.K., Eds.; Springer Nature: Cham, Switzerland, 2021; pp. 1–12. ISBN 978-3-030-61458-4.

57. Lerouge, S.; Wertheimer, M.R.; Yahia, L. Plasma sterilization: A review of parameters, mechanisms, and limitations. *Plasmas Polym.* **2001**, *6*, 175–188. [CrossRef]

58. Schnabel, U.; Niquet, R.; Schmidt, C.; Stachowiak, J.; Schlüter, O.; Andrasch, M.; Ehlbeck, J. Antimicrobial efficiency of non-thermal atmospheric pressure plasma processed water (PPW) against agricultural relevant bacteria suspensions. *Int. J. Environ. Agric. Res.* **2016**, *2*, 212–2014.

59. Uchida, G.; Nakajima, A.; Ito, T.; Takenaka, K.; Kawasaki, T.; Koga, K.; Shiratani, M.; Setsuhara, Y. Effects of nonthermal plasma jet irradiation on the selective production of H_2O_2 and NO_2- in liquid water. *J. Appl. Phys.* **2016**, *120*, 203302. [CrossRef]

60. Kawasaki, T.; Sato, A.; Kusumegi, S.; Kudo, A.; Sakanoshita, T.; Tsurumaru, T.; Uchida, G.; Koga, K.; Shiratani, M. Two-dimensional concentration distribution of reactive oxygen species transported through a tissue phantom by atmospheric-pressure plasma-jet irradiation. *Appl. Phys. Express* **2016**, *9*, 076202. [CrossRef]

61. Busco, G.; Omran, A.V.; Ridou, L.; Pouvesle, J.M.; Robert, E.; Grillon, C. Cold atmospheric plasma-induced acidification of tissue surface: Visualization and quantification using agarose gel models. *J. Phys. D Appl. Phys.* **2019**, *52*, 24LT01. [CrossRef]

62. Kaushik, N.K.; Ghimire, B.; Li, Y.; Adhikari, M.; Veerana, M.; Kaushik, N.; Jha, N.; Adhikari, B.; Lee, S.J.; Masur, K.; et al. Biological and medical applications of plasma-activated media, water and solutions. *Biol. Chem.* **2018**, *400*, 39–62. [CrossRef] [PubMed]

63. Dzimitrowicz, A.; Greda, K.; Lesniewicz, T.; Jamroz, P.; Nyk, M.; Pohl, P. Size-controlled synthesis of gold nanoparticles by a novel atmospheric pressure glow discharge system with a metallic pin electrode and a flowing liquid electrode. *RSC Adv.* **2016**, *6*, 80773–80783. [CrossRef]

64. Slawiak, M.; Łojkowska, E.; van der Wolf, J.M. First report of bacterial soft rot on potato caused by *Dickeya* sp. (syn. *Erwinia chrysanthemi*) in Poland. *Plant Pathol.* **2009**, *58*, 794. [CrossRef]

65. Potrykus, M.; Sledz, W.; Golanowska, M.; Slawiak, M.; Binek, A.; Motyka, A.; Zoledowska, S.; Czajkowski, R.; Lojkowska, E. Simultaneous detection of major blackleg and soft rot bacterial pathogens in potato by multiplex polymerase chain reaction. *Ann. Appl. Biol.* **2014**, *165*, 474–487. [CrossRef]

66. Cockerill, F.R.; Wikler, M.A.; Alder, J.; Dudley, M.N.; Eliopoulos, G.M.; Ferraro, M.J.; Hardy, D.J.; Hecht, D.W.; Hindler, J.A.; Patel, J.B.; et al. *Methods for Dilution Antimicrobial Susceptibility Tests for Bacteria That Grow Aerobically; Approved Standard*, 9th ed.; Clinical and Laboratory Standards Institute: Wayne, PA, USA, 2012.

International Journal of
Molecular Sciences

Article

The Quest to Quantify Selective and Synergistic Effects of Plasma for Cancer Treatment: Insights from Mathematical Modeling

Charlotta Bengtson * and Annemie Bogaerts 

Research Group PLASMANT, Department of Chemistry, University of Antwerp, Universiteitsplein 1, B-2610 Wilrijk-Antwerp, Belgium; annemie.bogaerts@uantwerpen.be
* Correspondence: charlotta.bengtson@uantwerpen.be

Abstract: Cold atmospheric plasma (CAP) and plasma-treated liquids (PTLs) have recently become a promising option for cancer treatment, but the underlying mechanisms of the anti-cancer effect are still to a large extent unknown. Although hydrogen peroxide (H_2O_2) has been recognized as the major anti-cancer agent of PTL and may enable selectivity in a certain concentration regime, the co-existence of nitrite can create a synergistic effect. We develop a mathematical model to describe the key species and features of the cellular response toward PTL. From the numerical solutions, we define a number of dependent variables, which represent feasible measures to quantify cell susceptibility in terms of the H_2O_2 membrane diffusion rate constant and the intracellular catalase concentration. For each of these dependent variables, we investigate the regimes of selective versus non-selective, and of synergistic versus non-synergistic effect to evaluate their potential role as a measure of cell susceptibility. Our results suggest that the maximal intracellular H_2O_2 concentration, which in the selective regime is almost four times greater for the most susceptible cells compared to the most resistant cells, could be used to quantify the cell susceptibility toward exogenous H_2O_2. We believe our theoretical approach brings novelty to the field of plasma oncology, and more broadly, to the field of redox biology, by proposing new ways to quantify the selective and synergistic anti-cancer effect of PTL in terms of inherent cell features.

Keywords: selective cancer treatment; cold atmospheric plasma; hydrogen peroxide; reaction network; mathematical modeling

check for
updates

Citation: Bengtson, C.; Bogaerts, A. The Quest to Quantify Selective and Synergistic Effects of Plasma for Cancer Treatment: Insights from Mathematical Modeling. *Int. J. Mol. Sci.* **2021**, *22*, 5033. https://doi.org/10.3390/ijms22095033

Academic Editor: Akikazu Sakudo

Received: 30 March 2021
Accepted: 6 May 2021
Published: 10 May 2021

Publisher's Note: MDPI stays neutral with regard to jurisdictional claims in published maps and institutional affiliations.

1. Introduction

In the last decade, the use of cold atmospheric plasma (CAP), which is an ionized gas near room temperature, has become a novel method to treat cancer. Both direct application of CAP (e.g., by the clinically approved kINPenMED® plasma jet) and indirect treatment by application of plasma-treated liquids (PTLs) have been shown to provide a significant anti-cancer effect [1]. Van Boxem et al. [2] showed that PTLs have an anti-cancer effect for a number of different CAP and liquid conditions, and Lin et al. [3] found that CAP can induce immunogenic cancer cell death. This mode of cell death induced by CAP was later attributed to the CAP generated short-lived reactive species [4]. Moreover, CAP and PTLs have been reported to cause a selective anti-cancer effect [5], although selectivity depends on the cell type, the type of cancer, and the culturing medium [6]. Bekeschus et al. [7] demonstrated, using an *in ovo* model, that CAP is a safe cancer treatment modality with respect to possible metastasis formation. A number of promising results of the clinical application of CAP for cancer treatment have also been published (see e.g., [8,9]).

It is widely believed that the processes leading to cancer cell death are initiated by reactive oxygen and/or nitrogen species (RONS), in particular hydrogen peroxide (H_2O_2), but the knowledge about the specific mechanisms underlying cell death induced by CAP and PTL is still very limited. The lack of understanding of the combined effect of RONS contained in CAP and PTL in terms of the cellular response to exposure is problematic in the development of CAP/PTL treatment as a standardized cancer therapy for clinical use. Ultimately, it should be possible to predict and quantify the susceptibility to CAP/PTL of a particular cell line in terms of features specifically associated with those cells.

So far, the vast majority of the literature in plasma oncology are experimental studies. As a complement, another approach to increase the understanding of complex biological systems such as the interaction between cells and PTL, is to develop a mathematical model that includes all the known information (of major importance in the given context) about the system and use it to investigate the system's response to various conditions. Especially, the system's response to a perturbation of the "normal" conditions can be analyzed. Furthermore, the development of the mathematical model itself can be seen as a way to summarize the current state of knowledge on the matter in a compact manner; it can be seen as the current "working hypothesis" of the mechanisms and processes governing the system dynamics.

Mathematical modeling has indeed proven to be a useful approach to increase our knowledge about the mechanisms of the cell's antioxidant defense and redox signaling. Some examples are the range of diffusion of H_2O_2 in the cytosol [10,11] and the cellular decomposition of exogenous H_2O_2 [12–14]. In the context of plasma oncology, two catalase-dependent apoptotic pathways associated with cancer cells, which possibly could be reactivated by CAP and thus explain the anti-cancer effect of CAP, have been investigated by mathematical modeling [15]. It was found that these pathways are unlikely to account for the anti-cancer effect of CAP and thus the underlying cause has to be studied further.

In the present study, we develop a mathematical model that includes the species and mechanisms of major importance in the context of a cell system exposed to PTL. The ultimate aim is to find a measure in terms of key features and characteristics of cells, which is able to quantify a particular cell system's susceptibility toward PTL and thus explain differences in response between normal cells and cancer cells. To the best of our knowledge, this is a completely novel approach in the field of plasma oncology, and we believe that our study will provide a new perspective and new insights as a complement to the experimental studies. An extensive summary of the background, leading to the more detailed research question, is provided in Section 2.

2. Experimentally Observed Cytotoxic Effects of CAP and PTL and Possible Features Determining Cancer Cell Susceptibility

An immediate effect of CAP treatment of cancer cells is an increase in intracellular RONS [16–20]. The significance of this RONS accumulation has been verified by the observation that the treatment does not succeed if the cancer cells have been pre-treated with intracellular RONS scavengers [17,21,22]. The origin of the increase in intracellular RONS after CAP treatment is still under investigation, but a hypothesis consistent with experimental observations is that it is caused by a diffusion of extracellular CAP-originated RONS across the cell membrane [17,18,23,24].

It has been demonstrated that the anti-cancer effect of CAP can also be induced by the species in PTL. In PTL, which mainly consists of H_2O_2, NO_2^- and NO_3^- [25–27], H_2O_2 has been shown to be of major importance [2,25–31]. It has been demonstrated that the H_2O_2 consumption rate, which is cell specific, of cancer cells after PTL treatment, is a key factor determining the specific susceptibility of cancer cell lines to PTL. More explicitly, it has been reported that the higher the H_2O_2 consumption rate of cancer cells, the lower the susceptibility toward CAP/PTL [32]. The susceptibility of cancer cells toward exogenous H_2O_2 has also been shown in [33–36]. However, it has been found that H_2O_2 alone cannot account for the total anti-cancer effect observed for PTL [27]. In this context, there are some reports of a synergistic effect of H_2O_2 and NO_2^- in PTL [25,26]. Thus, the cytotoxic effect of H_2O_2 seems to be enhanced in the presence of NO_2^-. The study in [26] found a *selective, synergistic* anti-cancer effect for H_2O_2 in the µM-range and NO_2^- in the mM-range, whereas in [25], a *non-selective, synergistic* anti-cancer effect was reported when H_2O_2 and NO_2^- were both in the mM-range. Since H_2O_2 and NO_2^- in PTL may react to form $ONOO^-$ [37], which is known to be highly toxic to cells, it has been speculated whether $ONOO^-$ is the species causing the synergistic effect of NO_2^- and H_2O_2. Its formation could thus potentially increase the cytotoxicity of PTL compared to an equal concentration of H_2O_2 only.

To summarize, some key points of the observed cytotoxic effects of CAP or PTL are:

- An intracellular increase of RONS, which is likely to be caused by diffusion of CAP-originated constituents through the cell membrane, is crucial for cell cytotoxicity.
- The key species in the anti-cancer effect of PTL is H_2O_2 (note that this may be different for direct CAP treatment, where short-lived RONS also play a crucial role [4,38]), and the corresponding cytotoxicity is inversely proportional to the extracellular consumption rate of H_2O_2.
- The effect of extracellular H_2O_2 is enhanced in the presence of NO_2^-, which can be a clue to understanding why PTL enables a more efficient treatment than a mock solution of H_2O_2 only.

Thus, from the information presented in literature, we can conclude that the cellular response to an addition of extracellular H_2O_2, with and without a simultaneous addition of NO_2^-, is crucial to understanding the anti-cancer effect of PTL. In Section 2.1, we introduce the key parameters to predict the response toward extracellular H_2O_2 of cells. We also relate this to general differences between normal cells and cancer cells. Based on this information and knowledge, we introduce our approach, and formulate our research question and aim in detail in Section 2.2.

2.1. Differences in Cellular Response to Exogenous Hydrogen Peroxide

In particular, two factors determine whether a certain cell line is susceptible to the exposure of exogenous H_2O_2:

- The plasma-membrane H_2O_2 diffusion rate constant and
- The intracellular expression of catalase.

Several cancer cell lines have shown a common phenotype of *decreased* catalase expression and *increased* aquaporin expression (that facilitates the transport of H_2O_2 through the cell membrane [39–42] and thus determines the H_2O_2 membrane diffusion rate) compared to normal cells. Hence, cancer cells in general can be assumed to be more susceptible to exogenous H_2O_2.

2.1.1. Membrane Diffusion Rate of Hydrogen Peroxide in Normal versus Cancer Cells

Aquaporins are proteins that form pores in the cell membrane. Primarily, they facilitate the transport of water between cells, but they also enable the trans-membrane diffusion of H_2O_2 (due to the chemical similarities between both molecules). Thus, the aquaporin expression in the cell membrane relates to the membrane diffusion rate of H_2O_2. Many aquaporins have been found to be overexpressed in tumors of different origins, especially in aggressive tumors [43]. Since different cancer cell lines express aquaporins to various extents [43,44], the different responses of H_2O_2-exposure by different cancer cell lines can at least partly be explained by the non-identical levels of aquaporin expression. In [45], it was found that aquaporin 3 accounted for nearly 80% of the membrane diffusion of H_2O_2 in a human pancreatic cancer cell line. For cells with a decreased aquaporin 3 expression, the rate of H_2O_2-uptake from the extracellular compartment was significantly decreased. It has furthermore been shown that for glioblastoma tumor cells, the anti-cancer effect of PTL as well as the increase of the concentration of intracellular RONS was significantly inhibited when aquaporin 8 was inhibited [46].

2.1.2. Catalase Activity in Normal versus Cancer Cells

Catalase is one of the main enzymes of the antioxidant defense system of cells of almost all aerobic organisms. The biological role of catalase is to regulate intracellular steady-state concentrations of H_2O_2, and experimental investigations and kinetic models using in vitro data have demonstrated that catalase is the major enzyme involved in the antioxidant defense against high concentrations of H_2O_2 [12,47–49]. In particular, catalase has been shown to be responsible for the clearance of *exogenous* H_2O_2 in vitro and in vivo [12,50–52].

Although catalase levels vary widely across cell lines, the total concentration of catalase (extracellular and intracellular) is frequently reported to be lower in cancer cells than in normal cells [36,53–60]. In [61], it was found that the catalase activity in various cancer cells is up to an order of magnitude lower compared to normal cells, and in [62], it was shown that normal cells had a better capacity to remove extracellular H_2O_2 than cancer cells; the rate constants for removal of extracellular H_2O_2 were on average two times higher in normal cells than in cancer cells. Furthermore, it was reported in [62] that the rate constants for H_2O_2 removal by different cell lines correlated with the number of active catalase monomers per cell.

However, while in general, the levels of catalase are low in cancer cells, catalase activity appears to vary greatly across different cancer cell lines [63]. In [34], it was found that three cancer cell lines (glioblastoma) that were extremely susceptible to H_2O_2 (generated by ascorbic acid) had reduced activity of intracellular catalase. Ascorbic acid-resistant cancer cell lines, on the other hand, exhibited significantly higher levels of catalase, but catalase knockdown sensitized these cell lines to extracellular H_2O_2.

An additional aspect of catalase that may be of interest in the context of cytotoxicity of CAP and PTL, is that it has been shown to decompose $ONOO^-$ [64]. Thus, if the synergistic effect of H_2O_2 and NO_2^- is to be found in the formation of $ONOO^-$, catalase might have a double function (i.e., as a protective factor toward exogenous exposure of both H_2O_2 and NO_2^-).

2.2. Approach and Research Question

In this study, we develop a mathematical model of the kinetics of the key species of PTL (i.e., H_2O_2 and NO_2^-) as well as of the processes governing the interaction with a cell system, which are given in terms of the H_2O_2 membrane diffusion rate constant and the intracellular catalase concentration. The system modeled is illustrated in Figure 1.

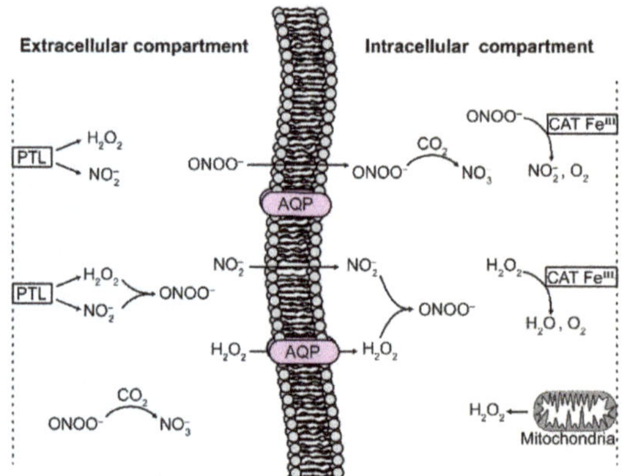

Figure 1. Illustration of the system representing a cell exposed to PTL.

As can be seen, the system consists of two compartments: the extracellular compartment (EC) and the intracellular compartment (IC). The two compartments are separated by the cell membrane, which some species in the system can diffuse through. Our mathematical model is explained in detail in Section 5, with references to all input data and assumptions made. Briefly, it takes into account (i) the diffusion of H_2O_2 and NO_2^- from the EC (where these species are supplied by PTL) to the IC, (ii) the formation of $ONOO^-$ from H_2O_2 and NO_2^- (in both the IC and EC), (iii) the mitochondrial production of H_2O_2 (in the IC), and (iv) the decomposition of H_2O_2 and $ONOO^-$ (in the IC). Furthermore, (v) since the CO_2-catalyzed consumption is considered to be the main route for $ONOO^-$-decay in biological systems (due to a high CO_2-concentration) [65–69], this reaction is also included.

There have already been attempts to capture the susceptibility toward exogenous H_2O_2 of different cell lines in terms of their H_2O_2 membrane diffusion rate constant and intracellular catalase concentration [13,14]. Two dependent variables that have been investigated recently are the intracellular steady-state concentration of H_2O_2 and the so-called latency (which describes the reduced average reaction rates for the observed decomposition of H_2O_2 due to the localization of encapsulated catalase in the peroxisomes). In [13], a lumped-parameter mathematical model, assuming that catalase is the major H_2O_2-removal enzyme, was developed and used to calculate the intracellular steady-state H_2O_2 concentration for several cell lines. The model was calibrated to the experimental values of the measured critical parameters, and the resulting intracellular steady-state H_2O_2 concentration was related to observed cell specific susceptibility to extracellular exposure of H_2O_2. The results showed that despite the fact that the experimental parameters including catalase concentration and H_2O_2 membrane diffusion rate constant, in particular, varied significantly across cell lines, the calculated steady-state intracellular-to-extracellular $[H_2O_2]$ ratio did not vary significantly across cell lines. In [14], it was investigated whether variations in the latency of peroxisomal catalase across cancer cell lines correlates with observed in vitro susceptibility to ascorbate at equivalent dosing of extracellular H_2O_2. The so-called effectiveness factor, which takes both the membrane diffusion rate and the overall reduced activity for encapsulated catalase into account, was used to quantify the effect of latency. The results suggest that latency alone is not a reliable parameter for predicting cell susceptibility to ascorbate (and hence, H_2O_2).

In this study, we explore new dependent variables that could possibly explain the difference in cell susceptibility to an external addition of H_2O_2, with and without a simultaneous addition of NO_2^-, and ultimately, quantify the effect in terms of the H_2O_2 membrane diffusion rate constant ($k_{D,1}$) and the intracellular catalase concentration ($[CATFe^{III}]_0$). Since we cannot distinguish a cancer cell from a normal cell solely by their H_2O_2 membrane diffusion rate constant and intracellular catalase concentration, we will have to work under the notations "cancer-like cells" (i.e., systems in the higher range of H_2O_2 membrane diffusion rate constant and the lower range of catalase concentration) and "normal-like cells" (i.e., systems in the lower range of H_2O_2 membrane diffusion rate constant and the higher range of catalase concentration). We investigate different regimes of the supplied extracellular H_2O_2- and NO_2^- concentrations according to experimental observations of the regimes of selective/non-selective and synergistic/non-synergistic anti-cancer effect of PTL [25,26]. The dependent variables that we investigate are:

1. **The temporal maximum of $[H_2O_2]$ and $[ONOO^-]$ in the IC.** As opposed to the steady-state value of the intracellular H_2O_2 concentration, the temporal maximum can be expected to be dependent on both $k_{D,1}$ and $[CATFe^{III}]_0$. These dependent variables may be related to the maximal intracellular oxidative power of the extracellularly added H_2O_2 (and NO_2^-), and thus it would be of interest to study whether a certain extracellularly added concentration of H_2O_2 (and NO_2^-) would result in a higher oxidative power in a more cancer-like cell than in a more normal-like cell.

2. **The system response time (i.e., the time out of equilibrium) with respect to $[H_2O_2]$ in the IC.** In order to achieve tumor progression, it is essential for cancer cells to optimize their RONS concentration and maintain the RONS equilibrium. For our mathematical model, this is translated into the question: does a more cancer-like cell have a longer response time compared to a more normal-like cell?

3. **The "load" of intracellular H_2O_2 and $ONOO^-$ (i.e., the time integral of $[H_2O_2]$ and $[ONOO^-]$ in the IC).** As the temporal maximum of $[H_2O_2]$ and $[ONOO^-]$ cannot capture any information about the total "load" of H_2O_2 and $ONOO^-$ (i.e., how much the intracellular $[H_2O_2]$ and $[ONOO^-]$ is increased over a period of time), it could be of interest to study such a dependent variable as a complement. The load can be seen as a measure that combines the temporal maximum concentration and the system

response time. Another possible way to define the load of intracellular H_2O_2 would be to only consider the concentration of H_2O_2 over a "baseline". Here, the steady-state intracellular $[H_2O_2]$, before the perturbation of an addition of extracellular H_2O_2 and at the upper limit of $[CATFe^{III}]_0$, is used as the baseline.

4.   **The inverse of the average and maximal rate of extracellular H_2O_2 consumption.** Since the cell susceptibility of CAP and PTL has been found to be inversely proportional to the (extracellular) consumption rate of H_2O_2, it is of interest to explore a dependent variable quantifying the system susceptibility in terms of the H_2O_2 consumption. We investigate two such candidates where one was defined in terms of the inverse of the average H_2O_2 consumption rate, and the other one in terms of the inverse of the maximal H_2O_2 consumption rate.

For all proposed dependent variables, we will analyze the dependence on $k_{D,1}$ and $[CATFe^{III}]_0$ and whether a more cancer-like cell is associated with a higher "response" than a more normal-like cell. Our main research question is thus: Can the difference in cell susceptibility toward PTL be understood, and even quantified, by one of these dependent variables?

To the best of our knowledge, this is the first study of its kind, and our aim is to take some initial steps in the direction of an increased understanding of the mechanisms underlying the selective and synergistic anti-cancer effect of PTL, and ultimately, be able to predict the response of different cells.

3. Results

As introduced in Section 2.2, in order to try to understand the combined role of the H_2O_2 membrane diffusion rate constant and the intracellular catalase concentration in determining the susceptibility of cells toward exogenous H_2O_2, we have to go beyond the steady-state value of the intracellular H_2O_2 concentration [13] (as well as latency [14]) and examine dependent variables that take the system's temporal response of a $[H_2O_2]$ perturbation in the EC into account. To be able to present the results in a more compact manner, the variables not yet introduced but of importance, and their denotations, are presented in Table 1. Details about the independent and dependent variables can be found in Section 5, where the mathematical model is presented. Likewise, details about the numerical calculations such as the values of the independent variables and parameters used in the model, can be found in Section 6.

Table 1. Denotations of variables used in the results analysis.

Variable	Meaning
$[H_2O_2]_0^{EC}$	Initial $[H_2O_2]$ in the EC
$[NO_2^-]_0^{EC}$	Initial $[NO_2^-]$ in the EC
$[H_2O_2]^{IC}$	$[H_2O_2]$ in the IC
$c_{1,\max}$	Temporal maximum of $[H_2O_2]$ in the IC

For the analysis and interpretation of the results, we mainly consider three important features of the dependent variable of interest:

- Does it account for selectivity with respect to different regimes of $[H_2O_2]_0^{EC}$?
- Does it account for a synergistic effect when NO_2^- is added to the system?
- Does it represent a feasible measure to quantify the susceptibility to exogenous H_2O_2 of a cell system in terms of $k_{D,1}$ and $[CATFe^{III}]_0$?

To qualify as a "measure" (i.e., as a quantification of the susceptibility in terms of $k_{D,1}$ and $[CATFe^{III}]_0$), the dependent variable should be associated with a higher value for cells with a higher susceptibility and a lower value for cells with a lower susceptibility. Thus, in accordance with experimental observations, a feasible measure should result in a higher value for more cancer-like cells than for more normal-like cells, at least in the expected regime of selectivity (that is, for $[H_2O_2]_0^{EC}$ in the µM-range [26]). However, it should be noted that in this study, we do not follow strict mathematical criteria for a function to be categorized as a measure.

3.1. The Temporal Maximum of the Intracellular Hydrogen Peroxide Concentration: A Possible Measure of the Cell Susceptibility to Exogenous Hydrogen Peroxide

Our calculation results suggest that the temporal maximum of $[H_2O_2]^{IC}$ (i.e., $c_{1,\max}$) is the dependent variable of major interest in terms of our requirements. Therefore, we focus our analysis on this variable. The results of the other dependent variables are presented in Appendix A.

Figure 2 shows $c_{1,\max}$ as a function of $k_{D,1}$ and $[CATFe^{III}]_0$ for $[H_2O_2]_0^{EC} = 1$ µM, with and without NO_2^-. The same results, but for $[H_2O_2]_0^{EC} = 1$ mM, are shown in Figure 3.

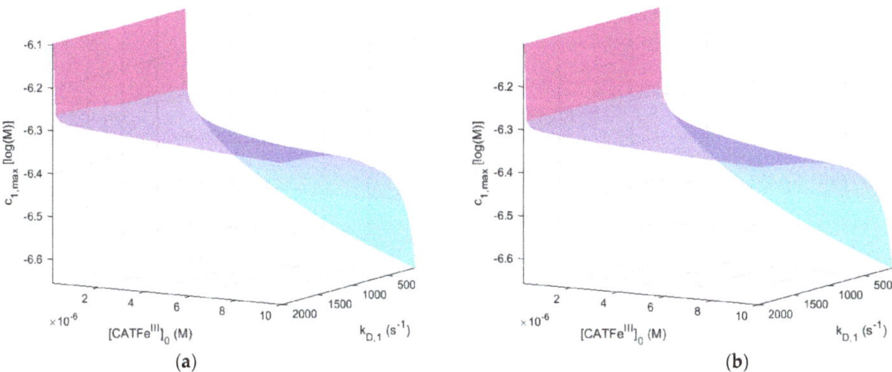

Figure 2. The dependent variable $c_{1,\max}$ (i.e., the temporal maximum of $[H_2O_2]$ in the IC) as a function of $k_{D,1}$ and $[CATFe^{III}]_0$ when $[H_2O_2]_0^{EC} = 1$ µM. $[NO_2^-]_0^{EC} = 0$ M (**a**) and $[NO_2^-]_0^{EC} = 1$ mM (**b**).

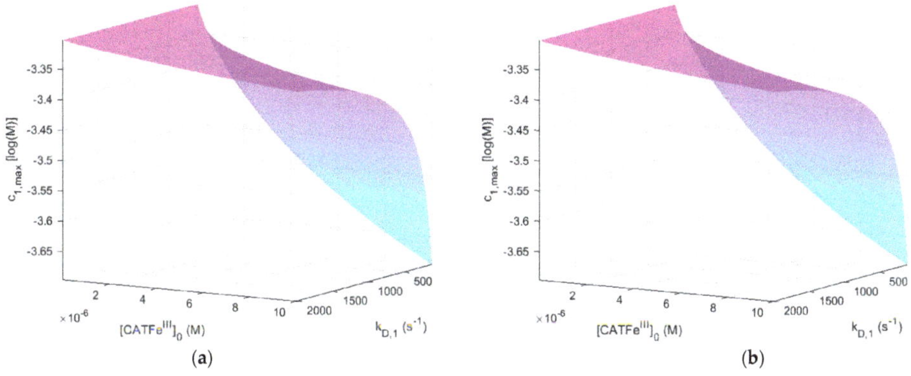

Figure 3. The dependent variable $c_{1,\max}$ (i.e., the temporal maximum of $[H_2O_2]$ in the IC) as a function of $k_{D,1}$ and $[CATFe^{III}]_0$ when $[H_2O_2]_0^{EC} = 1$ mM. $[NO_2^-]_0^{EC} = 0$ M (**a**) and $[NO_2^-]_0^{EC} = 1$ mM (**b**).

When comparing the result for the different $[H_2O_2]_0^{EC}$-regimes for $[NO_2^-]_0^{EC} = 0$ M (see Figures 2a and 3a), we see that $c_{1,\max}$ is also in different concentration regimes, which is logical. Indeed, for $[H_2O_2]_0^{EC} = 1$ mM, $c_{1,\max} \gtrsim 10^{-4}$ M, whereas for $[H_2O_2]_0^{EC} = 1$ µM, $c_{1,\max} < 10^{-6}$ M. Thus, by assuming that there exists a threshold value $c_{1,\max} > 10^{-6}$ M for which all types of cells undergo cell death, selectivity could be accounted for. However, there is no obvious synergetic effect; when comparing Figure 2a,b, $c_{1,\max}$ is almost identical.

Thus, the addition of NO_2^- does not change $c_{1,max}$ significantly. The same is true for Figure 3a,b.

For $[H_2O_2]_0^{EC} = 1$ µM, $c_{1,max}$ shows an increased $k_{D,1}$-dependence with increasing $[CATFe^{III}]_0$. The lowest value of $c_{1,max}$ is for the lowest values of $k_{D,1}$ and highest values of $[CATFe^{III}]_0$, as would be expected for a dependent variable that would qualify as a measure of the cell susceptibility in terms of $k_{D,1}$ and $[CATFe^{III}]_0$. In addition, the highest value of $c_{1,max}$ is associated with the lowest value of $[CATFe^{III}]_0$. However, in this regime, the dependence on $k_{D,1}$ is insignificant. Here, in contrast, there exists a significant $[CATFe^{III}]_0$-dependence and by changing the scale on the $[CATFe^{III}]_0$-axis to a log-scale (see Figure 4), we see that there are two distinct regimes with a clear shift from one regime to another at about $[CATFe^{III}]_0 \sim 10^{-7}$ M. The regimes of $k_{D,1}$ and $[CATFe^{III}]_0$ with the most profound difference between the value of $c_{1,max}$ is between cells with $[CATFe^{III}]_0 < 10^{-7}$ M (for all $k_{D,1}$) and cells with the lowest possible $k_{D,1}$ and highest possible $[CATFe^{III}]_0$. Thus, $c_{1,max}$ (i.e., the temporal maximum of $[H_2O_2]^{IC}$) is associated with a higher value for cancer-like cells than for normal-like cells. Indeed, $c_{1,max}$ is about four times greater for the most susceptible cells compared to the most resistant cells.

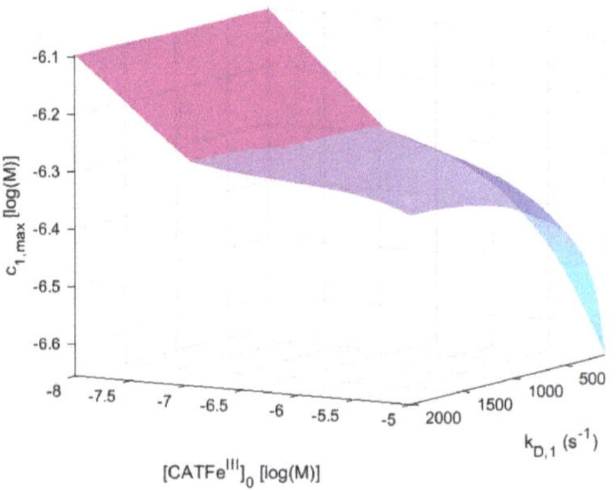

Figure 4. The dependent variable $c_{1,max}$ (i.e., the temporal maximum of $[H_2O_2]$ in the IC) as a function of $k_{D,1}$ and $\log([CATFe^{III}]_0)$ when $[H_2O_2]_0^{EC} = 1$ µM and $[NO_2^-]_0^{EC} = 0$ M.

In summary, $c_{1,max}$ does capture the dependence of $k_{D,1}$ and $[CATFe^{III}]_0$ in a manner that is consistent with experimental observations and could thus represent a feasible measure to quantify the susceptibility of different cells in terms of their H_2O_2 membrane diffusion rate constant and intracellular catalase concentration. However, in our model, it cannot yet account for the synergistic effect when NO_2^- is added.

To the best of our knowledge, there are not yet any experimental results to support our findings. We hope that our theoretical work will inspire future experimental studies. In the next section (Section 3.2), we discuss possible opportunities to experimentally quantify $c_{1,max}$.

3.2. Physical Interpretation and the Use of the Temporal Maximum of the Intracellular Hydrogen Peroxide Concentration as an Experimental Probe

Our model, with all the equations, is explained in detail in Section 5. Here we use the equations to better understand how we can use $c_{1,max}$ as a measure to quantify the response of different cells. In order to analyze and write the equations in a more compact manner, we first introduce some short notations as well as some new notations (see Table 2).

Table 2. Denotations of variables used in the results analysis.

Variable	Meaning
c_1^{IC}	$[H_2O_2]^{IC}$
c_1^{EC}	$[H_2O_2]^{EC}$
c_2	$\left[CATFe^{III}\right]$
c_3	$\left[CATFe^{IV}O^{\bullet+}\right]$
k_P	Rate of mitochondrial H_2O_2 production
k_1	Rate of H_2O_2 consumption by $\left[CATFe^{III}\right]$
k_2	Rate of H_2O_2 consumption by $\left[CATFe^{IV}O^{\bullet+}\right]$

The temporal maximum of c_1^{IC} (i.e., $c_{1,\max}$) occurs when the production and consumption of intracellular H_2O_2 are equal and $c_1^{EC} - c_1^{IC} \geq 0$ (as opposed to the steady-state value of $[H_2O_2]^{IC}$, which is governed by the same rate equation but for which $c_1^{EC} - c_1^{IC} = 0$). If we exclude in Equation (10) (see Section 5.2.2) the term representing the formation of $ONOOH$ from H_2O_2 and NO_2^- (since it is much smaller than the other terms), we have

$$- k_1 c_{1,\max} c_2 - k_2 c_{1,\max} c_3 + k_{D,1}\left(c_1^{EC} - c_{1,\max}\right) + k_P = 0.$$

Thus,

$$k_{D,1}\left(c_1^{EC} - c_{1,\max}\right) + k_P = k_1 c_{1,\max} c_2 + k c_{1,\max} c_3.$$

Here, we can furthermore use the constraint

$$c_3 = \left[CATFe^{III}\right]_0 - \left[CATFe^{III}\right],$$

since the total catalase concentration will be constant. By noting that $k_1 \sim k_2 = k$ (see Section 6.2.1), we can use the approximate expression

$$k_{D,1}\left(c_1^{EC} - c_{1,\max}\right) + k_P = k c_{1,\max}\left[CATFe^{III}\right]_0.$$

From our numerical calculations, we know that for low $\left[CATFe^{III}\right]_0$, $c_{1,\max}$ is independent on $k_{D,1}$, whereas for high $\left[CATFe^{III}\right]_0$, $c_{1,\max}$ is highly dependent on $k_{D,1}$. Furthermore (for $\left[CATFe^{III}\right]_0 > 10^{-7}$ M), for low $k_{D,1}$, $c_{1,\max}$ is highly dependent on $\left[CATFe^{III}\right]_0$, whereas for high $k_{D,1}$, $c_{1,\max}$ is independent on $\left[CATFe^{III}\right]_0$. The question is whether this behavior can be understood.

In the analysis, we first note that the implicit importance of $\left[CATFe^{III}\right]_0$ and $k_{D,1}$ in determining the value of c_1^{EC} at the time of $c_{1,\max}$, and thus $c_{1,\max}$, is hidden. The dependence on $\left[CATFe^{III}\right]_0$ originates from the fact that in our model, $c_{1,0}^{IC}$ is determined by $\left[CATFe^{III}\right]_0$. Equation (3) (in Section 5.2.1) can be approximated as:

$$\frac{dc_1^{EC}}{dt} = -k_{D,1}\left(c_1^{EC} - c_1^{IC}\right).$$

Hence, the initial rate, or driving force, of H_2O_2-consumption in the EC will crucially depend on $c_{1,0}^{IC}$, and thus, $\left[CATFe^{III}\right]_0$. In fact, $c_{1,0}^{IC} \sim 10^{-7}$ M for $\left[CATFe^{III}\right]_0 \sim 10^{-8}$ M, whereas $c_{1,0}^{IC} \sim 10^{-10}$ M for $\left[CATFe^{III}\right]_0 \sim 10^{-5}$ M (see Equation (18), Section 6). It means that the initial driving force is about ten times higher in the latter case compared to the former. This could explain why $c_{1,\max}$ is seemingly independent on $k_{D,1}$ at low values of $\left[CATFe^{III}\right]_0$; if $k_{D,1}\left(c_1^{EC} - c_{1,\max}\right) \ll k_P$ for all values of $k_{D,1}$, k_P will be the dominant factor of the build up of H_2O_2 in the IC. For higher values of $\left[CATFe^{III}\right]_0$, it seems like somewhat at $\left[CATFe^{III}\right]_0 > 10^{-7}$ M, $c_{1,\max}$ becomes increasingly dependent on $k_{D,1}$. It is thus reasonable to believe that the term $k_{D,1}\left(c_1^{EC} - c_{1,\max}\right)$ is becoming increasingly dominant and that the larger the value of $\left[CATFe^{III}\right]_0$, the larger the value of $\left(c_1^{EC} - c_{1,\max}\right)$. For a fixed value of $\left[CATFe^{III}\right]_0$, $c_{1,\max}$ will thus increase with increasing values of $k_{D,1}$.

In summary, this means that the lower the value of $\left[CATFe^{III}\right]_0$, the less important is the value of $k_{D,1}$, and the other way around. Thus, the susceptibility (toward exogenous H_2O_2) of cancer-like cells is not much influenced by the H_2O_2 membrane diffusion rate constant and this is due to their much higher level of intracellular H_2O_2 prior to the perturbation by the addition of exogenous H_2O_2. Normal-like cells, on the other hand, are more sensitive to the value of the H_2O_2 membrane diffusion rate constant, since the difference in concentration between the intracellular and extracellular H_2O_2 will be much larger.

Another aspect of $c_{1,\max}$ is whether it could provide an opportunity to extract information about different cell lines in terms of their H_2O_2 membrane diffusion rate constant and intracellular catalase concentration. By measuring $c_{1,\max}$ and the corresponding c_1^{EC} for different $c_{1,0}^{EC}$ it could be possible to roughly quantify $k_{D,1}$ and $\left[CATFe^{III}\right]_0$. There are many experimental techniques for the detection and quantification of the H_2O_2 concentration in vitro and in vivo. The intracellular H_2O_2 concentration has been detected and measured by a chemoselective fluorescent naphthylimide peroxide probe [70], by a genetically encoded red fluorescent sensor [71], and by fluorescent reporter proteins [72]. Thus, even if $c_{1,\max}$ does not represent a feasible measure of the cell susceptibility in terms of $k_{D,1}$ and $\left[CATFe^{III}\right]_0$, it could still possibly be used to gain more knowledge about the correlation between $k_{D,1}$ and $\left[CATFe^{III}\right]_0$ and cell susceptibility toward exogenous H_2O_2 and PTLs.

4. Discussion

In this study, we use a theoretical approach to increase the knowledge about possible underlying causes of the anti-cancer effect of PTL. Although the model is fairly simple, it does include the major pathways for species production and consumption relevant for such a cell system. It also puts emphasis on two important features (i.e., the H_2O_2 membrane diffusion rate constant and the intracellular catalase concentration), possibly explaining the different cell responses and cell susceptibility toward PTL when comparing normal cells to cancer cells, but also when comparing resistant vs. sensitive cancer cells. Nevertheless, it is important to keep in mind that in our model, different cells are only defined in terms of these two features, which are independent variables in our analysis, whereas in reality, there are countless of other features characteristic for different types of cells that could play an important role in the context of the anti-cancer effect of PTL. Here, we merely analyze the immediate cell response determined by the scavenging system active at high concentrations of H_2O_2. However, we do believe that our results contribute to a better understanding of some mechanisms probably underlying the anti-cancer effect of PTL. It brings novelty to the field of plasma oncology, and more broadly to the field of redox biology by using a theoretical approach and by proposing new ways to quantify the selective and synergistic anti-cancer effect of PTL in terms of inherent factors of cells. Here, we discuss each of our main findings and their potential implications. We also highlight what we believe are the most important limitations of the model.

As opposed to the steady-state intracellular concentration of H_2O_2, which has been evaluated in previous studies [13], our results suggest that the temporal maximal concentration of intracellular H_2O_2 could be a measure feasible to quantify the cell susceptibility toward exogenous H_2O_2 in terms of the H_2O_2 membrane diffusion rate constant and the intracellular catalase concentration. This result furthermore enables us to speculate whether the mode of action of H_2O_2 is as a signaling molecule rather than as a toxic substance causing necrosis. It is known that the intracellular concentration of a signaling molecule rises and falls within a short period. Indeed, whether a signaling molecule is effective or not is determined by how rapidly it is produced, how rapidly it is removed, and the concentration it must reach to alter the activity of its target effector. Of particular relevance in our context is that several reports have demonstrated that the rate of H_2O_2 generation and its concentration as a function of time play a key role in determining target cell damage or destruction [73–75]. RONS are regulators of signaling pathways such as the extracellular signal-regulated kinase (ERK) mitogen activated protein kinase (MAPK) pathway, which is important for cell proliferation, and a number of studies have demonstrated the ability of exogenous oxidants to activate the ERK MAPK pathway [76–80]. As in the general case, the duration and intensity of the ERK MAPK signal determine the outcome of the cellular response; there is a connection between the levels of ROS in a cell and the levels of MAPK signaling. In particular, MAPKs are activated in response to H_2O_2 [81–83].

Based on our modeling results (presented in the Appendix A; i.e., Figures A1 and A9), we do not think that the formation of $ONOO^-$ itself plays a major role in the explanation of the synergistic effect of H_2O_2 and NO_2^-. This is because although the overall intracellular concentration of $ONOO^-$ is increased with about one order of magnitude when NO_2^- is added to the system, the dependence on the H_2O_2 membrane diffusion rate constant is such that cells with a higher value of the H_2O_2 membrane diffusion rate constant (i.e., cancer-like cells) are associated with a lower maximal intracellular $ONOO^-$ concentration than more normal-like cells (i.e., cells with a lower value of the H_2O_2 membrane diffusion rate constant). In addition, the load of $ONOO^-$ is independent of the H_2O_2 membrane diffusion rate constant. However, an important aspect to keep in mind regarding our results for $ONOO^-$ and the choice to include CO_2-catalyzed consumption of $ONOO^-$ in our model, is that CO_2 redirects much of the $ONOO^-$ produced in vivo toward radical mechanisms [65]. Indeed, many of the reactions of $ONOO^-$ in vivo are more likely to be mediated by reactive intermediates derived from the reaction of $ONOO^-$ with CO_2 than by $ONOO^-$ itself [84,85]. Thus, if the production of such reactive intermediates were to be monitored instead of $ONOO^-$, our results might be different. In this context, especially the formation of $CO_3^{\bullet-}$ should be considered; a fraction (about 30%) of the formed $ONOOCOO^-$ will produce cage-escaped $^\bullet NO_2$ and $CO_3^{\bullet-}$ radicals according to [86–88]

$$ONOOCOO^- \rightleftharpoons CO_3^{\bullet-} + {}^\bullet NO_2,$$

where $k = 1.9 \times 10^9$ s^{-1} and $k' = 5 \times 10^8$ M^{-1}s^{-1} [89]. A possibly important target in the context of our study is catalase; catalase is so far the best known protein target for $CO_3^{\bullet-}$ and the rate constant of the reaction of bovine liver catalase with $CO_3^{\bullet-}$ is $(3.7 \pm 0.4) \times 10^9$ M^{-1}s^{-1} at $pH = 8.4$ [90]. Since the temporal maximum of intracellular $[H_2O_2]$ (i.e., $c_{1,\max}$) is inversely dependent on the catalase concentration (i.e., $[CATFe^{III}]_0$) with an increasingly steeper incline for lower catalase concentrations in the regime $10^{-8} \leq [CATFe^{III}]_0 \leq 10^{-7}$ M (see Figure 2), cancer-like cells would be more vulnerable to a decrease in the catalase concentration than normal-like cells, which are associated with higher values of $[CATFe^{III}]_0$. Thus, including these reaction pathways may possibly also make the dependent variable $c_{1,\max}$ able to account for the synergetic effect of NO_2^-. Such an extension of our model was out of the scope for this study, but would be highly interesting in a future model development.

In experiments, the consumption rate of extracellular H_2O_2 has been found to inversely correlate with the susceptibility of cancer cells toward exogenous H_2O_2 [32]. Thus, cancer cell lines with a high consumption rate were less susceptible. Our results cannot yet account for this correlation; when cells are defined in terms of their H_2O_2 membrane diffusion rate constant and their intracellular catalase concentration, susceptibility in terms of the inverse of the extracellular H_2O_2 consumption rate is not consistent with the experimental observations of cancer cells having a higher H_2O_2 membrane diffusion rate constant and a lower catalase concentration (see Section 2.1). The fact that our model does not reproduce these patterns leaves an open question of how to construct a dependent variable in terms of the inverse of the extracellular H_2O_2 consumption rate so that it corresponds to the experimental correlation.

The fact that our mathematical model, as well as our criteria for a dependent variable to represent a feasible measure of the cell susceptibility, does not select the system response time as a good candidate does not necessarily indicate that this variable, in general, cannot capture cell susceptibility toward exogenous H_2O_2. Indeed, in our definition of this dependent variable, we assume a tolerance of a 10% increase of the intracellular steady-state H_2O_2 concentration, and a different assumption of the tolerance might give a different result.

Except for the limitations of the model already mentioned in this discussion, some other model assumptions could hamper a realistic representation of a cell system in inter-action with PTL. One such limitation is that in our model, the rate of mitochondrial H_2O_2 production is constant. Although it can be argued that this assumption is a valid starting point, in a model development, it could be important to modify this aspect to represent a cancer cell in a more realistic manner. Indeed, it has been shown that in some cancer cells, the mitochondrial respiration is decreased (in favor of aerobic glycolysis), and moreover, this shift seems to be a dynamic process (see [91] and references therein). We believe that future models could benefit from trying to take such variation of the rate of mitochondrial H_2O_2 production into account, but this was out of the scope for this study.

Another aspect to take into account in a more realistic model is the fact that the H_2O_2 membrane diffusion rate constant is not a static but dynamic property. In [92], it was shown that cellular stress conditions reversibly inhibit the diffusion of H_2O_2 (and H_2O) of aquaporin 8. Thus, a more complex model taking the implicit time-dependence of the H_2O_2 membrane diffusion rate (caused by the increased intracellular H_2O_2 concentration after the addition of exogenous H_2O_2) could potentially produce results different from our model.

A third aspect to be aware of is that in our model, we assume that the addition of PTL does not affect the membrane diffusion rate constants. However, a number of studies have reported an enhanced cell membrane permeability (and thus, increased membrane diffusion rate constants) after CAP/PTL treatment [93–95]. For the aim and approach of our study, where the membrane diffusion rate constant of the key species H_2O_2 is varied within a range of possible values, we believe that our assumption is a valid starting point. Nevertheless, for future model extensions and developments, this aspect might be important to take into account.

Finally, it should be mentioned that the rate equations used to model the system are derived from information (collected from the literature) about rate constants and reaction orders for each reaction as they appear in the experiments. Most likely, the experimental conditions will deviate from the conditions of cells treated with PTL, which will affect the accuracy of the results produced by the model. However, for the purpose of our study, we believe that parameter values of the correct order of magnitude are sufficient at this stage.

5. Mathematical Model

Mathematical models of biological reaction networks such as the system considered in this study, can generally be divided into two categories: predictive and descriptive models. Since the experimental studies on which we build our model on are primarily in vitro studies, we construct a predictive model in this work. This means that we put together the information about each of the involved reactions (reaction orders, rate constants, etc.) as they appear in the experiments. From there, the result for a certain set of initial conditions is generated by solving the time-dependent equations of motion, representing the time evolution of the system.

In this section, we systematically present the species and reactions in the system considered (Section 5.1), how the system time evolution is modeled (Section 5.2), and we explicitly define the dependent variables that are analyzed (Section 5.3).

5.1. Species and Reactions in the System

The involved species of interest are H_2O_2, NOO^-, $ONOOH$, NO_2^-, CO_2, H^+, $CATFe^{III}$ and $CATFe^{IV}O^{\bullet+}$ (see Figure 1 in Section 2). The following reactions and interactions of the species H_2O_2, NO_2^- and (native) catalase $(CATFe^{III})$ in the system are taken into account.

- Decomposition of H_2O_2 by catalase:

$$CATFe^{III} + H_2O_2 \xrightarrow{k_1} CATFe^{IV}O^{\bullet+} + H_2O,$$

$$CATFe^{IV}O^{\bullet+} + H_2O_2 \xrightarrow{k_2} CATFe^{III} + O_2 + H_2O.$$

- Generation of $ONOO^-$ through reaction between H_2O_2 and NO_2^-:

$$NO_2^- + H_2O_2 + H^+ \xrightarrow{k_3} ONOOH + H_2O,$$

where the equilibrium between $ONOOH$ and $ONOO^-$ is described by:

$$ONOOH \underset{k_{-4}}{\overset{k_4}{\rightleftharpoons}} ONOO^- + H^+.$$

- Decomposition of $ONOO^-$ by catalase:

$$2ONOO^- \underset{CATFe^{III}}{\overset{k_5}{\rightarrow}} O_2 + 2NO_2^-.$$

- CO_2-catalyzed consumption of $ONOO^-$:

$$ONOO^- + CO_2 \xrightarrow{k_6} ONOOCOO^-.$$

The denotations of the time-dependent concentrations of the different species are shown in Table 3.

Table 3. Denotations of the time-dependent concentrations in the system.

Species	Denotation
$[H_2O_2]$	c_1
$[CATFe^{III}]$	c_2
$[CATFe^{IV}O^{\bullet+}]$	c_3
$[ONOO^-]$	c_4
$[NO_2^-]$	c_5
$[H^+]$	c_6
$[ONOOH]$	c_7
$[CO_2]$	c_8

5.2. Modeling the System

The mathematical model considers the kinetics of the reactions in the system composed of two subsystems (EC and IC), see Figure 1 in Section 2, as well as diffusion of certain species between the two subsystems. The equation governing the kinetics of each species i is given by the sum of the reaction rates (describing the rate of production and consumption of species i),

$$\frac{dc_i}{dt} = r_i, \tag{1}$$

and (in the case of species 1, 4, 5, and 7), the diffusion rate through the cell membrane, from the EC to the IC,

$$\frac{dc_i}{dt} = -k_{D,i}\left(c_i^{EC} - c_i^{IC}\right). \tag{2}$$

Equation (1) represents the resulting rate equation, derived from the rate constants and reaction orders for each reaction as they appear in the experiments. Equation (2) describes the rate of membrane diffusion of species i according to Fick's law of diffusion with a linear concentration gradient over the cell membrane. Here, $k_{D,i}$ is the rate of species i exchange through the membrane. We denote this as "membrane diffusion rate constant". More information about the derivation of Equation (2) can be found in Appendix B.

Explicitly, our mathematical model is used to analyze the behavior of a dependent variable $y(\bar{x})$, where $\bar{x}$ denotes the set of independent variables that are varied in the system. The independent variables in our model are:

- The H_2O_2 membrane diffusion rate constant through the cell membrane ($k_{D,1}$) and
- The initial intracellular catalase concentration ($[CATFe^{III}]_0$).

The species CO_2 and H^+ are assumed to be present in equal initial concentrations in both the EC and the IC (thus, $c_{8,0}^{EC} = c_{8,0}^{IC}$ and $c_{6,0}^{EC} = c_{6,0}^{IC}$). Since we do not explicitly study the kinetics of these species, we make such an assumption to reduce the complexity of the model.

Detailed information about the mathematical model is presented in the following sections.

5.2.1. Mathematical Model of the Reaction Kinetics in the Extracellular Compartment

At $t = 0$, H_2O_2 and NO_2^- in certain initial concentrations ($c_{1,0}^{EC} = [H_2O_2]_0^{EC}$ and $c_{5,0}^{EC} = [NO_2^-]_0^{EC}$) are inserted into the EC, representing treatment of the cell by PTL (as these species are the dominant RONS in PTLs), and their reactions as well as diffusion through the membrane into the IC is monitored. The reaction network and resulting set of differential equations are given below.

- Reaction network

$$NO_2^- + H_2O_2 + H^+ \xrightarrow{k_3} ONOOH + H_2O,$$

$$ONOOH \underset{k_{-4}}{\overset{k_4}{\rightleftharpoons}} ONOO^- + H^+,$$

$$ONOO^- + CO_2 \xrightarrow{k_6} ONOOCOO^-,$$

$$H_2O_2 \xrightarrow{k_{D,1}} IC,$$

$$ONOO^- \xrightarrow{k_{D,4}} IC,$$

$$NO_2^- \xrightarrow{k_{D,5}} IC,$$

$$ONOOH \xrightarrow{k_{D,7}} IC.$$

- Differential equations

$$\frac{dc_1^{EC}}{dt} = -k_3 c_1^{EC} c_5^{EC} c_6^{EC} - k_{D,1}\left(c_1^{EC} - c_1^{IC}\right), \tag{3}$$

$$\frac{dc_4^{EC}}{dt} = k_4 c_7^{EC} - k_{-4} c_4^{EC} c_6^{EC} - k_6 c_4^{EC} c_8^{EC} - k_{D,4}\left(c_4^{EC} - c_4^{IC}\right), \tag{4}$$

$$\frac{dc_5^{EC}}{dt} = -k_3 c_1^{EC} c_5^{EC} c_6^{EC} - k_{D,5}\left(c_5^{EC} - c_5^{IC}\right), \tag{5}$$

$$\frac{dc_6^{EC}}{dt} = -k_3 c_1^{EC} c_5^{EC} c_6^{EC} + k_4 c_7^{EC} - k_{-4} c_4^{EC} c_6^{EC},\tag{6}$$

$$\frac{dc_7^{EC}}{dt} = k_3 c_1^{EC} c_5^{EC} c_6^{EC} - k_4 c_7^{EC} + k_{-4} c_4^{EC} c_6^{EC} - k_{D,7}\left(c_7^{EC} - c_7^{IC}\right),\tag{7}$$

$$\frac{dc_8^{EC}}{dt} = -k_6 c_4^{EC} c_8^{EC}.\tag{8}$$

5.2.2. Mathematical Model of the Reaction Kinetics in the Intracellular Compartment

At $t = 0$, the concentration of H_2O_2 is at a certain steady-state value ($c_{1,0}^{IC} = [H_2O_2]_0^{IC}$) because it is continuously produced by the mitochondria at the rate

$$\frac{dc_1^{IC}}{dt} = k_P,\tag{9}$$

and decomposed by catalase, which exists in the IC, and is modeled as free in the solution. (More information can be found in Section 6.2.3). It is assumed that at $t = 0$, the total amount of catalase exists as $CATFe^{III}$ (i.e., $c_{3,0}^{IC} = 0$). The reaction network and resulting set of differential equations are given below.

- Reaction network

$$CATFe^{III} + H_2O_2 \xrightarrow{k_1} CATFe^{IV}O^{\bullet+} + H_2O,$$

$$CATFe^{IV}O^{\bullet+} + H_2O_2 \xrightarrow{k_2} CATFe^{III} + O_2 + H_2O,$$

$$NO_2^- + H_2O_2 + H^+ \xrightarrow{k_3} ONOOH + H_2O,$$

$$ONOOH \underset{k_{-4}}{\overset{k_4}{\rightleftharpoons}} ONOO^- + H^+,$$

$$2ONOO^- \xrightarrow{k_5} O_2 + 2NO_2^-,$$

$$ONOO^- + CO_2 \xrightarrow{k_6} ONOOCOO^-,$$

$$EC \xrightarrow{k_{D,1}} H_2O_2,$$

$$EC \xrightarrow{k_{D,4}} ONOO^-,$$

$$EC \xrightarrow{k_{D,5}} NO_2^-,$$

$$EC \xrightarrow{k_{D,7}} ONOOH.$$

- Differential equations

$$\frac{dc_1^{IC}}{dt} = -k_1 c_1^{IC} c_2 - k_2 c_1^{IC} c_3 - k_3 c_1^{IC} c_5^{IC} c_6^{IC} + k_{D,1}\left(c_1^{EC} - c_1^{IC}\right) + k_P,\tag{10}$$

$$\frac{dc_2}{dt} = -k_1 c_1^{IC} c_2 + k_2 c_1^{IC} c_3,\tag{11}$$

$$\frac{dc_3}{dt} = k_1 c_1^{IC} c_2 - k_2 c_1^{IC} c_3, \tag{12}$$

$$\frac{dc_4^{IC}}{dt} = k_4 c_7^{IC} - k_{-4} c_4^{IC} c_6^{IC} - k_5 c_2 c_4^{IC} - k_6 c_4^{IC} c_8^{IC} + k_{D,4}\left(c_4^{EC} - c_4^{IC}\right), \tag{13}$$

$$\frac{dc_5^{IC}}{dt} = -k_3 c_1^{IC} c_5^{IC} c_6^{IC} + k_5 c_2 c_4^{IC} + k_{D,5}\left(c_5^{EC} - c_5^{IC}\right), \tag{14}$$

$$\frac{dc_6^{IC}}{dt} = -k_3 c_1^{IC} c_5^{IC} c_6^{IC} + k_4 c_7^{IC} - k_{-4} c_4^{IC} c_6^{IC}, \tag{15}$$

$$\frac{dc_7^{IC}}{dt} = k_3 c_1^{IC} c_5^{IC} c_6^{IC} - k_4 c_7^{IC} + k_{-4} c_4^{IC} c_6^{IC} + k_{D,7}\left(c_7^{EC} - c_7^{IC}\right), \tag{16}$$

$$\frac{dc_8^{IC}}{dt} = -k_6 c_4^{IC} c_8^{IC}. \tag{17}$$

Equations (3)–(8) and (10)–(17) are solved numerically. The details about the numerical calculations can be found in Section 6.3.

5.3. Dependent Variables

In the following sections, we explicitly define the dependent variables analyzed in this study.

5.3.1. Temporal Maximum of the Intracellular Hydrogen Peroxide and Peroxynitrite Concentration

The dependent variables $c_{1,\text{max}}$ and $c_{4,\text{max}}$ are defined as

$$c_{1,\text{max}}\left(k_{D,1}, [CATFe^{III}]_0\right) = \max\left([H_2O_2]^{IC}\right),$$

and

$$c_{4,\text{max}}\left(k_{D,1}, [CATFe^{III}]_0\right) = \max\left([ONOO^-]^{IC}\right).$$

5.3.2. System Response Time of Intracellular Hydrogen Peroxide

Assuming that the system has a tolerance of an increase of 10% of the baseline H_2O_2 concentration (see Section 3.2), the dependent variable τ can be formulated

$$\tau\left(k_{D,1}, [CATFe^{III}]_0\right) = t \ni \frac{(100+10)}{100} \times \left[H_2O_2\right]_t^{IC} = [H_2O_2]_0^{IC}.$$

5.3.3. Load of Intracellular Hydrogen Peroxide and Peroxynitrite

The simplest way of creating a quantitative measure of the "load" of intracellular H_2O_2 and $ONOO^-$ is to use the time-integral over the whole time regime ($0 \leq t \leq t_f$) as the dependent variable, in other words,

$$l_1\left(k_{D,1}, [CATFe^{III}]_0\right) = \int_0^{t_f} [H_2O_2]^{IC},$$

and

$$l_4\left(k_{D,1}, [CATFe^{III}]_0\right) = \int_0^{t_f} [ONOO^-]^{IC}.$$

For the "load" over the baseline concentration of intracellular H_2O_2, if we denote this baseline constant $[H_2O_2]_{BS}^{IC}$, the dependent variable is defined as

$$l_{1,BS}\left(k_{D,1},\left[CATFe^{III}\right]_0\right) = \int_0^{t_f}\left([H_2O_2]^{IC} - [H_2O_2]^{IC}_{BS}\right).$$

5.3.4. Rate of Extracellular Hydrogen Peroxide Consumption

Here, we first define the dependent variable r as

$$r\left(k_{D,1},\left[CATFe^{III}\right]_0\right) = \frac{d[H_2O_2]^{EC}}{dt}.$$

The average extracellular consumption rate of H_2O_2 is then defined as

$$\bar{r} = \frac{1}{t_f}\int_0^{t_f} r\,dt,$$

and the maximal extracellular consumption rate of H_2O_2 as

$$r_{max} = \max(|r|).$$

In order to create a potential measure (i.e., a dependent variable where a more cancer-like cell is associated with a higher susceptibility), we use the variables

$$\bar{s} = \frac{1}{\bar{r}},$$

and

$$s_{max} = \frac{1}{r_{max}},$$

in our calculations.

6. Numerical Calculations

6.1. Independent Variables

The two independent variables in the system are $k_{D,1}$ and $c_{2,0} = [CATFe^{III}]_0$ (i.e., the diffusion rate constant of H_2O_2 through the cell membrane from the EC to the IC, and the initial catalase concentration in the IC). Furthermore, we use four different combinations of $c_{1,0}^{EC} = [H_2O_2]_0^{EC}$ and $c_{5,0}^{EC} = [NO_2^-]_0^{EC}$ in our calculations. The motivation and details of these variables are given in the following sections, and have also been introduced in Section 2.

6.1.1. Membrane Diffusion Rate Constant of Hydrogen Peroxide

In [96], the diffusion rate constant for H_2O crossing lipid bilayers was found to be 920 s^{-1}. Due to the chemical similarities between H_2O and H_2O_2, we use this value as a reference value for $k_{D,1}$ and we vary $k_{D,1}$ within the range $100 \le k_{D,1} \le 2000$ s^{-1}.

6.1.2. Initial Catalase Concentration in the Intracellular Compartment

The intracellular concentration of catalase is calculated from two different premises (see Appendix C). Considering the rough estimates in both approaches, it seems reasonable to use an effective catalase concentration in the range of 10^{-8}–10^{-5} M in our calculations. As a reference value, catalase concentration in human blood cells is about 2–3 μM [97,98].

6.1.3. Initial Hydrogen Peroxide and Peroxynitrite Concentration in the Extracellular Compartment

Several publications have shown that H_2O_2 and NO_2^- are formed at concentrations ranging from μM to mM in plasma-treated liquids (PTLs) [99–102]. In this study, we use the initial conditions for $[H_2O_2]^{EC}$ and $[NO_2^-]^{EC}$ shown in Table 4. The different regimes of these four combinations are specified in the last column. We assume that the selectivity is related to the concentration of extracellular H_2O_2 (i.e., selective cancer killing only occurs at low H_2O_2 concentrations (order of 1 µM [26]), while higher H_2O_2 concentrations (e.g., order of 1 mM) kill both cancer and normal cells) [25]. Based on this assumption, we want to compare the dependent variables for the selective versus non-selective regime. For both regimes (selective versus non-selective), we furthermore want to investigate whether a synergistic effect can be found (i.e., if the values of the dependent variables are enhanced when H_2O_2 and NO_2^- are added together) [25,26].

Table 4. Initial concentrations of H_2O_2 and NO_2^- in the extracellular compartment.

$[H_2O_2]^{EC}$ (M)	$[NO_2^-]^{EC}$ (M)	Regime
10^{-3}	10^{-3}	Non-selective, synergistic
10^{-3}	0	Non-selective, non-synergistic
10^{-6}	10^{-3}	Selective, synergistic
10^{-6}	0	Selective, non-synergistic

6.2. Parameter Values

6.2.1. Reaction Rate Constants

The used rate constants are summarized in Table 5, along with the references where the data is adopted from, and some remarks about the conditions for which these values were reported.

Table 5. Reaction rate constants.

Rate Constant	Parameter Value	Reference	Remark
k_1	$1.7 \times 10^7 \ M^{-1}s^{-1}$	[103]	Mammalian catalases
k_2	$2.6 \times 10^7 \ M^{-1}s^{-1}$	[103]	Mammalian catalases
k_3	$1.1 \times 10^3 \ M^{-2}s^{-1}$	[37]	At $pH = 3.3$ and $T = 25\,°C$
k_4	$K_a k_{-4} = 10^{-pK_a} k_{-4}$	[104,105]	The pK_a-value at $T = 25\,°C$ is 6.5–6.8
k_{-4}	$\sim 10^{10} \ M^{-1}s^{-1}$	[106]	
k_5	$1.7 \times 10^6 \ M^{-1}s^{-1}$	[64]	At $pH = 7.1$ and $T = 25\,°C$
k_6	$5.8 \times 10^4 \ M^{-1}s^{-1}$	[84]	At $T = 37\,°C$, pH-independent

6.2.2. Membrane Diffusion Rate Constants

NO_2^-, when protonated (i.e., as HNO_2), is reported to diffuse easily across biological membranes [107]. When not protonated, anionic channels have been shown to be permeable to NO_2^- [108]. It has furthermore been established that $ONOO^-$ is able to penetrate cell membranes [96,109]. In [96], using model phospholipid vesicular systems, it was demonstrated that $ONOO^-$ freely crosses phospholipid membranes. The diffusion rate constant for $ONOO^-$ crossing lipid bilayers was found to be $k_{D,4} = 320 \ s^{-1}$. Due to the acid–base equilibrium between $ONOO^-$ and its conjugated acid $ONOOH$, this is likely an average value for $ONOO^-$ and $ONOOH$. Thus, $k_{D,7} = k_{D,4} = 320 \ s^{-1}$. Since NO_2^- is an anion as well as similar in size, we assume the same value (i.e., $k_{D,4} = k_{D,5}$). The used diffusion rate constants are summarized in Table 6. Note that we do not consider the potential effect of the PTL on the membrane diffusion rate constants in our model.

Table 6. Membrane diffusion rate constants.

Rate Constant	Parameter Value	Reference	Remark
$k_{D,4}$	$320 \ s^{-1}$	[96]	
$k_{D,5}$	$320 \ s^{-1}$		Assigned
$k_{D,7}$	$320 \ s^{-1}$	[96]	

6.2.3. Initial Concentrations

The used initial concentrations in the EC and IC are summarized in Tables 7 and 8, respectively.

Table 7. Initial concentrations of the species in the extracellular compartment.

Species	Initial Concentration (*M*)	Reference	Remark
H_2O_2	Varied		
$ONOO^-$	0		Assigned
NO_2^-	Varied		
H^+	10^{-7}		Assigned
$ONOOH$	0		Assigned
CO_2	10^{-3}	[110]	

Table 8. Initial concentrations of the species in the intracellular compartment.

Species	Initial Concentration (*M*)	Reference	Remark
H_2O_2	Varied		
$CATFe^{III}$	Varied		
$CATFe^{IV}O^{\bullet+}$	0		Assigned
$ONOO^-$	0		Assigned
NO_2^-	10^{-4}	[111–113]	See Appendix D
H^+	10^{-7}		Assigned
$ONOOH$	0		Assigned
CO_2	10^{-3}	[110]	See Appendix D

The initial concentration of intracellular H_2O_2 (i.e., $c_{1,0}^{IC}$) is varied with the initial concentration of catalase in order to achieve the correct steady-state $c_{1,0}^{IC}$ for each $[CATFe^{III}]_0$. The H_2O_2 generation from mitochondria is in the range of 50 μmol kg^{-1} min^{-1} [114], which corresponded to $k_P = 1 \times 10^{-7}$ Ms^{-1} [115].

Thus, from Equation (10), at t $= 0$ (and thus, the term $k_{D,1}\left(c_1^{EC} - c_1^{IC}\right)$ vanishes),

$$\frac{dc_1^{IC}}{dt} = -k_1 c_1^{IC} c_2 - k_2 c_1^{IC} c_3 - k_3 c_1^{IC} c_5^{IC} c_6^{IC} + k_P = 0.$$

Assuming that $c_3 = 0$ at steady-state,

$$k_P = c_{1,ss}^{IC}\left(k_1 c_2 + k_3 c_5^{IC} c_6^{IC}\right) \Leftrightarrow$$

$$c_{1,ss}^{IC} = \frac{k_P}{\left(k_1 c_2 + k_3 c_5^{IC} c_6^{IC}\right)}. \tag{18}$$

Hence, $c_{1,ss}^{IC} = c_{1,0}^{IC}$ in our model.

6.3. Software and Details about the Calculations

The numerical calculations are performed in MATLAB. Due to significant differences in time scales, we use the solver ode23s to solve the set of rate equations.

The simulations are performed at time-scales covering the transient of the system's response. For the calculations, we use the time intervals and time steps shown in Table 9. We start with a very short time-step in the first 10 ms, which is then enlarged by a factor of 100 until 1 s, and again by a factor of 100 until the final time of 100 s.

Table 9. Time intervals and time steps.

Time	Value (s)	Time Step	Value (s)
t_1	10^{-2}	dt_1	10^{-7}
t_2	1	dt_2	10^{-5}
t_f	10^2	dt_3	10^{-3}

We furthermore vary the independent variables according to Table 10.

Table 10. Minimal and maximal values as well as number of steps of independent variables.

Independent Variable	Minimal Value	Maximal Value	Number of Steps
$\left[CATFe^{III}\right]$	10^{-8} M	10^{-5} M	100
$k_{D,1}$	$100\ s^{-1}$	$2000\ s^{-1}$	100

7. Conclusions

With this study, we aim to gain insights about the possible mechanisms underlying the anti-cancer effect of plasma-treated liquids (PTLs). In particular, we are interested in whether cell susceptibility toward PTL can be quantified in terms of cell-specific features, how selectivity arises, and why H_2O_2 combined with NO_2^- (as in PTL) offers a synergistic and thus enhanced anti-cancer effect compared with H_2O_2 only. By developing a mathematical model describing the kinetics of the species in PTL-treated cells, we analyze four different dependent variables as a function of the H_2O_2 membrane diffusion rate constant and the intracellular catalase concentration. Ultimately, one or more of these dependent variables could be used to quantify selective and synergistic effects of PTLs for different types of cells. In accordance with experimental observations, cancer cells are supposed to be associated with a higher H_2O_2 membrane diffusion rate constant and a lower intracellular catalase concentration compared to normal cells, and we use this knowledge in the evaluation of our proposed dependent variables.

The model is built up *ab initio* based on the species, reactions, and processes of major importance in the context of cell susceptibility toward PTL, and parameter values such as rate constants are extracted from the literature. Thus, the model itself summarizes the current state of knowledge on the matter in a compact and descriptive manner. This type of mathematical modeling to gain insight into the underlying mechanisms of the anti-cancer effect of PTL is novel and this study is the first of its kind in the field of plasma oncology. Furthermore, we propose new ways to quantify the selective and synergistic anti-cancer effect of PTL in terms of inherent cell features, which is also an innovative approach in the ongoing research on the mode of action of PTL.

We find that the temporal maximal intracellular H_2O_2 concentration shows a dependency of the H_2O_2 membrane diffusion rate constant and the intracellular catalase concentration, so that it could possibly be used to quantify the anti-cancer effect of exogenous H_2O_2, but it does not account for the synergistic effect of H_2O_2 and NO_2^- in PTL. However, by including the reactions where $CO_3^{\bullet-}$ is produced in the CO_2 catalyzed consumption of $ONOO^-$, and the interaction between $CO_3^{\bullet-}$ and catalase, the dependent variable $c_{1,max}$ could possibly also be able to account for the synergetic effect of NO_2^-.

We believe that our model is an important step to unveil the underlying mechanisms of the anti-cancer effect of CAP and PTLs, but more efforts are needed in order to understand the full picture of the causes and action. Here, both positive and negative results are important to share, in order to increase our collective knowledge of which clues may lead us forward in our search, and which clues we can leave behind, at least for now. Theoretical and experimental approaches to investigate possible key features of cells and their interaction with CAP and PTLs play complementary roles in our aim to push the limit of knowledge further. We hope, and believe, that our study contributes to the quest to quantify selective and synergistic effects of plasma for cancer treatment.

Author Contributions: Conceptualization, C.B.; Methodology, C.B.; Software, C.B.; Validation, C.B. and A.B.; Formal analysis, C.B.; Investigation, C.B.; Resources, A.B.; Data curation, C.B.; Writing—original draft preparation, C.B.; Writing—review and editing, C.B. and A.B.; Visualization, C.B.; Supervision, A.B.; Project administration, C.B.; Funding acquisition, A.B. All authors have read and agreed to the published version of the manuscript.

Funding: This research received no external funding.

Appendix A. Additional Results

Appendix A.1. Temporal Maximum of the Intracellular Peroxynitrite Concentration

Figure A1 shows the result for the dependent variable $c_{4,max}$ for $[H_2O_2]_0^{EC} = 1$ μM, with and without NO_2^-. The same results, but for $[H_2O_2]_0^{EC} = 1$ mM, are shown in Figure A2.

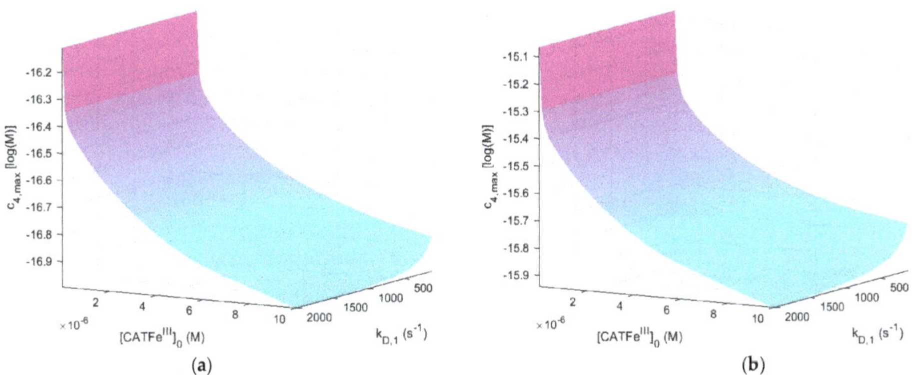

Figure A1. The dependent variable $c_{4,max}$ (i.e., the temporal maximum of $[ONOO^-]$ in the IC) as a function of $k_{D,1}$ and $[CATFe^{III}]_0$ when $[H_2O_2]_0^{EC} = 1$ μM. $[NO_2^-]_0^{EC} = 0$ M (**a**) and $[NO_2^-]_0^{EC} = 1$ mM (**b**).

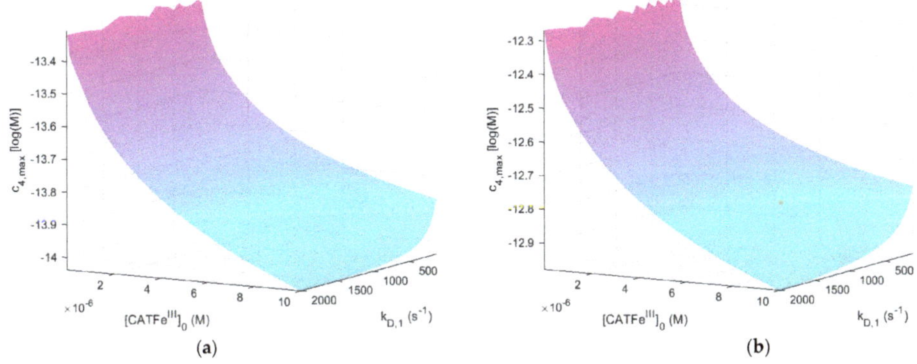

Figure A2. The dependent variable $c_{4,max}$ (i.e., the temporal maximum of $[ONOO^-]$ in the IC) as a function of $k_{D,1}$ and $[CATFe^{III}]_0$ when $[H_2O_2]_0^{EC} = 1$ mM. $[NO_2^-]_0^{EC} = 0$ M (**a**) and $[NO_2^-]_0^{EC} = 1$ mM (**b**).

The dependent variable $c_{4,max}$ could account for selectivity with respect to the different regimes of $[H_2O_2]_0^{EC}$ in the same manner as $c_{1,max}$ (see Figures A1a and A2a ($c_{4,max}$) and Figures 2a and 3a ($c_{1,max}$)). Indeed, there is a difference in $c_{4,max}$ of about three orders of magnitude for $[H_2O_2]_0^{EC} = 1$ mM (Figure A2a) compared with $[H_2O_2]_0^{EC} = 1$ µM (Figure A1a).

When comparing Figure A1a,b, we see that although the overall behavior of $c_{4,\max}$ is very similar, there is an order of magnitude difference in its value. In other words, the addition of NO_2^- increases the value of $c_{4,\max}$ for all $k_{D,1}$ and $[CATFe^{III}]_0$. Thus, $c_{4,\max}$ could account for the observed synergetic effect of PTL.

We see that for $[H_2O_2]_0^{EC} = 1$ µM (Figure A1a), $c_{4,\max}$ shows a clear $[CATFe^{III}]_0$-dependence and is relatively independent of $k_{D,1}$ for low $[CATFe^{III}]_0$. However, the $k_{D,1}$-dependence gradually increases with increasing $[CATFe^{III}]_0$. In this case, the dependence is such that $c_{4,\max}$ (for a given value of $[CATFe^{III}]_0$) is inversely dependent on $k_{D,1}$. This is not consistent with the pattern we are looking for and although the formation of intracellular $ONOO^-$ could play a role in the overall cytotoxicity of PTL, we do not believe that it plays the main role. In this context, it should also be noted that even the maximal amount formed corresponds to a very low concentration.

In summary, the dependent variable $c_{4,\max}$ could account for selectivity with respect to the concentration of H_2O_2 as well as the synergistic effect of PTLs. Indeed, the addition of NO_2^- in the extracellular compartment does increase $c_{4,\max}$ with about one order of magnitude. This could, however, be expected since the formation of $ONOO^-$ is directly proportional to the concentration of NO_2^- and with $[NO_2^-]_0^{EC} = 10[NO_2^-]_0^{IC}$, the intracellular concentration of NO_2^- will increase with about one order of magnitude compared to when no extracellular NO_2^- is added. Still, $c_{4,\max}$ does not show a dependency that is consistent with a measure of the cell susceptibility toward PTL. However, as we discuss in Sections 4 and 7, instead, by using the temporal maximum of intracellular $[CO_3^{\bullet-}]$, which is produced in the CO_2-catalyzed decomposition of $ONOO^-$, as a dependent variable, a feasible measure could possibly be found.

Appendix A.2. System Response Time of Intracellular Hydrogen Peroxide

Figure A3 shows the result for the dependent variable τ for $[H_2O_2]_0^{EC} = 1$ µM, with and without NO_2^-. The same results, but for $[H_2O_2]_0^{EC} = 1$ mM, are shown in Figure A4.

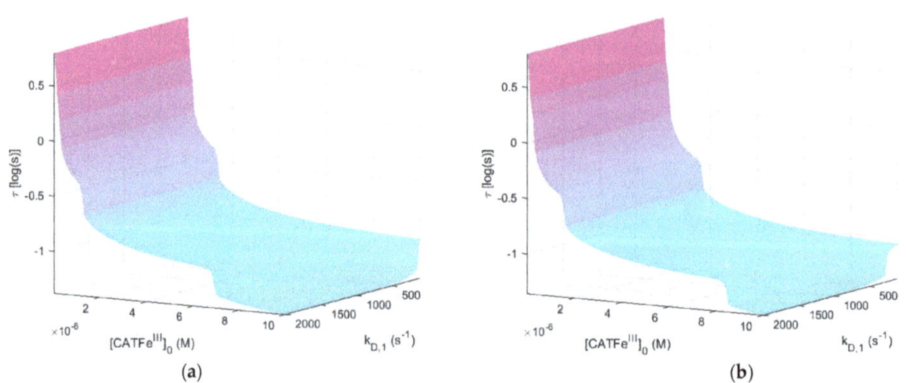

Figure A3. The dependent variable τ (i.e., the system response time of $[H_2O_2]$ in the IC) as a function of $k_{D,1}$ and $[CATFe^{III}]_0$ when $[H_2O_2]_0^{EC} = 1$ µM. $[NO_2^-]_0^{EC} = 0$ M (**a**) and $[NO_2^-]_0^{EC} = 1$ mM (**b**).

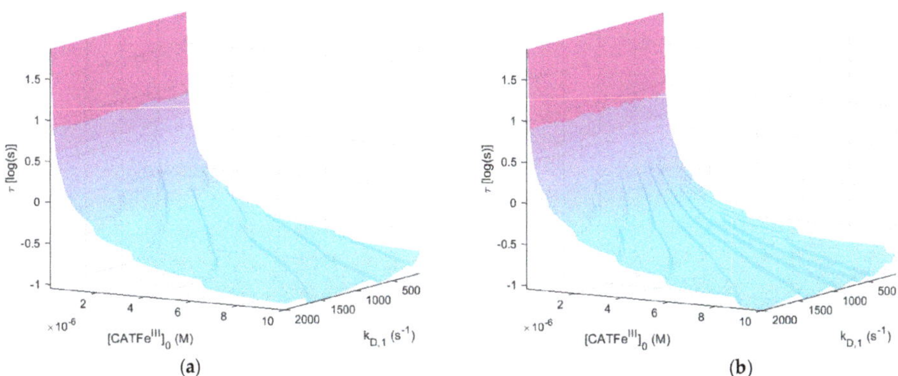

Figure A4. The dependent variable τ (i.e., the system response time of $[H_2O_2]$ in the IC) as a function of $k_{D,1}$ and $[CATFe^{III}]_0$ when $[H_2O_2]_0^{EC} = 1$ mM. $[NO_2^-]_0^{EC} = 0$ M (**a**) and $[NO_2^-]_0^{EC} = 1$ mM (**b**).

Here, the different concentration regimes of $[H_2O_2]_0^{EC}$ do not lead to well separated regimes of τ; in Figure A3a, we see that $-1.5 \lesssim \log(\tau) \lesssim 1$ whereas for $[H_2O_2]_0^{EC} = 1$ mM (Figure A4a), $-1 \lesssim \log(\tau) \lesssim 2$. Thus, τ is not a dependent variable that clearly takes into account selectivity with respect to $[H_2O_2]_0^{EC}$. Moreover, since the system response time is decreased with increased $[H_2O_2]_0^{EC}$, cells would then be less sensitive to higher concentrations of extracellular H_2O_2, which is not in accordance with the experimental observations.

However, there is a synergistic effect for a subspace of the total $k_{D,1}$, $[CATFe^{III}]_0$-space (see Figure A3a,b); in the region of high $[CATFe^{III}]_0$ and for approximately the whole $k_{D,1}$-regime, the addition of NO_2^- corresponds to an increased value of τ.

In Figure A3a, we see that τ is more or less independent of $k_{D,1}$; the overall dominant independent variable is $[CATFe^{III}]_0$. The $[CATFe^{III}]_0$-dependence is such that the decrease of τ for increased $[CATFe^{III}]_0$ has regions with distinct drops of τ in the overall exponential decrease of τ. Moreover, there is a region at high $[CATFe^{III}]_0$ and low $k_{D,1}$ where there is a slight $k_{D,1}$-dependence. This dependence is such that if a longer response time is associated with higher susceptibility, in a certain region of $k_{D,1}$, a higher value of $k_{D,1}$ has a protective effect compared to a lower value of $k_{D,1}$ (for a constant $[CATFe^{III}]_0$). This effect is increased when NO_2^- is added to the system. Since this does not correspond to the current state of knowledge (see Section 2.1), the system response time does not seem like a suitable dependent variable to quantify the cellular response to H_2O_2 and NO_2^-.

In summary, the system response time, τ, does not seem to be a suitable measure to quantify the cell susceptibility toward PTL.

Appendix A.3. Load of Intracellular Hydrogen Peroxide and Peroxynitrite

Figure A5 shows the result for the dependent variable l_1 for $[H_2O_2]_0^{EC} = 1$ µM, with and without NO_2^-. The same results, but for $[H_2O_2]_0^{EC} = 1$ mM, are shown in Figure A6.

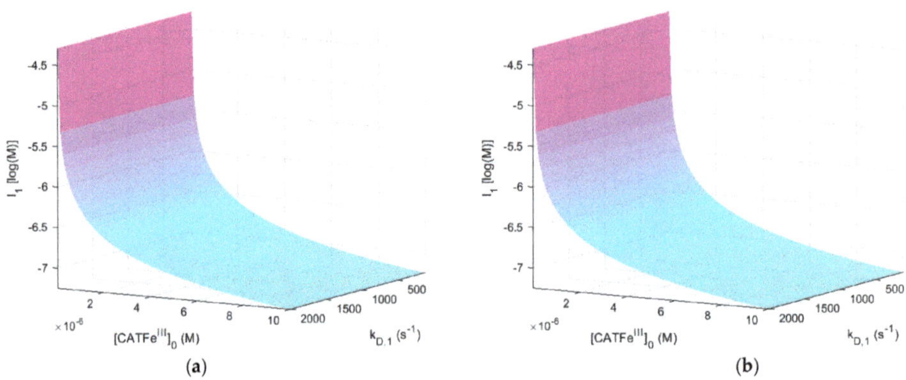

Figure A5. The dependent variable l_1 (i.e., the load of H_2O_2 in the IC) as a function of $k_{D,1}$ and $[CATFe^{III}]_0$ when $[H_2O_2]_0^{EC} = 1$ µM. $[NO_2^-]_0^{EC} = 0$ M (**a**) and $[NO_2^-]_0^{EC} = 1$ mM (**b**).

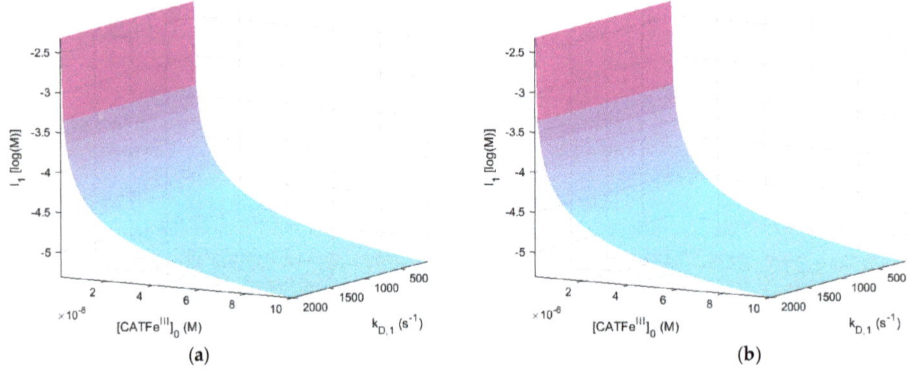

Figure A6. The dependent variable l_1 (i.e., the load of H_2O_2 in the IC) as a function of $k_{D,1}$ and $[CATFe^{III}]_0$ when $[H_2O_2]_0^{EC} = 1$ mM. $[NO_2^-]_0^{EC} = 0$ M (**a**) and $[NO_2^-]_0^{EC} = 1$ mM (**b**).

Likewise, Figure A7 shows the results for the dependent variable $l_{1,BS}$ for $[H_2O_2]_0^{EC} = 1$ µM, with and without NO_2^-. The same results, but for $[H_2O_2]_0^{EC} = 1$ mM, are shown in Figure A8.

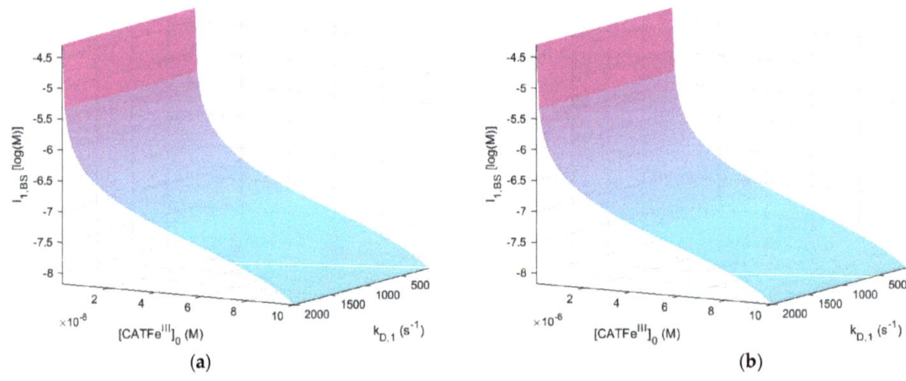

Figure A7. The dependent variable $l_{1,BS}$ (i.e., the load over the baseline of H_2O_2 in the IC) as a function of $k_{D,1}$ and $[CATFe^{III}]_0$ when $[H_2O_2]_0^{EC} = 1$ µM. $[NO_2^-]_0^{EC} = 0$ M (**a**) and $[NO_2^-]_0^{EC} = 1$ mM (**b**).

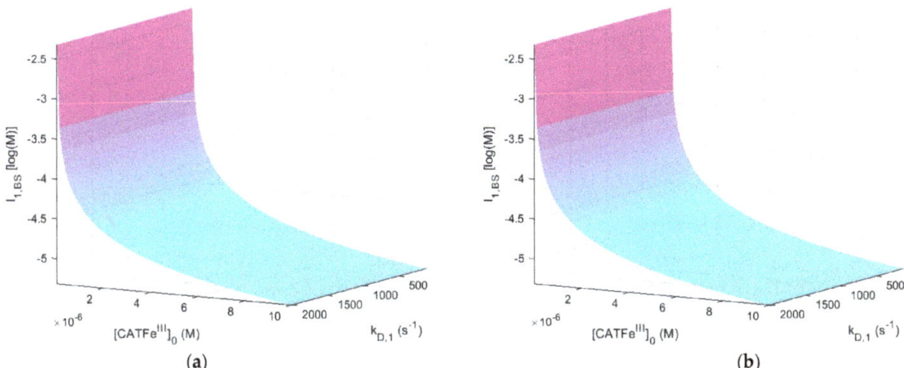

Figure A8. The dependent variable $l_{1,BS}$ (i.e., the load over the baseline of H_2O_2 in the IC) as a function of $k_{D,1}$ and $[CATFe^{III}]_0$ when $[H_2O_2]_0^{EC} = 1$ mM. $[NO_2^-]_0^{EC} = 0$ M (**a**) and $[NO_2^-]_0^{EC} = 1$ mM (**b**).

If one excludes the systems with the lowest levels of $[CATFe^{III}]_0$, l_1 could account for the selectivity with respect to different $[H_2O_2]_0^{EC}$; when comparing Figures A5a and A6a, there is a region for which $-7.5 \lesssim \log(l_1) \lesssim -5.5$ for $[H_2O_2]_0^{EC} = 1$ µM, whereas in the same region, $-5.5 \lesssim \log(l_1) \lesssim -3.5$ for $[H_2O_2]_0^{EC} = 1$ mM. The very same situation applies for $l_{1,BS}$; by excluding the lowest regime of $[CATFe^{III}]_0$, selectivity with respect to different $[H_2O_2]_0^{EC}$ could be taken into account (see Figures A7a and A8a).

When comparing Figure A5a,b, there is no observable effect on l_1 when adding NO_2^- into the system. Thus, l_1 does not seem to take the synergistic effect into account. The same situation applies for $l_{1,BS}$.

In Figure A5a, we see that l_1 is independent of $k_{D,1}$ and only dependent on $[CATFe^{III}]_0$. Thus, l_1 is not suitable as a measure to quantify cell susceptibility in terms of both $k_{D,1}$ and $[CATFe^{III}]_0$. From Figure A7a, we see that $l_{1,BS}$ differs a bit from l_1; $l_{1,BS}$ is less sensitive to an increased $[CATFe^{III}]_0$ at low $[CATFe^{III}]_0$ in particular. In addition, there is a point of inflection somewhere along the $[CATFe^{III}]_0$-axis (i.e., the graph is initially concave up and then shifts to concave down). This means that the rate of change of $l_{1,BS}$ with respect to $[CATFe^{III}]_0$ changes from increasing to decreasing somewhere on the $[CATFe^{III}]_0$-axis. Nevertheless, $l_{1,BS}$ is still independent of $k_{D,1}$, and thus it does not represent a feasible measure to quantify the cell susceptibility toward PTL.

Figure A9 shows the result for the dependent variable l_4 for $[H_2O_2]_0^{EC} = 1$ µM, with and without NO_2^-. The same results, but for $[H_2O_2]_0^{EC} = 1$ mM, are shown in Figure A10.

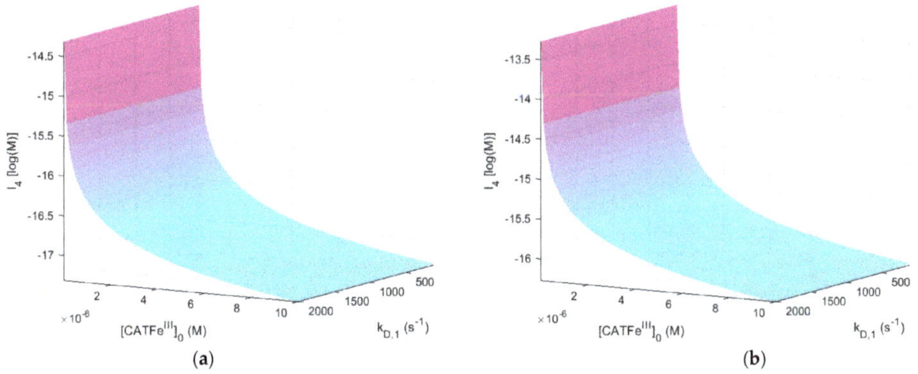

Figure A9. The dependent variable l_4 (i.e., the load of $ONOO^-$ in the IC) as a function of $k_{D,1}$ and $[CATFe^{III}]_0$ when $[H_2O_2]_0^{EC} = 1$ µM. $[NO_2^-]_0^{EC} = 0$ M (**a**) and $[NO_2^-]_0^{EC} = 1$ mM (**b**).

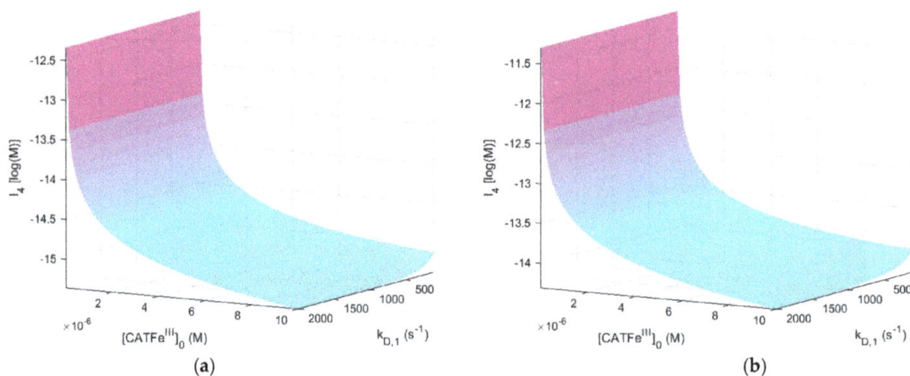

Figure A10. The dependent variable l_4 (i.e., the load of $ONOO^-$ in the IC) as a function of $k_{D,1}$ and $[CATFe^{III}]_0$ when $[H_2O_2]_0^{EC} = 1$ mM. $[NO_2^-]_0^{EC} = 0$ M (**a**) and $[NO_2^-]_0^{EC} = 1$ mM (**b**).

The pattern for l_4 more or less follows the same pattern as for l_1 selectivity with respect to $[H_2O_2]_0^{EC}$ is only taken into account if the $[CATFe^{III}]_0$-regime is modified by removing the lowest levels of $[CATFe^{III}]_0$. In addition, l_4 is independent of $k_{D,1}$ and thus, it does not represent a feasible measure for quantifying cell-susceptibility. However, there exists a synergistic effect; l_4 is about one order of magnitude higher at every point in the $(k_{D,1}, [CATFe^{III}]_0)$-space when NO_2^- is added.

In summary, looking for a variable that depends on both $[CATFe^{III}]_0$ and $k_{D,1}$, none of the dependent variables l_1, l_4, or $l_{1,BS}$ seem to be appropriate candidates for a dependent variable able to capture and quantify the cell susceptibility toward PTL.

Appendix A.4. Rate of Extracellular Hydrogen Peroxide Consumption

Figure A11 shows the result for the dependent variable $\bar{s}$ for $[H_2O_2]_0^{EC} = 1$ μM, with and without NO_2^-. The same results, but for $[H_2O_2]_0^{EC} = 1$ mM, are shown in Figure A12.

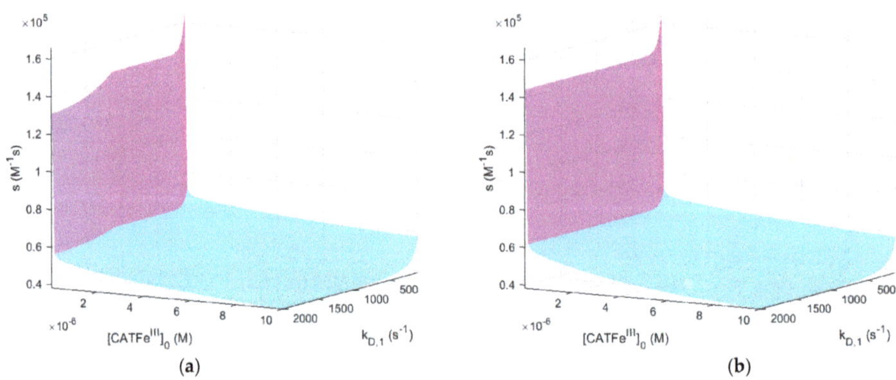

Figure A11. The dependent variable $\bar{s}$ (i.e., the inverse of the average rate of H_2O_2 consumption in the EC) as a function of $k_{D,1}$ and $[CATFe^{III}]_0$ when $[H_2O_2]_0^{EC} = 1$ μM. $[NO_2^-]_0^{EC} = 0$ M (**a**) and $[NO_2^-]_0^{EC} = 1$ mM (**b**).

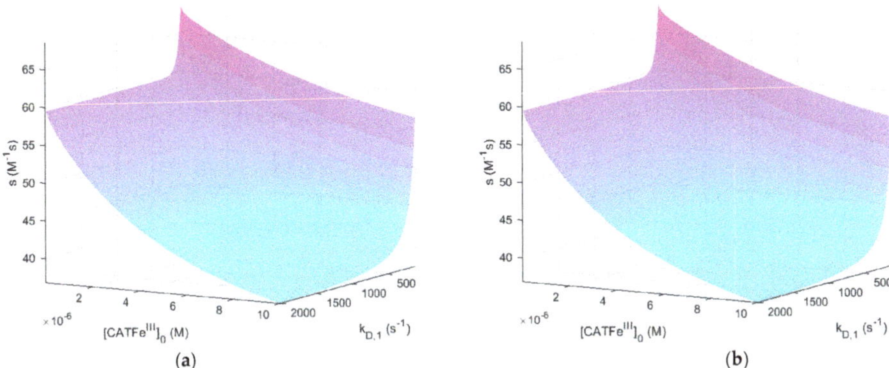

Figure A12. The dependent variable $\bar{s}$ (i.e., the inverse of the average rate of H_2O_2 consumption in the EC) as a function of $k_{D,1}$ and $[CATFe^{III}]_0$ when $[H_2O_2]_0^{EC} = 1$ mM. $[NO_2^-]_0^{EC} = 0$ M (**a**) and $[NO_2^-]_0^{EC} = 1$ mM (**b**).

Here, one has to be a bit careful to compare $\bar{s}$ for different regimes of $[H_2O_2]_0^{EC}$. Obviously, a larger value of $[H_2O_2]_0^{EC}$ will correspond to a higher membrane diffusion rate of H_2O_2 from the EC to the IC, which will affect the consumption rate so that it increases with increasing H_2O_2 membrane diffusion rate. When comparing $\bar{s}$ for $[H_2O_2]_0^{EC} = 1$ μM (Figure A11a) and for $[H_2O_2]_0^{EC} = 1$ mM (Figure A12a), we see that it is many orders of magnitude higher in the latter case, even though we know that $[H_2O_2]_0^{EC} = 1$ mM is associated with a higher cytotoxic effect than $[H_2O_2]_0^{EC} = 1$ μM. Thus, for this choice of dependent variable, it is not meaningful to compare the results for different regimes of $[H_2O_2]_0^{EC}$; in other words, selectivity with respect to $[H_2O_2]_0^{EC}$ cannot be verified.

For the potential synergistic effect, we see in Figure A11 that even though there is a slight deviation of Figure A11a compared to Figure A11b for very low values of $[CATFe^{III}]_0$, it is not enough to verify a synergistic effect when NO_2^- is added to the system.

The variable $\bar{s}$ is strongly dependent on $[CATFe^{III}]_0$ in the regime of very low values of $[CATFe^{III}]_0$. Here, the dependence is such that is could capture specific cell susceptibility in terms of $[CATFe^{III}]_0$. However, for larger values of $[CATFe^{III}]_0$, $\bar{s}$ is more or less constant and moreover, the very weak $k_{D,1}$-dependence is reverse to what would be expected (see Section 2.1). Thus, $\bar{s}$ does not seem to represent a feasible measure to quantify cell susceptibility in terms of $k_{D,1}$ and $[CATFe^{III}]_0$.

Finally, Figure A13 shows the result for the dependent variable $s_{\max}$ for $[H_2O_2]_0^{EC} = 1$ μM, with and without NO_2^-. The same results, but for $[H_2O_2]_0^{EC} = 1$ mM, are shown in Figure A14.

As for $\bar{s}$, we cannot compare $s_{\max}$ for different regimes of $[H_2O_2]_0^{EC}$ (see Figures A13a and A14a), so again the feature of selectivity with respect to $[H_2O_2]_0^{EC}$ cannot be verified. Furthermore, there is no observable synergistic effect when NO_2^- is added to the system (see Figure A13a,b).

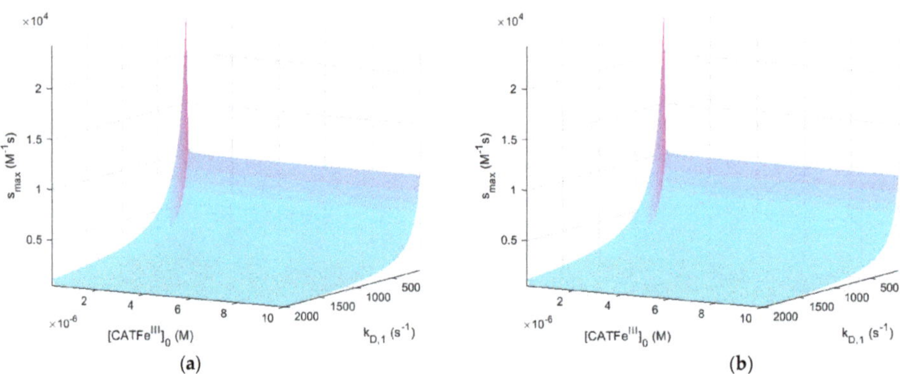

(a)

(b)

Figure A13. The dependent variable $s_{\max}$ (i.e., the inverse of the maximal rate of H_2O_2 consumption in the EC) as a function of $k_{D,1}$ and $[CATFe^{III}]_0$ when $[H_2O_2]_0^{EC} = 1\ \mu$M. $[NO_2^-]_0^{EC} = 0$ M (**a**) and $[NO_2^-]_0^{EC} = 1$ mM (**b**).

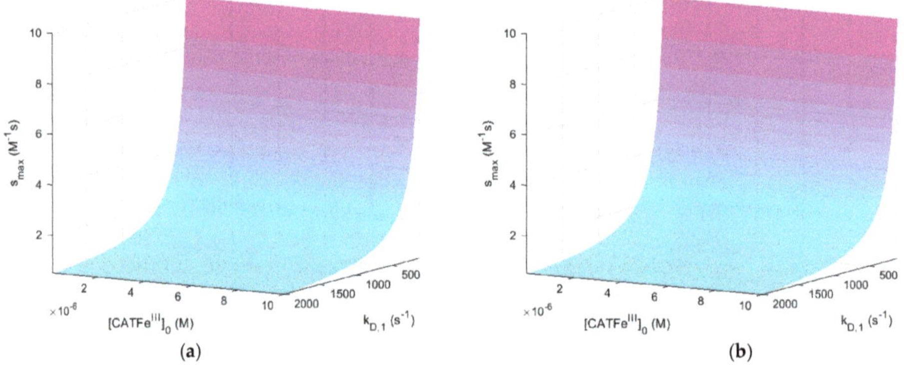

(a)

(b)

Figure A14. The dependent variable $s_{\max}$ (i.e., the inverse of the maximal rate of H_2O_2 consumption in the EC) as a function of $k_{D,1}$ and $[CATFe^{III}]_0$ when $[H_2O_2]_0^{EC} = 1$ mM. $[NO_2^-]_0^{EC} = 0$ M (**a**) and $[NO_2^-]_0^{EC} = 1$ mM (**b**).

Opposed to $\bar{s}$, $s_{\max}$ is more or less independent of $[CATFe^{III}]_0$ and only dependent of $k_{D,1}$. However, the dependence of $k_{D,1}$ is reverse to what would be expected (see Section 2.1); the lower value of $k_{D,1}$, the higher cell susceptibility toward PTL. Thus, $s_{\max}$ does not qualify as a measure to quantify the cell susceptibility in terms of $k_{D,1}$ and $[CATFe^{III}]_0$.

In summary, neither $\bar{s}$ or $s_{\max}$ seem to be appropriate candidates for a dependent variable able to capture and quantify the cell response to PTL.

Appendix B. Derivation of the Rate of Diffusion Equation

According to Fick's first law, the flow of a species through a membrane can be written as

$$J = \frac{1}{A}\frac{dn}{dt} = -D\frac{dc}{dx}, \tag{A1}$$

where J denotes the flow in mol m^{-2}s^{-1}; A is the surface area of the membrane (m^{-2}); n is the amount of the substance (mol); D is the diffusion coefficient (m^{2}s^{-1}); c is the concentration (M). By assuming a linear concentration gradient over the cell membrane, we can write

$$\frac{dc}{dx} = \frac{c_2 - c_1}{\Delta x}, \tag{A2}$$

where Δx is the width of the membrane, see Figure A15, Equations (A1) and (A2) yields

$$\frac{dn}{dt} = -\frac{DA}{\Delta x}(c_2 - c_1),\tag{A3}$$

$$\Leftrightarrow \frac{1}{V}\frac{dn}{dt} = -\frac{DA}{V\Delta x}(c_2 - c_1),\tag{A4}$$

where V denotes the encapsulated volume (m^3). Here, it is a custom to define the term

$$P = \frac{D}{\Delta x},\tag{A5}$$

as the permeability of the membrane. In these terms, Equation (A4) can be written as

$$\frac{dc}{dt} = \frac{-AP}{V}(c_2 - c_1).\tag{A6}$$

Thus, in our Equation (2) in the main paper,

$$k_D = \frac{-AP}{V}.\tag{A7}$$

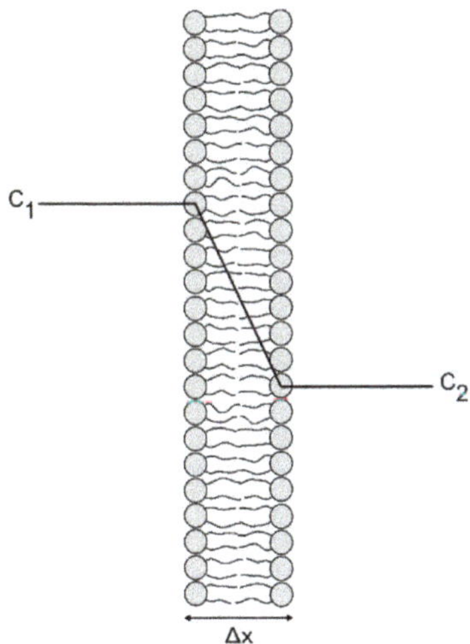

Figure A15. Illustration of a linear concentration gradient over a membrane.

Appendix C. Estimates of Intracellular Catalase Concentration

Appendix C.1. Catalase Concentration Corresponding to a Normal Steady-State Concentration of Hydrogen Peroxide

H_2O_2 is continuously produced in vivo [103] and remains in a quasi-steady-state. Normal intracellular steady-state concentrations of H_2O_2 are 10 nM or less [103,116].

Assuming that the mitochondria are the major source of intracellular H_2O_2, and that catalase is the only enzyme responsible for its decomposition, we estimate the average intracellular concentration of catalase. Denoting $x = [H_2O_2]$, $y = [CATFe^{III}]$, and $z = [CATFe^{IV}O^{\bullet+}]$, and including the constraint that $z = y_0 - y$, the rate equation for x is

$$\frac{dx}{dt} = -x(k_1 y + k_2(y_0 - y)) + k_P.\tag{A8}$$

Denoting the steady state concentration of H_2O_2 as x_{ss}, Equation (A8) yields

$$-x_{ss}(k_1 y + k_2(y_0 - y)) + k_P = 0 \Rightarrow$$

$$y_0 = -y\left(\frac{k_1}{k_2} - 1\right) + \frac{k_P}{k_2 x_{ss}}, \tag{A9}$$

in other words, the solutions are found on the straight line given by Equation (A9). For example, $y = y_0$ ($x_{ss} \sim 10^{-8}$ M),

$$y_0 = \frac{k_P}{k_1 x_{ss}} \sim 6 \times 10^{-7} \text{ M.}$$

Appendix C.2. Catalase Concentration from Detected Catalase Monomers per Cell

In [62], the effective number of fully active catalase monomers per cell for various cancer cell lines was detected. This number varied from 101×10^3 to 538×10^3 and there was a strong correlation between the rate constant of H_2O_2-decomposition and the number of fully active catalase monomers per cell. Since each catalase molecule consists of four monomers, the number of catalase molecules per cell is thus roughly $(25–125) \times 10^3$. The conclusions in [62] were that the rate constant for removal of extracellular H_2O_2 were on average two times higher in normal cells than in cancer cells, and that catalase activity is critical in removing this H_2O_2. If normal cells have a capacity that is twice as large to remove H_2O_2, the number of catalase molecules for the cancer cell lines' normal counterparts should be $(50–250) \times 10^3$. If N denotes the number of molecules per cell, the number of moles per cell is

$$n = \frac{N}{N_A},$$

where

$$N_A \sim 6.022 \times 10^{23} \text{ mol}^{-1},$$

is Avogadro's number. If each cell can be assumed to be a sphere of radius $r \sim 20$ μM (as has been estimated for HeLa-cells [117]), its volume is given by

$$V = \frac{4\pi}{3} r^3.$$

Thus, the average concentration of catalase per cell is

$$\left[CATFe^{III}\right] = \frac{n}{V}.$$

For $N = 100 \times 10^3$, $\left[CATFe^{III}\right] \sim 6 \times 10^{-6}$ M.

Appendix D. Concentration of Intracellular Nitrite and Carbon Dioxide

The NO_2^--levels measured in human physiological fluids are 0.5–210 μM [111–113].

The partial pressure in human alveolar has been found to be $p_{CO_2} = (4.0–9.3) \times 10^3$ Pa [110], which according to Henry's law yields $[CO_2] = Hp$, where $H = 3.4 \times 10^{-2}$ M atm^{-1} is Henry's constant for CO_2 in water at $T = 298.15$ K (i.e., $[CO_2] = (1.3–3.1) \times 10^{-3}$ M). Cell culture media (Eagle's) contained 2200 mgL^{-1} of $NaHCO_3$ (i.e., $\left[HCO_3^-\right] = 36$ mM). At $pH = 7.0$, this corresponds to $[CO_2] = 36$ mM.

References

1. Privat-Maldonado, A.; Bengtson, C.; Razzokov, J.; Smits, E.; Bogaerts, A. Modifying the tumour microenvironment: Challenges and Future Perspectives for Anticancer Plasma Treatments. *Cancers* **2019**, *11*, 1920. [CrossRef]
2. Van Boxem, W.; Van der Paal, J.; Gorbanev, Y.; Vanuytsel, S.; Smits, E.; Dewilde, S.; Bogaerts, A. Anti-cancer capacity of plasma-treated PBS: Effect of chemical composition on cancer cell cytotoxicity. *Sci. Rep.* **2017**, *7*, 15. [CrossRef]

3.	Lin, A.; Truong, B.; Patel, S.; Kaushik, N.; Choi, E.H.; Fridman, G.; Fridman, A.; Miller, V. Nanosecond-Pulsed DBD plasma-generated reactive oxygen species trigger immunogenic cell death in A549 lung carcinoma cells through intracellular oxidative stress. *Int. J. Mol. Sci.* **2017**, *18*, 966. [CrossRef] [PubMed]

4.	Lin, A.; Gorbanev, Y.; De Backer, J.; Van Loenhout, J.; Van Boxem, W.; Lemiere, F.; Cos, P.; Dewilde, S.; Smits, E.; Bogaerts, A. Non-thermal plasma as a unique delivery system of short-lived reactive oxygen and nitrogen species for immunogenic cell death in melanoma cells. *Adv. Sci.* **2019**, *6*, 15. [CrossRef] [PubMed]

5.	Bisag, A.; Bucci, C.; Coluccelli, S.; Girolimetti, G.; Laurita, R.; De Iaco, P.; Perrone, A.M.; Gherardi, M.; Marchio, L.; Porcelli, A.M.; et al. Plasma-activated Ringer's Lactate solution displays a selective cytotoxic effect on ovarian cancer cells. *Cancers* **2020**, *12*, 476. [CrossRef] [PubMed]

6.	Biscop, E.; Lin, A.; Van Boxem, W.; Van Loenhout, J.; De Backer, J.; Deben, C.; Dewilde, S.; Smits, E.; Bogaerts, A. The influence of cell type and culture medium on determining cancer selectivity of cold atmospheric plasma treatment. *Cancers* **2019**, *11*, 1287. [CrossRef]

7.	Bekeschus, S.; Freund, E.; Spadola, C.; Privat-Maldonado, A.; Hackbarth, C.; Bogaerts, A.; Schmidt, A.; Wende, K.; Weltmann, K.D.; von Woedtke, T.; et al. Risk Assessment of kINPen plasma treatment of four human pancreatic cancer cell lines with respect to metastasis. *Cancers* **2019**, *11*, 1237. [CrossRef] [PubMed]

8.	Schuster, M.; Seebauer, C.; Rutkowski, R.; Hauschild, A.; Podmelle, F.; Metelmann, C.; Metelmann, B.; von Woedtke, T.; Hasse, S.; Weltmann, K.D.; et al. Visible tumor surface response to physical plasma and apoptotic cell kill in head and neck cancer. *J. Cranio-Maxillo-Facial Surg.* **2016**, *44*, 1445–1452. [CrossRef] [PubMed]

9.	Metelmann, H.R.; Seebauer, C.; Miller, V.; Fridman, A.; Bauer, G.; Graves, D.B.; Pouvesle, J.M.; Rutkowski, R.; Schuster, M.; Bekeschus, S.; et al. Clinical experience with cold plasma in the treatment of locally advanced head and neck cancer. *Clin. Plasma Med.* **2018**, *9*, 6–13. [CrossRef]

10.	Lim, J.B.; Huang, B.K.; Deen, W.M.; Sikes, H.D. Analysis of the lifetime and spatial localization of hydrogen peroxide generated in the cytosol using a reduced kinetic model. *Free Radic. Biol. Med.* **2015**, *89*, 47–53. [CrossRef]

11.	Lim, J.B.; Langford, T.F.; Huang, B.K.; Deen, W.M.; Sikes, H.D. A reaction-diffusion model of cytosolic hydrogen peroxide. *Free Radic. Biol. Med.* **2016**, *90*, 85–90. [CrossRef]

12.	Komalapriya, C.; Kaloriti, D.; Tillmann, A.T.; Yin, Z.K.; Herrero-de-Dios, C.; Jacobsen, M.D.; Belmonte, R.C.; Cameron, G.; Haynes, K.; Grebogi, C.; et al. Integrative Model of Oxidative Stress Adaptation in the Fungal Pathogen Candida albicans. *PLoS ONE* **2015**, *10*, e0137750.

13.	Erudaitius, D.; Mantooth, J.; Huang, A.; Soliman, J.; Doskey, C.M.; Buettner, G.R.; Rodgers, V.G.J. Calculated cell-specific intracellular hydrogen peroxide concentration: Relevance in cancer cell susceptibility during ascorbate therapy. *Free Radic. Biol. Med.* **2018**, *120*, 356–367. [CrossRef]

14.	Erudaitius, D.T.; Buettner, G.R.; Rodgers, V.G.J. The latency of peroxisomal catalase in terms of effectiveness factor for pancreatic and glioblastoma cancer cell lines in the presence of high concentrations of H_2O_2: Implications for the use of pharmacological ascorbate in cancer therapy. *Free Radic. Biol. Med.* **2020**, *156*, 20–25. [CrossRef] [PubMed]

15.	Bengtson, C.; Bogaerts, A. On the anti-cancer effect of cold atmospheric plasma and the possible role of catalase-dependent apoptotic pathways. *Cells* **2020**, *9*, 2330. [CrossRef] [PubMed]

16.	Ahn, H.J.; Kim, K.I.; Kim, G.; Moon, E.; Yang, S.S.; Lee, J.S. Atmospheric-pressure plasma jet induces apoptosis involving mitochondria via generation of free radicals. *PLoS ONE* **2011**, *6*, e28154. [CrossRef]

17.	Arjunan, K.P.; Friedman, G.; Fridman, A.; Clyne, A.M. Non-thermal dielectric barrier discharge plasma induces angiogenesis through reactive oxygen species. *J. R. Soc. Interface* **2012**, *9*, 147–157. [CrossRef]

18.	Yan, X.; Xiong, Z.L.; Zou, F.; Zhao, S.S.; Lu, X.P.; Yang, G.X.; He, G.Y.; Ostrikov, K. Plasma-Induced Death of HepG2 cancer cells: Intracellular effects of reactive species. *Plasma Process. Polym.* **2012**, *9*, 59–66. [CrossRef]

19.	Ishaq, M.; Kumar, S.; Varinli, H.; Han, Z.J.; Rider, A.E.; Evans, M.D.M.; Murphy, A.B.; Ostrikov, K. Atmospheric gas plasma-induced ROS production activates TNF-ASK1 pathway for the induction of melanoma cancer cell apoptosis. *Mol. Biol. Cell* **2014**, *25*, 1523–1531. [CrossRef] [PubMed]

20.	Kim, S.J.; Chung, T.H. Cold atmospheric plasma jet-generated RONS and their selective effects on normal and carcinoma cells. *Sci. Rep.* **2016**, *6*, 14. [CrossRef]

21.	Kalghatgi, S.; Kelly, C.M.; Cerchar, E.; Torabi, B.; Alekseev, O.; Fridman, A.; Friedman, G.; Azizkhan-Clifford, J. Effects of non-thermal plasma on mammalian cells. *PLoS ONE* **2011**, *6*, e16270. [CrossRef]

22.	Vandamme, M.; Robert, E.; Lerondel, S.; Sarron, V.; Ries, D.; Dozias, S.; Sobilo, J.; Gosset, D.; Kieda, C.; Legrain, B.; et al. ROS implication in a new antitumor strategy based on non-thermal plasma. *Int. J. Cancer* **2012**, *130*, 2185–2194. [CrossRef]

23.	Van der Paal, J.; Neyts, E.C.; Verlackt, C.C.W.; Bogaerts, A. Effect of lipid peroxidation on membrane permeability of cancer and normal cells subjected to oxidative stress. *Chem. Sci.* **2016**, *7*, 489–498. [CrossRef] [PubMed]

24.	Van der Paal, J.; Verheyen, C.; Neyts, E.C.; Bogaerts, A. Hampering effect of cholesterol on the permeation of reactive oxygen species through phospholipids bilayer: Possible explanation for plasma cancer selectivity. *Sci. Rep.* **2017**, *7*, 11. [CrossRef] [PubMed]

25.	Girard, P.M.; Arbabian, A.; Fleury, M.; Bauville, G.; Puech, V.; Dutreix, M.; Sousa, J.S. Synergistic effect of H_2O_2 and NO_2 in cell death induced by cold atmospheric he plasma. *Sci. Rep.* **2016**, *6*, 17. [CrossRef] [PubMed]

26. Kurake, N.; Tanaka, H.; Ishikawa, K.; Kondo, T.; Sekine, M.; Nakamura, K.; Kajiyama, H.; Kikkawa, F.; Mizuno, M.; Hori, M. Cell survival of glioblastoma grown in medium containing hydrogen peroxide and/or nitrite, or in plasma-activated medium. *Arch. Biochem. Biophys.* **2016**, *605*, 102–108. [CrossRef] [PubMed]

27. Yan, D.Y.; Talbot, A.; Nourmohammadi, N.; Cheng, X.Q.; Canady, J.; Sherman, J.; Keidar, M. Principles of using Cold Atmospheric Plasma Stimulated Media for Cancer Treatment. *Sci. Rep.* **2015**, *5*, 17. [CrossRef]

28. Adachi, T.; Tanaka, H.; Nonomura, S.; Hara, H.; Kondo, S.; Hod, M. Plasma-activated medium induces A549 cell injury via a spiral apoptotic cascade involving the mitochondrial-nuclear network. *Free Radic. Biol. Med.* **2015**, *79*, 28–44. [CrossRef]

29. Adachi, T.; Nonomura, S.; Horiba, M.; Hirayama, T.; Kamiya, T.; Nagasawa, H.; Hara, H. Iron stimulates plasma-activated medium-induced A549 cell injury. *Sci. Rep.* **2016**, *6*, 12. [CrossRef]

30. Judee, F.; Fongia, C.; Ducommun, B.; Yousfi, M.; Lobjois, V.; Merbahi, N. Short and long time effects of low temperature Plasma Activated Media on 3D multicellular tumor spheroids. *Sci. Rep.* **2016**, *6*, 12. [CrossRef]

31. Yokoyama, M.; Johkura, K.; Sato, T. Gene expression responses of HeLa cells to chemical species generated by an atmospheric plasma flow. *Biochem. Biophys. Res. Commun.* **2014**, *450*, 1266–1271. [CrossRef] [PubMed]

32. Yan, D.Y.; Lin, L.; Sherman, J.H.; Canady, J.; Trink, B.; Keidar, M. The Correlation Between the Cytotoxicity of Cold Atmospheric Plasma and the Extracellular H_2O_2-Scavenging Rate. *IEEE Trans. Radiat. Plasma Med. Sci.* **2018**, *2*, 618–623. [CrossRef]

33. Antunes, F.; Cadenas, E. Cellular titration of apoptosis with steady state concentrations of H_2O_2: Submicromolar levels of H_2O_2 induce apoptosis through Fenton chemistry independent of the cellular thiol state. *Free Radic. Biol. Med.* **2001**, *30*, 1008–1018. [CrossRef]

34. Klingelhoeffer, C.; Kammerer, U.; Koospal, M.; Muhling, B.; Schneider, M.; Kapp, M.; Kubler, A.; Germer, C.T.; Otto, C. Natural resistance to ascorbic acid induced oxidative stress is mainly mediated by catalase activity in human cancer cells and catalase-silencing sensitizes to oxidative stress. *BMC Complement. Altern. Med.* **2012**, *12*, 10. [CrossRef]

35. Sestili, P.; Brandi, G.; Brambilla, L.; Cattabeni, F.; Cantoni, O. Hydrogen peroxide mediates the killing of U937 tumor cells elicited by pharmacologically attainable concentrations of ascorbic acid: Cell death prevention by extracellular catalase or catalase from cocultured erythrocytes or fibroblasts. *J. Pharmacol. Exp. Ther.* **1996**, *277*, 1719–1725.

36. Verrax, J.; Calderon, P.B. Pharmacologic concentrations of ascorbate are achieved by parenteral administration and exhibit antitumoral effects. *Free Radic. Biol. Med.* **2009**, *47*, 32–40. [CrossRef]

37. Lukes, P.; Dolezalova, E.; Sisrova, I.; Clupek, M. Aqueous-phase chemistry and bactericidal effects from an air discharge plasma in contact with water: Evidence for the formation of peroxynitrite through a pseudo-second-order post-discharge reaction of H_2O_2 and HNO_2. *Plasma Sources Sci. Technol.* **2014**, *23*, 15. [CrossRef]

38. Privat-Maldonado, A.; Gorbanev, Y.; Dewilde, S.; Smits, E.; Bogaerts, A. Reduction of human glioblastoma spheroids using cold atmospheric plasma: The combined effect of short- and long-lived reactive species. *Cancers* **2018**, *10*, 394. [CrossRef]

39. Almasalmeh, A.; Krenc, D.; Wu, B.H.; Beitz, E. Structural determinants of the hydrogen peroxide permeability of aquaporins. *FEBS J.* **2014**, *281*, 647–656. [CrossRef]

40. Bienert, G.P.; Chaumont, F. Aquaporin-facilitated transmembrane diffusion of hydrogen peroxide. *Biochim. Biophys. Acta Gen. Subj.* **2014**, *1840*, 1596–1604. [CrossRef] [PubMed]

41. Miller, E.W.; Dickinson, B.C.; Chang, C.J. Aquaporin-3 mediates hydrogen peroxide uptake to regulate downstream intracellular signaling. *Proc. Natl. Acad. Sci. USA* **2010**, *107*, 15681–15686. [CrossRef] [PubMed]

42. Yusupov, M.; Razzokov, J.; Cordeiro, R.M.; Bogaerts, A. Transport of Reactive Oxygen and Nitrogen Species across Aquaporin: A Molecular Level Picture. *Oxidative Med. Cell. Longev.* **2019**, *2019*, 2930504. [CrossRef]

43. Verkman, A.S.; Hara-Chikuma, M.; Papadopoulos, M.C. Aquaporins–New players in cancer biology. *J. Mol. Med.* **2008**, *86*, 523–529. [CrossRef] [PubMed]

44. Papadopoulos, M.C.; Saadoun, S. Key roles of aquaporins in tumor biology. *Biochim. Biophys. Acta Biomembr.* **2015**, *1848*, 2576–2583. [CrossRef] [PubMed]

45. Erudaitius, D.; Huang, A.; Kazmi, S.; Buettner, G.R.; Rodgers, V.G.J. Peroxiporin Expression Is an Important Factor for Cancer Cell Susceptibility to Therapeutic H_2O_2: Implications for Pharmacological Ascorbate Therapy. *PLoS ONE* **2017**, *12*, e0170442. [CrossRef] [PubMed]

46. Yan, D.Y.; Xiao, H.J.; Zhu, W.; Nourmohammadi, N.; Zhang, L.G.; Bian, K.; Keidar, M. The role of aquaporins in the anti-glioblastoma capacity of the cold plasma-stimulated medium. *J. Phys. D Appl. Phys.* **2017**, *50*, 8. [CrossRef]

47. Makino, N.; Mochizuki, Y.; Bannai, S.; Sugita, Y. Kinetic-studies on the removal of extracellular hydrogen-peroxide by cultured fibroblasts. *J. Biol. Chem.* **1994**, *269*, 1020–1025. [CrossRef]

48. Makino, N.; Sasaki, K.; Hashida, K.; Sakakura, Y. A metabolic model describing the H_2O_2 elimination by mammalian cells including H_2O_2 permeation through cytoplasmic and peroxisomal membranes: Comparison with experimental data. *Biochim. Biophys. Acta Gen. Subj.* **2004**, *1673*, 149–159. [CrossRef]

49. Sasaki, K.; Bannai, S.; Makino, N. Kinetics of hydrogen peroxide elimination by human umbilical vein endothelial cells in culture. *Biochim. Biophys. Acta Gen. Subj.* **1998**, *1380*, 275–288. [CrossRef]

50. Johnson, R.M.; Ho, Y.S.; Yu, D.Y.; Kuypers, F.A.; Ravindranath, Y.; Goyette, G.W. The effects of disruption of genes for peroxiredoxin-2, glutathione peroxidase-1, and catalase on erythrocyte oxidative metabolism. *Free Radic. Biol. Med.* **2010**, *48*, 519–525. [CrossRef]

51. Mueller, S.; Riedel, H.D.; Stremmel, W. Direct evidence for catalase as the predominant H_2O_2-removing enzyme in human erythrocytes. *Blood* **1997**, *90*, 4973–4978. [CrossRef]

52. Winterbourn, C.C.; Stern, A. Human red-cells scavenge extracellular hydrogen-peroxide and inhibit formation of hypochlorous acid and hydroxyl radical. *J. Clin. Investig.* **1987**, *80*, 1486–1491. [CrossRef]

53. Baker, A.M.; Oberley, L.W.; Cohen, M.B. Expression of antioxidant enzymes in human prostatic adenocarcinoma. *Prostate* **1997**, *32*, 229–233. [CrossRef]

54. Beck, R.; Pedrosa, R.C.; Dejeans, N.; Glorieux, C.; Leveque, P.; Gallez, B.; Taper, H.; Eeckhoudt, S.; Knoops, L.; Calderon, P.B.; et al. Ascorbate/menadione-induced oxidative stress kills cancer cells that express normal or mutated forms of the oncogenic protein Bcr-Abl. An in vitro and in vivo mechanistic study. *Investig. New Drugs* **2011**, *29*, 891–900. [CrossRef] [PubMed]

55. Bostwick, D.G.; Alexander, E.E.; Singh, R.; Shan, A.; Qian, J.Q.; Santella, R.M.; Oberley, L.W.; Yan, T.; Zhong, W.X.; Jiang, X.H.; et al. Antioxidant enzyme expression and reactive oxygen species damage in prostatic intraepithelial neoplasia and cancer. *Cancer* **2000**, *89*, 123–134. [CrossRef]

56. Ho, J.C.M.; Zheng, S.; Comhair, S.A.A.; Farver, C.; Erzurum, S.C. Differential expression of manganese superoxide dismutase and catalase in lung cancer. *Cancer Res.* **2001**, *61*, 8578–8585.

57. Kwei, K.A.; Finch, J.S.; Thompson, E.J.; Bowden, G.T. Transcriptional repression of catalase in mouse skin tumor progression. *Neoplasia* **2004**, *6*, 440–448. [CrossRef] [PubMed]

58. Lauer, C.; Volkl, A.; Riedl, S.; Fahimi, H.D.; Beier, K. Impairment of peroxisomal biogenesis in human colon carcinoma. *Carcinogenesis* **1999**, *20*, 985–989. [CrossRef] [PubMed]

59. Glorieux, C.; Sandoval, J.M.; Fattaccioli, A.; Dejeans, N.; Garbe, J.C.; Dieu, M.; Verrax, J.; Renard, P.; Huang, P.; Calderon, P.B. Chromatin remodeling regulates catalase expression during cancer cells adaptation to chronic oxidative stress. *Free Radic. Biol. Med.* **2016**, *99*, 436–450. [CrossRef]

60. Sato, K.; Ito, K.; Kohara, H.; Yamaguchi, Y.; Adachi, K.; Endo, H. Negative regulation of catalase gene-expression in hepatoma-cells. *Mol. Cell. Biol.* **1992**, *12*, 2525–2533. [CrossRef]

61. Guner, G.; Kokoglu, E.; Guner, A. Hydrogen-peroxide detoxication by catalase in subcellular-fractions of human-brain tumors and normal brain-tissues. *Cancer Lett.* **1985**, *27*, 221–224. [CrossRef]

62. Doskey, C.M.; Buranasudja, V.; Wagner, B.A.; Wilkes, J.G.; Du, J.; Cullen, J.J.; Buettner, G.R. Tumor cells have decreased ability to metabolize H_2O_2: Implications for pharmacological ascorbate in cancer therapy. *Redox Biol.* **2016**, *10*, 274–284. [CrossRef]

63. Marklund, S.L.; Westman, N.G.; Lundgren, E.; Roos, G. Copper-containing and zinc-containing superoxide-dismutase, manganese-containing superoxide-dismutase, catalase, and glutathione-peroxidase in normal and neoplastic human cell-lines and normal human-tissues. *Cancer Res.* **1982**, *42*, 1955–1961. [PubMed]

64. Gebicka, L.; Didik, J. Catalytic scavenging of peroxynitrite by catalase. *J. Inorg. Biochem.* **2009**, *103*, 1375–1379. [CrossRef]

65. Alvarez, B.; Radi, R. Peroxynitrite reactivity with amino acids and proteins. *Amino Acids* **2003**, *25*, 295–311. [CrossRef] [PubMed]

66. Augusto, O.; Goldstein, S.; Hurst, J.K.; Lind, J.; Lymar, S.V.; Merenyi, G.; Radi, R. Carbon dioxide-catalyzed peroxynitrite reactivity–The resilience of the radical mechanism after two decades of research. *Free Radic. Biol. Med.* **2019**, *135*, 210–215. [CrossRef]

67. Carballal, S.; Bartesaghi, S.; Radi, R. Kinetic and mechanistic considerations to assess the biological fate of peroxynitrite. *Biochim. Biophys. Acta Gen. Subj.* **2014**, *1840*, 768–780. [CrossRef]

68. Radi, R. Peroxynitrite, a Stealthy Biological Oxidant. *J. Biol. Chem.* **2013**, *288*, 26464–26472. [CrossRef]

69. Squadrito, G.L.; Pryor, W.A. Oxidative chemistry of nitric oxide: The roles of superoxide, peroxynitrite, and carbon dioxide. *Free Radic. Biol. Med.* **1998**, *25*, 392–403. [CrossRef]

70. Rong, L.; Zhang, C.; Lei, Q.; Hu, M.M.; Feng, J.; Shu, H.B.; Liu, Y.; Zhang, X.Z. Hydrogen peroxide detection with high specificity in living cells and inflamed tissues. *Regen. Biomater.* **2016**, *3*, 217–222. [CrossRef]

71. Ermakova, Y.G.; Bilan, D.S.; Matlashov, M.E.; Mishina, N.M.; Markvicheva, K.N.; Subach, O.M.; Subach, F.V.; Bogeski, I.; Hoth, M.; Enikolopov, G.; et al. Red fluorescent genetically encoded indicator for intracellular hydrogen peroxide. *Nat. Commun.* **2014**, *5*, 9. [CrossRef]

72. Enyedi, B.; Zana, M.; Donko, A.; Geiszt, M. Spatial and Temporal Analysis of NADPH Oxidase-Generated Hydrogen Peroxide Signals by Novel Fluorescent Reporter Proteins. *Antioxid. Redox Signal.* **2013**, *19*, 523–534. [CrossRef] [PubMed]

73. Nathan, C.F.; Silverstein, S.C.; Brukner, L.H.; Cohn, Z.A. Extracellular cytolysis by activated macrophages and granulocytes. 2. hydrogen-peroxide as a mediator of cytotoxicity. *J. Exp. Med.* **1979**, *149*, 100–113. [CrossRef]

74. Roos, D.; Weening, R.S.; Voetman, A.A.; Vanschaik, M.L.J.; Bot, A.A.M.; Meerhof, L.J.; Loos, J.A. Protection of phagocytic leukocytes by endogenous glutathione–studies in a family with glutathione-reductase deficiency. *Blood* **1979**, *53*, 851–866. [CrossRef] [PubMed]

75. Roos, D.; Weening, R.S.; Wyss, S.R.; Aebi, H.E. Protection of human-neutrophils by endogenous catalase–studies with cells from catalase-deficient individuals. *J. Clin. Investig.* **1980**, *65*, 1515–1522. [CrossRef] [PubMed]

76. Abe, M.K.; Chao, T.S.O.; Solway, J.; Rosner, M.R.; Hershenson, M.B. Hydrogen-peroxide stimulates mitogen-activated protein-kinase in bovine tracheal myocytes-implications for human airway disease. *Am. J. Respir. Cell Mol. Biol.* **1994**, *11*, 577–585. [CrossRef] [PubMed]

77. Bae, G.U.; Seo, D.W.; Kwon, H.K.; Lee, H.Y.; Hong, S.; Lee, Z.W.; Ha, K.S.; Lee, H.W.; Han, J.W. Hydrogen peroxide activates p70(S6k) signaling pathway. *J. Biol. Chem.* **1999**, *274*, 32596–32602. [CrossRef]

78. Milligan, S.A.; Owens, M.W.; Grisham, M.B. Differential regulation of extracellular signal-regulated kinase and nuclear factor-kappa B signal transduction pathways by hydrogen peroxide and tumor necrosis factor. *Arch. Biochem. Biophys.* **1998**, *352*, 255–262. [CrossRef] [PubMed]

79. Muller, J.M.; Cahill, M.A.; Rupec, R.A.; Baeuerle, P.A.; Nordheim, A. Antioxidants as well as oxidants activate c-fos via Ras-dependent activation of extracellular-signal-regulated kinase 2 and Elk-1. *Eur. J. Biochem.* **1997**, *244*, 45–52. [CrossRef]

80. Stevenson, M.A.; Pollock, S.S.; Coleman, C.N.; Calderwood, S.K. X-irradiation, phorbol esters, and H_2O_2 stimulate mitogen-activated protein-kinase activity in nih-3t3 cells through the formation of reactive oxygen intermediates. *Cancer Res.* **1994**, *54*, 12–15. [PubMed]

81. Bilsland, E.; Molin, C.; Swaminathan, S.; Ramne, A.; Sunnerhagen, P. Rck1 and Rck2 MAPKAP kinases and the HOG pathway are required for oxidative stress resistance. *Mol. Microbiol.* **2004**, *53*, 1743–1756. [CrossRef] [PubMed]

82. Park, B.G.; Yoo, C.I.; Kim, H.T.; Kwon, C.H.; Kim, Y.K. Role of mitogen-activated protein kinases in hydrogen peroxide-induced cell death in osteoblastic cells. *Toxicology* **2005**, *215*, 115–125. [CrossRef]

83. Staleva, L.; Hall, A.; Orlow, S.J. Oxidative stress activates FUS1 and RLM1 transcription in the yeast Saccharomyces cerevisiae in an oxidant-dependent manner. *Mol. Biol. Cell* **2004**, *15*, 5574–5582. [CrossRef] [PubMed]

84. Denicola, A.; Freeman, B.A.; Trujillo, M.; Radi, R. Peroxynitrite reaction with carbon dioxide/bicarbonate: Kinetics and influence on peroxynitrite-mediated oxidations. *Arch. Biochem. Biophys.* **1996**, *333*, 49–58. [CrossRef]

85. Uppu, R.M.; Squadrito, G.L.; Pryor, W.A. Acceleration of peroxynitrite oxidations by carbon dioxide. *Arch. Biochem. Biophys.* **1996**, *327*, 335–343. [CrossRef] [PubMed]

86. Bonini, M.G.; Radi, R.; Ferrer-Sueta, G.; Ferreira, A.M.D.; Augusto, O. Direct EPR detection of the carbonate radical anion produced from peroxynitrite and carbon dioxide. *J. Biol. Chem.* **1999**, *274*, 10802–10806. [CrossRef] [PubMed]

87. Goldstein, S.; Czapski, G. Viscosity effects on the reaction of peroxynitrite with CO_2: Evidence for radical formation in a solvent cage. *J. Am. Chem. Soc.* **1999**, *121*, 2444–2447. [CrossRef]

88. Hodges, G.R.; Ingold, K.U. Cage-escape of geminate radical pairs can produce peroxynitrate from peroxynitrite under a wide variety of experimental conditions. *J. Am. Chem. Soc.* **1999**, *121*, 10695–10701. [CrossRef]

89. Squadrito, G.L.; Pryor, W.A. Mapping the reaction of peroxynitrite with CO_2: Energetics, reactive species, and biological implications. *Chem. Res. Toxicol.* **2002**, *15*, 885–895. [CrossRef]

90. Gebicka, L.; Didik, J.; Gebicki, J. Reactions of heme proteins with carbonate radical anion. *Res. Chem. Intermed.* **2009**, *35*, 401–409. [CrossRef]

91. Movahed, Z.G.; Rastegari-Pouyani, M.; Mohammadi, M.H.; Mansouri, K. Cancer cells change their glucose metabolism to overcome increased ROS: One step from cancer cell to cancer stem cell? *Biomed. Pharmacother.* **2019**, *112*, 15.

92. Medrano-Fernandez, I.; Bestetti, S.; Bertolotti, M.; Bienert, G.P.; Bottino, C.; Laforenza, U.; Rubartelli, A.; Sitia, R. Stress Regulates Aquaporin-8 Permeability to Impact Cell Growth and Survival. *Antioxid. Redox Signal.* **2016**, *24*, 1031–1044. [CrossRef] [PubMed]

93. Leduc, M.; Guay, D.; Leask, R.L.; Coulombe, S. Cell permeabilization using a non-thermal plasma. *New J. Phys.* **2009**, *11*, 12. [CrossRef]

94. Sasaki, S.; Honda, R.; Hokari, Y.; Takashima, K.; Kanzaki, M.; Kaneko, T. Characterization of plasma-induced cell membrane permeabilization: Focus on OH radical distribution. *J. Phys. D Appl. Phys.* **2016**, *49*, 8. [CrossRef]

95. Chung, T.H.; Stancampiano, A.; Sklias, K.; Gazeli, K.; Andre, F.M.; Dozias, S.; Douat, C.; Pouvesle, J.M.; Sousa, J.S.; Robert, E.; et al. Cell Electropermeabilisation Enhancement by Non-Thermal-Plasma-Treated PBS. *Cancers* **2020**, *12*, 219. [CrossRef] [PubMed]

96. Marla, S.S.; Lee, J.; Groves, J.T. Peroxynitrite rapidly permeates phospholipid membranes. *Proc. Natl. Acad. Sci. USA* **1997**, *94*, 14243–14248. [CrossRef] [PubMed]

97. Kirkman, H.N.; Gaetani, G.F. Catalase–A tetrameric enzyme with 4 tightly bound molecules of nadph. *Proc. Natl. Acad. Sci. USA* **1984**, *81*, 4343–4347. [CrossRef]

98. Kirkman, H.N.; Rolfo, M.; Ferraris, A.M.; Gaetani, G.F. Mechanisms of protection of catalase by NADPH–Kinetics and stoichiometry. *J. Biol. Chem.* **1999**, *274*, 13908–13914. [CrossRef] [PubMed]

99. Ahn, H.J.; Kim, K.I.; Hoan, N.N.; Kim, C.H.; Moon, E.; Choi, K.S.; Yang, S.S.; Lee, J.S. Targeting cancer cells with reactive oxygen and nitrogen species generated by atmospheric-pressure air plasma. *PLoS ONE* **2014**, *9*, e86173. [CrossRef] [PubMed]

100. Hirst, A.M.; Simms, M.S.; Mann, V.M.; Maitland, N.J.; O'Connell, D.; Frame, F.M. Low-temperature plasma treatment induces DNA damage leading to necrotic cell death in primary prostate epithelial cells. *Br. J. Cancer* **2015**, *112*, 1536–1545. [CrossRef]

101. Mohades, S.; Laroussi, M.; Sears, J.; Barekzi, N.; Razavi, H. Evaluation of the effects of a plasma activated medium on cancer cells. *Phys. Plasmas* **2015**, *22*, 6. [CrossRef]

102. Yang, H.; Lu, R.; Xian, Y.; Gan, L.; Lu, X.; Yang, X. Effects of atmospheric pressure cold plasma on human hepatocarcinoma cell and its 5-fluorouracil resistant cell line. *Phys. Plasmas* **2015**, *22*, 9. [CrossRef]

103. Chance, B.; Sies, H.; Boveris, A. Hydroperoxide metabolism in mammalian organs. *Physiol. Rev.* **1979**, *59*, 527–605. [CrossRef] [PubMed]

104. Kissner, R.; Nauser, T.; Bugnon, P.; Lye, P.G.; Koppenol, W.H. Formation and properties of peroxynitrite as studied by laser flash photolysis, high-pressure stopped-flow technique, and pulse radiolysis. *Chem. Res. Toxicol.* **1997**, *10*, 1285–1292. [CrossRef]

105. Pryor, W.A.; Squadrito, G.L. The chemistry of peroxynitrite—A product from the reaction of nitric-oxide with superoxide. *Am. J. Physiol. Lung Cell. Mol. Physiol.* **1995**, *268*, L699–L722. [CrossRef] [PubMed]

106. Gutman, M.; Nachliel, E. The dynamic aspects of proton-transfer processes. *Biochim. Biophys. Acta* **1990**, *1015*, 391–414. [CrossRef]

107. Samouilov, A.; Woldman, Y.Y.; Zweier, J.L.; Khramtsov, V.V. Magnetic resonance study of the transmembrane nitrite diffusion. *Nitric Oxide Biol. Chem.* **2007**, *16*, 362–370. [CrossRef]

108. May, J.M.; Qu, Z.C.; Xia, L.; Cobb, C.E. Nitrite uptake and metabolism and oxidant stress in human erythrocytes. *Am. J. Physiol. Cell Physiol.* **2000**, *279*, C1946–C1954. [CrossRef]

109. Denicola, A.; Souza, J.M.; Radi, R. Diffusion of peroxynitrite across erythrocyte membranes. *Proc. Natl. Acad. Sci. USA* **1998**, *95*, 3566–3571. [CrossRef] [PubMed]

110. Jordanoglou, J.; Tatsis, G.; Danos, J.; Gougoulakis, S.; Orfanidou, D.; Gaga, M. Alveolar partial pressures of carbon-dioxide and oxygen measured by a helium washout technique. *Thorax* **1990**, *45*, 520–524. [CrossRef]

111. Gaston, B.; Reilly, J.; Drazen, J.M.; Fackler, J.; Ramdev, P.; Arnelle, D.; Mullins, M.E.; Sugarbaker, D.J.; Chee, C.; Singel, D.J.; et al. Endogenous nitrogen-oxides and bronchodilator s-nitrosothiols in human airways. *Proc. Natl. Acad. Sci. USA* **1993**, *90*, 10957–10961. [CrossRef] [PubMed]

112. Green, L.C.; Wagner, D.A.; Glogowski, J.; Skipper, P.L.; Wishnok, J.S.; Tannenbaum, S.R. Analysis of nitrate, nitrite, and n-15 -labeled nitrate in biological-fluids. *Anal. Biochem.* **1982**, *126*, 131–138. [CrossRef]

113. Leone, A.M.; Francis, P.L.; Rhodes, P.; Moncada, S. A rapid and simple method for the measurement of nitrite and nitrate in plasma by high-performance capillary electrophoresis. *Biochem. Biophys. Res. Commun.* **1994**, *200*, 951–957. [CrossRef] [PubMed]

114. Jones, D.P. Radical-free biology of oxidative stress. *Am. J. Physiol. Cell Physiol.* **2008**, *295*, C849–C868. [CrossRef] [PubMed]

115. Adimora, N.J.; Jones, D.P.; Kemp, M.L. A model of redox kinetics implicates the thiol proteome in cellular hydrogen peroxide responses. *Antioxid. Redox Signal.* **2010**, *13*, 731–743. [CrossRef]

116. Oshino, N.; Chance, B.; Sies, H.; Bucher, T. Role of H_2O_2 generation in perfused rat-liver and reaction of catalase compound 1 and hydrogen donors. *Arch. Biochem. Biophys.* **1973**, *154*, 117–131. [CrossRef]

117. Puck, T.T.; Marcus, P.I.; Cieciura, S.J. Clonal growth of mammalian cells invitro-growth characteristics of colonies from single hela cells with and without a feeder layer. *J. Exp. Med.* **1956**, *103*, 273–283. [CrossRef]

International Journal of
Molecular Sciences

Article

Argon Plasma Exposure Augments Costimulatory Ligands and Cytokine Release in Human Monocyte-Derived Dendritic Cells

Sander Bekeschus [1,*], Dorothee Meyer [1], Kevin Arlt [1], Thomas von Woedtke [1,2], Lea Miebach [1,3], Eric Freund [1,3] and Ramona Clemen [1]

[1] The Centre for Innovation Competence (ZIK) Plasmatis, Leibniz Institute for Plasma Science and Technology (INP), 17489 Greifswald, Germany; dorothee.j.meyer@gmail.com (D.M.); kevin.arlt@inp-greifswald.de (K.A.); woedtke@inp-greifswald.de (T.v.W.); lea.miebach@inp-greifswald.de (L.M.); eric.freund@inp-greifswald.de (E.F.); ramona.clemen@inp-greifswald.de (R.C.)

[2] Institute of Hygiene and Environmental Medicine, Greifswald University Medical Center, 17475 Greifswald, Germany

[3] Department of General, Visceral, Thoracic, and Vascular Surgery, Greifswald University Medical Center, 17475 Greifswald, Germany

* Correspondence: sander.bekeschus@inp-greifswald.de; Tel.: +49-38-34554-3948

Abstract: Cold physical plasma is a partially ionized gas expelling many reactive oxygen and nitrogen species (ROS/RNS). Several plasma devices have been licensed for medical use in dermatology, and recent experimental studies suggest their putative role in cancer treatment. In cancer therapies with an immunological dimension, successful antigen presentation and inflammation modulation is a key hallmark to elicit antitumor immunity. Dendritic cells (DCs) are critical for this task. However, the inflammatory consequences of DCs following plasma exposure are unknown. To this end, human monocyte-derived DCs (moDCs) were expanded from isolated human primary monocytes; exposed to plasma; and their metabolic activity, surface marker expression, and cytokine profiles were analyzed. As controls, hydrogen peroxide, hypochlorous acid, and peroxynitrite were used. Among all types of ROS/RNS-mediated treatments, plasma exposure exerted the most notable increase of activation markers at 24 h such as CD25, CD40, and CD83 known to be crucial for T cell costimulation. Moreover, the treatments increased interleukin (IL)-1α, IL-6, and IL-23. Altogether, this study suggests plasma treatment augmenting costimulatory ligand and cytokine expression in human moDCs, which might exert beneficial effects in the tumor microenvironment.

Keywords: CAP; cancer; cold atmospheric pressure plasma; hydrogen peroxide; hypochlorous acid; moDCs; peroxynitrite; reactive oxygen and nitrogen species; RNS; ROS

Citation: Bekeschus, S.; Meyer, D.; Arlt, K.; von Woedtke, T.; Miebach, L.; Freund, E.; Clemen, R. Argon Plasma Exposure Augments Costimulatory Ligands and Cytokine Release in Human Monocyte-Derived Dendritic Cells. *Int. J. Mol. Sci.* **2021**, *22*, 3790. https://doi.org/10.3390/ijms22073790

Academic Editors: Akikazu Sakudo and Yoshihito Yagyu

Received: 21 March 2021
Accepted: 30 March 2021
Published: 6 April 2021

Publisher's Note: MDPI stays neutral with regard to jurisdictional claims in published maps and institutional affiliations.

1. Introduction

The resolution of many diseases is controlled by precise modulation of inflammation [1–3]. Pro-inflammatory responses help to promote pathogen clearance and antigen presentation to engage adaptive immunity, while anti-inflammatory responses often counterbalance preceding inflammation, which—if left unchecked—lead to tissue damage and cellular dysfunction. Cells of the innate immune system critically modulate inflammatory responses and provide the link between a pathological condition, e.g., infection or cancer and adaptive immune responses that can, for instance, specifically target infected or malignant cells. Especially, dendritic cells (DCs) are the most prominent example of providing activating antigens for T cell stimulation, which is why these cells are a member of the professional antigen-presenting cells (APCs) family [4].

Two major DC subsets exist: conventional DCs (cDCs) and plasmacytoid DCs (pDCs). The former are characterized by toll-like receptor (TLR) 2 and 4 expression and can be further divided into cDC-1, which are more abundant and provide major stimulus for

107

T cells, and for anticancer immunity, and the less abundant cDC-2, which are critical in targeting infection, e.g., in wounds [5]. The latter (pDCs) mainly express TLR7, TLR8, and TLR9, and are also critical in antitumor effects due to their direct cytotoxic activity towards cancer cells, such as melanoma [6]. DCs can be differentiated from human monocytes via chemokine/cytokine stimulation that leads to the upregulation of several surface receptors crucial for antigen presentation to T cells, such as major histocompatibility complex class II (MHC-II or HLA-DR), and cluster of differentiation (CD) 40, CD80, CD83, and CD86, along with changes in their metabolic profiles [7]. Antigen presentation is done after maturation and migration in a CCR7-dependent manner to secondary lymphatic organs such as the lymph node or spleen, which are patrolled by naïve T cells searching for their cognate antigen [8]. Reactive oxygen and nitrogen species (ROS/RNS) also play vital roles in the biology of DCs as their intracellular levels correlate with the priming, function, and development of these cells, with implications in inflammatory responses [9].

ROS/RNS release is the prime hallmark of cold physical plasmas. This accredited technology has been established in many dermatological centers with the main focus on wound management in the last decade [10] and is based on evidence from several clinical trials [11–13]. Many in vivo models have supported the claim of physical plasma-induced healing being also promoted in sterile wounds, i.e., in the absence of infection [14–16]. Moreover, recent evidence found a role of physical plasma treatment in oncology [17]. Cancer patients benefited from this treatment during palliation [18], and several syngeneic animal models suggested an involvement of antitumor immunity unleashed by plasma-induced tumor cell killing and subsequent putative transport of tumor antigens by DCs promoting T cell activation [18–21]. The patients' results also suggest the role of the immune system [22], with DCs being among the key cells takiing up antigens from inactivated tumor cells. Ultimately, this suggests physical plasma-derived ROS/RNS, which are plentiful and diverse [22], directly affect ROS/RNS sensing and redox signaling and ultimately contributing to biologically relevant consequences [23].

While the tumor-toxic activity of physical plasma treatment has been shown numerous times [24], and it is clear that DCs are propagating antitumor immunity in the cancer context [25], it has not been investigated so far how plasma exposure affects DCs alone that—in an in vivo setting—undoubtedly are present in the tumor microenvironment (TME) [26]. To this end, we investigated DC activity, surface marker expression, and cytokine release in human-monocyte-derived DCs (moDCs) in vitro following physical plasma exposure and compared responses to treatments with single ROS/RNS and lipopolysaccharide (LPS). All treatments had an impact on DCs, while plasma treatment elicited notable changes potentially beneficial for their function.

2. Results

2.1. Toxicity towards Argon Plasma Treatment and ROS/RNS

In this study, the kINPen argon plasma jet was used (Figure 1a) for the treatment of monocyte-derived dendritic cells (moDCs). Analysis was conducted 24 h later (Figure 1b). To estimate the sensitivity of moDCs compared to lymphocytes, cell suspensions were treated together, and the number of dead cells was analyzed for each population separately (Figure 1c). It was found that lymphocytes were markedly more sensitive to argon plasma-induced cytotoxic effects compared to the moDCs (Figure 1d).

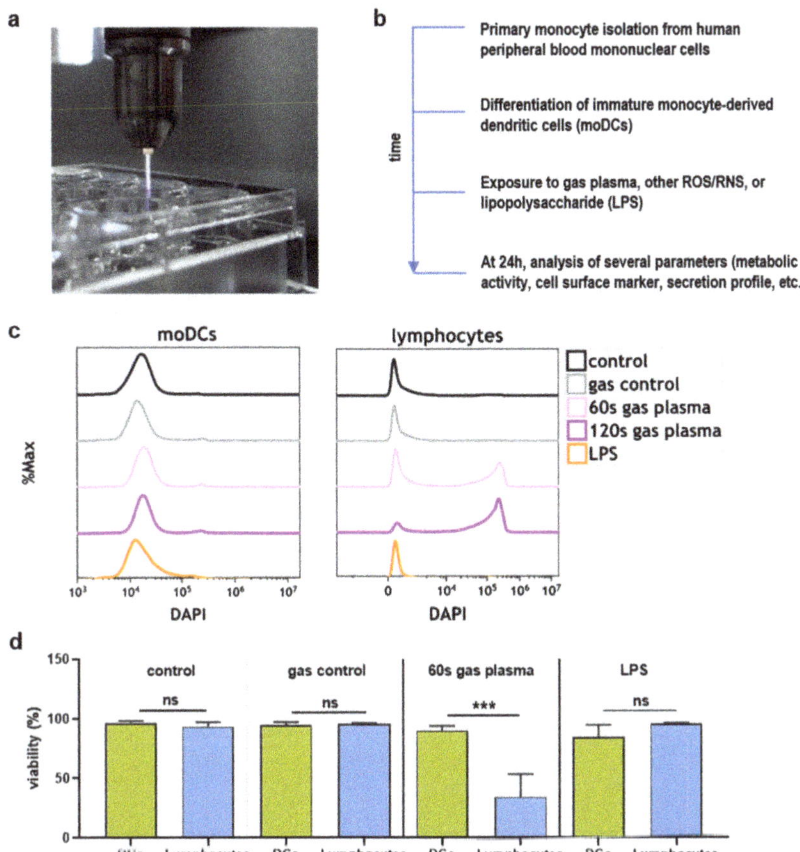

Figure 1. Study protocol and toxicity comparison. (**a**) Image of argon plasma treatment of cells in 24-well plates; (**b**) scheme of study protocol; (**c,d**) overlay 4′,6-diamidino-2-phenylindole (DAPI) histograms of monocyte-derived DCs (moDCs) and lymphocytes according to the indicated treatments (**c**) and quantification of the percentage of viable moDCs and lymphocytes (**d**) treated together in a single well. Data are mean and standard error of three experiments, and statistical analysis was performed using t-test with $p < 0.001$ (***) differing significantly or non-significantly (ns).

Next, the metabolic activity and viability of moDCs following exposure to argon plasma or ROS/RNS were investigated. Pronounced differences were observed for metabolic activity when assayed kinetically after exposure to pilot amounts of ROS/RNS or plasma treatment (Figure 2a), demonstrating the assay sensing differences in metabolic activity in stress inactivated cells. Subsequently, metabolic activity endpoint assays were conducted 24 h after exposure to different argon plasma treatment times (Figure 2b) or concentrations of hydrogen peroxide (H_2O_2, Figure 2c), hypochlorous acid (HOCl, Figure 2d), peroxynitrite (ONOO⁻, Figure 2e), and lipopolysaccharide (LPS, Figure 2f). Except for LPS, which is known to be non-toxic at lower concentrations, all agents showed a treatment time-dependent reduction in metabolic activity. These data were in agreement with the results obtained via flow cytometry assaying cell viability (Figure 2g–k), as illustrated by the heatmap (Figure 2l).

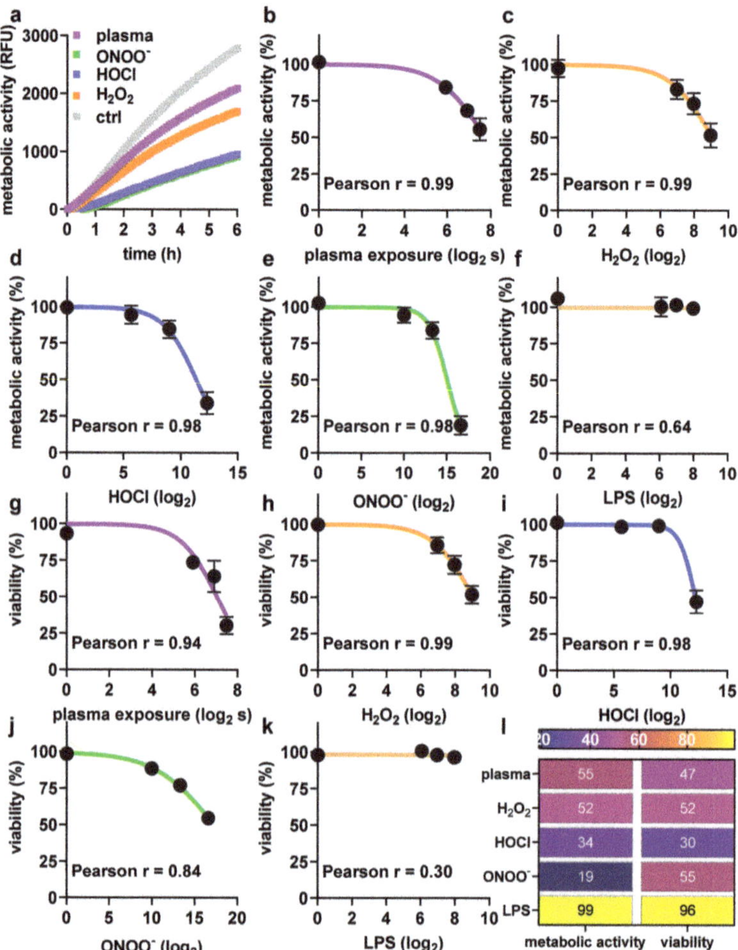

Figure 2. Argon plasma treatment and reactive oxygen and nitrogen species (ROS/RNS) have dose-dependent toxicity profiles. (**a**) Kinetic metabolic activity of moDCs treated as indicated over 6 h; (**b–f**) metabolic activity in response to several argon plasma treatments times or concentrations of ROS/RNS or LPS at 24 h; (**g–k**) viability in response to several argon plasma treatment times or concentrations of ROS/RNS or lipopolysaccharide (LPS) at 24 h; and (**l**) heatmap for comparison between reduction in metabolic activity and viability. Data are mean and standard error of 3–6 different donors.

2.2. Surface Marker Expression and Cytokine Release after Plasma or ROS/RNS Exposure

Next, the surface marker expression levels on moDCs were investigated. Pilot experiments using unstained, stained, and LPS-pulsed, and stained cells confirmed the known upregulation of several costimulatory molecules after activation, including CCR7, CD25, CD40, CD83, CD86, and HLA-DR (Figure 3a).

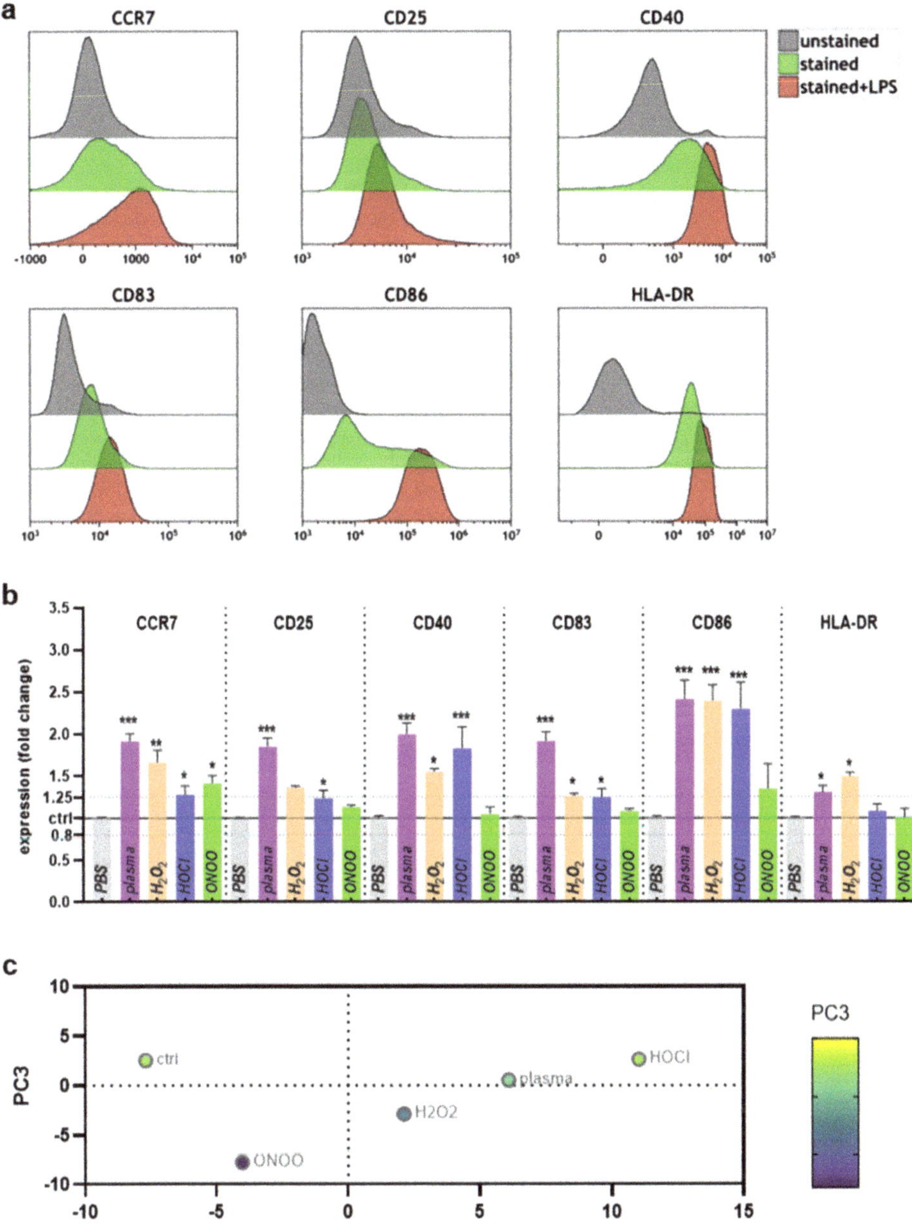

Figure 3. Argon plasma treatment and ROS/RNS modulate the surface marker expression profiles. (**a**) overlay histograms of several cell surface markers of unstained, stained, stained, and LPS-pulsed human moDCs; (**b**) quantification and fold-change differences in human moDCs treated as indicated and analyzed 24 h later; (**c**) data summary using principal-component analysis. Data are mean and standard error of 3–6 different donors. Statistical analysis was done using one-way analysis of variances with $p < 0.05$ (*), $p < 0.01$ (**), and $p < 0.001$ (***).

Next, moDCs were exposed to partially toxic argon plasma or ROS/RNS conditions, and the surface marker expression was investigated by multicolor flow cytometry and normalized to that of cells receiving vehicle only. Shown are also the boundary lines for ± 1.25 differences in fold-change, and a significant downregulation was not observed for any of the markers or conditions (Figure 3b). Argon plasma treatment led to a pronounced increase of expression across all markers investigated. Exposure to H_2O_2 produced similar but less pronounced results. The same findings were made for HOCl, except for lack of significant increase of CD25 and HLA-DR. ONOO⁻ exposure resulted in the most minor changes across the ROS/RNS treatments, with a significant increase observed for CCR7 only. The surface marker expression values were subsequently fed into a principal component analysis, showing that argon plasma treatment was more similar to H_2O_2 and HOCl than ONOO⁻ (Figure 3c). Altogether, the results suggest an increase of costimulatory surface marker expression in moDCs, which was pronounced for argon plasma treatment and observed for the other ROS/RNS regimens as well.

Next, the secretory profiles of the treatment conditions were examined for 10 different cytokines (Figure 4a–j). Argon plasma treatment led to significantly elevated levels of IL-1α, IL-2, IL-6, and IL-23. For H_2O_2, levels significantly increased from that of controls samples for IL-1α, IL-6, IL-10, IL-12p70, IL-23, and TGF-β. In the case of HOCl exposure, significant differences were found for IL-1α, IL-6, IL-10, IL-18, and IL-23. For ONOO⁻ treatments, significantly higher concentrations were found for IL-1α, IL-6, IL-10, IL-12p70, IL-23, and TGF-β. Despite the differences being significant, the amplitude of differences were subtle, overall. Consistent changes across all ROS/RNS treatments were found for IL-1α, IL-6, and IL-23. All data were fed into a principal component analysis, revealing the argon plasma condition being in between H_2O_2 and HOCl and more apart from ONOO⁻ conditions, which mimicked the results obtained for the surface marker expression. These data provided evidence of argon plasma treatment mildly increasing the inflammatory cytokine profile in human moDCs.

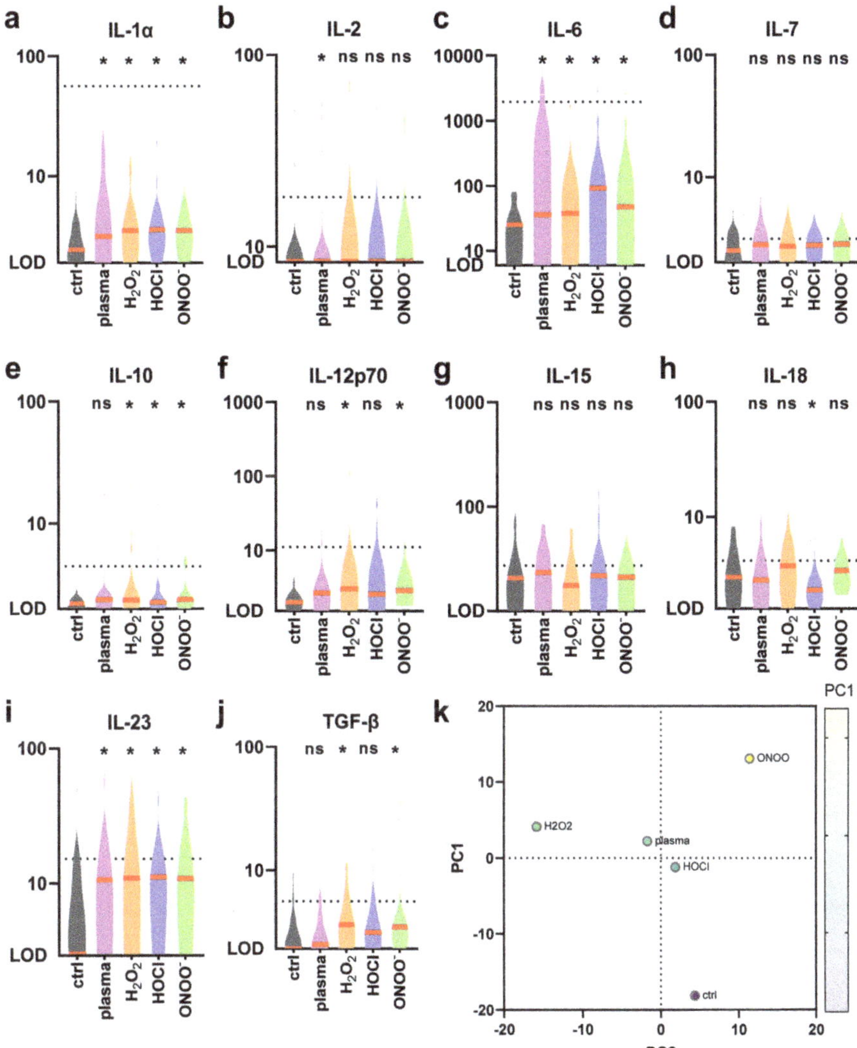

Figure 4. Argon plasma treatment and ROS/RNS modulate the cytokine release profiles. Cells were treated as indicated, supernatants were collected 24 h later, and absolute cytokine concentrations of 10 analytes (**a–j**) were assessed. All data were also related to each other using principal component analysis (**k**). Data are violin plots and median (red lines) of 3–6 different donors. Statistical analysis was done using one-way analysis of variances with $p < 0.05$ (*). Dashed lines show values of LPS-positive controls of moDCs included for the analytes.

3. Discussion

Dendritic cells (DCs) are critical for eliciting antitumor immunity and present in the tumor microenvironment (TME) that is envisaged to be targeted with cold physical plasma. While antitumor effects of this technology have been shown in many experimental models, the effects of argon plasma treatment on DCs have not been elucidated so far. This was the current study's aim.

We found that monocyte-derived DCs (moDCs) were much more resistant to argon plasma-induced cytotoxic effects compared to lymphocytes. This mirrors previous findings comparing lymphocyte survival to that of undifferentiated monocytes after argon plasma exposure and showing a survival advantage of the latter in both primary human cells as well as cell lines [27,28]. Mechanistically, previous studies suggested an association between the enhanced oxidative stress resistance of myeloid progenitor cells (stem-like cells) and increased intracellular expression of catalase, glutathione peroxidase, and manganese superoxide dismutase [29]. Vice versa, it has been shown that transduction of lymphocytes with catalase increased their viability in response to ROS treatment [30]. On the transcriptional level, FoxO was suggested to be associated with oxidative-stress resistance in hematopoietic cells [31]. In general, it is understood that monocytes and lineages deriving from them (e.g., DCs and macrophages) are less affected by ROS-induced cell death since these cells are capable of producing them themselves via activation of NADPH oxidases and myeloperoxidase at the cell membrane or in phagosomes, for instance [32]. Human moDCs, for instance, were previously found to generate millimolar concentrations of ROS in their phagosomes per second [33]. This, and the enhanced transcriptional regulation of redox-related genes in moDCs [34], underlines their enhanced oxidative stress resistance, which is greater in cDCs than pDCs [35].

DCs are present in the TME, and their role is to elicit T cell responses in secondary lymphatic organs via the presentation of tumor antigen [36]. In this process, DCs become activated and upregulate the surface expression of several costimulatory ligands, such as CD40, CD83, CD86, and MHC-II [37,38]. In our study, argon plasma and the exogenous addition of single ROS/RNS all significantly increased CD40, CD83, and CD86, which agrees with a previous study using H_2O_2 [35]. It should be noted that industrially produced H_2O_2 contains stabilizers, which might cause more individual effects than plasma and H_2O_2 derived from it. In professional antigen-presenting cells, CD40 ligation through CD154 activates DCs [39]. CD83 and CD86 are critical in providing costimulation for T cell activation, and especially elevated CD83 expression demarcates DC maturation [40]. The slight but significant increase of MHC-II in our study with argon plasma or H_2O_2 exposure underlines this notion, as MHC-II is upregulated in activated DCs, and ensures enhanced antigen presentation to T cells [41]. While many reports found H_2O_2 to be an important product for the physical plasma effects observed [42–45], other short-lived species and precursors might also be in place for reaction with biomolecules [46–49]. Another report also found an upregulation of TLR2 and TLR4 in response to H_2O_2 treatment [50], targets that were not investigated in our work. Hence, although functional T cell stimulation assays are lacking, our data suggest that especially argon plasma treatment and also ROS/RNS exposure, in general, may foster moDC activation as seen by their increased expression of costimulatory molecules. The effect of argon plasma treatment also was stronger than that of, e.g., H_2O_2 treatment alone. This might be due to the physical plasma treatment being a multimodality regimen generating a multitude of different ROS/RNS as well as other physical effects such as mild UV radiation and electric fields that may support the action of the plasma-derived ROS/RNS on the cells [51–54].

The altered cytokine secretion profiles support this. While the literature on moDC H_2O_2, HOCl, ONOO⁻ exposure, and multi-cytokine secretion of moDCs is scarce and for physical plasma treatment absent, there is ample knowledge on the role of the cytokines investigated in DC biology. IL-1α showed a slight but consistent increase with all treatments, and the cytokine can generate inflammatory DCs in an autocrine fashion [55]. IL-6 was also increased across all conditions, and its secretion has—similar to many other cytokines—pleiotropic effects. For instance, DC-derived IL-6 was found to promote colon cancer metastasis [56], drive TH17 differentiation in concert with IL-1 and TGF-β, and foster macrophage differentiation from monocytes [57]. Along similar lines, DC-derived IL-23, which was also found to be increased, is associated with inflammatory bowel disease [58], osteosarcoma progression [59], and autoinflammation and autoimmunity in general [60]. Nevertheless, it should be mentioned that the absolute changes in concentrations across all 10 cytokines investigated were very moderate, especially in the light of hundred-fold-changes observed in literature with activating agents such as TLR-agonists, Poly I:C, Interferon β, and TNF-α [61]. Therefore, a strong effect of the argon plasma or ROS/RNS induced cytokine changes found in our study might not be expectable. Notwithstanding the relatively small changes identified in our study, they served to consistently separate the different ROS/RNS treatments in principal component analysis in both surface marker expression and cytokine secretion profiles.

4. Materials and Methods

4.1. Dendritic Cells

Human PBMCs were isolated from donor blood from buffy coats dedicated for research purposes and obtained at the Institute of Transfusion Medicine (Greifswald University Medical Center, Greifswald, Germany) based on a density-gradient protocol as described before [62]. Subsequently, CD14$^+$ monocytes were separated using magnetic isolation (BioLegend, Amsterdam, The Netherlands), and viable cell counts and purity were assessed using an attune Nxt flow cytometer (Applied Biosystems, Darmstadt, Germany). In each well of a 24-well plate (Eppendorf, Hamburg, Germany), 2×10^5 cells were seeded in 500 µL of fully supplemented cell culture medium. The cell culture medium was Roswell-Park Memorial Institute (RPMI) 1640 medium supplemented with 10% fetal bovine serum, 5% glutamine, 0.1 mg/L penicillin, and 100 U/L streptomycin (all Sigma-Aldrich, Taufkirchen, Germany). Maturation was induced using human granulocyte-macrophage stimulating factor (GM-CSF, 800IU; PeproTech, Hamburg, Germany) and human interleukin (IL) 4 (IL-4, 500IU; PeproTech, Hamburg, Germany) to retrieve immature human monocytes-derived dendritic cells.

4.2. Argon Plasma Treatment and ROS Exposure

MoDCs were exposed to either hydrogen peroxide (H_2O_2 in water with 0.5 ppm stannate-containing compounds and 1 ppm phosphorus-containing compounds as stabilizers, final concentrations were 500 µM, 250 µM, and 125 µM; Sigma-Aldrich, Taufkirchen, Germany), hypochlorous acid (HOCl, final concentrations were 5000 µM, 500 µM, and 50 µM; Sigma-Aldrich, Taufkirchen, Germany), or peroxynitrite ($ONOO^-$, final concentrations were 10,000 µM, 1000 µM, and 100 µM; Sigma-Aldrich, Taufkirchen, Germany) at different concentrations to obtain dose-response toxicity relationships. Alternatively, cells were left untreated, exposed to argon gas alone (2 standard liters per minute; Air Liquide, Hamburg, Germany), or exposed to argon plasma at different treatment times (60 s, 120 s, and 180 s). The atmospheric pressure argon plasma jet kINPen (neoplas, Greifswald, Germany) was used. The jet and its physical properties have been reviewed recently [63], with the operation frequency being 1 MHz and the plasma-dissipated power and input power being about 1 W and 20 W, respectively. The distance between the nozzle of the head and the liquid was 15 mm. The evaporation was compensated for by adding a pre-determined amount of double-distilled water after argon plasma treatment was finished to restore isosmotic conditions for the cells. Significant changes in the temperature of the liquid were not observed and were about 5 °C in increase from 20 °C [64], while the medium in the incubator was 37 °C. As positive control, lipopolysaccharide (LPS, final concentrations were 250 ng/mL, 125 ng/mL, and 67 ng/mL; Sigma-Aldrich, Taufkirchen, Germany) was used.

4.3. Flow Cytometry

After incubation for 24 h, supernatants and cells were collected, and cells were washed. For analyzing the cells' viability, 4′,6-diamidino-2-phenylindole (DAPI, 1 µM; BioLegend, Amsterdam, The Netherlands) and CD11c conjugated to phycoerythrin (PE) cyanine (Cy) 7 (clone S-HCL-3; BioLegend, Amsterdam, The Netherlands) were added and incubated for 15 min at room temperature in the dark. Subsequently, cells were washed, resuspended in running buffer (Miltenyi Biotec, Bergisch-Gladbach, Germany), and cell data were acquired on a Gallios flow cytometer (Beckman-Coulter, Krefeld, Germany) equipped with an autosampler for FACS tubes. For surface marker analysis, cells were incubated with antibodies (Table 1), washed, and data were acquired on a CytoFLEX LX flow cytometer (Beckman-

Coulter, Krefeld, Germany) equipped with an autosampler for 96-well plates. Data analysis was performed using Kaluza 2.1.1 software (Beckman-Coulter, Krefeld, Germany).

Table 1. Antibodies used in this study.

Target	Clone	Conjugate
CD11c	S-HCL-3	PE-Cy7
CD25	BC96	Alexa Fluor 488
CD40	5C3	APC
CD83	HB15	PerCP-Cy5.5
CD86	IT2.2	PE
CD197	GO43H7	Brilliant Violet 785
HLA-DR	APC-Cy7	APC-Cy7

4.4. Supernatant Analysis

To assess metabolic activity 24 h after treatment of moDCs, 100 µM of resazurin (Alfa Aesar, Kandel, Germany) resolved in fully supplemented cell culture medium was added to each well. After 4 h of incubation at 37 °C in the incubator, the plate was read on a multimode plate reader (F200; Tecan, Männedorf, Switzerland) at λ_{ex} 535 nm and λ_{em} 590 nm. In some experiments, kinetic readings were performed on the device by preheating it up to 37 °C for 1 h, adding additional double-distilled water to the outer rim of the Eppendorf 24-well plate as evaporation shield protection, and continuously supplying 5% CO_2 to the plate reader using the Gas Control Module (GCM; Tecan, Männedorf, Switzerland). For analyzing the concentration of cytokines in cell culture supernatants, the LEGENDplex (BioLegend, Amsterdam, The Netherlands) multiplex bead technology was used as recently described [65]. Briefly, supernatants were collected in 96-well plates and stored at −20 °C for longitudinal analysis. After thawing, supernatants were mixed with capture beads and antibodies, washed, and analyzed using a CytoFLEX S (Beckman-Coulter, Krefeld, Germany) flow cytometry equipped with an autosampler for 96-well plates. Absolute concentrations were calculated against a 5 log fit from a 7-seven serial dilution series (Figure A1) using LEGENDplex software (BioLegend, Amsterdam, The Netherlands).

4.5. Statistical Analysis

Statistical analysis was performed using prism 9.02 (GraphPad Software, San Diego, CA, USA). For comparison between two groups, t-test was used. Regressions were calculated from log-2 transformed data and non-linear fit; Pearson's r was calculated to obtain the goodness of fit. For comparing more than two groups, one-way analysis of variances was performed. For principal component analysis (PCA), data were standardized and PCs were selected based on parallel analysis using Monte Carlo simulations on random data of equal dimensions to the input data and subsequent calculation of eigenvalues for all resulting PCs.

Author Contributions: Conceptualization, S.B.; methodology, S.B., E.F., and R.C.; software, S.B. and E.F.; formal analysis, R.C., K.A., D.M., L.M., and E.F.; investigation, D.M. and K.A.; resources, S.B. and T.v.W.; writing—original draft preparation, S.B.; visualization, S.B.; supervision, S.B. and T.v.W.; project administration, S.B.; funding acquisition, S.B. All authors have read and agreed to the published version of the manuscript.

Funding: This research was funded by the German Federal Ministry of Education and Research (BMBF), grant numbers 03Z22DN11 and 03Z22Di1. D.M. and L.M. were supported by the Gerhard-Domagk-Foundation (Greifswald, Germany).

Institutional Review Board Statement: Apheresis products (buffy coats) were collected from healthy donors according to the German guidelines for hemotherapy with written ethical consent and under approval of University Medicine Greifswald, approval number BB 014-14.

Informed Consent Statement: Blood donor consent for this specific study was waived due to the blood donors' general consent upon blood donation in the transfusion medicine that their blood, transfusion products, or parts of it may be used for research purposes.

Data Availability Statement: The data presented in this study are available on request from the corresponding author.

Acknowledgments: The authors gratefully acknowledge technical support by Felix Niessner.

Conflicts of Interest: The authors declare no conflict of interest. The founding sponsors had no role in the design of the study; in the collection, analyses, or interpretation of data; in the writing of the manuscript; or in the decision to publish the results.

Appendix A

The determination of absolute levels of cytokines was done against a fit of a seven-fold dilution series of each of the analytes (Figure A1). The goodness of fit for each of the analytes across this dilution series was always higher than $R^2 > 0.99$, providing the basis for accurate quantification for each sample and analyte. The coefficient of variation (CV) of each of the dilution series was always smaller than 1 (except for IL2 due to the last dilution step).

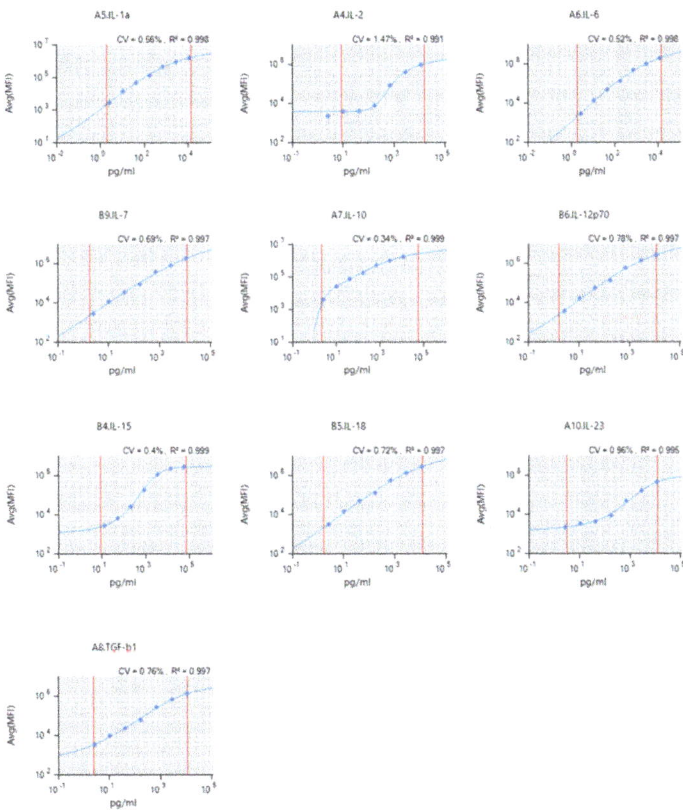

Figure A1. Standard curves from the multiplex cytokine analysis for each analyte across a range of concentrations shown together with the corresponding average mean fluorescence intensities (Avg(MFI)). All curve fittings had an R^2 of greater 0.99, providing great accuracy for the quantification of absolute cytokine levels in moDC culture supernatants.

References

1. Coussens, L.M.; Werb, Z. Inflammation and cancer. *Nature* **2002**, *420*, 860–867. [CrossRef] [PubMed]
2. Fujiwara, N.; Kobayashi, K. Macrophages in inflammation. *Curr. Drug Targets Inflamm. Allergy* **2005**, *4*, 281–286. [CrossRef] [PubMed]
3. Immenschuh, S.; Schroder, H. Heme oxygenase-1 and cardiovascular disease. *Histol. Histopathol.* **2006**, *21*, 679–685.
4. Nowarski, R.; Gagliani, N.; Huber, S.; Flavell, R.A. Innate immune cells in inflammation and cancer. *Cancer Immunol. Res.* **2013**, *1*, 77–84. [CrossRef]
5. Collin, M.; Bigley, V. Human dendritic cell subsets: An update. *Immunology* **2018**, *154*, 3–20. [CrossRef] [PubMed]
6. Drobits, B.; Holcmann, M.; Amberg, N.; Swiecki, M.; Grundtner, R.; Hammer, M.; Colonna, M.; Sibilia, M. Imiquimod clears tumors in mice independent of adaptive immunity by converting pDCs into tumor-killing effector cells. *J. Clin. Investig.* **2012**, *122*, 575–585. [CrossRef] [PubMed]
7. Kelly, B.; O'Neill, L.A. Metabolic reprogramming in macrophages and dendritic cells in innate immunity. *Cell Res.* **2015**, *25*, 771–784. [CrossRef]
8. Maverakis, E.; Kim, K.; Shimoda, M.; Gershwin, M.E.; Patel, F.; Wilken, R.; Raychaudhuri, S.; Ruhaak, L.R.; Lebrilla, C.B. Glycans in the immune system and The Altered Glycan Theory of Autoimmunity: A critical review. *J. Autoimmun.* **2015**, *57*, 1–13. [CrossRef]
9. Sheng, K.C.; Pietersz, G.A.; Tang, C.K.; Ramsland, P.A.; Apostolopoulos, V. Reactive oxygen species level defines two functionally distinctive stages of inflammatory dendritic cell development from mouse bone marrow. *J. Immunol.* **2010**, *184*, 2863–2872. [CrossRef]
10. Boeckmann, L.; Bernhardt, T.; Schafer, M.; Semmler, M.L.; Kordt, M.; Waldner, A.C.; Wendt, F.; Sagwal, S.; Bekeschus, S.; Berner, J.; et al. Current indications for plasma therapy in dermatology. *Hautarzt* **2020**, *71*, 109–113. [CrossRef]
11. Stratmann, B.; Costea, T.C.; Nolte, C.; Hiller, J.; Schmidt, J.; Reindel, J.; Masur, K.; Motz, W.; Timm, J.; Kerner, W.; et al. Effect of Cold Atmospheric Plasma Therapy vs. Standard Therapy Placebo on Wound Healing in Patients With Diabetic Foot Ulcers: A Randomized Clinical Trial. *JAMA Netw. Open* **2020**, *3*, e2010411. [CrossRef]
12. Ulrich, C.; Kluschke, F.; Patzelt, A.; Vandersee, S.; Czaika, V.A.; Richter, H.; Bob, A.; von Hutten, J.; Painsi, C.; Hugel, R.; et al. Clinical use of cold atmospheric pressure argon plasma in chronic leg ulcers: A pilot study. *J. Wound Care* **2015**, *24*, 196–203. [CrossRef] [PubMed]
13. Isbary, G.; Heinlin, J.; Shimizu, T.; Zimmermann, J.L.; Morfill, G.; Schmidt, H.U.; Monetti, R.; Steffes, B.; Bunk, W.; Li, Y.; et al. Successful and safe use of 2 min cold atmospheric argon plasma in chronic wounds: Results of a randomized controlled trial. *Br. J. Dermatol.* **2012**, *167*, 404–410. [CrossRef]
14. Schmidt, A.; von Woedtke, T.; Vollmar, B.; Hasse, S.; Bekeschus, S. Nrf2 signaling and inflammation are key events in physical plasma-spurred wound healing. *Theranostics* **2019**, *9*, 1066–1084. [CrossRef] [PubMed]
15. Schmidt, A.; Bekeschus, S.; Wende, K.; Vollmar, B.; von Woedtke, T. A cold plasma jet accelerates wound healing in a murine model of full-thickness skin wounds. *Exp. Dermatol.* **2017**, *26*, 156–162. [CrossRef]
16. Arndt, S.; Unger, P.; Wacker, E.; Shimizu, T.; Heinlin, J.; Li, Y.F.; Thomas, H.M.; Morfill, G.E.; Zimmermann, J.L.; Bosserhoff, A.K.; et al. Cold atmospheric plasma (CAP) changes gene expression of key molecules of the wound healing machinery and improves wound healing in vitro and in vivo. *PLoS ONE* **2013**, *8*, e79325.
17. Dai, X.; Bazaka, K.; Richard, D.J.; Thompson, E.R.W.; Ostrikov, K.K. The Emerging Role of Gas Plasma in Oncotherapy. *Trends Biotechnol.* **2018**, *36*, 1183–1198. [CrossRef]
18. Lin, A.G.; Xiang, B.; Merlino, D.J.; Baybutt, T.R.; Sahu, J.; Fridman, A.; Snook, A.E.; Miller, V. Non-thermal plasma induces immunogenic cell death in vivo in murine CT26 colorectal tumors. *Oncoimmunology* **2018**, *7*, e1484978. [CrossRef]
19. Bekeschus, S.; Clemen, R.; Niessner, F.; Sagwal, S.K.; Freund, E.; Schmidt, A. Medical Gas Plasma Jet Technology Targets Murine Melanoma in an Immunogenic Fashion. *Adv. Sci.* **2020**, *7*, 1903438. [CrossRef]
20. Mizuno, K.; Shirakawa, Y.; Sakamoto, T.; Ishizaki, H.; Nishijima, Y.; Ono, R. Plasma-Induced Suppression of Recurrent and Reinoculated Melanoma Tumors in Mice. *IEEE TRPMS* **2018**, *2*, 353–359. [CrossRef]
21. Mizuno, K.; Yonetamari, K.; Shirakawa, Y.; Akiyama, T.; Ono, R. Anti-tumor immune response induced by nanosecond pulsed streamer discharge in mice. *J. Phys. D Appl. Phys.* **2017**, *50*, 12LT01. [CrossRef]
22. Schmidt-Bleker, A.; Bansemer, R.; Reuter, S.; Weltmann, K.-D. How to produce an NOx- instead of Ox-based chemistry with a cold atmospheric plasma jet. *Plasma Process. Polym.* **2016**, *13*, 1120–1127. [CrossRef]
23. Privat-Maldonado, A.; Schmidt, A.; Lin, A.; Weltmann, K.D.; Wende, K.; Bogaerts, A.; Bekeschus, S. ROS from Physical Plasmas: Redox Chemistry for Biomedical Therapy. *Oxid. Med. Cell. Longev.* **2019**, *2019*, 9062098. [CrossRef]
24. Semmler, M.L.; Bekeschus, S.; Schafer, M.; Bernhardt, T.; Fischer, T.; Witzke, K.; Seebauer, C.; Rebl, H.; Grambow, E.; Vollmar, B.; et al. Molecular Mechanisms of the Efficacy of Cold Atmospheric Pressure Plasma (CAP) in Cancer Treatment. *Cancers* **2020**, *12*, 269. [CrossRef]
25. Binnewies, M.; Mujal, A.M.; Pollack, J.L.; Combes, A.J.; Hardison, E.A.; Barry, K.C.; Tsui, J.; Ruhland, M.K.; Kersten, K.; Abushawish, M.A.; et al. Unleashing Type-2 Dendritic Cells to Drive Protective Antitumor CD4(+) T Cell Immunity. *Cell* **2019**, *177*, 556.e16–571.e16. [CrossRef]
26. Fu, C.; Jiang, A. Dendritic Cells and CD8 T Cell Immunity in Tumor Microenvironment. *Front. Immunol.* **2018**, *9*, 3059. [CrossRef] [PubMed]

27. Bekeschus, S.; Kolata, J.; Muller, A.; Kramer, A.; Weltmann, K.-D.; Broker, B.; Masur, K. Differential Viability of Eight Human Blood Mononuclear Cell Subpopulations After Plasma Treatment. *Plasma Med.* **2013**, *3*, 1–13. [CrossRef]

28. Bundscherer, L.; Bekeschus, S.; Tresp, H.; Hasse, S.; Reuter, S.; Weltmann, K.-D.; Lindequist, U.; Masur, K. Viability of Human Blood Leukocytes Compared with Their Respective Cell Lines after Plasma Treatment. *Plasma Med.* **2013**, *3*, 71–80. [CrossRef]

29. Dernbach, E.; Urbich, C.; Brandes, R.P.; Hofmann, W.K.; Zeiher, A.M.; Dimmeler, S. Antioxidative stress-associated genes in circulating progenitor cells: Evidence for enhanced resistance against oxidative stress. *Blood* **2004**, *104*, 3591–3597. [CrossRef] [PubMed]

30. Ando, T.; Mimura, K.; Johansson, C.C.; Hanson, M.G.; Mougiakakos, D.; Larsson, C.; Martins da Palma, T.; Sakurai, D.; Norell, H.; Li, M.; et al. Transduction with the antioxidant enzyme catalase protects human T cells against oxidative stress. *J. Immunol.* **2008**, *181*, 8382–8390. [CrossRef]

31. Tothova, Z.; Kollipara, R.; Huntly, B.J.; Lee, B.H.; Castrillon, D.H.; Cullen, D.E.; McDowell, E.P.; Lazo-Kallanian, S.; Williams, I.R.; Sears, C.; et al. FoxOs are critical mediators of hematopoietic stem cell resistance to physiologic oxidative stress. *Cell* **2007**, *128*, 325–339. [CrossRef] [PubMed]

32. Gordon, S. The role of the macrophage in immune regulation. *Res. Immunol.* **1998**, *149*, 685–688. [CrossRef]

33. Paardekooper, L.M.; Dingjan, I.; Linders, P.T.A.; Staal, A.H.J.; Cristescu, S.M.; Verberk, W.; van den Bogaart, G. Human Monocyte-Derived Dendritic Cells Produce Millimolar Concentrations of ROS in Phagosomes Per Second. *Front. Immunol.* **2019**, *10*, 1216. [CrossRef] [PubMed]

34. Van Brussel, I.; Schrijvers, D.M.; Martinet, W.; Pintelon, I.; Deschacht, M.; Schnorbusch, K.; Maes, L.; Bosmans, J.M.; Vrints, C.J.; Adriaensen, D.; et al. Transcript and protein analysis reveals better survival skills of monocyte-derived dendritic cells compared to monocytes during oxidative stress. *PLoS ONE* **2012**, *7*, e43357. [CrossRef]

35. Pazmandi, K.; Magyarics, Z.; Boldogh, I.; Csillag, A.; Rajnavolgyi, E.; Bacsi, A. Modulatory effects of low-dose hydrogen peroxide on the function of human plasmacytoid dendritic cells. *Free Radic. Biol. Med.* **2012**, *52*, 635–645. [CrossRef]

36. Lapteva, N.; Aldrich, M.; Weksberg, D.; Rollins, L.; Goltsova, T.; Chen, S.Y.; Huang, X.F. Targeting the intratumoral dendritic cells by the oncolytic adenoviral vaccine expressing RANTES elicits potent antitumor immunity. *J. Immunother.* **2009**, *32*, 145–156. [CrossRef]

37. Gerlach, A.M.; Steimle, A.; Krampen, L.; Wittmann, A.; Gronbach, K.; Geisel, J.; Autenrieth, I.B.; Frick, J.S. Role of CD40 ligation in dendritic cell semimaturation. *BMC Immunol.* **2012**, *13*, 22. [CrossRef]

38. Obendorf, J.; Viveros, P.R.; Fehlings, M.; Klotz, C.; Aebischer, T.; Ignatius, R. Increased expression of CD25, CD83, and CD86, and secretion of IL-12, IL-23, and IL-10 by human dendritic cells incubated in the presence of Toll-like receptor 2 ligands and Giardia duodenalis. *Parasites Vectors* **2013**, *6*. [CrossRef]

39. Frentsch, M.; Arbach, O.; Kirchhoff, D.; Moewes, B.; Worm, M.; Rothe, M.; Scheffold, A.; Thiel, A. Direct access to CD4+ T cells specific for defined antigens according to CD154 expression. *Nat. Med.* **2005**, *11*, 1118–1124. [CrossRef]

40. Prechtel, A.T.; Steinkasserer, A. CD83: An update on functions and prospects of the maturation marker of dendritic cells. *Arch. Dermatol. Res.* **2007**, *299*, 59–69. [CrossRef]

41. Nakayama, M. Antigen Presentation by MHC-Dressed Cells. *Front. Immunol.* **2014**, *5*, 672. [CrossRef]

42. Bekeschus, S.; Kolata, J.; Winterbourn, C.; Kramer, A.; Turner, R.; Weltmann, K.D.; Broker, B.; Masur, K. Hydrogen peroxide: A central player in physical plasma-induced oxidative stress in human blood cells. *Free Radic. Res.* **2014**, *48*, 542–549. [CrossRef]

43. Adachi, T.; Tanaka, H.; Nonomura, S.; Hara, H.; Kondo, S.; Hori, M. Plasma-activated medium induces A549 cell injury via a spiral apoptotic cascade involving the mitochondrial-nuclear network. *Free Radic. Biol. Med.* **2015**, *79*, 28–44. [CrossRef] [PubMed]

44. Bauer, G. Intercellular singlet oxygen-mediated bystander signaling triggered by long-lived species of cold atmospheric plasma and plasma-activated medium. *Redox. Biol.* **2019**, *26*, 101301. [CrossRef] [PubMed]

45. Bekeschus, S.; Schmidt, A.; Jablonowski, H.; Bethge, L.; Hasse, S.; Wende, K.; Masur, K.; von Woedtke, T.; Weltmann, K.D. Environmental Control of an Argon Plasma Effluent and Its Role in THP-1 Monocyte Function. *IEEE Trans. Plasma Sci.* **2017**, *45*, 3336–3341. [CrossRef]

46. Bruno, G.; Heusler, T.; Lackmann, J.-W.; von Woedtke, T.; Weltmann, K.-D.; Wende, K. Cold physical plasma-induced oxidation of cysteine yields reactive sulfur species (RSS). *Clin. Plas. Med.* **2019**, *14*. [CrossRef]

47. Heusler, T.; Bruno, G.; Bekeschus, S.; Lackmann, J.-W.; von Woedtke, T.; Wende, K. Can the effect of cold physical plasma-derived oxidants be transported via thiol group oxidation? *Clin. Plas. Med.* **2019**, *14*. [CrossRef]

48. Tanaka, H.; Nakamura, K.; Mizuno, M.; Ishikawa, K.; Takeda, K.; Kajiyama, H.; Utsumi, F.; Kikkawa, F.; Hori, M. Non-thermal atmospheric pressure plasma activates lactate in Ringer's solution for anti-tumor effects. *Sci. Rep.* **2016**, *6*, 36282. [CrossRef]

49. Wende, K.; Bruno, G.; Lalk, M.; Weltmann, K.-D.; von Woedtke, T.; Bekeschus, S.; Lackmann, J.-W. On a heavy path–determining cold plasma-derived short-lived species chemistry using isotopic labelling. *RSC Adv.* **2020**, *10*, 11598–11607. [CrossRef]

50. Batal, I.; Azzi, J.; Mounayar, M.; Abdoli, R.; Moore, R.; Lee, J.Y.; Rosetti, F.; Wang, C.; Fiorina, P.; Sackstein, R.; et al. The mechanisms of up-regulation of dendritic cell activity by oxidative stress. *J. Leukoc. Biol.* **2014**, *96*, 283–293. [CrossRef]

51. Babaeva, N.Y.; Kushner, M.J. Intracellular electric fields produced by dielectric barrier discharge treatment of skin. *J. Phys. D Appl. Phys.* **2010**, *43*, 185206. [CrossRef]

52. Von Woedtke, T.; Schmidt, A.; Bekeschus, S.; Wende, K.; Weltmann, K.D. Plasma Medicine: A Field of Applied Redox Biology. *In Vivo* **2019**, *33*, 1011–1026. [CrossRef] [PubMed]

53. Chung, T.H.; Stancampiano, A.; Sklias, K.; Gazeli, K.; Andre, F.M.; Dozias, S.; Douat, C.; Pouvesle, J.M.; Santos Sousa, J.; Robert, E.; et al. Cell Electropermeabilisation Enhancement by Non-Thermal-Plasma-Treated PBS. *Cancers* **2020**, *12*, 219. [CrossRef] [PubMed]

54. Wolff, C.M.; Kolb, J.F.; Weltmann, K.D.; von Woedtke, T.; Bekeschus, S. Combination Treatment with Cold Physical Plasma and Pulsed Electric Fields Augments ROS Production and Cytotoxicity in Lymphoma. *Cancers* **2020**, *12*, 845. [CrossRef] [PubMed]

55. Eriksson, U.; Kurrer, M.O.; Sonderegger, I.; Iezzi, G.; Tafuri, A.; Hunziker, L.; Suzuki, S.; Bachmaier, K.; Bingisser, R.M.; Penninger, J.M.; et al. Activation of dendritic cells through the interleukin 1 receptor 1 is critical for the induction of autoimmune myocarditis. *J. Exp. Med.* **2003**, *197*, 323–331. [CrossRef]

56. Toyoshima, Y.; Kitamura, H.; Xiang, H.; Ohno, Y.; Homma, S.; Kawamura, H.; Takahashi, N.; Kamiyama, T.; Tanino, M.; Taketomi, A. IL6 Modulates the Immune Status of the Tumor Microenvironment to Facilitate Metastatic Colonization of Colorectal Cancer Cells. *Cancer Immunol Res.* **2019**, *7*, 1944–1957. [CrossRef]

57. Chomarat, P.; Banchereau, J.; Davoust, J.; Palucka, A.K. IL-6 switches the differentiation of monocytes from dendritic cells to macrophages. *Nat. Immunol.* **2000**, *1*, 510–514. [CrossRef]

58. McGovern, D.; Powrie, F. The IL23 axis plays a key role in the pathogenesis of IBD. *Gut* **2007**, *56*, 1333–1336. [CrossRef]

59. Kansara, M.; Thomson, K.; Pang, P.; Dutour, A.; Mirabello, L.; Acher, F.; Pin, J.P.; Demicco, E.G.; Yan, J.; Teng, M.W.L.; et al. Infiltrating Myeloid Cells Drive Osteosarcoma Progression via GRM4 Regulation of IL23. *Cancer Discov.* **2019**, *9*, 1511–1519. [CrossRef]

60. Jin, J.; Xie, X.; Xiao, Y.; Hu, H.; Zou, Q.; Cheng, X.; Sun, S.C. Epigenetic regulation of the expression of Il12 and Il23 and autoimmune inflammation by the deubiquitinase Trabid. *Nat. Immunol.* **2016**, *17*, 259–268. [CrossRef]

61. Jensen, S.S.; Gad, M. Differential induction of inflammatory cytokines by dendritic cells treated with novel TLR-agonist and cytokine based cocktails: Targeting dendritic cells in autoimmunity. *J. Inflamm.* **2010**, *7*, 37. [CrossRef] [PubMed]

62. Bekeschus, S.; Masur, K.; Kolata, J.; Wende, K.; Schmidt, A.; Bundscherer, L.; Barton, A.; Kramer, A.; Broker, B.; Weltmann, K.D. Human Mononuclear Cell Survival and Proliferation is Modulated by Cold Atmospheric Plasma Jet. *Plasma Process. Polym.* **2013**, *10*, 706–713. [CrossRef]

63. Reuter, S.; von Woedtke, T.; Weltmann, K.D. The kINPen-a review on physics and chemistry of the atmospheric pressure plasma jet and its applications. *J. Phys. D Appl. Phys.* **2018**, *51*. [CrossRef]

64. Clemen, R.; Freund, E.; Mrochen, D.; Miebach, L.; Schmidt, A.; Rauch, B.H.; Lackmann, J.W.; Martens, U.; Wende, K.; Lalk, M.; et al. Gas Plasma Technology Augments Ovalbumin Immunogenicity and OT-II T Cell Activation Conferring Tumor Protection in Mice. *Adv. Sci.* **2021**. [CrossRef]

65. Bekeschus, S.; Ressel, V.; Freund, E.; Gelbrich, N.; Mustea, A.; Stope, M.B. Gas Plasma-Treated Prostate Cancer Cells Augment Myeloid Cell Activity and Cytotoxicity. *Antioxidants* **2020**, *9*, 323. [CrossRef]

Article

Acidified Nitrite Contributes to the Antitumor Effect of Cold Atmospheric Plasma on Melanoma Cells

Tom Zimmermann [1], Lisa A. Gebhardt [2], Lucas Kreiss [3,4], Christin Schneider [1], Stephanie Arndt [5], Sigrid Karrer [5], Oliver Friedrich [4], Michael J. M. Fischer [2,6] and Anja-Katrin Bosserhoff [1,7,*]

[1] Emil-Fischer-Center, Institute of Biochemistry, University of Erlangen-Nuernberg, 91054 Erlangen, Germany; tom.zimmermann@fau.de (T.Z.); schneiderchristin@gmx.de (C.S.)

[2] Institute of Physiology and Pathophysiology, University of Erlangen-Nuernberg, 91054 Erlangen, Germany; lisa.a.gebhardt@fau.de (L.A.G.); michael.jm.fischer@meduniwien.ac.at (M.J.M.F.)

[3] Department of Medicine I, University Clinics Erlangen, 91054 Erlangen, Germany; lucas.kreiss@fau.de

[4] Institute of Medical Biotechnology, University of Erlangen-Nuernberg, 91052 Erlangen, Germany; oliver.friedrich@mbt.uni-erlangen.de

[5] Department of Dermatology, University Hospital of Regensburg, 93053 Regensburg, Germany; stephanie.arndt@ukr.de (S.A.); sigrid.karrer@ukr.de (S.K.)

[6] Institute of Physiology, Medical University of Vienna, 1090 Vienna, Austria

[7] Comprehensive Cancer Center (CCC) Erlangen-EMN, 91054 Erlangen, Germany

* Correspondence: anja.bosserhoff@fau.de

Citation: Zimmermann, T.; Gebhardt, L.A.; Kreiss, L.; Schneider, C.; Arndt, S.; Karrer, S.; Friedrich, O.; Fischer, M.J.M.; Bosserhoff, A.-K. Acidified Nitrite Contributes to the Antitumor Effect of Cold Atmospheric Plasma on Melanoma Cells. *Int. J. Mol. Sci.* **2021**, *22*, 3757. https://doi.org/10.3390/ijms22073757

Academic Editor: Akikazu Sakudo

Received: 11 March 2021
Accepted: 2 April 2021
Published: 4 April 2021

Publisher's Note: MDPI stays neutral with regard to jurisdictional claims in published maps and institutional affiliations.

Abstract: Cold atmospheric plasma (CAP) is partially ionized gas near room temperature with previously reported antitumor effects. Despite extensive research and growing interest in this technology, active components and molecular mechanisms of CAP are not fully understood to date. We used Raman spectroscopy and colorimetric assays to determine elevated nitrite and nitrate levels after treatment with a MiniFlatPlaster CAP device. Previously, we demonstrated CAP-induced acidification. Cellular effects of nitrite and strong extracellular acidification were assessed using live-cell imaging of intracellular Ca^{2+} levels, cell viability analysis as well as quantification of p21 and DNA damage. We further characterized these observations by analyzing established molecular effects of CAP treatment. A synergistic effect of nitrite and acidification was found, leading to strong cytotoxicity in melanoma cells. Interestingly, protein nitration and membrane damage were absent after treatment with acidified nitrite, thereby challenging their contribution to CAP-induced cytotoxicity. Further, phosphorylation of ERK1/2 was increased after treatment with both acidified nitrite and indirect CAP. This study characterizes the impact of acidified nitrite on melanoma cells and supports the importance of RNS during CAP treatment. Further, it defines and evaluates important molecular mechanisms that are involved in the cancer cell response to CAP.

Keywords: cold atmospheric plasma; malignant melanoma; acidification; nitrite; acidified nitrite; nitration; membrane damage

1. Introduction

Cold atmospheric plasma (CAP) consists of a heterogeneous mixture of reactive oxygen (ROS) and nitrogen (RNS) species, as well as other ions, uncharged particles, and small amounts of radiation in ultraviolet and infrared ranges. Manifold effects of CAP have been described in the past, mainly referring to strong antibacterial, antiviral, and antifungal action [1–3]. Significant efforts have been made to evaluate and use its positive impact on wound healing [4–6], dental health [7,8], and regenerative medicine [9]. Furthermore, beneficial effects of plasma treatment have been reported both in vitro and in vivo for numerous cancer types, including malignant melanoma [10–12], colon [13,14], and brain tumors [15–17]. Given the constantly high incidence and mortality of such malignancies, and their enormous burden on the patient as well as the health care system, CAP technology displays a promising approach to the development of novel therapeutic treatments.

The potential impact of an oncological application is highlighted by recent studies showing that chemo-resistance might be challenged directly via apoptosis [18,19] and indirectly by restoration of chemo-sensitivity [20]. Furthermore, it was reported that CAP shows strong selectivity against cancer cells, while healthy cells remain largely unaffected [21–23]. Such preferential killing of tumor cells, however, remains controversial as treatment conditions and cell culture media largely affect the potency of CAP [24]. Despite extensive research and ongoing advances in plasma medicine, the exact molecular mechanisms of CAP treatment are still unknown. In terms of malignant melanoma, multiple studies have contributed to a better understanding of CAP effects and underlying molecular mechanisms. For example, it was shown that melanoma cells enter apoptosis in response to DNA damage and mitochondrial dysfunction caused by CAP-induced ROS and RNS [25,26]. Our group previously reported dose-dependent effects of CAP ranging from senescence to apoptosis [11]. On a molecular level, the establishment of cellular senescence was tightly linked to an immediate elevation of cytoplasmic Ca^{2+} levels, mainly originating from intracellular stores [27]. In comparison to direct treatment of melanoma cells, indirect treatment by application of CAP-treated physiological buffers showed similar but attenuated results, indicating that many of the activating agents are able to dilute in aqueous solutions and remain fairly stable. This was supported by the finding that such CAP-treated buffers do not lose their biological effect if their application is delayed for one hour. Another study revealed strong extracellular acidification during CAP treatment to be essential for its effect on melanoma cells [28]. However, since the exact molecular and cellular mechanisms are still unknown, further research is required to enable the development of plasma-based tumor therapy and valid plasma devices for such approaches. The aim of this study was to characterize reactive species involved in the CAP effect on melanoma cells and to further understand the CAP-induced mechanisms as a basis for the generation of personalized plasma therapy.

2. Results

2.1. CAP Induces Production of Nitrate and Nitrite in Aqueous Solutions

Recently, we described the effects of surface micro discharge (SMD) generated CAP on tumor cells, linking these to the induction of reactive species. For an unbiased characterization of CAP-induced production of ROS and RNS, we used Raman spectroscopy. Even though this approach was not able to detect reactive species with a short lifetime, significantly elevated levels of nitrate ($1048\ cm^{-1}$) and nitrite ($817\ cm^{-1}$ and $1336\ cm^{-1}$) were found after 2 min CAP treatment (Figure 1A). The validity of these findings was confirmed using standard solutions of potassium nitrate and sodium nitrite (Figures S1 and S2). An established colorimetric assay based on the transnitration of salicylic acid [29] was then used to validate and quantify the observed nitrate production. We determined a dose-dependent increase in nitrate levels reaching approximately 1 mM nitrate after 2 min CAP treatment (Figure 1B). Nitrite levels were assessed using a colorimetric assay based on the Griess diazotization reaction. A similar dose-dependency was observed, resulting in nitrite levels of 2 mM after 2 min CAP (Figure 1C). In previous studies, we could show that CAP-treated solutions retained a substantial fraction of their cellular effects even 1 h after incubation [27]. We, therefore, assessed the stability of nitrate and nitrite and found high stability of the CAP-induced substances (Figure 1B,C).

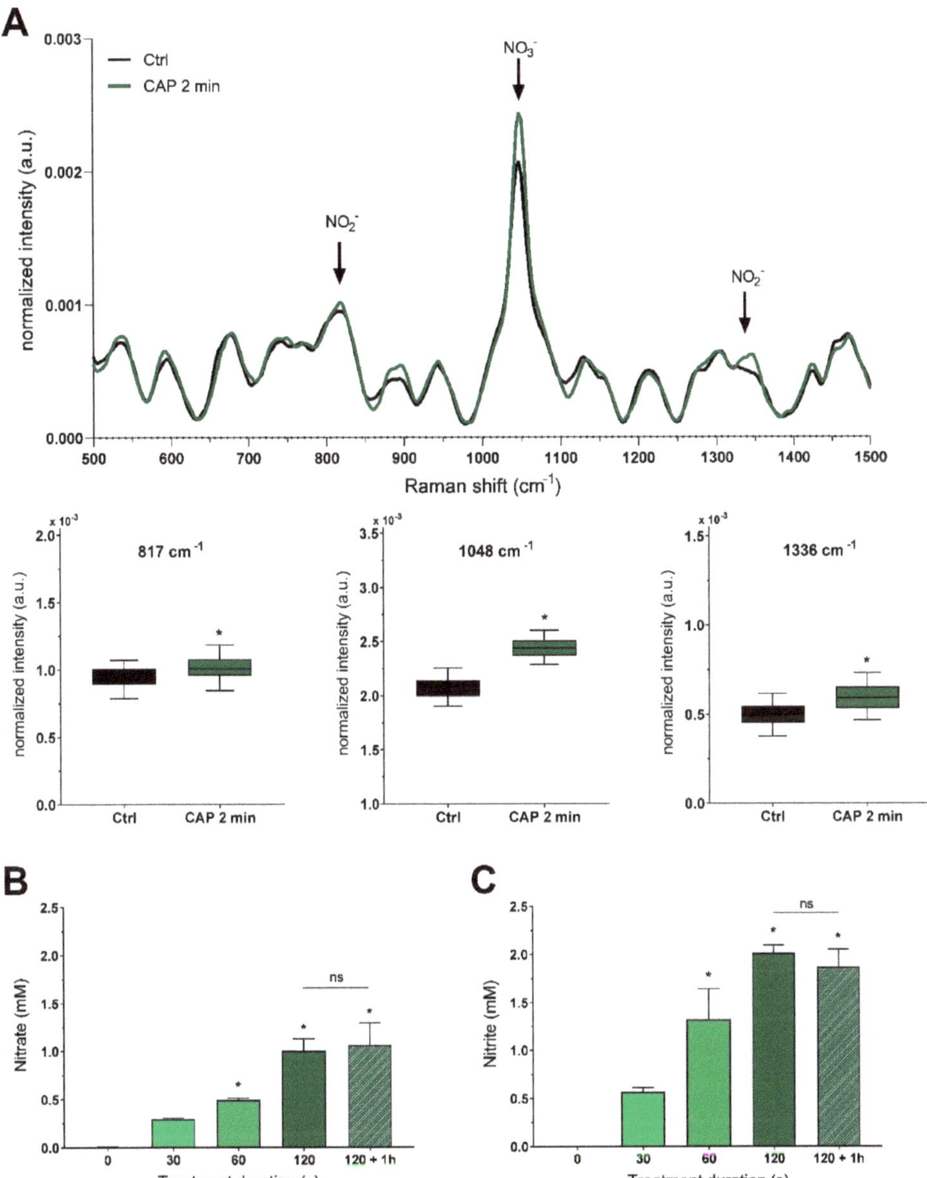

Figure 1. CAP treatment causes the production of nitrate and nitrite. (**A**) Raman spectroscopy of untreated extracellular solution (Ctrl) versus a solution treated with 2 min CAP. Traces are mean only, and box plots are mean with 95% confidence interval (Student's *t*-test, $n = 300$). (**B**) Photometric quantification of nitrate and (**C**) nitrite after different doses of CAP. Stability of these molecules was assessed 1 h after treatment ($F_{(4,10)} = 29.31$ and $F_{(4,10)} = 29.63$, both $p < 0.0001$). Bars are shown as mean $\pm$ SEM (ANOVA followed by Tukey's HSD post-hoc test vs. no treatment, $n = 3$, *: $p < 0.05$, ns: not significant ($p > 0.05$)).

2.2. Nitrite and Acidification Have Synergistic Effects on Ca²⁺ Influx and Cytotoxicity

Next, we assessed the effects of nitrate and nitrite on the cytoplasmic Ca^{2+} levels using the fluorescent calcium indicator fura-2 AM. Cells of the melanoma cell line Mel Im (derived from metastasis) were treated with 1 mM nitrate or 2 mM nitrite in a physiological HEPES-buffered solution. However, no alterations of cytoplasmic calcium concentrations were observed. Since previous studies revealed CAP-induced acidification to be essential for cellular effects of plasma-treated solutions, we combined acidic buffer solutions (pH 3.9) with nitrate and nitrite to resemble these aspects of CAP treatment. Interestingly, a synergistic effect of acidification and nitrite was observed, leading to a strong and immediate increase in cytoplasmic Ca^{2+} fluorescence (Figure 2A). Acidification alone, however, did not show significant effects. We also investigated the combination of nitrate and acidification but could not detect any increase in cytoplasmic calcium levels (Figure S3). Functional consequences of treatment with acidic nitrite solution were additionally analyzed using the melanoma cell line Mel Juso (derived from the primary tumor). Cell viability was assessed using the tetrazolium-based XTT assay, revealing significant cytotoxicity of a 5 min treatment in Mel Juso (Figure 2B) and Mel Im (Figure S4A). To exclude any contribution of HEPES to this effect, we repeated the experiment with a phosphate-buffered solution without HEPES. Similar cytotoxicity was observed, indicating that HEPES does not actively contribute to the effects of acidic nitrite solutions (Figure 2C and Figure S4B). Since phosphate-buffered solutions are not particularly suited for such low pH levels, all further experiments of this study used HEPES-buffered solutions. On a molecular level, cytotoxicity was accompanied by a strong induction of cell cycle inhibitor p21, which was found both on mRNA and protein levels (Figure 2D–F and Figure S4C–E). Immunofluorescent stainings of promyelocytic leukemia protein (PML) were used to evaluate DNA damage in Mel Juso. The number of PML nuclear bodies was significantly increased in response to treatment with acidic nitrite solution (Figure 2G,H). Finally, we assessed whether acidic nitrite solutions exhibited the same tumor selectivity that was previously reported for CAP. As expected, normal human fibroblasts showed no significant response to the treatment as compared to melanoma cell lines Mel Juso and Mel Im (Figure S4F).

2.3. Molecular Effects of Acidic Nitrite Solution Compared to CAP Treatment

Multiple studies proposed the generation of peroxynitrite (ONOO-) to be a major consequence of CAP treatment [30,31]. We aimed to estimate intracellular ONOO- generation indirectly via quantification of 3-nitrotyrosin, a marker that was previously reported to be a major consequence of peroxynitrite-dependent protein nitration [32,33]. Therefore, Western blot analysis of 3-nitrotyrosine was performed directly after a 5 min treatment with acidic nitrite solution (Figure 3A). Interestingly, we could not detect relevant amounts of 3-nitrotyrosine even after prolonged treatment periods of 1 h. A physiological buffer solution treated with 2 min CAP was used as a positive control and resulted in protein nitration after it was applied on the cells for 5 min and strong induction of 3-nitrotyrosine after 1 h incubation. Another established feature of CAP was the introduction of membrane damage, which was assessed next. We used a 5 min propidium iodide (PI) staining without fixation or permeabilization to detect ruptures in the cellular membrane. While there was a minor increase in PI signal after treatment with acidic nitrite solution, we could not find significant induction of membrane damage (Figure 3B). Again, a buffer solution treated with 2 min CAP was used as a positive control and led to a significant increase in PI signal. Next, we analyzed MAPK activity by Western blot analysis of ERK1/2 phosphorylation and found a significant elevation of pERK1/2 after 5 min treatment with acidic nitrite solution or indirect CAP treatment (Figure 3C).

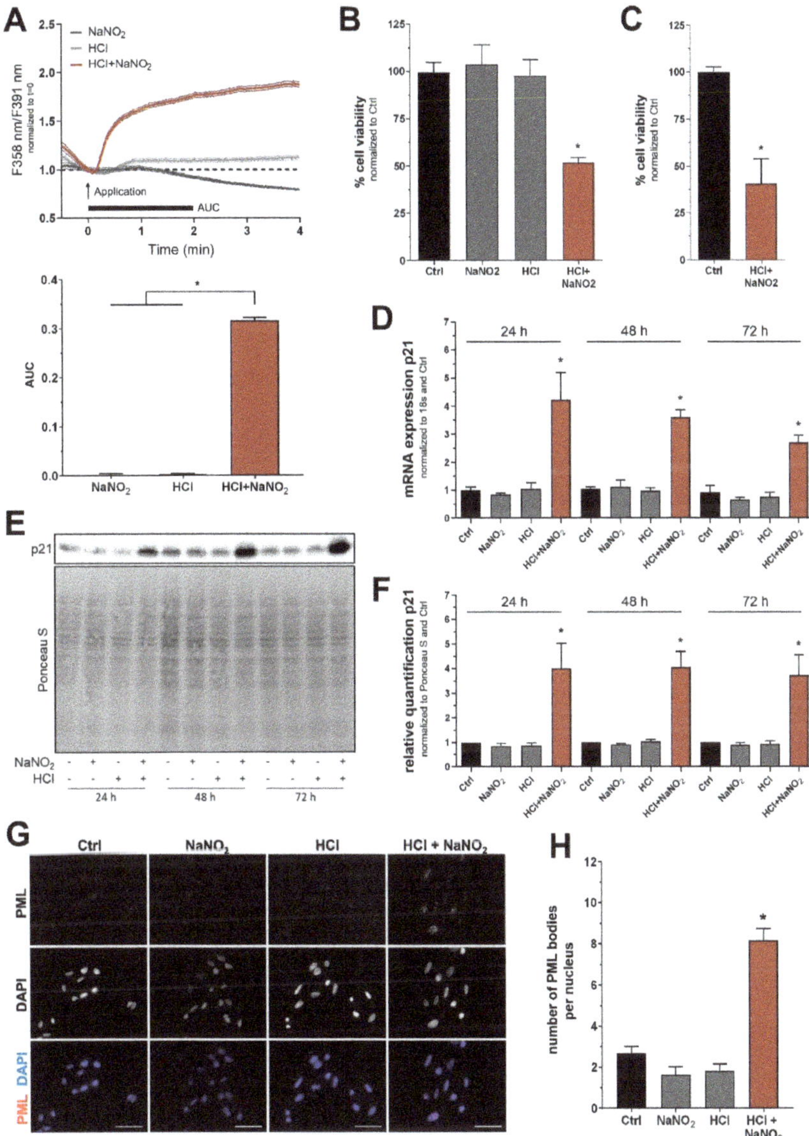

Figure 2. Nitrite and acidification have synergistic effects on melanoma cells. (**A**) Time course of cytoplasmic Ca^{2+} levels due to a 4-min treatment with acidic ECS (HCl), nitrite solution (NaNO$_2$), or a combination of both. Ca^{2+} levels were quantified by calculating the area under the curve (AUC) of the first 120 s after the start of the application ($F_{(2,828)} = 1891$, $p < 0.0001$, $n = 260–293$). (**B**) Cell viability analysis 24 h after a 5 min treatment with untreated ECS (Ctrl) or solutions described in (**A**) ($F_{(3,8)} = 10.93$, $p = 0.0033$). (**C**) Cell viability analysis 24 h after a 5 min treatment with phosphate-buffered ECS without HEPES (Ctrl) or a combination of acidic phosphate-buffered ECS without HEPES and nitrite (Student's *t*-test). (**D**) Expression analysis of p21 during the time span of 24–72 h after treatment ($F_{(11,24)} = 13.82$, $p < 0.0001$). (**E**,**F**) Western blot analysis and quantification of p21 protein levels with similar incubation time as (**D**) ($F_{(11,24)} = 10.44$, $p < 0.0001$). (**G**,**H**) Immunofluorescent stainings of PML and DAPI to assess DNA damage. The amount of PML nuclear bodies was quantified in the bar chart ($F_{(3,8)} = 55.13$, $p < 0.0001$). Scale bars: 50 μm. Traces are mean with 95% confidence interval, bars are mean ± SEM (ANOVA followed by Tukey's HSD post-hoc test vs. Ctrl, $n = 3$, *: $p < 0.05$).

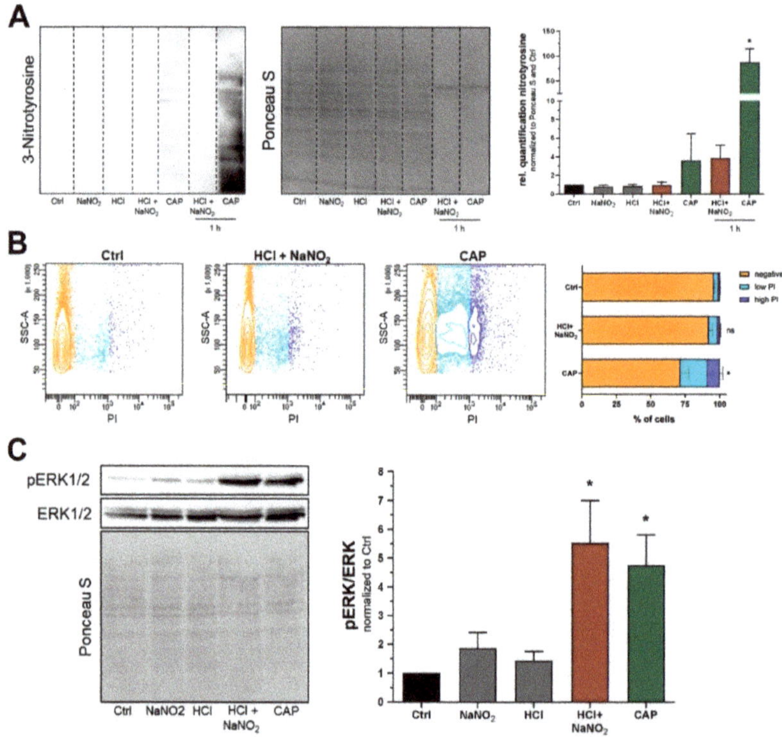

Figure 3. Molecular effects of acidic nitrite solution and CAP. (**A**) Western blot analysis of protein nitration directly after 5 min treatment with acidified nitrite solution or indirect CAP ($F_{(6,14)}$ = 9.437, p = 0.0003). Reference samples of 1 h treatment serve as a positive control. (**B**) Propidium iodide staining in combination with flow cytometry to assess membrane damage after treatment ($F_{(2,6)}$ = 11.62, p = 0.0086). (**C**) Western blot analysis of pERK1/2 and ERK1/2 after treatment with acidic nitrite solution or indirect CAP ($F_{(4,10)}$ = 7.404, p = 0.0049). Control treatment (Ctrl) refers to ECS without nitrite at pH 7.4. Bars are shown as mean ± SEM (ANOVA followed by Tukey's HSD post-hoc test vs. Ctrl, n = 3, *: $p < 0.05$).

3. Discussion

Generation of nitrite and nitrate after CAP treatment was addressed in some publications before, but their presence and quantity highly depend on the used plasma device and experimental conditions. Since we wanted to combine these molecules with the previously reported acidification after CAP treatment, detection and quantification of both molecules were an essential part of developing a valid RNS-based treatment comparable to CAP. We used Raman spectroscopy to identify long-lived reactive species and found increased levels of nitrate and nitrite, which were then quantified. The use of Raman spectroscopy has shown to be especially practicable due to the all-optical assessment allowing contact-free and label-free quantification of samples. Our results of up to mM ranges are supported by studies on plasma-treated aqueous solutions using other plasma devices [34,35] and display an important baseline for comparison of CAP effects.

When monitoring cytoplasmic calcium levels in melanoma cells, we detected a significant difference between treatments using acidic nitrite and nitrate solutions. It, therefore, seems important to differentiate between these two species when assessing RNS-related effects of CAP, which is not reliably done to date. While it is possible that nitrate contributes to the CAP effect on melanoma cells, most probably due to interaction with ROS and other components of the plasma [36], its role is most definitely minor in comparison to acidic nitrite solutions.

The combination of inorganic nitrite and acidification was previously referred to as *acidified nitrite* [37–39]. On a molecular level, such treatment mainly results in the generation of nitrous acid (HNO_2), an unstable compound that degrades to NO and NO_2. However, the strong acidification might result in further protonation and production of additional reactive species such as $H_2NO_2^+$ or N_2O_3 [40,41]. Identification and evaluation of these reactive species were not addressed in this study, but since the CAP and nitrite effects both depend on strong extracellular acidification, it is possible that their cytotoxicity is not solely based on HNO_2. Acidified nitrite solutions have been well studied due to their antimicrobial activity and positive effects on wound healing, including successful clinical trials using acidified nitrite creams [42,43]. The similarity of these effects with CAP indicates a potential role during plasma treatment. However, the antitumor effects of acidified nitrite are mostly unknown to date. To our knowledge, there is only one study by Morcos et al. [44] showing that 50 µM sodium nitrite inhibits human bladder tumor cells at pH 5.5 to 6 by interfering with DNA replication. However, due to the narrow methodological spectrum of this study and the focus on a physiological rather than therapeutic setting, their work was hardly proof of antitumor effects of acidified nitrite. Nevertheless, it supports our findings and indicates that such treatment might be able to inhibit a wide variety of cancer cells. In the present study, acidified nitrite caused a strong reduction of cell viability in human melanoma cells. Surviving cells were characterized by significant DNA damage and activation of cell cycle inhibitor p21, thereby indicating that this treatment causes lasting damage to melanoma cells. Due to these severe cytotoxic effects, we propose that acidification and nitrite are important components of CAP. Cytotoxicity in normal human fibroblasts was not significant, unlike both melanoma cell lines used in this study, indicating tumor selectivity similar to CAP. However, further comparative studies will be necessary to validate this observation. It was previously reported that HEPES might undergo chemical changes in response to reactive species [45,46]. Since the products of such reaction were found to be cytotoxic, it was necessary to rule out any contribution of modified HEPES to the effects observed in this study. The cytotoxicity of acidified nitrite, however, was still present after interchanging HEPES with a phosphate buffer. We, therefore, conclude that molecular changes of HEPES play a negligible role during our experiments. Thus far, molecular studies of CAP show a strong tendency towards ROS, mainly due to their superior reactivity and cytotoxicity in comparison to RNS. It is, therefore, not surprising that previous studies on CAP-induced nitrite only used it in combination with ROS (namely H_2O_2) to assess antitumor effects [47,48].

On a molecular level, CAP effects on tumor cells were previously linked to oxidative stress, such as the generation of peroxynitrite ($ONOO^-$) and the resulting formation of nitrated proteins. For example, we recently reported increased levels of 3-nitrotyrosine after short CAP treatment of melanoma cells [28]. Furthermore, several studies proposed a causative role of $ONOO^-$ during antimicrobial and even cytotoxic effects of CAP [49,50]. In the present work, we used an experimental setup solely based on RNS and could not detect significant amounts of protein nitration, most probably due to the absence of ROS. Nevertheless, acidified nitrite was found to be strongly cytotoxic, indicating that molecular mechanisms independent of peroxynitrite exist and might also be involved in CAP effects. Another common feature of CAP is the induction of membrane damage in prokaryotic and eukaryotic cells [51,52]. Such alterations of the plasma membrane were previously used to improve drug delivery to cells and through tissues [53] but also have a potential role in antimicrobial and antitumor effects. Interestingly, we could not detect increased membrane damage in response to treatment with acidified nitrite. A possible explanation can be found in the studies of He et al. [52,54], which proposed ROS-dependent lipid peroxidation as the main cause of CAP-induced membrane damage. Our findings of strong cytotoxicity without membrane damage suggest that this process might not be a major contributor to CAP-induced killing of cancer cells. At this stage, however, we are not able to draw a final conclusion on the importance of protein nitration and membrane damage during the antitumor effects of CAP. It is likely that CAP utilizes a broader range of molecular mechanisms to induce cell death, some of which may depend on protein nitration or membrane damage.

When assessing signaling pathways, we found phosphorylation of ERK1/2 to be increased in response to CAP treatment, which was not described before in tumor cells. However, a few articles reported such an activation of the MAPK pathway in normal cells, leading to diverse cellular effects [55–57]. Since a similar increase in pERK1/2 was found after treatment with acidified nitrite, the MAPK pathway might be involved in the observed antitumor effects. This hypothesis is supported by an already established linkage between MAPK activation and apoptosis [58,59].

In summary, this study highlights the importance of acidified nitrite during CAP treatment and calls for further research on CAP-induced RNS. We could show that acidified nitrite is a potent inhibitor of melanoma cells, although it represents only a fraction of all reactive species involved in CAP. Additionally, the comparison of acidified nitrite and CAP treatment is a useful approach for the identification of molecular mechanisms and their evaluation in the context of antitumor effects. Our observations, therefore, contribute to a better understanding of CAP action on tumor cells and facilitate development of CAP-based anti-cancer therapies.

4. Materials and Methods

4.1. Chemicals and Solutions

Extracellular solution (ECS) and phosphate-buffered ECS (pbECS) were prepared as previously described [28]. For ECS, the following chemicals were diluted in bi-distilled water: 145 mM NaCl, 5 mM KCl, 10 mM glucose, 1.25 mM $CaCl_2$, 1 mM $MgCl_2$, and 10 mM HEPES. For preparation of pbECS, 133 mM NaCl, 3.5 mM KCl, 10 mM glucose, 1.25 mM $CaCl_2$, 1 mM $MgCl_2$, 1.5 mM KH_2PO_4, and 8.1 mM Na_2HPO_4 were diluted in bi-distilled water. Both solutions were adjusted to pH 7.4. Sources for further chemicals: Fura-2 AM and pluronic F-127 (Biotium, Fremont, CA, USA), ionomycin (Enzo Life Sciences, Farmingdale, NY, USA), KNO_3 (Carl Roth, Karlsruhe, Germany), $NaNO_3$ (Acros Organics, Fair Lawn, NJ, USA), $NaNO_2$ (Sigma Aldrich, Steinheim, Germany), salicylic acid (Carl Roth), Sulfuric acid (Carl Roth, Karlsruhe, Germany), sulfanilamide (Sigma Aldrich, Steinheim, Germany), naphthylethylenediamine dihydrochloride (Sigma Aldrich, Steinheim, Germany).

4.2. CAP Treatment

CAP treatment of aqueous solutions was described previously [28]. Eight droplets of ECS (20 µL each) were distributed evenly inside a 35 mm petri dish. A MiniFlatPlaster device was placed directly above the dish, resulting in a distance of approximately 10 mm between electrode and sample. Air circulation was minimized by contact of the device to the plastic dish with careful application of pressure. After treatment was finished, all droplets were collected and transferred to reaction tubes for further processing. ECS treated with 2 min CAP was used for the indirect treatment of melanoma cells.

4.3. Raman Spectroscopy

A laser diode with a wavelength of 785 nm (Laser-785-LAB-ADJ-S, Ocean Optics, Dunedin, FL, USA) was used in combination with a low noise spectrometer (QE65000 Pro-Raman, Ocean Optics, Dunedin, FL, USA) and a Raman probe (General Purpose Raman, RIP-RPB-785-SMA-SMA, Ocean Optics, Dunedin, FL, USA). Before each measurement, a reference 'dark spectrum' was recorded and subtracted. An enclosed aluminum chamber was used to measure liquid samples (Figure S5). In this custom design, a 100 µL sample carrier was placed at the working distance of the Raman probe. The molecular concentration of each solution was quantified before the experiment, and a negative control sample of ultra-pure water (Merck-Millipore Chemicals GmbH, Darmstadt, Germany) served as negative control (0 mmol). From each solution, 100 subsequent spectra were recorded with 2 s integration time each. Data processing of the Raman spectra was performed as previously described [60]. The raw spectra were cropped to the spectral range 500–1500 cm^{-1}. Denoising was achieved by a median filter and discrete wavelet denoising (DWT) (k = 2, J_{max} = 2) [61]. The autofluorescence background was modeled by asymmetric least square [62] with λ = 73 and p = 0.001 and then subtracted. Finally, each spectrum was normalized to its entire area under the curve. The peak value was determined within the spectral resolution of 12 cm^{-1}.

4.4. Photometric Nitrate Assay

Quantification of nitrate was based on the transnitration of salicylic acid. A volume of 2 µL of the sample solution was combined with 8 µL of 5% salicylic acid in concentrated sulfuric acid and allowed to incubate for 20 min. Then, 200 µL of 8% NaOH was added to achieve basic pH. The solution was mixed thoroughly, measured at 410 nm using a Clariostar Plus Multiplate reader (BMG Labtech, Ortenberg, Germany), and quantified using a standard curve of $NaNO_3$.

4.5. Photometric Nitrite Assay

A modified Griess diazotization reaction was used to quantify nitrite levels. Briefly, 1.5 µL of the sample solution was transferred to a 96-well plate, followed by 100 µL 1% sulfanilamide in 1 M HCl and 100 µL 0.2% naphthylethylenediamine dihydrochloride (NED) in bi-distilled water. After 15 min incubation, the solution was mixed thoroughly, measured at 540 nm using a Clariostar Plus Multiplate reader, and quantified using a standard curve of $NaNO_2$.

4.6. Cell Culture

Melanoma cell line Mel Im and normal human fibroblasts were cultivated in DMEM low glucose, while the Mel Juso cell line required RPMI 1640 medium with 2% sodium bicarbonate. All media were supplemented with 10% FCS and 1% penicillin/streptomycin. Cells were incubated at 37 °C and 8% CO_2 until approximately 80% confluence. Following a washing step with PBS, a solution of 0.05% trypsin and 0.02% EDTA in PBS was applied to detach the cells. After centrifugation and removal of the trypsin solution, cells were counted using a Neubauer counting chamber. Mycoplasma contamination was regularly excluded for all cell lines. All cell culture chemicals and media were obtained from Sigma Aldrich.

4.7. Calcium Imaging

The experimental setup and procedures were described elsewhere [27]. Briefly, 200,000 cells were seeded in 35 mm cell culture dishes. On the next day, cells were stained with fura-2 AM (3 µM) in ECS with 0.02% pluronic for 30 min at 37 °C and 8% CO_2. After a 5 min washing step with ECS, the solution was removed before the dish was mounted on an inverted microscope, and the perfusion outlet was placed within 1 mm distance to the cells. The imaging procedure started with 1 min background measurement, followed by 4 min treatment with the sample solution. During this time, cells were alternatingly excited at 358 nm and 391 nm while recording fura-2 fluorescence. Intracellular calcium levels were evaluated by calculation of the F358 nm/F391 nm ratio. The area under the curve (AUC) refers to the first 120 s after treatment began, relative to the fluorescence 10 s before treatment. To ensure responsiveness of the cells and validate the staining, 2 µM ionomycin was applied after the treatment. Consequently, non-responsive or erratic cells were excluded from analysis. Areas of interest were placed on individual cells, and their fluorescence ratio time courses were calculated after background subtraction. Further information on data evaluation and imaging equipment can be found in a previous publication [63].

4.8. Nitrite Treatment

A stock solution of 50 mM $NaNO_2$ was prepared freshly using ECS or pbECS with the according pH. Cells were washed with PBS to remove all cell culture media and cell debris, followed by the addition of a 2 mM $NaNO_2$ solution. Unless otherwise specified, treatment duration was 5 min at 37 °C. The solution was removed afterward, and cells were cultivated for 24–72 h in their regular cell culture medium.

4.9. Cell Viability Assay

One day prior to treatment, 6000 cells/well were seeded in a 96-well plate. Following treatment and subsequent incubation for 24 h, cell viability was assessed using the Cell Proliferation Kit II (Roche, Basel, Switzerland) according to the manufacturer's instructions. Photometric detection was realized with a Clariostar Plus Multiplate reader. Absorbance values were normalized to control. The resulting ratios were visualized as % of control.

4.10. Analysis of mRNA Expression by Real-Time PCR

Approximately 150,000 cells/well were seeded in 6-well plates 1 day before treatment. RNA isolation was performed 24 h, 48 h, and 72 h after treatment using E.Z.N.A.® Total RNA Kit (Omega Bio-Tek, Norcross, GA, USA) according to manufacturer's instructions, followed by cDNA generation using reverse transcriptase reaction as previously described [64]. Real-time PCR was carried out in LightCycler® 480 II devices (Roche, Basel, Switzerland) with forward and reverse primers from Sigma-Aldrich: p21_for: 5′-CGAGGCACCGAGGCACTCAGAGG-3′; p21_rev: 5′-CCTGCCTCCTCCCAACTCATCCC-3′; 18s_for: 5′-TCTGTGATGCCCTTAGATGTCC-3′; 18s_rev: 5′-CCATCCAATCGGTAGTAGCG-3′.

4.11. Western Blot Protein Analysis

Approximately 150,000 cells/well were seeded in 6-well plates 1 day before treatment. Total protein isolation was performed 24 h, 48 h, and 72 h after treatment by addition of radio-immunoprecipitation assay buffer (Roche, Basel, Switzerland) as described elsewhere [65]. Detection of 3-nitrotyrosin required immediate protein isolation after treatment. 20 µg protein were loaded on a 10.00% or 12.75% SDS polyacrylamide gel for electrophoresis and subsequently blotted onto a PVDF membrane (Bio-Rad, Hercules, CA, USA). After a short incubation in methanol, Ponceau S staining was performed to quantify total protein load. Membranes were then washed with double distilled water and incubated in 5% non-fat dried milk/TBS-T for 1 h to block unspecific binding sites. Primary antibodies against p21 (1:5000 in 5% NFDM, Abcam, ab109199), 3-nitrotyrosine (1:1000 in TBST, Merck Millipore, 06-284), pERK and ERK (1:1000 in 5% BSA, Cell Signaling, 4370 and 9102) were incubated overnight shaking at 4 °C. Secondary antibodies conjugated to horseradish peroxidase (HRP, Cell Signaling, 7074) were applied for 1 h at room temperature. Visualization of HRP-conjugated antibodies was achieved by the addition of Clarity™ Western ECL Substrate (Bio-Rad) in combination with a Chemostar chemiluminescence imager (Intas, Goettingen, Germany). Signal intensity was then quantified using LabImage software version 4.2.3 (Kapelan Bio-Imaging GmbH, Leipzig, Germany).

4.12. Immunofluorescent Staining

Approximately 35,000 cells were seeded on 18 mm round coverslips the day before treatment. 24 h after treatment, cells were fixed and stained as previously described [66]. The following antibodies were used: anti-PML (1:200, Santa Cruz, Dallas, TX, USA), Cy3 anti-mouse (1:400, Thermo Fisher, Waltham, MA, USA). Cells were eventually stained with DAPI (1:10,000, Sigma Aldrich, Steinheim, Germany). Final stainings were stored at 4 °C and analyzed using an Olympus IX83 inverted microscope in combination with Olympus CellSens Dimension software (Olympus, Tokio, Japan).

4.13. Detection of Membrane Damage

Approximately 200,000 cells/well were seeded in 6-well plates. After cultivation for 24 h, cells were washed with PBS, treated with each sample solution, and washed again. Staining was achieved by the addition of 1 mL propidium iodide solution (10 µg/mL, PromoCell, Heidelberg, Germany) and 5 min incubation at room temperature. The staining solution was then removed, cells were washed with PBS and detached from the plate using trypsin. Following another washing step, cells were eventually resuspended in 1% BSA/PBS and analyzed by flow cytometry (LSRFortessa^{TM}, BD Biosciences, San Jose, CA, USA). Data analysis was done using FACSDiva 9.0 software (BD Biosciences).

4.14. Statistical Analysis

Experimental results were analyzed and visualized using GraphPad Prism 7 software (GraphPad Software Inc., San Diego, CA, USA). If not otherwise specified, at least 3 biological replicates were measured, and statistical analysis was performed by one-way ANOVA. A significant F-test was followed by Tukey's HSD post-hoc tests. A critical value of $p < 0.05$ was set for statistical significance. All results were given as mean $\pm$ standard error of the mean (SEM).

Supplementary Materials: The following are available online at https://www.mdpi.com/article/10.3390/ijms22073757/s1, Figure S1: Calibration of Raman spectroscopy to potassium nitrate solutions of known concentration, Figure S2: Calibration of Raman spectroscopy to sodium nitrite solutions of known concentration, Figure S3: Acidified nitrate solution does not induce a cytoplasmic Ca^{2+} release in melanoma cells, Figure S4: Effects of acidic nitrite solution in melanoma cell line Mel Im and normal human fibroblasts, Figure S5: Custom designed experimental chamber for Raman spectroscopy.

Author Contributions: Conceptualization, T.Z., A.-K.B.; methodology, T.Z., L.A.G., L.K., O.F., M.J.M.F., A.-K.B.; formal Analysis, T.Z., L.A.G., L.K.; investigation, T.Z., L.A.G., L.K., C.S.; resources, O.F., M.J.M.F., A.-K.B.; writing—original draft, T.Z., L.K., A.-K.B.; writing—review and editing, T.Z., L.A.G., L.K., C.S., S.A., S.K., O.F., M.J.M.F., A.-K.B.; visualization: T.Z., L.K.; supervision, O.F., M.J.M.F., A.-K.B.; funding acquisition, O.F., M.J.M.F., A.-K.B. All authors have read and agreed to the published version of the manuscript.

Funding: This work was funded by the German Research Foundation (DFG) (BO1573/23 and TRR241-C01) and the Interdisciplinary Center for Clinical Research (IZKF) Erlangen (D31).

Institutional Review Board Statement: Not applicable.

Informed Consent Statement: Not applicable.

Data Availability Statement: Not applicable.

Acknowledgments: We thank Michaela Pommer, Ingmar Henz, Sebastian Staebler, Chafia Chiheb and Ines Boehme for technical and methodological assistance and discussions.

Conflicts of Interest: The authors declare no conflict of interest.

References

1. Morfill, G.E.; Shimizu, T.; Steffes, B.; Schmidt, H.U. Nosocomial infections—A new approach towards preventive medicine using plasmas. *N. J. Phys.* **2009**, *11*, 115019. [CrossRef]
2. Zimmermann, J.L.; Dumler, K.; Shimizu, T.; Morfill, G.E.; Wolf, A.; Boxhammer, V.; Schlegel, J.; Gansbacher, B.; Anton, M. Effects of cold atmospheric plasmas on adenoviruses in solution. *J. Phys. D Appl. Phys.* **2011**, *44*, 505201. [CrossRef]
3. Maisch, T.; Shimizu, T.; Li, Y.F.; Heinlin, J.; Karrer, S.; Morfill, G.; Zimmermann, J.L. Decolonisation of MRSA, S. aureus and E. coli by cold-atmospheric plasma using a porcine skin model in vitro. *PLoS ONE* **2012**, *7*, e34610. [CrossRef]
4. Arndt, S.; Unger, P.; Wacker, E.; Shimizu, T.; Heinlin, J.; Li, Y.F.; Thomas, H.M.; Morfill, G.E.; Zimmermann, J.L.; Bosserhoff, A.K.; et al. Cold atmospheric plasma (CAP) changes gene expression of key molecules of the wound healing machinery and improves wound healing in vitro and in vivo. *PLoS ONE* **2013**, *8*, e79325. [CrossRef]
5. Chatraie, M.; Torkaman, G.; Khani, M.; Salehi, H.; Shokri, B. In vivo study of non-invasive effects of non-thermal plasma in pressure ulcer treatment. *Sci. Rep.* **2018**, *8*, 5621. [CrossRef] [PubMed]
6. Arndt, S.; Unger, P.; Berneburg, M.; Bosserhoff, A.K.; Karrer, S. Cold atmospheric plasma (CAP) activates angiogenesis-related molecules in skin keratinocytes, fibroblasts and endothelial cells and improves wound angiogenesis in an autocrine and paracrine mode. *J. Dermatol. Sci.* **2018**, *89*, 181–190. [CrossRef] [PubMed]
7. Pan, J.; Sun, K.; Liang, Y.; Sun, P.; Yang, X.; Wang, J.; Zhang, J.; Zhu, W.; Fang, J.; Becker, K.H. Cold plasma therapy of a tooth root canal infected with enterococcus faecalis biofilms in vitro. *J. Endod.* **2013**, *39*, 105–110. [CrossRef] [PubMed]
8. Aparecida Delben, J.; Evelin Zago, C.; Tyhovych, N.; Duarte, S.; Eduardo Vergani, C. Effect of atmospheric-pressure cold plasma on pathogenic oral biofilms and in vitro reconstituted oral epithelium. *PLoS ONE* **2016**, *11*, e0155427.

9.   Scharf, C.; Eymann, C.; Emicke, P.; Bernhardt, J.; Wilhelm, M.; Görries, F.; Winter, J.; Von Woedtke, T.; Darm, K.; Daeschlein, G.; et al. Improved wound healing of airway epithelial cells is mediated by cold atmospheric plasma: A time course-related proteome analysis. *Oxid. Med. Cell. Longev.* **2019**, *2019*. [CrossRef]
10. Binenbaum, Y.; Ben-David, G.; Gil, Z.; Slutsker, Y.Z.; Ryzhkov, M.A.; Felsteiner, J.; Krasik, Y.E.; Cohen, J.T. Cold atmospheric plasma, created at the tip of an elongated flexible capillary using low electric current, can slow the progression of Melanoma. *PLoS ONE* **2017**, *12*, e0169457. [CrossRef]
11. Arndt, S.; Wacker, E.; Li, Y.F.; Shimizu, T.; Thomas, H.M.; Morfill, G.E.; Karrer, S.; Zimmermann, J.L.; Bosserhoff, A.K. Cold atmospheric plasma, a new strategy to induce senescence in melanoma cells. *Exp. Dermatol.* **2013**, *22*, 284–289. [CrossRef]
12. Yajima, I.; Iida, M.; Kumasaka, M.Y.; Omata, Y.; Ohgami, N.; Chang, J.; Ichihara, S.; Hori, M.; Kato, M. Non-equilibrium atmospheric pressure plasmas modulate cell cycle-related gene expressions in melanocytic tumors of RET-transgenic mice. *Exp. Dermatol.* **2014**, *23*, 424–425. [CrossRef]
13. Schneider, C.; Arndt, S.; Zimmermann, J.L.; Li, Y.; Karrer, S.; Bosserhoff, A.K. Cold atmospheric plasma treatment inhibits growth in colorectal cancer cells. *Biol. Chem.* **2018**, *400*, 111–127. [CrossRef] [PubMed]
14. Tuhvatulin, A.I.; Sysolyatina, E.V.; Scheblyakov, D.V.; Logunov, D.Y.; Vasiliev, M.M.; Yurova, M.A.; Danilova, M.A.; Petrov, O.F.; Naroditsky, B.S.; Morfill, G.E.; et al. Non-thermal plasma causes P53-dependent apoptosis in human colon carcinoma cells. *Acta Nat.* **2012**, *4*, 82–87. [CrossRef]
15. Chen, Z.; Simonyan, H.; Cheng, X.; Gjika, E.; Lin, L.; Canady, J.; Sherman, J.H.; Young, C.; Keidar, M. A novel micro cold atmospheric plasma device for glioblastoma both in vitro and in vivo. *Cancers* **2017**, *9*, 61. [CrossRef]
16. Walk, R.M.; Snyder, J.A.; Srinivasan, P.; Kirsch, J.; Diaz, S.O.; Blanco, F.C.; Shashurin, A.; Keidar, M.; Sandler, A.D. Cold atmospheric plasma for the ablative treatment of neuroblastoma. *J. Pediatr. Surg.* **2013**, *48*, 67–73. [CrossRef]
17. Vandamme, M.; Robert, E.; Dozias, S.; Sobilo, J.; Lerondel, S.; Le Pape, A.; Pouvesle, J.M. Response of human glioma U87 xenografted on mice to non thermal plasma treatment. *Plasma Med.* **2011**, *1*, 27–43. [CrossRef]
18. Utsumi, F.; Kajiyama, H.; Nakamura, K.; Tanaka, H.; Mizuno, M.; Ishikawa, K.; Kondo, H.; Kano, H.; Hori, M.; Kikkawa, F. Effect of indirect nonequilibrium atmospheric pressure plasma on anti-proliferative activity against chronic chemo-resistant ovarian cancer cells in vitro and in vivo. *PLoS ONE* **2013**, *8*, e81576. [CrossRef] [PubMed]
19. Ishaq, M.; Han, Z.J.; Kumar, S.; Evans, M.D.M.; Ostrikov, K. Atmospheric-pressure plasma- and TRAIL-induced apoptosis in TRAIL-resistant colorectal cancer cells. *Plasma Process. Polym.* **2015**, *12*, 574–582. [CrossRef]
20. Köritzer, J.; Boxhammer, V.; Schäfer, A.; Shimizu, T.; Klämpfl, T.G.; Li, Y.F.; Welz, C.; Schwenk-Zieger, S.; Morfill, G.E.; Zimmermann, J.L.; et al. Restoration of sensitivity in chemo—Resistant glioma cells by cold atmospheric plasma. *PLoS ONE* **2013**, *8*, e64498. [CrossRef] [PubMed]
21. Zucker, S.N.; Zirnheld, J.; Bagati, A.; DiSanto, T.M.; Des Soye, B.; Wawrzyniak, J.A.; Etemadi, K.; Nikiforov, M.; Berezney, R. Preferential induction of apoptotic cell death in melanoma cells as compared with normal keratinocytes using a non-thermal plasma torch. *Cancer Biol. Ther.* **2012**, *13*, 1299–1306. [CrossRef]
22. Panngom, K.; Baik, K.Y.; Nam, M.K.; Han, J.H.; Rhim, H.; Choi, E.H. Preferential killing of human lung cancer cell lines with mitochondrial dysfunction by nonthermal dielectric barrier discharge plasma. *Cell Death Dis.* **2013**, *4*, e642. [CrossRef] [PubMed]
23. Alimohammadi, M.; Golpur, M.; Sohbatzadeh, F.; Hadavi, S.; Bekeschus, S.; Niaki, H.A.; Valadan, R.; Rafiei, A. Cold atmospheric plasma is a potent tool to improve chemotherapy in melanoma in vitro and in vivo. *Biomolecules* **2020**, *10*, 1011. [CrossRef] [PubMed]
24. Biscop, E.; Lin, A.; Van Boxem, W.; Van Loenhout, J.; De Backer, J.; Deben, C.; Dewilde, S.; Smits, E.; Bogaerts, A. Influence of cell type and culture medium on determining cancer selectivity of cold atmospheric plasma treatment. *Cancers* **2019**, *11*, 1287. [CrossRef] [PubMed]
25. Kim, G.J.; Kim, W.; Kim, K.T.; Lee, J.K. DNA damage and mitochondria dysfunction in cell apoptosis induced by nonthermal air plasma. *Appl. Phys. Lett.* **2010**, *96*, 021502. [CrossRef]
26. Ishaq, M.; Kumar, S.; Varinli, H.; Han, Z.J.; Rider, A.E.; Evans, M.D.M.; Murphy, A.B.; Ostrikov, K. Atmospheric gas plasma-induced ROS production activates TNF-ASK1 pathway for the induction of melanoma cancer cell apoptosis. *Mol. Biol. Cell* **2014**, *25*, 1523–1531. [CrossRef]
27. Schneider, C.; Gebhardt, L.; Arndt, S.; Karrer, S.; Zimmermann, J.L.; Fischer, M.J.M.; Bosserhoff, A.K. Cold atmospheric plasma causes a calcium influx in melanoma cells triggering CAP-induced senescence. *Sci. Rep.* **2018**, *8*, 10048. [CrossRef]
28. Schneider, C.; Gebhardt, L.; Arndt, S.; Karrer, S.; Zimmermann, J.L.; Fischer, M.J.M.; Bosserhoff, A.K. Acidification is an essential process of cold atmospheric plasma and promotes the anti-cancer effect on malignant melanoma cells. *Cancers* **2019**, *11*, 671. [CrossRef]
29. Cataldo, D.A.; Haroon, M.H.; Schrader, L.E.; Youngs, V.L. Rapid colorimetric determination of nitrate in plant tissue by nitration of salicylic acid. *Commun. Soil Sci. Plant Anal.* **1975**, *6*, 71–80. [CrossRef]
30. Tarabová, B.; Lukeš, P.; Hammer, M.U.; Jablonowski, H.; Von Woedtke, T.; Reuter, S.; Machala, Z. Fluorescence measurements of peroxynitrite/peroxynitrous acid in cold air plasma treated aqueous solutions. *Phys. Chem. Chem. Phys.* **2019**, *21*, 8883–8896. [CrossRef]
31. Lukes, P.; Dolezalova, E.; Sisrova, I.; Clupek, M. Aqueous-phase chemistry and bactericidal effects from an air discharge plasma in contact with water: Evidence for the formation of peroxynitrite through a pseudo-second-order post-discharge reaction of H_2O_2 and HNO_2. *Plasma Sources Sci. Technol.* **2014**, *23*, 015019. [CrossRef]

32. Ferrer-Sueta, G.; Campolo, N.; Trujillo, M.; Bartesaghi, S.; Carballal, S.; Romero, N.; Alvarez, B.; Radi, R. Biochemistry of Peroxynitrite and protein tyrosine nitration. *Chem. Rev.* **2018**, *118*, 1338–1408. [CrossRef] [PubMed]

33. Yamakura, F.; Taka, H.; Fujimura, T.; Murayama, K. Inactivation of human manganese-superoxide dismutase by peroxynitrite is caused by exclusive nitration of tyrosine 34 to 3-nitrotyrosine. *J. Biol. Chem.* **1998**, *273*, 14085–14089. [CrossRef]

34. Griseti, E.; Merbahi, N.; Golzio, M. Anti-cancer potential of two plasma-activated liquids: Implication of long-lived reactive oxygen and nitrogen species. *Cancers* **2020**, *12*, 721. [CrossRef]

35. Girard, P.M.; Arbabian, A.; Fleury, M.; Bauville, G.; Puech, V.; Dutreix, M.; Sousa, J.S. Synergistic effect of H_2O_2 and NO_2 in cell death induced by cold atmospheric he plasma. *Sci. Rep.* **2016**, *6*. [CrossRef]

36. Xu, D.; Cui, Q.; Xu, Y.; Liu, Z.; Chen, Z.; Xia, W.; Zhang, H.; Liu, D.; Chen, H.; Kong, M.G. NO^{2-} and NO^{3-} enhance cold atmospheric plasma induced cancer cell death by generation of $ONOO^-$. *AIP Adv.* **2018**, *8*, 105219. [CrossRef]

37. Weller, R.; Price, R.J.; Ormerod, A.D.; Benjamin, N. Antimicrobial effect of acidified nitrite on dermatophyte fungi, Candida and bacterial skin pathogens. *J. Appl. Microbiol.* **2001**, *90*, 648–652. [CrossRef] [PubMed]

38. Müller-Herbst, S.; Wüstner, S.; Kabisch, J.; Pichner, R.; Scherer, S. Acidified nitrite inhibits proliferation of Listeria monocytogenes—Transcriptional analysis of a preservation method. *Int. J. Food Microbiol.* **2016**, *226*, 33–41. [CrossRef] [PubMed]

39. Finnen, M.J.; Hennessy, A.; McLean, S.; Bisset, Y.; Mitchell, R.; Megson, I.L.; Weller, R. Topical application of acidified nitrite to the nail renders it antifungal and causes nitrosation of cysteine groups in the nail plate. *Br. J. Dermatol.* **2007**, *157*, 494–500. [CrossRef]

40. Nguyen, M.T.; Hegarty, A.F. Protonation of nitrous acid and formation of the nitrosating agent NO+: An ab initio study. *J. Chem. Soc. Perkin Trans. 2* **1984**, *12*, 2037–2041. [CrossRef]

41. Anastasio, C.; Liang, C. Photochemistry of nitrous acid (HONO) and nitrous acidium ion (H 2ONO +) in aqueous solution and ice. *Environ. Sci. Technol.* **2009**, *43*, 1108–1114. [CrossRef] [PubMed]

42. Ormerod, A.D.; Shah, A.A.J.; Li, H.; Benjamin, N.B.; Ferguson, G.P.; Leifert, C. An observational prospective study of topical acidified nitrite for killing methicillin-resistant Staphylococcus aureus (MRSA) in contaminated wounds. *BMC Res. Notes* **2011**, *4*, 458. [CrossRef] [PubMed]

43. Weller, R.; Ormerod, A.D.; Hobson, R.P.; Benjamin, N.J. A randomized trial of acidified nitrite cream in the treatment of tinea pedis. *J. Am. Acad. Dermatol.* **1998**, *38*, 559–563. [CrossRef]

44. Morcos, E.; Carlsson, S.; Weitzberg, E.; Wiklund, N.P.; Lundberg, J.O. Inhibition of cancer cell replication by inorganic nitrite. *Nutr. Cancer* **2010**, *62*, 501–504. [CrossRef]

45. Habib, A.; Tabata, M. Oxidative DNA damage induced by HEPES (2-[4-(2-hydroxyethyl)-1-piperazinyl] ethanesulfonic acid) buffer in the presence of Au(III). *J. Inorg. Biochem.* **2004**, *98*, 1696–1702. [CrossRef]

46. Zhao, G.; Chasteen, N.D. Oxidation of Good's buffers by hydrogen peroxide. *Anal. Biochem.* **2006**, *349*, 262–267. [CrossRef]

47. Bauer, G. Intercellular singlet oxygen-mediated bystander signaling triggered by long-lived species of cold atmospheric plasma and plasma-activated medium. *Redox Biol.* **2019**, *26*, 101301. [CrossRef]

48. Bauer, G. The synergistic effect between hydrogen peroxide and nitrite, two long-lived molecular species from cold atmospheric plasma, triggers tumor cells to induce their own cell death. *Redox Biol.* **2019**, *26*, 101291. [CrossRef]

49. Zhou, R.; Zhou, R.; Prasad, K.; Fang, Z.; Speight, R.; Bazaka, K.; Ostrikov, K. Cold atmospheric plasma activated water as a prospective disinfectant: The crucial role of peroxynitrite. *Green Chem.* **2018**, *20*, 5276–5284. [CrossRef]

50. Bauer, G.; Sersenová, D.; Graves, D.B.; Machala, Z. Cold Atmospheric plasma and plasma-activated medium trigger RONS-based tumor cell apoptosis. *Sci. Rep.* **2019**, *9*, 1–28. [CrossRef]

51. Xu, H.; Zhu, Y.; Du, M.; Wang, Y.; Ju, S.; Ma, R.; Jiao, Z. Subcellular mechanism of microbial inactivation during water disinfection by cold atmospheric-pressure plasma. *Water Res.* **2021**, *188*, 116513. [CrossRef]

52. He, Z.; Liu, K.; Scally, L.; Manaloto, E.; Gunes, S.; Ng, S.W.; Maher, M.; Tiwari, B.; Byrne, H.J.; Bourke, P.; et al. Low dose cold atmospheric plasma induces membrane oxidation, stimulates endocytosis and enhances uptake of nanomaterials in Glioblastoma multiforme cells. *BioRxiv* **2019**. [CrossRef]

53. Wen, X.; Xin, Y.; Hamblin, M.R.; Jiang, X. Applications of cold atmospheric plasma for transdermal drug delivery: A review. *Drug Deliv. Transl. Res.* **2020**, 1–7. [CrossRef]

54. He, Z.; Liu, K.; Scally, L.; Manaloto, E.; Gunes, S.; Ng, S.W.; Maher, M.; Tiwari, B.; Byrne, H.J.; Bourke, P.; et al. Cold Atmospheric plasma stimulates clathrin-dependent endocytosis to repair oxidised membrane and enhance uptake of nanomaterial in glioblastoma multiforme cells. *Sci. Rep.* **2020**. [CrossRef] [PubMed]

55. Jang, J.Y.; Hong, Y.J.; Lim, J.; Choi, J.S.; Choi, E.H.; Kang, S.; Rhim, H. Cold atmospheric plasma (CAP), a novel physicochemical source, induces neural differentiation through cross-talk between the specific RONS cascade and Trk/Ras/ERK signaling pathway. *Biomaterials* **2018**, *156*, 258–273. [CrossRef] [PubMed]

56. Schmidt, A.; Bekeschus, S.; Jarick, K.; Hasse, S.; Von Woedtke, T.; Wende, K. Cold physical plasma modulates p53 and mitogen-activated protein kinase signaling in keratinocytes. *Oxid. Med. Cell. Longev.* **2019**, *2019*. [CrossRef]

57. Bundscherer, L.; Nagel, S.; Hasse, S.; Tresp, H.; Wende, K.; Walther, R.; Reuter, S.; Weltmann, K.D.; Masur, K.; Lindequist, U. Non-thermal plasma treatment induces MAPK signaling in human monocytes. *Open Chem.* **2015**, *13*. [CrossRef]

58. Cagnol, S.; Chambard, J.C. ERK and cell death: Mechanisms of ERK-induced cell death—Apoptosis, autophagy and senescence. *FEBS J.* **2010**, *277*, 2–21. [CrossRef] [PubMed]

59. Yue, J.; López, J.M. Understanding MAPK signaling pathways in apoptosis. *Int. J. Mol. Sci.* **2020**, *21*, 2346. [CrossRef]

60. Kreiß, L.; Hohmann, M.; Klämpfl, F.; Schürmann, S.; Dehghani, F.; Schmidt, M.; Friedrich, O.; Büchler, L. Diffuse reflectance spectroscopy and Raman spectroscopy for label-free molecular characterization and automated detection of human cartilage and subchondral bone. *Sens. Actuators B Chem.* **2019**, *301*, 127121. [CrossRef]

61. Von Wegner, F.; Both, M.; Fink, R.H.A.; Friedrich, O. Fast XYT imaging of elementary calcium release events in muscle with multifocal multiphoton microscopy and wavelet denoising and detection. *IEEE Trans. Med. Imaging* **2007**, *26*, 925–934. [CrossRef]

62. Peng, J.; Peng, S.; Jiang, A.; Wei, J.; Li, C.; Tan, J. Asymmetric least squares for multiple spectra baseline correction. *Anal. Chim. Acta* **2010**, *683*, 63–68. [CrossRef] [PubMed]

63. Babes, A.; Sauer, S.K.; Moparthi, L.; Kichko, T.I.; Neacsu, C.; Namer, B.; Filipovic, M.; Zygmunt, P.M.; Reeh, P.W.; Fischer, M.J.M. Photosensitization in porphyrias and photodynamic therapy involves TRPA1 and TRPV1. *J. Neurosci.* **2016**, *36*, 5264–5278. [CrossRef] [PubMed]

64. Arndt, S.; Bosserhoff, A.K. TANGO is a tumor suppressor of malignant melanoma. *Int. J. Cancer* **2006**, *119*, 2812–2820. [CrossRef] [PubMed]

65. Stieglitz, D.; Lamm, S.; Braig, S.; Feuerer, L.; Kuphal, S.; Dietrich, P.; Arndt, S.; Echtenacher, B.; Hellerbrand, C.; Karrer, S.; et al. BMP6-induced modulation of the tumor micro-milieu. *Oncogene* **2019**, *38*, 609–621. [CrossRef]

66. Feuerer, L.; Lamm, S.; Henz, I.; Kappelmann-Fenzl, M.; Haferkamp, S.; Meierjohann, S.; Hellerbrand, C.; Kuphal, S.; Bosserhoff, A.K. Role of melanoma inhibitory activity in melanocyte senescence. *Pigment Cell Melanoma Res.* **2019**, *32*, 777–791. [CrossRef] [PubMed]

International Journal of
Molecular Sciences

Article

Genotypic and Phenotypic Changes in *Candida albicans* as a Result of Cold Plasma Treatment

Ewa Tyczkowska-Sieroń [1] , Tadeusz Kałużewski [2] , Magdalena Grabiec [2],
Bogdan Kałużewski [2] and Jacek Tyczkowski [3,*]

1   Department of Biology and Parasitology, Medical University of Lodz, Żeligowski Str. 7/9,
    90-752 Lodz, Poland; ewa.tyczkowska-sieron@umed.lodz.pl
2   Laboratory of Medical Genetics of the "Genos" Partnership - R&D Division, Inowrocławska Str. 9/132,
    91-033 Lodz, Poland; t.kaluzewski@genos.com.pl (T.K.); magrabiec@gmail.com (M.G.);
    b.kaluzewski42@gmail.com (B.K.)
3   Department of Molecular Engineering, Faculty of Process and Environmental Engineering,
    Lodz University of Technology, Wólczańska Str. 213, 90-924 Lodz, Poland
*   Correspondence: jacek.tyczkowski@p.lodz.pl

Received: 15 September 2020; Accepted: 27 October 2020; Published: 30 October 2020

Abstract: We treated *Candida albicans* cells with a sublethal dose of nonequilibrium (cold) atmospheric-pressure He plasma and studied alterations in the genome of this fungus as well as changes in the phenotypic traits, such as assimilation of carbon from carbohydrates, hydrolytic enzyme activity, and drug susceptibility. There is a general problem if we use cold plasma to kill microorganism cells and some of them survive the process—whether the genotypic and phenotypic features of the cells are significantly altered in this case, and, if so, whether these changes are environmentally hazardous. Our molecular genetic studies have identified six single nucleotide variants, six insertions, and five deletions, which are most likely significant changes after plasma treatment. It was also found that out of 19 tested hydrolytic enzymes, 10 revealed activity, of which nine temporarily decreased their activity and one (naphthol-AS-BI- phosphohydrolase) permanently increased activity as a result of the plasma treatment. In turn, carbon assimilation and drug susceptibility were not affected by plasma. Based on the performed studies, it can be concluded that the observed changes in *C. albicans* cells that survived the plasma action are not of significant importance to the environment, especially for the drug resistance and pathogenicity of this fungus.

Keywords: *Candida albicans*; cold plasma treatment; genome; hydrolytic enzyme activity; carbon assimilation; drug susceptibility

---

1. Introduction

The application of nonequilibrium (cold) atmospheric-pressure plasma for the inactivation of microorganisms in the areas of medicine, biotechnology, and food processing has attracted a rapidly growing interest in recent years. In particular, much attention has been paid to plasma medicine and healthcare [1–4]. Numerous attempts to use cold plasma in oncology [5], dermatology [6–8], and wound healing [9], as well as in disinfection and sterilization [10,11], are already quite advanced. The studies have long since gone beyond in vitro experiments and are conducted in vivo in animals as well as in humans as clinical trials [12–16]. A significant problem that has emerged as a result of these studies is the finding that plasma treatment does not necessarily kill all microorganisms in the area of plasma action. Some of them survive, and it is suspected that certain changes in their genome and phenotypic traits occur [17,18]. Particularly dangerous changes would be the increase in virulence and drug resistance, which would place the application of plasma technology in a disadvantageous position in medicine. Although researchers have just turned their attention to the plasma-induced sublethal effects in living organisms [19,20], also studying them on isolated DNA [21–23] or enzymes [24], the precise mechanism of this phenomenon, especially the types of DNA damage, is poorly characterized [25]. This problem, therefore, requires further and more advanced studies.

Following this research trend, in this paper, we present the results of studies on *Candida albicans* cells in the sublethal state, which survived plasma treatment, looking for genotypic and phenotypic changes that would be a consequence of such a treatment. *C. albicans* was chosen for two reasons. Firstly, it is a well-characterized microorganism [26], which is recognized as a model in the study of fungal pathogens [27], but most of all—secondly—it is the most important cause of fungal infections in humans [28,29], commonly known as candidiasis, which poses a serious therapeutic problem, among others, due to its constantly increasing resistance to antifungal agents [30–32]. It is not surprising then that new and effective strategies of fighting against candidiasis are being searched for. Considering that superficial candidiasis, mainly skin infections, are the most numerous and widespread group of all fungal infections [33,34], cold plasma, which is especially suitable for surface treatment, seems to be a very promising therapeutic method in this case. As this method is becoming increasingly popular and is beginning to be applied in practice, the determination of changes that may occur in *C. albicans* cells that survived plasma treatment is particularly justified.

Several tests with the use of cold plasma have already been carried out on *C. albicans*. Apart from determining the influence of plasma parameters and treatment time on the survival rate of the cells of this fungus using various types of plasma sources [35–39], a few preliminary studies have also been carried out on the changes that occur in sublethal cells. For example, Rahimi-Verki et al. [40] found a reduction in the activity of phospholipase and proteinase enzymes for plasma-treated *C. albicans* samples compared to untreated controls. A decrease in the amount of ergosterol after plasma treatment was also noticed. On the other hand, Borges et al. [41] did not observe the effect of plasma on exoenzyme (phospholipase and proteinase) production but found a promising impact of plasma treatment on morphogenesis, with an almost 40-fold reduction of the filamentation rate compared to the nonexposed group. As can be seen, just taking enzymes as an example, the response of *C. albicans* to plasma treatment is inconclusive and requires further detailed investigation.

2. Results and Discussion

Surface plasma treatment of *C. albicans* culture induced the growth inhibition zone with an elliptical-like shape, as shown in Figure 1. The zone edge has been defined as a place where cell survival reaches an average of 10 % compared to the plasma-untreated region [39]. Due to the appropriately selected plasma parameters (see Section 3.1) and the relatively short time of plasma action (see Section 3.2), not all *C. albicans* cells in the elliptical zone were killed. Some of them survived, which were revealed, after incubation, as spot colonies inside this zone (Figure 1). These colonies were collected for further investigations.

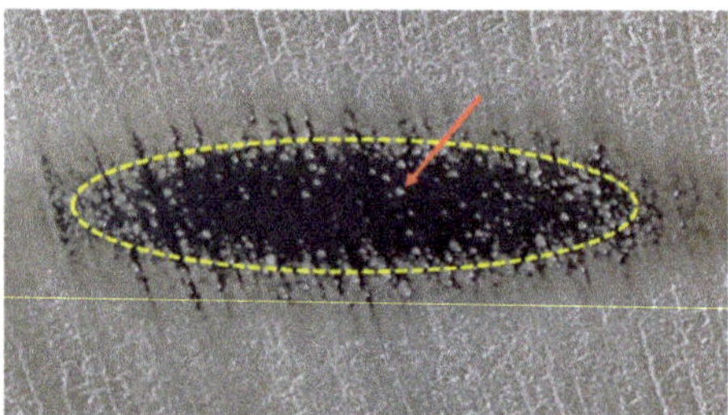

Figure 1. The growth inhibition zone induced by plasma that was generated by the linear microdischarge jet. The red arrow indicates one of the spot colonies, which grew after plasma treatment.

2.1. Genomic Alterations

The results of nuclear DNA sequencing are presented on a circular plot (Figure 2). The variant calling allows the identification of numerous genomic alterations in each evaluated sample. However, since no genomic changes affecting the mitochondrial genome have been recorded, mitochondrial DNA is not discussed further.

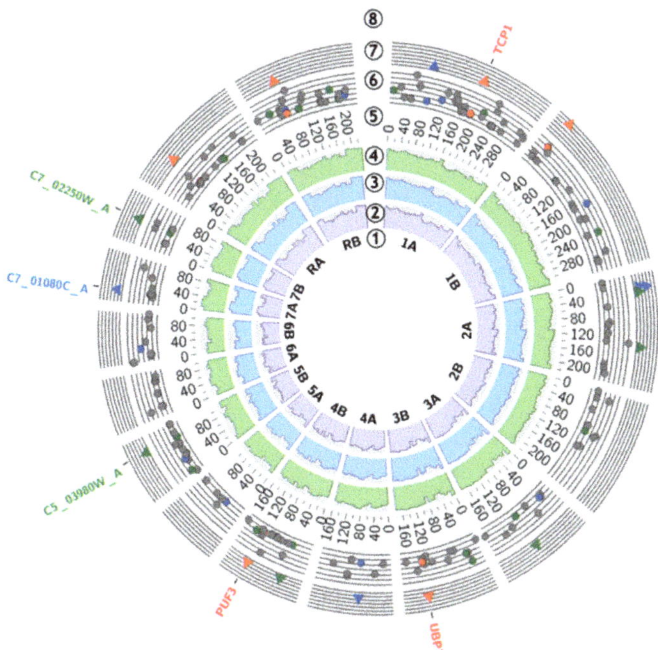

Figure 2. Genome diagram: (**1**) *C. albicans* SC5314 A22 ideogram; (**2**) control sample coverage; (**3**) 12_1 sample coverage; (**4**) 12_2 sample coverage; (**5**) chromosomal coordinates; (**6**) genomic alternations (grey—low confidence single nucleotide variants (SNVs), red—high confidence SNVs, blue—deletions, green—insertions); (**7**) genomic alternations statistically significant (red—high confidence SNVs, blue—deletions, green—insertions); (**8**) genes affected by high-confidence SNVs (red), deletion (blue), insertions (green).

In this study, we focused on the single nucleotide variants (SNVs), insertions, and deletions, which were present in two investigated samples after 12 cycles of plasma treatment (12_1 and 12_2) and were significantly more frequent than in the control sample. This approach led to the identification of six single nucleotide variants, six insertions, and five deletions. The following genes were affected by high-confidence SNVs: TCP1 (p.Gly>Arg and p.Gly>Glu), UBP1 (p.Gln>STOP), and PUF3 (p.Gly>Gly). Short insertions were present in the sequence of genes: C5_03980W_A and C7_02250W_A. The sequence of C7_01080C_A was interfered with by deletion. All of the abovementioned mutations were not exclusive to the tested samples. They were also present in the control sample, but they were significantly less frequent, according to the VarScan software evaluation. Particular attention was paid to genes known to alter the *C. albicans* physiology—ALS1−ALS12, HWP1, EAP1, ECM33, MP65, PHR1, SAP1−SAP10, SAP30, SAP98, SAP99, SAP155, SAP190, RSR1, BIG1, LIP1−LIP10, PMT1, PMT4, BCR1, TEC1, ACE2, EFG1, ZAP1, MDR1, CDR1, CDR2, EFG1, CPH1, CPH2, ECE1, TUP1, NRG1, RFG [42], and ERG3, ERG11 [43]). There were no significant changes in the sequence of those genes. A detailed summary of the molecular findings is presented in Table 1.

Table 1. Summary of statistically significant mutations detected in the tested samples.

Standard Name	Systematic Name	Localization	Mutation	Sample 12_1 *p*-Value	Sample 12_2 *p*-Value	Description [44]
				Single Nucleotide Variants		
TCP1	C1_08560W_A	Chr1A:1874103 Chr1A:1874104	G>C (p. Gly>Arg) G>A (p.Gly>Glu)	0.0369 0.0471	0.0095 0.0112	Chaperonin- containing T-complex subunit, induced by alpha pheromone in SpicerM medium; stationary phase enriched protein
Uncharacterized region		Chr1B:275481	T>A	0.0033	0.0104	–
UBP1	C3_05870C_A	Chr3B: 1315611	C>T (p.Gln>STOP)	0.0208	0.0119	Ortholog(s) have thiol-dependent ubiquitin-specific protease activity, role in negative regulation of protein autoubiquitination, protein deubiquitination and cytoplasm, endoplasmic reticulum localization
PUF3	C4_05370W_A	Chr4B:1172318	T>C (p.Gly>Gly)	0.0304	0.0304	RNA-binding protein involved in the regulation of mitochondrial biogenesis
Uncharacterized region		ChrRA:655094	A>G	0.0405	0.0082	–
Uncharacterized region		ChrRA:655095	C>G	0.0336	0.0031	–
				Insertions and Deletions		
Uncharacterized region		Chr1A:837098	-AAAAGAAAAG	0.0064	0.0428	–
Uncharacterized region		Chr2A:254001	-TAGAAGAAGAAT	0.0027	0.0345	–
Uncharacterized region		Chr2A:342691	+C	0.0028	0.0257	–
Uncharacterized region		Chr2A:239531	-T	0.0105	0.0337	–
Uncharacterized region		Chr2A:1421683	+GATATTTAAGG	0.0153	0.0256	–
Uncharacterized region		Chr3A:980388	+T	0.0338	0.0291	–
Uncharacterized region		Chr4A:686524	-T	0.0373	0.0346	–
Uncharacterized region		Chr4B:456519	+TAAAT	0.0185	0.0285	–
–	C5_03980W_A	Chr5B: 878800	+CAACAACAA	0.0403	0.0364	Protein of unknown function; Spider biofilm induced
–	C7_01080C_A	Chr7A:228708	-TC	0.0182	0.0231	Major repeat sequence (MRS) on Chromosome 7; MRS units are composed of variable numbers of RPS units flanked by HOK and RB2 sequences; found on most chromosomes; may serve as recombination hot spot
–	C7_02250W_A	Chr7B:490089	+TTCCAA	0.0022	0.0208	Ortholog of *C. dubliniensis* CD36: Cd36_72050, *C. parapsilosis* CDC317: CPAR2_301140, *C. tenuis* NRRL Y-1498: CANTEDRAFT_ 135055, and *Debaryomyces hansenii* CBS767: DEHA2E07678g

The genome of *C. albicans* has a length of 28,605,418 bp (including the nuclear and mitochondrial genome) and consists of eight pairs of chromosomes [44]. A characteristic feature of this organism is the occurrence of frequent genetic alternations (translocations, insertions, and deletions on both the chromosomal and nucleotide levels), which are part of its strategy to adapt to environmental conditions [45]. The abovementioned circumstances, along with the alternative codon usage (CUG translated into serine rather than leucine), make it a challenging organism to study.

In the performed analysis, we identified several changes in the nuclear genome of the examined strain. During the analysis, we proved the presence of thousands of SNVs, insertions, and deletions in each of the samples, including the control. The genomic alternations presented in the results were selected after a detailed statistical analysis, taking into account the methodology of the experiment. However, there is no evidence that any of them appeared as a result of plasma treatment, and it is not simply a matter of coincidence. To gain a deeper understanding of possible plasma-induced genetic changes in *C. albicans*, phenotypic characteristics such as carbon assimilation, enzyme activity, and susceptibility to antifungal drugs should be investigated, and, if changes occur due to the plasma treatment, to try to link them to the genetic changes. The following part of this paper presents the results of such studies.

2.2. Phenotypic Changes

2.2.1. Carbon Assimilation and Hydrolytic Enzyme Activity

Tests of the ability to assimilate carbohydrates as a carbon source did not show any changes in the investigated *C. albicans* strain after repeated sublethal treatments with the use of cold atmospheric plasma. The analyzes were performed for the untreated culture as well as for cultures after 1, 7, and 12 plasma treatments. Both before and after the plasma treatment, the studied fungal cells were characterized by the ability to assimilate carbon from the same 13 compounds. As an example, Table 2 shows the API 20C AUX test results for the untreated culture and after 12 cycles of plasma treatment. A plus sign indicates that *C. albicans* grows on this carbohydrate medium.

Table 2. The ability of the tested strain to absorb carbon from 19 carbohydrates, assessed using the API 20C AUX system.

Carbohydrate	Carbohydrate Symbol	Before Plasma Treatment	After 12 × Plasma Treatment
D-glucose	GLU	+	+
glycerol	GLY	+	+
2-keto-D-gluconate	2KG	+	+
L-arabinose	ARA	−	−
D-xylose	XYL	+	+
adonitol	ADO	+	+
xylitol	XLT	+	+
D-galactose	GAL	+	+
inositol	INO	−	−
D-sorbitol	SOR	+	+
methyl-αD-glucopyranoside	MDG	+	+
N-acetyl-D-glucosamine	NAG	+	+
D-cellobiose	CEL	−	−
D-lactose	LAC	−	−
D-maltose	MAL	+	+
D-sucrose	SAC	+	+
D-trehalose	TRE	+	+
D-melesitose	MLZ	−	−
D-raffinose	RAF	−	−

The situation is different, however, in the case of the activity of hydrolytic enzymes, which is associated with the virulence of *C. albicans* [46–48] and therefore requires more careful analysis. Among 19 tested enzymes (Table 3), 10 of them showed activity in the investigated strain of *C. albicans*. In nine cases, the activity decreased in the course of successive cycles of plasma treatment, even falling to zero, while in one case, it increased. Relative average activities for successive plasma treatment cycles of these enzymes are shown in Figure 3.

Table 3. List of hydrolytic enzymes tested, using the API ZYM system and the characteristic wavelengths, at which the absorbance of the products of the enzymatic reactions was measured.

No.	Enzyme	Wavelength [nm]
e2	Alkaline phosphatase	537
e3	Esterase (C4)	537
e4	Esterase Lipase (C8)	537
e5	Lipase (C14)	537
e6	Leucine arylamidase	494
e7	Valine arylamidase	494
e8	Cystine arylamidase	494
e9	Trypsin	494
e10	α-Chymotrypsin	494
e11	Acid phosphatase	537
e12	Naphthol-AS-BI-phosphohydrolase	582
e13	α-Galactosidase	547
e14	β-Galactosidase	547
e15	β-Glucuronidase	582
e16	α-Glucosidase	537
e17	β-Glucosidase	537
e18	N-Acetyl-β-glucosaminidase	454
e19	α-Mannosidase	537
e20	β-Fucosidase	537

After 12 cycles, further screening tests were performed without plasma to determine the persistence of the observed changes. As shown in Figure 4, the activity of the enzymes decreased due to the plasma action returning to the initial state, while the activity of the enzyme No. 12 (naphthol-AS-BI-phosphohydrolase), which increased, remained unchanged. The next 12 screenings did not change the activity of enzyme No. 12, indicating a permanent change.

Based on these results, we have two problems to solve. First, what is the reason for the unstable decrease in the activity of nine of the tested enzymes, which returned to their original state after the first screening without plasma treatment? Secondly, why did enzyme No. 12 permanently change its activity?

The reversible changes in enzyme activity, observed for nine enzymes, may be due to cell-to-cell transmission of information from dying cells about danger during the plasma action [49], for example, via cytoplasmic flow [50]. It can also be caused by stress induced by changes in the environment [51], for example, the byproducts of killed cells, which are constantly present during the growth of colonies after plasma treatment. The fact that virtually all cells growing in a given colony revealed reduced enzyme activity directly after plasma treatment and while being reinoculated onto a new medium, they forgot this change and returned to the state of original activity, indicates a stress effect that is not a reflection of genetic changes. To find out more precisely what could be causing the stress, the strain was inoculated on the medium that had been treated with plasma for 12 min. No changes in enzyme activity were observed in this case. This indicates that it is not the changes in the substrate due to the action of the plasma but the products derived from killed cells that are a source of stress.

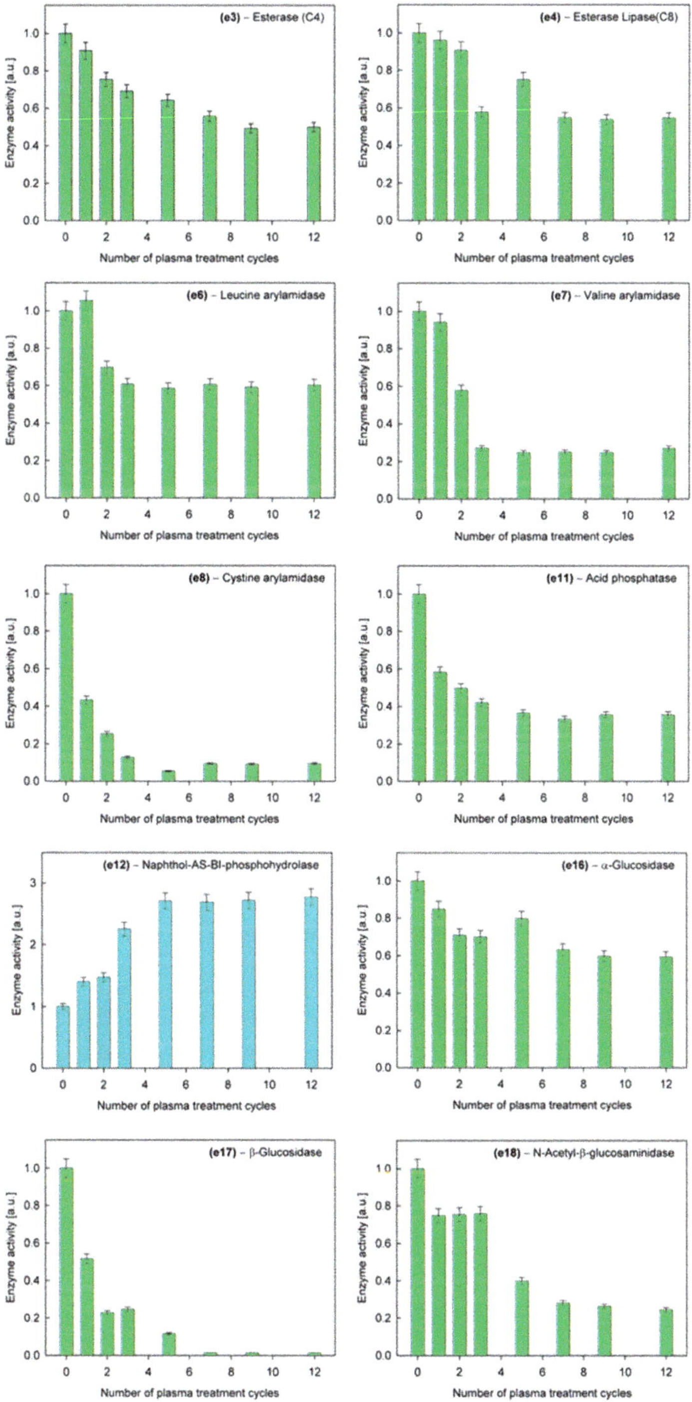

Figure 3. Activities of hydrolytic enzymes for C. albicans as a function of the number of plasma cycles.

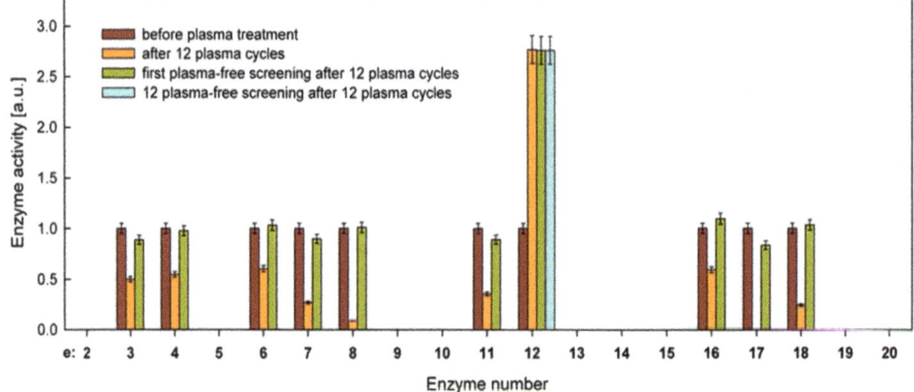

Figure 4. Summary of normalized activities of hydrolytic enzymes for *C. albicans*, before plasma treatment, after 12 plasma cycles, and after subsequent plasma-free screenings. All activity values for active enzymes before plasma treatment were normalized to 1.

On the other hand, the sustained increase in naphthol-AS-BI-phosphohydrolase activity after plasma treatment seems to have a genetic basis. Changes in the genome could be generated by both the cell-to-cell communication process during plasma operation and the stress induced by plasma products. Unfortunately, the lack of a connection in the literature between this enzyme and specific genetic sequences makes it difficult to determine the exact mechanism that caused these changes. However, it should be added that the naphthol-AS-BI-phosphohydrolase enzyme is not, so far, considered to be an important hydrolytic enzyme that affects the virulence of *C. albicans* [48,52].

2.2.2. Susceptibility to Antifungal Drugs

Susceptibility studies of *C. albicans* to antifungal drugs as a function of the number of cycles of plasma treatment did not reveal any differences in comparison to the untreated culture. Table 4 shows the MIC (minimum inhibitory concentration) values determined for the tested drugs for the untreated culture and the culture after 12 plasma treatment cycles, i.e., after the highest plasma exposure we have used for cells that survived such treatment. These results were confirmed by repeating the measurement series three times. Although the dose of plasma in each cycle was large enough to kill the vast majority of cells in its area of action (Figure 1), it did not alter drug susceptibility in this small number of surviving cells. No increase in drug resistance for cells that have survived the plasma action bodes well for the future progress of using plasma techniques in the fight against superficial candidiasis.

Table 4. Minimum inhibitory concentration (MIC) values (μg/mL) of antifungal agents for the investigated *C. albicans* strain.

Antifungal Agent	Before Plasma Treatment [MIC Value]	After 12 × Plasma Treatment [MIC Value]
Voriconazole	0.094	0.094
Fluconazole	2.0	2.0
Caspofungin	0.125	0.125
Amphotericin B	0.125	0.125
Micafungin	0.012	0.012
Anidulafungin	0.004	0.004

3. Materials and Methods

3.1. Plasma Source

As a source of nonequilibrium (cold) atmospheric plasma for the treatment of *Candida* cells, we used a linear microdischarge jet (produced in-house), sometimes called a plasma razor jet [53]. More details on the design of the plasma jet, its principle of operation, and process characteristics can be found in [39]. In this study, plasma was generated at 13.56 MHz in helium as a working gas. The helium flow rate was 1.9 L/min, while the discharge power was set at 17 W. The plasma beam had a cylindrical shape, with a length of 40 mm and a diameter of about 1.5 mm, and was aligned parallel to the treated surface at a distance of 5 mm. The experimental system is shown in Figure 5.

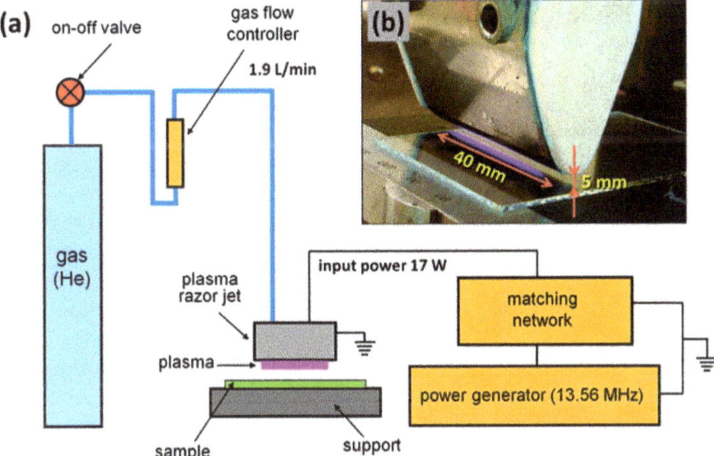

Figure 5. The experimental system: (**a**) schematic diagram of the setup; (**b**) photo of the plasma razor jet.

3.2. Microbiological and Plasma Treatment Procedures

The reference strain of the American Type Culture Collection (ATCC) *Candida albicans* ATCC® 10231 (National Collection of Pathogenic Fungi (NCPF), Salisbury, UK) was used in this study. A 24-h culture of the fungus for further studies was prepared by uniformly spreading 100 µL of phosphate buffer solution (PBS) containing 5×10^7 CFU/mL (5 units of the McFarland scale, determined with a DEN-1B type densitometer, Biosan, Riga, Latvia) onto a Petri dish with Sabouraud dextrose agar (bioMérieux, Marcy-l'Etoile, France), after which it was exposed to the plasma treatment for 1 min and then incubated at 37 °C for 24 h. *C. albicans* cells that were not killed in the plasma-induced growth inhibition zone were the protoplasts of growing spot colonies, which were the starting material for further studies. They also served as the basis for the preparation of the next culture and its plasma treatment, exactly as it was done the first time. This procedure was repeated 12 times in each measurement series. Then, subsequent screenings without plasma treatment were performed to determine the durability of occurring changes that had taken place under the influence of the plasma. As control samples, a 24-h culture of the fungus was used.

3.3. DNA Isolation and Genome Bioinformatics Analysis

The molecular genetic research was based on a comparison between the control sample ("control") and two test samples after 12 exposure periods ("12_1" and "12_2") of the selected *Candida albicans* strain. The analysis aims to identify possible genetic changes caused by cold plasma treatment.

The DNA isolation was performed with 50 mg of 24-h cultures of the *C. albicans*, before and after plasma treatment. The biological material was triturated in liquid nitrogen, in a sterile mortar, to which the CTAB extraction buffer (OPS Diagnostics LLC, Lebanon, NJ, USA), proteinase K (Promega, Madison, WI, USA), and RNase-A (Promega, Madison, WI, USA) were added. The prepared material was transferred to sterile Eppendorf tubes and incubated in a ThermoMixer C (Eppendorf, Hamburg, Germany) at 70 °C for 30 min. The tubes were centrifuged in an Eppendorf Centrifuge. The collected 200 μL of supernatant fluid was transferred to Maxwell® 16LEV Plant DNA Kit cartridges (Promega, Madison, WI, USA), and the isolation continued in a Maxwell® 16 machine (Promega, Madison, WI, USA). The amount and purity of DNA were assessed using a Nanodrop ND 2000C (ThermoFisher Scientific, Waltham, MA, USA) and a Qubit fluorimeter (ThermoFisher Scientific, Waltham, MA, USA). Sequencing libraries were prepared using the TruSeq DNA PCR-free reagent set (350 bp insert; Illumina, San Diego, CA, USA), and the type of prepared libraries was Illumina-Shotgun. Next-generation sequencing was carried out on the HiSeq X Ten platform (Illumina, San Diego, CA, USA) to generate 2 × 150 bp paired-end reads, assuming a path coverage in the flow cell for each sample of 20 %.

The quality of the obtained raw reads was checked using FastQC v0.11.4 software [54]. Then, the adapter sequences and low-quality regions of raw reads were trimmed using the Trimmomatic v0.36 tool [55], with the following operating parameters: initial and final regions quality >20, average read quality >30, minimum reads length = 90. The quality of processed reads has been confirmed by rechecking the samples with the FastQC program. The obtained reads were mapped using STAR mapper v2.7.3a [56] to the current, up-to-date version of the *C. albicans* reference genome (*C. albicans* SC5314 A22) taken from the Candida Genome Database [44], implementing the default operating parameters. To identify single nucleotide variations, as well as insertions and deletions in the nuclear sequences, the data was converted to mpileup format using SAMtools v1.10 software [57]. The variant calling and comparison between the samples was performed by VarScan v2.4.4 [58], with default parameters. The effect of selected SNVs on protein products was evaluated using IGV browser v2.7.2 [59]. The obtained data were visualized in the Perl environment with the usage of Circos v.0.69.9 [60]. The coverage data for visualization was obtained by deepTools v2.0 [61].

3.4. Carbohydrate Assimilation Test

To evaluate the changes in the biochemical properties of the strain after plasma treatment, investigations were carried out using the API 20C AUX system (bioMérieux, Marcy-l'Etoile, France). It is a biochemical identification series based on the assessment of the ability of fungi to absorb carbon (auxanogram) from 19 compounds, i.e., D-glucose (GLU), glycerol (GLY), 2-keto-D-gluconate (2KG), L-arabinose (ARA), D-xylose (XYL), adonitol (ADO), xylitol (XLT), D-galactose (GAL), inositol (INO), D-sorbitol (SOR), methyl-αD-glucopyranoside (MDG), N-acetyl-D-glucosamine (NAG), D-cellobiose (CEL), D-lactose (LAC), D-maltose (MAL), D-sucrose (SAC), D-trehalose (TRE), D-melesitose (MLZ), and D-raffinose (RAF). The API 20C AUX strip consists of 20 cupules containing the 19 dehydrated substrates and a place for the control sample, where assimilation tests are performed. A semiliquid starvation medium is introduced into the cupules. Yeast-like fungi grow if they can use a given substrate as the only carbon source. The API strips were prepared according to the manufacturer's instructions and incubated at 30 °C for 72 h, after which the carbohydrate assimilation patterns were read.

3.5. Hydrolytic Enzyme Activity

The hydrolytic enzyme activities for the investigated fungal strain were determined using an API ZYM system (bioMérieux, Marcy-l'Etoile, France). The system enables rapid determination of the activity of 19 enzymes (5 peptidases, 3 lipases, 3 phosphatases, and 8 carbohydrases) using very small amounts of unpurified samples. The list of the enzymes is presented in Table 3. Although it is a qualitative test based on visual comparison of the colors produced by enzymatic reactions, with a printed color standard to more quantitatively determine the enzyme activities, we used spectrophotometric analysis.

Based on the manufacturer's instructions, enzymatic activity measurements were carried out by suspending a given fungal sample in 2 mL of sterile distilled water with turbidity between 5 and 6 on the McFarland scale and then transferring 65 μL of this suspension to each cupule of the test strip. The strips were incubated for 4 h at 37 °C. After incubation, one drop of ZYM A reagent (trihydroxymethylaminomethane, 37% hydrochloric acid, lauryl sulfate, distilled water) and one drop of ZYM B reagent (fast blue 2 BB, methoxyethanol) were added to each cupule to stop the reaction. After 5 min of color development, the test strip was exposed to intense visible light for 10 s to eliminate a yellow tint due to an excess of unreacted fast blue 2 BB. Then, a 5-μL sample was taken from each cupule and placed in a microvolume Nano Stick-S for UV–vis spectrophotometry (PIKE Technologies, Fitchburg, WI, USA). The absorbance of each sample was measured using a spectrophotometer UV–VIS Jasco V-630 (ABL&E-JASCO Polska, Cracow, Poland) for the wavelength characteristic of the given enzyme reaction (Table 3) [62]. Two identical tests were performed at the same time for each sample, while each series of measurements was repeated three times. The maximum relative uncertainty in determining the activity of a given enzyme for a given sample is estimated to be ±5%.

3.6. Susceptibility Testing

The in-vitro activity of typical antifungal agents (voriconazole, fluconazole, amphotericin B, caspofungin, micafungin, and anidulafungin) was determined by the Etest strips (bioMérieux, Marcy- l'Etoile, France). This method consists of placing a narrow plastic strip soaked with a given agent of increasing concentration along this strip on a fungal culture. This allows us to estimate the minimum inhibitory concentration (MIC), which is a measure of drug activity. Inoculum suspensions in sterile saline (0.85 % NaCl) were prepared from primary and plasma-treated *C. albicans* cultures, with an optical density of 0.5 McFarland standard (approximately 5×10^6 CFU/mL). The suspensions were inoculated directly onto plates, with RPMI-1640 agar (bioMérieux, Marcy-l'Etoile, France) as the base medium on which the Etest strips were placed, according to the manufacturer's instructions. MIC values were read 24 h after incubation at 35 °C, accurate to scale on the Etest strips. Each series of measurements was repeated three times.

4. Conclusions

One of the problems associated with the development of plasma medicine is the risk that microorganisms surviving the plasma action may unfavorably change their characteristics; for example, their virulence or resistance to drugs may increase. The studies conducted on the *C. albicans* fungus have not confirmed these concerns. Although both genotypic and phenotypic changes were observed as a result of repeated plasma treatment, they did not have a significant effect on the virulence and drug susceptibility of the tested strain.

Among the observed changes, the alteration in enzyme activity is of particular interest. The results showed that 9 out of 19 tested enzymes reduced, some significantly, their activity after plasma treatment, which could be interpreted as a decrease in the virulence of this fungus. However, when such cells were inoculated onto a new medium devoid of plasma products, for example, killed cells, the activities of these enzymes returned to their original state. This most likely indicates the occurrence of a stress effect, which should be further investigated in more detail.

Another interesting result is the permanent increase in the activity of one of the enzymes (naphthol-AS-BI-phosphohydrolase) under plasma treatment. Multiple screening of cells on the medium, which had not been in contact with the plasma, did not change the elevated enzyme activity. This permanent shift in the activity must result from alterations in the genotype. Additionally, although we have identified some of the genetic alterations that were possibly plasma-induced in *C. albicans*, their link to the change in enzyme behavior, as well as explaining the genesis of this phenomenon, requires further research.

In summary, it should be noted that the performed studies, on the one hand, have shown changes in both the genotype and phenotype of *C. albicans* under the influence of plasma, which, however, are not dangerous in terms of virulence and drug resistance, and, on the other hand, have revealed some interesting effects related to the activity of hydrolytic enzymes.

Author Contributions: E.T.-S. contributed to manuscript writing, plasma treatment, measurements of carbon assimilation from carbohydrates, hydrolytic enzyme activity, and drug sensitivity; T.K. contributed to manuscript writing and genetic analysis; M.G. contributed to genetic analysis; B.K. was the consultant in the field of molecular genetic research; J.T. conceptualized, supervised, wrote, and edited the manuscript. All authors have read and agreed to the published version of the manuscript.

Funding: This work was financially supported by the Ministry of Science and Higher Education (MNiSW) as part of statutory research (No. 7189 and No. 7297 (2018)).

Acknowledgments: The authors would like to thank Jan Paweł Jastrzębski (Department of Plant Physiology, Genetics and Biotechnology University of Warmia and Mazury in Olsztyn, Poland) for valuable discussions and Ryszard Kapica and Justyna Markiewicz (Department of Molecular Engineering, Lodz University of Technology, Poland) for their technical assistance.

Abbreviations

C. albicans	*Candida albicans*
CFU/mL	Colony-forming unit per milliliter
CUG	Ttrinucleotide codon composed of cytosine, uracil and guanine
L/min	Liter per minute
MHz	Megahertz (frequency unit)
MIC	Minimum inhibitory concentration
SNVs	Single nucleotide variants
W	Watt (power unit)

References

1. Laroussi, M.; Kong, M.G.; Morfill, G.; Stolz, W. (Eds.) *Plasma Medicine—Applications of Low-Temperature Gas. Plasmas in Medicine and Biology*, 1st ed.; Cambridge University Press: Cambridge, UK, 2012.
2. Fridman, A.; Friedman, G. *Plasma Medicine*, 1st ed.; Wiley: Chichester, UK, 2013.
3. Shintani, H.; Sakudo, A. (Eds.) *Gas Plasma Sterilization in Microbiology—Theory, Applications, Pitfalls and New Perspectives*, 1st ed.; Caister Academic Press: Norfolk, UK, 2016.
4. Laroussi, M. Cold plasma in medicine and healthcare: The new frontier in low temperature plasma applications. *Front. Phys.* **2020**, *8*, 74. [CrossRef]
5. Dai, X.; Zhang, Z.; Zhang, J.; Ostrikov, K. Dosing: The key to precision plasma oncology. *Plasma Process. Polym.* **2020**, e1900178. [CrossRef]
6. Daeschlein, G.; Scholz, S.; Emmet, S.; von Podewils, S.; Haase, H.; von Woedtke, T.; Jüngen, M. Plasma medicine in dermatology: Basic antimicrobial efficacy testing as prerequisite to clinical plasma therapy. *Plasma Med.* **2012**, *2*, 33–69. [CrossRef]
7. Emmert, S.; Brehmer, F.; Hänßle, H.; Helmke, A.; Mertens, N.; Ahmed, R.; Simon, D.; Wandke, D.; Maus-Friedrichs, W.; Däschlein, G.; et al. Atmospheric pressure plasma in dermatology: Ulcus treatment and much more. *Clin. Plasma Med.* **2013**, *1*, 24–29. [CrossRef]
8. Liu, D.; Zhang, Y.; Xu, M.; Chen, H.; Lu, X.; Ostrikov, K. Cold atmospheric pressure plasmas in dermatology: Sources, reactive agents, and therapeutic effects. *Plasma Process. Polym.* **2020**, *17*, e1900218. [CrossRef]
9. Lloyd, G.; Friedman, G.; Jafri, S.; Schultz, G.; Fridman, A.; Harding, K. Gas plasma: Medical uses and developments in wound care. *Plasma Process. Polym.* **2010**, *7*, 194–211. [CrossRef]
10. Scholtz, V.; Pazlarova, J.; Souskova, H.; Khun, J.; Julak, J. Nonthermal plasma—A tool for decontamination and disinfection. *Biotechnol. Adv.* **2015**, *33*, 1108–1119. [CrossRef]
11. Sakudo, A.; Yagyu, Y.; Onodera, T. Disinfection and sterilization using plasma technology: Fundamentals and future perspectives for biological applications. *Int. J. Mol. Sci.* **2019**, *20*, 5216. [CrossRef]

12. Isbary, G.; Heinlin, J.; Shimizu, T.; Zimmermann, J.L.; Morfill, G.; Schmidt, H.U.; Monetti, R.; Steffes, B.; Bunk, W.; Li, Y.; et al. Successful and safe use of 2 min cold atmospheric argon plasma in chronic wounds: Results of a randomized controlled trial. *Br. J. Dermatol.* **2012**, *167*, 404–410. [CrossRef]
13. Park, G.Y.; Park, S.J.; Choi, M.Y.; Koo, I.G.; Byun, J.H.; Hong, J.W.; Sim, J.Y.; Collins, G.J.; Lee, J.K. Atmospheric-pressure plasma sources for biomedical applications. *Plasma Sources Sci. Technol.* **2012**, *21*, 043001. [CrossRef]
14. Isbary, G.; Zimmermann, J.L.; Shimizu, T.; Li, Y.F.; Morfill, G.E.; Thomas, H.M.; Steffes, B.; Heinlin, J.; Karrer, S.; Stolz, W. Non-thermal plasma—More than five years of clinical experience. *Clin. Plasma Med.* **2013**, *1*, 19–23. [CrossRef]
15. Ulrich, C.; Kluschke, F.; Patzelt, A.; Vandersee, S.; Czaika, V.A.; Richter, H.; Bob, A.; von Hutten, J.; Painsi, C.; Hügel, R.; et al. Clinical use of cold atmospheric pressure argon plasma in chronic leg ulcers: A pilot study. *J. Wound Care* **2015**, *24*, 196–203. [CrossRef]
16. He, M.; Duan, J.; Xu, J.; Ma, M.; Chai, B.; He, G.; Gan, L.; Zhang, S.; Duan, X.; Lu, X.; et al. *Candida albicans* biofilm inactivated by cold plasma treatment in vitro and in vivo. *Plasma Process. Polym.* **2020**, *17*, e1900068. [CrossRef]
17. Zhang, X.; Zhang, X.F.; Li, H.P.; Wang, L.Y.; Zhang, C.; Xing, X.H.; Bao, C.Y. Atmospheric and room temperature plasma (ARTP) as a new powerful mutagenesis tool. *App. Microbiol. Biotechnol.* **2014**, *98*, 5387–5396. [CrossRef] [PubMed]
18. Privat-Maldonado, A.; O'Connell, D.; Welch, E.; Vann, R.; van der Woude, M.W. Spatial dependence of DNA damage in bacteria due to low-temperature plasma application as assessed at the single cell level. *Sci. Rep.* **2016**, *6*, 35646. [CrossRef] [PubMed]
19. Wang, L.Y.; Huang, Z.L.; Li, G.; Zhao, H.X.; Xing, X.H.; Sun, W.T.; Li, H.P.; Gou, Z.X.; Bao, C.Y. Novel mutation breeding method for *Streptomyces avermitilis* using an atmospheric pressure glow discharge plasma. *J. Appl. Microbiol.* **2010**, *108*, 851–858. [CrossRef]
20. Mai-Prochnow, A.; Murphy, A.B.; McLean, K.M.; Kong, M.G.; Ostrikov, K. Atmospheric pressure plasmas: Infection control and bacterial responses. *Int. J. Antimicrob. Agents* **2014**, *43*, 508–517. [CrossRef]
21. Li, G.; Li, H.P.; Wang, L.Y.; Wang, S.; Zhao, H.X.; Sun, W.T.; Xing, X.H.; Bao, C.Y. Genetic effects of radio-frequency, atmospheric-pressure glow discharges with helium. *Appl. Phys. Lett.* **2008**, *92*, 221504. [CrossRef]
22. Alkawareek, M.Y.; Alshraiedeh, N.H.; Higginbotham., S.; Flynn, P.B.; Algwari, Q.T.; Gorman, S.P.; Graham, W.G.; Gilmore, B.F. Plasmid DNA damage following exposure to atmospheric pressure nonthermal plasma: Kinetics and influence of oxygen admixture. *Plasma Med.* **2014**, *4*, 211–219. [CrossRef]
23. Edengeiser, E.; Lackmann, J.W.; Bründermann, E.; Schneider, S.; Benedikt, J.; Bandow, J.E.; Havenith, M. Synergistic effects of atmospheric pressure plasma-emitted components on DNA oligomers: A Raman spectroscopic study. *J. Biophotonics* **2015**, *8*, 918–924. [CrossRef]
24. Li, H.P.; Wang, L.Y.; Li, G.; Jin, L.H.; Le, P.S.; Zhao, H.X.; Xing, X.H.; Bao, C.Y. Manipulation of lipase activity by the helium radio-frequency, atmospheric-pressure glow discharge plasma jet. *Plasma Process. Polym.* **2011**, *8*, 224–229. [CrossRef]
25. Guo, L.; Zhao, Y.; Liu, D.; Liu, Z.; Chen, C.; Xu, R.; Tian, M.; Wang, X.; Chen, H.; Kong, M.G. Cold atmospheric-pressure plasma induces DNA–protein crosslinks through protein oxidation. *Free Radic. Res.* **2018**, *52*, 783–798. [CrossRef] [PubMed]
26. Merz, W.G. *Candida albicans* strain delineation. *Clin. Microbiol. Rev.* **1990**, *3*, 321–334. [CrossRef] [PubMed]
27. Kabir, M.A.; Hussain, M.A.; Ahmed, Z. *Candida albicans*: A model organism for studying fungal pathogens. *ISRN Microbiol.* **2012**, 538694. [CrossRef] [PubMed]
28. Yapar, N. Epidemiology and risk factors for invasive candidiasis. *Ther. Clin. Risk Manag.* **2014**, *10*, 95–105. [CrossRef] [PubMed]
29. Pappas, P.G.; Kauffman, C.A.; Andes, D.R.; Clancy, C.J.; Marr, K.A.; Ostrosky-Zeichner, L.; Reboli, A.C.; Schuster, M.G.; Vazquez, J.A.; Walsh, T.J.; et al. Clinical practice guideline for the management of candidiasis: 2016 update by the infectious diseases society of America. *Clin. Infect. Dis.* **2016**, *62*, e1–e50. [CrossRef]
30. Pfaller, M.A. Antifungal drug resistance: Mechanisms, epidemiology, and consequences for treatment. *Am. J. Med.* **2012**, *125*, S3–S13. [CrossRef]

31. Pfaller, M.A.; Diekema, D.J.; Turnidge, J.D.; Castanheira, M.; Jones, R.N. Twenty years of the SENTRY antifungal surveillance program: Results for *Candida* species from 1997–2016. *Open Forum Infect. Dis.* **2019**, *6*, S79–S94. [CrossRef]

32. Roemer, T.; Krysan, D.J. Antifungal drug development: Challenges, unmet clinical needs, and new approaches. *Cold Spring Harb. Perspect. Med.* **2014**, *4*, a019703. [CrossRef]

33. Havlickova, B.; Czaika, V.A.; Friedrich, M. Epidemiological trends in skin mycoses worldwide. *Mycoses* **2008**, *51*, 2–15. [CrossRef]

34. Raiesi, O.; Siavash, M.; Mohammadi, F.; Chabavizadeh, J.; Mahaki, B.; Maherolnaghsh, M.; Dehghan, P. Frequency of cutaneous fungal infections and azole resistance of the isolates in patients with diabetes mellitus. *Adv. Biomed. Res.* **2017**, *6*, 71. [CrossRef] [PubMed]

35. Shi, X.M.; Zhang, G.J.; Yuan, Y.K.; Ma, Y.; Xu, G.M.; Yang, Y. Research on the inactivation effect of low-temperature plasma on *Candida albicans*. *IEEE Trans. Plasma Sci.* **2008**, *36*, 498–503. [CrossRef]

36. Song, Y.; Liu, D.; Ji, L.; Wang, W.; Zhao, P.; Quan, C.; Niu, J.; Zhang, X. The inactivation of resistant *Candida albicans* in a sealed package by cold atmospheric pressure plasmas. *Plasma Process. Polym.* **2012**, *9*, 17–21. [CrossRef]

37. Song, Y.; Liu, D.; Ji, L.; Wang, W.; Niu, J.; Zhang, X. Plasma inactivation of *Candida albicans* by an atmospheric cold plasma brush composed of hollow fibers. *IEEE Trans. Plasma Sci.* **2012**, *40*, 1098–1102. [CrossRef]

38. Kostov, K.G.; Borges, A.C.; Koga-Ito, Y.; Nishime, T.M.C.; Prysiazhnyi, V.; Honda, R.Y. Inactivation of *Candida albicans* by cold atmospheric pressure plasma jet. *IEEE Trans. Plasma Sci.* **2015**, *43*, 770–775. [CrossRef]

39. Tyczkowska-Sieroń, E.; Kapica, R.; Markiewicz, J.; Tyczkowski, J. Linear microdischarge jet for microbiological applications. *Plasma Med.* **2018**, *8*, 57–71. [CrossRef]

40. Rahimi-Verki, N.; Shapoorzadeh, A.; Razzaghi-Abyaneh, M.; Atyabi, S.M.; Shams-Ghahfarokhi, M.; Jahanshiri, Z.; Gholami-Shabani, M. Cold atmospheric plasma inhibits the growth of *Candida albicans* by affecting ergosterol biosynthesis and suppresses the fungal virulence factors in vitro. *Photodiagnosis Photodyn. Ther.* **2016**, *13*, 66–72. [CrossRef]

41. Borges, A.C.; Nishime, T.M.C.; Kostov, K.G.; de Morais Gouvêa Lima, G.; Gontijo, A.V.L.; de Carvalho, J.N.M.M.; Honda, R.Y.; Koga-Ito, C.Y. Cold atmospheric pressure plasma jet modulates *Candida albicans* virulence traits. *Clin. Plasma Med.* **2017**, *7/8*, 9–15. [CrossRef]

42. Yan, L.; Yang, C.; Tang, J. Disruption of the intestinal mucosal barrier in *Candida albicans* infections. *Microbiol. Res.* **2013**, *168*, 389–395. [CrossRef]

43. Zhou, Y.; Liao, M.; Zhu, C.; Hu, Y.; Tong, T.; Peng, X.; Li, M.; Feng, M.; Cheng, L.; Ren, B.; et al. *ERG3* and *ERG11* genes are critical for the pathogenesis of *Candida albicans* during the oral mucosal infection. *Int. J. Oral Sci.* **2018**, *10*, 9. [CrossRef]

44. Skrzypek, M.S.; Binkley, J.; Binkley, G.; Miyasato, S.R.; Simison, M.; Sherlock, G. The *Candida* Genome Database (CGD): Incorporation of Assembly 22, systematic identifiers and visualization of high throughput sequencing data. *Nucleic Acids Res.* **2017**, *45*, D592–D596. [CrossRef]

45. Selmecki, A.; Forche, A.; Berman, J. Genomic plasticity of the human fungal pathogen *Candida albicans*. *Eukaryot. Cell* **2010**, *9*, 991–1008. [CrossRef] [PubMed]

46. Schaller, M.; Borelli, C.; Korting, H.C.; Hube, B. Hydrolytic enzymes as virulence factors of *Candida albicans*. *Mycoses* **2005**, *48*, 365–377. [CrossRef]

47. Silva, S.; Negri, M.; Henriques, M.; Oliviera, R.; Williams, D.W.; Azeredo, J. *Candida glabrata*, *Candida parapsilosis* and *Candida tropicalis*: Biology, epidemiology, pathogenicity and antifungal resistance. *FEMS Microbiol. Rev.* **2012**, *36*, 288–305. [CrossRef] [PubMed]

48. Staniszewska, M.; Gizińska, M.; Kurządkowski, W. The role of leucine arylamidase in the virulence of Candida albicans. *Postep. Microbiol.* **2013**, *52*, 373–380.

49. Majumdar, S.; Pal, S. Information transmission in microbial and fungal communication: From classical to quantum. *J. Cell Commun. Signal.* **2018**, *12*, 491–502. [CrossRef] [PubMed]

50. Bloemendal, S.; Kück, U. Cell-to-cell communication in plants, animals, and fungi: A comparative review. *Sci. Nat.* **2013**, *100*, 3–19. [CrossRef] [PubMed]

51. Brown, A.J.P.; Budge, S.; Kaloriti, D.; Tillmann, A.; Jacobsen, M.D.; Yin, Z.; Ene, I.V.; Bohovych, I.; Sandai, D.; Kastora, S.; et al. Stress adaptation in a pathogenic fungus. *J. Exp. Biol.* **2014**, *217*, 144–155. [CrossRef]

52. Mohammadi, F.; Ghasemi, Z.; Familsatarian, B.; Salehi, E.; Sharifynia, S.; Marikani, A.; Mirzadeh, M.; Hosseini, M.A. Relationship between antifungal susceptibility profile and virulence factors in *Candida albicans* isolated from nail specimens. *Rev. Soc. Bras. Med. Trop.* **2020**, *53*, e20190214. [CrossRef]

53. Tyczkowski, J.; Kazimierski, P.; Zieliński, J. Microplasma Electrode Reactor for the Surface Treatment under Atmospheric Pressure. Patent PL 212569, 31 October 2012.

54. Babraham Bioinformatics; FastQC A Quality Control Tool for High Throughput Sequence Data. Available online: http://www.bioinformatics.babraham.ac.uk/projects/fastqc/ (accessed on 26 August 2020).

55. Bolger, A.M.; Lohse, M.; Usadel, B. Trimmomatic: A flexible trimmer for Illumina sequence data. *Bioinformatics* **2014**, *30*, 2114–2120. [CrossRef]

56. Dobin, A.; Davis, C.A.; Schlesinger, F.; Drenkow, J.; Zaleski, C.; Jha, S.; Batut, P.; Chaisson, M.; Gingeras, T.R. STAR: Ultrafast universal RNA-seq aligner. *Bioinformatics* **2013**, *29*, 15–21. [CrossRef]

57. Li, H.; Handsaker, B.; Wysoker, A.; Fennell, T.; Ruan, J.; Homer, N.; Marth, G.; Abecasis, G.; Durbin, R. The sequence alignment/map format and SAMtools. *Bioinformatics* **2009**, *25*, 2078–2079. [CrossRef]

58. Koboldt, D.C.; Chen, K.; Wylie, T.; Larson, D.E.; McLellan, M.D.; Mardis, E.R.; Weinstock, G.M.; Wilson, R.K.; Ding, L. VarScan: Variant detection in massively parallel sequencing of individual and pooled samples. *Bioinformatics* **2009**, *25*, 2283–2285. [CrossRef] [PubMed]

59. Robinson, J.T.; Thorvaldsdóttir, H.; Winckler, W.; Guttman, M.; Lander, E.S.; Getz, G.; Mesirov, J.P. Integrative genomics viewer. *Nat. Biotechnol.* **2011**, *29*, 24–26. [CrossRef]

60. Krzywinski, M.; Schein, J.; Birol, I.; Connors, J.; Gascoyne, R.; Horsman, D.; Jones, S.J.; Marra, M.A. Circos: An information aesthetic for comparative genomics. *Genome Res.* **2009**, *19*, 1639–1645. [CrossRef]

61. Ramírez, F.; Ryan, D.P.; Grüning, B.; Bhardwaj, V.; Kilpert, F.; Richter, A.S.; Heyne, S.; Dündar, F.; Manke, T. DeepTools2: A next generation web server for deep-sequencing data analysis. *Nucleic Acids Res.* **2016**, *44*, W160–W165. [CrossRef]

62. Collin, R.; Starr, M.J. Comparative ontogenetic changes in enzyme activity during embryonic development of calyptraeid gastropods. *Biol. Bull.* **2013**, *225*, 8–17. [CrossRef] [PubMed]

International Journal of
Molecular Sciences

Article

Inhibition of Angiogenesis by Treatment with Cold Atmospheric Plasma as a Promising Therapeutic Approach in Oncology

Lyubomir Haralambiev [1,2,*,†], Ole Neuffer [1,†], Andreas Nitsch [1], Nele C. Kross [1], Sander Bekeschus [3], Peter Hinz [1], Alexander Mustea [4], Axel Ekkernkamp [1,2], Denis Gümbel [1,2] and Matthias B. Stope [4]

[1] Department of Trauma, Reconstructive Surgery and Rehabilitation Medicine, University Medicine Greifswald, Ferdinand-Sauerbruch-Straße, 17475 Greifswald, Germany; oleneuffer@aol.de (O.N.); an124100@uni-greifswald.de (A.N.); nk144013@uni-greifswald.de (N.C.K.); peter.hinz@med.uni-greifswald.de (P.H.); ekkernkamp@ukb.de (A.E.); denis.guembel@uni-greifswald.de (D.G.)

[2] Department of Trauma and Orthopaedic Surgery, BG Klinikum Unfallkrankenhaus Berlin gGmbH, Warener Straße 7, 12683 Berlin, Germany

[3] ZIK plasmatis, Leibniz Institute for Plasma Science and Technology (INP), Felix-Hausdorff-Straße 2, 17489 Greifswald, Germany; sander.bekeschus@inp-greifswald.de

[4] Department of Gynecology and Gynecological Oncology, University Hospital Bonn, Venusberg-Campus 1, 53127 Bonn, Germany; alexander.mustea@ukbonn.de (A.M.); matthias.stope@ukbonn.de (M.B.S.)

* Correspondence: lyubomir.haralambiev@uni-greifswald.de; Tel.: +49-3834-8622541

† These authors contributed equally to this work.

Received: 15 September 2020; Accepted: 24 September 2020; Published: 26 September 2020

Abstract: Background: Cold atmospheric plasma (CAP) is increasingly used in the field of oncology. Many of the mechanisms of action of CAP, such as inhibiting proliferation, DNA breakage, or the destruction of cell membrane integrity, have been investigated in many different types of tumors. In this regard, data are available from both in vivo and in vitro studies. Not only the direct treatment of a tumor but also the influence on its blood supply play a decisive role in the success of the therapy and the patient's further prognosis. Whether the CAP influences this process is unknown, and the first indications in this regard are addressed in this study. Methods: Two different devices, kINPen and MiniJet, were used as CAP sources. Human endothelial cell line HDMEC were treated directly and indirectly with CAP, and growth kinetics were performed. To indicate apoptotic processes, caspase-3/7 assay and TUNEL assay were used. The influence of CAP on cellular metabolism was examined using the MTT and glucose assay. After CAP exposure, tube formation assay was performed to examine the capillary tube formation abilities of HDMEC and their migration was messured in separate assays. To investigate in a possible mutagenic effect of CAP treatment, a hypoxanthine-guanine-phosphoribosyl-transferase assay with non malignant cell (CCL-93) line was performed. Results: The direct CAP treatment of the HDMEC showed a robust growth-inhibiting effect, but the indirect one did not. The MMT assay showed an apparent reduction in cell metabolism in the first 24 h after CAP treatment, which appeared to normalize 48 h and 72 h after CAP application. These results were also confirmed by the glucose assay. The caspase 3/7 assay and TUNEL assay showed a significant increase in apoptotic processes in the HDMEC after CAP treatment. These results were independent of the CAP device. Both the migration and tube formation of HDMEC were significant inhibited after CAP-treatment. No malignant effects could be demonstrated by the CAP treatment on a non-malignant cell line.

Keywords: apoptosis; cell migration; cold atmospheric plasma; endothelial cells VEGF

---

1. Introduction

Cancer is one of the most common diseases worldwide [1] and represents a major challenge for the healthcare system [2]. The treatment costs caused by malignant neoplasms in Germany rose from 11.7 billion euros in 2002 to 19.9 billion euros in 2015, illustrating the growing social importance of cancer [3].

The differentiation and spread of malignant tumors at the time of diagnosis is crucial for the treatment so that knowledge of the tumor extent is essential for the prognosis of cancer [4]. Cancer can often be treated by the surgical removal of the tumor, chemotherapy, radiation, or a combination of these treatment options [5]. However, curative therapy is often not possible for many types of cancers due to a lack of surgical resectability, tumor remission, or possible metastases. Chemotherapeutic agents are usually not specific for tumor cells, which is why severe adverse drug reactions often occur [5,6]. In addition, even if the initial response to chemotherapy is good, development of resistance plays an important role in therapy failure and tumor recurrence [7].

Regardless of the type of cancer, cancer cells of solid tumors lose their organ-specific functionality and destruct the surrounding healthy tissue through uncontrolled and invasive growth [8]. Tumor cells can invade surrounding blood and lymphatic vessels spread throughout the body, and form metastases in other organs [9]. Therefore, tumor-associated angiogenesis is a critical issue in anti-oncological therapy. This complex process is controlled by soluble angiogenic factors, such as VEGF (vascular endothelial growth factor), bFGF (basic fibroblast growth factor), PDGF (platelet derivated growth factor), angiopoietin, prostaglandins and other proangiogenetic factors [10,11]. Their production by tumor cells is primarily induced in hypoxia [12,13]. When VEGF-A, in particular, binds to the tyrosine kinase receptors KDR (kinase insert domain receptor), and Flt-1 (fms related tyrosine kinase 1) in endothelial cells, these become activated [14,15], and endothelial surface proteins, such as PECAM 1 (platelet endothelial adhesion molecule 1), and VE-Cadherin (vascular endothelial cadherin) are induced [16–18], as well as VEGF [19]. As a result, endothelial cells gain mobility, can overcome physical barriers and can migrate into the interstitium [20]. Elevated concentrations of VEGF also lead to a local increase in blood vessel permeability [15]. The release of angiogenic factors exerts a chemotactic effect on endothelial cells so that the sprouting of the cells occurs specifically along the VEGF gradient [21,22]. Such vascular networks are characteristic of tumor angiogenesis, which is primarily mediated by VEGF secreted by cancer cells [19,23–25]. As a result, intense vascularization is a characteristic of aggressively growing tumors and facilitates the formation of metastases [26]. Antibodies directed against VEGF have been used in cancer therapy for several years to inhibit tumor progression and subsequent metastasis [24,27]. VEGF-specific antibodies, such as bevacizumab, can reduce vascularization of tumors; however, a variety of side effects and the development of resistance to these antibodies limit their use. In order to improve and expand the existing treatment options in oncology, the application of cold atmospheric plasma (CAP) is becoming increasingly important.

CAP is a highly reactive gas mixture consisting of ions, neutral particles, such as atoms and molecules, radicals, free electrons, photons, and electromagnetic radiation [28,29]. The medical application of CAP has already been established in wound therapy, the sterilization of medical implants and devices, and the disinfection of air [30,31]. Another promising application of CAP is its use in the field of oncology, since an anti-proliferative effect of CAP on cancer cells of various cancers has already been demonstrated [32–35]. It is believed that, among other effects, the induction of apoptosis is the main reason for the anti-proliferative effect of CAP on cancer cells [34]. Different components of CAP, such as oxygen and nitrogen species—e.g., hydrogen peroxide—are responsible for the induction of programmed cell death [36–38].

The influence of CAP on endothelial cell proliferation and thus on tumor-associated angiogenesis is underexplored and hence investigated in this study.

2. Results

Human Dermal Microvascular Endothelial Cells (HDMEC) were directly treated with cold atmospheric plasma (CAP) and compared to cells treated with argon carrier gas only (mock control). Direct treatment of HDMEC was performed with the two different CAP devices: kINPen (neoplas tools, Greifswald, Germany) and Mini Jet (Heuermann HF-Technik GmbH). Anti-proliferative effects were observed with both CAP devices, which led to statistically significant growth inhibition from an incubation period of 48 h (Figure 1A,B). Growth inhibitory effects were treatment time-dependent (data not shown). For both CAP devices, a treatment time was determined, in which the cell number of CAP-treated cells was reduced to approximately half that of the control group treated with argon. In the case of kINPen, this was achieved by 15 s treatment time. After 120 h, the cell number of CAP-treated cells was 50.3 ± 6.0% compared to the control group. Treatment time had to be increased for the Mini Jet to obtain an anti-proliferative effect comparable to the treatment with kINPen. This was achieved by a 30 s treatment time. After 120 h, the cell number of CAP-treated cells showed 49.8 ± 7.8% compared to the control group. These treatment times were applied in all subsequent experiments.

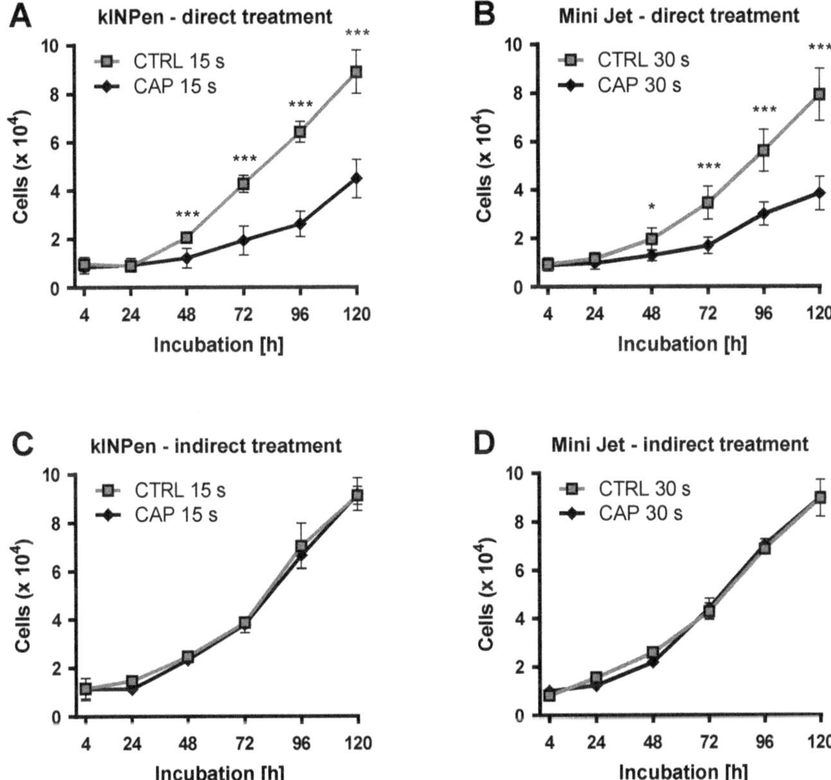

Figure 1. Curbing the HDMEC growth after direct and indirect CAP-treatment. HDMEC were treated with cold atmospheric plasma (CAP) generated by a kINPen (**A**) or Mini Jet (**B**) device. HDMEC were treated with CAP-treated cell culture media by kINPen (**C**) and by Mini Jet (**D**). As control (CTRL) cells were treated with the carrier gas argon. The number of viable cells was examined at the indicated times. Four independent experiments were performed. Results are presented as means ± SD. Two-way RM ANOVA was performed. Student's *t*-test was used with the following significance levels. The results of the Sidak's post-hoc tests were indicated as follows: $p < 0.05$ (*), $p < 0.001$ (***).

The anti-proliferative effect of CAP treatment on cells could possibly be influenced by components of the cell culture medium. In particular, chemical modifications of these components—e.g., amino acids—could influence cellular metabolism. To investigate this indirect influence, untreated HDMEC were incubated with CAP- and argon-treated cell culture media for 120 h. The number of viable cells was examined 4, 24, 48, 72, 96, and 120 h after treatment. For both CAP devices, no statistically significant difference in growth between the indirectly treated HDMEC compared to control cells was found (Figure 1C,D).

After inhibition of HDMEC proliferation after CAP-treatment had been demonstrated, the influence of CAP on cellular metabolism was examined using MTT assay. This is an established assay to assess the general metabolic activity of viable cells. HDMEC were treated with CAP or argon gas and analyzed by MTT assay after 4, 24, 48, and 72 h of incubation. For the first two measurement points, a statistically significant reduction in cell metabolism was detected after treatment with both CAP devices (Figure 2). This reduction was more pronounced after CAP treatment with the kINPen than after treatment with the Mini Jet. After 4 h, metabolic activities in CAP-treated HDMEC were reduced to 72.7% ± 9.0% ($p = 0.041$) (kINPen) and 80.2% ± 7.4% ($p < 0.001$) (Mini Jet) compared to the control. After 24 h, reductions to 70.2% ± 8.2% ($p = 0.002$) (kINPen) and 89.7% ± 6.1% ($p = 0.040$) (Mini Jet) were obtained. The inhibitory effect of CAP treatment was not statistically significant after longer incubation periods of 48 and 72 h.

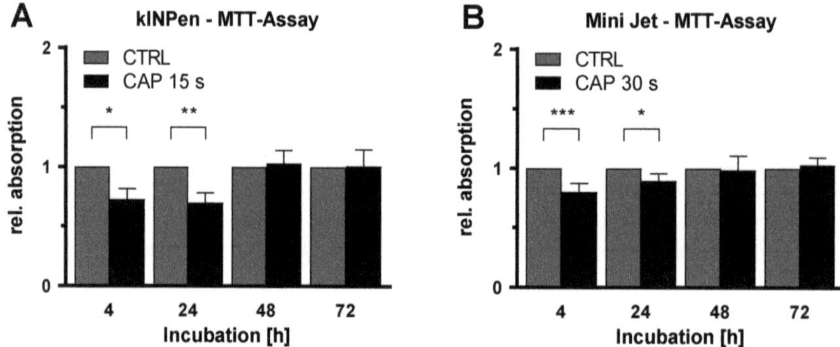

Figure 2. Inhibition of metabolism of endothelial cells after CAP-treatment. HDMEC were treated with cold atmospheric plasma (CAP) generated by a kINPen (**A**) or Mini Jet (**B**) device and incubated up to 72 h. MTT assays were performed at the indicated time points. Absorption was measured and normalized to the number of viable cells. Data were normalized to the argon-treated control cells and are presented as mean ± SD. At least 5 independent experiments, each with three replicates, was performed. Two-way RM ANOVA was used before normalization. The results of the Sidak's post-hoc tests are indicated as follows: $p < 0.05$ (*), $p < 0.01$ (**), $p < 0.001$ (***).

The cell glucose uptake after CAP treatment was measured to further investigate the effects of CAP treatment on cell metabolism. For this purpose, the glucose concentration in the culture medium was determined 4, 24, 48 and 72 h after the treatment. The measured concentration was normalized to the number of living cells. The glucose concentration of the cell culture medium per cell differed significantly ($F_{(1, 10)} = 9.897$) between CAP and the control (Figure 3). Then, 48 and 72 h after treatment the glucose concentration per cell was significantly (48 h: $p = 0.009$, 72 h: $p = 0.007$) higher in the CAP treated group (48 h: CAP: 0.1634 ± 0.0167, CTRL: 0.1206 ± 0.0168, 72 h: CAP: 0.1405 ± 0.0257, CTRL: 0.0783 ± 0.0116).

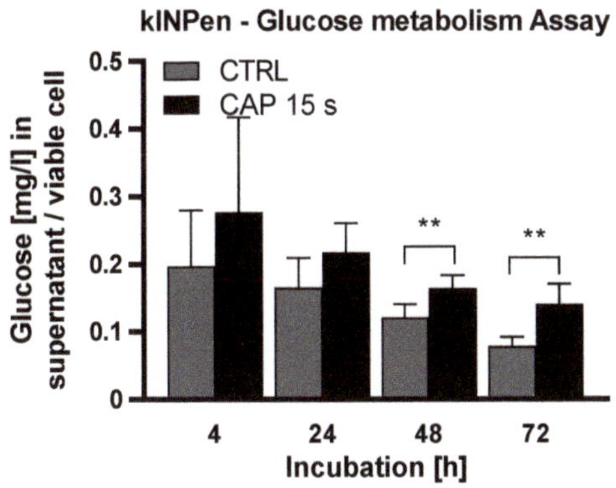

kINPen - Glucose metabolism Assay

Figure 3. CAP treatment reduces glucose metabolism in HDMEC. HDMEC were treated with cold atmospheric plasma (CAP) generated by a kINPen and incubated up to 72 h. The glucose concentration in the culture medium was measured at the indicated time points and normalized to the number of viable cells. Six independent experiments were performed, each with three replicates. The experiment was statistically evaluated with a two-way RM ANOVA. The results of the Sidak's post-hoc tests are indicated as follows: $p < 0.01$ (**).

To verify whether the anti-proliferative effect of CAP treatment was based on the induction of apoptotic processes, caspase-3/caspase-7 activity assays (caspase-3/7 assay) were performed. HDMEC were treated with CAP or carrier gas argon as a control and incubated for 24 h, 48 h, and 72 h. At each time, CAP-treated HDMEC demonstrated a significantly increased caspase-3 and caspase-7 signal, with lower apoptosis induction after Mini Jet treatment than after kINPen treatment (Figure 4A,B). The activity of both caspases increased with prolonged incubation time. After 72 h, the signal strengths of activated caspase-3 and caspase-7 in CAP-treated HDMEC were 2.5 fold ($p < 0.001$) (kINPen) and 2.3 fold ($p < 0.001$) (Mini Jet) compared to control.

TUNEL assays confirmed the findings obtained by the caspase-3/7 assays. HDMEC were treated with CAP or argon gas and incubated for 24, 48, and 72 h. For treatment with both CAP devices, an increased signal of DNA fragmentation was observed (Figure 4C,D). The TUNEL signals increased 1.4 fold ($p = 0.006$) (kINPen) and 1.4 fold ($p < 0.001$) (Mini Jet) 72 h after CAP-treatment.

The migration of endothelial cells into surrounding tissues is a key event of angiogenesis and is a directed process that can be stimulated primarily by the angiogenic factor VEGF. After it was shown that CAP-treatment inhibits the proliferation of HDMEC, the influence of CAP on the migration behavior of HDMEC was also investigated. For this purpose, a migration assay was established in which HDMEC migrated through the membrane of a FluoroBlok Transwell insert (Corning, New York, NY, USA; pore size 8 µm). The number of migrated cells was determined by fluorescence microscopy. In the presence of VEGF, the migration of HDMEC was statistically significantly increased. The number of migrated HDMEC increased 9.5-fold ($p < 0.001$) after an incubation of 6 h (VEGF–negative: 449 ± 157, VEGF–positive: 4250 ± 495 (Figure 5A).

Following this, the effect of CAP treatment on the migration behavior of HDMEC was examined in this setup. For this purpose, the HDMEC were treated with CAP or argon gas (control) before the migration assay. To stimulate the migration of HDMEC, VEGF was present in both approaches. After an incubation period of 6 h, a statistically significant migration inhibition in CAP-treated HDMEC was detected (Figure 5B). The number of migrated CAP-treated HDMEC after 6 h was 3.2-fold lower than the control cells ($p < 0.001$) (CAP: 1287 ± 286, Control: 4129 ± 533).

To investigate the influence of CAP on the ability of HDMEC to form tubes in Matrigel, tube formation assays was performed. The total tube length after an incubation period of 6 h was analyzed. The total tube length of CAP treated HDMEC was significantly shorter than the control treated cells ($p = 0.005$, CTRL: 151.1 ± 48.5, CAP: 97.9 ± 42.2) (Figure 6).

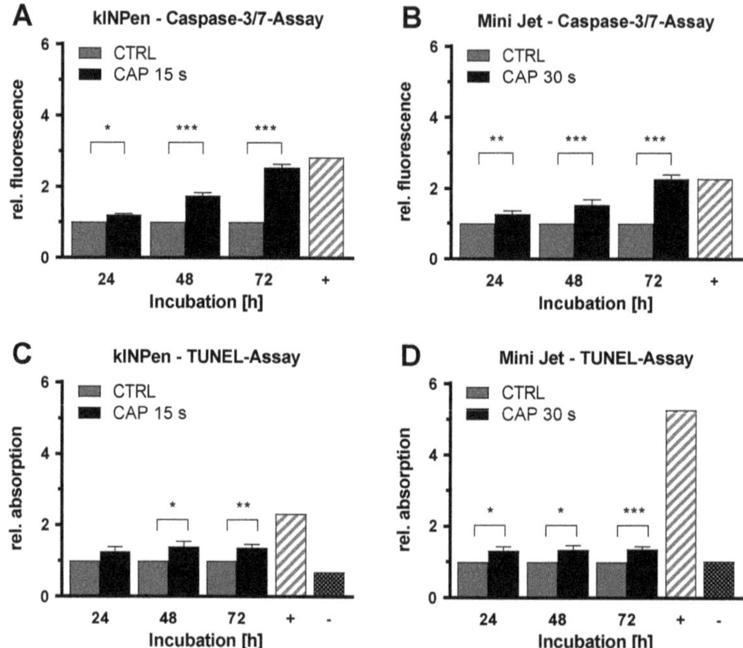

Figure 4. CAP treatment increases the apoptotic signals independent of CAP device. Endothelial cells were treated with cold atmospheric plasma (CAP) or argon carrier gas as control (CTRL), released by the kINPen (**A,C**) or Mini Jet (**B,D**) device. Caspase-3/7 and TUNEL assays were performed at the indicated times. Positive controls (+) were carried out by treatment with Cycloheximide (**A,B**) or nuclease treatment (**C,D**). For TUNEL assay, a negative control (-) without labeling was performed. Five independent experiments were performed. The results were normalized to control cells and are presented as mean ± SD. All experiments were statistically evaluated with two-way RM ANOVA with Sidak's post hoc tests before normalization. The following significance levels were defined: $p < 0.05$ (*), $p < 0.01$ (**), $p < 0.001$ (***).

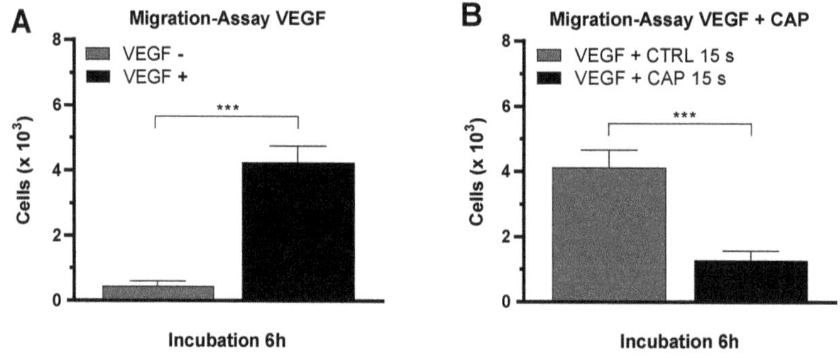

Figure 5. CAP inhibits the migration of endothelial cells. HDMEC were seeded in cell culture inserts with 8 μm pores. A VEGF gradient was established over the membrane. The number of migrated cells was determined after 6 h of incubation (**A**). HDMEC were treated with CAP or argon carrier gas before performing the migration assay with VEGF (**B**). Five experiments were performed. The results are presented as means ± SD. All experiments were statistically evaluated with the Student's *t*-test: $p < 0.001$ (***).

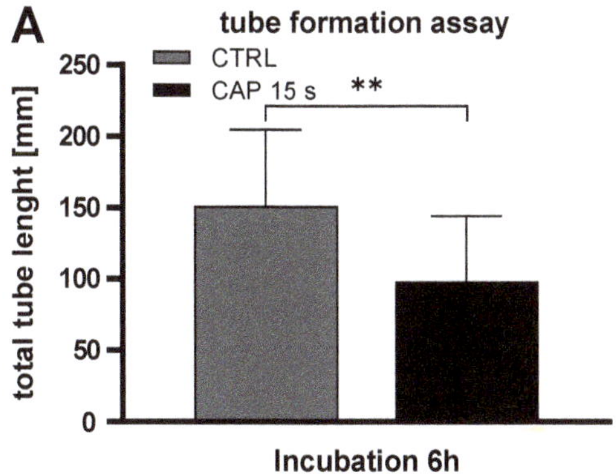

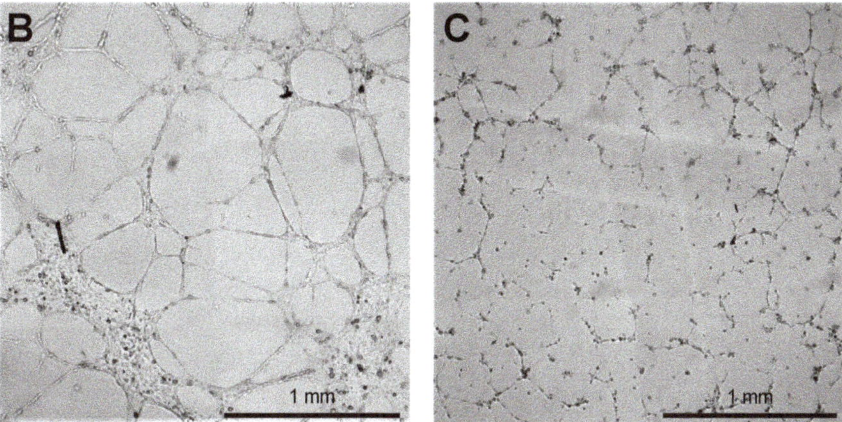

Figure 6. CAP inhibits the tube formation of HDMEC. HDMEC were treated with CAP or carrier gas Argon and seeded in µ-slides pre-coated with Matrigel/Medium. After an incubation period of 6 h, pictures were captured and the total tube length were analyzed with imageJ. The total tube length of the control treated HDMEC was significantly longer than the CAP treated cells (**A**). Six independent experiments were performed. Data are given as means ± SD. The data were statistically evaluated with the paired *t*-test: $p < 0.01$ (**). (**B**,**C**) show representative capture of control-(**B**) or CAP-(**C**) treated cells.

To examine a possible mutagenic effect of the CAP treatment, a hypoxanthine-guanine-phosphoribosyl-transferase (HPRT) assay was carried out [39]. For this purpose, CCL-93 cells were treated with CAP or the carrier gas. Growth kinetics were carried out over 72 h. This showed a significant reduction in the number of cells after CAP treatment after 48 ($p < 0.001$) and 72 h ($p < 0.001$). After, treatment cells were incubated in medium containing 6-thioguanine (TG). TG should inhibit cell growth in non-mutated cells (Figure 7).

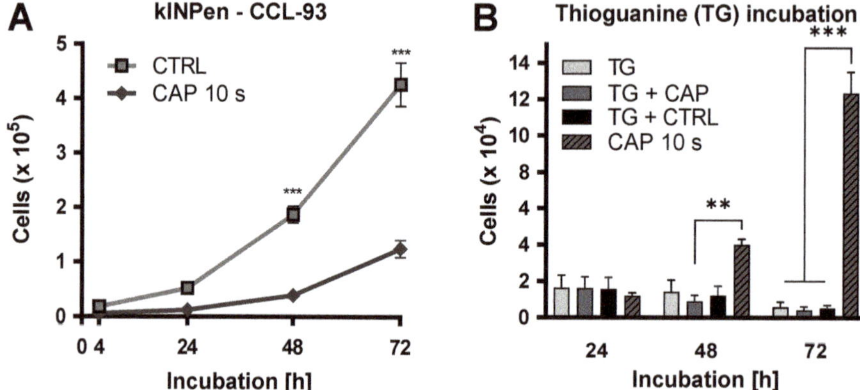

Figure 7. CAP treatment does not lead to mutagenicity. CCL-93 cells were treated with CAP or carrier gas. Growth kinetics were performed over 72 h (**A**). CAP- and argon-treated cells were propagated in medium containing 6-thioguanine (TG) (**B**). Number of viable cells was measured after 24, 48, and 72 h. Untreated cells propagated in medium containing TG and CAP treated cells propagated in medium without TG were used as controls. Three independent experiments were performed. Data are given as means ± SD. All experiments were statistically evaluated with two-way RM ANOVA with Sidak's post hoc tests. The following significance levels were defined: $p < 0.01$ (**), $p < 0.001$ (***).

3. Discussion

A high degree of vascularization is required to satisfy the high oxygen and nutrient requirement of tumor cells and is the most critical prerequisite for rapid and aggressive tumor growth [26]. Therefore, tumor-associated angiogenesis is a major factor in the initiation and progression of malignant tumors [25]. Tumor cells can induce angiogenesis by secreting cytokines, such as VEGF, thereby stimulating endothelial cells to proliferate and migrate [25,26,40,41]. This confirms the cell culture model established in the present study and demonstrated that the presence of VEGF significantly increased the invasiveness of HDMEC.

In addition to radiation, chemotherapy and surgical debridement, the treatment of malignant tumors can be complemented by the inhibition of tumor-induced angiogenesis [26,42]. Several drugs are currently available to suppress angiogenesis, such as the VEGF-specific antibody, bevacizumab. However, bevacizumab therapy can cause severe systemic side effects or development of resistance, which may restrict its application [43–46]. Against this background, CAP treatment can be considered a new alternative in anti-oncological therapy. It has been shown that CAP treatment can inhibit both the proliferation and the cell mobility of in vitro propagated tumor cells [35,47–49].

CAP treatment of HDMEC inhibited their migration, depending on the duration of treatment and the CAP device. The inhibitory effects of the CAP treatment also have a direct influence on the tube formations of HDMEC. This suggests that CAP treatment can inhibit not only the tumor cells directly but also tumor-associated cells, such as endothelial cells. One such example is the inhibitory effect of CAP on fibroblasts [50,51]. These effects were also shown in the current study. In contrast to systemic therapies, CAP only has a locally limited effect. Other studies also indicate that CAP treatment has fewer systemic side effects than conventional therapies [52–54]. Another favorable property of CAP is that no resistance to CAP treatment has been observed in previous studies [52,55]. Over the entire study, no reduced CAP efficacy could be detected in the present study. Indirectly mediated anti-proliferative CAP effects have not been demonstrated. A possible explanation can be explained as the inclusion of ascorbic acid in the cell medium. This is known to reduce reactive oxygen species (ROS) and thus prevents apoptosis [56] and thus can weaken the CAP effects [51]. In addition to the migration-inhibiting effect of CAP on endothelial cells, an anti-proliferative effect on human microvascular endothelial cells was also demonstrated for the first time for both CAP devices. The selection of the appropriate treatment time is crucial, since short treatment times have hardly any anti-proliferative effects, but a sharply increased CAP treatment can damage adjacent tissue [57]. Similar to the use of pharmacological compounds, a dose–response relationship should, therefore, be defined for CAP treatment. It is known that the anti-proliferative effect of both CAP sources depends on the treatment time [34]. However, device-specific parameters are also conceivable with which a dose–response relationship may be defined, such as the voltage between the electrodes and the flow rate of the carrier gas [58]. These findings help to understand why different treatment times with both CAP devices were required to achieve an almost identical anti-proliferative effect. With the Mini Jet, the gas flow rate (1.5 L/min) was only half as high as with the kINPen (3 L/min), while the treatment time required to achieve a comparable effect was twice as long. The carrier gas and the CAP device itself represent other critical parameters on which the composition of the cold atmospheric plasma depends [55]. CAP includes charged particles and radicals as well as electromagnetic radiation [59,60]. Due to this heterogeneity of CAP composition, treatment parameters for each CAP source must be defined individually for each cell type and in clinical practice for each tumor type [34].

The anti-proliferative effect of CAP in tumor cells is mainly due to the increase in intracellular concentrations of ROS and nitrogen species (RNS) with corresponding cell biological consequences [61]. The glucose assay showed inhibition of the glucose metabolism of the HDMEC after CAP treatment, since the consumption of glucose by the CAP-treated cells was significantly suppressed up to 72 h after the treatment. Additionally, the MTT assay demonstrated that the metabolic activity [62,63] of endothelial cells was reduced up to 24 h after CAP treatment. However, the metabolism of HDMEC does not appear to be permanently restricted, as the surviving HDMEC were able to regenerate and showed a metabolic activity comparable to that of the control group after 48 h. In other endothelial cells (HUVEC), the restoration of physiological cell metabolism after CAP treatment has already been shown [64]. Comparable metabolic CAP effects have also been described in squamous epithelium cells and fibroblasts [65,66]. This demonstrated regeneration of cell metabolism could not be observed in the tumor cells examined [58,66–68]. The different effects of CAP on the metabolic activity of different cell types may be explained by cell type-specific redox-protective mechanisms. In contrast to non-malignant cells, tumor cells are believed to be highly susceptible to oxidative stress [58,65,66]. It is known that oxidative stress triggers apoptosis in epithelial cells [69]. Although CAP-induced apoptosis has been demonstrated in various tumor entities [34,35,70,71], the effect in endothelial cells is still largely unexplored. Analysis of apoptotic processes revealed both increased DNA fragmentation and increased activity of caspase 3 and caspase 7 in HDMEC directly treated with CAP. Indirectly mediated anti-proliferative CAP effects were not observed in HDMEC.

Discussing the use of CAP for medical treatment, the question of the safety of this procedure always arises. This is particularly relevant for a possible intraoperative use of CAP, since non-malignant areas of the surrounding tissue can also be affected by the CAP treatment.

The reports by other authors about the impact of CAP on normal, non-malignant cells show no undesirable effects to date [72]. The control tests of this study with a non-malignant cell line (CCL-93) also showed that the CAP treatment does not have any mutagenic effects on normal cells [73].

4. Materials and Methods

4.1. Cell Culture

Human endothelial cell line HDMEC (Promocell, Heidelberg, Germany) were propagated in a humidified atmosphere at 5% CO_2 and 37 °C in ECGM-MV medium (Endothelial Cell Growth Medium MV 2; Promocell, Heidelberg, Germany) with the following medium supplements: Fetal Calf Serum 0.05 mL/mL; Epidermal Growth Factor 5 ng/mL, Basic Fibroblast Growth Factor 10 ng/mL; Insulin-like Growth Factor (Long R3 IGF) 20 ng/mL; Vascular Endothelial Growth Factor 165 0.5 ng/mL;

Ascorbic Acid 1 µg/mL; Hydrocortisone 0.2 µg/mL. Further, 1% *v/v* penicillin/streptomycin (PAN Biotech, Aidenbach, Germany) was then added to this medium.

CCL-93 cells (ATCC, Manassas, VA, USA), the non-malignant fibroblast cells of cricetulus griseus, were propagated in a humidified atmosphere at 5% CO_2 and 37 °C in DMEM containing 4.5 g/L glucose with 10% *v/v* FCS and 1% *v/v* penicillin/streptomycin (all from PAN Biotech, Aidenbach, Germany).

4.2. Proliferation Assay after CAP Exposure

Cell growth was determined after 4, 24, 48, 72, 96, and 120 h using a CASY cell counter and analyzer model TT (Roche Applied Science, Mannheim, Germany) with a 150 µm capillary. For this purpose, 1×10^4 endothelial cells were suspended in 200 µL culture media and treated with CAP or carrier gas argon (control group) for 10 s (kINPen) or 30 s (Mini Jet). After treatment, the cell suspension was transferred to another 24-Well cell culture plate. The treatment well was rinsed with 200 µL fresh media, which was also transferred to the culture plate. Then, 800 µL fresh media was added, and cells were incubated in a humidified atmosphere at 5% CO_2 and 37 °C. Cell count was determined by suspending the cells by trypsin/EDTA (Ethylenediaminetetraacetic acid) treatment and diluting 100 µL cell suspension in 10,000 µL CASYton (Roche Applied Science, Mannheim, Germany). The measurement was performed three times with 400 µL each of this dilution and was performed in triplicate. To discriminate between cell debris, dead cells, and living cells, gates of 5.88 µm/11.13 µm were used.

4.3. Proliferation Assay after Indirect Exposure

1×10^4 endothelial cells were suspended in 800 µL culture media and transferred to the wells of a 24 well plate. Then, 200 µL culture medium was treated with CAP or carrier gas argon for 15 s (kINPen) or 30 s (MiniJet) in a separate 24-well plate and added to the cell suspension. Cells counts were performed at 4, 24, 48, 72, 96, and 120 h after the indirect CAP-exposure, as described in proliferation assay after direct CAP treatment.

4.4. MTT-Assay

2.0×10^4 (4 h), 1.5×10^4 (24 h), $1.0 \times\times 10^4$ (48 h) and 6.5×10^3 (72 h) cells were treated with CAP or carrier gas argon by kINPen (15 s) or Mini Jet (30 s) and were incubated for 4, 24, 48, and 72 h. To be able to normalize the metabolized MTT to the cell number, a second plate was planted out parallel. For performing the MTT-Assay, 20 µL MTT-Solution (MTT (3-(4,5-dimethylthiazol-2-yl)-2,5-diphenyltetrazolium bromide; Carl Roth, Karlsruhe, Germany) in a. bidest (5mg/mL) was added and incubated over 3 h. After incubation, the culture media were removed, and 120 µL lysis reagent (DMSO (Carl Roth, Karlsruhe, Germany) with 2.5% w/w SDS (Sodium Dodecyl Sulfate) (Carl Roth, Karlsruhe, Germany) and 150 µL 37% HCl (Carl Roth, Karlsruhe, Germany) were added and incubated over 10 min. Absorption at 565 nm was measured using an Infinite M200 plate reader (Tecan, Männedorf, Switzerland).

4.5. Glucose Metabolism-Assay

2.0×10^4 cells were treated with 15 s CAP or carrier gas argon by kINPen and were incubated for 4, 24, 48, and 72 h. The glucose concentration of the cell-free supernatant was measured with a glucose detection kit (r-biopharm, Darmstadt, Germany) after the incubation. The number of viable cells was examined with CASY cell counter and analyzer. The glucose concentration was normalized to the cell count.

4.6. Caspase-3/7-Assay

1.5×10^4 (24 h), 1.0×10^4 (48 h) and 6.5×10^3 (72 h) cells were treated with CAP or argon for 15 s (kINPen) and 30 s (Mini Jet). As positive control cells were treated with cycloheximide (15 µM in cell culture medium; Carl Roth, Karlsruhe, Germany). To be able to normalize the measured fluorescence intensity to the cell number, a second plate was planted out parallel. After the incubation period, the used medium was removed, and 100 µL of Caspase 3/7 detection solution (CellEvent[TM] Caspase 3/7 Green Detection Reagent (Thermo Fisher Scientific, Waltham, MA, USA) and 10 µM in DPBS with 5% FCS *v/v*) were incubated for 45 min. The fluorescence (535 nm) was measured after excitation (495 nm) using a plate reader.

4.7. TUNEL-Assay

1.5×10^4 (24 h), 1.0×10^4 (48 h) and 6.5×10^3 (72 h) cells were treated with CAP or argon for 15 s (kINPen) and 30 s (Mini Jet). Additionally, negative controls lacking fluorescent labeling and a nuclease-treated positive control were included. To be able to normalize the measured absorption to the cell number later, a second cell culture plate was simultaneously performed. The TiterTACS™ Colorimetric Apoptosis Detection Kit (Trevigen, Gaithersburg, MD, USA) was used according to the manufacturer's instructions. Absorption was measured using the Infinite M200 plate reader (Tecan, Männedorf, Switzerland).

4.8. Migration-Assay

FluoroBlok Transwell inserts (Corning, New York, NY, USA; pore size 8 µm) were inserted into the wells of a 24-well cell culture plate. For the assay, 5.0×10^4 cells were suspended in 200 µL culture media. The wells of the cell culture plate were filled with medium with or without VEGF (26.6 ng/mL). After 6 h of incubation (37 °C, 5% CO_2), the insets were removed, washed twice with DPBS, and fixated with 100% Methanol. Cells were stained with DAPI (1 µg/mL) over 15 min at room temperature. The bottom membrane was cut out and transferred to microscope slides. The cover slips were mounted with 33% Glycerol. Edges were sealed with transparent nail varnish. With the BZ-II Analyzer software, the number of cells was determined.

To investigate the influence of CAP on migration, cell suspensions were treated with CAP or argon carrier gas released by the kINPen over 15 s before seeding. The subsequent assay procedure remained unchanged. The assay setup is shown in Figure 8.

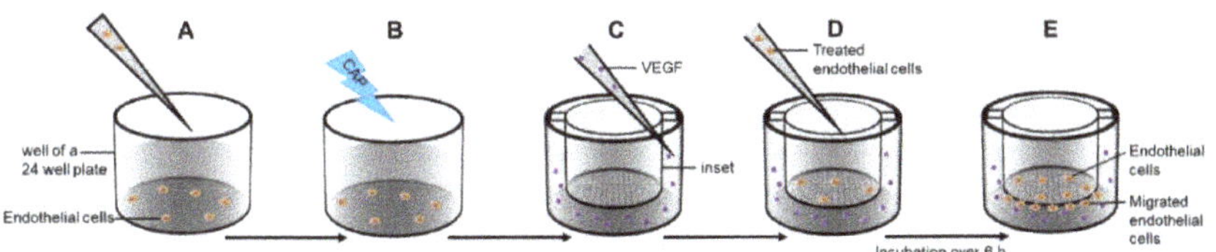

Figure 8. Migration assay. Endothelial cells were transferred in a well of a 24-well plate (**A**) and treated with cold atmospheric plasma (CAP) or argon carrier gas (**B**). Cell culture inserts were placed in another 24 well plates and a VEGF gradient was established over the membrane (**C**). Treated cells were transferred into the inset (**D**) and incubated over 6 h. Cells migrated through the membrane pores (**F**)

4.9. Tube Formation-Assay

µ-Slides (ibidi, Gräfelfing, Germany) were coated with 10 µL Matrigel/Medium (5 mg/mL Matrigel (Corning, New York, NY, USA)/RPMI 1640 (PAN Biotech, Aidenbach, Germany) 1:1, *v/v*) per Well. In total, 4.0×10^4 HDMEC were suspended in 200 µL medium and were treated with 15 s CAP or carrier gas argon only (as control treatment). Then, 50 µL cell suspension was transferred into the wells of the µ slides in triplicates. After an incubation period of 6 h at 37 C and 5% CO_2 pictures were captured with Observer Z1 (Zeiss, Jena, Germany) with the software Zen 2012 pro (blue version, Zeiss, Jena, Germany). We analyzed the total tube length with imageJ (version 1.53e, NIH, Bethesda, MD, USA).

4.10. Hypoxanthine-Guanine-Phosphoribosyl-Transferase (HPRT) Assay

5×10^4 CCL-93 cells were suspended in 200 µL media and treated with 10 s CAP or carrier gas argon. The cells were propagated in medium containing 6-thioguanin (10 µg/mL). As control CAP and carrier gas treated cells were also propagated in medium without 6-thioguanin 24, 48 and 72 h after CAP exposure, the number of viable cells were measured with CASY cell counter and analyzer, as described under point 4.2.

5. Conclusions

This study demonstrated the anti-proliferative effects of CAP on endothelial HDMEC, which depend on the duration of treatment and are due to the induction of apoptosis by CAP. In addition, it was shown that the migration of endothelial cells could be inhibited directly by CAP. The results of this work show that CAP is used to suppress angiogenesis and that tumor growth and the metastatic spread of malignant tumors can be directly influenced.

Author Contributions: Conceptualization, D.G., M.B.S. and A.E.; methodology, L.H. and O.N.; software, A.N.; validation, P.H., A.N., M.B.S. and S.B.; formal analysis, D.G., A.N. and L.H.; investigation, O.N., L.H., N.C.K. and S.B.; resources, A.M., A.E. and P.H.; data curation, M.B.S., N.C.K. and O.N.; writing—original draft preparation, L.H., A.N. and O.N.; writing—review and editing, L.H., N.C.K. and M.B.S.; visualization, O.N., N.C.K. and A.N.; supervision, D.G., A.E., A.M. and P.H.; project administration, A.M., P.H. and A.E.; All authors have read and agreed to the published version of the manuscript.

Funding: No external funding.

Conflicts of Interest: The authors declare no conflict of interest.

Abbreviations

bFGF	basic fibroblast growth factor
CAP	cold atmospheric plasma
CTRL	control
ECGM-MV2	Endothelial Cell Growth Medium MV 2
Flt-1	fms related tyrosine kinase 1
HDMEC	Human Dermal Microvascular Endothelial Cells
KDR	kinase insert domain receptor
PDGF	platelet derivated growth factor
PECAM 1	platelet endothelial adhesion molecule 1
SDS	Sodium Dodecyl Sulfate
VE Cadherin	vascular endothelial cadherin
VEGF	vascular endothelial growth factor

References

1. Fitzmaurice, C.; Akinyemiju, T.F.; Al Lami, F.H.; Alam, T.; Alizadeh-Navaei, R.; Allen, C.; Alsharif, U.; Alvis-Guzman, N.; Amini, E.; Anderson, B.O. Global, regional, and national cancer incidence, mortality, years of life lost, years lived with disability, and disability-adjusted life-years for 29 cancer groups, 1990 to 2016: A systematic analysis for the global burden of disease study. *JAMA Oncol.* **2018**, *4*, 1553–1568. [PubMed]

2. Hall, P.; Hamilton, P.; Hulme, C.; Meads, D.; Jones, H.; Newsham, A.; Marti, J.; Smith, A.; Mason, H.; Velikova, G. Costs of cancer care for use in economic evaluation: A UK analysis of patient-level routine health system data. *Br. J. Cancer* **2015**, *112*, 948–956. [CrossRef] [PubMed]

3. Statistisches Bundesamt. Krankheitskosten 2002, 2004, 2006 und 2008 7.2 ed. 2015. Available online: https://www.destatis.de/DE/Themen/Gesellschaft-Umwelt/Gesundheit/Krankheitskosten/_inhalt.html (accessed on 12 May 2020).

4. Gress, D.M.; Edge, S.B.; Greene, F.L.; Washington, M.K.; Asare, E.A.; Brierley, J.D.; Byrd, D.R.; Compton, C.C.; Jessup, J.M.; Winchester, D.P. Principles of cancer staging. *AJCC Cancer Staging Man.* **2017**, *8*, 3–30.

5. Robert Koch-Institut. *Krebs in Deutschland für 2013/2014*; Robert Koch-Institut: Berlin, Germany, 2017.

6. Pagani, M. The complex clinical picture of presumably allergic side effects to cytostatic drugs: Symptoms, pathomechanism, reexposure, and desensitization. *Med Clin.* **2010**, *94*, 835–852. [CrossRef]

7. Gottesman, M.M.; Fojo, T.; Bates, S.E. Multidrug resistance in cancer: Role of ATP–dependent transporters. *Nat. Rev. Cancer* **2002**, *2*, 48–58. [CrossRef]

8. Jones, P.A.; De Clerck, Y.A. Extracellular matrix destruction by invasive tumor cells. *Cancer Metastasis Rev.* **1982**, *1*, 289–317. [CrossRef]

9. Lorentzen, A.; Becker, P.F.; Kosla, J.; Saini, M.; Weidele, K.; Ronchi, P.; Klein, C.; Wolf, M.J.; Geist, F.; Seubert, B. Single cell polarity in liquid phase facilitates tumour metastasis. *Nat. Commun.* **2018**, *9*, 1–20. [CrossRef]

10. Ferrara, N. Vascular endothelial growth factor and the regulation of angiogenesis. *Recent Prog. Horm. Res.* **2000**, *55*, 15–35, Discussion 35–16.

11. Holmes, D.I.; Zachary, I. The vascular endothelial growth factor (VEGF) family: Angiogenic factors in health and disease. *Genome Biol.* **2005**, *6*, 1–10. [CrossRef]

12. Fraisl, P.; Mazzone, M.; Schmidt, T.; Carmeliet, P. Regulation of angiogenesis by oxygen and metabolism. *Dev. Cell* **2009**, *16*, 167–179. [CrossRef] [PubMed]

13. Cramer, T.; Johnson, R.S. A novel role for the hypoxia inducible transcription factor HIF-1alpha: Critical regulation of inflammatory cell function. *Cell Cycle* **2003**, *2*, 191–192. [CrossRef]

14. Kroll, J.; Waltenberger, J. Regulation der Endothelfunktion und der Angiogenese durch den vaskulären endothelialen Wachstumsfaktor-A (VEGF-A). *Z. Kardiol.* **2000**, *89*, 206–218. [CrossRef]

15. Gille, H.; Kowalski, J.; Li, B.; LeCouter, J.; Moffat, B.; Zioncheck, T.F.; Pelletier, N.; Ferrara, N. Analysis of biological effects and signaling properties of Flt-1 and KDR: A reassessment using novel highly receptor-specific VEGF mutants. *J. Biol. Chem.* **2000**. [CrossRef]

16. Cao, G.; O'Brien, C.D.; Zhou, Z.; Sanders, S.M.; Greenbaum, J.N.; Makrigiannakis, A.; DeLisser, H.M. Involvement of human PECAM-1 in angiogenesis and in vitro endothelial cell migration. *Am. J. Physiol. Cell Physiol.* **2002**, *282*, C1181–C1190. [CrossRef] [PubMed]

17. Chang, S.-H.; Kanasaki, K.; Gocheva, V.; Blum, G.; Harper, J.; Moses, M.A.; Shih, S.-C.; Nagy, J.A.; Joyce, J.; Bogyo, M. VEGF-A induces angiogenesis by perturbing the cathepsin-cysteine protease inhibitor balance in venules, causing basement membrane degradation and mother vessel formation. *Cancer Res.* **2009**, *69*, 4537–4544. [CrossRef]

18. Li, H.; Shi, X.; Liu, J.; Hu, C.; Zhang, X.; Liu, H.; Jin, J.; Opolon, P.; Vannier, J.; Perricaudet, M. The soluble fragment of VE-cadherin inhibits angiogenesis by reducing endothelial cell proliferation and tube capillary formation. *Cancer Gene Ther.* **2010**, *17*, 700–707. [CrossRef]

19. Lee, S.; Jilani, S.M.; Nikolova, G.V.; Carpizo, D.; Iruela-Arispe, M.L. Processing of VEGF-A by matrix metalloproteinases regulates bioavailability and vascular patterning in tumors. *J. Cell Biol.* **2005**, *169*, 681–691. [CrossRef]

20. Vempati, P.; Mac Gabhann, F.; Popel, A.S. Quantifying the proteolytic release of extracellular matrix-sequestered VEGF with a computational model. *PLoS ONE* **2010**, *5*, e11860. [CrossRef]

21. Leung, D.W.; Cachianes, G.; Kuang, W.-J.; Goeddel, D.V.; Ferrara, N. Vascular endothelial growth factor is a secreted angiogenic mitogen. *Science* **1989**, *246*, 1306–1309. [CrossRef]

22. Eilken, H.M.; Adams, R.H. Dynamics of endothelial cell behavior in sprouting angiogenesis. *Curr. Opin. Cell Biol.* **2010**, *22*, 617–625. [CrossRef] [PubMed]

23. Bates, D.O.; Cui, T.-G.; Doughty, J.M.; Winkler, M.; Sugiono, M.; Shields, J.D.; Peat, D.; Gillatt, D.; Harper, S.J. VEGF165b, an inhibitory splice variant of vascular endothelial growth factor, is down-regulated in renal cell carcinoma. *Cancer Res.* **2002**, *62*, 4123–4131. [PubMed]

24. Kerbel, R.; Folkman, J. Clinical translation of angiogenesis inhibitors. *Nat. Rev. Cancer* **2002**, *2*, 727–739. [CrossRef] [PubMed]

25. Ping, Y.F.; Bian, X.W. Consice review: Contribution of cancer stem cells to neovascularization. *Stem Cells (Dayt., Ohio)* **2011**, *29*, 888–894. [CrossRef]

26. Klagsbrun, M. Angiogenesis and cancer: AACR special conference in cancer research. American association for cancer research. *Cancer Res.* **1999**, *59*, 487–490. [PubMed]

27. Folkman, J. Angiogenesis-dependent diseases. *Semin. Oncol.* **2001**, *28*, 536–542. [CrossRef]

28. Adhikari, B.; Pangomm, K.; Veerana, M.; Mitra, S.; Park, G. Plant disease control by non-thermal atmospheric-pressure plasma. *Front. Plant Sci.* **2020**, *11*. [CrossRef]

29. Gay-Mimbrera, J.; García, M.; Tejera, B.; Rodero, A.; Antonio Vélez García, N.; Ruano, J. Clinical and biological principles of cold atmospheric plasma application in skin cancer. *Adv. Ther.* **2016**, *2016*, 1–16. [CrossRef]

30. Ulrich, C.; Kluschke, F.; Patzelt, A.; Vandersee, S.; Czaika, V.A.; Richter, H.; Bob, A.; Hutten, J.; Painsi, C.; Hüge, R.; et al. Clinical use of cold atmospheric pressure argon plasma in chronic leg ulcers: A pilot study. *J. Wound Care* **2015**, *24*, 196–203. [CrossRef]

31. Von Woedtke, T.; Schmidt, A.; Bekeschus, S.; Wende, K.; Weltmann, K.-D. Plasma medicine: A field of applied redox biology. *In Vivo* **2019**, *33*, 1011–1026. [CrossRef]

32. Semmler, M.L.; Bekeschus, S.; Schäfer, M.; Bernhardt, T.; Fischer, T.; Witzke, K.; Seebauer, C.; Rebl, H.; Grambow, E.; Vollmar, B.; et al. Molecular mechanisms of the efficacy of Cold Atmospheric Pressure Plasma (CAP) in cancer treatment. *Cancers* **2020**, *12*, 269. [CrossRef] [PubMed]

33. Arndt, S.; Wacker, E.; Li, Y.-F.; Shimizu, T.; Thomas, H.M.; Morfill, G.E.; Karrer, S.; Zimmermann, J.L.; Bosserhoff, A.-K. Cold atmospheric plasma, a new strategy to induce senescence in melanoma cells. *Exp. Dermatol.* **2013**, *22*, 284–289. [CrossRef] [PubMed]

34. Haralambiev, L.; Wien, L.; Gelbrich, N.; Lange, J.; Bakir, S.; Kramer, A.; Burchardt, M.; Ekkernkamp, A.; Gümbel, D.; Stope, M.B. Cold atmospheric plasma inhibits the growth of osteosarcoma cells by inducing apoptosis, independent of the device used. *Oncol. Lett.* **2020**, *19*, 283–290. [CrossRef] [PubMed]

35. Koensgen, D.; Besic, I.; Gumbel, D.; Kaul, A.; Weiss, M.; Diesing, K.; Kramer, A.; Bekeschus, S.; Mustea, A.; Stope, M.B. Cold Atmospheric Plasma (CAP) and CAP-stimulated cell culture media suppress ovarian cancer cell growth—A putative treatment option in ovarian cancer therapy. *Anticancer Res.* **2017**, *37*, 6739–6744. [CrossRef]

36. Yan, X.; Xiong, Z.; Zou, F.; Zhao, S.; Lu, X.; Yang, G.; He, G.; Ostrikov, K. Plasma-induced death of HepG2 cancer cells: Intracellular effects of reactive species. *Plasma Process. Polym.* **2012**, *9*, 59–66. [CrossRef]

37. Zamzami, N.; Marchetti, P.; Castedo, M.; Decaudin, D.; Macho, A.; Hirsch, T.; Susin, S.A.; Petit, P.X.; Mignotte, B.; Kroemer, G. Sequential reduction of mitochondrial transmembrane potential and generation of reactive oxygen species in early programmed cell death. *J. Exp. Med.* **1995**, *182*, 367–377. [CrossRef] [PubMed]

38. Graves, D.B. Reactive species from cold atmospheric plasma: Implications for cancer therapy. *Plasma Process. Polym.* **2014**, *11*, 1120–1127. [CrossRef]

39. Bradley, M.O.; Bhuyan, B.; Francis, M.C.; Langenbach, R.; Peterson, A.; Huberman, E. Mutagenesis by chemical agents in V79 chinese hamster cells: A review and analysis of the literature. A report of the Gene-Tox Program. *Mutat. Res.* **1981**, *87*, 81–142. [CrossRef]

40. Melincovici, C.S.; Boşca, A.B.; Şuşman, S.; Mărginean, M.; Mihu, C.; Istrate, M.; Moldovan, I.M.; Roman, A.L.; Mihu, C.M. Vascular endothelial growth factor (VEGF)—Key factor in normal and pathological angiogenesis. *Rom. J. Morphol. Embryol. Rev. Roum. Morphol. Embryol.* **2018**, *59*, 455–467.

41. Kunstfeld, R.; Hawighorst, T.; Streit, M.; Hong, Y.-K.; Nguyen, L.; Brown, L.F.; Detmar, M. Thrombospondin-2 overexpression in the skin of transgenic mice reduces the susceptibility to chemically induced multistep skin carcinogenesis. *J. Dermatol. Sci.* **2014**, *74*, 106–115. [CrossRef]

42. Xie, L.; Ji, T.; Guo, W. Anti-angiogenesis target therapy for advanced osteosarcoma (Review). *Oncol. Rep.* **2017**, *38*, 625–636. [CrossRef] [PubMed]

43. Deng, Z.; Zhou, J.; Han, X.; Li, X. TCEB2 confers resistance to VEGF-targeted therapy in ovarian cancer. *Oncol. Rep.* **2016**, *35*, 359–365. [CrossRef] [PubMed]

44. Guerrouahen, B.S.; Pasquier, J.; Abu Kaoud, N.; Maleki, M.; Beauchamp, M.-C.; Yasmeen, A.; Ghiabi, P.; Lis, R.; Vidal, F.; Saleh, A.; et al. Akt-activated endothelium constitute the niche for residual disease and resistance to bevacizumab in ovarian cancer. *Mol. Cancer Ther.* **2014**. [CrossRef] [PubMed]

45. Lubner, S.J.; Uboha, N.V.; Deming, D.A. Primary and acquired resistance to biologic therapies in gastrointestinal cancers. *J. Gastrointest. Oncol.* **2017**, *8*, 499–512. [CrossRef] [PubMed]

46. Mitamura, T.; Pradeep, S.; McGuire, M.; Wu, S.Y.; Ma, S.; Hatakeyama, H.; Lyons, Y.A.; Hisamatsu, T.; Noh, K.; Villar-Prados, A.; et al. Induction of anti-VEGF therapy resistance by upregulated expression of microseminoprotein (MSMP). *Oncogene* **2018**, *37*, 722–731. [CrossRef] [PubMed]

47. Gümbel, D.; Gelbrich, N.; Napp, M.; Daeschlein, G.; Kramer, A.; Sckell, A.; Burchardt, M.; Ekkernkamp, A.; Stope, M.B. Peroxiredoxin expression of human osteosarcoma cells is influenced by cold atmospheric plasma treatment. *Anticancer Res.* **2017**, *37*, 1031–1038. [CrossRef]

48. Gümbel, D.; Daeschlein, G.; Ekkernkamp, A.; Kramer, A.; Stope, M.B. Cold atmospheric plasma in orthopaedic and urologic tumor therapy. *GMS Hyg. Infect. Control* **2017**, *12*, Doc10. [CrossRef]

49. Kramer, A.; Bekeschus, S.; Matthes, R.; Bender, C.; Stope, M.B.; Napp, M.; Lademann, O.; Lademann, J.; Weltmann, K.-D.; Schauer, F. Cold physical plasmas in the field of hygiene—Relevance, significance, and future applications. *Plasma Process. Polym.* **2015**, *12*, 1410–1422. [CrossRef]

50. Balzer, J.; Heuer, K.; Demir, E.; Hoffmanns, M.A.; Baldus, S.; Fuchs, P.C.; Awakowicz, P.; Suschek, C.V.; Opländer, C. Non-thermal Dielectric Barrier Discharge (DBD) effects on proliferation and differentiation of human fibroblasts are primary mediated by hydrogen peroxide. *PLoS ONE* **2015**, *10*, e0144968. [CrossRef]

51. Maisch, T.; Bosserhoff, A.K.; Unger, P.; Heider, J.; Shimizu, T.; Zimmermann, J.L.; Morfill, G.E.; Landthaler, M.; Karrer, S. Investigation of toxicity and mutagenicity of cold atmospheric argon plasma. *Environ. Mol. Mutagenesis* **2017**, *58*, 172–177. [CrossRef]

52. Bauer, G.; Sersenová, D.; Graves, D.B.; Machala, Z. Cold atmospheric plasma and plasma-activated medium trigger RONS-based tumor cell apoptosis. *Sci. Rep.* **2019**, *9*, 14210. [CrossRef] [PubMed]

53. Haertel, B.; Woedtke, T.; Weltmann, K.-D.; Lindequist, U. Non-thermal atmospheric-pressure plasma possible application in wound healing. *Biomol. Ther.* **2014**, *22*, 477–490. [CrossRef] [PubMed]

54. Jablonowski, L.; Kocher, T.; Schindler, A.; Müller, K.; Dombrowski, F.; von Woedtke, T.; Arnold, T.; Lehmann, A.; Rupf, S.; Evert, M.; et al. Side effects by oral application of atmospheric pressure plasma on the mucosa in mice. *PLoS ONE* **2019**, *14*, e0215099. [CrossRef]

55. Dubuc, A.; Monsarrat, P.; Virard, F.; Merbahi, N.; Sarrette, J.-P.; Laurencin-Dalicieux, S.; Cousty, S. Use of cold-atmospheric plasma in oncology: A concise systematic review. *Ther. Adv. Med. Oncol.* **2018**, *10*, 1–12. [CrossRef] [PubMed]

56. Liu, Y.; Hong, H.; Lu, X.; Wang, W.; Liu, F.; Yang, H. L-ascorbic acid protected against extrinsic and intrinsic apoptosis induced by cobalt nanoparticles through ROS attenuation. *Biol. Trace Elem. Res.* **2017**, *175*, 428–439. [CrossRef] [PubMed]

57. Wang, M.; Holmes, B.; Cheng, X.; Zhu, W.; Keidar, M.; Zhang, L.G. Cold atmospheric plasma for selectively ablating metastatic breast cancer cells. *PLoS ONE* **2013**, *8*, e73741. [CrossRef]

58. Cheng, X.; Sherman, J.; Murphy, W.; Ratovitski, E.; Canady, J.; Keidar, M. The effect of tuning cold plasma composition on glioblastoma cell viability. *PLoS ONE* **2014**, *9*, e98652. [CrossRef]

59. Bernhardt, T.; Semmler, M.L.; Schäfer, M.; Bekeschus, S.; Emmert, S.; Boeckmann, L. Plasma medicine: Applications of cold atmospheric pressure plasma in dermatology. *Oxid. Med. Cell. Longev.* **2019**, *2019*, 3873928. [CrossRef]

60. Kalghatgı, S.; Kelly, C.M.; Cerchar, E.; Torabi, B.; Alekseev, O.; Fridman, A.; Friedman, G.; Azizkhan-Clifford, J. Effects of non-thermal plasma on mammalian cells. *PLoS ONE* **2011**, *6*, e16270. [CrossRef]

61. Ishaq, M.; Evans, M.; Ostrikov, K. Effect of atmospheric gas plasmas on cancer cell signaling. *Int. J. Cancer* **2014**, *134*, 1517–1528. [CrossRef]

62. Sieuwerts, A.M.; Klijn, J.G.; Peters, H.A.; Foekens, J.A. The MTT tetrazolium salt assay scrutinized: How to use this assay reliably to measure metabolie activity of cell cultures in vitro for the assessment of growth characteristics, IC50-values and cell survival. *Clin. Chem. Lab. Med.* **1995**, *33*, 813–824. [CrossRef] [PubMed]

63. Dias, N.; Nicolau, A.; Carvalho, G.S.; Mota, M.; Lima, N. Miniaturization and application of the MTT assay to evaluate metabolic activity of protozoa in the presence of toxicants. *J. Basic Microbiol.* **1999**, *39*, 103–108. [CrossRef]

64. Siu, A.; Volotskova, O.; Cheng, X.; Khalsa, S.S.; Bian, K.; Murad, F.; Keidar, M.; Sherman, J.H. Differential effects of cold atmospheric plasma in the treatment of malignant glioma. *PLoS ONE* **2015**, *10*, e0126313. [CrossRef]

65. Kim, S.J.; Chung, T.H. Cold atmospheric plasma jet-generated RONS and their selective effects on normal and carcinoma cells. *Sci. Rep.* **2016**, *6*, 20332. [CrossRef] [PubMed]

66. Guerrero-Preston, R.; Ogawa, T.; Uemura, M.; Shumulinsky, G.; Valle, B.L.; Pirini, F.; Ravi, R.; Sidransky, D.; Keidar, M.; Trink, B. Cold atmospheric plasma treatment selectively targets head and neck squamous cell carcinoma cells. *Int. J. Mol. Med.* **2014**, *34*, 941–946. [CrossRef] [PubMed]

67. Jalili, A.; Irani, S.; Mirfakhraie, R. Combination of cold atmospheric plasma and iron nanoparticles in breast cancer: Gene expression and apoptosis study. *OncoTargets Ther.* **2016**, *9*, 5911.

68. Steuer, A.; Wolff, C.M.; von Woedtke, T.; Weltmann, K.-D.; Kolb, J.F. Cell stimulation versus cell death induced by sequential treatments with pulsed electric fields and cold atmospheric pressure plasma. *PLoS ONE* **2018**, *13*, e0204916. [CrossRef]

69. Zhou, J.; Wang, S.; Ye, Y.; Yang, S.; Cui, Z. Change in apoptosis and its mechanism in intestinal epithelial cells under oxidative stress. *Chin. Crit. Care Med.* **2005**, *17*, 268–272.

70. Jacoby, J.M.; Strakeljahn, S.; Nitsch, A.; Bekeschus, S.; Hinz, P.; Mustea, A.; Ekkernkamp, A.; Tzvetkov, M.V.; Haralambiev, L.; Stope, M.B. An innovative therapeutic option for the treatment of skeletal sarcomas: Elimination of Osteo-and Ewing's Sarcoma cells using physical gas plasma. *Int. J. Mol. Sci.* **2020**, *21*, 4460. [CrossRef]

71. Weiss, M.; Gümbel, D.; Hanschmann, E.-M.; Mandelkow, R.; Gelbrich, N.; Zimmermann, U.; Walther, R.; Ekkernkamp, A.; Sckell, A.; Kramer, A.; et al. Cold atmospheric plasma treatment induces anti-proliferative effects in prostate cancer cells by redox and apoptotic signaling pathways. *PLoS ONE* **2015**, *10*, e0130350. [CrossRef]

72. Canal, C.; Fontelo, R.; Hamouda, I.; Guillem-Marti, J.; Cvelbar, U.; Ginebra, M.-P. Plasma-induced selectivity in bone cancer cells death. *Free Radic. Biol. Med.* **2017**, *110*, 72–80. [CrossRef] [PubMed]

73. Yan, D.; Sherman, J.H.; Keidar, M. Cold atmospheric plasma, a novel promising anti-cancer treatment modality. *Oncotarget* **2017**, *8*, 15977–15995. [CrossRef] [PubMed]

Article

Antibiotic-Resistant and Non-Resistant Bacteria Display Similar Susceptibility to Dielectric Barrier Discharge Plasma

Akikazu Sakudo [1,2,*] and Tatsuya Misawa [3]

1   School of Veterinary Medicine, Okayama University of Science, Imabari, Ehime 794-8555, Japan
2   Laboratory of Biometabolic Chemistry, School of Health Sciences, University of the Ryukyus, Nishihara, Okinawa 903-0215, Japan
3   Department of Electrical and Electronic Engineering, Faculty of Science and Engineering, Saga University, Saga 840-8502, Japan; misawa@cc.saga-u.ac.jp
*   Correspondence: akikazusakudo@gmail.com; Fax: +81-898-52-9198

Received: 30 July 2020; Accepted: 31 August 2020; Published: 31 August 2020

Abstract: Here, we examined whether antibiotic-resistant and non-resistant bacteria show a differential susceptibility to plasma treatment. *Escherichia coli* DH5α were transformed with pPRO-EX-HT-CAT, which encodes an ampicillin resistance gene and chloramphenicol acetyltransferase (CAT) gene, and then treated with a dielectric barrier discharge (DBD) plasma torch. Plasma treatment reduced the viable cell count of *E. coli* after transformation/selection and further cultured in ampicillin-containing and ampicillin-free medium. However, there was no significant difference in viable cell count between the transformed and untransformed *E. coli* after 1 min- and 2 min-plasma treatment. Furthermore, the enzyme-linked immunosorbent assay (ELISA) and acetyltransferase activity assay showed that the CAT activity was reduced after plasma treatment in both transformed and selected *E. coli* grown in ampicillin-containing or ampicillin-free medium. Loss of lipopolysaccharide and DNA damage caused by plasma treatment were confirmed by a Limulus test and polymerase chain reaction, respectively. Taken together, these findings suggest the plasma acts to degrade components of the bacteria and is therefore unlikely to display a differential affect against antibiotic-resistant and non-resistant bacteria. Therefore, the plasma method may be useful in eliminating bacteria that are recalcitrant to conventional antibiotic therapy.

Keywords: antibiotic resistant bacteria; antibiotic resistance gene; disinfection; *E. coli*; inactivation; plasma; sterilization

---

1. Introduction

In recent years, the emergence of antibiotic-resistant bacteria has become a global problem. One of the main drivers for the emergence of antibiotic-resistant bacteria is the excessive or inappropriate use of antibiotics [1,2].

Clinical isolates of antibiotic-resistant *Escherichia coli* were first isolated in the 1970s [3], soon followed by strains with extended-spectrum β-lactamase (ESBL) genes, which confer resistance to third generation cephalosporins [4], which are the most commonly prescribed classes of antibiotics with bactericidal activity [5]. Multiple resistant bacteria are now a major problem in human and veterinary medicine, especially for nosocomial infections [3,6]. These antibiotic resistance genes are found on plasmids that also carry genes conferring resistance to other non-β-lactam antibiotics such as aminoglycosides and trimethoprim-sulfamethoxazole [4]. *E. coli* harboring these plasmids are only susceptible to carbapenems and colistin.

Antibiotic-resistant bacteria are commonly found in agricultural settings, hospital environments, factory farms, and contaminated foodstuffs [7]. Regardless of the source of antibiotic-resistant bacteria, development of efficient cleaning and disinfection methods are urgently needed to prevent their spread [8].

Several technologies to eliminate antibiotic-resistant bacteria and/or antibiotic resistance genes have been employed including ultraviolet (UV) light [9], solar photo-Fenton process [10], photocatalysis [11], photoelectrocatalysis [12], and nanoparticle treatment [13]. However, some of these technologies have given disappointing levels of disinfection [14,15].

Recently, plasma technologies have been shown to effectively inactivate various bacteria including antibiotic-resistant bacteria [16,17]. Currently, however, it is not known whether the sensitivity against plasma differs between antibiotic-resistant and non-resistant bacteria. Previous studies were conducted in a way that may have masked possible differences arising from sensitivity due to antibiotic resistance. Here, we compared antibiotic-resistant bacteria harboring a plasmid that confers antibiotic resistance with its non-transformed counterpart in terms of susceptibility against plasma generated by a dielectric barrier discharge (DBD) plasma torch. In addition, we analyzed biochemical changes in the bacteria after exposure to the plasma to better understand the mechanism of inactivation.

2. Results

To obtain antibiotic-resistant bacteria, *Escherichia coli* DH5α was transformed with pPRO-EX-HT -CAT, which contains an ampicillin resistance gene and chloramphenicol acetyltransferase (CAT) gene. Colonies were picked after overnight selection on Luria-Bertani (LB) agar medium containing ampicillin. The selected colonies were subsequently cultured in LB liquid medium with or without ampicillin for 24 h. Aliquots of the resultant *E. coli* were then treated with a dielectric barrier discharge (DBD) plasma torch (Figure 1). As a control, untransformed *E. coli* were also analyzed. The viable cell counts of all *E. coli* samples decreased after plasma treatment. The viable cell number for *E. coli* transformed/selected with pPRO-EX-HT-CAT and cultured in LB liquid medium supplemented with ampicillin was lower than the control ($2.40 \times 10^6 \pm 1.38 \times 10^6$ CFU/mL, 0 min) (Figure 2). Plasma treatment for 1 min reduced viable cell count to $1.38 \times 10^5 \pm 0.80 \times 10^5$ CFU/mL. However, cultures of the transformed and selected *E. coli* in ampicillin-free LB liquid medium were $2.27 \times 10^6 \pm 1.31 \times 10^6$ CFU/mL (0 min) and $1.06 \times 10^5 \pm 0.61 \times 10^5$ CFU/mL (1 min). In summary, no significant difference in viable cell count was observed between the ampicillin-containing culture and ampicillin-free culture. This result suggests that selection pressure with ampicillin does not change the susceptibility of the bacteria to plasma treatment.

Similarly, *E. coli* transformed/selected with pPRO-EX-HT-CAT and non-transformed *E. coli* were cultured in ampicillin-free LB liquid medium (Figure 3) and then subjected to plasma treatment. Viable cell number was $2.13 \times 10^7 \pm 0.20 \times 10^7$ CFU/mL (0 min) and $1.80 \times 10^4 \pm 1.80 \times 10^4$ CFU/mL (1 min) for the transformed *E. coli*, and $2.59 \times 10^7 \pm 0.47 \times 10^7$ CFU/mL (0 min) and $7.83 \times 10^4 \pm 5.17 \times 10^4$ CFU/mL (1 min) for the non-transformed *E. coli*, indicating no significant difference between the two groups. In all cases, the number of viable bacteria was below the detection limit after 2 min treatment (Figures 2 and 3).

Next, we aimed to examine the effect of plasma treatment on components of *E. coli* using the DBD plasma torch. To this end, we performed DNA analysis by polymerase chain reaction (PCR), CAT quantification by ELISA and CAT activity using an acetyltransferase activity assay as well as lipopolysaccharide (LPS) analysis by the Limulus test. PCR amplification using sequence-specific primers for bacterial 16S rDNA followed by agarose gel electrophoresis gave a discrete band in the untreated *E. coli* sample (0 min). DNA sequencing of the amplified band verified that it corresponded to *E. coli* DH5α 16S rDNA (i.e., $99.25 \pm 0.35\%$ [$5' \rightarrow 3'$ direction, $N = 4$] and $99.48 \pm 0.34\%$ [$3' \rightarrow 5'$ direction, $N = 4$] identical to Genbank accession number NZ_CP026085.1) (Figure 4). A band was detected in *E. coli* samples (0 min) in DH5α Amp($-$) (non-transformed *E. coli* proliferated in an ampicillin-free LB liquid medium) and pPRO-EX-CAT DH5α Amp($-$) (*E. coli* transformed with pPRO-EX-CAT, selected in ampicillin-containing LB agar, and proliferated in ampicillin-free LB liquid medium), although the levels of intact DNA was low at 1 min and undetectable at 2 min for both *E. coli* samples.

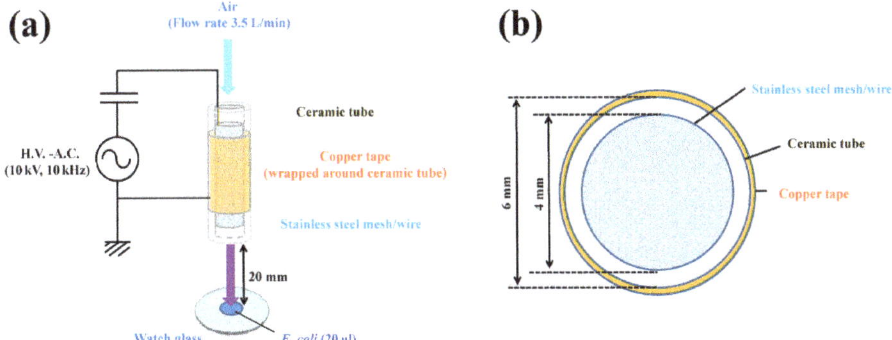

Figure 1. Schematic of a dielectric barrier discharge (DBD) plasma torch. (**a**) The DBD plasma torch comprises a ceramic tube (Al_2O_3) (length, 100 mm) containing a stainless-steel mesh (SUS304) and covered with copper tape (thickness, 80 µm) on the outside. The copper tape and stainless steel mesh with stainless steel wire were connected to a power supply (10 kV, 10 kHz); (**b**) Cross-sectional view of the torch, which is made up of a ceramic tube (inner diameter, 4 mm; outer diameter, 6 mm). During gas plasma generation, air flow was maintained at 3.5 L/min using an air pump (Suishin SSPP-2S; Suisaku Co., Sakai, Japan). A suspension (20 µL) of *Escherichia coli* was dropped onto a watch glass. The distance from the top of the plasma torch to the liquid surface on the watch glass was fixed at 20 mm.

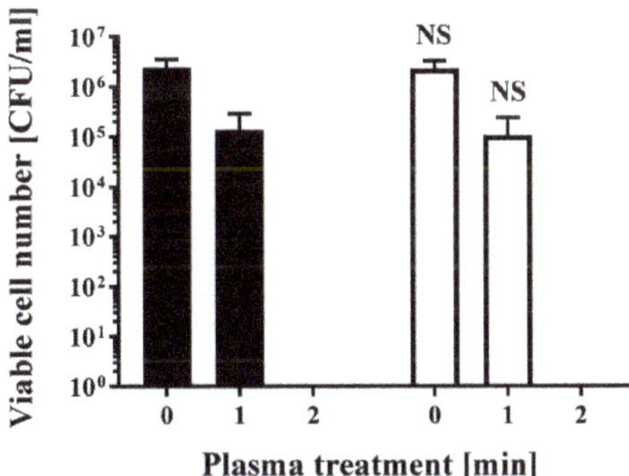

Figure 2. Treatment with a DBD plasma torch reduced the viable cell number of *Escherichia coli* transformed with a plasmid conferring ampicillin resistance, which was selected by spreading on ampicillin-containing agar medium and proliferated in ampicillin-containing liquid medium (■) by a similar amount as plasmid-transformed *E. coli* proliferated in ampicillin-free liquid medium (□). A suspension of *E. coli* transformed with an ampicillin resistance gene-containing plasmid pPRO-EX-HT-CAT, selected in ampicillin-containing agar medium, and proliferated either in ampicillin-containing liquid medium (■) or ampicillin-free liquid medium (□) was exposed to a DBD plasma torch for the indicated time (min). Viable cell count (colony forming units (CFU)/mL) was then determined before and after treatment. NS means no significant difference between ■ and □ at each time when verified by the non-repeated measured ANOVA (analysis of variance) followed by the Bonferroni correction. ANOVA for non-repeated measures was used for comparing the intragroup, while the Bonferroni correction was used as post-hoc analysis.

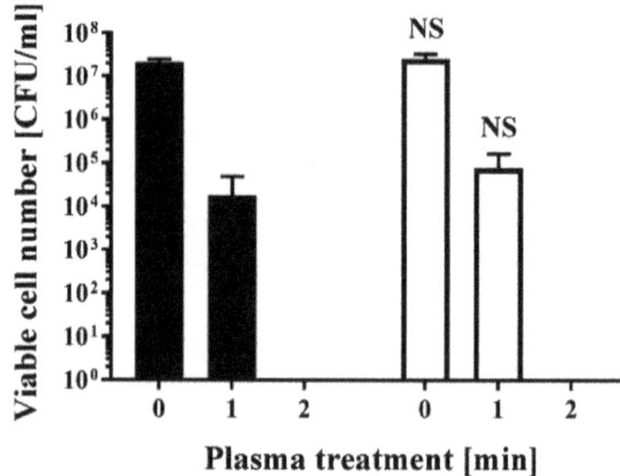

Figure 3. Treatment with a DBD plasma torch reduced the viable cell number of *E. coli* transformed with a plasmid conferring ampicillin resistance, which was selected by spreading on ampicillin-containing agar medium and proliferated in ampicillin-free medium (■), compared with non-transformed *E. coli* proliferated in ampicillin-free medium (□). A suspension of *E. coli* transformed with an ampicillin resistance gene-containing plasmid pPRO-EX-HT-CAT and selected in ampicillin-containing agar medium and proliferated in ampicillin-free liquid medium (■) as well as non-transformed *E. coli* proliferated in ampicillin-free liquid medium (□) was exposed to a DBD plasma torch for the indicated time (min). Viable cell count (colony forming units (CFU)/mL) was then determined before and after treatment. NS means no significant difference between ■ and □ at each time when verified by the non-repeated measured ANOVA followed by the Bonferroni correction.

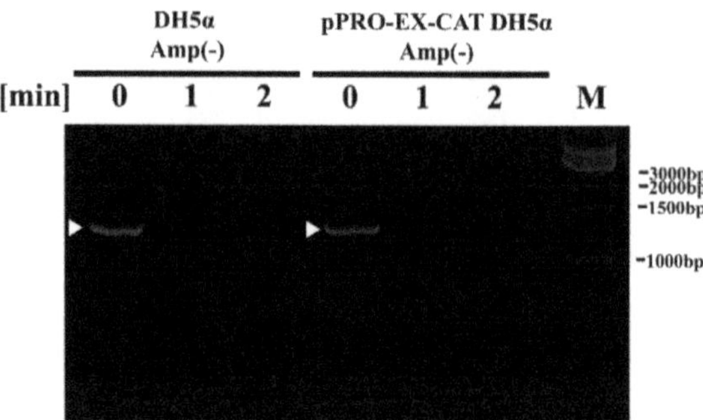

Figure 4. Genomic DNA (16S rDNA) of non-transformed *E. coli* proliferated in ampicillin-free medium (DH5α Amp(−)) and *E. coli* transformed with a plasmid conferring ampicillin resistance, selected in ampicillin-containing agar medium and proliferated in ampicillin-free medium (pPRO-EX-CAT DH5α Amp(−)), was degraded by DBD plasma torch treatment. A suspension of non-transformed *E. coli* proliferated in ampicillin-free medium (DH5α Amp(−)) and *E. coli* transformed with an ampicillin resistance gene-containing pPRO-EX-CAT plasmid, selected in ampicillin-containing agar medium and proliferated in ampicillin-free liquid medium (pPRO-EX-CAT DH5α Amp(−)), was exposed to a DBD plasma torch for the indicated time (min). Viable cell count (colony forming units (CFU)/mL) was then determined before and after treatment. Bands corresponding to amplified *E. coli* 16S rDNA are indicated by arrowheads. Bands corresponding to a DNA size ladder (M) are labelled on the right-hand side of the gel.

Next, to examine the effect of plasma on the quantity of protein, *E. coli* transformed with pPRO-EX-HT-CAT containing CAT gene was treated with the DBD plasma torch and analyzed by CAT ELISA (Figure 5). After exposure to the plasma torch, CAT was reduced in *E. coli* for both transformed and selected *E. coli* cultured in LB medium with or without ampicillin. For *E. coli* transformed with the plasmid and proliferated in the ampicillin-containing medium, CAT concentration before gas plasma treatment was 3147.7302 ± 224.9762 ng/mL (0 min), but was significantly reduced to 0.0325 ± 0.0001 ng/mL and 0.0335 ± 0.0009 ng/mL after plasma treatment for 1 min and 2 min, respectively. In addition, significant reduction of CAT concentration to 0.0344 ± 0.0004 ng/mL at 1 min and 0.0339 ± 0.0004 ng/mL at 2 min was also observed in transformed *E. coli* cultured in ampicillin-free medium compared to the control of 0.3961 ± 0.0182 ng/mL at 0 min. In summary, a significant decrease in the amount of CAT for the transformed *E. coli* was observed at each time point compared to 0 min.

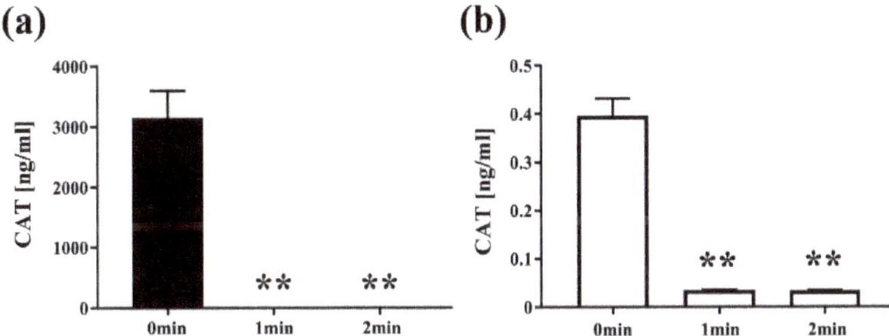

Figure 5. The amount of chloramphenicol acetyltransferase (CAT) in *E. coli* transformed with a plasmid conferring ampicillin resistance, selected on ampicillin-containing agar medium and then proliferated in liquid medium with ampicillin (■) or without ampicillin (□), was reduced by DBD plasma torch treatment. *E. coli* transformed with an ampicillin resistance gene- and CAT gene-containing plasmid pPRO-EX-HT-CAT and selected on ampicillin-containing LB agar were proliferated in ampicillin-containing liquid LB medium (Amp(+)) (■) (a) or ampicillin-free liquid LB medium (Amp(−)) (□) (b) and then subjected to DBD plasma torch treatment for 0–2 min. The concentration of CAT (ng/mL) was subsequently measured by enzyme-linked immunosorbent assay (ELISA). Differences where $p < 0.01$ (**) versus control (0 min) were significant when verified by the non-repeated measured ANOVA followed by the Bonferroni correction.

Next, to examine the effect of plasma on protein function, the activity of CAT in transformed *E. coli* was measured by an acetyltransferase activity assay (Figure 6). We speculated that plasma might interfere with the activity of CAT. Specifically, the activity of CAT in *E. coli* transformed with pPR-EX-HT-CAT, selected in ampicillin-containing LB agar, and then proliferated in ampicillin-containing LB liquid medium was 5617.49 ± 92.17 RFU (0 min), but decreased to 322.22 ± 1.07 RFU and 345.11 ± 10.88 RFU after DBD plasma torch treatment for 1 min and 2 min, respectively. In addition, 1991.12 ± 27.76 RFU (0 min) and 343.10 ± 7.47 RFU (1 min) and 343.27 ± 5.74 RFU (2 min) were also observed in cultured *E. coli* after transformation/selection of pPRO-EX-HT-CAT in the ampicillin-free LB liquid medium. Thus, a significant decrease in CAT activity of the transformed *E. coli* was observed at each time compared to 0 min.

Finally, a chromogenic Limulus test was performed to measure the amount of intact LPS in non-transformed *E. coli* following DBD plasma torch treatment (Figure 7). The results showed that intact LPS in *E. coli* after plasma treatment significantly decreased from 218.94 ± 16.38 EU/mL at 0 min to 4.14 ± 0.22 EU/mL at 1 min and 1.53 ± 0.24 EU/mL at 2 min. These findings indicate that *E. coli* LPS lipid A, which is located on the outer surface of the bacteria, is degraded after plasma treatment. In conclusion, plasma treatment may degrade the integrity of the cell wall of *E. coli*.

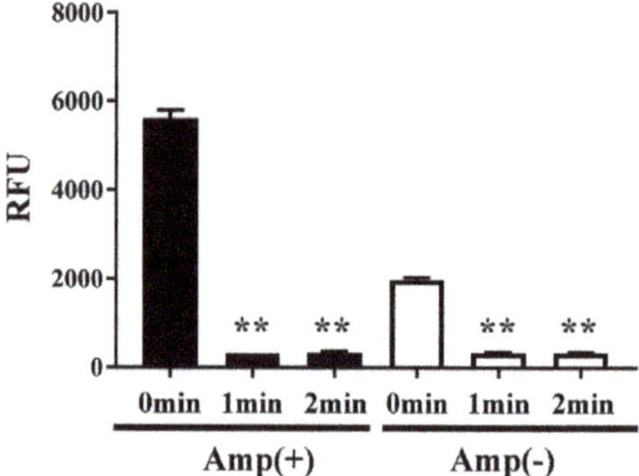

Figure 6. CAT activity in *E. coli* transformed with an ampicillin resistance gene- and CAT gene-containing plasmid, selected on ampicillin-containing agar medium, and proliferated in liquid medium with ampicillin (■) or without ampicillin (□) was decreased by DBD plasma torch treatment. *E. coli* transformed with an ampicillin resistance gene- and CAT gene-containing plasmid pPRO-EX-HT-CAT and selected on ampicillin-containing medium were proliferated in ampicillin-containing liquid medium (Amp(+)) (■) or ampicillin-free liquid medium (Amp(−)) (□) and then subjected to DBD plasma torch treatment for 0–2 min. The activity of CAT in the bacteria was subsequently measured using an acetyltransferase activity assay kit (Enzo Life Sciences, Inc.) with an excitation wavelength of 400 nm and emission wavelength of 490 nm. Differences where $p < 0.01$ (**) versus control (0 min) were significant when verified by the non-repeated measured ANOVA followed by the Bonferroni correction.

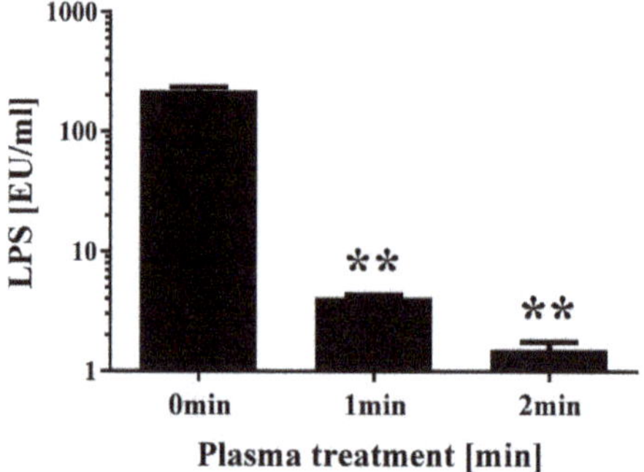

Figure 7. Changes to the level of lipopolysaccharides (LPS) in *E. coli* following DBD plasma torch treatment. The concentration of LPS measured by the Limulus test (Limulus-color KY test, Wako Pure Chemical Industries Ltd., Osaka, Japan) is expressed as endotoxin units (EU) per mL. LPS level was reduced after plasma treatment for 0–2 min. Differences where $p < 0.01$ (**) versus control (0 min) were considered significant when verified by the non-repeated measured ANOVA followed by the Bonferroni correction.

3. Discussion

The growing risk of disease from antibiotic-resistant bacteria is now a global public health concern [18]. The spread of antibiotic resistance is caused, in part, from the extensive and indiscriminate use of antibiotics, which leads to the emergence of antibiotic-resistant bacteria. To restrict the increase of antibiotic-resistant bacteria, efficient inactivation methods that do not rely on antibiotics are required.

Previous studies have shown that exposure to plasma, which is a state of matter predominantly composed of ions and electrons, can inactivate a variety of antibiotic-resistant bacteria. For example, glow discharge plasma has been shown to inactivate antibiotic resistant *E. coli* [16].

Surface micro-discharge (SMD) plasma treatment was demonstrated to inactivate antibiotic-resistant bacteria including *Yersinia enterocolitica*, *Staphylococcus aureus*, *Klebsiella*, *E. coli*, *Acinetobacter*, and *Enterococcus faecium* [17]. Moreover, a floating-electrode DBD plasma device has successfully inactivated methicillin-resistant *S. aureus* (MRSA) [19]. In addition, suspensions and biofilms of MRSA were inactivated by RF (radiofrequency) plasma [20] and floating-electrode DBD [21]. Therefore, plasma treatment provides a robust and efficient way of eliminating antibiotic-resistant bacteria. However, to date, there are no reports showing whether antibiotic-resistant and non-resistant bacteria display the same or different level of sensitivity to plasma treatment.

In this study, we have clarified that antibiotic-resistant and non-resistant bacteria show no significant difference in susceptibility to treatment with plasma. Furthermore, this study suggests that the emergence of plasma-resistant bacteria following plasma treatment is low.

Our analysis showed a reduction in the levels of CAT, DNA, and LPS in *E. coli* after plasma treatment. Previous studies have analyzed gas plasma [22–25], which led to oxidation/modification of DNA and protein as well as their degradation. LPS, also termed as endotoxin, is a component of the outer membrane of Gram-negative bacteria and is known to be highly resistant to both physical and chemical treatment [26]. However, a previous study has shown that the level of endotoxin can be reduced by exposure of bacteria to nitrogen gas plasma [27]. Here, we found that LPS is degraded by DBD plasma torch treatment. These findings are consistent with a previous study, in which the authors proposed that cell surface components are important in the mechanism of inactivation because the thickness of the polysaccharide membrane influences sensitivity to plasma [17]. Furthermore, as previously determined [28], the temperature of the liquid surface was $38.40 \pm 0.12\ °C$ at the 2 min treatment time using the same DBD plasma torch under identical conditions. *E. coli* can survive at around $38\ °C$ [29]. Thus, heating during operation of the plasma torch cannot have been the cause of *E. coli* inactivation.

To clarify the likely inactivation mechanism, analysis of plasma components by electron spin resonance (ESR) was also performed (Supplementary Figure S1). These experiments showed that OH radicals (OH·) and H radicals (H·) were present in the plasma. Thus, reactive chemical species such as OH and H radicals may act to eliminate bacteria by oxidizing biomolecules. Indeed, these findings are consistent with a previous study that showed that the OH radical is the most important factor for bacterial inactivation during plasma treatment [16,30]. Thus, oxidation and degradation of biomolecules may be caused by reactive chemical species such as reactive nitrogen species (RNS) and reactive oxygen species (ROS) generated by the plasma. Nevertheless, it is possible that the reactive chemical species may act on other biomolecules. In addition, as antioxidants have antimicrobial activity against antibiotic-resistant bacteria [31], modulation of the oxidative conditions might change the antibacterial activity. Thus, further studies on the potential relationship between oxidative modulation by plasma and its antibacterial activity would be necessary to perform.

Regardless of the underlying mechanism of inactivation, the DBD plasma torch was found to be highly effective at eliminating antibiotic-resistant bacteria. Presumably, this was because changes to cell components induced by the plasma treatment such as modifications to the proteins and DNA were independent of antibiotic resistance. Further studies are also needed to clarify the inactivation mechanism of the DBD plasma torch including detailed analysis of bactericidal factors generated by the plasma that induce damage to DNA and proteins as well as cell surface degradation including LPS, which is a component of the outer membrane of Gram-negative bacteria. Antibiotic resistant bacteria employ multiple mechanisms to confer resistance including antibiotic inactivation, target-site modification, and reduction of cytoplasmic antibiotic concentration [32]. The specific type of antibiotic resistance mechanism might be related to plasma susceptibility. Therefore, further studies on the relationship between the various mechanisms of antibiotic resistance and plasma susceptibility are required.

Finally, the development of process compatible technology design is required to convert the DBD plasma device into a commercial and practical technology for disinfection in agricultural settings.

To approach this issue, we have recently designed a new type of DBD plasma device (a roller conveyer plasma device) that is well suited to the disinfection of agricultural products such as vegetables and fruits during the sorting process on rollers [33]. To achieve broad applicability for the DBD plasma device across a range of agricultural food products, further development and improvement such as scale-up and better cost-performance will be needed. Thus, further optimization of plasma generation to maximize disinfection efficiency will be required before the device can be widely applied in a commercial setting.

4. Materials and Methods

4.1. DBD Plasma Torch

The DBD plasma torch used in the present study was the same as described in a previous report [34]. Briefly, a ceramic tube (Al_2O_3) (inner diameter, 4 mm; outer diameter, 6 mm; length, 100 mm) was wound around the tube by a copper tape (thickness, 80 μm; length, 60 mm) that was used as an earth electrode (Figure 1). A stainless steel mesh (SUS304) with stainless steel wires was placed inside the tube to act as a high voltage electrode. Next, a low frequency and high-voltage power supply (10 kV peak-to-peak, 10 kHz) was connected to the two electrodes. During plasma generation, an air pump (Suishin SSPP-2S; Suisaku Co., Sakai, Japan) was used at an air flow rate of 3.5 L/min. The distance from the torch top to the liquid surface was set at 20 mm. Aliquots (20 μL) of *E. coli* suspension were dropped on a watch glass and subjected to plasma treatment.

4.2. Plasma Treatment of Transformed Bacteria and Colony Counting

E. coli (DH5α Competent Cells, Takara Bio, Shiga, Japan) was used for the plasma treatment. Transformation of pPRO-EX-HT-CAT (Thermo Fisher Scientific, Waltham, MA, USA) with a plasmid carrying an ampicillin-resistance gene and chloramphenicol acetyltransferase (CAT) gene was performed according to a conventional methodology [35]. *E. coli* obtained after selection on LB (Luria-Bertani) agar medium supplemented with ampicillin (50 mg/L) were picked and then cultured in LB liquid medium with or without ampicillin (50 mg/L) at 37 °C for 24 h. The resultant *E. coli* were then subjected to plasma treatment. As a control, untransformed *E. coli* were also exposed to plasma. Viable cell count after plasma treatment was performed by spreading onto LB agar medium with or without ampicillin (50 mg/L). After overnight culture at 37 °C, colony counting was performed to determine the colony forming units per mL (CFU/mL).

4.3. Enzyme-Linked Immunosorbent Assay (ELISA)

CAT was quantified by ELISA using a CAT ELISA Kit (cat no. #11 363 727 001; Roche, Rotkreuz, Switzerland) in the accordance with the manufacturer's instructions.

4.4. Acetyltransferase Activity Assay

CAT activity was measured using an acetyltransferase activity assay kit (Enzo Life Sciences Inc., Farmingdale, NY, USA) as the index of RFU (relative fluorescence units) with an excitation wavelength of 400 nm and an emission wavelength of 490 nm.

4.5. PCR

The level of intact *E. coli* genomic DNA was assessed using a Bacterial 16S rDNA PCR Kit (Takara Bio Inc., Shiga, Japan) according to the manufacturer's instructions. Bacterial DNA samples were subjected to 30 cycles of PCR amplification conditions (94 °C for 0.5 min, 55 °C for 0.5 min, 72 °C for 1 min) using a PC320 thermal cycler (Astec Co. Ltd., Fukuoka, Japan). Amplified DNA was analyzed by agarose gel electrophoresis (1% gel) and the DNA was visualized using a WSE-5200 Printgraph 2M device (ATTO Corporation, Tokyo, Japan). The amplified products were extracted using a QIAquick Gel Extraction Kit (QIAGEN, Hilden, Germany) and directly subjected to DNA sequencing using primers F1 and R1 (*bacterial*) of the Bacterial 16S rDNA PCR Kit with an ABI 373 OXL Genetic Analyzer (Applied Biosystems, Foster City, CA, USA).

4.6. Measurement of LPS

DBD plasma-treated or untreated bacterial suspensions were resuspended in distilled water (Otsuka Pharmaceutical Co. Ltd., Tokyo, Japan) and subjected to the chromogenic Limulus test (Limulus-color KY test; Wako Pure Chemical Industries Ltd., Osaka, Japan) for quantifying LPS. Absorbance at 415 nm by reference to that at 655 nm was compared with a standard curve obtained using a LPS solution (Wako Pure Chemical Industries Ltd., Osaka, Japan). LPS concentrations (Endotoxin Units (EU)/mL) were then estimated from the index of absorbance.

4.7. Statistical Analysis

GraphPad Prism 7.02 software (GraphPad Prism Software Inc., La Jolla, CA, USA) was used for the statistical analysis. Mean ± standard deviation of experiments carried out at least in triplicate are shown. Non-repeated analysis of variance (ANOVA) followed by Bonferroni's multiple comparison test was applied to the statistical analysis of significant difference.

5. Conclusions

The present study established that there was no difference in susceptibility between antibiotic-resistant and non-resistant bacteria to plasma treatment using an DBD plasma torch. In addition, ESR has shown that the plasma torch generates reactive chemical species including OH and H radicals, which may contribute to the inactivation of DNA, proteins, and LPS. However, OH and H radicals may react with various other bacterial components to generate oxidation products. Therefore, the bactericidal mechanism needs to be further clarified by the detailed analysis of changes to the components of the bacteria induced by radicals generated by the plasma.

Nonetheless, the mechanism of inactivation is unlikely to differ between antibiotic-resistant and non-resistant bacteria. Therefore, the plasma method may be especially useful for eliminating antibiotic-resistant bacteria from the environment. Data on the likely inactivation mechanism may contribute to improving the efficiency of the plasma system by fine-tuning plasma generation to maximize bactericidal action. Indeed, future optimization of the plasma system based on knowledge gained concerning the inactivation mechanism(s) may facilitate increased inactivation efficiency.

Supplementary Materials: The following are available online at http://www.mdpi.com/1422-0067/21/17/6326/s1.

Author Contributions: Conceptualization, A.S. and T.M.; Resources, T.M.; Data curation, A.S.; Writing—original draft, A.S.; Writing—review and editing, A.S. and T.M. All authors have read and agreed to the published version of the manuscript.

Funding: This work was supported in part by Grant-in-Aid from The Ito Foundation. This work was also supported by the JSPS (Japan Society for the Promotion of Science) KAKENHI grant numbers JP20K03919, JP16K04997, JP22110514, and JP24110717, and the Science and Technology Research Promotion Program for Agriculture, Forestry, Fisheries, and Food Industry (Grant number 26015A) as well as the Promotion of Basic Research Activities for Innovative Biosciences from Bio-oriented Technology Research Advancement Institution (BRAIN).

Acknowledgments: The authors would like to acknowledge Risa Yamashiro (University of the Ryukyus, Japan) and Kunihiko Tajima (Kyoto Institute of Technology) for their technical assistance.

Conflicts of Interest: The authors declare no conflict of interest.

Int. J. Mol. Sci. **2020**, *21*, 6326

Abbreviations

ANOVA	Analysis of variance
CAT	Chloramphenicol acetyltransferase
CFU	Colony forming units
DBD	Dielectric barrier discharge
ELISA	Enzyme-linked immunosorbent assay
ESBL	Extended-Spectrum β-lactamase
ESR	Electron spin resonance
EU	Endotoxin Units
H·	H radical
LB	Luria-Bertani
LPS	Lipopolysaccharides
MRSA	Methicillin-resistant *S. aureus*
OH·	OH radical
PCR	Polymerase chain reaction
RFU	Relative fluorescence units
SMD	Surface micro-discharge
UV	Ultraviolet

References

1. Wen, Q.; Yang, L.; Zhao, Y.; Huang, L.; Chen, Z. Insight into effects of antibiotics on reactor performance and evolutions of antibiotic resistance genes and microbial community in a membrane reactor. *Chemosphere* **2018**, *197*, 420–429. [CrossRef] [PubMed]

2. Koskiniemi, S.; Virtanen, P. Selective killing of antibiotic-resistant bacteria from within. *Nature* **2019**, *570*, 449–450. [CrossRef] [PubMed]

3. Dunn, S.J.; Connor, C.; McNally, A. The evolution and transmission of multi-drug resistant *Escherichia coli* and *Klebsiella pneumoniae*: The complexity of clones and plasmids. *Curr. Opin. Microbiol.* **2019**, *51*, 51–56. [CrossRef] [PubMed]

4. Mathers, A.J.; Peirano, G.; Pitout, J.D. The role of epidemic resistance plasmids and international high-risk clones in the spread of multidrug-resistant *Enterobacteriaceae*. *Clin. Microbiol. Rev.* **2015**, *28*, 565–591. [CrossRef] [PubMed]

5. Chaudhry, S.B.; Veve, M.P.; Wagner, J.L. Cephalosporins: A focus on side chains and beta-lactam cross-reactivity. *Pharmacy* **2019**, *7*, 103. [CrossRef]

6. Guenther, S.; Ewers, C.; Wieler, L.H. Extended-spectrum beta-lactamases producing *E. coli* in wildlife, yet another form of environmental pollution? *Front. Microbiol.* **2011**, *2*, 246. [CrossRef]

7. Landers, T.F.; Cohen, B.; Wittum, T.E.; Larson, E.L. A review of antibiotic use in food animals: Perspective, policy, and potential. *Public Health Rep.* **2012**, *127*, 4–22. [CrossRef]

8. Van Breda, L.K.; Ward, M.P. Evidence of antimicrobial and disinfectant resistance in a remote, isolated wild pig population. *Prev. Vet. Med.* **2017**, *147*, 209–212. [CrossRef]

9. Chang, P.H.; Juhrend, B.; Olson, T.M.; Marrs, C.F.; Wigginton, K.R. Degradation of extracellular antibiotic resistance genes with UV254 treatment. *Environ. Sci Technol.* **2017**, *51*, 6185–6192. [CrossRef]

10. De la Obra Jimenez, I.; Lopez, J.L.C.; Ibanez, G.R.; Garcia, B.E.; Perez, J.A.S. Kinetic assessment of antibiotic resistant bacteria inactivation by solar photo-Fenton in batch and continuous flow mode for wastewater reuse. *Water Res.* **2019**, *159*, 184–191. [CrossRef]

11. Guo, C.; Wang, K.; Hou, S.; Wan, L.; Lv, J.; Zhang, Y.; Qu, X.; Chen, S.; Xu, J. H_2O_2 and/or TiO_2 photocatalysis under UV irradiation for the removal of antibiotic resistant bacteria and their antibiotic resistance genes. *J. Hazard. Mater.* **2017**, *323*(Pt. B), 710–718. [CrossRef]

12. Jiang, Q.; Yin, H.; Li, G.; Liu, H.; An, T.; Wong, P.K.; Zhao, H. Elimination of antibiotic-resistance bacterium and its associated/dissociative bla_{TEM-1} and *aac(3)-II* antibiotic-resistance genes in aqueous system via photoelectrocatalytic process. *Water Res.* **2017**, *125*, 219–226. [CrossRef]

13. Tung le, M.; Cong, N.X.; Huy le, T.; Lan, N.T.; Phan, V.N.; Hoa, N.Q.; Vinh le, K.; Thinh, N.V.; Tai le, T.; Ngo, D.T.; et al. Synthesis, Characterizations of superparamagnetic Fe_3O_4-Ag hybrid nanoparticles and their application for highly effective bacteria inactivation. *J. Nanosci. Nanotechnol.* **2016**, *16*, 5902–5912. [CrossRef]

14. Chen, H.; Zhang, M. Effects of advanced treatment systems on the removal of antibiotic resistance genes in wastewater treatment plants from Hangzhou, China. *Environ. Sci. Technol.* **2013**, *47*, 8157–8163. [CrossRef]

15. Sousa, J.M.; Macedo, G.; Pedrosa, M.; Becerra-Castro, C.; Castro-Silva, S.; Pereira, M.F.R.; Silva, A.M.T.; Nunes, O.C.; Manaia, C.M. Ozonation and UV_{254nm} radiation for the removal of microorganisms and antibiotic resistance genes from urban wastewater. *J. Hazard. Mater.* **2017**, *323*, 434–441. [CrossRef] [PubMed]

16. Yang, Y.; Wan, K.; Yang, Z.; Li, D.; Li, G.; Zhang, S.; Wang, L.; Yu, X. Inactivation of antibiotic resistant *Escherichia coli* and degradation of its resistance genes by glow discharge plasma in an aqueous solution. *Chemosphere* **2020**, *252*, 126476. [CrossRef]

17. Lis, K.A.; Kehrenberg, C.; Boulaaba, A.; von Kockritz-Blickwede, M.; Binder, S.; Li, Y.; Zimmermann, J.L.; Pfeifer, Y.; Ahlfeld, B. Inactivation of multidrug-resistant pathogens and *Yersinia enterocolitica* with cold atmospheric-pressure plasma on stainless-steel surfaces. *Int. J. Antimicrob. Agents* **2018**, *52*, 811–818. [CrossRef] [PubMed]

18. Serra-Burriel, M.; Keys, M.; Campillo-Artero, C.; Agodi, A.; Barchitta, M.; Gikas, A.; Palos, C.; Lopez-Casasnovas, G. Impact of multi-drug resistant bacteria on economic and clinical outcomes of healthcare-associated infections in adults: Systematic review and meta-analysis. *PLoS ONE* **2020**, *15*, e0227139. [CrossRef] [PubMed]

19. Kvam, E.; Davis, B.; Mondello, F.; Garner, A.L. Nonthermal atmospheric plasma rapidly disinfects multidrug-resistant microbes by inducing cell surface damage. *Antimicrob. Agents Chemother.* **2012**, *56*, 2028–2036. [CrossRef]

20. Brun, P.; Bernabe, G.; Marchiori, C.; Scarpa, M.; Zuin, M.; Cavazzana, R.; Zaniol, B.; Martines, E. Antibacterial efficacy and mechanisms of action of low power atmospheric pressure cold plasma: Membrane permeability, biofilm penetration and antimicrobial sensitization. *J. Appl. Microbiol.* **2018**, *125*, 398–408. [CrossRef]

21. Joshi, S.G.; Paff, M.; Friedman, G.; Fridman, G.; Fridman, A.; Brooks, A.D. Control of methicillin-resistant *Staphylococcus aureus* in planktonic form and biofilms: A biocidal efficacy study of nonthermal dielectric-barrier discharge plasma. *Am. J. Infect. Control* **2010**, *38*, 293–301. [CrossRef] [PubMed]

22. Kim, S.M.; Kim, J.I. Decomposition of biological macromolecules by plasma generated with helium and oxygen. *J. Microbiol.* **2006**, *44*, 466–471. [PubMed]

23. Maeda, K.; Toyokawa, Y.; Shimizu, N.; Imanishi, Y.; Sakudo, A. Inactivation of *Salmonella* by nitrogen gas plasma generated by a static induction thyristor as a pulsed power supply. *Food Control* **2015**, *52*, 54–59. [CrossRef]

24. Modic, M.; McLeod, N.P.; Sutton, J.M.; Walsh, J.L. Cold atmospheric pressure plasma elimination of clinically important single- and mixed-species biofilms. *Int. J. Antimicrob. Agents* **2017**, *49*, 375–378. [CrossRef] [PubMed]

25. Sakudo, A.; Toyokawa, Y.; Nakamura, T.; Yagyu, Y.; Imanishi, Y. Nitrogen gas plasma treatment of bacterial spores induces oxidative stress that damages the genomic DNA. *Mol. Med. Rep.* **2017**, *15*, 396–402. [CrossRef]

26. Shintani, H.; Sakudo, A.; Burke, P.; McDonnell, G. Gas plasma sterilization of microorganisms and mechanisms of action. *Exp. Ther. Med.* **2010**, *1*, 731–738. [CrossRef]

27. Shintani, H.; Shimizu, N.; Imanishi, Y.; Sekiya, T.; Tamazawa, K.; Taniguchi, A.; Kido, N. Inactivation of microorganisms and endotoxins by low temperature nitrogen gas plasma exposure. *Biocontrol Sci.* **2007**, *12*, 131–143. [CrossRef]

28. Yamashiro, R.; Misawa, T.; Sakudo, A. Key role of singlet oxygen and peroxynitrite in viral RNA damage during virucidal effect of plasma torch on feline calicivirus. *Sci. Rep.* **2018**, *8*, 17947. [CrossRef]

29. Sakudo, A.; Yamashiro, R.; Haritani, M.; Furusaki, K.; Onishi, R.; Onodera, T. Inactivation of non-enveloped viruses and bacteria by an electrically charged disinfectant containing meso-structure nanoparticles via modification of the genome. *Int. J. Nanomedicine* **2020**, *15*, 1387–1395. [CrossRef]

30. Sakudo, A.; Yagyu, Y.; Onodera, T. Disinfection and sterilization using plasma technology: Fundamentals and future perspectives for biological applications. *Int. J. Mol. Sci.* **2019**, *20*, 5216. [CrossRef]

31. Chaves-Lopez, C.; Usai, D.; Donadu, M.G.; Serio, A.; Gonzalez-Mina, R.T.; Simeoni, M.C.; Molicotti, P.; Zanetti, S.; Pinna, A.; Paparella, A. potential of *Borojoa patinoi* Cuatrecasas water extract to inhibit nosocomial antibiotic resistant bacteria and cancer cell proliferation in vitro. *Food Funct.* **2018**, *9*, 2725–2734. [CrossRef] [PubMed]

32. Usai, D.; Donadu, M.; Bua, A.; Molicotti, P.; Zanetti, S.; Piras, S.; Corona, P.; Ibba, R.; Carta, A. Enhancement of antimicrobial activity of pump inhibitors associating drugs. *J. Infect. Dev. Ctries.* **2019**, *13*, 162–164. [CrossRef] [PubMed]
33. Toyokawa, Y.; Yagyu, Y.; Misawa, T.; Sakudo, A. A new roller conveyer system of non-thermal gas plasma as a potential control measure of plant pathogenic bacteria in primary food production. *Food Control* **2017**, *72*, 62–72. [CrossRef]
34. Sakudo, A.; Miyagi, H.; Horikawa, T.; Yamashiro, R.; Misawa, T. Treatment of *Helicobacter pylori* with dielectric barrier discharge plasma causes UV induced damage to genomic DNA leading to cell death. *Chemosphere* **2018**, *200*, 366–372. [CrossRef]
35. Green, M.R.; Sambrook, J. *Molecular Cloning: A Laboratory Manual*, 4th ed.; Cold Spring Harbor Laboratory Press: Cold Spring Harbor, NY, USA, 2012.

International Journal of
Molecular Sciences

Article

Enhanced Osteogenic Differentiation of Human Mesenchymal Stem Cells on Amine-Functionalized Titanium Using Humidified Ammonia Supplied Nonthermal Atmospheric Pressure Plasma

Jae-Sung Kwon [1,2,*], Sung-Hwan Choi [2,3], Eun Ha Choi [4], Kwang-Mahn Kim [1,2] and Paul K. Chu [5,*]

[1]   Department and Research Institute of Dental Biomaterials and Bioengineering, Yonsei University College of Dentistry, Seoul 03722, Korea; kmkim@yuhs.ac
[2]   BK21 PLUS Project, Yonsei University College of Dentistry, Seoul 03722, Korea; selfexam@yuhs.ac
[3]   Department of Orthodontics, Institute of Craniofacial Deformity, Yonsei University College of Dentistry, Seoul 03722, Korea
[4]   Plasma Bioscience Research Center, Kwangwoon University, Seoul 01897, Korea; ehchoi@kw.ac.kr
[5]   Department of Physics, Department of Materials Science and Engineering, and Department of Biomedical Engineering, City University of Hong Kong, Kowloon, Hong Kong, China
*   Correspondence: jkwon@yuhs.ac (J.-S.K.); paul.chu@cityu.edu.hk (P.K.C.)

Received: 22 July 2020; Accepted: 19 August 2020; Published: 24 August 2020

Abstract: The surface molecular chemistry, such as amine functionality, of biomaterials plays a crucial role in the osteogenic activity of relevant cells and tissues during hard tissue regeneration. Here, we examined the possibilities of creating amine functionalities on the surface of titanium by using the nonthermal atmospheric pressure plasma jet (NTAPPJ) method with humidified ammonia, and the effects on human mesenchymal stem cell (hMSC) were investigated. Titanium samples were subjected to NTAPPJ treatments using nitrogen (N-P), air (A-P), or humidified ammonia (NA-P) as the plasma gas, while control (C-P) samples were not subjected to plasma treatment. After plasma exposure, all treatment groups showed increased hydrophilicity and had more attached cells than the C-P. Among the plasma-treated samples, the A-P and NA-P showed surface oxygen functionalities and exhibited greater cell proliferation than the C-P and N-P. The NA-P additionally showed surface amine-related functionalities and exhibited a higher level of alkaline phosphatase activity and osteocalcin expression than the other samples. The results can be explained by increases in fibronectin absorption and focal adhesion kinase gene expression on the NA-P samples. These findings suggest that NTAPPJ technology with humidified ammonia as a gas source has clinical potential for hard tissue generation.

Keywords: atmospheric-pressure plasma; titanium; amine; osteogenic differentiation; mesenchymal stem cells

1. Introduction

Osteoporosis, trauma, and cancer cause damage to and loss of hard tissue, and such conditions are increasing as the baby boomer generation ages [1–5]. Consequently, therapeutic strategies based on hard tissue regeneration that can replace or repair damaged hard tissues are in increasing demand. Over two million hard tissue replacement procedures are currently conducted annually worldwide [3–5]. Biomaterials are commonly used in hard tissue regeneration, and titanium (Ti) and Ti alloys are commonly used in orthopedics and dentistry [6,7]. Because of their good biocompatibility, corrosion resistance, and mechanical properties, such as elastic modulus, density, and surface hardness,

Ti and Ti alloys have several advantages over other biomaterials in hard tissue regeneration [8]. However, despite recent advances in Ti technology, rapid and firm integration of Ti with surrounding tissue, which is necessary for dental and orthopedic hard tissue regeneration, remains challenging [9]. The integration of biomaterials such as Ti with surrounding bone tissues is commonly known as osseointegration. Osseointegration involves osteoconduction, which enables existing osteoblasts to attach onto the material and proliferate, and osteoinduction, which induces stem cells to differentiate into the osteoblastic lineage [10,11]. Failure or impairment of these two processes results in soft tissue formation surrounding the implant, which causes fibrous encapsulation of the material and prevents achievement of firm fixation over a short period of time [12].

Because of their potential for unlimited proliferation, stem cells are of interest in the development of hard tissue regeneration therapies [2]. Among the various types of stem cells, human mesenchymal stem cells (hMSCs) can be obtained from the bone marrow of patients. These cells are more abundant than embryonic stem cells and present fewer ethical concerns with their extraction. Furthermore, hMSCs represent the precursors of the osteoblastic lineage [13,14] and can also differentiate into other cell lineages, such as chondrocytes [15], myocytes [16], and adipocytes [17]. Differentiation of hMSCs on the surface of biomaterials may be influenced by the properties of the specific biomaterial, such as its elasticity [18] and surface topography [12,19]. Chemical inducing agents, such as dexamethasone and glycerophosphate, have well-known effects on the osteogenic differentiation of hMSCs; however, these agents do not naturally occur in the human body [20]. Recently, the formation of chemical functional groups, such as amines, on biomaterials has been shown to influence hMSC differentiation into the osteoblastic cell lineage [21–23]. For example, Curran et al. [22] demonstrated that amino functionalization on a glass coverslip resulted in osteogenic hMSC differentiation, and Wang et al. [23] utilized plasma immersion ion implantation with ammonia gas to modify polytetrafluoroethylene and to induce hMSC differentiation into osteoblasts.

Recently, we reported that the application of NTAPPJ could improve the selected properties of biomaterials without causing thermal damage [24,25]. Compared with low-pressure plasma devices, NTAPPJ technology does not require a vacuum, and it is also cost effective, portable, and easy to use. The effects of NTAPPJ-treated Ti result in the removal of hydrocarbons and enhanced osteogenic activity, similar to the effects achieved by ultraviolet treatment of Ti [26]. Nevertheless, current NTAPPJ technology with supplies of air, oxygen, nitrogen, and argon, etc. is limited in terms of its ability to form specific osteogenic functionalities such as amines.

Herein, we demonstrate the feasibility of guiding hMSC differentiation on Ti via the use of nonthermal atmospheric pressure plasma jet (NTAPPJ) technology, which produces electrons, ions, and free radicals at atmospheric pressure [27]. We developed a specific way to form amine functionalities on the surface of Ti using NTAPPJ supplied with humidified ammonia (Figure 1) following pilot testing with different humidified sources, where samples of titanium were exposed to NTAPPJ with different conditions as outlined in Table 1. We hypothesized that the NTAPPJ-promoted formation of various chemical functionalities on the surface of Ti would affect the differentiation of hMSCs, whereby the amine functionalized surface of Ti by humidified ammonia supplied by NTAPPJ would further enhance the osteogenic activity of hMSCs.

Table 1. Nonthermal atmospheric pressure plasma jet exposure conditions and assignment of sample codes.

	Gas Sources	Gas Flow Rate (L/min)	Voltage (kV)	Current (mA)	Treatment Duration (min)	Sample Code
Control	No gas sources	N/A	N/A	N/A	0	C-P
Test Groups	Air (compressed)	1	2.24	1.08	4	A-P
	Nitrogen	1	2.24	1.08	4	N-P
	Nitrogen/Ammonia (humidified)	1	2.24	1.08	4	NA-P

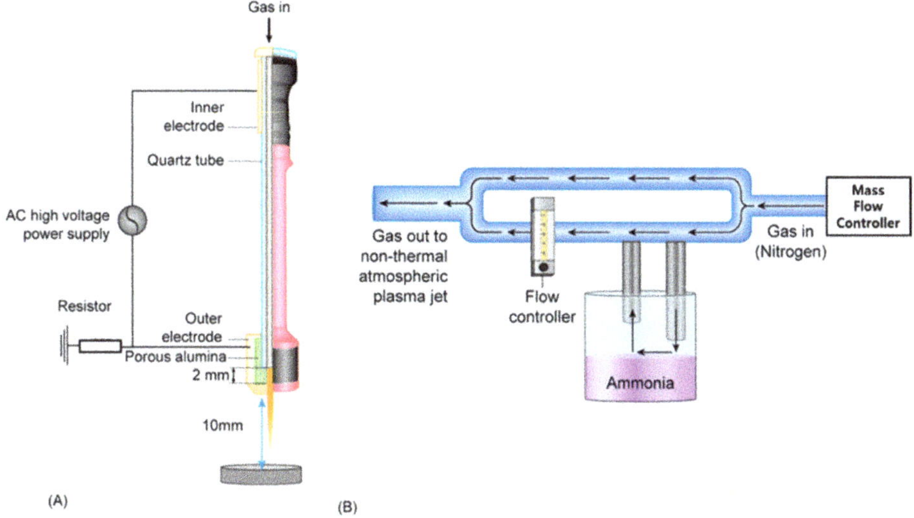

Figure 1. (**A**) Schematic diagram of the nonthermal atmospheric pressure plasma jet and (**B**). the design for a customized device that would supply nitrogen mixed with humidified ammonia.

2. Results

2.1. Morphology and Hydrophilicity of NTAPPJ-Treated Ti Samples

The morphology of the samples was examined by scanning electron microscopy (SEM), and the results showed that a smooth surface with a morphology typical of machined-cut Ti was evident in the control (C-P) group and that the morphology was preserved after plasma exposure for the nitrogen (N-P), air (A-P), and humidified ammonia (NA-P) groups.

Despite the preserved surface morphology, the hydrophilicity of the Ti changed after NTAPPJ exposure. The contact angle on the C-P samples is approximately 75°, indicating a relatively hydrophobic surface, whereas that on the N-P sample is not measurable because of its superhydrophilicity. The A-P and NA-P samples are also relatively hydrophilic compared with the C-P and present contact angles of approximately 25° and 20°, respectively.

Additionally, the changes in contact angle on the Ti surface with respect to time after NTAPPJ exposure have been considered in order to consider reversion of the hydrophilic state of the NTAPPJ-exposed surface (Figure 2). The results showed that N-P, A-P, and NA-P showed reversion in the hydrophilic state with respect to time after initial exposure, though the contact angle values were still lower than C-P at 24 h.

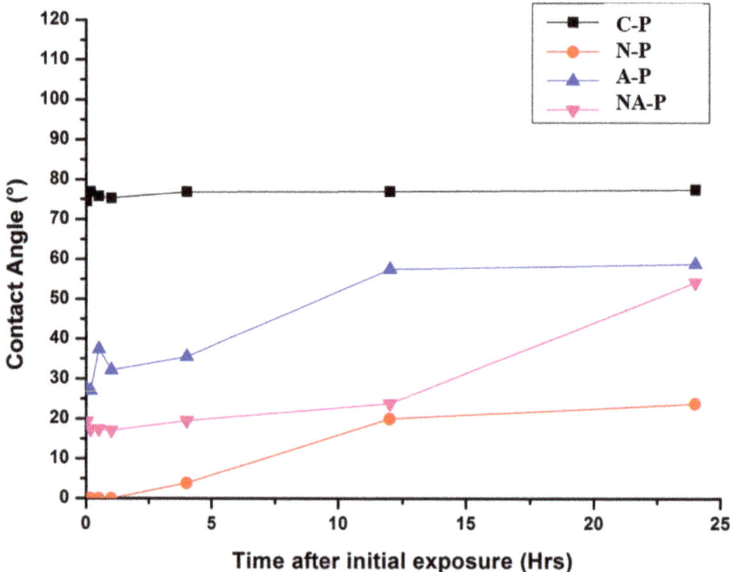

Figure 2. Water contact angles on the samples with time after exposure to nonthermal atmospheric pressure plasma jet.

2.2. Surface Chemistry of NTAPPJ-Treated Ti Samples

The chemical shifts and subsequent changes in chemical composition induced by the formation of new functional groups on the Ti surface were examined via XPS and are shown in Figure 3. Figure 3a–c displays high-resolution XPS spectra for C1s, O1s, and N1s, respectively, on all samples. Figure 3d shows a histogram of the surface chemistry atomic percentage on each sample calculated from the broad spectrum. Figure 3d shows that O is the most abundant element on the surface of all samples. The C-P treatment showed C as the next most abundant element followed by Ti and very minimal N. In terms of N-P, a decrease in C content was noticeable compared with that of C-P, and the reduction in the C-H/C-C peak (C_2) is shown in Figure 3a. Although a decrease in the peak corresponding to C-H/C-C (C_2) was also found for the A-P and NA-P treatments (Figure 3a), the atomic percentage of C was only slightly reduced (Figure 3d) compared with that of the N-P treatment because the A-P and NA-P treatments showed a peak of high binding energy corresponding to the C=O (C_1) functional group. This shift to a higher binding energy for the two groups was also identified in the O1s spectra, as a shift of the peak from a lower binding energy for the O-Ti (O_2) functional group to a higher binding energy for the O-H (O_1) group in both A-P and NA-P is shown in Figure 3b.

In the N1s spectra (Figure 3c), a single peak spectrum corresponding to the N-H (N_2) functional group was observed for the C-P treatment, and similar spectra were observed for the N-P and A-P treatments. However, a significantly higher intensity of the peak corresponding to the N-H (N_2) functional group was observed in the NA-P treatment, and peaks exhibiting a high binding energy for N-O (N_1) and lower binding energy for N-Ti (N_3) were observed. Indeed, the composition of elemental N was much higher in the NA-P samples compared with any of the other samples (Figure 3d) as the composition has been calculated from the area under curve of XPS results.

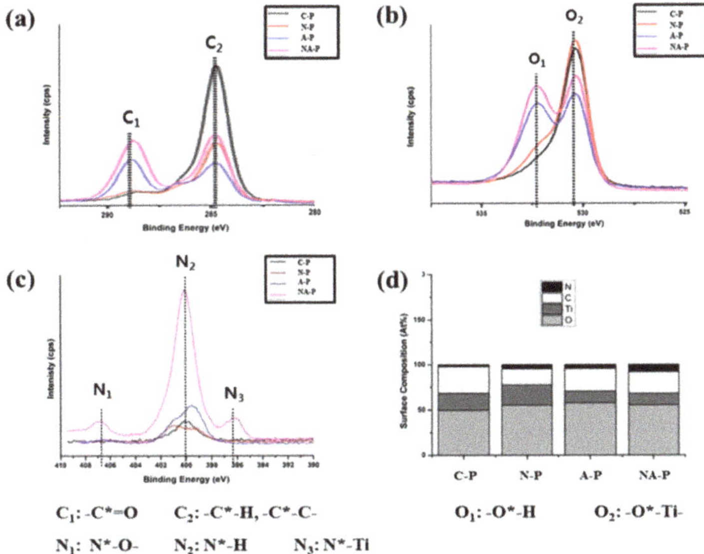

C_1: -C*=O $\quad$ C_2: -C*-H, -C*-C-

N_1: N*-O- $\quad$ N_2: N*-H $\quad$ N_3: N*-Ti

O_1: -O*-H $\quad$ O_2: -O*-Ti-

Figure 3. Surface chemistry of the samples as indicated by high-resolution spectra: (**a**) C1s, (**b**) O1s, and (**c**) N1s. The text below the figure refers to the contribution of each peak. (**d**) Atomic percentages. Each relevant element analyzed is marked with *.

2.3. *Number of Viable Cells and Cell Morphology on the Titanium before and after NTAPPJ Exposure*

The viability of the hMSCs in the C-P and test groups was examined under a confocal laser microscope, and the live cells appeared green and the dead cells appeared red (Figure 4a–d). The results indicated that dead red cells were not visible on the surface of either the C-P or test groups, indicating biocompatibility of both the C-P specimen and NTAPPJ-treated Ti. A significantly greater number of green cells were visible for all of the test groups (Figure 4b–d) compared with the C-P (Figure 4a).

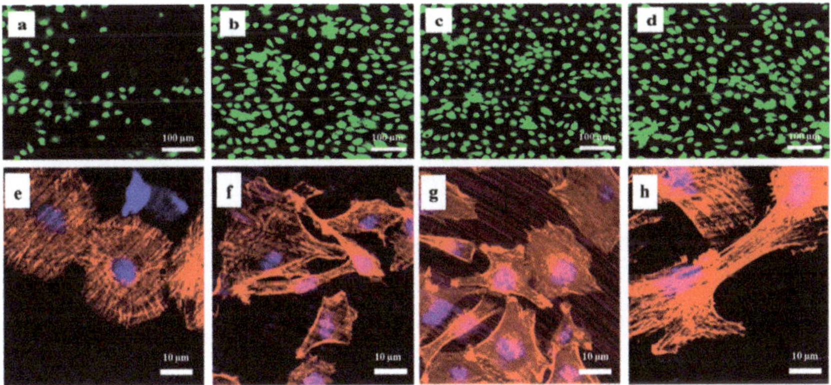

Figure 4. Immunofluorescence images of attached live mesenchymal stem cells: (**a**) control (C-P), (**b**) nitrogen (N-P), (**c**) air (A-P), and (**d**) humidified ammonia (NA-P) (scale bar: 100 μm). Immunofluorescence images of the cytoskeleton of the mesenchymal stem cells (MSCs): (**e**) C-P, (**f**) N-P, (**g**) A-P, and (**h**) NA-P (scale bar: 10 μm). The red color indicates actin filaments, and the blue color shows the nuclei.

The morphology of cells after 4 h of incubation was examined using immunofluorescent images and is shown in Figure 4e–h. A rounded cell morphology with relatively undeveloped actin filaments was evident for the C-P in Figure 4e, whereas stretched cells with actin filaments were evident for N-P (Figure 4f), A-P (Figure 4g), and NA-P (Figure 4h). In terms of quantitative analyses, the results showed that there were no significant differences in number of cell attachments between N-P, A-P, and NA-P ($p > 0.05$) (Figure 5).

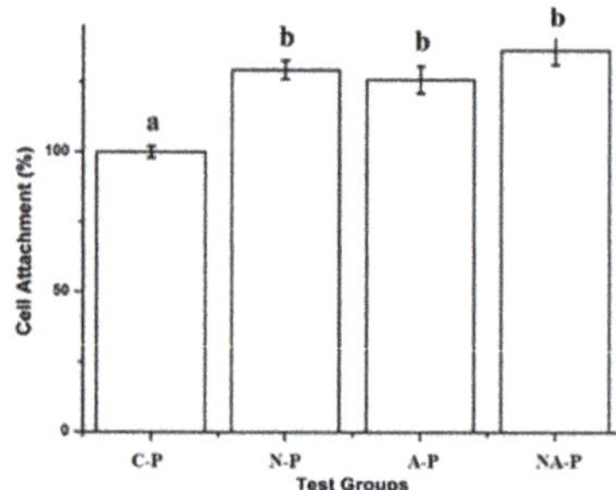

Figure 5. Cell attachment on control and test groups after 4 h: The same lowercase letter indicates no significant differences ($p > 0.05$), and different lowercase letters indicate significant differences ($p < 0.05$), where the greater cellular attachment is higher in alphabetical order (i.e., "b" is significantly greater than "a" and so on).

2.4. Cell Proliferation, Alkaline Phosphatase Acitivity, and Osteogenic Differentiation on Titanium before and after NTAPPJ Exposure

The cell proliferation rate is shown in Figure 6a. Similar to cell attachment, a significantly higher rate of hMSC proliferation was observed for the N-P, A-P, and NA-P groups compared with C-P, although in terms of the differences among the test groups, both A-P and NA-P showed significantly higher levels of proliferation than N-P.

Alkaline Phosphatase (ALP) activity is shown in Figure 6b, and its trend was similar to that for cell proliferation, with significantly higher ALP activity levels observed for the hMSCs in all test groups compared with C-P. In addition, the hMSCs in both the A-P and NA-P groups showed significantly higher ALP activity than the hMSCs in the N-P group. Finally, the NA-P group showed a significantly higher level of ALP activity than the A-P group.

The results of the qRT-PCR analysis for osteogenic markers are shown in Figure 6c–f. The results for ALP gene expression are shown in Figure 6c, and they were similar to the ALP activity results; the N-P, A-P, and NA-P groups all showed significantly higher levels of relative ALP gene expression than C-P, and A-P and NA-P showed significantly higher levels of ALP gene expression than N-P. However, compared with the ALP activity test results, the A-P and NA-P groups did not show differences in terms of the level of relative gene expression. Although significant differences in the relative gene expression of Bone Sialoprotein (BSP) were not observed among the groups (Figure 6d), significantly higher levels of Osteopontin (OPN) expression were observed in N-P, A-P, and NA-P than in C-P and significant differences in expression were not observed among the test groups (Figure 6e). Finally, only NA-P showed a significantly higher level of Osteocalcin (OCN) gene expression compared with C-P (Figure 6f).

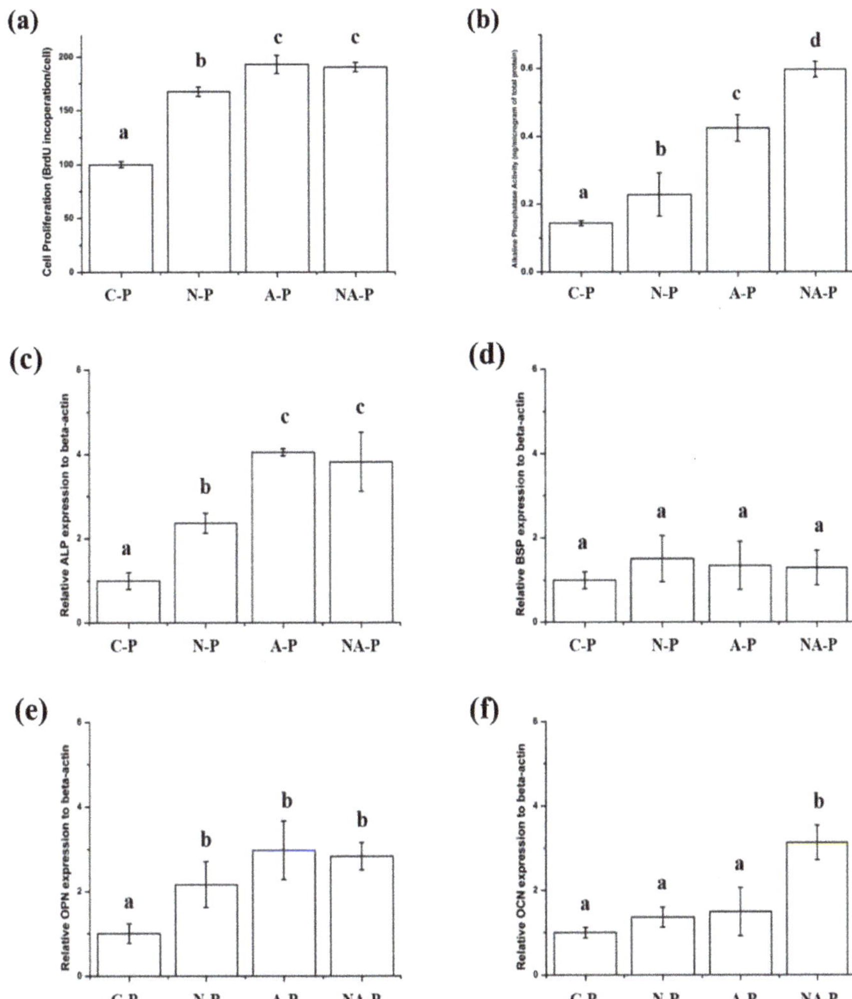

Figure 6. (**a**) Cell proliferation rate and (**b**) alkaline phosphatase activity of human mesenchymal stem cells (hMSCs) cultured on the samples and osteogenic differentiation of hMSCs on the surface of the control and test groups analyzed by qPCR for osteogenic markers: (**c**) alkaline phosphatase (ALP), (**d**) bone sialoprotein (BSP), (**e**) osteopontin (OPN), and (**f**) osteocalcin (OCN) after culturing for 14 days. All the results are expressed relative to the gene expression of C-P as the control. The same lowercase letter indicates no significant difference ($p > 0.05$), and different lowercase letters indicate significant differences ($p < 0.05$). The greater the cellular proliferation or alkaline phosphatase activity or relative gene expression, the higher the letter in alphabetical order (i.e., "b" is significantly greater than "a", and so on).

2.5. Protein Absorption of Titanium and FAK Gene Expression of hMSCs

The amount of Bovine Serum Albumin (BSA) absorption on the surface of the C-P and test groups was measured, and the results for the C-P group were not significantly different compared to those of the test groups (Figure 7a). However, the results for fibronectin absorption were inconsistent with the results for BSA absorption, which was significantly higher in the A-P and NA-P groups than in the C-P and N-P groups (Figure 7b).

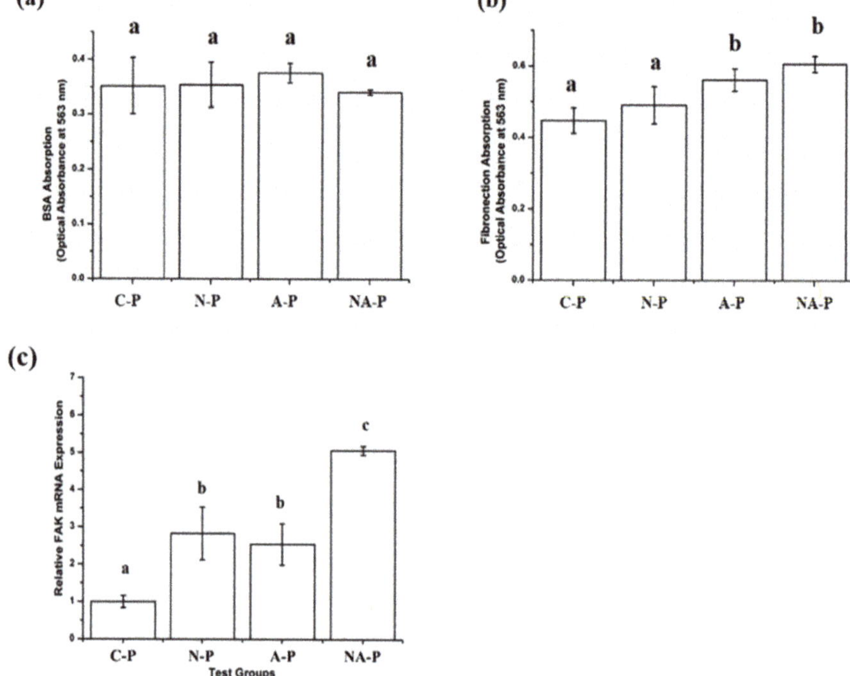

Figure 7. Protein absorption in the control and test groups for two different proteins: (**a**) bovine serum albumin and (**b**) fibronectin. (**c**) Relative gene expression level of focal adhesion kinase (FAK) in hMSCs cultured on the control and test groups where all the results are expressed as relative to gene expression of C-P as the control: The same lowercase letter indicates no significant difference ($p > 0.05$), and different lowercase letters indicate significant differences ($p < 0.05$). The higher the protein absorption or higher the relative gene expression, the higher the letter in alphabetical order (i.e., "b" is significantly greater than "a" and so on).

The relative expression of Focal Adhesion Kinase (FAK) by hMSCs cultured on the C-P and test samples was measured, and significantly higher levels of expression were observed for the hMSCs cultured on the NA-P sample than those observed in the remaining groups (Figure 7c). Although the level of FAK gene expression was significantly higher for N-P and A-P compared with C-P, significant differences were not observed between the N-P and A-P test groups.

3. Discussion

Various molecular functionalities were formed on the surface of the Ti samples, and they influenced on the osteogenic differentiation of hMSCs. Similar results have been reported by other authors who reported the osteogenic guidance of hMSCs on various biomaterials using different chemical functionalities [22,23]. However, this is the first study that considered the use of humid ammonia-supplied NTAPPJ, which forms various chemical functionalities on the surface of Ti that included amine-related functionalities.

Compared with other methods capable of forming osteogenesis-promoting functionalities on biomaterial, the use of NTAPPJ preserved the topographical features of the Ti biomaterial. Topographical changes have been considered a limitation with previous methods of chemical functionalities formation, as such changes would have an additional influence on cellular behavior and tissue formation [28]. This has been shown to be a problem with methods such as magnetron sputtering, physical vapor deposition, or anodization, as the results will cause changes in surface topographical features [7,12,23].

The first change observed after the exposure of Ti to NTAPPJ was an improvement of hydrophilicity. Hydrophilic biomaterials may be favored in dental and medical implants for hard tissue regeneration because the access to blood, which contains necessary cells and proteins, is enhanced [29,30]. The hydrophilicity of Ti changes over time, and aging produces a progressively more hydrophobic surface (Figure 2). This limitation is not solely associated with NTAPPJ technology because other chemical modification techniques exhibit similar time-dependent effects [31,32]. However, this issue is not prohibitive as long as the effects are well understood and addressed during clinical applications; moreover, the portability of the technology offers a unique advantage over other conventional surface treatment techniques because surface treatments via NTAPPJ could occur immediately prior to implant placement. Nonetheless, the samples exposed to NTAPPJ were more hydrophilic than the C-P samples, even after 24 h. Our previous studies also considered samples after NTAPPJ treatment longer than 24 h [33]. The results showed eventual reversion of hydrophilicity to its original state, though the reversions were indicated to be minimized by storage in solution such as distilled water as the method prevented hydrocarbon contamination [33].

Chemical changes to biomaterials often alter the surface hydrophilicity of the material [24]; however, changes in hydrophilicity do not necessarily cause corresponding variations in osteogenic differentiation by hMSCs because osteogenicity is also related to chemical features other than wettability [29]. Accordingly, the chemical changes after plasma exposure were monitored. As described, NTAPPJ produces electrons, radicals, and ions that interact directly or indirectly with the Ti surface. The chemical species produced by NTAPPJ modified the material in terms of chemical composition and formation of new functional groups as indicated by XPS (Figure 3). The formation of chemical functionalities on the surface of Ti can be varied using different gases supplied for NTAPPJ. First, the N-P treatment resulted in a simple reduction of the carbon content compared with C-P. Although this decrease was also observed for the A-P and NA-P treatments, the atomic percentage of carbon was not greatly reduced compared with that of the N-P treatment. The extreme hydrophilicity of N-P and relatively lower hydrophilicity of A-P and NA-P may have been caused by the removal of hydrocarbons and the reduced carbon content on the samples [24,34] because similar results have been achieved by other methods, such as ultraviolet exposure of Ti [29,31,34]. Surface hydrocarbons are formed as a reaction between surface titanium dioxide and the surrounding atmosphere including carbon dioxide [29,31,33]. However, the N-P treatment showed a simple reduction of carbon content without formation of any new functional groups, whereas the A-P and NA-P treatments both formed oxygen-related functional groups, such as C=O or O-H. Finally, the A-P and NA-P samples showed differences in the N1s spectra and N concentration as indicated by a significantly high intensity peak corresponding to the functional group N-H and peaks of high binding energy for the N-O group and lower binding energy for the N-Ti group in the NA-P samples, whereas these peaks were not observed for the A-P samples. It is interesting to note that N-P treatment that used nitrogen as the source of NTAPPJ caused very little changes in surface nitrogen of Ti (Figure 3c). This is perhaps linked to the previously reported study that optical emission spectra analyses of N-P NTAPPJ showed very small amounts of reactive nitrogen species [25,30].

Biological experiments were performed to assess cell viability, attachment, proliferation, and differentiation. The hMSCs showed high levels of cell attachment (Figures 4 and 5) and little to no evidence of cytotoxicity, which are critical characteristics for dental and medical applications [35,36] and consistent with previous studies [24,37,38]. More substantial cell attachment was observed in the N-P, A-P, and NA-P treatments than in C-P (Figures 4 and 5). hMSC attachment is an early event that is essential to the proliferation and differentiation required for bone formation [39]. As previously discussed, topographical differences were not observed between the C-P and test samples; thus, any differences should have been caused by chemical effects introduced by NTAPPJ exposure. Despite the superior hydrophilicity of the N-P samples compared with the A-P and NA-P samples, significant differences ($p > 0.05$) were not observed between the test groups in terms of the number of attached cells, which may have been related to limitations of the in vitro tests. However, this finding provides evidence that cellular attachment is not as strongly correlated with hydrophilic surfaces as previously demonstrated [29].

Similar trends in cell morphology were observed after 4 h (Figure 4), and actin filament formation and stretched cells were observed for all of the test samples. Cell morphology is an important parameter in cell–biomaterial interactions [40]. For example, the high level of actin cytoskeleton formation on the surface of the test groups exposed to NTAPPJ indicates superior cellular perception by the material [41]. However, despite the different chemical changes, obvious differences in cell viability, attachment, or morphology were not observed between the groups exposed to NTAPPJ.

Differences were observed in the proliferation rate and ALP activity (Figure 6a–b) of hMSCs on the plasma-treated samples. Both the A-P and NA-P treatments showed a higher level ($p < 0.05$) of proliferation than the N-P treatment, and the NA-P showed higher levels ($p < 0.05$) of ALP activity than A-P. Proliferation of hMSCs may indicate osteoconductive properties of the material because it is consistent with the number of mature or progenitor osteoblast cells derived from hMSCs [42]. These features may be modified by changes in the surface chemistry induced by NTAPPJ or similar treatments [30,37,38,43]. The observed chemical changes improved the cell proliferation of the test groups relative to C-P, and the presence of chemical functionality, such as C=O and O-H groups, further enhanced the effects as shown for the A-P and NA-P treatments. This finding highlights the advantage of NTAPPJ over other technologies, such as ultraviolet treatments for dental implants, because these treatments reduce only the carbon content but do not form other functional groups [29,31,34]. A reduction in the carbon content enhances cell attachment and morphological transformation but has limited effects on the proliferation rates in the absence of the necessary bioactive chemical functional groups.

ALP is a key enzyme involved in osteogenic cell formation, and it regulates phosphatase metabolism via the hydrolysis of phosphate esters [44]. The improved ALP activity therefore indicates increased osteogenic activity. Here, greater osteogenic activity was observed from all test groups, with the largest enhancement shown by the NA-P group, which possesses more nitrogen-related functionalities, such as N-H, N-O, and N-Ti; is linked to the osteogenic differentiation of hMSCs on the Ti samples; and is related to successful osteoinduction and early osseointegration. Four primers were considered in the quantitative real-time polymerase chain reaction (qPCR) amplification of ALP, BSP, OPN, and OCN (Figure 6c–f). High BSP expression indicates the formation of proteins that bind strongly to hydroxyapatite via a negatively charged domain, thereby resulting in bone formation [45]. OPN expression is related to the formation of a phosphoprotein that serves as the bridge between osteoblasts and hydroxyapatite [46]. OCN is a calcium-binding protein that regulates bone crystal growth and serves as the most specific marker for osteoblast maturation [47,48]. To investigate the effects of NTAPPJ exposure on the chemical functional groups while minimizing other influences, the cells were cultured in basic medium without osteogenic supplements such as dexamethasone or glycerophosphate. A higher level of ALP and OPN gene expression was observed in each test group than in C-P ($p < 0.05$). However, only NA-P showed a significantly higher level ($p < 0.05$) of OCN gene expression than C-P. Because OCN represents the most specific marker in osteoblast maturation, these findings indicate that the NA-P samples were the most potent in terms of osteogenic differentiation of hMSCs. The difference between the NA-P and the C-P and other test groups results from the presence of nitrogen-related functional groups, such as N-O, N-H, and N-Ti, which favor osteogenic differentiation of stem cells [49] because the positive charge of the chemical functional groups increases affinity for fibronectin [50,51].

To study the possible mechanisms underlying osteogenic differentiation and variations in hMSC differentiation, the absorption of fibronectin and BSA was examined (Figure 7a,b). The results show that greater fibronectin absorption ($p < 0.05$) occurred on the NA-P samples than on the C-P and N-P samples, whereas differences in BSA absorption ($p > 0.05$) were not observed between the C-P and test groups. BSA carries a neutral charge and is less influenced by the charge on the surface of the biomaterial [52]. However, because fibronectin carries negative charges, absorption of this protein is influenced by the positive charge on the biomaterial generated by the various functional groups [52].

The presence of fibronectin and the subsequent exposure of receptors that bind to specific integrin receptors increase the expression of osteogenic differentiation genes via the FAK-ERK pathway [53,54]. Therefore, the absorption of fibronectin is an important initial step in the osteogenic differentiation of hMSCs, and it was investigated by qRT-PCR analyses of FAK gene expression in the hMSCs cultured on the samples for 6 h in a basic medium (Figure 7c). The results show a significantly higher level ($p < 0.05$) of FAK gene expression in the hMSCs cultured on the NA-P samples than in those cultured on the other samples, which corroborates the hypothesis. However, the data do not explain all observed phenomena. As observed from the NA-P group, a substantially higher level ($p < 0.05$) of fibronectin absorption occurred on the A-P samples than on the C-P or N-P samples, and significant differences ($p > 0.05$) were not observed between the NA-P and A-P samples. Nonetheless, the hMSCs cultured on the A-P samples showed a significantly lower level ($p < 0.05$) of FAK gene expression than those cultured on the NA-P samples.

The possible mechanism underlying these differences is summarized in Figure 8. The C-P samples that were not subjected to NTAPPJ treatment possessing hydrocarbons exhibited a lower degree of cell attachment and a more rounded cellular morphology relative to the hMSCs on the plasma-exposed samples. NTAPPJ removes hydrocarbons and increases cell attachment, and the attached cells appear stretched. Oxygen-related chemical functional groups, such as OH, were formed on the A-P and NA-P samples, which resulted in more absorption of fibronectin and consequently higher levels of hMSC proliferation and ALP activity on the A-P and NA-P samples than on the N-P samples. Differences between the A-P and NA-P groups were observed because nitrogen-related functional groups may play an additional role in enhancing FAK gene expression and osteogenic differentiation.

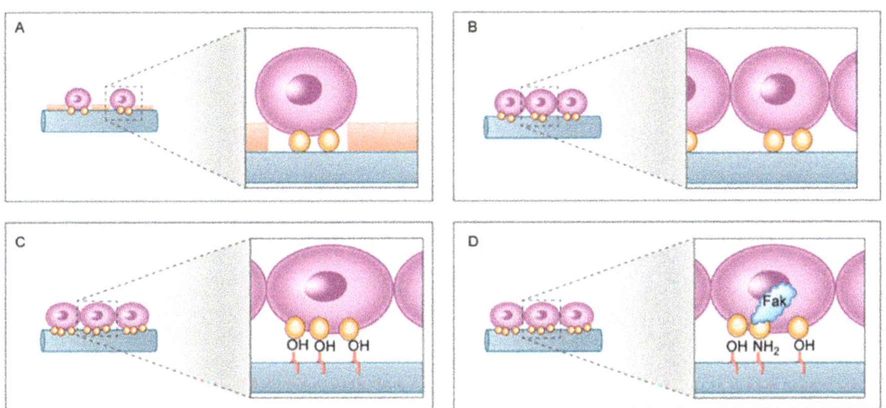

Figure 8. Schematic drawing of the putative mechanism underlying interactions between hMSCs and the (**A**) C-P, (**B**) N-P, (**C**) A-P, and (**D**) NA-P groups: For the N-P, A-P, and NA-P groups, more cell attachment was observed due to removal of hydrocarbon molecules from C-P. Higher levels of fibronectin absorption were observed on the A-P and NA-P groups, likely due to the presence of oxygen-related functional groups, which enhance osteogenic differentiation. The presence of nitrogen-related functional groups is considered to increase focal adhesion kinase (FAK) gene expression, further enhancing osteogenic differentiation on the NA-P samples.

Despite the limitations associated with in vitro studies and the need for further in vivo or other related research, the present findings may provide new insights for developing Ti implants with a higher level of osseointegration using humid ammonia-supplied NTAPPJ. Future studies are currently planned to determine possible clinical applications of the technology as a bioactive surface treatment for biomaterials, including Ti.

4. Materials and Methods

4.1. Sample Preparation

Machine-cut, grade IV commercially pure titanium (cp-Ti) was used. Round specimens with a diameter of 12 mm and a thickness of 1 mm were cleaned ultrasonically with acetone (99.5%, Duksan Pure Chemical, Ansan, Korea), ethanol (99.9%, Duksan Pure Chemical, Ansan, Korea), and distilled water sequentially for 5 min each at room temperature. The samples were dried naturally in air at room temperature before exposure to NTAPPJ. The specimens for the cellular experiments were sterilized in an autoclave at 121 °C for 15 min.

4.2. NTAPPJ Treatment

NTAPPJ was developed by the Plasma Bioscience Research Center (PBRC, Kwangwoon University, Seoul, Korea). The maximum discharge voltage and discharge current were 2.24 kV and 1.08 mA, respectively. Discharge occurred between the inner and outer electrodes of the NTAPPJ system, with porous alumina being the dielectric between the two electrodes (Figure 1a). The gases were supplied to the NTAPPJ system through mass flow controllers (MFC; AFC500, ATOVAC, Yongin-si, Korea) at a flow rate of 1 L/min (Figure 1b). Three different gases were used: nitrogen, air (compressed air), and humidified ammonia. In the case of humidified ammonia, special settings were considered involving the passage of nitrogen through the 0.1 M ammonia solution (Duksan Pure Chemical, Ansan, Korea) at a rate of 1 L/min (Figure 1b). The gas flow was controlled such that the ratio of nitrogen to humidified ammonia was 1 to 9. The control group was not exposed to NTAPPJ. The titanium specimens were placed 10 mm away from the NTAPPJ system (Figure 1a,b), and the plasma exposure time was 4 min. The NTAPPJ operation conditions are summarized in Table 1.

4.3. Surface Characterization

The surface morphology of the control and test samples was examined by SEM (JSM-6700F, JEOL, Japan) at 20 kV and 5000× magnification. The specimens were coated with platinum prior to SEM examination. The wettability of the test and control groups was measured by monitoring the water contact angles in accordance with previous studies [24,30,33]. A volume of 8 μL of distilled water was used for measurements, and the contact angle was measured after 30 s at room temperature on a video contact angle measurement system (Phoenix 300, SEO, Suwon, Korea). The change in wettability at 1, 2, 4, 12, and 24 h after NTAPPJ exposure was monitored. The surface chemical composition was determined by X-ray photoelectron spectroscopy (XPS; K-alpha, Thermo VG Scientific, Waltham, MA, USA). Monochromatic Al Kα was used as the X-ray source (Al Kα line: 1486.6 eV), and the sampling area was 400 μm. The spectra were recorded with a pass energy of 200 eV (step size of 1.0 eV) in survey mode and 50 eV (step size of 0.1 eV) in the high-resolution mode to acquire the C1s, O1s, N1s, and Ti2p spectra with a resolution of 0.78 eV measured from the Ag 3d5/2 peaks. The binding energies were referenced to the C1s peak at 284.8 eV.

4.4. hMSC Cell Culture

The hMSCs were purchased commercially (Lonza, Allendale, NJ, USA). Cells were passaged three to five times to maintain multilineage capabilities. The cells were cultured on a 100-mm-diameter culture dish (SPL, Daegu, Korea) at 37 °C and 5% CO_2 in an incubator. Dulbecco's modified Eagle's medium (DMEM, Gibco, Grand Island, NY, USA) supplemented with 10% human fetal bovine serum

(FBS; Gibco, Grand Island, NY, USA) and 1% antibiotic-antimycotic (Gibco, Grand Island, NY, USA) were used without the addition of any osteogenic differentiation-inducing supplements.

4.5. Cell Viability Assay

The viability of hMSCs cultured on the test and control samples was examined using a calcein and ethidium homodimer-1-based staining kit (LIVE/DEADTM Viability/Cytotoxicity Kit, Invitrogen Co., Eugene, OR, USA). A total of 5×10^4 cells was placed on the samples, transferred to 12-well culture plates (SPL, Daegu, Korea), and cultured in an incubator at 37 °C and 5% CO_2 for 24 h. The cell culture medium was removed from each well, and the cells were washed with phosphate-buffered saline (PBS; Gibco, Grand Island, NY, USA). The stock solution mixed with 10 mL of PBS was applied to the cells for 45 min at room temperature. The cells on each specimen were examined under a confocal laser microscope (LSM700, Carl-Zeiss, Thornwood, NY, USA) with live cells appearing green and dead cells appearing red.

4.6. Cell Morphology

A total of 5×10^4 cells was placed on the samples, transferred to a 12-well culture plate (SPL, Daegu, Korea), and cultured in an incubator at 37 °C and 5% CO_2 for 4 h. The unattached cells were washed away with phosphate buffered saline (PBS; Gibco, Grand Island, NY, USA). The morphology of the attached cells was assessed using fluorescent dyes and a confocal laser microscope. After culturing for 4 h under the previously described conditions, the cells were stained with DAPI (4′,6-diamidino-2-phenylindole, blue coloration for nuclei, Invitrogen, Grand Island, NY, USA) and rhodamine phalloidin (red coloration for actin filament, Invitrogen, Grand Island, NY, USA) and visualized under a confocal laser microscope (LSM700, Carl-Zeiss, Thornwood, NY, USA).

4.7. Cell Proliferation Assay

The proliferation of hMSCs was evaluated using BrdU incorporation during DNA synthesis. A total of 5×10^4 cells was placed on each sample, transferred to a 12-well culture plate (SPL, Daegu, Korea), and cultured in an incubator at 37 °C and 5% CO_2. After 3 days, 100 μL of 100 mM BrdU solution (Roche Applied Science, Penzberg, Germany) was added to each culture well, and the cells with incorporated BrdU were further incubated for 12 h. The cells with incorporated BrdU reacted with anti-BrdU conjugated with peroxidase (Roche Applied Science, Penzberg, Germany) for 90 min. Finally, tetramethylbenzidine (Roche Applied Science, Penzberg, Germany) was added to enable color development. The optical absorbance was measured on a reader (Epoch, BioTek Instruments, Winooski, VT, USA) at 370 nm, and the rate of proliferation was expressed as the percentage of that of the control, C-P.

4.8. Alkaline Phosphatase Activity Assay

The ALP activity of hMSCs cultured on each sample for 14 days was examined using a Sensolyte®® p-nitrophenyl phosphate (pNPP) ALP assay kit (AnaSpec, San Jose, CA, USA). Briefly, 1×10^4 cells were placed on the surface of each sample, placed in a 12-well culture plate (SPL, Daegu, Korea), and cultured in an incubator at 37 °C and 5% CO_2. After 14 days, the cells on each sample were washed twice with PBS and lysed with 1 mL of 0.2% Triton X-100 (AnaSpec, San Jose, CA, USA). The lysed products were further incubated at 4 °C for 10 min under agitation, and the cell suspension was centrifuged at $2500 \times g$ for 10 min at 4 °C. Fifty microliters of the supernatant weree reacted with 50 μL of the pNPP assay buffer, and the absorbance was determined at 405 nm on an absorbance reader (Epoch, BioTek Instruments, Winooski, VT, USA) after 2 h. The quantity of ALP was determined by allowing 50 μL of ALP with the standard concentration to react with 50 μL of the pNPP assay buffer, and a calibration curve was derived. Total protein was quantified using a microbicinchoninic acid (MicroBCA)-based total protein assay kit (Thermo Fisher Scientific, Rockford, IL, USA) according to the manufacturer's instructions. Briefly, the supernatant (150 μL) obtained from the previous centrifugation step was mixed with 150 µL of the working solution from the kit and incubated at 37 °C for 2 h. The optical absorbance was measured at 562 nm on a plate reader (Epoch, BioTek Instruments, Winooski, VT, USA). As described previously, the quantity of total protein was determined by reacting 150 µL of bovine serum albumin with 150 µL of the working solution and a calibration curve was generated. The ALP activity was expressed as the amount of quantified ALP per the amount of quantified total protein.

4.9. Gene Expression Analysis

The osteogenic gene expression of hMSCs on each sample was determined by qRT-PCR. A total of 1×10^4 cells was placed on each sample, transferred to each well of a 12-well culture plate (SPL, Daegu, Korea), and cultured in an incubator at 37 °C and 5% CO_2 for 14 days. To analyze FAK gene expression, 5×10^4 cells were placed on each sample in a 12-well culture plate (SPL, Daegu, Korea). The cells were cultured in an incubator at 37 °C and 5% CO_2 for 12 h. After the incubation periods (14 days for osteogenic gene expression and 12 h for FAK gene expression), the culture media were removed and the cells were washed with PBS (Gibco, Grand Island, NY, USA). Total RNA was extracted from the cells using TrizolTM reagent (Sigma-Aldrich, St. Louis, MO, USA), and cDNA was prepared using a high capacity RNA-to-cDNA kit (Applied Biosystems, Foster City, CA, USA). qRT-PCR was performed with cDNA using SYBR green dye (Applied Biosystems, Foster City, CA, USA), with primers corresponding to ALP, BSP, OCN, and OPN for osteogenic gene expression and to FAK for FAK gene expression [30]. Glyceraldehyde-3-phosphate dehydrogenase (GAPDH) was used as a housekeeping reference gene in both experiments. Quantification was carried out on the PCR system (7300 Real-Time PCR System, Applied Biosystems, Foster City, CA, USA) using the following cycle protocol: 1 cycle at 50 °C for 2 min, 1 cycle at 95 °C for 10 min, 40 cycles at 95 °C for 15 s, and 1 cycle at 60 °C for 1 min. The expression of each osteogenic gene was expressed as the CT (cycle threshold) value relative to that of the control (C-P). All experiments were repeated three times.

4.10. Protein Absorption Assay

The absorption of two proteins, BSA and fibronectin, on each sample was investigated. BSA (Pierce Biotechnology, Inc., Rockford, IL, USA) and fibronectin (Sigma-Aldrich, St. Louis, MO, USA) solutions were prepared by mixing the corresponding protein powders with PBS (1 mg of protein per 1 mL of PBS). In each group, 300 µL of the protein solution was pipetted and spread over a titanium disk. After incubation for 24 h at 37 °C, the non-adherent proteins were removed and the amounts of each protein were measured by adding 300 µL of the microbicinchoninic acid (MicroBCA)-based total protein assay kit (Thermo Fisher Scientific, Rockford, IL, USA). The samples were further incubated at 37 °C for 2 h, and the optical absorbance was measured at 562 nm on a plate reader (Epoch, BioTek Instruments, Winooski, VT, USA).

4.11. Statistical Analysis

All statistical analyses were performed using IBM SPSS soft-ware, version 23.0 (IBM Korea Inc., Seoul, Korea) for Windows. The biological data are expressed as the mean ± S.D. of at least three independent experiments. Statistical significance was evaluated by one-way analysis of variance with Tukey's post hoc test. Values of $p < 0.05$ were considered statistically significant.

5. Conclusions

The present work demonstrates the advantages of using NTAPPJ technology with respect to the generation of bioactive functional groups on Ti. In particular, the use of humid ammonia-supplied NTAPPJ resulted in the formation of amine chemical functional groups on Ti without changing the topography of the material. Despite the limitations associated with in vitro studies and the need for further in vivo or other related research, the present findings provide new insights for developing Ti implants with a higher level of osseointegration. Future studies are currently planned to determine possible clinical applications of the technology as a bioactive surface treatment for biomaterials, including Ti. Nevertheless, it was evident that different plasma gases had different effects on osteogenic differentiation. NA-P treatment generated nitrogen-related functional groups and led to a higher level of osteogenic differentiation in the corresponding hMSCs. Thus, we discovered a novel surface treatment method for Ti using NTAPPJ with NA-P that can enhance the osteoconductive and osteoinductive activities of Ti.

Author Contributions: Conceptualization, J.-S.K. and K.-M.K.; methodology, J.-S.K., K.-M.K., S.-H.C. and E.H.C.; validation, K.-M.K., E.H.C. and P.K.C.; formal analysis, J.-S.K.; investigation, J.-S.K.; resources, K.-M.K. and E.H.C.; data curation, J.-S.K.; writing—original draft preparation, J.-S.K.; writing—review and editing, S.-H.C., E.H.C., K.-M.K. and P.K.C.; visualization, J.-S.K.; supervision, K.-M.K. and E.H.C.; project administration, J.-S.K. and S.-H.C.; funding acquisition, J.-S.K. and E.H.C. All authors have read and agreed to the published version of the manuscript.

Funding: This work was supported by the Yonsei University College of Dentistry (6-2019-0021) and by the program through the National Research Foundation of Korea (NRF) funded by the Korea government (MSIT) (NRF-2016K1A4A3914113).

Acknowledgments: Plasma device has been supported by Kwangwoon University. Special thanks to Yong Hee Kim for his support with plasma device.

Conflicts of Interest: The authors declare no conflict of interest.

Abbreviations

NTAPPJ	Nonthermal Atmospheric Pressure Plasma Jet
hMSC	Human Mesenchymal Stem Cells
XPS	X-ray Photoelectron Spectroscopy
Ti	Titanium
qRT-PCR	Quantitative Real-Time Polymerase Chain Reaction
ALP	Alkaline Phosphatase
OPN	Osteopontin
OCN	Osteocalcin
BSP	Bone Sialoprotein
FAK	Focal Adhesion Kinase
BSA	Bovine Serum Albumin

References

1. Gaharwar, A.K.; Mihaila, S.M.; Swami, A.; Patel, A.; Sant, S.; Reis, R.L.; Marques, A.P.; Gomes, M.E.; Khademhosseini, A. Bioactive silicate nanoplatelets for osteogenic differentiation of human mesenchymal stem cells. *Adv. Mater.* **2013**, *25*, 3329–3336. [CrossRef] [PubMed]

2. Gentleman, E.; Swain, R.J.; Evans, N.D.; Boonrungsiman, S.; Jell, G.; Ball, M.D.; Shean, T.A.V.; Oyen, M.L.; Porter, A.; Stevens, M.M. Comparative materials differences revealed in engineered bone as a function of cell-specific differentiation. *Nat. Mater.* **2009**, *8*, 763–770. [CrossRef] [PubMed]

3. Khademhosseini, A.; Langer, R.; Borenstein, J.; Vacanti, J.P. Microscale technologies for tissue engineering and biology. *Proc. Natl. Acad. Sci. USA* **2006**, *103*, 2480–2487. [CrossRef] [PubMed]

4. Langer, R.; Tirrell, D.A. Designing materials for biology and medicine. *Nature* **2004**, *428*, 487–492. [CrossRef] [PubMed]

5. Peppas, N.A.; Langer, R. New challenges in biomaterials. *Science* **1994**, *263*, 1715–1720. [CrossRef] [PubMed]

6. Hamlekhan, A.; Moztarzadeh, F.; Mozafari, M.; Azami, M.; Nezafati, N. Preparation of laminated poly(epsilon-caprolactone)-gelatin-hydroxyapatite nanocomposite scaffold bioengineered via compound techniques for bone substitution. *Biomatter* **2011**, *1*, 91–101. [CrossRef] [PubMed]

7. Minagar, S.; Berndt, C.C.; Wang, J.; Ivanova, E.; Wen, C. A review of the application of anodization for the fabrication of nanotubes on metal implant surfaces. *Acta Biomater.* **2012**, *8*, 2875–2888. [CrossRef]

8. Sousa, S.R.; Lamghari, M.; Sampaio, P.; Moradas-Ferreira, P.; Barbosa, M.A. Osteoblast adhesion and morphology on TiO2 depends on the competitive preadsorption of albumin and fibronectin. *J. Biomed. Mater. Res.* **2008**, *84*, 281–290. [CrossRef]

9. Puleo, D.A.; Nanci, A. Understanding and controlling the bone-implant interface. *Biomaterials* **1999**, *20*, 2311–2321. [CrossRef]

10. Urist, M.R. Bone: Formation by autoinduction. *Science* **1965**, *150*, 893–899. [CrossRef]

11. Yuan, H.; Fernandes, H.; Habibovic, P.; de Boer, J.; Barradas, A.M.; de Ruiter, A.; Walsh, W.R.; van Blitterswijk, C.A.; de Bruijn, J.D. Osteoinductive ceramics as a synthetic alternative to autologous bone grafting. *Proc. Natl. Acad. Sci. USA* **2010**, *107*, 13614–13619. [CrossRef] [PubMed]

12. Dalby, M.J.; Gadegaard, N.; Tare, R.; Andar, A.; Riehle, M.O.; Herzyk, P.; Wilkinson, C.D.W.; Oreffo, R.O.C. The control of human mesenchymal cell differentiation using nanoscale symmetry and disorder. *Nat. Mater.* **2007**, *6*, 997–1003. [CrossRef] [PubMed]

13. Cancedda, R.; Bianchi, G.; Derubeis, A.; Quarto, R. Cell therapy for bone disease: A review of current status. *Stem Cells* **2003**, *21*, 610–619. [CrossRef] [PubMed]

14. Derubeis, A.R.; Cancedda, R. Bone marrow stromal cells (BMSCs) in bone engineering: Limitations and recent advances. *Ann. Biomed. Eng.* **2004**, *32*, 160–165. [CrossRef]

15. Mackay, A.M.; Beck, S.C.; Murphy, J.M.; Barry, F.P.; Chichester, C.O.; Pittenger, M.F. Chondrogenic differentiation of cultured human mesenchymal stem cells from marrow. *Tissue Eng.* **1998**, *4*, 415–428. [CrossRef]

16. Toma, C.; Pittenger, M.F.; Cahill, K.S.; Byrne, B.J.; Kessler, P.D. Human mesenchymal stem cells differentiate to a cardiomyocyte phenotype in the adult murine heart. *Circulation* **2002**, *105*, 93–98. [CrossRef]

17. Matsushita, K.; Wu, Y.J.; Okamoto, Y.; Pratt, R.E.; Dzau, V.J. Local renin angiotensin expression regulates human mesenchymal stem cell differentiation to adipocytes. *Hypertension* **2006**, *48*, 1095–1102. [CrossRef]

18. Engler, A.J.; Sen, S.; Sweeney, H.L.; Discher, D.E. Matrix elasticity directs stem cell lineage specification. *Cell* **2006**, *126*, 677–689. [CrossRef]

19. Oh, S.; Brammer, K.S.; Li, Y.S.J.; Teng, D.; Engler, A.J.; Chien, S.; Jin, S. Stem cell fate dictated solely by altered nanotube dimension. *Proc. Natl. Acad. Sci. USA* **2009**, *106*, 2130–2135. [CrossRef]

20. Kaur, G.; Valarmathi, M.T.; Potts, J.D.; Jabbari, E.; Sabo-Attwood, T.; Wang, Q. Regulation of osteogenic differentiation of rat bone marrow stromal cells on 2D nanorod substrates. *Biomaterials* **2010**, *31*, 1732–1741. [CrossRef]

21. Benoit, D.S.W.; Schwartz, M.P.; Durney, A.R.; Anseth, K.S. Small functional groups for controlled differentiation of hydrogel-encapsulated human mesenchymal stem cells. *Nat. Mater.* **2008**, *7*, 816–823. [CrossRef] [PubMed]

22. Curran, J.M.; Chen, R.; Hunt, J.A. The guidance of human mesenchymal stem cell differentiation in vitro by controlled modifications to the cell substrate. *Biomaterials* **2006**, *27*, 478–4793. [CrossRef] [PubMed]

23. Wang, H.Y.; Kwok, D.T.K.; Xu, M.; Shi, H.G.; Wu, Z.W.; Zhang, W.; Chu, P.K. Tailoring of mesenchymal stem cells behavior on plasma-modified polytetrafluoroethylene. *Adv. Mater.* **2012**, *24*, 3315–3324. [CrossRef] [PubMed]

24. Kwon, J.S.; Kim, Y.H.; Choi, E.H.; Kim, K.N. Development of ultra-hydrophilic and non-cytotoxic dental vinyl polysiloxane impression materials using a non-thermal atmospheric-pressure plasma jet. *J. Phys. D Appl Phys.* **2013**, *46*, 195201. [CrossRef]

25. Yoo, E.M.; Uhm, S.H.; Kwon, J.S.; Choi, H.S.; Choi, E.H.; Kim, K.M.; Kim, K.N. The study on inhibition of planktonic bacterial growth by non-thermal atmospheric pressure plasma jet treated surfaces for dental application. *J. Biomed. Nanotechnol.* **2015**, *11*, 334–341. [CrossRef]

26. Aguiar, C.; Mendes, D.; Camara, A.; Figueiredo, J. Endodontic treatment of a mandibular second premolar with three root canals. *J. Contemp. Dent. Pract.* **2010**, *11*, 78–84. [CrossRef]

27. Kalghatgi, S.; Kelly, C.M.; Cerchar, E.; Torabi, B.; Alekseev, O.; Fridman, A.; Friedman, G.; Azizkhan-Clifford, J. Effects of non-thermal plasma on mammalian cells. *PLoS ONE* **2011**, *6*, e16270. [CrossRef]

28. Huang, H.H.; Ho, C.T.; Lee, T.H.; Lee, T.L.; Liao, K.K.; Chen, F.L. Effect of surface roughness of ground titanium on initial cell adhesion. *Biomol. Eng.* **2004**, *21*, 93–97. [CrossRef]

29. Aita, H.; Hori, N.; Takeuchi, M.; Suzuki, T.; Yamada, M.; Anpo, M.; Ogawa, T. The effect of ultraviolet functionalization of titanium on integration with bone. *Biomaterials* **2009**, *30*, 1015–1025. [CrossRef]

30. Lee, E.J.; Kwon, J.S.; Uhm, S.H.; Song, D.H.; Kim, Y.H.; Choi, E.H.; Kim, K.N. The effects of non-thermal atmospheric pressure plasma jet on cellular activity at SLA-treated titanium surfaces. *Curr. Appl. Phys.* **2013**, *13*, S36–S41. [CrossRef]

31. Att, W.; Hori, N.; Iwasa, F.; Yamada, M.; Ueno, T.; Ogawa, T. The effect of UV-photofunctionalization on the time-related bioactivity of titanium and chromium-cobalt alloys. *Biomaterials* **2009**, *30*, 4268–4276. [CrossRef] [PubMed]

32. Att, W.; Hori, N.; Takeuchi, M.; Ouyang, J.; Yang, Y.; Anpo, M.; Ogawa, T. Time-dependent degradation of titanium osteoconductivity: An implication of biological aging of implant materials. *Biomaterials* **2009**, *30*, 5352–5363. [CrossRef] [PubMed]

33. Choi, S.H.; Ryu, J.H.; Kwon, J.S.; Kim, J.E.; Cha, J.Y.; Lee, K.J.; Yu, H.S.; Cho, E.H.; Kim, K.M.; Hwang, C.J. Effect of wet storage on the bioactivity of ultraviolet light- and non-thermal atmospheric pressure plasma-treated titanium and zirconia implant surfaces. *Mater. Sci. Eng. C* **2019**, *105*, 110049. [CrossRef] [PubMed]

34. Iwasa, F.; Hori, N.; Ueno, T.; Minamikawa, H.; Yamada, M.; Ogawa, T. Enhancement of osteoblast adhesion to UV-photofunctionalized titanium via an electrostatic mechanism. *Biomaterials* **2010**, *31*, 2717–2727. [CrossRef]

35. Mazzanti, G.; Daniele, C.; Tita, B.; Vitali, F.; Signore, A. Biological evaluation of a polyvinyl siloxane impression material. *Dent. Mater.* **2005**, *21*, 371–374. [CrossRef]

36. Wataha, J.C. Principles of biocompatibility for dental practitioners. *J. Prosthet. Dent.* **2001**, *86*, 203–209. [CrossRef]

37. Choi, Y.R.; Kwon, J.S.; Song, D.H.; Choi, E.H.; Lee, Y.K.; Kim, K.N.; Kim, K.M. Surface modification of biphasic calcium phosphate scaffolds by non-thermal atmospheric pressure nitrogen and air plasma treatment for improving osteoblast attachment and proliferation. *Thin Solid Film.* **2013**, *547*, 235–240. [CrossRef]

38. Seo, H.Y.; Kwon, J.S.; Choi, Y.R.; Kim, K.M.; Choi, E.H.; Kim, K.N. Cellular attachment and differentiation on titania nanotubes exposed to air- or nitrogen-based non-thermal atmospheric pressure plasma. *PLoS ONE* **2014**, *9*, e113477. [CrossRef]

39. Anselme, K. Osteoblast adhesion on biomaterials. *Biomaterials* **2000**, *21*, 667–681. [CrossRef]

40. Kim, M.J.; Kim, C.W.; Lim, Y.J.; Heo, S.J. Microrough titanium surface affects biologic response in MG63 osteoblast-like cells. *J. Biomed. Mater. Res.* **2006**, *79*, 1023–1032. [CrossRef]

41. Geiger, B.; Bershadsky, A.; Pankov, R.; Yamada, K.M. Transmembrane extracellular matrix-cytoskeleton crosstalk. *Nat. Rev. Mol. Cell Biol.* **2001**, *2*, 793–805. [CrossRef]

42. Lewandrowski, K.U.; Gresser, J.D.; Wise, D.L.; Trantol, D.J. Bioresorbable bone graft substitutes of different osteoconductivities: A histologic evaluation of osteointegration of poly(propylene glycol-co-fumaric acid)-based cement implants in rats. *Biomaterials* **2000**, *21*, 757–764. [CrossRef]

43. Chen, M.; Zhang, Y.; Sky Driver, M.; Caruso, A.N.; Yu, Q.; Wang, Y. Surface modification of several dental substrates by non-thermal, atmospheric plasma brush. *Dent. Mater.* **2013**, *29*, 871–880. [CrossRef] [PubMed]

44. Liu, Y.; Cooper, P.R.; Barralet, J.E.; Shelton, R.M. Influence of calcium phosphate crystal assemblies on the proliferation and osteogenic gene expression of rat bone marrow stromal cells. *Biomaterials* **2007**, *28*, 1393–1403. [CrossRef] [PubMed]

45. Ogata, Y. Bone sialoprotein and its transcriptional regulatory mechanism. *J. Periodontal Res.* **2008**, *43*, 127–135. [CrossRef]

46. Denhardt, D.T.; Guo, X.J. Osteopontin: A Protein with Diverse Functions. *FASEB J.* **1993**, *7*, 1475–1482. [CrossRef]

47. Ducy, P.; Desbois, C.; Boyce, B.; Pinero, G.; Story, B.; Dunstan, C.; Smith, E.; Bonadio, J.; Goldstein, S.; Gundberg, C.; et al. Increased bone formation in osteocalcin-deficient mice. *Nature* **1996**, *382*, 448–452. [CrossRef]

48. Owen, T.A.; Aronow, M.; Shalhoub, V.; Barone, L.M.; Wilming, L.; Tassinari, M.S.; Kennedy, M.B.; Pockwinse, S.; Lian, J.D.; Stein, G.S. Progressive development of the rat osteoblast phenotype invitro: Reciprocal relationships in expression of genes associated with osteoblast proliferation and differentiation during formation of the bone extracellular-matrix. *J. Cell. Physiol.* **1990**, *143*, 420–430. [CrossRef]

49. Kaivosoja, E.; Barreto, G.; Levon, K.; Virtanen, S.; Ainola, M.; Konttinen, Y.T. Chemical and physical properties of regenerative medicine materials controlling stem cell fate. *Ann. Med.* **2012**, *44*, 635–650. [CrossRef]

50. Liu, L.Y.; Chen, S.F.; Giachelli, C.M.; Ratner, B.D.; Jiang, S.Y. Controlling osteopontin orientation on surfaces to modulate endothelial cell adhesion. *J. Biomed. Mater. Res.* **2005**, *74*, 23–31. [CrossRef]

51. Phillips, J.E.; Petrie, T.A.; Creighton, F.P.; Garcia, A.J. Human mesenchymal stem cell differentiation on self-assembled monolayers presenting different surface chemistries. *Acta Biomater.* **2010**, *6*, 12–20. [CrossRef] [PubMed]

52. Han, I.; Vagaska, B.; Park, B.J.; Lee, M.H.; Lee, S.J.; Park, J.C. Selective fibronectin adsorption against albumin and enhanced stem cell attachment on helium atmospheric pressure glow discharge treated titanium. *J. Appl. Phys.* **2011**, *109*, 124701. [CrossRef]

53. Keselowsky, B.G.; Collard, D.M.; Garcia, A.J. Surface chemistry modulates focal adhesion composition and signaling through changes in integrin binding. *Biomaterials* **2004**, *25*, 5947–5954. [CrossRef] [PubMed]

54. Keselowsky, B.G.; Wang, L.; Schwartz, Z.; Garcia, A.J.; Boyan, B.D. Integrin alpha(5) controls osteoblastic proliferation and differentiation responses to titanium substrates presenting different roughness characteristics in a roughness independent manner. *J. Biomed. Mater. Res.* **2007**, *80*, 700–710. [CrossRef]

International Journal of
Molecular Sciences

Article

Positive Effect of Cold Atmospheric Nitrogen Plasma on the Behavior of Mesenchymal Stem Cells Cultured on a Bone Scaffold Containing Iron Oxide-Loaded Silica Nanoparticles Catalyst

Agata Przekora [1,*,†], Maïté Audemar [2], Joanna Pawlat [3], Cristina Canal [4,5,6], Jean-Sébastien Thomann [7], Cédric Labay [4,5,6], Michal Wojcik [1], Michal Kwiatkowski [3], Piotr Terebun [3], Grazyna Ginalska [1], Sophie Hermans [2,*,†] and David Duday [7,*,†]

[1] Chair and Department of Biochemistry and Biotechnology, Medical University of Lublin, Chodzki 1 Street, 20-093 Lublin, Poland; michal.wojcik@umlub.pl (M.W.); g.ginalska@umlub.pl (G.G.)
[2] IMCN Institute, Université catholique de Louvain, Place Louis Pasteur 1, 1348 Louvain-la-Neuve, Belgium; maite.audemar@uclouvain.be
[3] Chair of Electrical Engineering and Electrotechnologies, Lublin University of Technology, Nadbystrzycka 38a, 20-618 Lublin, Poland; askmik@hotmail.com (J.P.); m.kwiatkowski@pollub.pl (M.K.); p.terebun@pollub.pl (P.T.)
[4] Biomaterials, Biomechanics and Tissue Engineering Group, Department of Materials Science and Engineering, Universitat Politècnica de Catalunya (UPC), Av. Eduard Maristany 14, 08930 Barcelona, Spain; cristina.canal@upc.edu (C.C.); cedric.labay@upc.edu (C.L.)
[5] Barcelona Research Center in Multiscale Science and Engineering, UPC, 08019 Barcelona, Spain
[6] Research Centre for Biomedical Engineering (CREB), UPC, 08019 Barcelona, Spain
[7] Material Research and Technology (MRT) Department, Luxembourg Institute of Science and Technology (LIST), 41 rue du Brill, L-4422 Belvaux, Luxembourg; jean-sebastien.thomann@list.lu
[*] Correspondence: agata.przekora@umlub.pl (A.P.); sophie.hermans@uclouvain.be (S.H.); david.duday@list.lu (D.D.); Tel.: +48-814487026 (A.P.)
[†] These authors contributed equally to this work.

Received: 9 June 2020; Accepted: 1 July 2020; Published: 3 July 2020

Abstract: Low-temperature atmospheric pressure plasma was demonstrated to have an ability to generate different reactive oxygen and nitrogen species (RONS), showing wide biological actions. Within this study, mesoporous silica nanoparticles (NPs) and Fe_xO_y/NPs catalysts were produced and embedded in the polysaccharide matrix of chitosan/curdlan/hydroxyapatite biomaterial. Then, basic physicochemical and structural characterization of the NPs and biomaterials was performed. The primary aim of this work was to evaluate the impact of the combined action of cold nitrogen plasma and the materials produced on proliferation and osteogenic differentiation of human adipose tissue-derived mesenchymal stem cells (ADSCs), which were seeded onto the bone scaffolds containing NPs or Fe_xO_y/NPs catalysts. Incorporation of catalysts into the structure of the biomaterial was expected to enhance the formation of plasma-induced RONS, thereby improving stem cell behavior. The results obtained clearly demonstrated that short-time (16s) exposure of ADSCs to nitrogen plasma accelerated proliferation of cells grown on the biomaterial containing Fe_xO_y/NPs catalysts and increased osteocalcin production by the cells cultured on the scaffold containing pure NPs. Plasma activation of Fe_xO_y/NPs-loaded biomaterial resulted in the formation of appropriate amounts of oxygen-based reactive species that had positive impact on stem cell proliferation and at the same time did not negatively affect their osteogenic differentiation. Therefore, plasma-activated Fe_xO_y/NPs-loaded biomaterial is characterized by improved biocompatibility and has great clinical potential to be used in regenerative medicine applications to improve bone healing process.

Keywords: reactive oxygen species; cold atmospheric plasma; mesoporous silica nanoparticles; biomaterials; bone regeneration; cytotoxicity; proliferation; osteogenic differentiation

1. Introduction

Plasma is an ionized gas that comprises various molecules, electrons, ions, excited species, and radicals. In the field of engineering of biomaterials and regenerative medicine, special attention has been paid to the application of non-thermal atmospheric pressure plasma, also known as cold plasma, for improving biological properties of the biomaterials and acceleration of healing processes [1–9]. In the field of biomaterials engineering, cold plasma technology is primarily used for surface functionalization with hydrophilic chemical groups to enhance cell adhesion and proliferation on the implants [10]. Atmospheric pressure plasma is also used to improve surface properties of the biomaterials, especially wettability and roughness that are known to significantly influence biocompatibility of the implants [11,12]. Materials scientists frequently use atmospheric pressure plasma combined with oxygen, argon, air, ammonia, or nitrogen gas for biocompatibility improvement of various polymeric biomaterials [5–8]. Atmospheric plasma processes are also used to graft molecules on the surface of biomaterials [13].

Since reactive species formed upon plasma treatment may have a positive effect on cell proliferation and differentiation, cold plasma has also attracted great attention in the regenerative medicine field. Possible applications of atmospheric pressure plasma have been reported for acceleration of wound healing [1–4] and bone regeneration [14,15]. The effectiveness of plasma technology application in the regenerative medicine field highly depends on the reactor design, the gas nature, the plasma power, and other characteristics of the treatment [11]. It was demonstrated that cold oxygen or nitrogen plasma may generate different reactive oxygen and nitrogen species (RONS) that may significantly affect cell metabolism, proliferation, and differentiation [16]. Importantly, the biological activity of RONS is mediated by the liquid environment surrounding the cells, indicating that reactive species, either short-lived e.g., singlet oxygen ($^1O_2$), hydroxyl (OH) or nitric oxide (NO) radicals, or long-lived i.e., H_2O_2 or NO_2, formed upon plasma treatment in a liquid environment (e.g., phosphate buffered saline, culture medium), have equivalent or superior activities over these generated in the gas phase [17]. Furthermore, comparative studies on the formation of reactive species ($^1O_2$, H radicals, OH radicals, and NO radicals) upon treatment of the phosphate buffered saline (PBS) with oxygen, air, helium, argon, nitrogen, and carbon dioxide plasmas, proved that nitrogen plasma has the ability to produce greater amounts of NO radicals and the largest amount of OH radicals among all tested gas species [17]. Therefore, within our studies, it was decided to use nitrogen as a source gas during plasma treatment of cell-seeded biomaterials.

It is also known that OH radicals may be formed in the liquid phase by the heterogeneous Fenton-like reaction [18] using metal or metal oxides (e.g., Au, Mn, Cu, Co, or Fe oxide) [19,20]. Therefore, herein, catalysts based on two types of silica nanoparticles (NPs) decorated with iron oxide (noted Fe_xO_y to denote magnetite + maghemite phase) as active phase were synthesized to enhance cold plasma-induced generation of RONS. The NPs produced with and without Fe_xO_y decoration were immobilized by the entrapment method within the polysaccharide matrix of a bone scaffold made of hybrid chitosan/curdlan matrix and hydroxyapatite (HA) granules. Then, modified biomaterials were seeded with human adipose tissue-derived mesenchymal stem cells (ADSCs) and subjected to cold atmospheric nitrogen plasma treatment to determine its effect on cell proliferation and osteogenic differentiation. The effect of embedded NPs on the production of OH radicals upon plasma treatment as well as basic mechanical and microstructural properties of the modified biomaterials were also evaluated. The primary goal of the presented study was to assess biomedical potential in regenerative medicine applications of Fe_xO_y/NPs-loaded biomaterial pre-seeded with human stem cells and activated by cold atmospheric nitrogen plasma.

2. Results and Discussion

2.1. Characterization of Produced NPs

Within this study, two types of silica NPs were produced to enhance reactive oxygen species (ROS) generation upon plasma treatment of NPs-loaded bone scaffolds: (1) Mobil Composition of Matter No. 48 (MCM-48) and (2) wormhole mesoporous silica nanoparticles (MSNPs). Produced NPs were subjected to basic characterization of their physicochemical and microstructural properties. The MCM-48 support exhibited a high specific surface area (SSA) of 1440 ± 46 m^2/g. MSNPs were characterized by lower SSA equal to 930 ± 10 m^2/g. Obtained nitrogen adsorption/desorption isotherms indicated a mesoporous structure of both NPs with a type IV isotherm. MSNPs were characterized by pore size of 2.8 ± 0.2 nm with a pore volume of 0.61 ± 0.05 cm^3/g, whereas MCM-48 particles had a pore size distribution centered on an average of 2.7 nm with a pore volume of 0.73 ± 0.05 cm^3/g (Supplementary Figure S1). These mesoporous silica materials were loaded with iron oxide by dry impregnation with iron nitrate in ethanol followed by reduction at 500 °C under H_2:Ar (1:4 *V:V*) after a calcination step. The reduction temperature was chosen to obtain a mix of Fe(II) and Fe(III), which is known to be the active phase in Fenton chemistry. The iron loading of NPs was measured by inductively coupled plasma (ICP), which gave a value of 4.7 wt.% and 4.9 wt.% for the MCM-48 and MSNPs materials, respectively.

The MSNPs preparation resulted in non-aggregated NPs with a diameter of 55 ± 5 nm (Supplementary Figure S2). MCM-48 NPs presented significantly higher size than MSNPs (around 400 nm). Zeta potential for MSNPs was -30 ± 5 mV, allowing a repulsion between the MSNPs in water. With the dry impregnation method, the iron oxide introduced within the pores of these materials was unsurprisingly not visible on the TEM images (Figure 1a) neither with the MCM-48 nor the MSNPs. Although TEM images of these materials showed no visible iron oxide particles decorating the silica nanoparticles, SEM-EDX mapping of Si, Fe, O showed a homogeneous dispersion of iron within the two different silica nanoparticles, confirming very small iron oxide particles sizes distributed throughout the entire silica (Figure 1b). X-ray photoelectron spectroscopy (XPS) analysis also revealed the presence of iron at the surface of the decorated silica samples (Table 1). However, no signal of iron oxide phases was visible by X-ray diffraction (XRD) for the Fe_xO_y/MCM-48 catalyst, due to very small or amorphous iron oxide particles (Figure 2). For the Fe_xO_y/MSNPs sample, a small peak at 35.5° was visible on the diffractogram, matching with the most intense diffraction peak of magnetite Fe_3O_4 or maghemite γ-Fe_2O_3. This suggests that the iron oxide active sites in the case of Fe_xO_y/MSNPs were at least partially crystalline and/or slightly bigger particles than within the MCM-48 support.

Table 1. Atomic composition (at. %) and atomic ratio from XPS for the Fe_xO_y/MCM-48 and Fe_xO_y/MSNPs samples, compared with pure MCM-48 and MSNPs.

	Fe_xO_y/MCM-48	Fe_xO_y/MSNPs	MCM-48	MSNPs
O 1s	64.24	68.80	68.99	60.50
Si 2p	33.41	29.18	29.74	30.87
O/Si	1.92	2.36	2.32	1.96
Fe 2p3/2	0.32	0.61	/	/
Fe/Si	0.01	0.02	/	/
C 1s	2.03	1.41	1.27	8.63

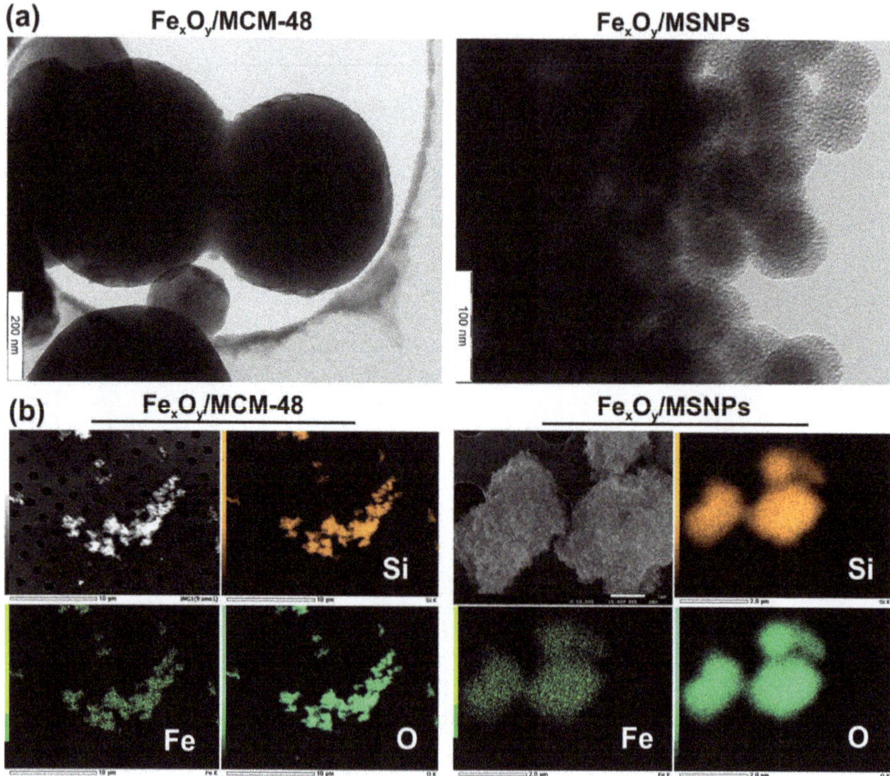

Figure 1. Visualization of produced NPs: (**a**) TEM images of Fe$_x$O$_y$/MCM-48 and Fe$_x$O$_y$/MSNPs catalysts obtained by dry impregnation; (**b**) SEM-EDX mapping of Si, Fe, O elements in the Fe$_x$O$_y$/MCM-48 and Fe$_x$O$_y$/MSNPs samples.

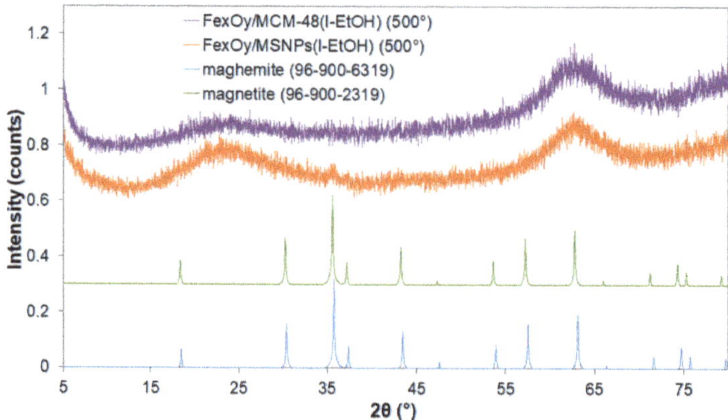

Figure 2. XRD diffractograms of the Fe$_x$O$_y$/MCM-48 and the Fe$_x$O$_y$/MNPs materials obtained by dry impregnation and reduced at 500 °C (bump at $2\theta = 63$–$64°$ is an artefact due to the sample holder) with reference data for magnetite and maghemite phases.

2.2. Biocompatibility Screening Tests on the NPs-Loaded Biomaterials

The produced NPs were incorporated into the polysaccharide matrix of chitosan/curdlan/HA scaffold to improve biocompatibility of the resultant biomaterial by enhancement of ROS formation upon exposure to cold nitrogen plasma. Before performing experiments making use of human ADSCs, aiming at determining the effect of plasma-induced ROS generation on proliferation and osteogenic differentiation of stem cells, biocompatibility screening tests were carried out using mouse calvarial preosteoblasts (MC3T3-E1 Subclone 4) to select a NPs–biomaterial combination that was non-toxic and the most supportive to osteoblast growth. For this purpose, biomaterials comprising three different concentrations (0.25, 0.125, and 0.05 wt.%) of MCM-48 and MSNPs with and without Fe_xO_y were tested.

MTT cytotoxicity test performed according to ISO 10993-5 (with the use of the extracts of the biomaterials) revealed that all NPs-loaded biomaterials were non-toxic to MC3T3-E1 preosteoblasts throughout the full length of the experiment (Figure 3a). Interestingly, biomaterials comprising MSNPs (without the Fe_xO_y active phase) at the concentrations of 0.25 wt.% and 0.125 wt.% significantly increased cell metabolism (thus viability) compared to other samples. Based on MTT test results, biomaterials containing NPs at the highest concentration (0.25 wt.%) were selected for further experiments with the use of confocal laser scanning microscope (CLSM). Live/dead staining and observation by confocal microscopy of the MC3T3-E1 preosteoblasts seeded at high concentration onto the surface of the samples confirmed non-toxicity of all tested biomaterials (Figure 3b). Preosteoblasts were well spread on the surface of the scaffolds and emitted only green fluorescence, indicating their high viability. Nevertheless, there were noticeably fewer cells on the biomaterial containing MCM-48 particles.

To select the variant of the biomaterial that is the most supportive to osteoblast growth and proliferation, preosteoblasts were seeded onto the biomaterials at low concentration and cultured for 3 days. Then, CLSM observation upon fluorescent staining of cytoskeleton was performed. CLSM images clearly demonstrated that biomaterials comprising both tested NPs without Fe_xO_y decoration negatively affected cell growth since there were meaningfully fewer cells on their surfaces compared to the control biomaterial without NPs (Figure 3c). Nevertheless, although biomaterials containing NPs without Fe_xO_y inhibited cell growth, higher magnification images showed that surfaces of all scaffolds allowed for good cell adhesion and spreading. Interestingly, addition of Fe_xO_y to the NPs overcame the negative effect of NPs without iron oxide decoration. The surface of the biomaterial with incorporated Fe_xO_y/MCM-48 was characterized by similar cell number to control biomaterial (without any NPs), whereas biomaterial containing Fe_xO_y/MSNPs noticeably improved cell growth and spreading. Based on the results obtained with the screening biocompatibility tests, bone scaffolds comprising Fe_xO_y/MSNPs and pure MSNPs (as a reference sample) at the concentration of 0.25 wt.% were selected for further experiments.

2.3. Characterization of the Selected NPs-Loaded Biomaterials

Based on the screening biocompatibility tests, MSNPs-loaded biomaterials were selected as the most promising. However, bone scaffolds for regenerative medicine applications should possess some key microstructural features, such as high stability, good compressive strength, or high porosity (at least 40%), allowing for new blood vessel formation and bone ingrowth deep into the implant [21–24]. Therefore, basic microstructural characterization of the scaffolds was performed to check whether incorporation of MSNPs and Fe_xO_y/MSNPs catalyst into the biomaterial structure affected its mechanical properties and porosity. Moreover, distribution of MSNPs within the polysaccharide matrix of the scaffolds was visualized by SEM.

The control scaffold (without any MSNPs) was composed of biopolymers and 80 wt.% HA granules, and on the macroscopic scale displayed the same morphology as biomaterials with the addition of MSNPs or MSNPs decorated with Fe_xO_y (Figure 4a). SEM observation made it possible to characterize the HA granules according to the structure of typical sintered ceramics, which were bound by a continuous polymer phase (Figure 4b) with a high degree of porosity that was estimated to be approximately 50% in all samples, as measured by Mercury Intrusion Porosimetry (MIP) (Table 2).

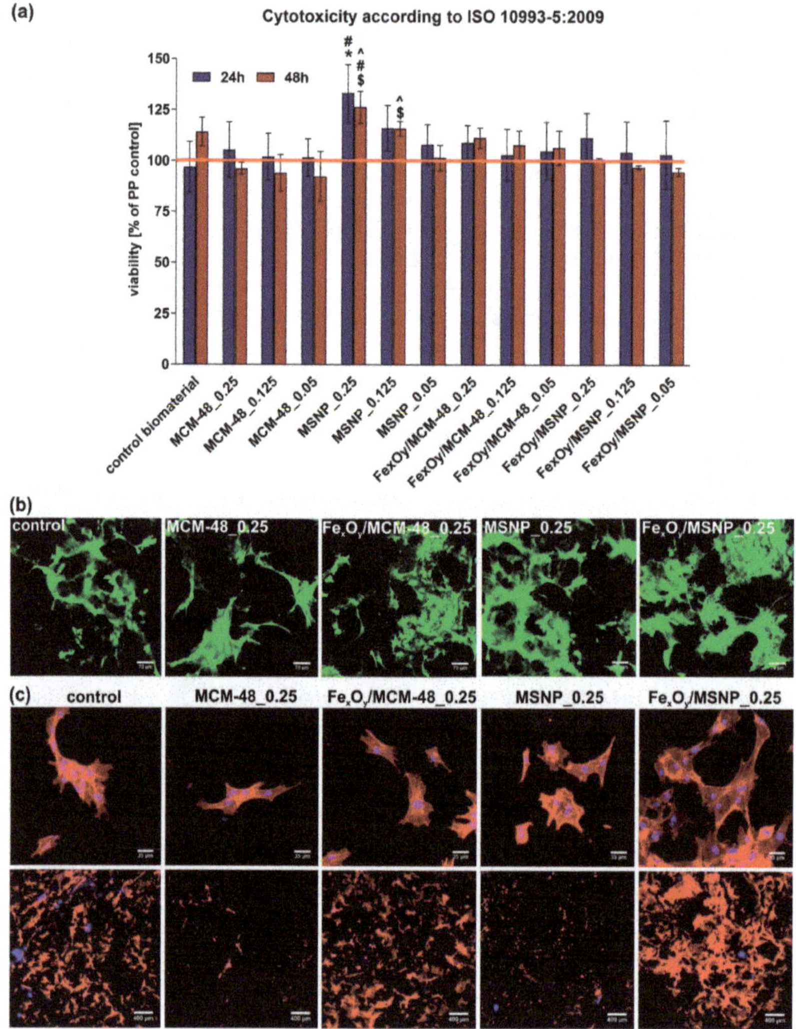

Figure 3. Biocompatibility screening tests on the NPs-loaded biomaterials: (**a**) MTT cytotoxicity test with the use of extracts of the biomaterials (PP control—cells exposed to the extract of polypropylene, revealing 100% viability; control biomaterial—extract of the biomaterial without any NPs; 0.25, 0.125, 0.05—concentration (wt.%) of NPs within the structure of the biomaterial; * statistically significant differences compared to the control biomaterial, # compared to the biomaterial containing Fe_xO_y/MCM-48 at corresponding concentration; $ compared to the biomaterial containing MCM at a corresponding concentration, and ^ compared to the biomaterial containing Fe_xO_y/MSNPs at a corresponding concentration, according to One-way Anova followed by Tukey's test, $p < 0.05$); (**b**) CLSM images of 3-day culture of preosteoblasts (MC3T3-E1 cells at high concentration were seeded) on the surface of the biomaterials upon fluorescent live/dead staining (control—biomaterial without any NPs; 0.25—concentration (wt.%) of NPs within the structure of the biomaterial; green fluorescence—viable cells, red fluorescence—nuclei of dead cells, magn. 200×); (**c**) CLSM images of 3-day culture of preosteoblasts (MC3T3-E1 cells at low concentration were seeded) on the surface of the biomaterials upon fluorescent staining of cytoskeleton and nuclei (red fluorescence—cytoskeleton, blue fluorescence—nuclei, upper images—magn. 400×, lower images—magn. 40×).

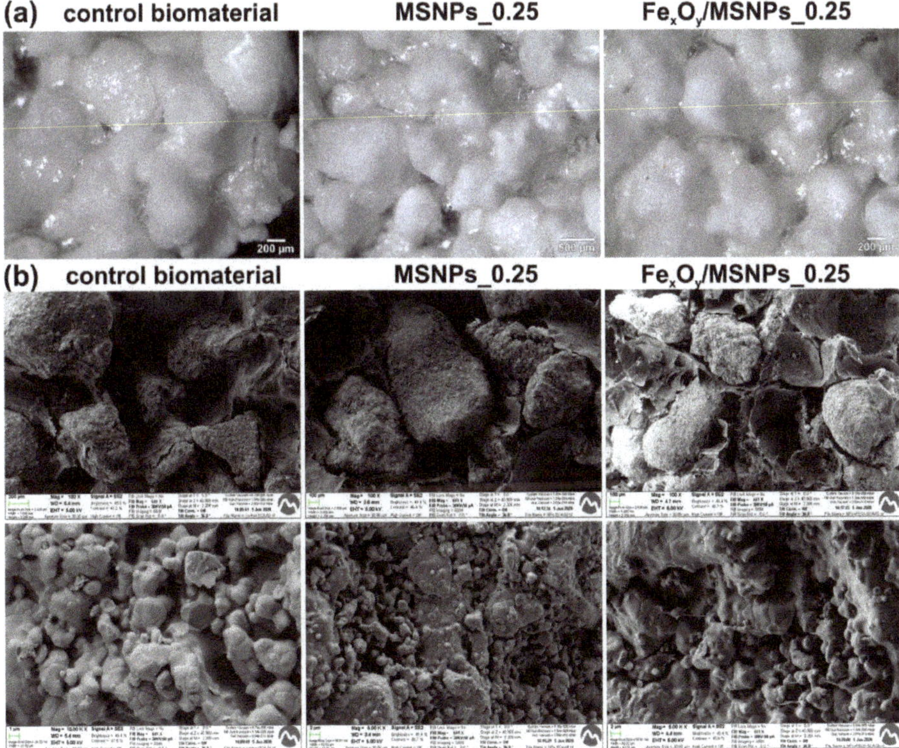

Figure 4. Visualization of the produced MSNPs-loaded scaffolds: (**a**) stereoscopic microscope images of the materials; (**b**) SEM micrographs of the scaffolds (upper images—magn. 100×, lower images—magn. 10000× for control and 5000× for NPs-loaded biomaterials).

Table 2. Total porosity as measured by Mercury Intrusion Porosimetry together with the elastic modulus and the work of fracture (WOF) of the different materials obtained from the compression tests.

Reference	Total Porosity (%)	Young's Modulus (GPa)	WOF (MPa.m$^{1/2}$)
Control Biomaterial	50.1	3.92 ± 0.75	1.72 ± 0.24
MSNPs_0.25	50.4	3.68 ± 0.78	1.63 ± 0.39
Fe$_x$O$_y$/MSNPs_0.25	50.2	3.18 ± 0.10	1.42 ± 0.21

As expected, the total porosity of the material remained unaltered with the incorporation of NPs and the Fe$_x$O$_y$ decoration at a concentration of 0.25 wt.% (Table 2), while the pore size distribution was slightly affected, displaying a shift of the main peak to larger pore sizes (Figure 5a). Thus, the Fe$_x$O$_y$/MSNPs_0.25 presented a slightly lower volume of pores but at larger entrance sizes compared to the control biomaterial.

Compression testing revealed that the three materials displayed strain hardening curve, characteristic of ductile composites, in agreement with the SEM images (Figure 5) showing a continuous polymeric phase with embedded HA. The fracture toughness or work of fracture (WOF), obtained from compression tests presented in Figure 5b, tended to decrease with the incorporation of the NPs and Fe$_x$O$_y$; however, no statistical differences can be pointed out between the different samples. The elastic modulus of the materials was also slightly decreasing with the incorporation of the different NPs; however, despite the known importance of biomechanics, no change should be expected in cell behavior from this low variation (Table 2).

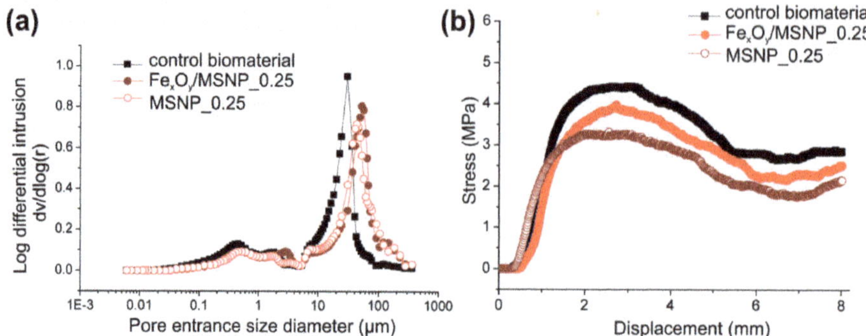

Figure 5. Basic characteristics of NPs-loaded biomaterial: (**a**) pore size distribution of the three different materials as determined by MIP; (**b**) stress–strain curves obtained with compression testing for the control biomaterial and scaffolds loaded with NPs and Fe_xO_y/NPs.

2.4. Plasma-Induced Production of OH radicals by NPs-Loaded Biomaterials

The main concept of this study was to incorporate Fe_xO_y-based catalysts into the structure of the biomaterials to enhance short-lived ROS and RNS formation upon nitrogen plasma treatment that would subsequently positively affect stem cell behavior. During our pilot studies, the effect of air and nitrogen plasma on viability of mouse preosteoblasts (MC3T3-E1 cell line) was determined. It was observed that longer exposure time (>16 s) had a negative effect on cell viability. Furthermore, cells revealed higher viability after 16 s exposure to nitrogen plasma compared to the air plasma. In our another study [15], it was demonstrated that 16 s nitrogen plasma treatment promoted preosteoblasts proliferation and enhanced their osteogenic differentiation. Importantly, the best results were obtained when MC3T3-E1 preosteoblasts were left in Hanks' Balanced Salt solution (HBSS) for 3 h after plasma exposure. Thus, on the basis of our previous experiments, in this study it was decided to use nitrogen as a source gas for plasma treatment and 16 s exposure time, followed by 3 h maintenance of the cells in HBSS after plasma treatment.

Based on the biocompatibility tests, Fe_xO_y/MSNPs catalyst was selected as the most promising material for bone therapies. Presence of Fe_xO_y in biomaterial suspended in water solution and treated by non-thermal plasma should enhance formation of oxygen-based active species, especially OH radicals in accordance to Fenton reaction mechanism [25]. OH radical production upon GlidArc (GAD) plasma treatment of the control scaffold (without nanoparticle addition) and scaffolds loaded with MSNPs and Fe_xO_y/MSNPs was determined by spectrofluorimetric measurements of the highly fluorescent probe umbelliferone, which is the product of coumarin and OH radical reaction [26]. Scaffolds suspended in HBSS were treated with plasma for 16 s in an analogous manner as cell-seeded biomaterials in the subsequent tests. Gliding arc is a versatile source, which allows working with pure nitrogen gas and renders relatively large area in comparison to helium/argon jet sources. For such a short treatment time, the control unit for the power supply with the timer was developed. First, gas flow was set using gas flow regulator, then the electrical discharge was ignited via control unit and it lasted for 16 s, then the power supply was shut down. The emission intensities of tested samples are presented in Figure 6a. The characteristic peak at 452 nm related to the presence of umbelliferone after short plasma treatment was clearly observed for the HBSS solutions with samples loaded with NPs. In the case of control scaffold (without any NPs), extremely low concentrations of generated active species were probably instantly consumed on-site by the surface reactions with organic part of biomaterial itself and they were almost below the detection limit in the liquid. Figure 6b depicts concentrations of umbelliferone in HBSS solution after 16 s of plasma treatment. The amount of generated OH radicals was low, but as expected, the umbelliferone concentrations obtained for Fe_xO_y/MSNPs-loaded biomaterial were slightly higher than for the MSNPs-loaded sample and control material.

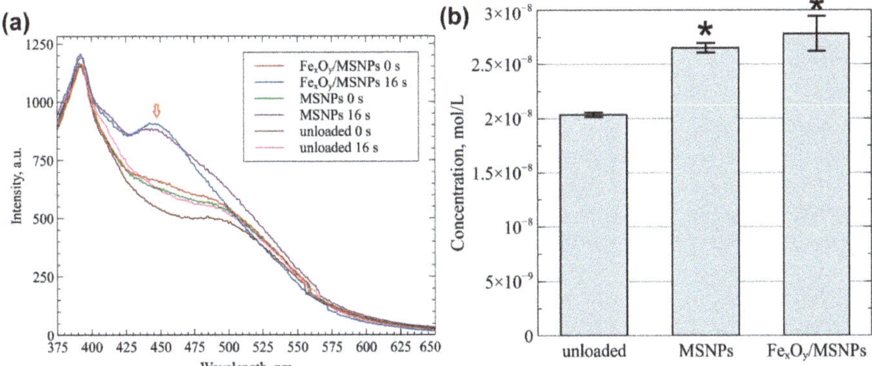

Figure 6. Plasma-induced OH production by biomaterials: (**a**) emission intensities of tested scaffolds (an arrow indicates emission peak of umbelliferone); (**b**) umbelliferone concentrations in HBSS solutions after 16 s GAD plasma treatment (* significantly different results compared to the control biomaterial (unloaded), according to One-way Anova followed by Tukey's test, $p < 0.05$).

It should be noted that NPs were incorporated by the entrapment method within the hydrogel matrix of the biomaterial. It is very likely that most OH radicals formed upon plasma exposure were entrapped within the hydrogel matrix and their release to the HBSS was hindered. Therefore, very low concentrations of OH radicals were detected in HBSS liquid medium. Nevertheless, it is assumed that high concentrations of reactive species could be present within the hydrogel matrix of the biomaterial directly affecting behavior of cells cultured on its surface. Because the plasma was not touching the HBSS and N_2 plasma gas was used for the treatment, it may also be assumed that a large quantity of NO_2/NO could be formed in our conditions [27]. As OH radicals are generated on Fe_xO_y nanocatalyst more efficiently than NO, both types of radicals are expected to be present at similar and high quantities in our conditions.

2.5. Plasma Effect on Stem Cell Behavior Cultured on the NPs-Loaded Biomaterials

Plasma treatment of the MSNPs- and Fe_xO_y/MSNPs-loaded biomaterials seeded with the human ADSCs was performed in Hanks' Balanced Salt solution (HBSS) to avoid unexpected results due to the high complexity of the culture medium [28]. Stem cells-seeded biomaterials were exposed to cold nitrogen plasma for 16 s and left for 3 h in HBSS after plasma treatment. These conditions were demonstrated in our previous studies to be very effective in stimulation of proliferation and ECM synthesis by MC3T3-E1 preosteoblasts [15].

To determine the effect of nitrogen plasma on ADSC proliferation on the biomaterials, stem cells were seeded directly on the samples at extremely low concentration. Within this study, it was clearly demonstrated that ADSCs proliferated significantly faster on the surface of plasma-activated biomaterials with incorporated MSNPs and Fe_xO_y/MSNPs catalyst compared to the plasma-activated control biomaterial (without any NPs) (Figure 7a).

Importantly, 6 days after plasma exposure, there were 3-fold more cells on the biomaterials with Fe_xO_y/MSNPs compared to the MSNPs-loaded scaffolds regardless of the plasma activation, indicating that the presence of the active phase (Fe_xO_y) itself was a key factor responsible for the enhancement of cell proliferation. This indicates that Fe_xO_y/MSNPs catalyst incorporated into the structure of the biomaterial had by itself the ability (without plasma activation) to induce radical formation by the heterogeneous Fenton-like reaction at sufficient level to improve cell proliferation. CLSM images of ADSCs grown on the biomaterials confirmed meaningfully greater number of cells on the biomaterials comprising Fe_xO_y active phase compared to other samples (Figure 7b). Plasma exposure of ADSCs grown on the Fe_xO_y/MSNPs-loaded biomaterial resulted in significant increase in cell number compared to the untreated Fe_xO_y/MSNPs-loaded sample (doubling time = 38.96 h for plasma-activated cells and doubling time = 60.74 h for untreated cells) (Figure 7a). Therefore, oxygen-based active species, more particularly OH radicals, generated upon plasma activation of Fe_xO_y/MSNPs-loaded biomaterial reached higher concentration compared to untreated sample, thereby significantly accelerating proliferation of mesenchymal stem cells. This is in agreement with previous works where other multipotent cells (i.e., human bone marrow-derived mesenchymal stem cells) displayed enhanced proliferation under controlled amounts of ROS generated by plasmas [29], and the differences observed after 6 days (Figure 7a) are clearly in agreement with the observed "delayed" effect of cold plasmas, where an initial dose of ROS is able to trigger long term effects on cells [28,29].

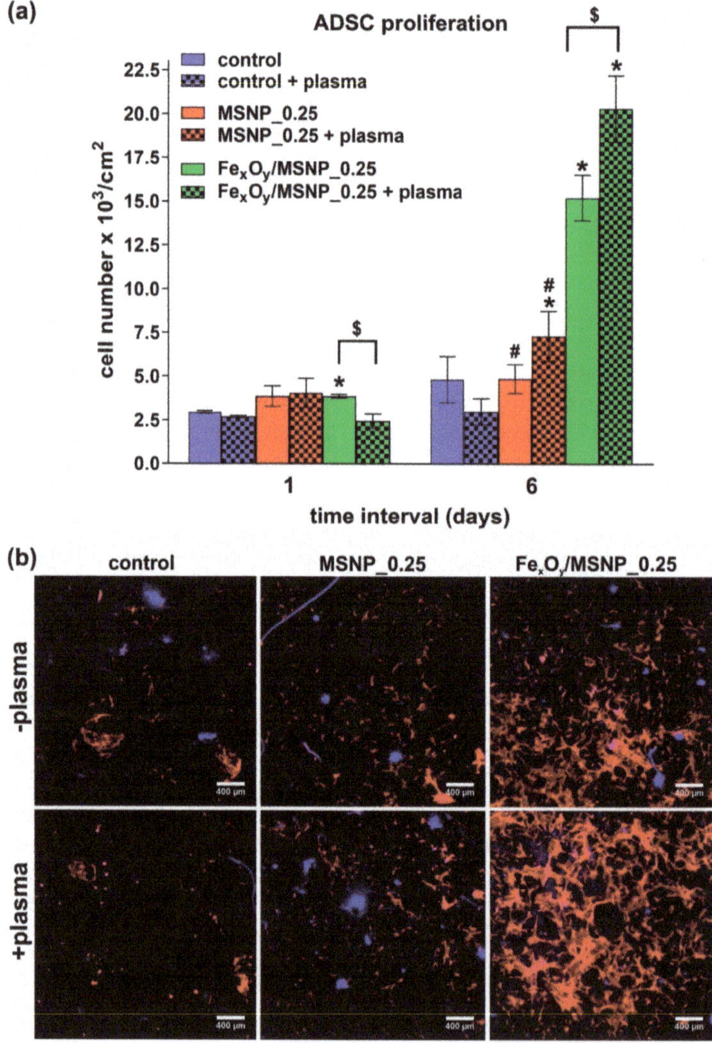

Figure 7. Proliferation of ADSCs on the surface of the NPs-loaded scaffolds upon nitrogen plasma exposure (16 s in HBSS; cells left in HBSS for 3 h after plasma treatment): (**a**) cell number increase with time determined by nuclei counting (control—biomaterial without any NPs; 0.25—concentration (wt.%)

of NPs within the structure of the biomaterial; * statistically significant results compared to the corresponding (with or without plasma activation) control biomaterial, # compared to the corresponding (with or without plasma activation) biomaterial containing Fe_xO_y/MSNPs; $ statistically significant results between plasma-treated and untreated biomaterials, according to One-way Anova followed by Tukey's test, $p < 0.05$; (**b**) CLSM images of ADSCs on the surface of the biomaterials (cells at extremely low concentration were seeded) 6 days after plasma treatment upon fluorescent staining of cytoskeleton and nuclei (red fluorescence—cytoskeleton, blue fluorescence—nuclei, magn. 40×).

Osteogenic differentiation of ADSCs cultured on the plasma-activated scaffolds was determined by detection of typical markers of bone formation process: bone alkaline phosphatase (bALP), type I collagen (Col I), and osteocalcin (OC). The experiment revealed that incorporation of MSNPs into the biomaterial structure significantly inhibited bALP production by ADSCs regardless of plasma activation (Figure 8). However, addition of Fe_xO_y overcame this negative effect of pure MSNPs, thus stem cells grown on the Fe_xO_y/MSNPs-loaded scaffold showed similar bALP level to control samples. Col I synthesis by ADSCs was significantly reduced in the presence of both MSNPs and Fe_xO_y/MSNPs compared to the control biomaterial. However, this effect was slightly overcome by plasma activation of the Fe_xO_y/MSNPs-loaded scaffold, whereas plasma treatment of MSNPs-loaded biomaterial did not influence Col I production. Similarly, OC production was significantly decreased by both MSNPs- and Fe_xO_y/MSNPs-loaded biomaterials. In this case, plasma activation significantly increased OC synthesis by ADSCs on both mentioned biomaterials compared to untreated samples. Surprisingly, plasma treatment of Fe_xO_y/MSNPs-loaded biomaterial only overcame negative effect of the Fe_xO_y/MSNPs catalyst on the OC production, whereas plasma activation of MSNPs-loaded scaffold resulted in 5-fold increase in OC level compared to untreated MSNPs-loaded biomaterial and 2-fold increase compared to the control biomaterial (plasma-treated and untreated). Importantly, no statistically significant differences between plasma-treated and untreated control samples—with respect to bALP, Col I, and OC production—were observed.

In our previous studies performed on MC3T3-E1 preosteoblasts, it was demonstrated that short-time (16 s) nitrogen plasma treatment of the cells significantly promoted cell proliferation and enhanced bALP and OC synthesis [15]. In this study, nitrogen plasma activation of control biomaterial (without NPs) seeded with ADSCs did not influence neither cell proliferation nor osteogenic differentiation. It indicates that ADSCs are less sensitive to plasma long-lived reactive species (e.g., H_2O_2) than preosteoblast cells. Nevertheless, plasma activation of Fe_xO_y/MSNPs-loaded biomaterial resulted in significantly accelerated cell proliferation (Figure 7). Thus, the proliferation of ADSCs seems to be more sensitive to short-lived reactive species (like OH radicals) provided by the presence of Fe_xO_y nanocatalysts on the MSNPs. Enhanced proliferation of ADSCs on the scaffold comprising Fe_xO_y/MSNPs upon exposure to nitrogen plasma may be explained by short-lived reactive species-mediated induction of fibroblast growth factor-2 (FGF-2) release by the cells. This mechanism of plasma-induced accelerated cell proliferation was proposed by Kalghatgi et al., who suggested that O_3, NO, H_2O_2, or OH generated upon plasma activation may lead to increased FGF-2 release [30]. Here, it was shown that short-lived reactive species and more probably OH radicals were produced in higher quantities, due to the combination of plasma activation and the presence of Fenton supported nanocatalysts, and were the key players for this mechanism. In turn, increased levels of FGF-2 in the microenvironment was an autocrine or paracrine signal for mesenchymal stem cells, influencing their proliferation [31–33]. Therefore, it indicates that accelerated proliferation of ADSCs observed in our studies resulted from enhanced production of active species, inducing local formation of OH radicals on the Fe_xO_y catalysts upon plasma activation of Fe_xO_y/MSNPs-loaded biomaterial.

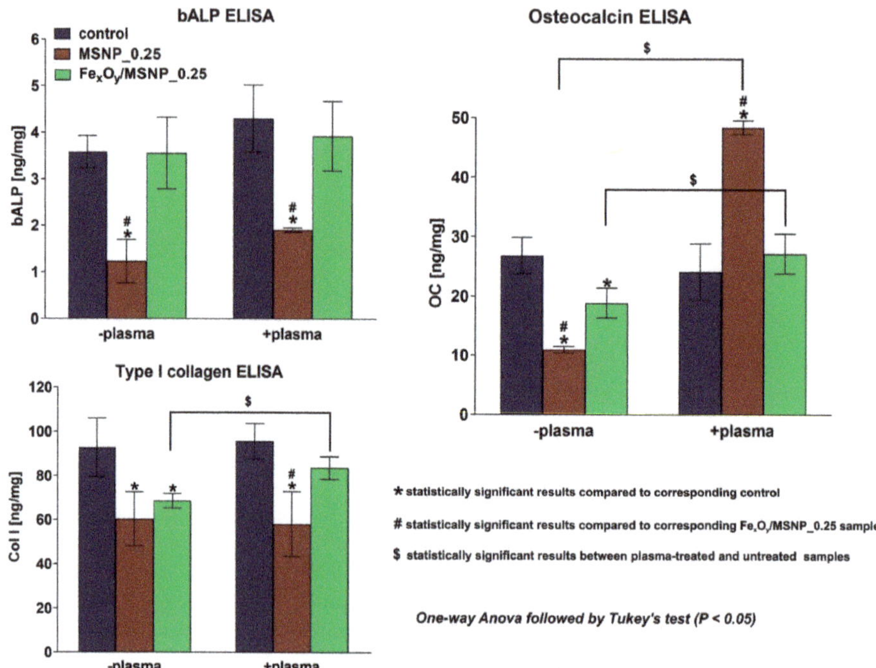

Figure 8. Osteogenic differentiation of ADSCs on the surface of the NPs-loaded scaffolds determined 20 days after nitrogen plasma exposure (16 s in HBSS; cells left in HBSS for 3 h after plasma treatment); control—biomaterial without any NPs; 0.25—concentration (wt.%) of NPs within the structure of the biomaterial.

It is known that Si plays an important role in several bone biochemical processes occurring in the living organism, including promotion of osteogenic differentiation [34]. However, incorporation of MSNPs with and without Fe$_x$O$_y$ decoration had a negative effect on osteogenic differentiation of ADSCs. This effect could be the result of the release of a certain amount of MSNPs and Fe$_x$O$_y$/MSNPs when the scaffolds were swelling in the aqueous environment (typical behavior of hydrogel type biomaterials) [23]. The attachment of free NPs on the cell membrane could alter the cell differentiation activity. To solve this issue, MSNPs could be covalently attached (instead of immobilization by entrapment) to the glucan matrix or hydroxyapatite granules to better control their release and avoid a burst release effect.

ADSCs cultured on the Fe$_x$O$_y$/MSNPs-loaded biomaterial showed slightly better osteogenic potential than cells on MSNPs-biomaterial (except OC synthesis upon plasma exposure) (Figure 8). Thus, the generation of short-lived species, mainly OH radicals, from the intrinsic H$_2$O$_2$ and at a lower extent the possible slow release of Fe ions from the Fe$_x$O$_y$/MSNPs-loaded biomaterial were shown to overcome negative effect of MSNPs on osteogenic differentiation. It is known that Fe is crucial for many biochemical reactions and processes occurring in the living organism, including oxygen transport and enzymatic reactions. Interestingly, recent studies have also demonstrated close correlation between iron supply and bone metabolism [35–37].

Without plasma treatment, ADSCs cultured on the NPs-loaded biomaterials generally revealed reduced osteogenic activity compared to the control biomaterial. However, upon plasma treatment of cells grown on the Fe$_x$O$_y$/MSNPs-loaded biomaterial, osteogenic markers (bALP, Col I, and OC) were at the same level as in the control sample. In the case of plasma-treated ADSCs cultured on the MSNPs-loaded biomaterial, bALP and Col I markers were reduced, whereas OC increased compared to the control biomaterial (Figure 8). Importantly, Tominami et al. demonstrated that activation of MC3T3-E1 preosteoblats with cold atmospheric helium plasma, which generates mainly H_2O_2, OH, and superoxide anion radicals, led to significantly increased bALP level and enhanced mineralization activity of the cells [14]. This indicates that the quantity of reactive species needs to be higher to induce an increase of bALP and Col I markers. However, surprisingly, a large increase of OC production by ADSCs was observed when plasma activation was combined with MSNPs free of Fe_xO_y. It seems that the quantity of long-lived reactive species generated in our conditions were close to the optimal one for OC production. Arai et al. suggested that H_2O_2 may up-regulate antioxidant system affecting expression of osteogenic genes, including enhancement of OC production [16]. Thus, a lower amount of reactive species seems to be needed to activate the expression of OC. Reduced OC synthesis by ADSCs grown on the Fe_xO_y/MSNPs-loaded scaffold compared to catalyst-free biomaterial (in the presence of plasma activation, Figure 8) was even more surprising result. In this case, the quantity of reactive species (short-lived and long-lived) was probably too high and the OC expression was disturbed. Other processes, which were in competition with the OC synthesis, were probably activated due to the high level of oxidative stress reached by combining plasma, MSNPs, and Fe_xO_y catalyst. Thus, a shorter plasma irradiation is needed to stay at the optimum dose of reactive species generated.

3. Materials and Methods

3.1. Silica NPs Synthesis

Spherical mesoporous Mobil Composition of Matter No. 48 (MCM-48) support was synthesized according to the procedure described by Schumacher et al. [38]. The template n-hexadecyltrimethylammonium bromide (2.4 g) (Sigma-Aldrich Chemicals, Steinheim, Germany) was dissolved under stirring in a mixture of deionized water/ethanol (50 mL/50 mL), followed by the addition of aqueous ammonia (28–30 wt.%, 12 mL) (ChemLab, Zedelgem, Belgium) and this solution was left under stirring for 10 min. Then, the silica source tetraethyl orthosilicate (3.4 g) (Sigma-Aldrich Chemicals, Steinheim, Germany) was added and the mixture was stirred for 2 h under constant stirring. The resulting solid was filtered out, washed with distilled water and dried in air at room temperature. The MCM-48 support was obtained by calcination at 823 K in a muffle furnace under static air, for 6 h, to remove the template.

Wormhole-like mesoporous silica nanoparticles (MSNPs) with interconnected porosity were synthesized using surfactant directed, base-catalyzed condensation of silica precursors with a sol–gel approach proposed earlier by Bein and co-workers [39,40]. The stock solution was prepared by mixing 13.75 mL (762.8 mmol) of milliQ water, 2.23 mL (38.2 mmol) of absolute ethanol, and 2.23 mL (1.69 mmol) of 25% CetylTrimethylAmmonium chloride (CTAC) by stirring in Radleys Mya 4® station for 10 min under Argon atmosphere. Then, TriEthanolAmine (TEA) (1.78 mL; 13.37 mmol) was added and mixed with the stock solution until complete dissolution. When TEA was fully dissolved, the mixture was heated at 60 °C, and then tetraethyl orthosilicate (TEOS) (1.454 mL; 6.5 mmol) was added drop by drop. The reaction was further stirred for 2 h, under Argon atmosphere. The molar ratio of this reaction was: TEOS/CTAC/TEA/H2O/EtOH 1/0.26/2/117.35/5.88. Extraction of the CTACl surfactant was achieved by applying several cycles of hydrochloric acid wash/centrifugation. Each extraction cycle included a dispersion of MSNPs in a solution of ethanol and hydrochloric acid at 2%, and an ultrasonication treatment for 40 min at 40 °C followed by centrifugation for 20 min at 45,000× g. After surfactant extraction, MSNPs were either dispersed in absolute ethanol or freeze-dried to obtain a dry powder. All reagents needed for the synthesis of MSNPs were purchased from Sigma-Aldrich Chemicals (Steinheim, Germany).

3.1.1. Preparation of Iron Oxide-Loaded NPs by Dry Impregnation

The NPs were impregnated with an ethanolic solution of $Fe(NO_3)_3$ $9H_2O$ (Alfa Aesar, Kandel, Germany), the concentration of which being adjusted to obtain a metal loading of 5 wt.%. Under these conditions, the volume of impregnation solution ensures the complete wetting of the support (by the process known as Incipient Wetness Impregnation, IWI). The impregnated powder was dried at 373 K for 12 h in stagnant air and then calcined at 773 K in a muffle oven. The iron oxide-based catalysts were then reduced at the desired temperature (773 K) under an H_2 5%/Ar flow.

3.1.2. Characterization of NPs

The nitrogen adsorption–desorption isotherms were obtained at 77 K with an ASAP-2020 Micrometrics instrument (Micromeritics, Norcross, Georgia, USA), allowing determination of specific surface area (SSA). Before analysis, the samples (0.02–0.10 g) were degassed under 0.133 Pa pressure for 2 h at 200 °C (10 °C/min). The specific surface areas were provided by the analysis of the isotherms with the Brunauer–Emmett–Teller (BET) equation, while the pores' average diameter was estimated using the DFT model. Zeta potential of nanoparticles were obtained with a Zetasizer Nano (Malvern® Panalytical, Malvern, UK) in deionized water.

Fe_xO_y/NPs were visualized by TEM. Images were obtained on a LEO 922 Omega Energy Filter Transmission Electron Microscope (LEO Electron Microscopy Inc., Cambridge, UK, now: Carl Zeiss NTS GmbH) operating at 120 kV. The samples were first suspended in acetone under ultrasonic treatment. A drop of the suspension was deposited on a holey carbon film supported on a copper grid (Holey Carbon Film 300 Mesh Cu, Electron Microscopy Sciences, Hatfield, PA, USA), which was dried overnight under vacuum at room temperature, before introduction in the microscope. The iron loading was measured by ICP on an ICAP 6500 instrument (Thermo Scientific, Cambridge, UK). Before analysis, the samples (known amount) were prepared by acid digestion process.

XRD was performed on a Bruker D8 advanced diffractometer with a Bragg Brentano geometry, using a LinkEye XE-T detector with Cu Kα radiation ($\lambda = 0.15418$ nm) and a power of 1200 W (40 kV, 30 mA). The samples were scanned from 0.8 to 10 (2θ range) at a scanning rate of 1.5 °C/min.

X-ray photoelectron spectroscopy (XPS) analyses were carried out at room temperature with a SSI-Xprobe (SSX 100/206) photoelectron spectrometer from Surface Science Instruments (Mountain View, California, USA), equipped with a monochromatized microfocus Al X-ray source. Samples were stuck onto small sample holders with double-face adhesive tape and then placed on an insulating Macor® ceramic carousel. Charge effects were avoided by placing a nickel grid above the samples and using a flood gun set at 8 eV. The binding energies were calculated with respect to the C-(C, H) component of the C1s peak fixed at 284.8 eV. Data treatment was performed using the CasaXPS program (Version 2.3.17dev6.0b, Casa Software Ltd., Teignmouth, UK). The peaks were decomposed into a sum of Gaussian/Lorentzian (85/15) after subtraction of a Shirley-type baseline.

SEM and SEM-EDX (Energy Dispersive X-ray Spectrometry analysis) analyses were performed using a JEOL FEG SEM 7600F (JEOL, Tokyo, Japan) equipped with an EDX system (Jeol JSM2300 with a resolution < 129 eV) operating at 15 keV with a working distance of about 8 mm. The acquisition time for the chemical spectra lasted 300 s with a probe current of 1 nA. The quantitative analysis of the atomic elements was performed with the integrated Analysis Station software. A two-step analysis procedure was applied in order to obtain quantitatively reliable results for the elements profiles over the entire cross sections: (i) the subtraction of the bremsstrahlung done with the classical "Top Hat Filter" method [41,42] and (ii) the quantification of the area under each atomic peak determined by the φ (ρz) model [43–45].

3.2. Fabrication of NPs-Loaded Biomaterials

The chitosan/curdlan/HA composite was produced according to the procedure described previously by [23,46,47] with some modifications. Krill chitosan (1174 kDa molecular weight, 73% deacetylation degree) was obtained from National Marine Fisheries Research Institute (Gdynia, Poland), whereas curdlan (β-1,3-glucan) was purchased from Wako Pure Chemicals Industries (Osaka, Japan). Suspension of various NPs (with and without Fe_xO_y) was prepared in distilled water at the following concentrations: 0.50 wt.%, 0.25 wt.%, and 0.10 wt.%. Then, 16 wt.% curdlan suspension was prepared in the appropriate NPs suspension or in distilled water (control biomaterial) and mixed 1:1 with 4 wt.% chitosan solution prepared in 1% acetic acid solution (Avantor Performance Materials, Gliwice, Poland). The final concentrations of the individual components in the blend were as follow: 8 wt.% curdlan, 2 wt.% chitosan, and 0.25 wt.%, 0.125 wt.% or 0.05 wt.% NPs. The 80 wt.% (*w/v*) hydroxyapatite granules (HA BIOCER, Chema Elektromet, Rzeszow, Poland) were added to the blend and the resultant paste was subjected to thermal gelation at 95 °C for 20 min., followed by neutralization in sodium hydroxide (Avantor Performance Materials, Gliwice, Poland).

3.3. Characterization of NPs-Loaded Biomaterials

Fabricated biomaterials containing MSNPs or Fe_xO_y/MSNPs were visualized by SEM and optical microscope. Pictures of the samples were taken with a stereoscopic microscope (Olympus SZ61TR, Olympus Polska Sp. z o. o., Warsaw, Poland). To study the microstructure of the scaffolds, cross-sections of the samples were made by cutting the cylinder-shaped samples in their middle. A Zeiss Neon 40 cross-beam workstation with Gemini SEM column (Carl Zeiss Iberia, S.L, Madrid, Spain) was used for SEM observation at 5 keV. Prior C-coating of the samples was performed using an EMITECH K950X Turbo Evaporator (Quorum Technologies Ltd., Lewes, UK). SEM images were recorded at magnification range from 100× to 10000× using a secondary electron detector.

Mechanical properties of the biomaterials were assessed by compression testing that was performed using a Bionix^{TM} Test System Model 858 (MTS, Eden Prairie, Minnesota, USA) with an axial load cell of 2.5 kN and a MTS FlexTest Model 40 controller. Compression rate was of 1 mm/min with a maximum piston extension (compression) of 8 mm, considering d = 0 when touching the sample. Monitoring of the system was done by Station Manager software provided by MTS. Three replicates were performed for each kind of samples set in vertical position, and results were presented by plotting the stress (MPa) as a function of the extension (mm).

Mercury Intrusion Porosimetry (MIP, AutoPore IV, Micromeritics, Norcross, Georgia, USA) was carried out to assess the pore entrance size distribution (PESD) within the biomaterials. For each MIP experiment 3-4 cylindrical samples were measured.

3.4. Plasma-Induced Production of OH Radicals by NPs-Loaded Biomaterials

For determination of OH radicals' presence, 40 mg of control scaffold (chitosan/curdlan/ hydroxyapatite) and biomaterials loaded with MSNPs or Fe_xO_y/MSNPs were immersed in 5 mL of 1×10^{-3} mol/L coumarin solution in HBSS. 99% purity coumarin and HBSS were purchased from ACROS Organics™ (Geel, Belgium) and Sigma-Aldrich Chemicals (Warsaw, Poland), respectively. Umbelliferone for calibration was purchased from Sigma-Aldrich (Steinheim, Germany). Spectrofluorimetric measurements of umbelliferone as fluorescent probe were performed with Shimadzu RF-6000 spectrofluorimeter with excitation and emission wavelengths at 346 nm and 452 nm, respectively.

Atmospheric pressure plasma was generated in GAD reactor presented in Figure 9 developed at Lublin University of Technology (LUT), consisting of two copper electrodes (1.5 mm thick, and 10 cm long with 12° angle between them), which has been described in detail in other related publications [15,48]. The smallest inter-electrode distance was 3 mm. The AC high-voltage power supply was operated at 50 Hz frequency. Maximum apparent power was 52 VA and 25 VA on primary and secondary side of transformer, respectively. The RMS voltage and current on the secondary side ranged 688 V and 36 mA, respectively. Current/voltage waveforms may be seen in Supplementary Figure S3. The reactor was supplied with nitrogen as a substrate gas via gas flow regulator at flow rate of 7.33 dm^3/min. The electrode tips were positioned at 3 cm distance from the surface of the liquid with immersed treated samples. Operation of power supply within selected treatment time of 16 s was controlled using control unit with timer designed at LUT. Measurements were performed in triplicates at room temperature and atmospheric pressure. The same conditions of plasma activation were applied in all cell culture tests.

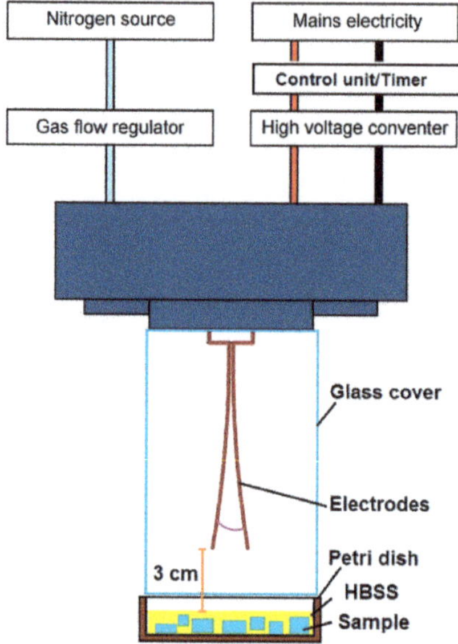

Figure 9. Scheme presenting plasma treatment set-up.

3.5. Cell Culture Experiments

Screening cell culture tests aiming at selecting the most promising NPs-loaded biomaterials were performed using mouse calvarial preosteoblast cell line (MC3T3-E1 Subclone 4, ATCC-LGC standards, Teddington, UK). These were cultured in alpha MEM medium (Gibco, Life Technologies, Carlsbad, California, USA) supplemented with 10% fetal bovine serum (FBS, Pan-Biotech GmbH, Aidenbach, Bavaria, Germany) and mixture of antibiotics (0.1 mg/mL streptomycin/100 U/mL penicillin) purchased from Sigma-Aldrich Chemicals (Warsaw, Poland).

The effect of cold plasma on cell behavior was assessed using human adipose tissue-derived mesenchymal stem cells (ADSCs, ATCC-LGC standards, Teddington, UK) that were cultured in Mesenchymal Stem Cell Basal Medium with the addition of the components of Adipose-Derived Mesenchymal Stem Cell Growth Kit Low Serum (ATCC-LGC Standards, Teddington, UK) and antibiotics (penicillin/streptomycin). All cells were maintained at 37 °C, 95% of air humidity and 5% CO_2.

3.5.1. Cytotoxicity of the Biomaterials

The cytotoxicity of the biomaterials containing various concentrations of NPs was determined according to ISO 10993-5 using 24 h extracts of the scaffolds prepared as it was described earlier [49]. MC3T3-E1 cells were seeded into 96-multiwell plates in 100 μL of a complete culture medium at a concentration of 2.5×10^5 cells/mL. After 24 h incubation, the culture medium was discarded and 100 μL of appropriate extracts of the biomaterials were added. Polypropylene extract was a negative control of cytotoxicity. The MC3T3-E1 cells were exposed to the extract for 24 h and 48 h and MTT colorimetric assay (Sigma-Aldrich Chemicals, Warsaw, Poland) was carried out to assess cell viability.

Cytotoxicity of the biomaterials was also determined by live/dead fluorescent staining of preosteoblasts cultured on the samples. Before the experiment, biomaterial samples (6 mm × 6 mm) were placed in the wells of a 48-multiwell plate and preincubated for 12 h in the complete culture medium. The preosteoblasts were seeded directly on the biomaterials at high concentration of 1×10^5 cells/sample. After 72 h of culture, MC3T3-E1 cells on the surface of the biomaterials were stained with the use of Live/Dead Double Staining Kit (Sigma-Aldrich Chemicals, Warsaw, Poland) according to the manufacturer protocol. The stained cells were analyzed using confocal laser scanning microscope (CLSM, Olympus Fluoview equipped with FV1000, Olympus Polska Sp. z o. o., Warsaw, Poland).

3.5.2. Osteoblast Growth on the Biomaterials

Ability of the biomaterials to support cell proliferation and growth was assessed by seeding MC3T3-E1 preosteoblasts directly on the biomaterials at low concentration of 3×10^4 cells/sample. After 72 h of culture, the MC3T3-E1 cells were fixed as described earlier [49] and F-actin cytoskeletal filaments were stained with AlexaFluor635-conjugated phallotoxin (Invitrogen, Carlsbad, California, USA). Cell nuclei were stained using 0.5 µg/mL DAPI (Sigma-Aldrich Chemicals, Warsaw, Poland). Stained cells were observed using CLSM.

3.5.3. Plasma Effect on Proliferation of Stem Cells on the Biomaterials

Human ADSCs were seeded directly on the biomaterials at extremely low concentration of 1.5×10^4 cells/sample. After 24 h of culture (when the cells were well attached to the biomaterials), the culture medium was discarded and replaced with Hanks' Balanced Salt solution (HBSS, Sigma-Aldrich Chemicals, Warsaw, Poland). Plasma treatment of the samples was performed using GlidArc reactor operated at the atmospheric pressure with the use of nitrogen as a substrate gas. The electrode tips were positioned at a 3 cm distance from the surface of the biomaterials and 7.33 dm^3/min flow-rate of nitrogen was applied. ADSCs were exposed to nitrogen plasma for 16 s and left for 3 h in HBSS after treatment. Stem cells cultured on the biomaterials and maintained for 3 h in HBSS without plasma treatment served as control samples. Then, HBSS was discarded, fresh complete culture medium was added, and the cells were cultured for further 6 days with medium renewal on the 3rd day. On the 1st and 6th day, stem cells were fixed and stained as described in Section 3.5.2. The number of cells on the surface of the biomaterials was determined by nuclei counting using ImageJ software version 1.52a (Wayne Rasband, National Institutes of Health, Bethesda, Maryland, USA). The doubling time for the stem cells grown on the samples was calculated with the use of Doubling Time Computing software version 3.1.0.

3.5.4. Plasma Effect on Osteogenic Differentiation of Stem Cells on the Biomaterials

Human ADSCs were seeded directly on the biomaterials at high concentration of 5×10^4 cells/sample. After 24 h of culture, the cells were treated with cold nitrogen plasma as described in Section 3.5.3. Three hours after plasma treatment, HBSS was replaced with osteogenic medium (Osteocyte Differentiation Tool, ATCC-LGC standards, Teddington, UK) and the cells were cultured for a further 20 days with half of a medium renewal every 3-4 days. Markers typical of the osteogenic differentiation (Col I, bALP, and OC) were assessed in the cell lysates that were prepared according to the procedure described earlier [46]. The levels of the osteogenic markers were evaluated using human-specific ELISA kits (bALP ELISA Kit, FineTest, Wuhan, China; Collagen alpha-1(I) chain ELISA Kit, EIAab, Wuhan, China; OC ELISA Kit, EIAab, Wuhan, China). The total protein content was also determined for each lysate using BCA Protein Assay Kit (Thermo Fisher Scientific, Waltham, Massachusetts, USA) to normalize the amount of osteogenic markers (ng) per mg of total cellular proteins.

3.6. Statistical Analysis

Cell culture tests were performed in at least three independent experiments ($n = 3$). Statistical significance was considered at $p < 0.05$ and determined using One-way ANOVA followed by Tukey's test (GraphPad Prism 8.0.0 Software, GraphPad Software Inc., California, CA, USA).

4. Conclusions

Within this study it was observed that extrinsic H_2O_2 generated by nitrogen plasma and intrinsic H_2O_2 generated by the presence of Fe_xO_y-free MSNPs in the scaffold positively affected expression of

OC gene, but did not compensate the negative effect of MSNPs presence for the expression of bALP and Col I genes. The addition of Fe_xO_y catalysts in MSNPs always lead to a significant increase of the quantity of reactive species (mainly short-lived species like OH radicals). The presence of these additional short-lived species had a positive effect on bALP and Col I markers but a negative effect on OC gene expression, where the concentration of reactive species was already at its optimum without the presence of the Fe_xO_y catalysts.

Presented results clearly demonstrated that short-time (16 s) exposure of ADSCs to nitrogen plasma was non-toxic, accelerated proliferation of cells grown on the biomaterial containing Fe_xO_y/MSNPs catalyst, and increased OC production by the cells cultured on the scaffold containing MSNPs without Fe_xO_y decoration. Plasma activation of the biomaterial containing Fe_xO_y/MSNPs catalyst resulted in the formation of sufficient amounts of active long-lived species with enhanced local generation of OH radicals, thanks to the Fe_xO_y catalysts, that had positive impact on stem cell proliferation and at the same time did not negatively affect their osteogenic differentiation. Therefore, plasma-activated Fe_xO_y/MSNPs-loaded biomaterial is characterized by improved biocompatibility and has great clinical potential to be used in regenerative medicine applications to improve bone healing process.

Supplementary Materials: Supplementary materials can be found at http://www.mdpi.com/1422-0067/21/13/4738/s1.

Author Contributions: Conceptualization, A.P., J.P., C.C., S.H. and D.D.; methodology, A.P., J.P., C.C., S.H. and D.D.; validation, A.P., J.P., M.A., J.-S.T., C.L., M.K. and P.T.; formal analysis, A.P., J.P., M.A., J.-S.T., C.C., S.H. and D.D.; investigation, A.P., J.P., M.A., J.-S.T., C.C., C.L., M.W., M.K. and P.T. and; resources, A.P., J.P., C.C., S.H., D.D. and G.G.; data curation, A.P., J.P., C.C., S.H., D.D. and M.W.; writing—original draft preparation, A.P., J.P., M.A., J.-S.T and C.L.; writing—review and editing, C.C., S.H., and D.D.; visualization, A.P., J.P., M.A., J.-S.T., C.L. and M.W.; supervision, A.P., J.P., C.C., S.H. and D.D.; project administration, C.C., S.H., D.D. and G.G.; funding acquisition, J.P., C.C., S.H., D.D. and G.G. All authors have read and agreed to the published version of the manuscript.

Funding: Financial assistance was provided within M-Era.Net 2 transnational research program by National Science Centre in Poland (NCN, project no. UMO-2016/22/Z/ST8/00694), and partially by Fonds National de la Recherche Luxembourg (FNR, Project No. INTER/MERA/16/11454672) and the Belgian Fonds de la Recherche Scientifique-FNRS (F.R.S.-FNRS, Convention No. R.50.13.17.F). The authors acknowledge also the Spanish Government for financial support through Project PCIN-2017-128. CC and CL belong to SGR2017 1165. The paper was developed using the equipment purchased within agreement no. POPW.01.03.00-06-010/09-00 Operational Program Development of Eastern Poland 2007–2013, Priority Axis I, Modern Economy, Operations 1.3. Innovations Promotion.

Acknowledgments: We would like to thank for technical support provided by Jean-Francois Statsyns, François Devred (XRD) and Delphine Magnin (SEM-EDX mapping) in Belgium.

Conflicts of Interest: The authors declare no conflict of interest.

Abbreviations

ADSCs	Adipose tissue-derived mesenchymal stem cells
bALP	Bone alkaline phosphatase
Col I	Type I collagen
CTAC	CetylTrimethylAmmonium chloride
FGF-2	Fibroblast growth factor-2
HA	Hydroxyapatite
HBSS	Hanks' Balanced Salt solution
ICP	Inductively coupled plasma
GAD	GlidArc plasma reactor
MCM-48	Mobil Composition of Matter No. 48
MIP	Mercury Intrusion Porosimetry
MSNPs	Mesoporous silica nanoparticles
NPs	Nanoparticles
OC	Osteocalcin

PBS	Phosphate buffered saline
PP	Polypropylene
RONS	Reactive oxygen and nitrogen species
ROS	Reactive oxygen species
SEM	Scanning electron microscopy
SSA	Specific surface area
TEA	TriEthanolAmine
TEM	Transmission electron microscopy
TEOS	Tetraethyl orthosilicate
XPS	X-ray photoelectron spectroscopy
XRD	X-ray diffraction

References

1. Arndt, S.; Unger, P.; Wacker, E.; Shimizu, T.; Heinlin, J.; Li, Y.F.; Thomas, H.M.; Morfill, G.E.; Zimmermann, J.L.; Bosserhoff, A.K.; et al. Cold atmospheric plasma (CAP) changes gene expression of key molecules of the wound healing machinery and improves wound healing in vitro and in vivo. *PLoS ONE* **2013**, *8*, e79325. [CrossRef]
2. Lin, Z.H.; Cheng, K.Y.; Cheng, Y.P.; Tschang, C.Y.T.; Chiu, H.Y.; Yeh, N.L.; Liao, K.C.; Gu, B.R.; Wu, J.S. Acute rat cutaneous wound healing for small and large wounds using Ar/O2 atmospheric-pressure plasma jet treatment. *Plasma Med.* **2017**, *7*, 227–244. [CrossRef]
3. Grigoras, C.; Topala, I.; Nastuta, A.V.; Jitaru, D.; Florea, I.; Badescu, L.; Ungureanu, D.; Badescu, M.; Dumitrascu, N. Influence of atmospheric pressure plasma treatment on epithelial regeneration process. *Rom. Rep. Phys.* **2011**, *56*, 54–61.
4. Haertel, B.; von Woedtke, T.; Weltmann, K.D.; Lindequist, U. Non-thermal atmospheric-pressure plasma possible application in wound healing. *Biomol. Ther.* **2014**, *22*, 477–490. [CrossRef]
5. Cheng, Q.; Lee, B.L.-P.; Komvopoulos, K.; Yan, Z.; Li, S. Plasma Surface Chemical Treatment of Electrospun Poly(L-Lactide) Microfibrous Scaffolds for Enhanced Cell Adhesion, Growth, and Infiltration. *Tissue Eng. Part A* **2013**, *19*, 1188–1198. [CrossRef]
6. Chen, T.F.; Siow, K.S.; Ng, P.Y.; Majlis, B.Y. Enhancing the biocompatibility of the polyurethane methacrylate and off-stoichiometry thiol-ene polymers by argon and nitrogen plasma treatment. *Mater. Sci. Eng. C* **2017**, *79*, 613–621. [CrossRef] [PubMed]
7. Griffin, M.; Palgrave, R.; Baldovino, V.G.; Butler, P.E.; Kalaskar, D.M. Argon plasma improves the tissue integration and angiogenesis of subcutaneous implants by modifying surface chemistry and topography. *Int. J. Nanomed.* **2018**, *13*, 6123–6141. [CrossRef]
8. Sagbas, B. Argon/Oxygen Plasma Surface Modification of Biopolymers for Improvement of Wettability and Wear Resistance. *Int. J. Chem. Mol. Nucl. Mater. Metall. Eng.* **2016**, *10*, 889–894.
9. Canal, C.; Gaboriau, F.; Villeger, S.; Cvelbar, U.; Ricard, A. Studies on antibacterial dressings obtained by fluorinated post-discharge plasma. *Int. J. Pharm.* **2009**, *367*, 155–161. [CrossRef]
10. Dal'Maz Silva, W.; Belmonte, T.; Duday, D.; Frache, G.; Noël, C.; Choquet, P.; Migeon, H.N.; Maliska, A.M. Interaction mechanisms between Ar-O$_2$ post-discharge and biphenyl. *Plasma Process. Polym.* **2012**, *9*, 207–216. [CrossRef]
11. Przekora, A. Current Trends in Fabrication of Biomaterials for Bone and Cartilage Regeneration: Materials Modifications and Biophysical Stimulations. *Int. J. Mol. Sci.* **2019**, *20*, 435. [CrossRef] [PubMed]
12. Labay, C.; Canal, J.M.; Modic, M.; Cvelbar, U.; Quiles, M.; Armengol, M.; Arbos, M.A.; Gil, F.J.; Canal, C. Antibiotic-loaded polypropylene surgical meshes with suitable biological behaviour by plasma functionalization and polymerization. *Biomaterials* **2015**, *71*, 132–144. [CrossRef] [PubMed]
13. Kudryavtseva, V.; Stankevich, K.; Gudima, A.; Kibler, E.; Zhukov, Y.; Bolbasov, E.; Malashicheva, A.; Zhuravlev, M.; Riabov, V.; Liu, T.; et al. Atmospheric pressure plasma assisted immobilization of hyaluronic acid on tissue engineering PLA-based scaffolds and its effect on primary human macrophages. *Mater. Des.* **2017**, *127*, 261–271. [CrossRef]
14. Tominami, K.; Kanetaka, H.; Sasaki, S.; Mokudai, T.; Kaneko, T.; Niwano, Y. Cold atmospheric plasma enhances osteoblast differentiation. *PLoS ONE* **2017**, *12*, e0180507. [CrossRef] [PubMed]

15. Przekora, A.; Pawlat, J.; Terebun, P.; Duday, D.; Canal, C.; Hermans, S.; Audemar, M.; Labay, C.; Thomann, J.S.; Ginalska, G. The effect of low temperature atmospheric nitrogen plasma on MC3T3-E1 preosteoblast proliferation and differentiation in vitro. *J. Phys. D. Appl. Phys.* **2019**, *52*, 275401. [CrossRef]

16. Arai, M.; Shibata, Y.; Pugdee, K.; Abiko, Y.; Ogata, Y. Effects of reactive oxygen species (ROS) on antioxidant system and osteoblastic differentiation in MC3T3-E1 cells. *IUBMB Life* **2007**, *59*, 27–33. [CrossRef]

17. Takamatsu, T.; Uehara, K.; Sasaki, Y.; Miyahara, H.; Matsumura, Y.; Iwasawa, A.; Ito, N.; Azuma, T.; Kohno, M.; Okino, A. Investigation of reactive species using various gas plasmas. *RSC Adv.* **2014**, *4*, 39901–39905. [CrossRef]

18. Bello, M.; Abdul Raman, A.A.; Asghar, A. A review on approaches for addressing the limitations of Fenton oxidation for recalcitrant wastewater treatment. *Process. Saf. Environ. Prot.* **2019**, *126*, 119–140. [CrossRef]

19. Palas, B.; Ersöz, G.; Atalay, S. Bioinspired metal oxide particles as efficient wet air oxidation and photocatalytic oxidation catalysts for the degradation of acetaminophen in aqueous phase. *Ecotoxicol. Environ. Saf.* **2019**, *182*, 109367. [CrossRef]

20. Dhakshinamoorthy, A.; Navalon, S.; Alvaro, M.; Garcia, H. Metal nanoparticles as heterogeneous fenton catalysts. *ChemSusChem* **2012**, *5*, 46–64. [CrossRef]

21. Kazimierczak, P.; Benko, A.; Palka, K.; Canal, C.; Kolodynska, D.; Przekora, A. Novel synthesis method combining a foaming agent with freeze-drying to obtain hybrid highly macroporous bone scaffolds. *J. Mater. Sci. Technol.* **2020**, *43*, 52–63. [CrossRef]

22. Przekora, A. The summary of the most important cell-biomaterial interactions that need to be considered during in vitro biocompatibility testing of bone scaffolds for tissue engineering applications. *Mater. Sci. Eng. C* **2019**, *97*, 1036–1051. [CrossRef] [PubMed]

23. Przekora, A.; Palka, K.; Ginalska, G. Biomedical potential of chitosan/HA and chitosan/β-1,3-glucan/HA biomaterials as scaffolds for bone regeneration—A comparative study. *Mater. Sci. Eng. C* **2016**, *58*, 891–899. [CrossRef]

24. Hannink, G.; Arts, J.J.C. Bioresorbability, porosity and mechanical strength of bone substitutes: What is optimal for bone regeneration? *Injury* **2011**, *42* (Suppl. 2), S22–S25. [CrossRef] [PubMed]

25. Bruggeman, P.; Kushner, M.; Locke, B.; Gardeniers, J.; Graham, W.; Graves, D.; Hofman-Caris, R.; Maric, D.; Reid, J.; Ceriani, E.; et al. Plasma-liquid interactions: A review and roadmap. *Plasma Sources Sci. Technol.* **2016**, *25*, 053002. [CrossRef]

26. Louit, G.; Foley, S.; Cabillic, J.; Coffigny, H.; Taran, F.; Valleix, A.; Renault, J.P.; Pin, S. The reaction of coumarin with the OH radical revisited: Hydroxylation product analysis determined by fluorescence and chromatography. *Radiat. Phys. Chem.* **2005**, *72*, 119–124. [CrossRef]

27. Uchida, G.; Takenaka, K.; Takeda, K.; Ishikawa, K.; Hori, M.; Setsuhara, Y. Selective production of reactive oxygen and nitrogen species in the plasma-treated water by using a nonthermal high-frequency plasma jet. *Jpn. J. Appl. Phys.* **2018**, *57*, 0102B4. [CrossRef]

28. Canal, C.; Fontelo, R.; Hamouda, I.; Guillem-Marti, J.; Cvelbar, U.; Ginebra, M.P. Plasma-induced selectivity in bone cancer cells death. *Free Radic. Biol. Med.* **2017**, *110*, 72–80. [CrossRef]

29. Tornin, J.; Mateu-Sanz, M.; Rodríguez, A.; Labay, C.; Rodríguez, R.; Canal, C. Pyruvate Plays a Main Role in the Antitumoral Selectivity of Cold Atmospheric Plasma in Osteosarcoma. *Sci. Rep.* **2019**, *9*, 10681. [CrossRef]

30. Kalghatgi, S.; Friedman, G.; Fridman, A.; Clyne, A.M. Endothelial cell proliferation is enhanced by low dose non-thermal plasma through fibroblast growth factor-2 release. *Ann. Biomed. Eng.* **2010**, *38*, 748–757. [CrossRef]

31. Shimoaka, T.; Ogasawara, T.; Yonamine, A.; Chikazu, D.; Kawano, H.; Nakamura, K.; Itoh, N.; Kawaguchi, H. Regulation of osteoblast, chondrocyte, and osteoclast functions by fibroblast growth factor (FGF)-18 in comparison with FGF-2 and FGF-10. *J. Biol. Chem.* **2002**, *277*, 7493–7500. [CrossRef]

32. Chikazu, D.; Katagiri, M.; Ogasawara, T.; Ogata, N.; Shimoaka, T.; Takato, T.; Nakamura, K.; Kawaguchi, H. Regulation of osteoclast differentiation by fibroblast growth factor 2: Stimulation of receptor activator of nuclear factor κB ligand/osteoclast differentiation factor expression in osteoblasts and inhibition of macrophage colony-stimulating factor functi. *J. Bone Miner. Res.* **2001**, *16*, 2074–2081. [CrossRef] [PubMed]

33. Dupree, M.A.; Pollack, S.R.; Levine, E.M.; Laurencin, C.T. Fibroblast growth factor 2 induced proliferation in osteoblasts and bone marrow stromal cells: A whole cell model. *Biophys. J.* **2006**, *91*, 3097–3112. [CrossRef] [PubMed]

34. Mao, L.; Xia, L.; Chang, J.; Liu, J.; Jiang, L.; Wu, C.; Fang, B. The synergistic effects of Sr and Si bioactive ions on osteogenesis, osteoclastogenesis and angiogenesis for osteoporotic bone regeneration. *Acta Biomater.* **2017**, *61*, 217–232. [CrossRef] [PubMed]

35. Jeney, V. Clinical impact and cellular mechanisms of iron overload-associated bone loss. *Front. Pharm.* **2017**, *8*, 77. [CrossRef]

36. Yamasaki, K.; Hagiwara, H. Excess iron inhibits osteoblast metabolism. *Toxicol. Lett.* **2009**, *191*, 211–215. [CrossRef] [PubMed]

37. Yang, J.; Zhang, J.; Ding, C.; Dong, D.; Shang, P. Regulation of Osteoblast Differentiation and Iron Content in MC3T3-E1 Cells by Static Magnetic Field with Different Intensities. *Biol. Trace Elem. Res.* **2018**, *184*, 214–225. [CrossRef]

38. Schumacher, K.; Grün, M.; Unger, K.K. Novel synthesis of spherical MCM-48. *Microporous Mesoporous Mater.* **1999**, *27*, 201–206. [CrossRef]

39. Möller, K.; Kobler, J.; Bein, T. Colloidal suspensions of nanometer-sized mesoporous silica. *Adv. Funct. Mater.* **2007**, *17*, 605–612. [CrossRef]

40. Kobler, J.; Möller, K.; Bein, T. Colloidal suspensions of functionalized mesoporous silica nanoparticles. *ACS Nano* **2008**, *2*, 791–799. [CrossRef]

41. Van Espen, P.; Van Grieken, R.; Markowicz, A. *Handbook of X-Ray Spectrometry: Methods and Techniques*; Marcel Dekker: New York, NY, USA, 1993.

42. Osán, J.; De Hoog, J.; Van Espen, P.; Szalóki, I.; Ro, C.U.; Van Grieken, R. Evaluation of energy-dispersive x-ray spectra of low-Z elements from electron-probe microanalysis of individual particles. *X-Ray Spectrom.* **2001**, *30*, 419–426. [CrossRef]

43. Kyotani, T.; Koshimizu, S. Identification of Individual Si-Rich Particles Derived from Kosa Aerosol by the Alkali Elemental Composition. *Bull. Chem. Soc. Jpn.* **2001**, *74*, 723–729. [CrossRef]

44. Boon, G.; Bastin, G. Quantitative analysis of thin specimens in the TEM using a $\varphi(\rho z)$-model. *Microchim. Acta* **2004**, *147*, 125–133. [CrossRef]

45. Bastin, G.F.; Dijkstra, J.M.; Heijligers, H.J.M. PROZA96: An Improved Matrix Correction Program for Electron Probe Microanalysis, Based on a Double Gaussian $\varphi(\rho z)$ Approach. *X-ray Spectrom.* **1998**, *27*, 3–10. [CrossRef]

46. Przekora, A.; Ginalska, G. Enhanced differentiation of osteoblastic cells on novel chitosan/β-1,3-glucan/bioceramic scaffolds for bone tissue regeneration. *Biomed. Mater.* **2015**, *10*. [CrossRef]

47. Przekora, A.; Palka, K.; Ginalska, G. Chitosan/β-1,3-glucan/calcium phosphate ceramics composites-Novel cell scaffolds for bone tissue engineering application. *J. Biotechnol.* **2014**, *182–183*, 46–53. [CrossRef]

48. Pawłat, J.; Terebun, P.; Kwiatkowski, M.; Tarabová, B.; Kovaľová, Z.; Kučerová, K.; Machala, Z.; Janda, M.; Hensel, K. Evaluation of Oxidative Species in Gaseous and Liquid Phase Generated by Mini-Gliding Arc Discharge. *Plasma Chem. Plasma Process.* **2019**, *39*, 627–642. [CrossRef]

49. Kazimierczak, P.; Kolmas, J.; Przekora, A. Biological response to macroporous chitosan-agarose bone scaffolds comprising Mg-and Zn-doped nano-hydroxyapatite. *Int. J. Mol. Sci.* **2019**, *20*, 3835. [CrossRef]

International Journal of
Molecular Sciences

Article

Large-Scale Image Analysis for Investigating Spatio-Temporal Changes in Nuclear DNA Damage Caused by Nitrogen Atmospheric Pressure Plasma Jets

Xu Han [1,2,†], James Kapaldo [1,2,†] , Yueying Liu [3], M. Sharon Stack [3,4], Elahe Alizadeh [5] and Sylwia Ptasinska [1,2,*]

[1] Radiation Laboratory, University of Notre Dame, Notre Dame, IN 46556, USA; xkapaldo@gmail.com (X.H.); james.kapaldo@gmail.com (J.K.)
[2] Department of Physics, University of Notre Dame, Notre Dame, IN 46556, USA
[3] Harper Cancer Research Institute, University of Notre Dame, Notre Dame, IN 46556, USA; yliu12@nd.edu (Y.L.); Sharon.Stack.11@nd.edu (M.S.S.)
[4] Department of Chemistry and Biochemistry, University of Notre Dame, Notre Dame, IN 46556, USA
[5] Queen's CardioPulmonary Unit (QCPU), Department of Medicine, Queen's University, Kingston, ON K7L 3J9, Canada; elahe.alizadeh@queensu.ca
* Correspondence: sptasins@nd.edu
† These authors contributed equally to this work.

Received: 9 May 2020; Accepted: 8 June 2020; Published: 10 June 2020

Abstract: The effective clinical application of atmospheric pressure plasma jet (APPJ) treatments requires a well-founded methodology that can describe the interactions between the plasma jet and a treated sample and the temporal and spatial changes that result from the treatment. In this study, we developed a large-scale image analysis method to identify the cell-cycle stage and quantify damage to nuclear DNA in single cells. The method was then tested and used to examine spatio-temporal distributions of nuclear DNA damage in two cell lines from the same anatomic location, namely the oral cavity, after treatment with a nitrogen APPJ. One cell line was malignant, and the other, nonmalignant. The results showed that DNA damage in cancer cells was maximized at the plasma jet treatment region, where the APPJ directly contacted the sample, and declined radially outward. As incubation continued, DNA damage in cancer cells decreased slightly over the first 4 h before rapidly decreasing by approximately 60% at 8 h post-treatment. In nonmalignant cells, no damage was observed within 1 h after treatment, but damage was detected 2 h after treatment. Notably, the damage was 5-fold less than that detected in irradiated cancer cells. Moreover, examining damage with respect to the cell cycle showed that S phase cells were more susceptible to DNA damage than either G1 or G2 phase cells. The proposed methodology for large-scale image analysis is not limited to APPJ post-treatment applications and can be utilized to evaluate biological samples affected by any type of radiation, and, more so, the cell-cycle classification can be used on any cell type with any nuclear DNA staining.

Keywords: atmospheric pressure plasma jets; large-scale imaging; machine learning; cancer treatment; cellular imaging

---

1. Introduction

In recent years, numerous in vitro studies have shown the considerable anticancer effects of nonthermal atmospheric pressure plasmas in approximately 20 types of malignant cell lines, including lung cancer [1], prostate cancer [2], ovarian cancer [3], osteosarcoma [4], and oral cancer [5]. Furthermore, several in vivo investigations using tumor models of pancreatic cancer [6], glioblastoma [7], melanoma [8,9], ovarian cancer [10], and breast cancer [11] have demonstrated the significant inhibition of cellular growth and tumor damage following atmospheric pressure plasma treatment. The ability of atmospheric pressure plasma jets (APPJs) to inactivate or kill malignant cells relies strongly on the production of a variety of plasma reactive species [12,13]. APPJs synergistically provide free electrons, positive ions, radicals, photons, and electromagnetic fields, which can damage biological targets without elevating the temperature of the treated area [14]. More importantly, plasma treatments in animal models have been reported to selectively damage targeted cancer cells, without affecting surrounding healthy tissues [15,16]. These features suggest that nonthermal atmospheric pressure plasmas may represent a promising alternative to conventional cancer treatments [14,17].

Although some primary clinical studies have previously been performed [18–20], the extensive clinical applications of APPJs require more detailed investigations to examine their effects on a variety of cancer cell lines, both in vitro and in vivo [21,22]. There is concern regarding the potential carcinogenic risk and side effects of prolonged clinical use due to the formation of free radicals. These can cause adverse and acute impacts that can present safety risks in long-term APPJ applications [14,23,24]. Also, technical issues, such as the optimal plasma dosage inside tissues, the penetration depth of reactive species, and the distribution of cellular damage, remain poorly understood and require further investigations. A variety of bioanalytical tools and imaging techniques have been used to quantify the induced damage and cellular responses following plasma irradiation, including fluorescence microscopy [25–27] and flow cytometry [28]. While these techniques can be utilized to perform routine cellular analyses, each possess both advantages and limitations, in terms of sample preparation requirements, sensitivity, measurable parameters, throughput, and costs. For example, fluorescence microscopy can capture images of small sample regions with high spatial resolution, facilitating the assessment of quantitative morphology [29]. In contrast, flow cytometry can facilitate the analysis of cellular kinetics and cell-cycle phases, but cannot provide spatial information; however, highly sensitive multicolor phenotypic data can be obtained from populations of different cells, within minutes [30].

In the current study, first we explored two dimensional (2D) spatial distributions of damage to deoxyribonucleic acid (DNA) induced by the APPJ treatment of cancer and nonmalignant cells. DNA damage was assessed by measuring double-strand break (DSB) formation in cell nuclei. In the cellular environment, DSBs trigger the phosphorylation of histone H2AX near the break site, resulting in the appearance of γH2AX foci and leading to local changes in the chromatin structure. These modifications are macroscopic structures that can be directly visualized with the assistance of antibody staining inside the cell nuclei.

Second, we developed a large-scale image analysis technique, using machine learning-based cell-cycle classifications, requiring only one staining dye. Generally, the cell cycle is divided into two major phases: interphase, including gap 1 (G1), DNA synthesis (S), and gap 2 (G2), and mitotic (M) phase. During the G1 phase, the cell grows in size at a high biosynthetic rate, producing proteins and copying organelles such as mitochondria and ribosomes to prepare for DNA synthesis (S phase). After DNA duplication, cells enter the second gap phase, G2, during which they grow rapidly and synthesize proteins and organelles in preparation for mitosis. As cells enter the M phase, they stop growing and synthesizing proteins to focus their energy on the complex and highly regulated cell division process. During standard analyses, different dyes are used to stain nuclear DNA in each cell phase. In this study, we used only 4',6-diamidino-2-phenylindole (DAPI) fluorescence stain to image and classify the cycle of every cell on a coverslip. Although we employed DAPI staining to build a correlation between nuclear DNA versus nucleus size for cell classification, our methodology can be applied to other types of nuclear DNA staining. However, some dyes can also stain mitochondrial DNA outside of the nucleus, and therefore will not be eligible for this method, because this method uses segmentation of nuclei due to their ellipse-like shape.

By combining microscopy images with a machine learning tool, we were able to study the spatial distribution of plasma-induced DNA damage in nuclei within each cell-cycle phase. By mapping the damage distribution over various post-treatment (incubation) times, in both malignant and nonmalignant oral cells, we revealed the spatio-temporal dependence of cellular responses to plasma-influenced regions. These results improve our understanding of APPJ effects on biological targets and their applications in plasma medicine. Furthermore, our proposed methodology for analyzing DNA damage in a large number of irradiated cells could facilitate the quantitative evaluation of the DNA damage caused by any other sources of cellular radiation, with widespread application in radiobiology research.

2. Results and Discussion

2.1. Computational Analysis

Figure 1a shows representative fluorescence images of nuclear DNA (denoted as the DAPI channel) and DSBs in nuclear DNA (denoted as the γH2AX channel) on a large scale. The slide scanner's movements followed a zig-zag pattern, first scanning the entire length of the coverslip along one direction (called the fast axis), then taking a short step to the side (called the slow axis), before again scanning the entire length of the coverslip in the opposite direction, until the whole surface was scanned.

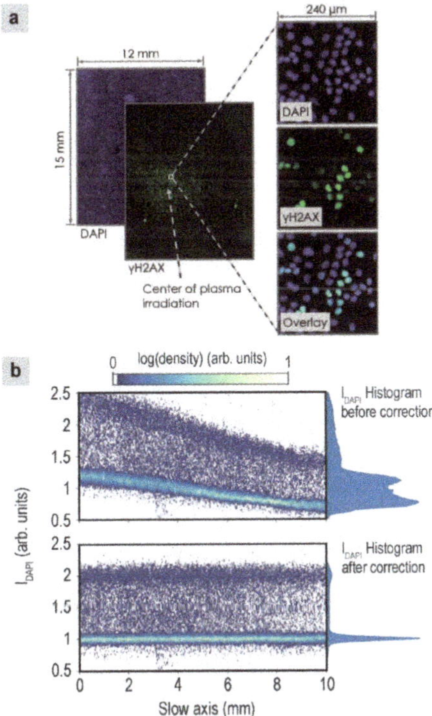

Figure 1. (a) Representative fluorescence images of cells, after 2-min plasma treatment and 1-h incubation. Green and blue fluorescence signals correspond to γH2AX and DAPI, respectively. (b) Effects of background correction. Plots show the nuclear DAPI intensity along the slow axis of the fluorescence slide scanner. Colors encode the nuclear density (along the slide scanner's fast axis). A histogram of the nuclear DAPI intensity is indicated to the right of each plot. The top and bottom plots show data before background correction and after flattening and stripe artifact removal, respectively.

The image analysis is comprised of four primary steps, including nuclei segmentation and background/foreground correction, feature extraction, machine learning to classify the cell phase, and quantification of DNA damage.

2.1.1. Nuclei Segmentation and Image Correction

Nuclei segmentation was achieved by filtering the DAPI images with a Gaussian blur filter ($\sigma = 1$) and then using an adaptive, log-weighted Otsu threshold [31]. This method successfully segmented all nuclei that were not in contact with other nuclei; however, in our studies, the initial confluence of the cells was 90%, and confluence increased during the incubation time, resulting in many overlapping nuclei. These overlapping nuclei needed to be further segmented for additional computational analyses, which was accomplished by applying a short-range attraction and a long-range repulsion (SALR) clustering algorithm [32], in which we identified the center of each nucleus following a geometric partitioning algorithm to segment the overlapping nuclei [33].

In addition, intensity correction of the image background and foreground is crucial for successful analysis. The image foreground represents all regions (pixels) containing nuclei; these regions were determined by the segmentation in the previous step. Likewise, the image background represents all regions not containing nuclei. Three primary causes of uneven background/foreground intensities were identified across our images: uneven fluorescence staining, which results in slowly changing background/foreground intensities; differing amounts of microscope illumination/collection, which can be caused by improper focusing across the whole coverslip; microscope calibration errors, which led to the appearance of bright stripes along the fast axis of the scanned images.

Using the background region, we computed the spatially variable background intensity for each channel and removed it from the image. After this background flattening, we computed the background stripe artifacts and removed them. A representative stripe artifact can be observed in the large γH2AX image in Figure 1a. DAPI channel foreground correction was performed before the foreground correction of other channels. Two bands of DAPI intensities are observed in Figure 1b, which likely correspond to the G1 and G2 phases, since cells in the G2 phase contain twice as much DNA as cells in the G1 phase and because cells spend the most time in the G1 and G2 phases. We fitted these two bands with locally weighted linear fits (lowess) to flatten the G1 and G2 bands and position them at DAPI intensity values of 1 and 2, respectively. Figure 1b shows the nuclear DAPI intensities across the slow axis of the scanner, before (top) and after (bottom) correction. After the correction, the G1 and G2 DAPI bands are flat and positioned correctly. The foreground correction for the γH2AX channel was performed with caution, as the intensity values should not be flat because plasma treatment causes localized DSBs in the nucleus. We first qualitatively selected all G1 cells, using the corrected DAPI intensities from above (cells with DAPI values in the range of 0.7–1.3). Using these cells, we fitted a surface to the locally varying data in the bottom 2%–4% of the γH2AX intensities and then subtracted this surface from the data. This process will correctly flatten the γH2AX channel as long as some of the cells in a large region of the image (~1 mm^2) are not influenced by the plasma, which is true in our images based on the fact that the maximum damage ratio of G1 cells is 50% (as described below) and therefore, the bottom 2%–4% of γH2AX intensities represent undamaged cells.

2.1.2. Feature Extraction

From each nucleus, we extracted a set of features from the DAPI channel for classifying cell-cycle phases. The features describe the nucleus shape, intensity, radial (shell) intensity, Haralick texture [34], and granularity. The Haralick and granularity features were computed using the same method as the commonly used CellProfiler software (version 2.1.1) [35]. The nuclear shapes were described, in part, using Fourier descriptors [36]. A list of the extracted features can be found in Appendix A. From the γH2AX channel, we extracted intensity features only. The integrated intensity of the γH2AX channel for each nucleus is proportional to the quantity of DSBs in each nucleus.

The code for feature extraction was written to compute the features of multiple nuclei in parallel on a graphical processing unit (GPU), in contrast with CellProfiler, which processes images in parallel on central processing units (CPUs) and processes nuclei features in serial. Extracting the nuclei features on a GPU results in a significant decrease in processing time when a large number of CPUs are not available. When applying our code during some basic tests, the entire processing time for the images (including segmentation and correction, which also used GPU acceleration), including feature extraction, required approximately 40–60 min per image, for both the DAPI and γH2AX channels, based on approximately 3 GB per channel. When using four CPUs, the same processing in CellProfiler took 2–3 h, without the application of SALR clustering for the accurate partitioning of nuclei clumps [35]. In addition, when using a simple watershed-based partitioning of overlapping cells, which is more similar to the method used by CellProfiler, our processing time is shortened to 20–30 min per image. The code was written in MATLAB (The MathWorks, Inc., Natick, MA, USA), and the hardware used for comparisons were an Nvidia GTX770, with 2 GB of memory, and an Intel 4 core i7-4790, 3.6 GHz.

2.1.3. Cell-Cycle Classification

The cell-cycle phase was classified using two steps. During the first step, supervised machine learning was used to classify five visually distinct classes. Class 1 included interphase (G1, S, and G2), and the other four classes were derived from the different mitosis stages, including prometaphase (class 2), metaphase (class 3), early anaphase (class 4), and late anaphase, telophase, and early G1 (class 5). Examples of these classes are presented in Figure 2a. We manually classified approximately 500 nuclei in each class to create a balanced dataset, which was divided into training, validation, and test sets at a 0.7/0.15/0.15 ratio. The employed learning method was a shallow neural network, with two hidden layers (sizes 30 and 10), and a softmax layer to output the final classifications. The training used early stopping with the validation data to prevent overtraining. The confusion matrix on the test dataset is shown in Figure 2c, indicating that the overall classification accuracy was 97%. The network creation and training were all performed using MATLAB built-in routines, namely the "patternnet" and "train" functions.

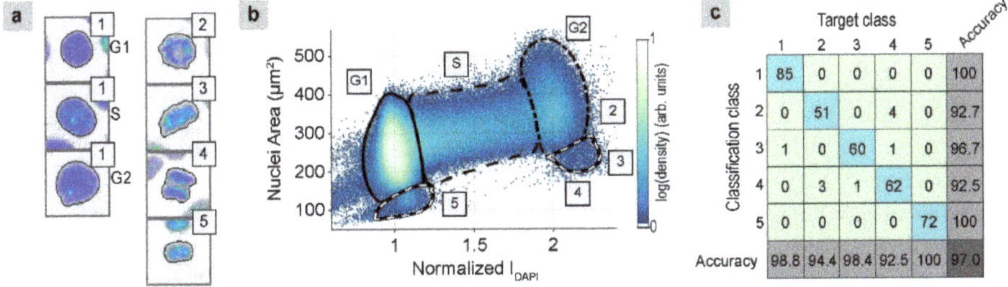

Figure 2. (a) Examples of nuclei from each classification group; 1: interphase, 2: prometaphase, 3: metaphase, 4: early anaphase, 5: late anaphase, telophase, and early G1. The images are 34 μm × 34 μm, and they are modified to show a white background, for ease of viewing. (b) Nuclear area versus DAPI intensity (color gives nuclear density), with classification contours that enclose regions that contain primarily G1 (solid), S (dashed), G2 (dot-dashed), and M (black/white dash) phase cells. The right M contour primarily contains classes 3 and 4, whereas the left M contour primarily contains class 5. (c) Confusion matrix of the test set, based on the classification of different cell classes.

During the second step, the interphase cells were classified using a mixture-of-Gaussians model, with a uniform background [37], using the nuclear area and DAPI intensity (Figure 2b). The classification used seven Gaussians, equally spaced along the line connecting the center of the G1 and G2 peaks (Figure 2b), and an eighth Gaussian, located on the G1 peak. This classification of interphases was confirmed by experiments using an anti- chromatin licensing and DNA replication factor 1 (CDT1) antibody to label G1 phase cells and 5-ethynyl-2′-deoxyuridine (EdU) to label S phase cells (results not shown). Figure 2b shows the distribution of the nuclear area versus the DAPI intensity, along with contours that denote regions containing cells in the G1, S, G2, and M phases. From this plot, we visualized the cell progression through the entire cell cycle. After the G1 phase ($I_{DAPI} = 1$), cells enter the S phase, during which they duplicate their DNA ($I_{DAPI} = 1 \rightarrow 2$). Then, they enter the G2 phase and prepare for mitosis ($I_{DAPI} = 2$). As they enter mitosis, they begin to condense, and their areas become smaller (I_{DAPI} approximately 2.3, area approximately 250 μm^2). During anaphase, the image segmentation splits the two halves into distinct objects, resulting in an area and a DAPI intensity equal to half of the previous values (I_{DAPI} approximately 1.1, area approximately 120 μm^2). The nuclei then start to grow and reenter the G1 phase.

2.1.4. Damage Quantification

During the final step, the damage to each nucleus is defined as a fraction of the total damaged DNA, which can be computed as the ratio between the γH2AX and DAPI intensities, $I_{\gamma H2AX}/I_{DAPI}$. To quantify the susceptibility of each cell phase to damage induced by the plasma jet, we used the damage ratio (DR), which indicates the number of cells with damage above a specified threshold divided by the total number of cells.

2.2. APPJ Irradiation

The above-described computational procedure for large-scale image analysis, with machine learning, was used to acquire the spatio-temporal distributions of DNA damage in malignant and nonmalignant cells after treatment with a nitrogen APPJ. The obtained images for two cell lines are presented, and the biological implications of plasma treatment are briefly discussed in the following subsections. These results are presented to illustrate the usefulness of the procedure for assessing the biological effects after plasma treatments, and certain conclusions have been derived; however, exploring the biological mechanisms induced by APPJs is not the focus of this study.

2.2.1. APPJ-Irradiated Malignant Cells

The first row of Figure 3a shows the 2D distributions of cancer cell damage after different incubation times. These distributions were obtained after subtracting the median damage value of the flow control cells, which did not show any effects following nitrogen flow treatment alone. After 1-h incubation, cells were observed to be damaged near the plasma jet treatment region, and the damage decreased radially outward. Damage that extends beyond the plasma jet treatment region may be caused by the following: (1) the diffusion of the plasma species above the liquid surface; (2) the diffusion of the reactive species in the liquid, induced by the plasma treatment; (3) cell−cell communications, which may contribute to the damage of bystander cells near the treated cells [38]. Similarly, an enlarged affected area has been observed in previous studies [5,15,39].

More importantly, the maximum damage was observed at locations approximately 1 mm away from the treatment center, as shown in cancer cells after 2 h of incubation, resulting in a ring-shape pattern centered on the treatment location. The radius of this ring was similar to the inner radius of the quartz tube orifice, which was 1 mm (Figure 4), suggesting that the most prominent damaging effects were caused by species located at the interface of the plasma jet and the surrounding air. At this interface, highly reactive species have formed that interact with cells, causing DSBs. For example, one highly reactive species produced at the interface is nitric oxide (NO) [5,40], which can cause DNA strand breaks via the production of other reactive nitrogen species (RNS), such as $ONOO^-$, HNO_2, and N_2O_3 [41]. NO can also trigger the production of intracellular reactive oxygen species (ROS), which can initiate various pathways, including apoptosis. Similar observations of ring-shaped regions have also been reported in previous studies, such as the inactivation patterns of bacteria, the distributions of reactive species, theoretical modeling [42], and for a gas-shield, helium-based APPJ, during interactions with cancer cells [43]. After 8 h of incubation, the measured damage decreased, which was likely due to cell detachment. Those cells that suffered from severe damage could not recover through repair processes and, consequently, undergo cell death. As a result, the dead cells detach from the coverslip and are removed during the washing steps of immunofluorescence procedures. Similar cell detachment after plasma treatment has been observed in other studies [44,45].

Furthermore, cancer cells in different cell-cycle phases have been found to respond with different sensitivities to plasma treatments. As shown in Figure 3a, after 1 h of incubation, almost 100% of S phase cells were damaged at the treatment region, whereas a much smaller proportion of G1 and G2 phase cells showed damage. In addition, the damaged area for S phase cells was found to be much larger than those for cells in the G1 and G2 phases. These results imply that S phase cells are more sensitive (susceptible) to damage induced by plasma treatment than cells in the G1 and G2 phases. During the S phase, DNA replication requires the exposure of a single-stranded portion of DNA near the replication fork, which results in an increased vulnerability to external attacks compared with other phases. Such attacks can be due to reactive species that are solvated/generated in the liquid above the cells during the treatment [46], which generally extend radially outwards beyond the location where the plasma jet makes direct contact with the liquid, through diffusion in the liquid phase or the transport of the gas phase plasma species across the liquid surface by the nitrogen gas flow. The damage caused during DNA replication may also be associated with the production of intracellular ROS/RNS, as a cellular response to plasma treatment [47].

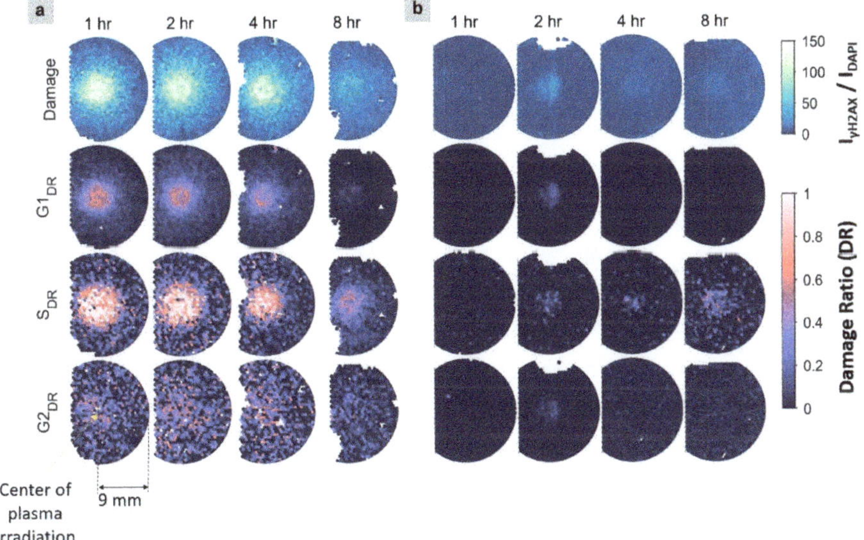

Figure 3. 2D distributions of the damaged DNA fraction in each nucleus (top row) and the damage ratios (DRs) for cells in G1, S, and G2 phases (rows 2–4), based on γH2AX staining in malignant cells (**a**) and nonmalignant cells (**b**) grown on coverslips. The damage threshold used to determine the DR was 75 (noting the damage range is 0–150). Each circular distribution, with a diameter of 18 mm, is centered on the treatment locations. Locations with white patches indicate regions without cells or blurry regions of the image, which were discarded. The circles for the cancer cells are cut off on the left side, due to a limitation of the slide scanner.

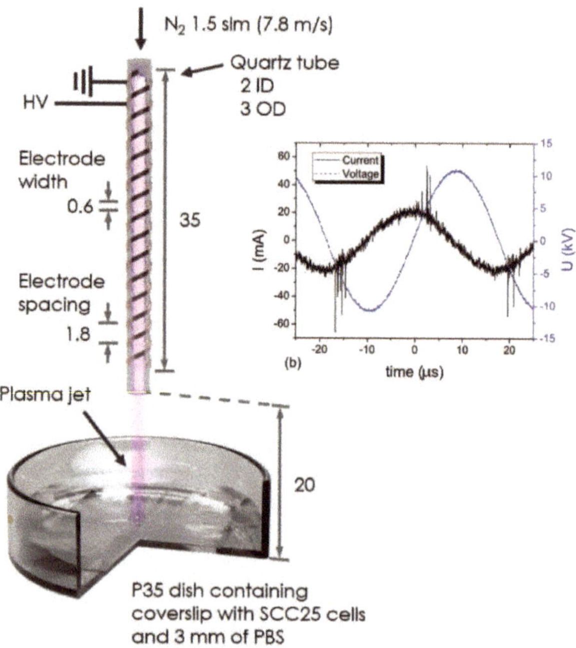

Figure 4. A schematic diagram of the nitrogen APPJ source used to treat cultured cells on coverslips. All dimensions are in mm. The inset shows the high voltage (HV) and current of alternating current (AC) waveforms of the nitrogen APPJ discharge.

Additionally, the communication between treated cells and adjacent cells (i.e., bystander cells) can initiate cellular damage pathways in the bystander cells, similar to those initiated in treated cells [38]. Therefore, DNA damage in a cell layer can propagate radially outwards from the treatment region. Notably, G1 phase cells also showed circular damage patterns, with a damage ratio of approximately 50% after 1 h of incubation, whereas the damage ratio observed for G2 phase cells was much lower and not prominently localized at the plasma treatment region. These results are different from the observations reported previously [28], in which no significant γH2AX signals were found for either the G1 or G2 phases. These differences may be due to the different plasma sources used (i.e., helium plasma jet in the previous study [28], compared with the nitrogen plasma jet in our study), as the composition and distribution of plasma species may differ, resulting in different levels of damaging effects.

2.2.2. APPJ-Irradiated Nonmalignant Cells

In contrast with the malignant cell line, plasma-treated nonmalignant cells did not show damage after 1 h of incubation (Figure 3b). After 2 h of incubation, the distribution of total damage in nonmalignant cells showed a circular pattern, with a much lower intensity than that observed in cancer cells (<1/5 the maximum value). This result suggested that under our experimental conditions, nonmalignant cells were minimally influenced by the plasma treatment. There are several factors: biological (e.g., gene expression, member structure, tolerance to oxidative stress) and experimental (e.g., dose and type of delivered radicals produced in APPJ from different plasma sources) factors reported that could contribute to different responses in various cell lines [14]. Because experimental conditions for both cell lines in this study are the same, the massive difference in damage susceptibility between two cell lines following plasma treatment is due to their biological differences. The observation that plasma-treated nonmalignant cells displayed mild damage at a later time (2 h of incubation compared with 1 h of incubation for cancer cells) suggested that the two types of cells respond at different rates. Therefore, to accurately compare the plasma-induced effects on different types of cells, the cellular responses must be monitored over time, instead of using a single time point after treatment. However, more studies are needed to assess the APPJ effectiveness in biological targets and to determine the mechanisms through which plasma interacts with cells, which are outside of the scope of this study.

3. Materials and Methods

A schematic diagram of the APPJ source and the experimental setup is illustrated in Figure 4. A detailed description of the plasma source used in this study has been previously reported [5]. The APPJ was ignited with a 22.4-kV peak-to-peak voltage and a 59-mA peak-to-peak leaking current, at 28 kHz (Figure 4). Ultrahigh, pure, 5.0-grade nitrogen (purity of 99.999%, Airgas, Radnor, PA, USA) was used as the feed gas, with a flow rate of 1.5 standard liters per minute (slm), corresponding to a gas speed of approximately 7.8 m/s. The gas flowed through a quartz tube, with an inner orifice of 2 mm. Oral cancer cells originally derived from squamous cell carcinoma of the tongue (SCC25) was obtained from American Type Culture Collection (ATCC, Rockville, MD, USA) and grown to approximately 90% confluence (approximately 10^5 cells·cm^{-2}), on coverslips, in p35 dishes. Prior to plasma exposure, the cell culture medium in the p35 dish was replaced with 2.4 mL phosphate-buffered saline (PBS, 1 X, Mediatech. Inc., Manassas, VA, USA), forming a 3-mm-thick liquid layer above the cells. The detailed recipes for the cell culture medium used with cells in this study are listed in Appendix B. During the treatment, the dish was placed 2 cm below the quartz tube orifice (Figure 4). A plasma treatment time of 2 min was selected based on our previous study as the optimal treatment time for the detection of DNA damage without causing excessive buffer evaporation [5]. After 2 min of treatment, the PBS was replaced with 2 mL fresh culture medium, and the dishes were transferred to an incubator for 1, 2, 4, and 8 h. The samples were incubated at 37 °C, in an atmosphere of > 95% humidity and 5% CO_2. Before image analysis, cells were fixed, permeabilized, and blocked, including PBS washes between each step, as described previously [5]. For imaging purposes, anti-phospho-histone H2AX (Ser139) antibody (Mouse, EMD Millipore Corp.) and goat anti-mouse IgG (H+L) (Alexa Fluor 488 from Thermo Fisher Scientific Inc., Waltham, MA, USA) were applied, to evaluate the DSBs in nuclear DNA. Fluorescent DAPI Mounting Solution (Vector Laboratories Inc., Burlingame, CA, USA) was used to stain nuclear DNA, and then the DNA contents were visualized and quantified.

We developed this large-scale image analysis method to facilitate cell-cycle classifications using only DAPI fluorescence. Cell-cycle classifications were confirmed by experiments using anti- CDT1 (Abcam plc.) antibody, which label G1 phase cells, and EdU (Thermo Fisher Scientific Inc.), which label S phase cells.

To compare the effects of plasma treatments between two cell lines, the same preparation and treatment procedures were performed for oral nonmalignant cells. Telomerase reverse transcriptase-immortalized normal oral keratiocytes (OKF6/T) cell were the generous gift of James Rhinewald, Brigham and Women's Hospital, Harvard Institutes of Medicine (Boston, MA, USA). Additionally, in the case of SCC25 and OKF6/T cells, two groups of samples were prepared using the same incubation times, including one group treated with nitrogen flow but no plasma irradiation and one group that received no treatment, which were used as control groups.

When the cells were not actively in culture, the cells were frozen and preserved for long-term storage. The procedures for cell culture and for freezing and thawing cells conducted in this work are provided in Appendices C and D, respectively.

To obtain cellular data, the coverslips were scanned with an Aperio fluorescence slide scanner (Leica Microsystems, Wetzlar, Germany), using the DAPI and Alexa Fluor 488 (denoted as γH2AX) channels, at 20× magnification. The immunofluorescence staining procedures are described in Appendix E.

4. Conclusions

We introduced a large-scale image analysis method, combined with machine learning, to study the spatial distributions of plasma jet-induced nuclear DNA damage over time in two cell lines from the oral cavity. The analysis included nuclear segmentation utilizing SALR clustering, the background and foreground correction of images, nuclei feature extraction, and cell-cycle classification. We demonstrated that the application of fluorescence microscopy using only DAPI staining (or any other dye for nuclear DNA staining) could achieve successful cell-cycle classification. This classification system not only preserves spatial information but also allows the discrimination of M phase cells from interphase cells, which flow cytometry often fails to accomplish.

By using this analysis method, we were able to visualize that the 2D DNA damage distributions depend on the different cell-cycle stages at various incubation times. Therefore, to realize the efficacy of plasma treatments, post-treatment time-dependent assessments are necessary to monitor the cellular responses in various irradiated cancerous cell lines and their nonmalignant counterparts. Determining the spatial and temporal distributions of nuclear DNA damage in plasma-irradiated cells can have significant impacts on revealing the molecular mechanisms of plasma effects. The robustness and versatility of this method will be beneficial for a more systematic and rigorous assessment of the strengths, such as selective targeting features and weaknesses, such as side effects of plasma clinical applications. Furthermore, our proposed methodology can be used to quantitatively examine the effects of other types of radiation on biological samples.

Author Contributions: Conceptualization, methodology, investigation, validation, software, visualization, writing—original draft preparation, J.K. and X.H.; resources, M.S.S. and S.P; review and editing, E.A. and S.P; supervision, M.S.S., Y.L. and S.P. All authors have read and agreed to the published version of the manuscript.

Funding: This material is based upon work supported by the U.S. Department of Energy Office of Science, Office of Basic Energy Science under Award Number DE-FC02-04ER15533. This is contribution number NDRL 5282 from the Notre Dame Radiation Laboratory.

Acknowledgments: The authors would like to acknowledge Guillermina Garcia from Sanford-Burnham Medical Research Institute (www.sbpdiscovery.org) for providing the whole slide fluorescence imaging.

Conflicts of Interest: The authors declare no conflict of interest.

Appendix A. Extracted Features

A list of the extracted features is given in Table A1, following a few notes. The number of Fourier descriptors for description nuclei shape was 20. Haralick textures [34] were computed using periods of 3 and 7 pixels and the features were averaged over the four computation directions (0°, 45°, 90°, and 135°). The granularity spectrum, length of 7 pixels, was computed after removing the background (morphological opening using a disk with radius of 10 pixels) using a background structuring element with radius of 10 pixels. The radial intensity features used the distance transform of a nucleus segmentation mask to extract intensity features from rings around the nucleus with different widths and radii. We computed the radial intensity values for three rings with a width of 4 pixels, as well as the center region of the nuclei as shown in Figure A1. The only features extracted from the γH2AX channels were Intensity features.

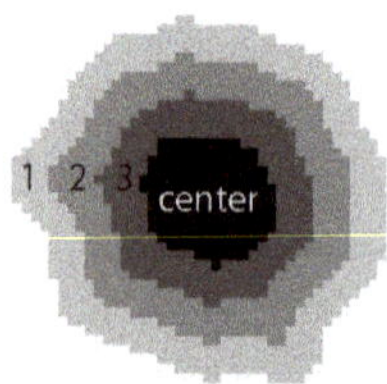

Figure A1. Radial intensity features. Example of the 3 rings with a width of 4 pixels and the center region. In each of these four regions, the mean and standard deviation of the intensity are computed.

Table A1. List of Extracted Features.

Feature Group	Feature Name
Shape	Area
	Major axis length
	Minor axis length
	Eccentricity solidity form factor
	Fourier descriptors (20)
Intensity	Integrated mean
	deviation
	Kurtosis
Radial intensity (N3 W4)	Mean
	Standard deviation
Haralick texture (3, 7)	Contrast
	Correlation
	Difference entropy
	Difference variance
	Energy
	Entropy
	Information measures correlation 1
	Information measures correlation 2
	Inverse difference moment
	Sum average
	Sum entropy
	Sum of squares variance
	Sum variance
Granularity	----

Appendix B. Cell Culture Medium Recipes

The preparation of cell culture medium is required to be carried out in a laminar flow hood in a sterile condition. All the components of the medium are required to be filtered. The filter used in this study is with a polyethersulfone microporous membrane which has a pore size of 0.2 μm.

Appendix B.1. SCC25 Cell Culture Medium: ~500 mL

- Dulbecco's modified Eagle medium (DMEM): 250 mL;
- Ham's nutrient mixture F-12 (F-12): 250 mL;
- Fetal bovine serum (FBS): 50 mL;
- Penicillin streptomycin (Pen Strep): 5 mL;
- Amphotericin B: 0.5 mL.

Appendix B.2. OKF6/T Cell Culture Medium: ~500 mL

- Keratinocyte-serum free medium (SFM): 500 mL;
- Epidermal growth factor 1-52 (EGF 1-53, 0.03 μg/μL): 3.3 μL;
- Bovine pituitary extract (BPE, 10 mg/mL): 1.25 mL;
- Calcium chloride (CaCl$_2$, 2M): 77.5 μL;
- Penicillin streptomycin (Pen Strep): 5 mL.

Appendix C. Cell Culture Procedure

The cell culture procedures are required to be carried out in a laminar flow hood in a sterile condition.

Appendix C.1. Primary Culture

This process is usually carried out every other day before the cells reach confluence. Fresh medium is needed to be prewarmed (37 °C) before use.

- Remove the cell culture medium in the p100 dish and wash the cells with PBS (1X) twice.
- Add 12 mL fresh medium in the dish.
- Return the dish to the incubator.

Appendix C.2. Subculture

This process is required when the cells are over 90% confluence. Fresh medium is needed to be prewarmed (37 °C) before use.

- Remove cell culture medium in the p100 dish and wash the cells with PBS (1X) twice.
- Add trypsin 2 mL (2 mL is enough for covering the surface of a p100 dish) and transfer the cell dish into an incubator for 3 min.
- Observe the cells under a microscope to check if the cells are detached from the dish surface. If not, prolong the incubation time or tab the side of the dish gently to expedite cell detachment.
- Add 4 mL fresh medium to the dish (make sure this amount is at least twice the volume of the added trypsin). Disperse the medium by pipetting the cell suspension over the cell layer surface several times.
- Harvest all the cell suspension into a 15 mL centrifuge tube.
- Centrifuge the tube at 1200 rpm for 2 min.
- Remove the supernatant above the cell pellet. Add 4 mL fresh medium, resuspend the cell pellet by pipetting and mixing.
- Add 11.5 mL fresh medium in a new p100 dish.
- Distribute 0.5 mL cells suspension into the new dish. As a result, the cells are seeded with 1:8 ratio in 12 mL medium in the p100 dish. The seeding ratio can be varied according to experimental needs. With 1:8 ratio dilution, SCC25 cells usually become confluent in approximately 4 days, and OKF6/T in approximately 6 days.
- Move the dish in four directions (forward, backward, left, and right) horizontally several times before transferring it to the incubator. This procedure will facilitate the cells to be seeded uniformly throughout the whole dish area.

Appendix D. Freezing and Thawing Cell Procedures

Appendix D.1. Freezing

This process can be carried out when the cells are over 90% confluence in p100 dishes. Freezing medium is needed to be prepared at 2–8 °C before use. Due to freezing medium contains dimethyl sulfoxide (DMSO), which is known to facilitate the entry of organic compound into human tissues, it is necessary to handle freezing medium carefully with proper personal protective equipment (PPE). A freezing container with isopropanol is needed to be prepared in room temperature. Once the isopropanol chamber is used five times for freezing, the isopropanol in the container must be replaced with fresh isopropanol.

- Conduct the first 6 steps in Appendix C.2.
- Remove the supernatant above the cell pellet. Add 1.2 mL prewarmed freezing medium, resuspend the cell pellet by pipetting and mixing.
- Dispense the cell suspension in two cryogenic storage vials with 0.5 mL each.
- Place the vials in the isopropanol chamber and store it at −80 °C overnight.
- Transfer the vials to a storage unit in a liquid nitrogen tank.

Appendix D.2. Thawing

The newly added medium should be changed the next day after the cells being thawed following the steps below. This is to remove the hazard of cryoprotective agents such as DMSO that present in the freezing medium. As mentioned in Appendix D.1, the freezing medium should be handled with care.

- Prepare a p100 dish with 10 mL fresh prewarmed cell culture medium in a sterilized laminar flow hood.
- Thaw the frozen cells rapidly (<1 min) by gently swirling the vial in a 37 °C water bath until the ice remains a little bit in the vial. At this step, a face mask or goggles are required in addition to gloves for PPE for the hazard of vial explosion.
- Sterilize the vial surface with 70% ethanol before transferring to the hood.
- Transfer the cells into the prepared dish filled with fresh medium with micropipette. Add 10 mL more fresh medium to the dish and gently mix with the cells.
- Move the dish in four directions horizontally several times before transferring it to the incubator.

Appendix E. Immunofluorescence Staining Procedures

Appendix E.1. Single Antibody (AB) Staining

Single antibody staining uses one primary antibody and one secondary antibody to stain the cells. The cells are grown on square coverslips in p35 culture dishes. Paraformaldehyde (PFA, from Electron Microscopy Sciences, PA, USA), Triton X-100 (from American Bioanalytical, MA, USA), and bovine serum albumin (BSA, from Gold Biotechnology, MO, USA) are previously prepared with desired concentrations in 1X PBS.

Day 1:

(1) Wash cells with PBS (1X) twice.
(2) Fix the cells by adding 1 mL PFA (4%) to the dish for 20 min.
(3) Remove PFA to a biowaste container. Since PFA is toxic, it should not be discarded with other conventional biowaste solutions.
(4) Rinse the cells with PBS (1X) twice.
(5) Permeabilize the cells with Triton X-100 (0.3%) for 5 min.
(6) Remove the Triton X-100 and wash the cells with PBS (1X) three times.
(7) Apply blocking solution BSA (3%) to the cells for 1 h.
(8) Prepare primary AB solution. For example, anti-phospho-histone H2AX needs to be diluted with a ratio of 1:250 in 3% BSA. The amount of antibody needed can be calculated as follows:

$$V_{AB}(\mu L) = \frac{V_{total}}{dilution\ factor} = \frac{75\mu L \times N_{sample} + 30\mu L}{250} \tag{A1}$$

$$V_{BSA} = V_{total} - V_{AB} \tag{A2}$$

where V_{AB} is the volume of antibody from the stock, V_{total} is the volume of the final solution, V_{BSA} is the volume of the 3% BSA as a solvent. The volume of the final solution applied to each coverslip for antibody staining is 75 μL. There is an extra volume of 30 μL in case of bubble formation inside the solution during pipetting and mixing.

(9) Prepare a flat parafilm sheet as a platform for AB incubation. The film can be placed at the bottom inside a square plastic container (e.g., 10 × 10 × 15 mm square petri dish with lid, SKS Science Products). The inside of the lid is covered with a wet Kimwipe tissue, which will facilitate maintaining the moisture during AB incubation process.
(10) Carefully add 75 μL primary AB each at a set of dispersed locations (the number of locations equals to N_{sample}) on the parafilm avoiding bubbles while pipetting.

(11) Take out the coverslips immersed in BSA from the dishes.

(12) Flip the coverslips enabling the face of the coverslip that having cells facing down towards the parafilm. Rest one side edge of the coverslip on the film, then carefully lower down the other side of the coverslip, till the cells on the entire coverslip are in contact with the solution. Make sure there are no bubbles in between the coverslip and the parafilm.

(13) After finishing the previous step for all the samples, put on the wet-tissue-covered lid. Transfer the container to a 4 °C room overnight.

Day 2:

(14) Fill coverglass staining jars with 1X PBS. Each jar can contain four coverslips vertically.

(15) Transfer the coverslips from the container to the staining jars. Make sure to record the orientation of the coverslip faces that have cells.

(16) Remove the PBS inside the jars. Fill the jars with fresh PBS and put the jars on a shaker (Orbit shaker, Barnstead Lab-Line, 600 rpm) for 5 min. Repeat this step one more time.

(17) Prepare secondary AB solution. For example, Alexa Fluor 488 goat anti-mouse needs to be diluted with a ratio of 1:400 in 3% BSA. The amount of the antibody needed can be calculated in the same method as Equation (E1) with a dilution factor of 400. During preparation, light exposure to the secondary AB must to be kept as short as possible.

(18) Repeat step (9), (10) with secondary AB, (11) with taking out coverslips immersed in PBS from the jars, and (12).

(19) After finishing the previous step for all samples, close the container with the wet-tissue-covered lid. Keep the container in dark for 20 min.

(20) Repeat step (15). Similar as the step (16), wash the coverslips in PBS on a shaker four times with 5 min each. Repeat washing the slides once with reverse osmosis (RO) water on a shaker for 5 min. Keep the entire washing process in dark.

(21) Take out the coverslips from the jars. Place the coverslips on a Kimwipe tissue with the cells facing up. Air dry the coverslips in dark.

(22) Prepare glass microscope slides with 20 μL DAPI droplets on top. One DAPI drop per slide is preferred. Be careful to protect DAPI from light.

(23) Repeat step (12) for staining the cells with DAPI solution. Press the coverslips on the slides vertically to remove extra DAPI with Kimwipe tissues.

(24) If the DAPI solution used is a soft-mounting medium, apply clear nail polish on four edges of the coverslips to seal the slides. If the DAPI used is a hard-mounting medium, no further sealing process is needed.

(25) Store the slides in a slide storage box in a 4 °C room in dark.

Appendix E.2. Double Antibody Staining

Double staining (DS) uses two primary antibodies and two secondary antibodies with different fluorescence emission ranges to stain two different parts of a cell. The procedures of double antibody staining are the same as the single antibody staining except following modifications of the step (8) and the step (17) to be (26) and (27), respectively:

(26) Prepare AB solution with two primary ABs. For example, anti-phospho-histone H2AX and cleaved caspase-3 antibody need to be diluted in 3% BSA with a ratio of 1:250 and 1:400, respectively. The amount of antibody can be calculated as follows:

$$V_{AB1}(\mu L) = \frac{V_{total}}{dilution\ factor\ 1} = \frac{75\mu L \times N_{sample} + 30\mu L}{250} \tag{A3}$$

$$V_{AB2}(\mu L) = \frac{V_{total}}{dilution\ factor\ 2} = \frac{75\mu L \times N_{sample} + 30\mu L}{400} \tag{A4}$$

$$V_{BSA} = V_{total} - V_{AB1} - V_{AB2} \tag{A5}$$

where V_{AB1} and V_{AB2} are the volumes of the two antibody solutions needed.

(27) Prepare AB solution with two secondary ABs. For example, both Alexa Fluor 488 goat anti-mouse and Alexa Fluor 594 goat anti-rabbit need to be diluted in 3% BSA with the ratio of 1:400. The amount of each antibody can be calculated in the same method as Equations (4) and (5) with a dilution factor of 400.

Appendix E.3. S Phase Cell Staining with Click-iT EdU Alexa Fluor 647 Imaging Kit and Single AB

Cells are grown on square coverslips in p35 culture dishes. PFA, Triton X-100, and BSA are previously prepared with desired concentrations in 1X PBS. Different stock solutions are previously prepared according to the manufacture protocol [48].

(28) Incubate cells with fresh medium containing 10 μM EdU for 2.5 h.
(29) Wash cells with PBS (1X) twice.
(30) Fix the cells by adding 1 mL PFA (4%) for 15 min.
(31) Remove PFA to a biowaste container. Since PFA is toxic, it should not be discarded with other conventional biowaste solutions.
(32) Wash cells with 1 mL 3% BSA twice.
(33) Permeabilize the cells with Triton X-100 (0.3%) for 20 min.
(34) Prepare the reaction buffer additive.
(35) Prepare the reaction cocktail according to the manufacture protocol [48] by adding the ingredients in the listed order. This cocktail solution should be used within 15 min of preparation.
(36) Wash cells with 1 mL 3% BSA twice.
(37) Add 500 μL reaction cocktail to each cell dish, incubate for 30 min in dark.
(38) Wash cells with 1 mL 3% BSA once.
(39) Follow single cell staining steps (8)–(25). If no AB staining needed (only EdU staining), proceed with steps (21)–(25).

References

1. Hou, J.; Ma, J.; Yu, K.N.; Li, W.; Cheng, C.; Bao, L.; Han, W. Non-thermal plasma treatment altered gene expression profiling in non-small-cell lung cancer A549 cells. *BMC Genom.* **2015**, *16*, 435. [CrossRef]
2. Hirst, A.M.; Simms, M.S.; Mann, V.M.; Maitland, N.J.; O'Connell, D.; Frame, F.M. Low-temperature plasma treatment induces DNA damage leading to necrotic cell death in primary prostate epithelial cells. *Br. J. Cancer* **2015**, *112*, 1536–1545. [CrossRef] [PubMed]
3. Iseki, S.; Nakamura, K.; Hayashi, M.; Tanaka, H.; Kondo, H.; Kajiyama, H.; Kano, H.; Kikkawa, F.; Hori, M. Selective killing of ovarian cancer cells through induction of apoptosis by nonequilibrium atmospheric pressure plasma. *Appl. Phys. Lett.* **2012**, *100*, 113702. [CrossRef]
4. Gumbel, D.; Bekeschus, S.; Gelbrich, N.; Napp, M.; Ekkernkamp, A.; Kramer, A.; Stope, M.B. Cold Atmospheric Plasma in the Treatment of Osteosarcoma. *Int. J. Mol. Sci.* **2017**, *18*, 2004. [CrossRef] [PubMed]
5. Han, X.; Klas, M.; Liu, Y.; Stack, M.S.; Ptasinska, S. DNA damage in oral cancer cells induced by nitrogen atmospheric pressure plasma jets. *Appl. Phys. Lett.* **2013**, *102*, 233703. [CrossRef]
6. Hattori, N.; Yamada, S.; Torii, K.; Takeda, S.; Nakamura, K.; Tanaka, H.; Kajiyama, H.; Kanda, M.; Fujii, T.; Nakayama, G.; et al. Effectiveness of plasma treatment on pancreatic cancer cells. *Int. J. Oncol.* **2015**, *47*, 1655–1662. [CrossRef]
7. Akter, M.; Jangra, A.; Choi, A.S.; Choi, E.H.; Han, I. Non-Thermal Atmospheric Pressure Bio-Compatible Plasma Stimulates Apoptosis via p38/MAPK Mechanism in U87 Malignant Glioblastoma. *Cancers* **2020**, *12*, 245. [CrossRef]
8. Liu, J.-R.; Wu, Y.-M.; Xu, G.-M.; Gao, L.-G.; Ma, Y.; Shi, X.-M.; Zhang, G.-J. Low-temperature plasma induced melanoma apoptosis by triggering a p53/PIGs/caspase-dependent pathway in vivo and in vitro. *J. Phys. D Appl. Phys.* **2019**, *52*, 315204. [CrossRef]

9. Chernets, N.; Kurpad, D.S.; Alexeev, V.; Rodrigues, D.B.; Freeman, T.A. Reaction Chemistry Generated by Nanosecond Pulsed Dielectric Barrier Discharge Treatment is Responsible for the Tumor Eradication in the B16 Melanoma Mouse Model. *Plasma Process. Polym.* **2015**, *12*, 1400–1409. [CrossRef]

10. Utsumi, F.; Kajiyama, H.; Nakamura, K.; Tanaka, H.; Mizuno, M.; Ishikawa, K.; Kondo, H.; Kano, H.; Hori, M.; Kikkawa, F. Effect of Indirect Nonequilibrium Atmospheric Pressure Plasma on Anti-Proliferative Activity against Chronic Chemo-Resistant Ovarian Cancer Cells in Vitro and in Vivo. *PLoS ONE* **2013**, *8*, e81576. [CrossRef] [PubMed]

11. Mirpour, S.; Piroozmand, S.; Soleimani, N.; Jalali Faharani, N.; Ghomi, H.; Fotovat Eskandari, H.; Sharifi, A.M.; Mirpour, S.; Eftekhari, M.; Nikkhah, M. Utilizing the micron sized non-thermal atmospheric pressure plasma inside the animal body for the tumor treatment application. *Sci. Rep.* **2016**, *6*, 29048. [CrossRef]

12. Khlyustova, A.; Labay, C.; Machala, Z.; Ginebra, M.-P.; Canal, C. Important parameters in plasma jets for the production of RONS in liquids for plasma medicine: A brief review. *Front. Chem. Sci. Eng.* **2019**, *13*, 238–252. [CrossRef]

13. von Woedtke, T.; Schmidt, A.; Bekeschus, S.; Wende, K.; Weltmann, K.D. Plasma Medicine: A Field of Applied Redox Biology. *In Vivo* **2019**, *33*, 1011–1026. [CrossRef]

14. Semmler, M.L.; Bekeschus, S.; Schäfer, M.; Bernhardt, T.; Fischer, T.; Witzke, K.; Seebauer, C.; Rebl, H.; Grambow, E.; Vollmar, B.; et al. Molecular Mechanisms of the Efficacy of Cold Atmospheric Pressure Plasma (CAP) in Cancer Treatment. *Cancers* **2020**, *12*, 269. [CrossRef]

15. Keidar, M.; Walk, R.; Shashurin, A.; Srinivasan, P.; Sandler, A.; Dasgupta, S.; Ravi, R.; Guerrero-Preston, R.; Trink, B. Cold plasma selectivity and the possibility of a paradigm shift in cancer therapy. *Br. J. Cancer* **2011**, *105*, 1295–1301. [CrossRef] [PubMed]

16. Yan, D.; Sherman, J.H.; Keidar, M. Cold atmospheric plasma, a novel promising anti-cancer treatment modality. *Oncotarget* **2017**, *8*, 15977–15995. [CrossRef] [PubMed]

17. Laroussi, M.; Lu, X.; Keidar, M. Perspective: The physics, diagnostics, and applications of atmospheric pressure low temperature plasma sources used in plasma medicine. *J. Appl. Phys.* **2017**, *122*, 020901. [CrossRef]

18. Metelmann, H.R.; Nedrelow, D.S.; Seebauer, C.; Schuster, M.; von Woedtke, T.; Weltmann, K.D.; Kindler, S.; Metelmann, P.H.; Finkelstein, S.E.; Von Hoff, D.D.; et al. Head and neck cancer treatment and physical plasma. *Clin. Plasma Med.* **2015**, *3*, 17–23. [CrossRef]

19. Dai, X.; Bazaka, K.; Richard, D.J.; Thompson, E.R.W.; Ostrikov, K.K. The Emerging Role of Gas Plasma in Oncotherapy. *Trends Biotechnol.* **2018**, *36*, 1183–1198. [CrossRef]

20. Schuster, M.; Seebauer, C.; Rutkowski, R.; Hauschild, A.; Podmelle, F.; Metelmann, C.; Metelmann, B.; von Woedtke, T.; Hasse, S.; Weltmann, K.D.; et al. Visible tumor surface response to physical plasma and apoptotic cell kill in head and neck cancer. *J. Craniomaxillofac. Surg.* **2016**, *44*, 1445–1452. [CrossRef]

21. Pranda, M.A.; Murugesan, B.J.; Knoll, A.J.; Oehrlein, G.S.; Stroka, K.M. Sensitivity of tumor versus normal cell migration and morphology to cold atmospheric plasma-treated media in varying culture conditions. *Plasma Process. Polym.* **2020**, *17*, 1900103. [CrossRef]

22. Boehm, D.; Bourke, P. Safety implications of plasma-induced effects in living cells—a review of in vitro and in vivo findings. *Biol. Chem.* **2018**, *400*, 3–17. [CrossRef]

23. Schuster, M.; Rutkowski, R.; Hauschild, A.; Shojaei, R.K.; von Woedtke, T.; Rana, A.; Bauere, G.; Metelmanng, P.; Seebauer, C. Side Effects in Cold Plasma Treatment of Advanced Oral Cancer—Clinical Data and Biological Interpretation. *Clin. Plasma Med.* **2018**, *10*, 9–15. [CrossRef]

24. Jablonowski, L.; Kocher, T.; Schindler, A.; Muller, K.; Dombrowski, F.; von Woedtke, T.; Arnold, T.; Lehmann, A.; Rupf, S.; Evert, M.; et al. Side Effects by Oral Application of Atmospheric Pressure Plasma on the Mucosa in Mice. *PLoS ONE* **2019**, *14*, e0215099. [CrossRef]

25. Itooka, K.; Takahashi, K.; Izawa, S. Fluorescence microscopic analysis of antifungal effects of cold atmospheric pressure plasma in Saccharomyces cerevisiae. *Appl. Microbiol. Biotechnol.* **2016**, *100*, 9295–9304. [CrossRef]

26. Delben, J.A.; Zago, C.E.; Tyhovych, N.; Duarle, S.; Vergani, C.E. Effect of Atmospheric-Pressure Cold Plasma on Pathogenic Oral Biofilms and In Vitro Reconstituted Oral Epithelium. *PLoS ONE* **2016**, *11*, e0155427. [CrossRef]

27. Smolkova, B.; Lunova, M.; Lynnyk, A.; Uzhytchak, M.; Churpita, O.; Jirsa, M.; Kubinova, S.; Lunov, O.; Dejneka, A. Non-Thermal Plasma, as a New Physicochemical Source, to Induce Redox Imbalance and Subsequent Cell Death in Liver Cancer Cell Line. *Cell Physiol. Biochem.* **2019**, *52*, 119–140. [CrossRef] [PubMed]

28. Volotskova, O.; Hawley, T.S.; Stepp, M.A.; Keidar, M. Targeting the cancer cell cycle by cold atmospheric plasma. *Sci. Rep.* **2012**, *2*, 636. [CrossRef] [PubMed]

29. Sanderson, M.J.; Smith, I.; Parker, I.; Bootman, M.D. Fluorescence microscopy. *Cold Spring Harb. Protoc.* **2014**, *10*. [CrossRef] [PubMed]

30. Blasi, T.; Hennig, H.; Summers, H.D.; Theis, F.J.; Cerveira, J.; Patterson, J.O.; Davies, D.; Filby, A.; Carpenter, A.E.; Rees, P. Label-free cell cycle analysis for high-throughput imaging flow cytometry. *Nat. Commun.* **2016**, *7*, 1–9. [CrossRef] [PubMed]

31. Otsu, N. A Threshold Selection Method from Gray-Level Histograms. *IEEE Trans. Syst. Man Cybern.* **1979**, *9*, 62–66. [CrossRef]

32. Kapaldo, J.; Han, X.; Mery, D. Seed-Point Detection of Clumped Convex Objects by Short-Range Attractive Long-Range Repulsive Particle Clustering. *arXiv* **2018**, arXiv:1804.04071.

33. Kapaldo, J. Seed-Point Based Geometric Partitioning of Nuclei Clumps. *arXiv* **2018**, arXiv:1804.04549.

34. Haralick, R.M.; Shanmugam, K.; Dinstein, I. Textural Features for Image Classification. *IEEE Trans. Syst. Man Cybern.* **1973**, *6*, 610–621. [CrossRef]

35. Carpenter, A.E.; Jones, T.R.; Lamprecht, M.R.; Clarke, C.; Kang, I.H.; Friman, O.; Guertin, D.A.; Chang, J.H.; Lindquist, R.A.; Moffat, J.; et al. CellProfiler: Image analysis software for identifying and quantifying cell phenotypes. *Genome Biol.* **2006**, *7*, R100. [CrossRef]

36. Zahn, C.T.; Roskies, R.Z. Fourier Descriptors for Plane Closed Curves. *IEEE Trans. Comput.* **1972**, *100*, 269–281. [CrossRef]

37. Connolly, A.J.; Genovese, C.; Moore, A.W.; Nichol, R.C.; Schneider, J.; Wasserman, L. Fast Algorithms and Efficient Statistics: Density Estimation in Large Astronomical Datasets. *arXiv* **2000**, arXiv:astro-ph/0008187.

38. Graves, D.B. Oxy-nitroso shielding burst model of cold atmospheric plasma therapeutics. *Clin. Plasma Med.* **2014**, *2*, 38–49. [CrossRef]

39. Han, X.; Liu, Y.; Stack, M.S.; Ptasinska, S. 3D Mapping of plasma effective areas via detection of cancer cell damage induced by atmospheric pressure plasma jets. *J. Phys. Conf. Ser.* **2014**, *565*, 012011. [CrossRef]

40. Klas, M.; Ptasinska, S. Characteristics of N2 and N2/O2 atmospheric pressure glow discharges. *Plasma Sources Sci. Technol.* **2013**, *22*, 025013. [CrossRef]

41. Arjunan, K.P.; Sharma, V.K.; Ptasinska, S. Effects of atmospheric pressure plasmas on isolated and cellular DNA-a review. *Int. J. Mol. Sci.* **2015**, *16*, 2971–3016. [CrossRef]

42. Arjunan, K.P.; Obrusnik, A.; Jones, B.T.; Zajickova, L.; Ptasinska, S. Effect of Additive Oxygen on the Reactive Species Profile and Microbicidal Property of a Helium Atmospheric Pressure Plasma Jet. *Plasma Process. Polym.* **2016**, *13*, 1089–1105. [CrossRef]

43. Kapaldo, J.; Han, X.; Ptasinska, S. Shielding-gas-controlled atmospheric pressure plasma jets: Optical emission, reactive oxygen species, and the effect on cancer cells. *Plasma Process. Polym.* **2019**, *16*, 1800169. [CrossRef]

44. Gweon, B.; Kim, M.; Bee Kim, D.; Kim, D.; Kim, H.; Jung, H.; Shin, J.H.; Choe, W. Differential responses of human liver cancer and normal cells to atmospheric pressure plasma. *Appl. Phys. Lett.* **2011**, *99*, 063701. [CrossRef]

45. Panngom, K.; Baik, K.Y.; Nam, M.K.; Han, J.H.; Rhim, H.; Choi, E.H. Preferential killing of human lung cancer cell lines with mitochondrial dysfunction by nonthermal dielectric barrier discharge plasma. *Cell Death Dis.* **2013**, *4*, e642. [CrossRef] [PubMed]

46. Weinberg, R.A. *The Biology of Cancer*, 2nd ed.; Garland Science, Taylor & Francis Group, LLC.: New York, NY, USA, 2014.

47. Lu, X.; Naidis, G.V.; Laroussi, M.; Reuter, S.; Graves, D.B.; Ostrikov, K. Reactive species in non-equilibrium atmospheric-pressure plasmas: Generation, transport, and biological effects. *Phys. Rep.* **2016**, *630*, 1–84. [CrossRef]

48. Life Technologies Corporation. *Click-iT EdU Imaging Kits (Mannual)*; Life Technologies Corp.: Carlsbad, CA, USA, 2011.

International Journal of
Molecular Sciences

Article

Cold Atmospheric Plasma and Silymarin Nanoemulsion Activate Autophagy in Human Melanoma Cells

Manish Adhikari [1,*,†], Bhawana Adhikari [1,†], Bhagirath Ghimire [1], Sanjula Baboota [2] and Eun Ha Choi [1,*]

[1] Plasma Bioscience Research Center, Applied Plasma Medicine Center, Department of Electrical and Biological Physics, Kwangwoon University, Seoul 01897, Korea; bnegi87@gmail.com (B.A.); ghimirebhagi@hotmail.com (B.G.)
[2] Department of Pharmaceutics, School of Pharmaceutical Education and Research, Jamia Hamdard, Delhi 110062, India; sbaboota@rediffmail.com
* Correspondence: manishadhikari85@gmail.com (M.A.); ehchoi@kw.ac.kr (E.H.C.)
† These authors contributed equally to this work and are joint first authors.

Received: 12 February 2020; Accepted: 11 March 2020; Published: 12 March 2020

Abstract: Background: Autophagy is reported as a survival or death-promoting pathway that is highly debatable in different kinds of cancer. Here, we examined the co-effect of cold atmospheric plasma (CAP) and silymarin nanoemulsion (SN) treatment on G-361 human melanoma cells via autophagy induction. Methods: The temperature and pH of the media, along with the cell number, were evaluated. The intracellular glucose level and *PI3K/mTOR* and *EGFR* downstream pathways were assessed. Autophagy-related genes, related transcriptional factors, and autophagy induction were estimated using confocal microscopy, flow cytometry, and ELISA. Results: CAP treatment increased the temperature and pH of the media, while its combination with SN resulted in a decrease in intracellular ATP with the downregulation of *PI3K/AKT/mTOR* survival and *RAS/MEK* transcriptional pathways. Co-treatment blocked downstream paths of survival pathways and reduced *PI3K* (2 times), *mTOR* (10 times), *EGFR* (5 times), *HRAS* (5 times), and *MEK* (10 times). CAP and SN co-treated treatment modulates transcriptional factor expressions (*ZKSCAN3, TFEB, FOXO1, CRTC2,* and *CREBBP*) and specific genes (*BECN-1, AMBRA-1, MAP1LC3A,* and *SQSTM*) related to autophagy induction. Conclusion: CAP and SN together activate autophagy in G-361 cells by activating *PI3K/mTOR* and *EGFR* pathways, expressing autophagy-related transcription factors and genes.

Keywords: cold atmospheric plasma; autophagy; silymarin nanoemulsion; *PI3K/mTOR* pathway

1. Introduction

Malignant melanoma arises from the irregular dysfunction of human skin pigment-producing cells (melanocytes), which results in the overproduction of skin color [1]. Around half of melanoma carry a mutation in the *BRAF* gene, which results in the dysregulation of several molecular signaling pathways [2]. Till now, various signaling pathways involved in resistance, as well as the spectrum of therapy-acquired mutations, have been known, while the cellular aspects of therapy resistance have not [3]. Antecedently, remarkable progress was achieved in both immunotherapy and molecular-targeted therapy as the standard of care for terminal melanoma patients [4,5]. Cold atmospheric plasma (CAP) is an evolving biomedical technique that has been used in multifarious ways, like in skin disorders, wound healing, dentistry, hair treatment, various kinds of cancers, etc., in past years [6–10]. It is well-known that CAP generates reactive oxygen and nitrogen species

(RONS), UV rays, and charged particles that physically and chemically change the biological surfaces by inducing oxidative stress, which results in a change in certain gene expression and epigenetic changes [11,12]. Our previous publication revealed the generation of RONS using a micro-dielectric barrier discharge (μ-DBD) device (same device used for this study), which aided in the reduction of melanoma cells by apoptosis [11]. In the last few years, nanotechnology has also played a key role in cancer treatment [13]. Our group recently published work demonstrating the synergistic effects of gold nanoparticles with CAP and silymarin nanoemulsion with CAP in which glioblastoma and human melanoma were killed by inhibiting the *PI3K/AKT* pathway and *HGF/c-MET* pathway, respectively [11,14]. Using a nanotechnology and pharmacology approach, we prepared, characterized, and performed experiments on silymarin nanoemulsion (SN), which was used in our previous publication [11]. The protein expression levels and associated signaling of the *PI3K/MEK* and *HRAS* pathways using CAP and SN have already been described and presented in our previous publication [11]. Silymarin is widely well-accepted as a heptoprotectant and reported in the treatment of different kinds of cancer [15,16]. SN has been shown to decrease silymarin hydrophobicity by its heterogeneous dispersion of two immiscible liquids (oil-in-water) and hence improve the bioavailability. In the past few years, nanotechnology combined with CAP technology has exhibited some promising aspects in cancer research by the activation/deactivation of cellular pathways [17]. Additionally, it has been proposed by some research groups that CAP can help in selective cell membrane permeability, which helps in the invasion of CAP-generated direct and indirect RONS species, leading to the intracellular invasion of nanoparticles towards applied sites [18–20]. However, at the clinical level, the outcome of the survival of melanoma patients has been very limited due to the targeted mutation in molecular targeted therapy [21]. Cell growth inhibition and cell death induction are the main objectives of every successful cancer treatment.

Melanoma seems to be a specific type of cancer that dysregulates apoptosis and hence evolves as a cancer that is resistant to programmed cell death [22]. The cellular death (apoptosis) of G-361 human melanoma by treatment with CAP and SN was evaluated using the annexin V-PI staining method [11]. Therefore, the induction of another form of cellular death is necessary and fundamental to conquering this resistance [23,24]. Recent studies on various types of cancers have demonstrated that in premature tumors, cancer cells can upregulate a related stress response process called autophagy [25]. Autophagy is a vibrant cellular self-digestion cum destruction process and in normal cells, occurs at constitutive stages to maintain internal cellular homeostasis [26,27]. Research has presented substantial outcomes showing that the autophagy process plays a vital role against various diseases, cancer, aging, and neurodegenerative disease [28]. Hence, the initiation of other death mechanisms, such as autophagy, provides a critical defensive approach to guarantee the elimination of potential cancer cells [29]. These data, for the first time, highlight that CAP, along with SN treatment, leads to the accumulation of autophagosomes, which is the signal of autophagy induction that occurs by inhibiting the classical autophagy specific *mTOR* survival pathway and *MEK* pathway, respectively. Our data also suggests an increase in autophagy-specific genes and their related transcriptional factors, which could be developed as a new strategy for melanoma treatment.

Therefore, the current study is focused on manipulating autophagy in melanoma cells by a μ-dielectric barrier discharge (μ-DBD) air CAP device using air as feeder gas with SN by evaluating autophagy influx, transcriptional factors, and the activation of autophagy-specific genes. Hence, these results present evidence of a plausible role of the air CAP with SN delivery in inhibiting melanoma progression by the induction of autophagy, which may improve clinical outcomes of patients in the future.

2. Results

2.1. Electrical and Optical Characteristics of the μ-DBD Plasma Instrument

The electrical and optical characteristics of discharge are shown in Figure 1c–e. These waveforms were recorded using a Lecroy wave surfer 434, 350 MHz oscilloscope with a Tektronix P6015A high-voltage probe and Tektronix P6022 current probe. The discharge was operated in dimming mode by using a DC-AC inverter whose operational time (T_{on}) and shutting time (T_{off}) were set to 18.77 ms and 232.82 ms, respectively. These waveforms are shown in Figure 1c. The duty percentage was $\approx$7% and the repetition frequency was $\approx$4 Hz. A longer shutting off time between consecutive discharges of short durations could enable longer operation of the source, without sufficient heating of the electrodes. The operational time of the discharge for the μ-DBD plasma source is characterized by repetitive sinusoidal waveforms that have a frequency of $\approx$60 kHz. These waveforms for voltage and current are shown in Figure 1d. Every positive half cycle of the applied voltage, a number of micro-discharges are initiated, which causes the positive polarity current. During this process, charges are accumulated on the surface of dielectric material and cause a negative polarity current during the negative half cycle. The values of the applied voltage (rms) and total current (rms) were found to be 1.40 kV and 20 mA, respectively. The dissipated power in one duty cycle was calculated by integrating the current and voltage waveforms over one cycle of the discharge period ($P = duty - ratio \times \frac{1}{T} \int_0^T IVdt$) [30] and was found to be 0.40 J/s.

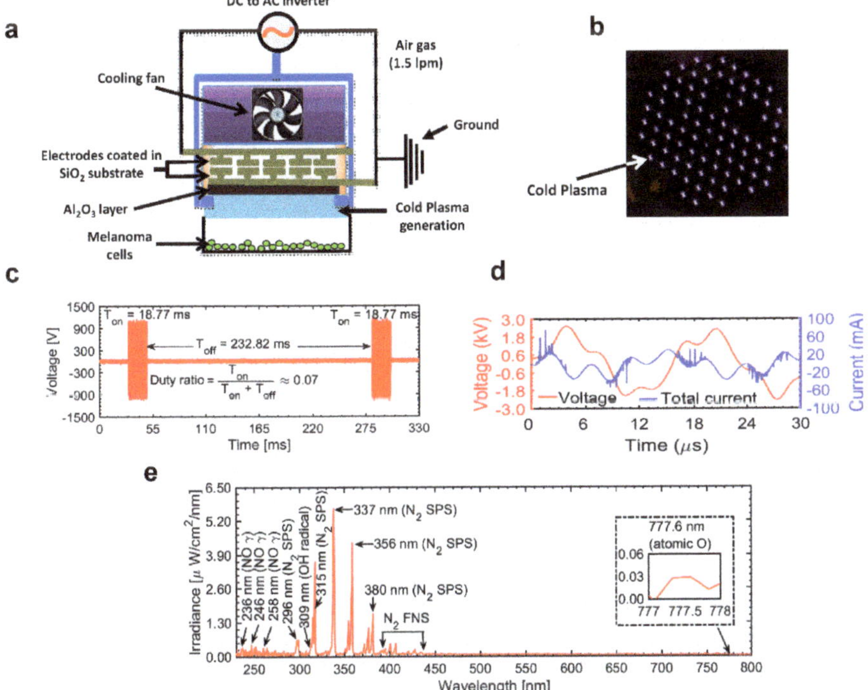

Figure 1. (**a**) Schematic diagram of the non-thermal cold atmospheric plasma (CAP) micro-dielectric barrier discharge (μ-DBD) plasma source with a distance of 2 mm from cell culture media. (**b**) Photograph of the μ-DBD plasma device showing plasma discharge using air as the feeder gas. (**c**) The μ-DBD device on time (T_{on}) and off-time (T_{off}). (**d**) Voltage and current waveforms for one cycle of the T_{on} period of the μ-DBD air CAP device. (**e**) Optical emission spectrum (OES) composition of the air μ-DBD CAP source between 200 nm and 800 nm.

The optical emission spectra (OES) of the μ-DBD air discharge recorded in the wavelength range of 200–900 nm is shown in Figure 1e. This spectrum was measured with an HR4000 spectrometer (Ocean Optics Corporation, Orlando, FL, USA). The conversion to absolute units (μW/cm²/nm) was performed by using a deuterium halogen lamp (Ocean Optics Corporation, Model: DH-3P-BAL-CAL).

In order to record the OES spectra of the inhomogeneous discharge, the integration time of the spectrometer was set to 500 ms, with an average of 4. In the OES, it can be observed that there are weak emission signals from the nitric oxide gamma band (NO γ) at 236, 246, and 258 nm, etc. [31]. These species originated from the collision of energetic electrons/metastable atoms with air molecules. A weak emission signal from the hydroxyl radical (OH) can be observed in the range of 306–309 nm. OH radicals are formed through the dissociation of water molecules which might be present in the feeding gas and ambient environment [30]. Strong emissions from the nitrogen second positive system (N_2 SPS) can be observed at 296, 315, 337, 356, and 380 nm, etc. There are also emissions from the nitrogen first negative system (N_2 FNS) in the range of 390–440 nm [32]. The origin of excited nitrogen species can be attributed to the dissociation of nitrogen molecules which may be present in the feeding gas, as well as the ambient environment. In addition to these, there is also emission from the atomic oxygen (O) at 777 nm.

2.2. Effect of CAP on Physical Parameters (Extracellular pH and Temperature)

Studies on different plasma devices revealed changes in the extracellular pH within cell culture media (RPMI) and its temperature. Here, the device (μ-DBD) shoots plasma in the form of dielectric barrier discharge which was prepared and characterized at the Plasma Bioscience Research Center, Seoul, South Korea. The temperature of RPMI-1640 cell culture media slightly increased when it received CAP for different time intervals. The CAP treatment time used for the extracellular pH and temperature estimation was 30, 60, 120, and 180 s, respectively. Initially, the recorded temperature was 32.63 °C (30 s), but it increased gradually up to 32.75 °C when treated with CAP for 60 s. Furthermore, treatment increased the extracellular temperature from 32.88 °C (120 s) to 33.05 °C (180 s) (Figure 2a).

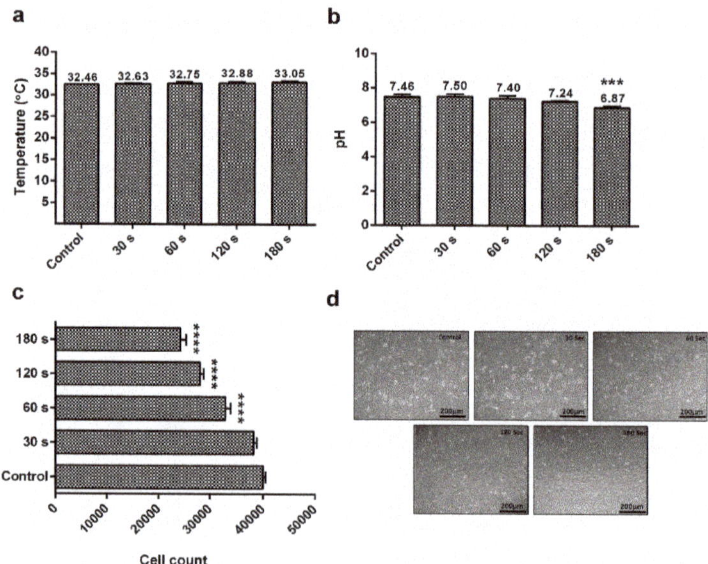

Figure 2. (**a**) Estimation of temperature within cell media (RPMI-1640) after the exposure of air CAP at different time intervals; (**b**) effect of different CAP doses on the pH level of cell media; (**c**) assessment of the G-361 human melanoma cell count using a hemocytometer counter after the exposure of CAP at different time intervals; (**d**) microscopic evaluation of G-361 human melanoma cells at different CAP doses (scale bar = 200 μm). Data are the mean ± SE of three experiments. Statistical analysis was performed using a one-way ANOVA test followed by Dunnett's test for comparisons. Each asterisk represents statistical differences between the treatment and control (*** $p < 0.001$ and **** $p < 0.0001$).

With ambient air as the feeder gas for CAP generation, the extracellular pH (cell culture media) was 7.5 for 30 s, which was decreased by every plasma dose setup. The pH level was 7.40 (60 s), 7.24 (120 s), and 6.87 (180 s), respectively (Figure 2b). To check the reduction in G-361 cells, the CAP was treated at different time intervals (Figure 2c) and the cell count was recorded 24 h after the treatment. The cell count in control samples was 40,000, which kept on reducing while receiving CAP treatment and decreased to the minimum at 180 s treatment (24,300 counts), which was also evident from microscopy images of different CAP-treated groups. CAP treatment progressively killed the G-361 cells as the dose increased, which resulted in detachment and a round structure, with a reduction of cells at higher CAP doses (Figure 2d).

2.3. Intracellular ATP and Glucose Estimation

CAP induces an extracellular acidic environment which may lead to acidic stress within G-361 cells. This will propagate to intracellular metabolic stress and a limited nutrient supply. Cancer cells need more energy (ATP) and hence produce high glucose channels on the cell surface, which helps them to activate aerobic glycolysis for their proper functioning, as compared to normal adjacent healthy cells. Therefore, we assessed glucose uptake in G-361 melanoma cells after exposure to SN, CAP, and CAP + SN. We used 2-NBDG (2-deoxy-2-[(7-nitro-2,1,3-benzoxadiazol-4-yl)amino]-D-glucose), which is a fluorescent non-metabolizable glucose analog, to measure glucose uptake within melanoma cells. The flow cytometry histogram analysis-calculated bar graph showed that the level of glucose in SN and CAP slightly decreased (~80%) compared to the control. Additionally, the combination of CAP + SN drastically significantly reduced the glucose uptake level (~35%) (Figure 3a,b).

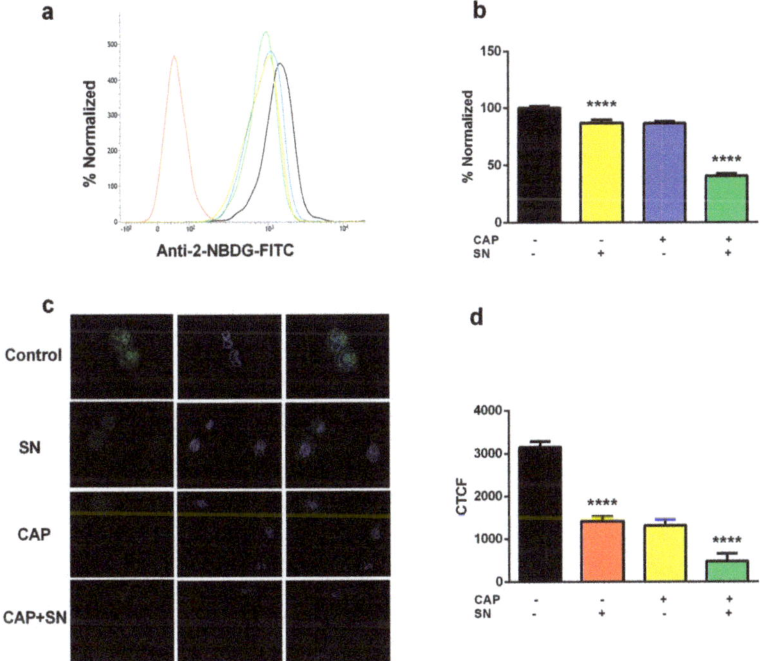

Figure 3. Estimation of intracellular glucose (ATP level) using fluorescent glucose analog 2-NBDG (**a**) employing flow cytometry in a bar graph; (**b**) calculated values in terms of the bar graph; (**c**) images were taken using confocal microscopy (intensity of green color represents the presence of glucose and blue indicates the nucleus); (**d**) captured images were expressed terms of Corrected Total Cell Fluorescence (CTCF) in a bar graph form.

Data are the mean ± SE of three experiments. Statistical analysis was performed using a one-way ANOVA test followed by Dunnett's test for comparisons. Each asterisk represents statistical differences between the treatment and control (**** $p < 0.0001$).

The glucose uptake was also represented by confocal microscopy imaging, where the expression of glucose was shown in green and the nucleus was counterstained with blue (4′,6-diamidino-2-phenylindole (DAPI) stained). The signal of glucose (Green) was recorded and the Corrected Total Cell Fluorescence (CTCF) was recorded using Image J software (Figure 3c). The imaging data showed a significant reduction of glucose uptake in SN and CAP samples compared to the control. The combination group-calculated fluorescence value further declined significantly, which proves that a CAP and SN combination can decrease the glucose uptake level in melanoma (Figure 3d).

2.4. PI3K Lead mTOR and EGF Signaling Arrest

mTOR is an important gene with a key role in the regulation of metabolism and cell growth. It is present in two different complexes: *mTORC1* (sensitive to rapamycin) and *mTORC2* (phosphorylates AKT and less sensitive to rapamycin). It also acts as a sensor for the availability of nutrients and glucose [33].

According to our proposed hypothesis, CAP and SN co-treatment can block the *GFR* and *EGF* receptor, which eventually results in autophagy activation (Figure 4a). In normal conditions, *GFR* induces the classical *PI3K* pathway and leads to *mTOR*-mediated cell survival by inhibiting apoptosis. Another key receptor is the Epidermal Growth factor (EGF) receptor, which is known to be required for cell survival. *EGF* induces the expression of transcriptional factors, which further modulates autophagy.

To evaluate the molecular signaling events involved in the *mTOR* pathway, we analyzed the upstream events that regulate the *mTOR* pathway in melanoma cells exposed to CAP, SN, and CAP and SN co-treatment conditions. Our findings suggest that autophagy was well-activated by the blocking of these two classical pathways when co-treated with CAP and SN. We analyzed the lesser expression of downstream genes of both pathways when treated with CAP and SN. However, the SN-treated samples increased the level of all the *mTOR* pathway signaling genes, which means that SN treatment helps in autophagy regression and hence promotes melanoma progression. However, the level of *MEK* and *mTOR* in the SN-treated group showed a lower expression compared to the control. The level of the *PI3K* gene was significantly increased (5 times) in the SN-treated group compared to all other groups. CAP treatment reduced the level of *MEK* (0.1 times), *PI3K* (0.6 times), and *mTOR* (0.4 times) compared to the control (Figure 4b,c). However, CAP and SN treatment further decreased the expression levels of all mentioned genes, which led to the activation of autophagy. Other genes responsible for the activation of autophagy-specific transcriptional factors *EGFR* and *HRAS* were also increased significantly when they received SN treatment (Figure 4d–f). However, both genes were reduced significantly when co-treated with CAP and SN. The gene expression for *HRAS* was 0.2 fold and for *EGFR* was 0.3 fold, respectively, for the CAP and SN co-treated group compared to the control, suggesting that the activation of transcriptional factors is responsible for autophagy (Figure 4e,f).

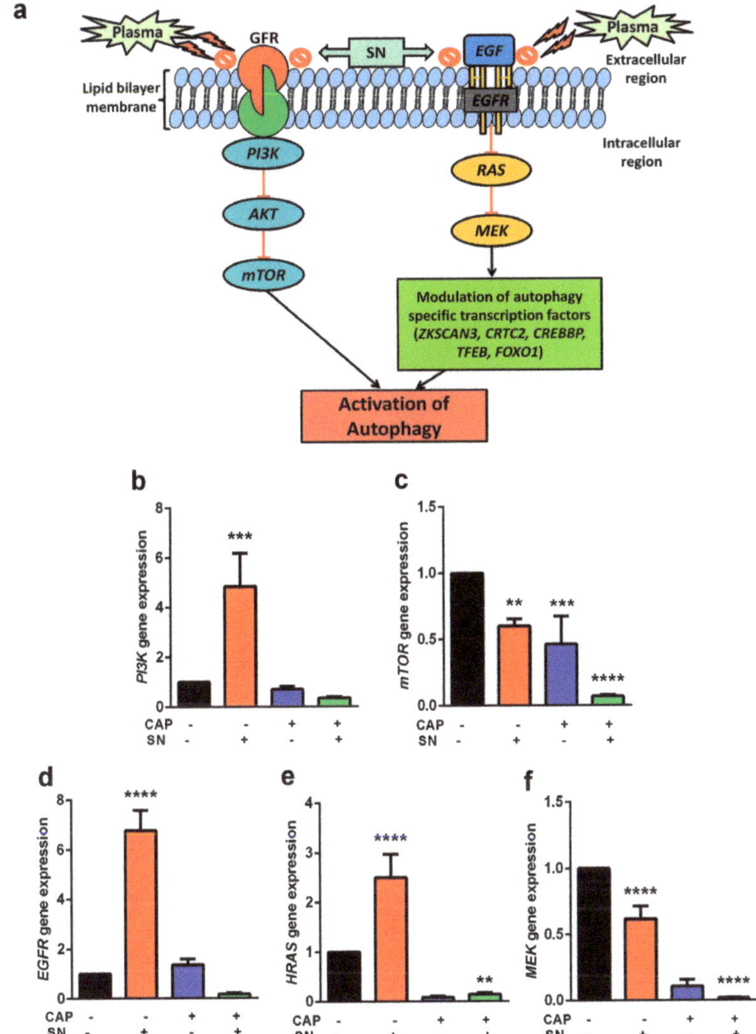

Figure 4. (**a**) Illustration of the CAP and silymarin nanoemulsion (SN)-mediated hypothesis about the blockage of *PI3K/AKT/mTOR* and *RAS/MEK* pathways for the activation of autophagy. Gene analysis of the *GFR*-mediated downstream expression of the (**b**) *PI3K* gene and (**c**) *mTOR* gene, and gene analysis of the Epidermal Growth factor (*EGF*)-mediated downstream expression of the (**d**) *EGFR* gene, (**e**) *HRAS* gene, and (**f**) *MEK* gene. Data are the mean ± SE of three experiments. Statistical analysis was performed using a one-way ANOVA test followed by Dunnett's test for comparisons. Each asterisk represents statistical differences between the treatment and control (** $p < 0.01$, *** $p < 0.001$, and **** $p < 0.0001$).

2.5. Increase in Autophagic-Related Gene Expressions and Related Transcriptional Factors

In addition to a decrease in the glucose uptake and *mTOR* and *EGFR* pathway component expressions, we also detected the expression level of genes responsible for autophagy. Autophagy-inducing gene expression was increased when receiving CAP and SN co-treatment. Autophagy is a regular process that is necessary for the normal functioning of the cell. The transcription factors responsible for autophagy were also evaluated. *ZKSCAN3* and *TFEB* are two important transcriptional factors required for autophagy [34]. The *ZKSCAN3* expression was alleviated 3.5 fold when treated with CAP alone; however, SN treatment non-significantly changed its expression level. The CAP and SN co-treated groups decreased the *ZKSCAN3* level significantly (Figure 5a). Additionally, *TFEB* gene expression revealed an increased expression level while treated with SN only, whilst CAP treatment decreased its expression level (2 times) compared to SN only treatment (7 times) (Figure 5b). The *FOXOs* family (acetylated *FOXO1*) responsible for autophagic induction by binding with *ATG7* and its downregulation is controlled by the autophagy process. Our results suggested that CAP treatment increased the *FOXO1* level 2-fold, while CAP treatment increased it 4-fold. Surprisingly, the CAP and SN co-treatment decreased the expression level up to half of the control, which is an indicator of the progression of autophagy (Figure 5c). The other transcriptional factors are *CRTC2* and *CREBBP*, which also increased while shifting to autophagy. In both transcriptional genes, SN showed a decrease (0.5 times) in the expression level compared to the control. However, CAP treatment increased the *CRTC2* level by 1.5 times and *CREBBP* by 2.2 times, respectively. CAP and SN co-treatment induced its effects and the expression level in this group exceeded 2 times in *CRTC2* and 2.5 times in the *CREBBP* gene, respectively (Figure 5d,e).

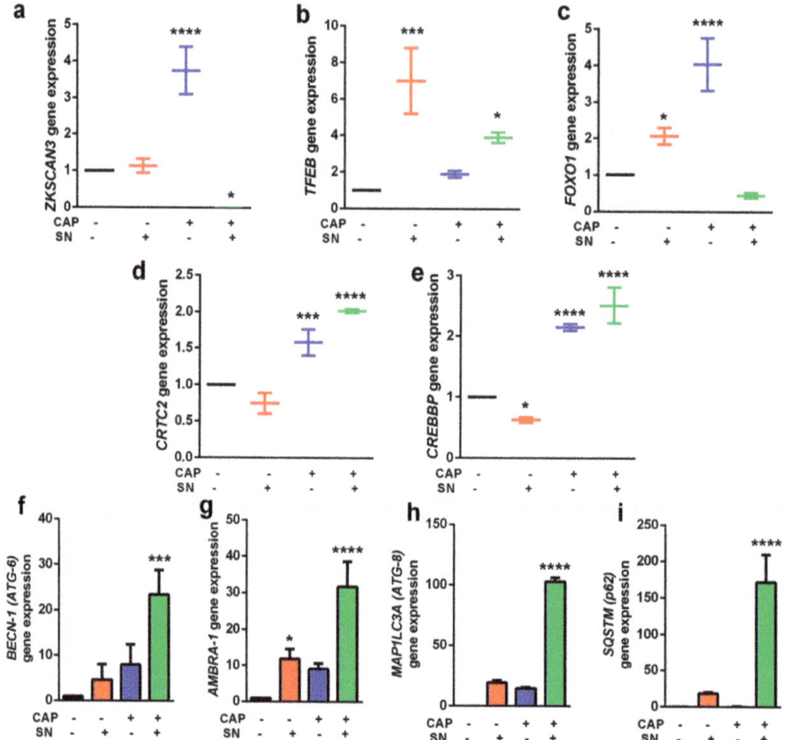

Figure 5. Gene expression levels of autophagy-related transcriptional factors: (**a**) *ZKSCAN3* gene; (**b**) *TFEB* gene; (**c**) *FOXO1* gene; (**d**) *CRTC2* gene; and (**e**) *CREBBP* gene. Expression of various gene levels directly responsible for autophagy: (**f**) *BECN-1 (ATG-6)*; (**g**) *AMBRA-1*; (**h**) *MAP1LC3A (ATG-8)*; and (**i**) *SQSTM (p62)*. Data are the mean ± SE of three experiments. Statistical analysis was performed using a one-way ANOVA test followed by Dunnett's test for comparisons. Each asterisk represents statistical differences between the treatment and control (* $p < 0.05$, *** $p < 0.001$, and **** $p < 0.0001$).

We also analyzed the expression status of autophagy-specific genes following exposure by CAP, SN, and CAP + SN in G-361 melanoma cells. *ATG-6/BECLIN-1*, which is one of the key genes, functions as *PI3K* complexes and helps catalyze autophagosomes in the process of autophagy. The SN alone and CAP alone groups increased level from 5 times to 8 times, while the combination of CAP and SN co-treatment significantly increased the *ATG-6* expression level (25 times) compared to the control (Figure 5f).

The protein associated with *ATG-6/BECLIN-1* is known as the activating molecule in *BECLIN 1*-regulated autophagy protein 1 (*AMBRA-1*) and is an important factor at the crossroad between autophagy and apoptosis. The expression level of *AMBRA-1* is directly related to the balance and conversion between autophagy and apoptosis. The CAP alone treatment increased the level of *AMBRA-1* (10 times), while the combination treatment significantly increased its level (31 times), respectively (Figure 5g). The *ATG8/LC3* family is essential for autophagosome biogenesis/ maturation and it also functions as an adaptor protein for selective autophagy. The *MAP1LC3* (microtubule-associated protein 1 light chain 3, hereafter referred to as *LC3*) is a homolog of *ATG8*, which is one of the core proteins present in the conjugation system that helps in elongation and maturation of the autophagosome. The level of *MAP1LC3* is negligible in the control group; however, the CAP and SN group significantly increased the level of the gene by up to 100 times (Figure 5h). Another classical receptor for autophagy is *p62/ SQSTM1*, which is involved in the proteasomal degradation of ubiquitinated proteins. The gene expression level of *SQSTM1* showed a significant increase (~160 times) in the CAP and SN co-treatment group, while it was completely undetectable in the control and CAP only group samples (Figure 5i).

2.6. Autophagy Induction in Human G-361 Melanoma Cells

Whether the activation of functional autophagy represents beneficial or destructive strategies for cells under metabolic stress is still questionable [35–37].

One of the classical approaches to inducing autophagy is the inhibition of *mTOR* by rapamycin or other signaling pathways [38,39], which may be regulated by the pH [40]. The autophagy flux was determined by immunocytochemistry (ICC) using a confocal microscope, where we kept the positive control-treated G-361 cells with Rapamycin and Chloroquine. The ICC image showed an increase in image intensity in the positive control group, which confirmed autophagy, and a decrease in the control group (Figure 6a,b). The CTCF values calculated by Image J exhibited maximum fluorescence in the positive control group (825.90), confirming maximum autophagy. The control group revealed a small amount of autophagy (219.57), which was increased in the SN only (291.87) and CAP only (378.94) groups. The autophagic image expression of CAP and SN co-treatment revealed the most fluorescence values (508.39), which confirmed the autophagy process. The autophagy was also investigated using flow cytometry using FITC-labeled anti-LC3B dye which specifically binds autolysosomes (LC3B only) and this revealed the same pattern of an increase in autophagy. The quantitative data from flow cytometry was converted into a histogram (Figure 6c,d) and showed an increase in autophagy when receiving SN only (2 times). The CAP only group showed comparative results that were slightly increased. Additionally, the CAP and SN co-treatment revealed a 2.5 times increase in autophagy compared to the control.

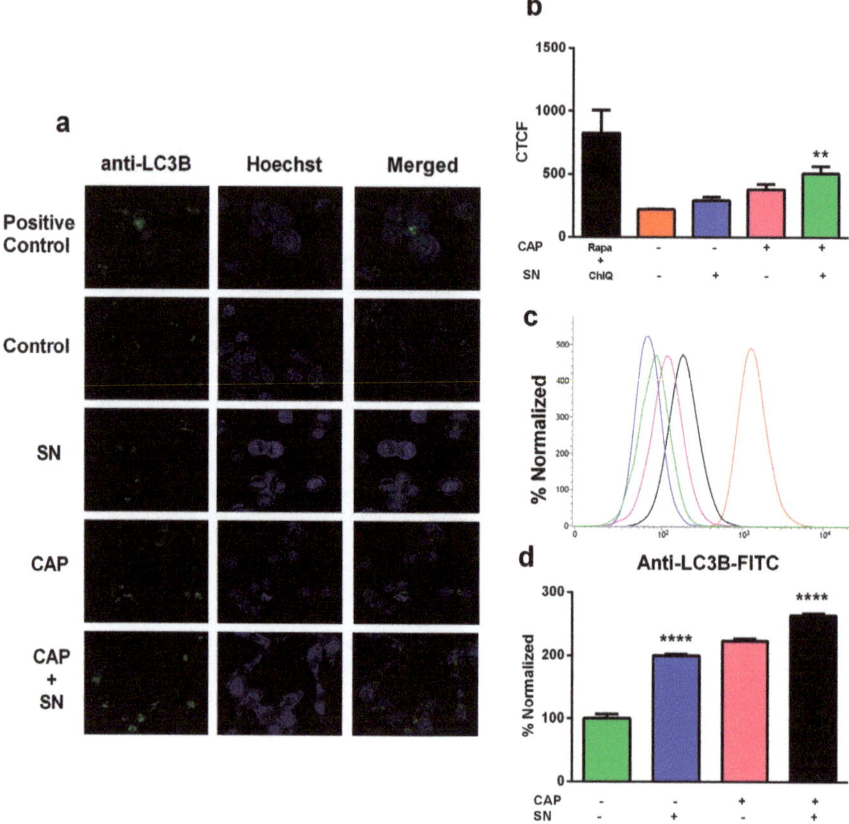

Figure 6. Interpretation of autophagic flux using a kit (Cyto ID®autophagy detection kit) (**a**) by confocal microscopy (green dots represent autophagic sites within G-361 cells and blue stains signify nuclei); (**b**) autophagy image intensities were expressed in terms of CTCF values using bar graphs; (**c**) flow cytometry evaluation of autophagy determination using anti-LC3B FITC-labeled dye; (**d**) FITC expression for autophagy was expressed in terms of a bar graph. Data are the mean ± SE of three experiments. Statistical analysis was performed using a one-way ANOVA test followed by Dunnett's test for comparisons. Each asterisk represents statistical differences between the treatment and control (** $p < 0.01$ and **** $p < 0.0001$).

3. Discussion

Autophagy is one of the most important mechanisms for the breakdown and reprocessing of long-lived proteins and their aggregates, along with organelles and intracellular pathogens [41]. It helps in maintaining cellular homeostasis in normal conditions by "flushing-out" unwanted/damaged intracellular moieties with the help of lysosomes, providing a potent tool for protection against the accumulation of toxic cellular components [42]. Hence, autophagy is considered as having an important role in various stress stimuli, including several human malignancies and cancer [39,43]. It is recognized that autophagy has a dual role in cancer; the dynamics by which this catabolic process either suppresses or promotes cancer still remain complex and are being debated [35,44]. Efforts to unravel cellular and molecular events that might induce autophagy processes in cancer cells have started to emerge [45,46].

In this article, we have reported that human melanoma G-361 cells treated and incubated with SN (100 nM) for 4 h prior to 180 s CAP treatment showed an increase in cellular stress by a decrease in extracellular pH and intracellular ATP (glucose uptake). The main reason for this is because the decrease in *AKT* phosphorylation due to the CAP and SN co-treated-induced blockage of the *PI3K/AKT* pathway leads to a low pH and reduction in glucose uptake [47], which was evident from a reduction in the 2-NBDG level in G-361 cells [48,49]. A major consequence of abnormal tumor metabolism is the accumulation of metabolic acids within the tumor microenvironment, characterized by a low extracellular pH which is distributed within the tumor mass [50,51]. Figure 2a,b shows that CAP treatment did not alter the extracellular temperature and ranged from 32.46 °C to 33.05 °C (180 s treatment), while the level of extracellular pH decreased as the CAP treatment time increased from 7.46 to 6.87 (180 s treatment). A decrease in extracellular pH helps in the glucose deprivation and de-activation of some crucial cellular pathways and hence cell death [52].

Although autophagy induction mainly occurs in the cytoplasm, it is virtually initiated in the nucleus as part of a transcriptional program controlling lysosome biogenesis/function, mediated by the *MIT/TFE* transcription factors [53]. Our study indicates that the CAP and SN combination dysregulated autophagy-related genes, including *TFEB* and *ZKSCAN3*, which have important functions in connecting *BRAF* signaling to autophagy-lysosome-mediated catabolism in melanoma [54,55]. We have presented evidence for the inactivation of *TFEB* and *ZKSCAN3* by the co-effect of CAP and SN and blockage of the *mTOR/EGFR* pathway that provides a compelling model to explain the suppression of melanoma and resultant autophagy-lysosome activation. In a broader context, the proposed model suggests a novel mechanism by which the loss of signaling through *ZKSCAN*, *TFEB*, and *FOXO1* results in activation of the autophagy-lysosome pathway in tumor suppression. *ZKSCAN3* is present inside the nucleus and translocates to the cytoplasm in the case of autophagy induction and increases its expression by alleviating its repression of *ATG* genes, which leads to a decreased autophagy process. It activates the transcriptional factor *TFEB*, which later dephosphorylates and translocates back to the nucleus [34]. Hence, the *TFEB* decrease expression level alleviates the autophagy process, which is also evident from our results, where the combination group expressed their levels significantly. Approximately 40–60% of melanoma is due to a mutation in the *BRAF* gene that promotes *MEK-MAPK* and *PI3K-AKT/mTOR* pathway activation and melanoma proliferation [56,57]. Our hypothesis for this study suggested that CAP and SN co-effectively blocked the classical *PI3K/mTOR* and *EGF/MEK* survival pathways, which will eventually lead to the activation of autophagy in G-361 human melanoma. As the results suggested (Figure 4a,b), the gene expression level of *PI3K* in the SN-treated group showed a significant increase, while the CAP alone and combination group exhibited a decreased *PI3K* expression. *mTOR* suppression is a classical pathway inhibition for autophagy induction [58] (Figure 4c) and is also evident from the results. Another pathway (*EGF/MEK*) also showed the significantly highest inhibition in its downstream process (Figure 4a) while receiving CAP and SN in combination. A plausible reason for this is that CAP generates multiple RONS and when it reacts with cell media and reaches cells, it forms many secondary reactive species that react and damage DNA and cytoplasmic *ATM*. This follows the activation of *p53* and *PARP*, which leads to autophagy induction [59].

It was proposed in our previous publication [11] that CAP and SN co-effectively inhibited *BRAF* gene expression in G-361 melanoma cells and hence repressed the mentioned survival pathways which lead to cell death. To address the above mechanism, we also checked the critical role of the other two transcriptional factors *CRTC2* and *CREBBP* [60,61], which were upregulated when receiving CAP and SN and directly related to autophagy induction. To confirm our results, we focused on autophagy-related biomarkers and checked the gene expression in all four treated groups. *AMBRA1* is a main key component of *BECN-1 (ATG-6)* and primarily involved in the formation of *PI3K*-rich membranes during the nucleation phase of autophagy [62,63]. As an initiating regulatory protein of autophagy, *AMBRA1* represents a potential marker of autophagy induction [64]. Our results indicated that treatment with SN and CAP individually expressed the *BECN-1* and *AMBRA-1* expression level, but CAP with SN treatment in combination increased *BECN-1* (25 times) and *AMBRA-1* (22 times) significantly compared to the control sample. Hence, combination treatment of CAP and SN helps in autophagic initiation via the activation of these two genes. *AMBRA-1, LC-3A,* and *SQSTM (p62)* are major biologically distinct markers of autophagic expression [65,66]. The lysosome is a fused and formed autophagosome which results in the transition of *LC3-I* to *LC3-II*, which is a gold standard for autophagy detection [67] and was checked by autophagy in our present work. The gene expression level of *MAP1LC3A*, which also stands for *ATG-8* or *LC3-II*, showed an increase in its expression level while receiving CAP and SN individually, while CAP and SN combination treatment significantly elevated the level (100 times) and clearly indicated the presence of autophagy induction in combination-treated melanoma groups. This can be attributed due to the activation of autophagy transcriptional factors (i.e., *FOXO*), which upregulated the genes *NIX* and *BNIP3* that are directly related to autophagy induction [34]. Additionally, the blocking of survival pathways *AKT* and *ERK* leads to a lower expression level of *mTOR*, which helps in autophagy induction [47].

4. Materials and Methods

4.1. Chemicals and Antibodies

RPMI-1640, phosphate-buffered saline, and a penicillin-streptomycin antibody cocktail solution were obtained from Welgene, Daegu, Korea. Isoropanol and 4′,6-diamidino-2-phenylindole (DAPI) were purchased from Thermofisher, Seoul, Korea. Diethyl pyrocarbonate (DEPC) was obtained from Biosesang, Gyeonggi, Korea. Trypsin (0.25%) with EDTA and fetal bovine serum was obtained from Hyclone, Seoul, Korea (GE Healthcare Life Sciences, Pittsburgh, PA, USA). The ReverTra Ace®qPCR RT Master Mix c-DNA synthesis kit was purchased from Toyobo, Japan. The SYBR®Green Master mix was procured from Bio-Rad, Seoul, Korea. All the primers for q-PCR studies were obtained from Searchbio, Gyeonggi, Korea. 2-(*N*-(7-Nitrobenz-2-oxa-1,3-diazol-4-yl)amino)-2-deoxyglucose (2-NBDG) was purchased from Biovision, Milpitas, CA, USA and 2′,7′-bis-(2-carboxyethyl)-5-(and-6)-carboxyfluorescein (BCECF-AM) was procured from Santacruz biotechnology, Dallas, TX, USA. The Cyto-ID®autophagy detection kit was purchased from Enzo lifesciences, Farmingdale, NY, USA.

4.2. Cell Culture

G-361 human melanoma cells were cultured in RPMI-1640 cell culture media supplemented with 10% fetal bovine serum, 100 U/mL penicillin, and 100 mg/mL streptomycin cocktail and were maintained at 37 °C in a 5% humidified CO_2 environment. The cells were routinely tested for mycoplasma (MycoAlert™ Mycoplasma Detection Kit, Lonza, Basel, Switzerland). The passage number for G-361 cells was kept constant (P-42) at the time of experiments. The experiments were repeated three times ($n = 3$) using identical conditions.

4.3. Characterization of the μ-Dielectric Barrier Discharge (μ-DBD) Plasma System Device

A schematic of the μ-DBD source used in this experiment is shown in Figure 1. The μ-DBD plasma source was fabricated using silver electrodes, which were placed in a coplanar configuration. These electrodes, with a width = 100 μm and thickness = 5 μm, were fabricated above the circular glass substrate (SiO_2) with a diameter and thickness of 35 mm and 1.8 mm, respectively. The spacing between adjacent silver electrodes on each plane was maintained at 2 mm. In order to prevent the arcing, an additional dielectric layer made up of SiO_2 (thickness: 50 μm) was coated above the electrodes. Moreover, a hydration prevention layer made up of alumina (thickness: 1 μm) was coated with SiO_2 to prevent the deposition of water molecules above the dielectric surface. This fabrication process was done by using the technique of photolithography. More details on a similar plasma device can be found elsewhere. One part of the electrode system was connected to the high-voltage power supply and the other part was grounded. In order to operate the discharge for a long time and prevent the heating of electrodes, the discharge was operated in dimming mode [on time (T_{on}) = 18.77 ms and off time (T_{off}) = 232.82 ms] by using a DC-AC inverter. The working gas was air [flow rate: 1.5 liters per minute (lpm)] and it was injected from two directions with a cooling fan to reduce the heating.

A photograph of the μ-DBD discharge operated by air is shown in Figure 1a, b. This plasma source was used to treat G-361 cells for 180 s.

4.4. Estimation of Extracellular pH and Temperature and Number of Melanoma Cells

RPMI-1640 serum-free media were placed in 48-well plates (1 mL: per well) and exposed to μ-DBD plasma using air as the feeder gas for 30, 60, 180, and 300 s. After exposure, the temperature and pH of the media were measured immediately in triplicate with an infrared (IR) camera (Fluke Ti100 Series Thermal Imaging Cameras, Everett, WA, USA) and pH meter (Eutech Instruments, Singapore). The G-361 cells were counted and DIC images (Leica, DMi8, Wetzlar, Germany) were taken after different CAP exposure times.

4.5. Glucose Uptake

G-361 cells after culture were treated with SN (100 nM), CAP (180 s), and a combination of SN + CAP and kept for the next 24 h in a CO_2 incubator. After incubation, G-361 cells were incubated with 100 μM 2-NBDG, which is a fluorescent derivative of 2-deoxy-D-glucose, for 45 min. The cells were trypsinized, collected, and washed with PBS and then analyzed by flow cytometry to collect green fluorescence. In a separate experiment, G-361 cells were fixed and permeabilized separately using 10% ethanol and 3.7% paraformaldehyde. They were then incubated with DAPI for 10 min to counterstain the nucleus, along with 2-NBDG. The fluorescence staining intensity and intercellular locations were examined using an Olympus IX83-FP confocal microscope (Tokyo, Japan).

4.6. qPCR Gene Analysis (Autophagy Related and Transcriptional Factor)

Any alterations in the cellular level are associated with changes in the gene expression at the mRNA level. To clarify this at molecular levels, we identified genes responsible for the induction of *mTOR*-mediated autophagy and related transcriptional factors and checked their levels by the production of plasma-generated reactive species. The desired primers for *mTOR* and *RAS/MEK*-mediated autophagy and its relevant transcriptional factors were designed and synthesized. G-361 melanoma cells were grown in Petri plates with a size of 35mm^2 and treated with CAP, SN, and CAP + SN, respectively, as described before. After incubation for 24 h, the cells were trypsinized and collected for RNA isolation using Trizol (Invitrogen, Carlsbad, CA, USA), after which q-PCR was performed with a Biorad 2X SYBR green mix (Biorad, Seoul, Korea). Reactions were carried out in a Biorad thermal cycler (Biorad, Seoul, Korea), and the results were expressed as the fold change calculated with the $2^{-\Delta\Delta Ct}$ method relative to a control sample. Meanwhile, *18S r-RNA* was used as an internal normalization control. All primers were purchased from Searchbio, Gyeonggi, Korea. Quantitative real-time PCR was performed according to the forward and reverse primer sequences listed in Supplementary Tables S1 and S2.

4.7. Evaluation of Autophagy by Immunocytochemistry (ICC), Flow Cytometry, and ELISA

G-361 melanoma cells were used to monitor autophagy fluorescence (a cyto-ID®autophagy detection kit specifically measures autophagic vacuoles and autophagic flux). To get insight into the functional role of autophagy in the mentioned groups, we analyzed autophagy in human melanoma using the Cyto ID autophagy detection kit, which specifically tags the formation of autophagosomes (LC3B formation). This kit contains the anti-LC3B (anti LC-3II) antibody which stains autophagic sites and CTCF values were determined using image-J software. The melanoma cells were plated in 35 mm^2 Petri dishes and treated with CAP, SN, and CAP + SN for 24 h, respectively, as per the standard experimental procedures mentioned above. Rapamycin (500 nM) and chloroquine (10 μM) were added to positive control samples and acted as autophagy inducers. After 18 h with autophagy inducers and 24 h with CAP, SN, and CAP + SN, the media was replaced and washed with 1× assay buffer (in 5% FBS), which was further kept in 1 mL 1× assay buffer (without FBS). The cells were stained with Cyto-ID®green detection reagent and counterstained with Hoechst 33342. The autophagy was determined by using an Olympus IX83-FP confocal microscope (Tokyo, Japan) and flow cytometry was performed by BD FACSVerse using the FACS suite software (Becton Dickinson and Co., Franklin Lakes, NJ, USA).

4.8. Statistical Analysis

All of the data are expressed as the mean ± SE of triplicates. The statistical software GraphPad Prism 6.0 (Microsoft Corporation, Redmond, WA, USA) was used for statistical analysis via one-way analysis of variance (ANOVA) followed by Dunnett's test.

5. Conclusions

Our present study delivers further support to therapeutic strategies targeting autophagy to kill cancer cells. This article suggests a plausible method of killing human melanoma cells using the same conditions and shows that autophagy plays a vital role in the eradication of melanoma by using CAP and SN. G-361 melanoma cell reduction was produced because of a decrease in the extracellular pH, which results in a reduction in the intracellular glucose level with the inhibition of *mTOR* and *EGF* survival pathways and leads to autophagy activation. However, more studies are needed to dissect the mechanisms that regulate autophagy in aggressive and different melanoma cells for the development of therapeutic strategies that will benefit patient outcomes.

Supplementary Materials: Supplementary materials can be found at http://www.mdpi.com/1422-0067/21/6/1939/s1.

Author Contributions: Conceptualization, methodology, writing—original draft preparation, investigation, and validation, M.A. and B.A.; software, M.A.; formal analysis, S.B.; resources, E.H.C.; data curation, M.A., B.A., and B.G.; writing—review and editing, M.A., B.A., and S.B.; visualization, B.A.; supervision, S.B. and E.H.C.; project administration, E.H.C.; funding acquisition, E.H.C. All authors have read and agreed to the published version of the manuscript.

Funding: This research was funded by the National Research Foundation of Korea (NRF), grant number ICT NRF-2016K1A4A3914113.

Acknowledgments: This work was also supported by Kwangwoon University in 2019-2020. The authors' give special thanks to the Department of Pharmaceutics, Jamia Hamdard, Delhi, for providing silymarin nanoemulsion (SN).

Conflicts of Interest: The authors declare no conflict of interest.

Abbreviations

CAP	Cold Atmospheric Plasma
SN	Silymarin Nanoemulsion
RONS	Reactive Oxygen and Nitrogen Species
PI3K	Phosphatidyl Inositol 3 Kinase
mTOR	Mammalian Target of Rapamycin
MEK	Mitogen-Activated Protein Kinase
ATP	Adenosine Triphosphate
ZKSCAN3	Zinc Finger with KRAB And SCAN Domains 3
TFEB	Transcription Factor EB
FOXO1	Forkhead Box Protein O1
CRTC2	cAMP-Response Element Binding Protein Regulated Transcription Coactivator 2
CREBBP	cAMP-Response Element Binding Protein
BECN-1	Beclin 1
AMBRA-1	Autophagy and Beclin 1 Regulator 1
MAP1LC3A	Microtubule Associated Protein 1 Light Chain 3 Alpha
SQSTM	Sequestosome-1
EGFR	Epidermal Growth Factor Receptor

References

1. Yamaguchi, Y.; Hearing, V.J. Melanocytes and their Diseases. *Cold Spring Harb. Perspect. Med.* **2014**, *4*. [CrossRef]
2. Ascierto, P.A.; Kirkwood, J.M.; Grob, J.J.; Simeone, E.; Grimaldi, A.M.; Maio, M.; Palmieri, G.; Testori, A.; Marincola, F.M.; Mozzillo, N. The Role of BRAF V600 Mutation in Melanoma. *J. Transl. Med.* **2012**, *10*, 85. [CrossRef]
3. Ndoye, A.; Weeraratna, A.T. Autophagy—An Emerging Target for Melanoma Therapy. *F1000Res* **2016**, *5*. [CrossRef]
4. Lugowska, I.; Teterycz, P.; Rutkowski, P. Immunotherapy of Melanoma. *Contemp. Oncol. (Pozn)* **2018**, *22*, 61–67. [CrossRef]
5. Chakraborty, R.; Wieland, C.N.; Comfere, N.I. Molecular Targeted Therapies in Metastatic Melanoma. *Pharmgenomics Pers. Med.* **2013**, *6*, 49–56.
6. Gay-Mimbrera, J.; Garcia, M.C.; Isla-Tejera, B.; Rodero-Serrano, A.; Garcia-Nieto, A.V.; Ruano, J. Clinical and Biological Principles of Cold Atmospheric Plasma Application in Skin Cancer. *Adv. Ther.* **2016**, *33*, 894–909. [CrossRef]
7. Zhang, J.; Guo, L.; Chen, Q.; Zhang, K.; Wang, T.; An, G.; Zhang, X.; Li, H.; Ding, G. Effects and Mechanisms of Cold Atmospheric Plasma on Skin Wound Healing of Rats. *Contrib. Plasma Phy.* **2019**, *59*, 92–101. [CrossRef]
8. Ranjan, R.; Krishnamraju, P.V.; Shankar, T.; Gowd, S. Nonthermal Plasma in Dentistry: An Update. *J. Int. Soc. Prev. Community Dent.* **2017**, *7*, 71–75.
9. Öngel, C.; Keleş, M.; Acar, E.; Birer, Ö. Atmospheric Pressure Plasma Jet Treatment of Human Hair Fibers. *J. Bio. Tribo. Corros.* **2015**, *1*, 1–10. [CrossRef]
10. Yan, D.; Sherman, J.H.; Keidar, M. Cold Atmospheric Plasma, a Novel Promising Anti-Cancer Treatment Modality. *Oncotarget* **2017**, *8*, 15977–15995. [CrossRef]
11. Adhikari, M.; Kaushik, N.; Ghimire, B.; Adhikari, B.; Baboota, S.; Al-Khedhairy, A.A.; Wahab, R.; Lee, S.J.; Kaushik, N.K.; Choi, E.H. Cold Atmospheric Plasma and Silymarin Nanoemulsion Synergistically Inhibits Human Melanoma Tumorigenesis Via Targeting HGF/C-MET Downstream Pathway. *Cell. Commun. Signal.* **2019**, *17*, 52–54. [CrossRef] [PubMed]
12. Schmidt, A.; Bekeschus, S.; Jablonowski, H.; Barton, A.; Weltmann, K.D.; Wende, K. Role of Ambient Gas Composition on Cold Physical Plasma-Elicited Cell Signaling in Keratinocytes. *Biophys. J.* **2017**, *112*, 2397–2407. [CrossRef] [PubMed]
13. Klochkov, S.G.; Neganova, M.E.; Nikolenko, V.N.; Chen, K.; Somasundaram, S.G.; Kirkland, C.E.; Aliev, G. Implications of Nanotechnology for the Treatment of Cancer: Recent Advances. *Semin. Cancer Biol.* **2019**. [CrossRef] [PubMed]
14. Kaushik, N.K.; Kaushik, N.; Yoo, K.C.; Uddin, N.; Kim, J.S.; Lee, S.J.; Choi, E.H. Low Doses of PEG-Coated Gold Nanoparticles Sensitize Solid Tumors to Cold Plasma by Blocking the PI3K/AKT-Driven Signaling Axis to Suppress Cellular Transformation by Inhibiting Growth and EMT. *Biomaterials* **2016**, *87*, 118–130. [CrossRef]
15. Vargas-Mendoza, N.; Madrigal-Santillan, E.; Morales-Gonzalez, A.; Esquivel-Soto, J.; Esquivel-Chirino, C.; Garcia-Luna Y Gonzalez-Rubio, M.; Gayosso-de-Lucio, J.A.; Morales-Gonzalez, J.A. Hepatoprotective Effect of Silymarin. *World J. Hepatol.* **2014**, *6*, 144–149. [CrossRef]
16. Ramasamy, K.; Agarwal, R. Multitargeted Therapy of Cancer by Silymarin. *Cancer Lett.* **2008**, *269*, 352–362. [CrossRef]
17. Kaushik, N.K.; Kaushik, N.; Linh, N.N.; Ghimire, B.; Pengkit, A.; Sornsakdanuphap, J.; Lee, S.J.; Choi, E.H. Plasma and Nanomaterials: Fabrication and Biomedical Applications. *Nanomaterials (Basel)* **2019**, *9*, 98. [CrossRef]
18. Bogaerts, A.; Yusupov, M.; Razzokov, J.; Van der Paal, J. Plasma for Cancer Treatment: How can RONS Penetrate through the Cell Membrane? Answers from Computer Modeling. *Front. Chem. Sci. Eng.* **2019**, *13*, 253–263. [CrossRef]
19. Pai, K.; Timmons, C.; Roehm, K.D.; Ngo, A.; Narayanan, S.S.; Ramachandran, A.; Jacob, J.D.; Ma, L.M.; Madihally, S.V. Investigation of the Roles of Plasma Species Generated by Surface Dielectric Barrier Discharge. *Sci. Rep.* **2018**, *8*, 16674. [CrossRef]

20. Cheng, X.; Rajjoub, K.; Sherman, J.; Canady, J.; Recek, N.; Yan, D.; Bian, K.; Murad, F.; Keidar, M. Cold Plasma Accelerates the Uptake of Gold Nanoparticles into Glioblastoma Cells. *Plasma Processes Polym.* **2015**, *12*, 1364–1369. [CrossRef]

21. Wong, D.J.; Ribas, A. Targeted Therapy for Melanoma. *Cancer Treat. Res.* **2016**, *167*, 251–262.

22. Grossman, D.; Altieri, D. Drug Resistance in Melanoma: Mechanisms, Apoptosis, and New Potential Therapeutic Targets. *Cancer Metastasis Rev.* **2001**, *20*, 3–11. [CrossRef]

23. Green, D.R.; Llambi, F. Cell Death Signaling. *Cold Spring Harb. Perspect. Biol.* **2015**, *7*. [CrossRef]

24. Mak, T.W.; Okada, H. Pathways of Apoptotic and Non-Apoptotic Death in Tumour Cells. *Nat. Rev. Cancer* **2004**, *4*, 592–603.

25. Yun, C.W.; Lee, S.H. The Roles of Autophagy in Cancer. *Int. J. Mol. Sci.* **2018**, *19*, 3466. [CrossRef]

26. Lorin, S.; Hamai, A.; Mehrpour, M.; Codogno, P. Autophagy Regulation and its Role in Cancer. *Semin. Cancer Biol.* **2013**, *23*, 361–379. [CrossRef]

27. Das, G.; Shravage, B.V.; Baehrecke, E.H. Regulation and Function of Autophagy during Cell Survival and Cell Death. *Cold Spring Harb. Perspect. Biol.* **2012**, *4*. [CrossRef]

28. Yen, W.L.; Klionsky, D.J. How to Live Long and Prosper: Autophagy, Mitochondria, and Aging. *Physiology (Bethesda)* **2008**, *23*, 248–262. [CrossRef]

29. Mathew, R.; Karantza-Wadsworth, V.; White, E. Role of Autophagy in Cancer. *Nat. Rev. Cancer.* **2007**, *7*, 961–967. [CrossRef]

30. Ghimire, B.; Szili, E.J.; Lamichhane, P.; Short, R.D.; Lim, J.S.; Attri, P.; Masur, K.; Weltmann, K.; Hong, S.; Choi, E.H. The Role of UV Photolysis and Molecular Transport in the Generation of Reactive Species in a Tissue Model with a Cold Atmospheric Pressure Plasma Jet. *App. Phys. Lett.* **2019**, *114*, 93701. [CrossRef]

31. Ghimire, B.; Sornsakdanuphap, J.; Hong, Y.J.; Uhm, H.S.; Weltmann, K.; Choi, E.H. The Effect of the Gap Distance between an Atmospheric-Pressure Plasma Jet Nozzle and Liquid Surface on OH and N2 Species Concentrations. *Phys. Plasmas* **2017**, *24*, 73502. [CrossRef]

32. Ghimire, B.; Lamichhane, P.; Lim, J.S.; Min, B.; Paneru, R.; Weltmann, K.; Choi, E.H. An Atmospheric Pressure Plasma Jet Operated by Injecting Natural Air. *Appl. Phys. Lett.* **2018**, *113*, 194101. [CrossRef]

33. Papadopoli, D.; Boulay, K.; Kazak, L.; Pollak, M.; Mallette, F.; Topisirovic, I.; Hulea, L. mTOR as a Central Regulator of Lifespan and Aging. *F1000Resesrch* **2019**, *8*. [CrossRef]

34. Füllgrabe, J.; Ghislat, G.; Cho, D.; Rubinsztein, D.C. Transcriptional Regulation of Mammalian Autophagy at a Glance. *J. Cell Sci.* **2016**, *129*, 3059–3066. [CrossRef]

35. Mathew, R.; White, E. Autophagy in Tumorigenesis and Energy Metabolism: Friend by Day, Foe by Night. *Curr. Opin. Genet. Dev.* **2011**, *21*, 113–119. [CrossRef]

36. Rabinowitz, J.D.; White, E. Autophagy and Metabolism. *Science* **2010**, *330*, 1344–1348. [CrossRef]

37. Song, J.; Guo, X.; Xie, X.; Zhao, X.; Li, D.; Deng, W.; Song, Y.; Shen, F.; Wu, M.; Wei, L. Autophagy in Hypoxia Protects Cancer Cells Against Apoptosis Induced by Nutrient Deprivation through a Beclin1-Dependent Way in Hepatocellular Carcinoma. *J. Cell. Biochem.* **2011**, *112*, 3406–3420. [CrossRef]

38. Meijer, A.J. Amino Acid Regulation of Autophagosome Formation. *Methods Mol. Biol.* **2008**, *445*, 89–109.

39. Meijer, A.J.; Codogno, P. Autophagy: Regulation and Role in Disease. *Crit. Rev. Clin. Lab. Sci.* **2009**, *46*, 210–240. [CrossRef]

40. Balgi, A.D.; Diering, G.H.; Donohue, E.; Lam, K.K.Y.; Fonseca, B.D.; Zimmerman, C.; Numata, M.; Roberge, M. Regulation of mTORC1 Signaling by pH. *PLoS ONE* **2011**, *6*, e21549. [CrossRef]

41. Orvedahl, A.; Levine, B. Eating the Enemy within: Autophagy in Infectious Diseases. *Cell Death Differentiation* **2009**, *16*, 57–69. [CrossRef]

42. Mizushima, N. Autophagy: Process and Function. *Genes Dev.* **2007**, *21*, 2861–2873. [CrossRef]

43. Thorburn, A. Autophagy and its Effects: Making Sense of Double-Edged Swords. *PLoS Biol.* **2014**, *12*, e1001967. [CrossRef]

44. Singh, S.S.; Vats, S.; Chia, A.Y.; Tan, T.Z.; Deng, S.; Ong, M.S.; Arfuso, F.; Yap, C.T.; Goh, B.C.; Sethi, G.; et al. Dual Role of Autophagy in Hallmarks of Cancer. *Oncogene* **2018**, *37*, 1142–1158. [CrossRef]

45.] Mowers, E.E.; Sharifi, M.N.; Macleod, K.F. Autophagy in Cancer Metastasis. *Oncogene* **2017**, *36*, 1619–1630. [CrossRef]

46. Dower, C.M.; Wills, C.A.; Frisch, S.M.; Wang, H.G. Mechanisms and Context Underlying the Role of Autophagy in Cancer Metastasis. *Autophagy* **2018**, *14*, 1110–1128. [CrossRef]

47. Shepherd, R.M.; Henquin, J.C. The Role of Metabolism, Cytoplasmic Ca2+, and pH-Regulating Exchangers in Glucose-Induced Rise of Cytoplasmic pH in Normal Mouse Pancreatic Islets. *J. Biol. Chem.* **1995**, *270*, 7915–7921. [CrossRef]

48. Zheng, J. Energy Metabolism of Cancer: Glycolysis Versus Oxidative Phosphorylation (Review). *Oncol. Lett.* **2012**, *4*, 1151–1157. [CrossRef]

49. Zou, C.; Wang, Y.; Shen, Z. 2-NBDG as a Fluorescent Indicator for Direct Glucose Uptake Measurement. *J. Biochem. Biophys. Methods* **2005**, *64*, 207–215. [CrossRef]

50. Gatenby, R.A.; Gillies, R.J. Why do Cancers have High Aerobic Glycolysis? *Nat. Rev. Cancer* **2004**, *4*, 891–899. [CrossRef]

51. Gillies, R.J.; Raghunand, N.; Karczmar, G.S.; Bhujwalla, Z.M. MRI of the Tumor Microenvironment. *J. Magnetic Res. Imag.* **2002**, *16*, 430–450. [CrossRef]

52. Rofstad, E.K.; Mathiesen, B.; Kindem, K.; Galappathi, K. Acidic Extracellular pH Promotes Experimental Metastasis of Human Melanoma Cells in Athymic Nude Mice. *Cancer Res.* **2006**, *66*, 6699–6707. [CrossRef]

53. Nezich, C.L.; Wang, C.; Fogel, A.I.; Youle, R.J. MiT/TFE Transcription Factors are Activated during Mitophagy Downstream of Parkin and Atg5. *J. Cell Biol.* **2015**, *210*, 435–450. [CrossRef]

54. Di Malta, C.; Cinque, L.; Settembre, C. Transcriptional Regulation of Autophagy: Mechanisms and Diseases. *Front. Cell. Dev. Biol.* **2019**, *7*, 114. [CrossRef]

55. Li, S.; Song, Y.; Quach, C.; Guo, H.; Jang, G.B.; Maazi, H.; Zhao, S.; Sands, N.A.; Liu, Q.; In, G.K.; et al. Transcriptional Regulation of Autophagy-Lysosomal Function in BRAF-Driven Melanoma Progression and Chemoresistance. *Nat. Commun.* **2019**, *10*, 1693–1698. [CrossRef]

56. McCain, J. The MAPK (ERK) Pathway: Investigational Combinations for the Treatment of BRAF-Mutated Metastatic Melanoma. *P T A Peer-Rev. J. Formul. Manag.* **2013**, *38*, 96–108.

57. Caporali, S.; Alvino, E.; Lacal, P.M.; Levati, L.; Giurato, G.; Memoli, D.; Caprini, E.; Antonini Cappellini, G.C.; D'Atri, S. Targeting the PI3K/AKT/mTOR Pathway Overcomes the Stimulating Effect of Dabrafenib on the Invasive Behavior of Melanoma Cells with Acquired Resistance to the BRAF Inhibitor. *Int. J. Oncol.* **2016**, *49*, 1164–1174. [CrossRef]

58. Kapuy, O.; Vinod, P.K.; Banhegyi, G. mTOR Inhibition Increases Cell Viability Via Autophagy Induction during Endoplasmic Reticulum Stress—An Experimental and Modeling Study. *FEBS Open Bio.* **2014**, *4*, 704–713. [CrossRef]

59. Kroemer, G.; Mariño, G.; Levine, B. Autophagy and the Integrated Stress Response. *Mol. Cell* **2010**, *40*, 280–293. [CrossRef]

60. Steven, A.; Seliger, B. Control of CREB Expression in Tumors: From Molecular Mechanisms and Signal Transduction Pathways to Therapeutic Target. *Oncotarget* **2016**, *7*, 35454–35465. [CrossRef]

61. Henriksson, E.; Sall, J.; Gormand, A.; Wasserstrom, S.; Morrice, N.A.; Fritzen, A.M.; Foretz, M.; Campbell, D.G.; Sakamoto, K.; Ekelund, M.; et al. SIK2 Regulates CRTCs, HDAC4 and Glucose Uptake in Adipocytes. *J. Cell. Sci.* **2015**, *128*, 472–486. [CrossRef]

62. Sun, W.L. Ambra1 in Autophagy and Apoptosis: Implications for Cell Survival and Chemotherapy Resistance. *Oncol. Lett.* **2016**, *12*, 367–374. [CrossRef]

63. Menon, M.B.; Dhamija, S. Beclin 1 Phosphorylation—At the Center of Autophagy Regulation. *Front. Cell. Dev. Biol.* **2018**, *6*, 137. [CrossRef]

64. Tang, D.Y.; Ellis, R.A.; Lovat, P.E. Prognostic Impact of Autophagy Biomarkers for Cutaneous Melanoma. *Front. Oncol.* **2016**, *6*, 236. [CrossRef]

65. Tanida, I.; Ueno, T.; Kominami, E. LC3 and Autophagy. *Methods Mol. Biol.* **2008**, *445*, 77–88.

66. Bjørkøy, G.; Lamark, T.; Pankiv, S.; Øvervatn, A.; Brech, A.; Johansen, T. Monitoring Autophagic Degradation of p62/SQSTM1. *Method. Enzymol.* **2009**, *452*, 181–197.

67. Runwal, G.; Stamatakou, E.; Siddiqi, F.H.; Puri, C.; Zhu, Y.; Rubinsztein, D.C. LC3-Positive Structures are Prominent in Autophagy-Deficient Cells. *Sci. Rep.* **2019**, *9*, 1–14. [CrossRef]

Article

Effects of Cold Jet Atmospheric Pressure Plasma on the Structural Characteristics and Immunoreactivity of Celiac-Toxic Peptides and Wheat Storage Proteins

Fusheng Sun [1], Xiaoxue Xie [1], Yufan Zhang [1], Jiangwei Duan [2], Mingyu Ma [2], Yaqiong Wang [1], Ding Qiu [1], Xinpei Lu [2], Guangxiao Yang [1,*] and Guangyuan He [1,*]

[1] The Key Laboratory of Molecular Biophysics of Chinese Ministry of Education, The Genetic Engineering International Cooperation Base of Chinese Ministry of Science and Technology, College of Life Science and Technology, Huazhong University of Science and Technology, Wuhan 430074, China; fufu4567@126.com (F.S.); xiexiaoxue_Cecilia@163.com (X.X.); zhangyufan1108@126.com (Y.Z.); wangyaqiongyou@126.com (Y.W.); qiuding1989@outlook.com (D.Q.)

[2] State Key Laboratory of Advanced Electromagnetic Engineering and Technology, School of Electrical and Electronic Engineering, Huazhong University of Science and Technology, Wuhan 430074, China; duanjiangwei@foxmail.com (J.D.); mmyhfut@163.com (M.M.); luxinpei@hotmail.com (X.L.)

* Correspondence: ygx@hust.edu.cn (G.Y.); hegy@hust.edu.cn (G.H.); Tel.: +86-27-87792271 (G.Y. & G.H.); Fax: +86-27-87792272 (G.Y. & G.H.)

Received: 13 December 2019; Accepted: 20 January 2020; Published: 4 February 2020

Abstract: The present research reported the effects of structural properties and immunoreactivity of celiac-toxic peptides and wheat storage proteins modified by cold jet atmospheric pressure (CJAP) plasma. It could generate numerous high-energy excited atoms, photons, electrons, and reactive oxygen and nitrogen species, including O_3, H_2O_2, •OH, NO_2^- and NO_3^- etc., to modify two model peptides and wheat storage proteins. The Orbitrap HR-LC-MS/MS was utilized to identify and quantify CJAP plasma-modified model peptide products. Backbone cleavage of QQPFP and PQPQLPY at specific proline and glutamine residues, accompanied by hydroxylation at the aromatic ring of phenylalanine and tyrosine residues, contributed to the reduction and modification of celiac-toxic peptides. Apart from fragmentation, oxidation, and agglomeration states were evaluated, including carbonyl formation and the decline of γ-gliadin. The immunoreactivity of gliadin extract declined over time, demonstrating a significant decrease by 51.95% after 60 min of CJAP plasma treatment in vitro. The CJAP plasma could initiate depolymerization of gluten polymer, thereby reducing the amounts of large-sized polymers. In conclusion, CJAP plasma could be employed as a potential technique in the modification and reduction of celiac-toxic peptides and wheat storage proteins.

Keywords: cold jet atmospheric pressure plasma; reactive oxygen and nitrogen species; backbone cleavage; hydroxylation; carbonyl formation

---

1. Introduction

Gluten proteins in wheat (*Triticum aestivum* L.) are a type of storage protein-containing gliadins and glutenins, which are responsible for the rheological and viscoelastic properties of wheat dough [1]. Gluten proteins contain abundant proline (Pro) and glutamine (Gln) residues, which can cause high resistance of gluten peptides to be hydrolyzed, a property that contributes to the gluten-related immunogenic nature of celiac disease (CD) patients [2]. The degradation of gluten proteins can decrease their immunological and toxic properties. To date, some biological, chemical, and physical approaches have been applied to reduce toxicity of these proteins to patients with CD. Biological strategies include downregulation of gliadin expression by CRISPR-Cas9 to produce low-gluten wheat lines with an average reduction of 66.7% and 61.7% with regard to R5 and G12 antibody reactivity [3]. For biochemical processing, usage of food-grade microbial transglutaminase (mTG) with lysine ethyl ester is able to modify gliadin peptides, which inhibits the ability of gluten to trigger the specific human immune response in vitro [4]. Huang et al. (2017) found a metal-catalyzed oxidation system can cause the modification and elimination of hordein, leading to the decrease of immunoreactivity of hordein [5].

Although enzymatic and chemical methods possess high efficiency in the modification of macromolecular compound, shortcomings of these techniques are concerned in points of higher cost, wastewater pollution and food safety etc. Physical techniques could make up for these deficiencies of enzymatic and chemical modifications [6]. The high temperature treatment modifies protein through denaturation and formation of inter/intramolecular bonds, which consequently resulted in specific epitopes unrecognizable by antibody [7]. Also, it has been reported that ozone, excited atoms, ions and electrons can be considered as efficient species because they can act as strong oxidizing agents for the object without leaving any residues [8]. According to this principle, cold plasma is regarded as an emerging non-thermal technique to modify the characteristics of gluten in wheat flour [9]. Cold plasma, regarded as the source of reactive oxygen and nitrogen species (RONS), is an ionized gas consisting of a variety of types of active species, such as singlet oxygen, ozone, excited molecular nitrogen, and other electrons or ions [10]. Exploitation of plasma has been widely researched in nonfood areas, such as the inactivation of bacteria, wound healing, tooth bleaching, and cancer therapy [11–14]. However, whether it can modify wheat gluten proteins to reduce the toxicity of celiac disease has rarely been investigated.

At present, the common method for quantification of food prolamins is based on antibodies against the epitopes Gln-Gln-Pro-Phe-Pro (QQPFP) and Pro-Gln-Pro-Gln-Leu-Pro-Tyr (PQPQLPY) [5]. Therefore, two model celiac-toxic peptides, QQPFP-repetitive domain in the γ-type, α-type and ω-type of prolamins in barley, rye and wheat, PQPQLPY-repetitive sequence in 33-mer peptide, were selected for plasma modification in our study [15]. Arentz-Hansen et al. (2002) previously stated that the toxicity mechanism of QQPFP sequence to CD patients may be due to its involvement in triggering T-cell activation after the deamidation of specific glutamine residues [16]. Furthermore, three previously identified celiac-specific T-cell epitopes play an important role in T-cell proliferation, namely, PQPQLPYPQ, PFPQPQLPY and PYPQPQLPY, all of which contain heptapeptide PQPQLPY [5]. In addition, exposure of Pro, Gln and phenylalanine (Phe) provides an important recognition site for G12 or R5 antibody, or stimulates the T-cell response [17]. Consequently, wheat gliadins in food containing those epitopes should be modified before their consumption in CD patients [18].

The objective of this study was to investigate the understanding of impacts of CJAP plasma on the structural properties and immunoreactivity of celiac-toxic peptides and wheat storage proteins in vitro. The effects of fragmentation at Pro and Gln and hydroxylation at Phe and tyrosine (Tyr) on celiac peptides-QQPFP and PQPQLPY, modifications and depolymerization behavior of gluten proteins, along with their immunoreactivity against R5 antibody after CJAP plasma treatment, were examined in this research.

2. Results

2.1. Experiment Setup, Discharge, Current and Optical Emission Spectrum (OES) Measurements

The schematic diagram of experiment setup is shown in Figure 1. The celiac-toxic peptides and wheat storage protein samples in 96-well plate (200 µL per sample) were modified by the special CJAP plasma. The discharge gas consisted of helium at a flow rate of 2 L/min and air at a flow rate of 10 mL/min. The distance from the surface of slurry to jet nozzle was 10 mm. The samples were modified by CJAP plasma at different time points (0, 5, 10, 20, 30, and 60 min).

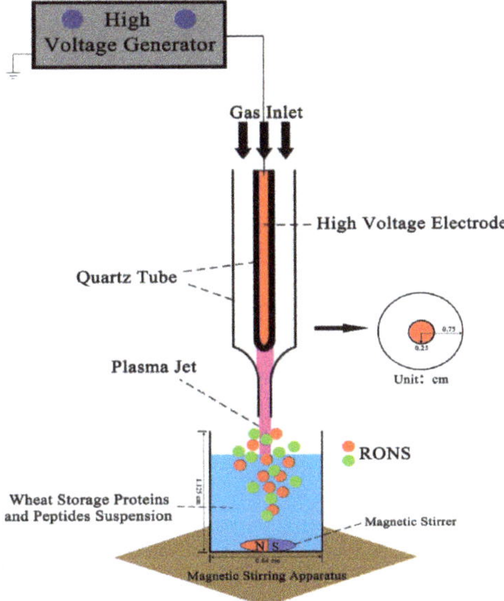

Figure 1. Schematic diagram of the plasma setup. The model peptides and gluten proteins in the 96-well plate were treated using a cold jet atmospheric pressure (CJAP) plasma jet.

The wave-forms of set voltage and single current of CJAP plasma were shown in Figure 2A. The peak-to-peak set voltage was 35 kV and frequency was 1 kHz. The OES of plasma were presented in Figure 2B, which were detected using an Ocean optics spectrometer. The emission lines of N_2, N_2^+, NO, •OH, O, and He can be found when the discharge gas is air [19].

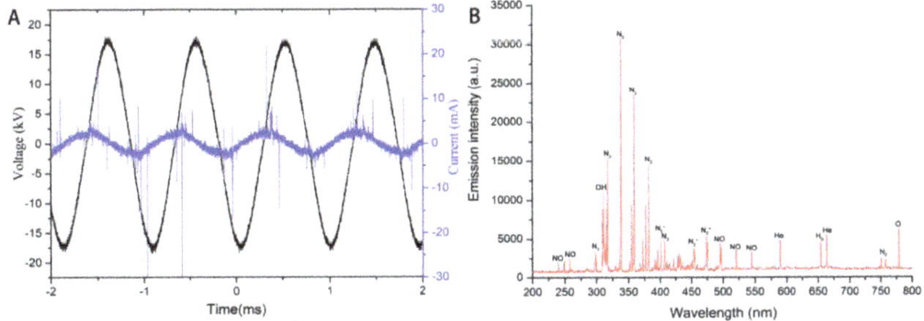

Figure 2. The waveforms of applied voltage and current (**A**) and the optical emissions spectra (OES) of the CJAP plasma jet (**B**).

2.2. Generation of Several RONS Induced by CJAP Plasma

In plasma processes, active species such as radicals and ions induced by the plasma play a dominant role. To study the effects of various RONS on plasma modification of two model peptides and gliadin, the contents of H_2O_2, OH radicals, O_3 and $NO_2^- + NO_3^-$ were determined in this part.

In order to understand the acidity and alkalinity of samples after treating the plasma, the pH value was determined. The pH value results showed a longer plasma treatment time correlated with a greater decline in pH value. From Figure 3A, it can be seen that the pH value decreased from 6.7 at 0 min to 2.2 after 60 min.

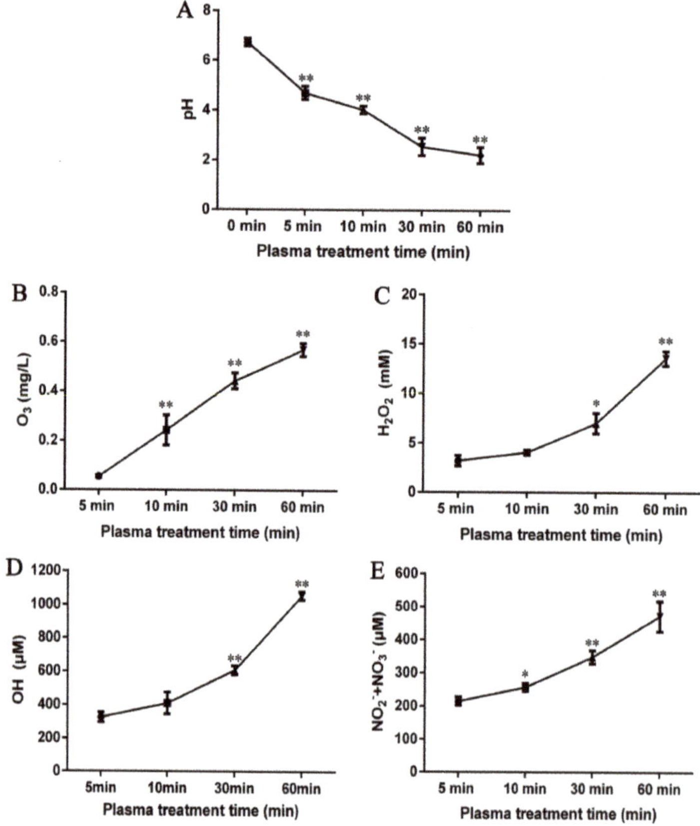

Figure 3. The pH value (**A**) and concentrations of O_3 (**B**), H_2O_2 (**C**), •OH (**D**), and total NO_2^- and NO_3^- (**E**) in CJAP plasma with treatment times of 5 min, 10 min, 30 min, and 60 min. Data are presented as means ± SE ($n = 3$) in the pH and reactive oxygen and nitrogen species (RONS) detection assays. Asterisks represent statistically significant differences from 5 min (* $p < 0.05$, ** $p < 0.01$).

Ozone is reported to selectively react with substrates by electrophilic attack and cycloaddition, while decomposing to produce OH radicals as another non-selective oxidant [20,21]. For a better understanding of the changes in ozone content following plasma initiation, results of the O_3 content were measured (Figure 3B), which presented a time-dependent increase, with significant increase ($p < 0.01$) at 10 min over control. It was found that the concentration of O_3 of the treated samples are 0.06 mg/L, 0.24 mg/L, 0.45 mg/L, and 0.58 mg/L at 5 min, 10 min, 30 min, and 60 min of plasma treatment, respectively.

H_2O_2 was considered to be another oxidizing agent and it was selected to be detected in this paper [22]. From Figure 3C, increase in H_2O_2 concentration was dependent on treatment time within 60 min, with the biggest increase from 30 to 60 min. We can see that the H_2O_2 content of the treated samples expanded from 3.23 mM at 5 min to 13.60 mM at 60 min.

For the most reactive form of ROS, OH radicals can undergo various types of reactions with amino acid side chains and peptide backbone [23]. Figure 3D shows that the pattern of change in •OH content was similar to that of H_2O_2. A significant ($p < 0.01$) increase occurred at 30 min, followed by a further increase at 60 min ($p < 0.01$). The •OH content was 325.53 μM at 5 min, 410.31 μM at 10 min, 608.30 μM at 30 min and 1055.10 μM at 60 min.

High oxidation features could also be possessed by reactive nitrogen species. To demonstrate whether nitrogen oxides were generated during plasma treatment to aid in modification and oxidation of proteins and peptides, the total NO_2^- and NO_3^- contents were measured. Figure 3E shows that the total contents of NO_2^- and NO_3^- increased rapidly to 215.21 μM at 5 min and then further increased at 60 min, reaching 472.83 μM.

2.3. Oxidation and Hydroxylation Gives Rise to P1~P7 after CJAP Plasma Treatment

Under non-plasma treatment conditions, the major peaks on the chromatograms were two original peptides, QQPFP (Supplementary Figure S1A) and PQPQLPY (Supplementary Figure S1B). After CJAP plasma modification, the peak intensities of two model peptides decreased and were almost undetectable after 60 min of treatment. With plasma processing, several new peaks appeared, mainly P1~P7. Fragmentation continued with continued treatment, but after 60 min, most of peaks had disappeared and less peptide was detected in columns.

2.4. Modification of Specific Pro, Gln, Phe, and Tyr Residues Leads to Degradation of the Model Peptides

The MS/MS spectra of QQPFP (616.27 *m/z*) and PQPQLPY (842.44 *m/z*) exhibited several fragmentation ions relating to b, x, y and z ions in the process of collision-induced dissociation (Figures 4A and 5A). For oxidation products, including P1, P2, and P6, there was a mass shift of −30 *m/z* for several y and b ions, corresponding to the formation of 2-pyrolidone at specific Pro residues on peptides with the loss of –COOH (−45 *m/z*) and subsequent addition of O (+16 *m/z*) linked to a double bond (−1 *m/z*) (Figure 4B,C and Figure 5C). In their hydroxylation products, namely, P3 and P5, the y2, y3, and y5 (only for P5) fragment ions showed an increase of 16 *m/z*, relevant to the addition of a hydroxyl group on the aromatic ring of peptides (Figures 4D and 5B). Similarly, P4 displayed a mass shift of +32 *m/z* for the y1, y2 and y3 fragment ions, suggesting that two hydroxyl groups were added to the benzene ring of specific pentapeptide (Figure 4E).

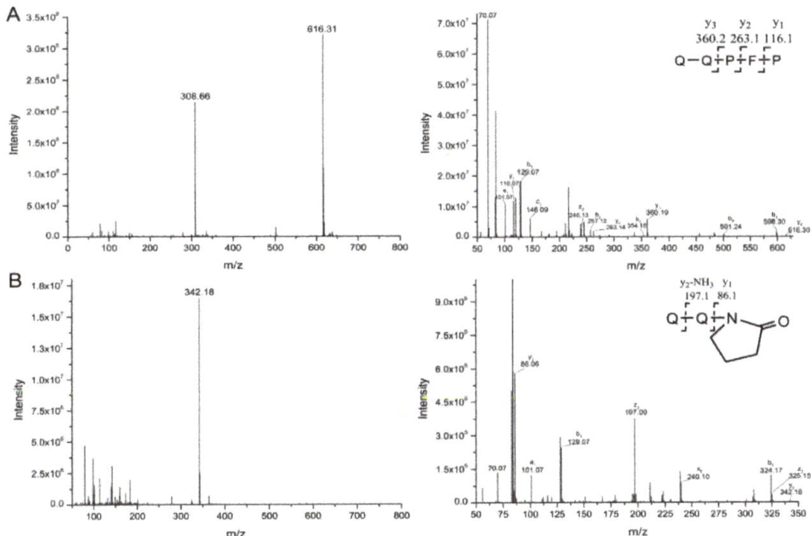

Figure 4. *Cont.*

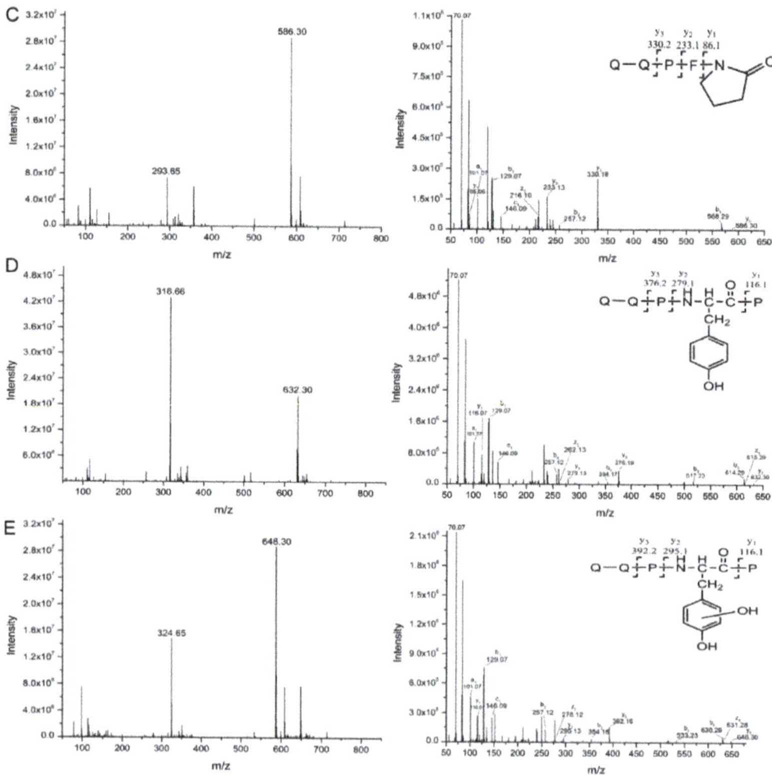

Figure 4. Full-scan mass spectrometry (MS) and representative MS/MS spectra of QQPFP and its products (P1~P4) during plasma treatment, showing evidence for •OH-mediated modification. In each panel, the left spectrum shows the full scan, while the right spectrum shows the fragment ions. The b and y ions originated from preferential cleavage at the N-terminus and C-terminus of specific amino acid residues during collision-induced dissociation, respectively. (**A**) Original peptide (616.3 *m/z*); (**B**) product 1 (342.2 *m/z*); (**C**) product 2 (586.3 *m/z*); (**D**) product 3 (632.3 *m/z*); (**E**) product 4 (648.3 *m/z*).

After plasma processing of original peptides, the Orbitrap HR-LC-MS/MS distinguished products P1, P2, P6 (Figure 6B,C and Figure 7C), in which Pro residues were attacked, as well as P3, P4 and P5 (Figures 6D and 7B), with the addition of hydroxyl groups on phenyl ring of Phe, leading to one or two +16-Da mass shifts. QQPFP gave rise to CJAP plasma-induced oxidation products at *m/z* 342.12 (P1) and *m/z* 586.30 (P2), as well as hydroxylation products at *m/z* 632.30 (P3, hydroxyphenylalanine) and *m/z* 648.30 (P4, dihydrophenelalanine, DOPA), while PQPQLPY generated products at *m/z* 858.43 (P5), *m/z* 311.17 (P6) and *m/z* 214.11 (P7) (Figures 4 and 5).

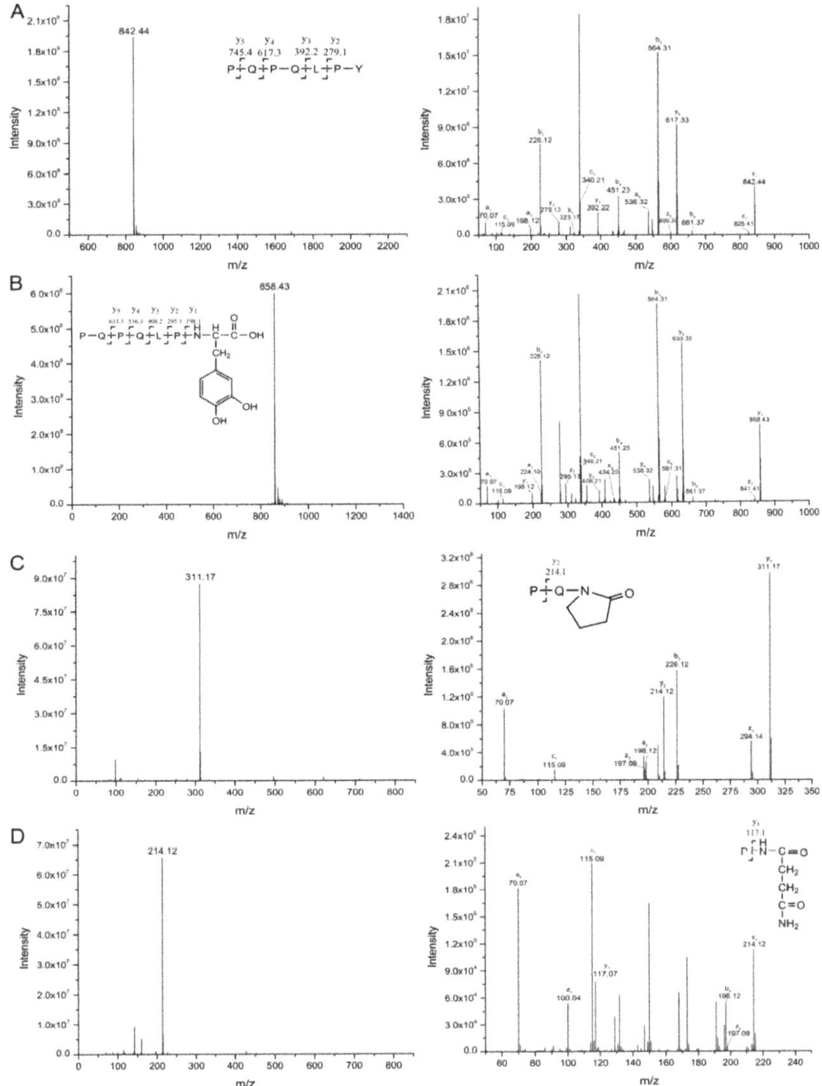

Figure 5. Full-scan MS and representative MS/MS spectra of PQPQLPY and its products (P5~P7) during plasma treatment, showing evidence for •OH-mediated modification. In each panel, the left spectrum shows the full scan, while the right spectrum shows the fragment ions. The b and y ions originated from preferential cleavage at the N-terminus and C-terminus of specific amino acid residues during collision-induced dissociation, respectively. (**A**) Original peptide (842.4 *m/z*); (**B**) product 5 (858.4 *m/z*); (**C**) product 6 (311.7 *m/z*); (**D**) product 7 (214.1 *m/z*).

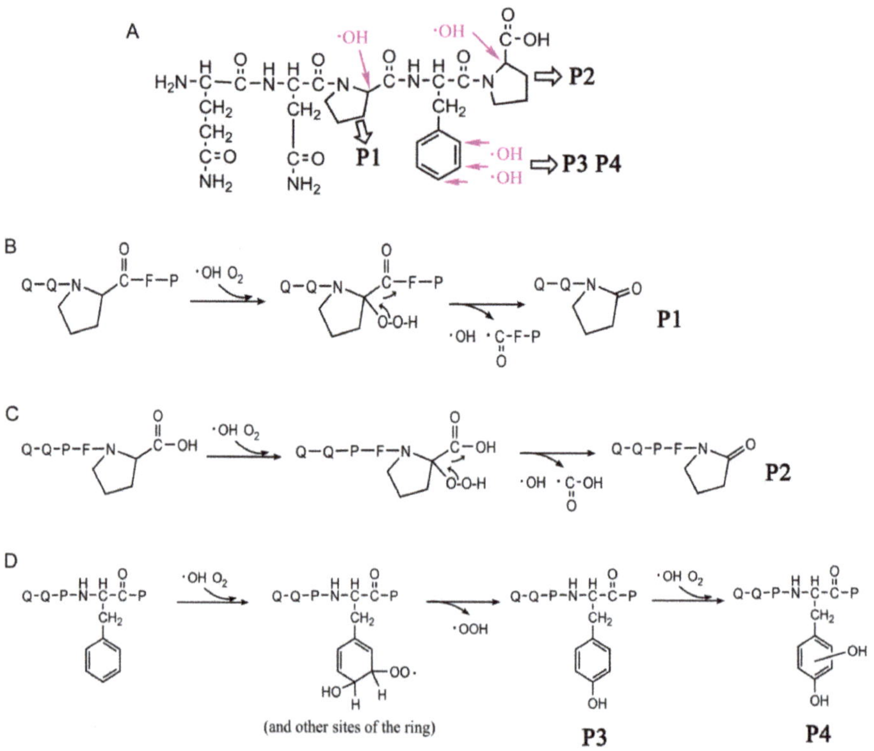

Figure 6. Structures of the celiac-toxic peptide QQPFP and its CJAP plasma-modified products referred to in this research (P1~P4), which were detected by Orbitrap HR-LC-MS/MS. (**A**) Structural changes in QQPFP during CJAP plasma treatment. The arrow shows the sites that were attacked at the intermediate region and C-terminus of the specific pentapeptide in this study. (**B**) The reaction pathway of P1 formation involved in •OH attack at the Pro residue in the intermediate region. (**C**) The reaction pathway of P2 formation associated with •OH attack at the Pro residue in the C-terminus. (**D**) P3 and P4 were formed through the reaction involved in •OH attack at the benzene ring of Phe, followed by a reaction with O_2 and, subsequently, elimination of •OOH.

As shown in Supplementary Figure S2A, plasma treatment exceeding 10 min resulted in detectable Pro fragmentation and hydroxylation of Phe in the peptide. After 10 min of plasma treatment, 40.21% of QQPFP and 66.73% of PQPQLPY remained intact, and further fragmentation was detected after a longer period of treatment, causing a decrease in QQPFP and PQPQLPY, leaving 1.43% and 15.03% at 30 min, and 0.16% and 0.08% at 60 min, respectively. Moreover, a shorter fragment, P7, was also detected and exhibited a tendency to increase over time (Supplementary Figure S2C). Extract ion chromatograms (EIC) opting for certain fragmentation ions were utilized to certify the intensity changes in P3, P4, and P5, which contained the different stereoisomers (Supplementary Figure S3).

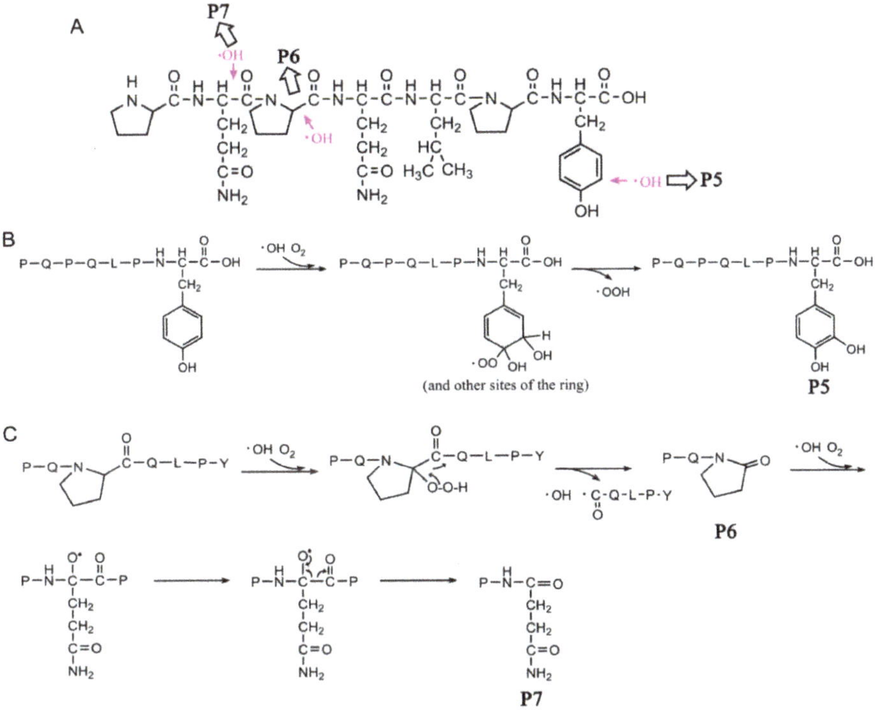

Figure 7. Structures of the celiac-toxic peptide PQPQLPY and its CJAP plasma-modified fragmentation products referred to in this research (P5~P7), which were detected by Orbitrap HR-LC-MS/MS. (**A**) Structural changes in PQPQLPY during CJAP plasma treatment. The arrow shows the sites attacked in the intermediate region and C-terminus of the specific pentapeptide in this study. (**B**) P5 was formed through the reaction involving •OH attack at the benzene ring of Tyr, followed by a reaction with O$_2$ and subsequent elimination of •OOH to form a mixture of DOPA stereoisomers. (**C**) The reaction pathway of P6 and P7 formation associated with •OH attack at the Pro residue and the formation of an alkoxy group on the α-carbon of Gln.

2.5. Agglomeration State of Wheat Gliadins Analyzed by SDS-PAGE

To display size changes in an observable way, gliadin proteins with and without plasma treatment were separated on the 12% SDS-PAGE. Comparing plasma-treated with untreated samples (Supplementary Figure S4), a decrease in intensity of stripe S1 was observed at the bottom of polyacrylamide gel. Additionally, another visible decrease in intensity of stripes S2 and S3 could also be seen in the plasma-treated samples. These observations demonstrated larger-sized proteins might undergo fragmentation to generate smaller-sized proteins under plasma conditions due to the depolymerization behavior of wheat gliadins during their exposure to the CJAP plasma.

2.6. Decrease in Quantity of Wheat Gliadins Treated with CJAP Plasma

To explore the CJAP plasma-treated specific gliadins changes in more detail, the 2-DE gel of gliadin extract is shown in Figure 8, which revealed that some protein spots disappeared or were reduced in quantity with longer CJAP plasma treatment. By comparison with the untreated sample, particular attention was directed toward three protein spots (G1, G2, and G3) with obviously reduced intensity in the treated samples. They were subjected to LC-MS/MS to clarify the type of gliadin.

To clarify the type of gliadin susceptible to CJAP plasma, those three protein spots arrowed in Figure 8 were subjected to LC-MS/MS experiment for further detection.

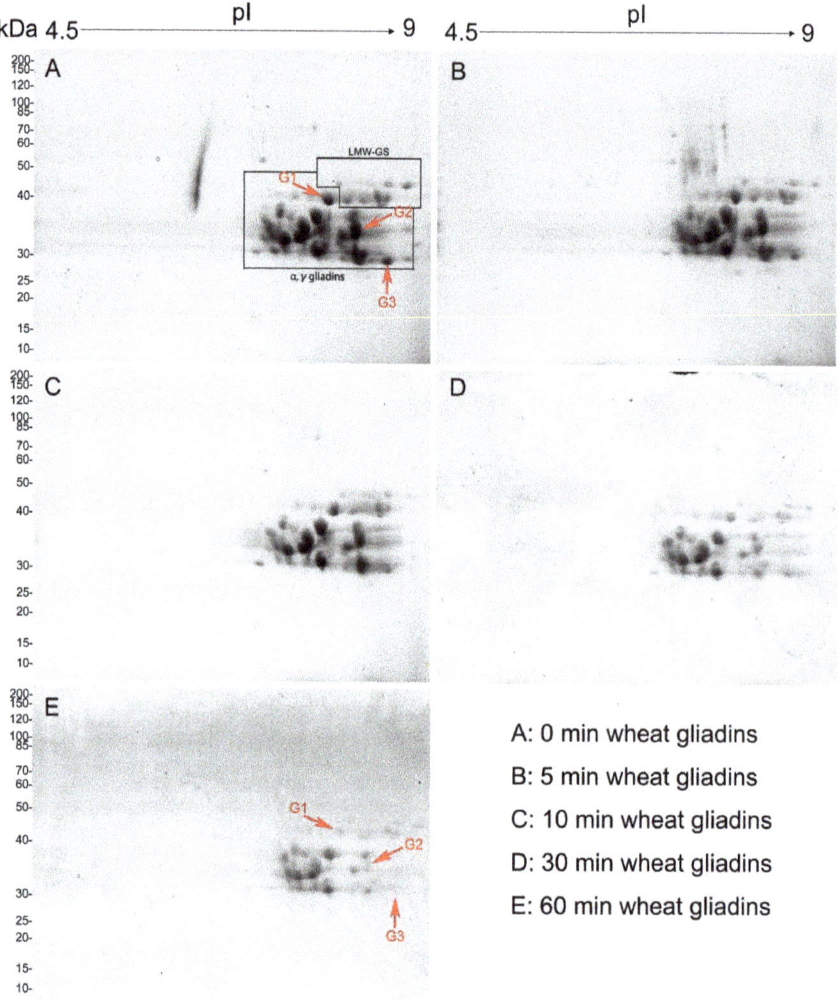

Figure 8. 2-DE detection of protein species in native and CJAP plasma-modified wheat gliadins. The major groups of wheat gliadins are indicated. The molecular weight is indicated on the vertical axis, and isoelectric point is shown on the horizontal axis. The A-E are the wheat gliadin samples treated with CJAP plasma for 0, 5, 10, 30 and 60 min, respectively.

2.7. LC-MS Identification of Proteins in 2-DE Spots

Proteins in 2-DE spots indicated by arrows were identified via LC-MS. Table 1 shows that γ-type gliadins were detected as the predominant protein in those three spots. In these spots, the predominant protein of G1 was a traditional gamma gliadin with an MW of 37,564 Daltons and pI of 8.51. An additional two protein spots (G2 and G3) with an MW of 34,502 and 33,067 Daltons with corresponding pI of 8.57 and 8.88 were also detected as gamma gliadin. More detailed information regarding these spots is presented in Table 1 below.

Table 1. LC-MS identification of wheat gliadin spots in 2-DE. The protein name, accession number, mass, pI, protein score, protein match, protein coverage, peptide sequence and mass error (ppm) were presented.

Spot	G1	G2	G3
Predominant protein	Gamma gliadin	Gamma gliadin	Gamma gliadin
Accession number	F2XAR5	Q30DX7	K7WVC4
MW	37,564	34,502	33,067
pI	8.51	8.57	8.88
Protein score	94	1762	2831
Protein match	2	91 (54)	62 (52)
Protein coverage	-	25.5	15.8
Sequence	R.APFASIVAGIGGQ. (ppm: 5.18) R.SLVLQTLPSMCSVYVPPECSIMR.A (ppm: −44.56)	R.TTTSVPFGVGTGVGAY. (ppm: 8.86) R.ILPTMCSVNVPLYR.T (ppm: 6.64) R.ILPTMCSVNVPLYR.T (ppm: 6.85) K.VFLQQQCSPVAMPQR.L (ppm: 9.19) R.SQMLQQSSCHVMQQQCCQQLPQ IPQQSR.Y (ppm: 8.80)	R.SDCQVMQQQCCQQLAQIPR.Q (ppm: 6.96) R.SDCQVMQQQCCQQLAQIPR.Q (ppm: 0.35) R.QPQQPFYQQPQQTFPQPQQ AFPHQPK.Q (ppm: 5.81)

The UniProt database was utilized to the spectral datasets searching.

2.8. Increment of Carbonyl Groups in Plasma-Treated Wheat Gliadins

Carbonyl groups are frequently used as biomarkers to demonstrate the protein oxidation level in biological samples [24]. As shown in Figure 9A, the separation gel presented new aggregates resulting from CJAP plasma-induced oxidation of wheat gliadins. Although carbonyl groups were formed in both plasma-treated and untreated samples, dissimilar blot intensities were observed in the samples, with the strongest intensity apparent at 30 and 60 min (Figure 9B). The quantification results (Figure 9C) revealed an increase in carbonyl groups with plasma treatment from 32.36 nmol/mL at 5 min to 86.30 nmol/mL at 60 min, with a significant increase ($p < 0.05$) first appearing at 10 min and a highly significant increase ($p < 0.01$) at 60 min compared with the untreated sample. These results showed that the CJAP plasma-treated wheat gliadins formed more carbonyl groups.

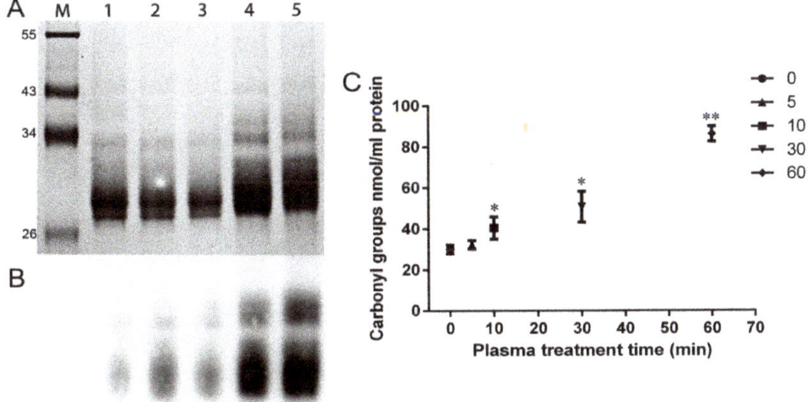

Figure 9. Qualitative and quantitative analysis of carbonyl groups of the wheat gliadin samples treated with CJAP plasma for 0 min, 5 min, 10 min, 30 min and 60 min, respectively. (**A**) SDS-PAGE gel stained with CBB. M: Protein molar mass standard, lane 1: 0 min, lane 2: 5 min, lane 3: 10 min, lane 4: 30 min, lane 5: 60 min. (**B**) Western blot of carbonyl groups using biotin-hydrazide. (**C**) Quantification of carbonyl groups using the 4-dinitrophenyl-hydrazine (DNPH) method. Data are presented as means ± SE ($n = 3$) in the quantitative detection assays. Asterisks represent statistically significant differences from 0 min (* $p < 0.05$, ** $p < 0.01$).

2.9. Immunological Activity Evaluation of Wheat Gliadins in ELISA

The R5 antibody can specifically recognize the epitope QQPFP, which exists in nearly all parts of gliadins [5]. The immunological activities of CJAP plasma-treated wheat gliadins were determined by ELISA. As shown in Figure 10, R5 antibody immunoreactivity in all treated samples exposed for different durations of plasma treatment were lower than in the untreated sample. After 5 min of plasma treatment, the prolamin concentration declined to 86.12% of its original level, and thereafter 84.23% of its initial content remained after 10 min of treatment. The lowest remaining prolamin concentration, at 48.05% of its initial level, was detected in samples treated with plasma for 60 min ($p < 0.05$). The decline in prolamin content implied that the epitopes for R5 recognition were modified by plasma treatment, which caused a reduction of immunological activity of wheat gliadins in vitro.

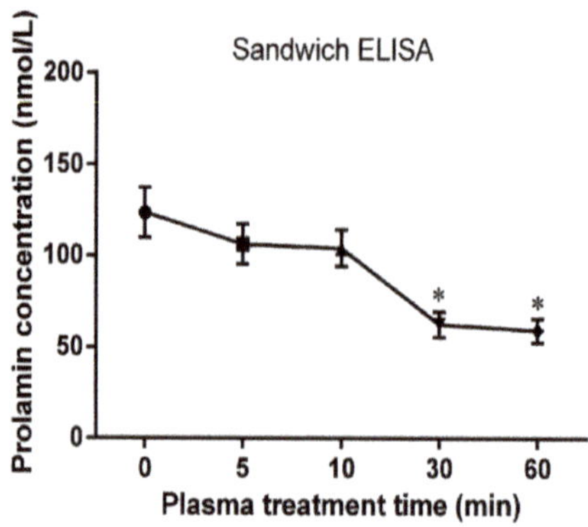

Figure 10. Sandwich ELISA of native and CJAP plasma-modified wheat gliadins. Data are presented as means ± SE ($n = 3$) in the ELISA system. Asterisks represent statistically significant differences from 0 min (* $p < 0.05$).

2.10. Size Distribution Change Analysis of Gluten Proteins before and after CJAP Plasma Treatment

SE-HPLC was a brilliant method for comparison of the molecular size distribution of native and plasma-treated gluten proteins by quantisation of the elution profiles and the size fractionation of those proteins with plasma treatment gave a good intimation of the degree of protein oxidation and modification. In the presence of RONS generated by CJAP plasma, an explicit decline of the peak height of SE-HPLC profiles was observed in the treated samples compared with untreated sample (Figure 11A). As shown in Figure 11B, lower values of F1% were observed at 5 min and 10 min compared with the untreated sample, with the biggest decrease observed at 60 min ($p < 0.05$). Besides, the ratios of F1%/F2% of CJAP plasma-treated samples gradually decreased and the most significant decline was achieved at 60 min ($p < 0.01$). The sample treated for 30 min had significantly higher ratios of (F3% + F4%)/F1% and a further increase appeared at 60 min ($p < 0.05$). These results implied that there was a time-dependent decrease in the amounts of large-sized polymers in plasma-treated samples when compared to the untreated one, suggesting that depolymerization of macropolymers probably occurred during plasma treatment.

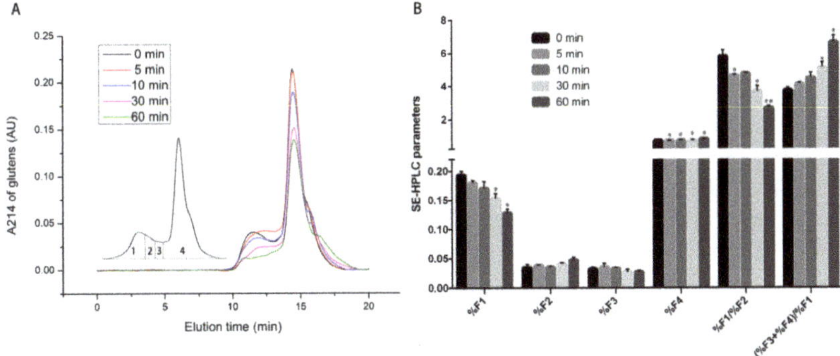

Figure 11. The influence of CJAP plasma on the aggregate formation of wheat gluten proteins. (**A**) SE-HPLC chromatogram curve of native and CJAP plasma-modified wheat gluten proteins. The inset refers to the area division of the wheat storage protein distribution, which corresponded to large-sized polymers (F1), medium-sized polymers (F2), monomeric and oligomeric proteins (F3) and monomeric gliadins and non-gluten proteins (F4). (**B**) with corresponding comparisons of six parameters (%F1, %F2, %F3, %F4, %F1/%F2, and (%F3 + %F4)/%F1) analyzed statistically. Data are presented as means ± SE (n = 3) in the assays. Asterisks represent statistically significant differences from 0 min (* $p < 0.05$, ** $p < 0.01$).

3. Discussion

In this context, an assortment of reactive oxygen and nitrogen species such as O_3, •OH, H_2O_2, NO_2^-, and NO_3^- generated during plasma treatment entailed that samples were subjected to an environment with high possibility of oxidation and modification. Moreover, the immunoreactivity of gliadin in wheat was successfully attenuated via CJAP-plasma modification, with oxidation and hydroxylation being of the major importance. Meanwhile, an Orbitrap HR-LC-MS/MS method was utilized to demonstrate the degree of oxidation at specific Pro and Gln and hydroxylation at Phe and Tyr through analyzing the products originated from plasma-treated peptides, QQPFP, and PQPQLPY.

3.1. CJAP Plasma can Generate a Variety of RONS

A diversity of RONS such as O_3, •OH, H_2O_2, NO_2^- and NO_3^- generated during CJAP-plasma treatment, indicating samples were subjected to an environment with a high possibility of oxidation and hydroxylation modification. RONS is in a dynamic process of mutual transformation. The OH radicals were found to be of great significance for oxidation and hydroxylation reactions. There are mainly two kinds of reaction pathways leading to the generation of OH radicals, which may explain the gradual increase of •OH density during CJAP-plasma treatment [25,26]. One is the dissociation of radicals and metastables (Reactions 1 and 2), and the other is the electron dissociation of H_2O (Reaction 3). Furthermore, it is well-documented that O_3 can preferentially produce in DBD, and O_3 mainly emerged from the reaction (Reaction 4) between the dissolved atomic oxygen and molecular oxygen [27,28].

$$e + O_2 \rightarrow O(^3P) + O(^1D) \tag{1}$$

$$O(^1D) + H_2O \rightarrow 2OH \tag{2}$$

$$e^- + H_2O \rightarrow OH + H + e^- \tag{3}$$

$$O + O_2 \leftrightarrow O_3 \tag{4}$$

According to Liu et al. (2016), the interaction (Reaction 5) between HO_2 and HO_3 is responsible for H_2O_2 generation [28]. It is worth mentioning that highly reactive OH radicals with a high density within a small diffusion length should be balanced via the recombination reaction (Reaction 6), which also contributes to the increase in H_2O_2.

$$HO_2 + HO_3 \rightarrow H_2O_2 + O_2 \tag{5}$$

$$OH + OH \rightarrow H_2O_2 \tag{6}$$

During the course of plasma treatment, various nitrogen oxides such as NO, NO_2^- and NO_3^- were detected, which presumably accounted for the decrease in pH value and increase in the density of NO_2^- and NO_3^- in our experiment [29].

3.2. CJAP Plasma-Induced Oxidation and Hydroxylation at Specific Pro, Gln, Phe, and Tyr Residues Lead to Reduction and Modification of Celiac-Toxic Peptides

Considering that the principal toxic components were not randomly scattered but clustered in Pro- and Gln-rich regions of gliadin, the modification and degradation of celiac-toxic peptides with application of CJAP plasma became our principle objective. Free radicals are available to attack substrates in various media, and the •OH accessibility of Pro, Gln, Phe and Tyr residues is critical for the formation of oxidation and hydroxylation products under CJAP plasma conditions. The higher Pro content in peptide PQPQLPY than in peptide QQPFP resulted in larger amounts of 2-pyrolidone products under RONS attack. A larger number of hydroxylation products was found in PQPQLPY than in QQPFP (Supplementary Figure S2). This finding may be attributed to the location of Tyr residue at the C-terminus of PQPQLPY, which was more attractive for OH radical attack [30]. Besides, the hydroxyl group in the benzene ring of Tyr residue could also increase reactivity of ortho-/meta-carbon compared with that of Phe residue.

The peak area of P1~P7 first appeared to increase, possibly owing to bond cleavage in peptide side chain, which was due to the addition of an oxygen atom to alpha carbon to generate a carbonyl and the introduction of a hydroxyl group resulting from the addition reaction on an aromatic ring of Phe [31,32]. Importantly, the P1, P2 and P6 detected by LC-MS/MS indicated that •OH-mediated cleavage of the peptide played an important role in attack of Pro residue at the C-terminus of specific pentapeptide and heptapeptide in 2-pyrrolidone formation in the presence of O_2 [33]. For P3 and P4, •OH rapidly attacked the benzene ring of Phe to generate a hydroxyl cyclohexadienyl radical, followed by a reaction with O_2 and subsequent elimination of •OOH to form o-/m-/p-Tyr (P3) and DOPA stereoisomers (P4). The substituent containing an unshared electron pair on the benzene ring can form a p-π-conjugated system with the benzene ring, resulting in an increase in the electron cloud density of the benzene ring, especially at ortho and para positions, which favors attack of the electrophilic group •OH generated by the plasma treatment process to form a mixture of o-/m-/p-Tyr (P3), which was responsible for three peaks in extract ion chromatograms as shown in Supplementary Figure S3A [34]. Attack of the second hydroxyl group has more positional selectivity due to the strong guiding action of the first hydroxyl substituent, leading to generation of the DOPA products (P4) [21].

Backbone cleavage mediated by the highly accumulated OH radical population induced by CJAP-plasma appears to occur at multiple sites, with positions adjacent to Pro residues at the C-terminus being of major importance [35]. With an open structure and water-soluble properties, two model peptides show greater availability for attack by OH radicals. Encouragingly, plasma treatment of peptides was more efficient in terms of structural changes of peptide than Fe/EDTA/AA treatment of QQPFP, for which 22% of the original peptide remained intact after 24 h of oxidation [5]. In contrast, 0.16% of peptide remained after 60 min of plasma processing in our experiment.

3.3. Decreased Immunoreactivity of Wheat Gliadins Modified by the CJAP Plasma

The chief epitope for R5 antibody binding is specific pentapeptide (QQPFP) in gliadins. Modification of the FP motif was a key factor in reducing the immunoreactivity of gliadin in R5-antibody recognition [30]. The Pro and ring structure of Phe can be oxidized more readily by OH radicals generated by CJAP plasma in our work. Numerous Pro and Gln are present in the primary structure of gliadin, and smaller fragments of gliadins could be produced at the α-carbon side of Pro and Gln when it was attacked by the RONS induced by CJAP plasma, making it easier for OH radicals to approach the exposed Pro sites of gliadins. Considering the fragmentation and side-chain modifications of the gliadin epitopes, a decreased ability of R5 antibody for detection of its binding sites was achieved in vitro. The 48.05% of original immunoreactivity against the R5 antibody was retained in gliadin treated with plasma for 60 min, compared with approximately 60% following 2-h metal-catalyzed oxidation of hordein [5]. The CJAP plasma could bring about the backbone cleavage and degradation of gluten protein (Figure 11), which resulted in the exposure of more epitope for •OH. Besides, it could produce the higher density of OH radicals at high voltage. These both contribute to the more decreased immunoreactivity than that of metal catalysis.

It is worth mentioning that although we have proved that CJAP plasma can reduce the immune activity of gliadin in vitro, more studies need to be done to verify the effectiveness of cold plasma technology. The gluten-specific CD4$^+$ T-cells could be activated inappropriately, triggered by gluten peptides bound to DQ2 and DQ8 heterodimers in vivo [4]. To date, 24 celiac disease-related epitopes have been identified from wheat gluten that induced T-cell response in patients with celiac disease [36]. Hence, it is important that using specific T-cell clones to evaluate celiac immunological activity [37]. Furthermore, it is reported that epitopes rich in Glu and Pro are derived from gluten across all different classes and the celiac disease is triggered by very small amounts of gluten [38]. Therefore, optimization of the discharge parameters of cold plasma is necessary to improve the efficiency of decreasing the immune activity. The development of stronger energy plasma could encourage to eliminate the celiac sensitivity of gluten in the future. Regarding that gluten is embed in complex matrix of lipids, starch and proteins, the region exposed to the digestive enzymes is closely related to those matrices. This accounts for the problematic quantification and identification of digestible peptides, which is significant for the gluten epitopes related to celiac disease [39]. Plasma modification target should be extended to the complex and diverse foods in the practical application.

3.4. Decrease in Large-Sized Polymers through Longer Exposure to CJAP Plasma

The agglomeration state with reduction of large-sized proteins and the appearance of small-sized proteins were observed by SDS-PAGE and SE-HPLC. We attributed the increase of small-sized aggregates to bond cleavage at Pro and Gln of the original protein due to the oxidation of RONS generated by CJAP plasma [40]. Based on the LC-MS results, it was found that plasma modification could cause quantitative decrease in gliadin with selectivity for γ-gliadin, and the longer plasma treatment, the greater was degree of protein reduction (Figure 8). The N-terminus comprising mostly highly repetitive sequences rich in Pro and Gln is distinct for each type of gliadin. For γ-gliadin, the typical unit, QPQQPFP, is repeated up to 16 times, which is more than that of the other gliadins [41]. γ-gliadin contained lower single-bond energy amino acid residues and its arrangement was less regular and less compact than that of α/β gliadin, which could also facilitate its attack by RONS [30,42].

3.5. Proposed Mechanism of the Model Peptide and Wheat Storage Protein Modified by CJAP Plasma

Based on the analytical results described above, possible mechanisms of two model peptide and wheat storage protein modified by CJAP plasma are shown in Figure 12. By performing cold jet plasma treatment under atmospheric pressure, large quantities of RONS could be produced. These RONS could cause the depolymerization of gluten protein and quickly attacked specific Pro and Gln residues to give rise to the 2-pyrolidone (P1, P2 and P6) and alkoxylate products (P7), respectively. In addition, aromatic amino acids such as Phe and Tyr were hydroxylated, and monohydroxylated Phe (P3), DOPA (P4) and monohydroxylated Tyr (P5) products are formed in the presence of RONS. Amino acid modifications and peptide cleavage, incorporated with gluten protein fragments induced by plasma treatment, generated a variety of smaller-sized aggregates and reduced the immunoreactivity of wheat storage protein.

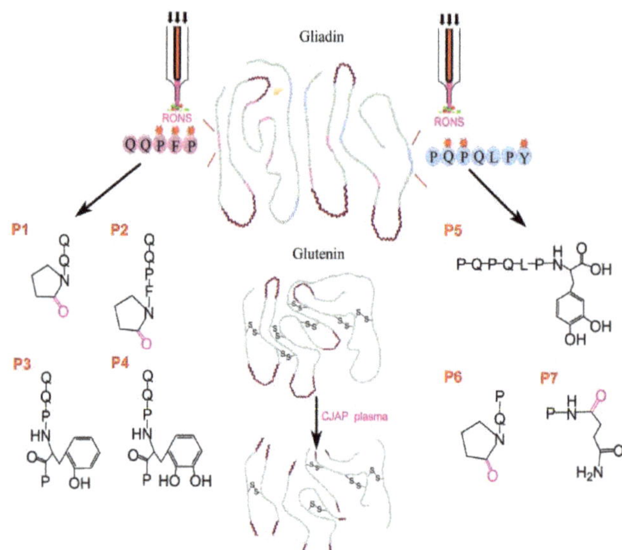

Figure 12. Schematic mechanism for wheat storage proteins and model celiac-toxic peptides modified by CJAP plasma.

4. Materials and Methods

4.1. Materials

The celiac-toxic peptides QQPFP (molecular weight, MW = 615.68 g/mol, purity 85.1%) and PQPQLPY (MW = 841.95 g/mol, purity 89.5%) were synthesized by Bioyeargene Biotechnology Inc. (Wuhan, China). The gliadins and storage proteins in wheat were extracted from flour of Zhengmai 9023 (*Triticum aestivum* L.), which was planted in the experimental field at Huazhong University of Science and Technology (HUST) (Wuhan, China).

4.2. Discharge, Current and Optical Emission Spectra (OES) Detection

Set voltage and discharge current were simultaneously measured by a Tektronix MSO 3045 mixed signal oscilloscope with a Tektronix P6015A high-voltage (H.V.) probe and a Pearson current monitor (Pearson Electronics, Palo Alto, CA, USA).

4.3. Extraction of Gliadins from Wheat

The gliadins in wheat were extracted as described by Marsh (2000) [42]. A fraction of 100 g flour was added to 1 L of water-saturated butanol solution and stirred constantly for 1 h at 20 °C After centrifugation (5000× g, 10 min, 20 °C) the precipitate was mixed with 1 L of 0.5 M NaCl for 1 h at 20 °C. This step was repeated twice. Next, pellet was washed with 1 L of distilled water for 10 min and centrifuged (5000× g, 10 min, 20 °C) again. The precipitate was dissolved in 1 L of 70% ethanol (*v/v*) for 1 h at 20 °C and centrifuged at 50,000× g in 20 °C for 10 min to collect supernatant. The supernatant containing gliadins was obtained during this procedure.

4.4. O_3 Measurement

The content of O_3 was detected using N,N-diethyl-p-phenyl-enediamine (DPD) method. A commercial kit (HuanKai Biotechnology, Guangzhou, China) was used in this experiment. Treated samples were supplemented with DPD reagent, which could react with O_3 to induce a color change of solution. The color depth of solution is positively correlated with amount of O_3.

4.5. H₂O₂ and pH Measurement

The concentration of H_2O_2 was measured using a Hydrogen Peroxide assay kit (Beyotime, Shanghai, China). The testing solution was mixed with sample at a ratio of 1:1 (*v/v*). 200 μL of mixed solution was allowed to remain at room temperature for 30 min and measured at a wavelength of 570 nm using microplate reader. pH test strips were used to measure the pH value.

4.6. •OH Measurement

The terephthalic acid (TA) method was employed for determination of hydrogen groups. TA was dissolved in 1.4 mM NaOH solution to obtain a 0.2 mM TA solution. The fluorescence microplate reader (FlexStation3, Molecular Devices, San Francisco, CA, USA) was used to measure the absorbance value of hydroxyl group at an excitation wavelength of 310 nm and an emission wavelength of 425 nm.

4.7. Total Content of NO₂⁻ and NO₃⁻ Determination

The total concentrations of NO_2^- and NO_3^- were analyzed with Nitric Oxide (NO) assay kit (Nitrate reductase method, Nanjing Jiancheng Bioengineering Institute, Nanjing, China). The reaction between NO_3^- and nitrate reductase leads to the formation of NO_2^-.

4.8. Identification of CJAP Plasma Treatment Fragments by Orbitrap HR-LC-MS/MS

The Q. Exactive Orbitrap HR-LC-MS/MS was applied to analyze the obtained CJAP plasma treatment peptide mixtures. An Ultimate 3000 UPLC system equipped with an Accucore aQ column (2.1 × 150 mm, 2.6 μm Thermo Scientific, Waltham, MA, USA) was coupled to the quadrupole electrostatic field orbital trap mass spectrometer (Q Exactive, Thermo Scientific, Waltham, MA, USA) in positive mode. The ESI pattern was used as ion source with a scanning range from 50 to 2400 *m/z*. Solvent A and solvent B consisted of an aqueous solution of 0.1% formic acid in water and of 0.1% formic acid in acetonitrile, respectively. A gradient of acetonitrile from 99% solvent A to 99% solvent B was generated for 20 min at a flow rate of 0.2 mL/min. The Xcalibur software (Thermo Scientific, Waltham, MA, USA) was applied for analysis of MS result.

4.9. Quantitative 2-DE and LC-MS Analysis

The CJAP plasma-treated wheat gliadins were separated by 2-DE as described by Nagib et al. (2010) [43]. First, 200 μL of the gliadin sample (1.25 μg/μL) was mixed with hydration loading buffer containing 8 M urea, 0.2% Bio-Lyte (*w/v*), 4% CHAPS (*w/v*) and 65 mM dithiothreitol (DTT). Each sample was subjected to IPG prefabricated strips pH 3–10 (7 cm, ReadyStrip, Bio-Rad, Irvine, CA, USA) with 16 h rehydration. IEF was applied using a PROTEAN IEF Cell (Bio-Rad, Irvine, CA, USA) at 17 °C with a series of voltages: 250 V and 500 V for desalting, 500–4000 V linear boost and 4000 V for focus. After IEF, IPG prefabricated strips were equilibrated in an equilibration buffer containing 6 M urea, 2% SDS (*w/v*), 0.375 M Tris-HCl (pH 8.8) and 20% (*v/v*) glycerol including 2% (*w/v*) DTT for 15 min, followed by a strip equilibration buffer containing 2.5% iodoacetamide (*w/v*) for another 15 min. In the second dimension, SDS-PAGE of a 12% separation gel at 20 mA was applied until blue dye line reached the end of gel.

Three sets of selected protein spots from 2-D gels were excised for trypsin digestion (37 °C, 16 h), according to the literature reported by Martinez-Esteso et al. (2016) [44]. The obtained peptide samples from digested protein spots were dissolved (0.1% formic acid, 5% acetonitrile) and loaded into the Q Exactive mass spectrometer equipped with a 75 μm i.d. × 150 mm, packed with an Acclaim PepMap RSLC C18, 2 μm, 100 Å, nanoViper column for sequence identification (Thermo Scientific, Waltham, MA, USA). Solvent A and solvent B consisted of an aqueous solution of 0.1% formic acid in water and of 0.1% formic acid, 80% ACN, respectively. The MS original file was processed and converted using MM File Conversion software to obtain the MGF format file, and the NCBI database and UniProt sapiens database were retrieved using MASCOT.

4.10. Qualitative and Quantitative Detection of Carbonyl Groups

The content of carbonyl groups was measured by 2,4-dinitrophenyl -hydrazine (DNPH) method [24]. First, 50-µL gliadin sample was supplemented with 800 µL of 10 mM DNPH in 2.5 M HCl, with 2.5 M HCl as a control. Two tubes were placed at room temperature for 1 h. The samples were precipitated with 1 mL of 20% trichloroacetic acid (TCA) (*w/v*) and left in an ice box for 5 min. After centrifugation (10,000× *g*, 10 min, 4 °C), the precipitate was washed with 1 mL of 10% TCA and centrifuged (10,000× *g*, 10 min, 4 °C). The precipitates were washed three times with 1 mL ethanol-ethyl acetate (1:1, *v/v*). The obtained pellets were dissolved in 500 µL of 6 M guaninide hydrochloride with general vortexing and again centrifuged (10,000× *g*, 10 min, 4 °C). Finally, 220 µL of supernatant from each tube was collected and the carbonyl content was calculated based on the absorption at 375 nm.

4.11. Sandwich Enzyme-Linked Immunosorbent Assay (ELISA)

A commercial ELISA kit (Gliadin, ml058393-2, Shanghai Enzyme-linked Biotechnology Co. Ltd., Shanghai, China) was applied to measure the concentration of prolamin. In the first step, 10 µL of CJAP-treated and untreated samples were added to 40 µL of sample diluent. All samples were then incubated at 37 °C for 30 min. The samples were washed five times with a washing solution. The samples were mixed with 50 µL of HRP-Conjugate reagent, incubated again at 37 °C for 30 min and then incubated and washed as described above. Thereafter, 50 µL each of reagent A and reagent B were mixed in the dark at 37 °C for 10 min. Finally, 50 µL of stop solution was used to terminate reaction. The concentration of prolamin can be calculated by absorption curve measured at a wavelength of 450 nm. Samples without addition of HRP-Conjugate reagents served as a blank control.

4.12. SE-HPLC Analysis

The gluten proteins were collected with 50 mM of sodium phosphate buffer with 0.5% SDS (*w/v*, pH 6.9) from flour of Zhengmai 9023 (*T. aestivum* L.), based on the documentation reported by Li et al. (2017) [45]. A 20-µL portion of supernatant filtered from a nylon membrane (pore size 0.45 µm) was poured into a Waters 1525 binary HPLC pump, fractionated on a Phenomenex Biosep-SEC-s4000 column (20 min, 0.5 mL min^{-1}) and then further tested at 214 nm using a Waters 2998 photodiode array detector (Waters Corp., Milford, CT, USA).

4.13. Statistical Analysis

The statistical software GraphPad Prism 6.0 (Microsoft Corporation, Redmond, WA, USA) was used for statistical analysis via one-way analysis of variance (ANOVA) followed by Tukey mean-comparison procedure. The 5% significant differences ($p < 0.05$) and 1% highly significant differences ($p < 0.01$) were evaluated using Origin 8.1 (OriginLab Corporation, Northampton, MA, USA).

5. Conclusions

In conclusions, fragmentation at Pro residues, along with modifications at Phe and Tyr residues of two CD-related peptides—QQPFP and PQPQLPY—could be achieved with the application of CAJP plasma. A variety of RONS such as OH, O_3, H_2O_2, NO_2^-, and NO_3^- were observed to increase during plasma processing. In addition, the immunoreactivity of gliadin using the R5 antibody appeared to be reduced because a large amount of recognition epitopes were modified after plasma treatment. Furthermore, the longer the plasma treatment, the greater was the formation of carbonyl and hydroxylation products and, consequently, the smaller was the size of the formed aggregates. The structural changes in the two model peptides, as well as the quantitative changes in gliadin after plasma treatment, favored the preparation of gluten-related product for the celiac disease patients.

Supplementary Materials: Supplementary materials can be found at http://www.mdpi.com/1422-0067/21/3/1012/s1. Figure S1. HR-LC chromatogram analysis of CJAP plasma-modified QQPFP and PQPQLPY identified by Orbitrap HR-LC-MS/MS at treatment times of 0 min, 5 min, 10 min, 30 min and 60 min, respectively. (A) QQPFP, (B) PQPQLPY. The black arrow indicates CJAP plasma-modified products of two celiac-toxic peptides. Figure S2. Two model peptides QQPFP and PQPQLPY modified by CJAP plasma. (A) The changes in the HR-LC-MS/MS peak area of QQPFP and PQPQLPY with the different treatment time. (B) The changes in HR-LC-MS/MS peak area of products P1~P4 from QQPFP with the different treatment time. (C) The changes in HR-LC-MS/MS peak area of products P5~P7 from PQPQLPY with different treatment time. Figure S3. Extract ion chromatograms (EIC) selected for certain fragmentation ions were utilized to certify the intensity changes of P3 (A), P4 (B) and P5 (C). Figure S4. SDS–PAGE analysis of the gliadins in wheat with different CJAP plasma treatment times. M: Protein molar mass standard, lane 1: 0 min treatment, lane 2: 5 min treatment, lane 3: 10 min treatment, lane 4: 30 min treatment and lane 5: 60 min treatment.

Author Contributions: G.H., G.Y. and F.S. conceived and designed the study. F.S., X.X. and Y.Z. performed all the biological experiments and analyzed the data. F.S., J.D., and M.M. participated in the physics experiments. Y.W. provided some help in the SE-HPLC experiment and its data analysis. D.Q. and X.L. proposed some valuable suggestions to the study. F.S. and X.X. wrote the draft manuscript. G.H. and G.Y. revised and finalized the manuscript. All authors have read and agreed to the published version of the manuscript.

Funding: The work was supported by National Genetically Modified New Varieties of Major Projects of China (2016ZX08010004-004) and the National Natural Science Foundation of China (Nos. 31771418, 31570261), and Key Project of Hubei Province (2017AHB041).

Acknowledgments: The authors acknowledge the Analytical and Testing Center of Huazhong University of Science and Technology (HUST) for technical assistance in the fragments identification of celiac-toxic peptides by used of Orbitrap HR-LC-MS/MS.

Conflicts of Interest: The authors declare that the research was accomplished without any commercial or financial relationships that could be identified as a potential conflict of interest.

Abbreviations

CJAP	cold jet atmospheric pressure plasma
RONS	reactive oxygen and nitrogen species
CD	celiac disease
QQPFP	Gln-Gln-Pro-Phe-Pro
PQPQLPY	Pro-Gln-Pro-Gln-Leu-Pro-Tyr
MW	molecular weight
kDa	kiloDalton
SDS	Sodium dodecyl sulfate
DNPH 2	4-dinitrophenyl-hydrazine
EDTA	ethylenediamine-tetraacetic acid
DTT	dithiothreitol
AC	alternating current
OES	optical emission spectra
H.V.	high-voltage
DPD	N,N-diethyl-p-phenyl-enediamine
TA	terephthalic acid
HR-LC-MS/MS	high resolution-liquid chromatography-mass spectrometry/mass spectrometry
EIC	extract ion chromatograms
SDS-PAGE	sodium dodecyl sulfate-polyacrylamide gel electrophoresis
2-DE	two dimensional electrophoresis
IEF	isoelectric focusing
ELISA	sandwich enzyme-linked immunosorbent assay
SE-HPLC	size exclusion high performance liquid phase chromatography
HMW	high molecular weight; LMW, low-molecular-weight

References

1. Shewry, P. Wheat. *J. Exp. Bot.* **2009**, *60*, 1537–1553. [CrossRef] [PubMed]
2. Valerii, M.; Ricci, C.; Spisni, E.; Silvestro, R.; Fazio, L.; Cavazza, E.; Lanzini, A.; Campieri, M.; Dalpiaz, A.; Pavan, B. Responses of peripheral blood mononucleated cells from non-celiac gluten sensitive patients to various cereal sources. *Food Chem.* **2015**, *176*, 167–174. [CrossRef] [PubMed]
3. Sánchez-León, S.; Gil-Humanes, J.; Ozuna, C.; Giménez, M.; Sousa, C.; Voytas, D.; Barro, F. Low-gluten, non-transgenic wheat engineered with CRISPR/Cas9. *Plant Biotechnol. J.* **2017**, *16*, 902–910. [CrossRef] [PubMed]
4. Gianfrani, C.; Siciliano, R.; Facchiano, A.; Camarca, A.; Mazzeo, M.; Costantini, S.; Salvati, V.; Maurano, F.; Mazzarella, G.; Iaquinto, G.; et al. Transamidation of wheat flour inhibits the response to gliadin of intestinal T cells in celiac disease. *Gastroenterology* **2007**, *133*, 780–789. [CrossRef] [PubMed]
5. Huang, X.; Sontag-Strohm, T.; Stoddard, F.; Kato, Y. Oxidation of proline decreases immunoreactivity and alters structure of barley prolamin. *Food Chem.* **2017**, *214*, 597–605. [CrossRef]
6. Wongsagonsup, R.; Deeyai, P.; Chaiwat, W.; Horrungsiwat, S.; Leejariensuk, K.; Suphantharika, M.; Fuongfuchat, A.; Dangtip, S. Modification of tapioca starch by non-chemical route using jet atmospheric argon plasma. *Carbohydr. Polym.* **2014**, *102*, 790–798. [CrossRef]
7. Cuadrado, C.; Cabanillas, B.; Pedrosa, M.M.; Varela, A.; Guillamon, E.; Muzquiz, M.; Crespo, J.; Rodriguez, J.; Burbano, C. Influence of thermal processing on IgE reactivity to lentil and chickpea proteins. *Mol. Nutr. Food Res.* **2010**, *53*, 1462–1468. [CrossRef]
8. Sandhu, H.; Manthey, F.; Simsek, S. Ozone gas affects physical and chemical properties of wheat (*Triticum aestivum* L.) starch. *Carbohydr. Polym.* **2012**, *87*, 1261–1268. [CrossRef]
9. Misra, N.; Kaur, S.; Tiwari, B.; Kaur, A.; Singh, N.; Cullen, P. Atmospheric pressure cold plasma (ACP) treatment of wheat flour. *Food Hydrocoll.* **2015**, *44*, 115–121. [CrossRef]
10. Bogaerts, A.; Neyts, E.; Gijbels, R.; Van der, J. Gas discharge plasmas and their applications. *Spectrochim. Acta Part B At. Spectrosc.* **2002**, *57*, 609–658. [CrossRef]
11. Ikawa, S.; Kitano, K.; Hamaguchi, S. Effects of pH on bacterial inactivation in aqueous solutions due to low-temperature atmospheric pressure plasma application. *Plasma Process. Polym.* **2010**, *7*, 33–42. [CrossRef]
12. Keidar, M.; Shashurin, A.; Volotskova, O.; Ann, M.; Srinivasan, P.; Sandler, A.; Trink, B. Cold atmospheric plasma in cancer therapy. *Phys. Plasmas* **2013**, *20*, 057101. [CrossRef]
13. Lee, H.; Kim, G.; Kim, J.; Park, J.; Lee, J.; Kim, G. Tooth bleaching with nonthermal atmospheric pressure plasma. *J. Endod.* **2009**, *35*, 587–591. [CrossRef]
14. Heinlin, J.; Isbary, G.; Stolz, W.; Morfill, G.; Landthaler, M.; Shimizu, T.; Steffes, B.; Nosenko, T.; Zimmermann, J.; Karrer, S. Plasma applications in medicine with a special focus on dermatology. *J. Eur. Acad. Dermatol. Venereol.* **2011**, *25*, 1–11. [CrossRef]
15. Osman, A.; Uhlig, H.; Valdes, I.; Amin, M.; Méndez, E.; Mothes, T. A monoclonal antibody that recognizes a potential coeliac-toxic repetitive pentapeptide epitope in gliadins. *Eur. J. Gastroenterol. Hepatol.* **2001**, *13*, 1189–1193. [CrossRef]
16. Arentz–Hansen, H.; Mcadam, S.; Molberg, Ø.; Fleckenstein, B.; Lundin, K.; Jørgensen, T.; Jung, G.; Roepstorff, P.; Sollid, L. Celiac lesion T cells recognize epitopes that cluster in regions of gliadins rich in proline residues. *Gastroenterology* **2002**, *123*, 803–809. [CrossRef]
17. Morón, B.; Cebolla, A.; Manyani, H.; Alvarez-Maqueda, M.; Megías, M.; Thomas, M.; López, M.; Sousa, C. Sensitive detection of cereal fractions that are toxic to celiac disease patients by using monoclonal antibodies to a main immunogenic wheat peptide. *Am. J. Clin. Nutr.* **2008**, *87*, 405–414. [CrossRef]
18. Vader, L.; Stepniak, D.; Bunnik, E.; Kooy, Y.; Haan, W.; Drijfhout, J.; Veelen, P.; Koning, F. Characterization of cereal toxicity for celiac disease patients based on protein homology in grains. *Gastroenterology* **2003**, *125*, 1105–1113. [CrossRef]
19. Liu, Z.; Xu, D.; Liu, D.; Cui, Q.; Cai, H.; Li, Q.; Chen, H.; Kong, M. Production of simplex RNS and ROS by nanosecond pulse N_2/O_2 plasma jets with homogeneous shielding gas for inducing myeloma cell apoptosis. *J. Phys. D Appl. Phys.* **2017**, *50*, 195204. [CrossRef]
20. Shad, A.; Li, C.; Zuo, J.; Liu, J.; Dar, A.; Wang, Z. Understanding the ozonated degradation of sulfadimethoxine, exploration of reaction site, and classification of degradation products. *Chemosphere* **2018**, *212*, 228–236. [CrossRef]

21. Li, Y.; Wang, X.; Yang, H.; Wang, X.; Xie, Y. Oxidation of isoprothiolane by ozone and chlorine: Reaction kinetics and mechanism. *Chemosphere* **2019**, *232*, 516–525. [CrossRef] [PubMed]

22. Ghimire, B.; Sornsakdanuphap, J.; Hong, Y.; Uhm, H.; Weltmann, K.; Choi, E. The effect of the gap distance between an atmospheric-pressure plasma jet nozzle and liquid surface on OH and N_2 species concentrations. *Phys. Plasmas* **2017**, *24*, 073502. [CrossRef]

23. Liu, F.; Lai, S.; Tong, H.; Lakey, P.; Shiraiwa, M.; Weller, M.; Pöschl, U.; Kampf, C. Release of free amino acids upon oxidation of peptides and proteins by hydroxyl radicals. *Anal. Bioanal. Chem.* **2017**, *409*, 2411–2420. [CrossRef] [PubMed]

24. Huang, X.; Kanerva, P.; Salovaara, H.; Sontag-Strohm, T. Degradation of C-hordein by metal-catalysed oxidation. *Food Chem.* **2016**, *196*, 1256–1263. [CrossRef] [PubMed]

25. Bruggeman, P.; Schram, D. On OH production in water containing atmospheric pressure plasmas. *Plasma Sources Sci. Technol.* **2010**, *19*, 045025. [CrossRef]

26. Pei, X.; Lu, Y.; Wu, S.; Xiong, Q.; Lu, X. A study on the temporally and spatially resolved OH radical distribution of a room-temperature atmospheric-pressure plasma jet by laser-induced fluorescence imaging. *Plasma Sources Sci. Technol.* **2013**, *22*, 025023. [CrossRef]

27. Capitelli, M.; Ferreira, C.; Gordiets, B.; Osipov, A. Plasma kinetics in atmospheric gases. *Vacuum* **2001**, *62*, 388–389.

28. Liu, D.; Liu, Z.; Chen, C.; Yang, A.; Li, D.; Rong, M.; Chen, H.; Kong, M. Aqueous reactive species induced by a surface air discharge: Heterogeneous mass transfer and liquid chemistry pathways. *Sci. Rep.* **2016**, *6*, 23737. [CrossRef]

29. Zangouei, M.; Haynes, B. The Role of Atomic Oxygen and Ozone in the Plasma and Post-plasma Catalytic Removal of N_2O. *Plasma Chem. Plasma Process.* **2019**, *39*, 89–108. [CrossRef]

30. Xu, G.; Chance, M. Hydroxyl radical-mediated modification of proteins as probes for structural proteomics. *Chem. Rev.* **2007**, *107*, 3514–3543. [CrossRef]

31. Jones, B.; Vergne, M.; Bunk, D.; Locascio, L.; Hayes, M. Cleavage of peptides and proteins using light-generated radicals from titanium dioxide. *Anal. Chem.* **2007**, *79*, 1327–1332. [CrossRef] [PubMed]

32. Kato, Y.; Uchida, K.; Kawakishi, S. Oxidative fragmentation of collagen and prolyl peptide by Cu (II)/H_2O_2. Conversion of proline residue to 2-pyrrolidone. *J. Biol. Chem.* **1992**, *267*, 23646–23651. [PubMed]

33. Uchida, K.; Kato, Y.; Kawakishi, S. A novel mechanism for oxidative cleavage of prolyl peptides induced by the hydroxyl radical. *Biochem. Biophys. Res. Commun.* **1990**, *169*, 265–271. [CrossRef]

34. Stadtman, E.; Levine, R. Free radical-mediated oxidation of free amino acids and amino acid residues in proteins. *Amino Acids* **2003**, *25*, 207–218. [CrossRef] [PubMed]

35. Morgan, P.; Pattison, D.; Davies, M. Quantification of hydroxyl radical-derived oxidation products in peptides containing glycine, alanine, valine, and proline. *Free Radic. Biol. Med.* **2012**, *52*, 328–339. [CrossRef] [PubMed]

36. Broeck, H.; Cordewener, J.; Nessen, M.; America, A.; Meer, I. Label free targeted detection and quantification of celiac disease immunogenic epitopes by mass spectrometry. *J. Chromatogr. A* **2015**, *1391*, 60–71. [CrossRef]

37. Zhou, L.; Kooy-Winkelaar, Y.M.C.; Cordfunke, R.A.; Dragan, I.; Thompson, A.; Drijfhout, J.W.; Van Veelen, P.A.; Chen, H.; Koning, F. Abrogation of immunogenic properties of gliadin peptides through transamidation by microbial transglutaminase is acyl-acceptor dependent. *J. Agric. Food Chem.* **2017**, *65*, 7542–7552. [CrossRef]

38. Peter, H.; Benjamin, L.; Ruby, G. Celiac disease. *J. Allergy Clin. Immunol.* **2015**, *135*, 1099–1106.

39. Fatma, B.; Barbara, P.; Andrea, F.; Stefano, S. A complete mass spectrometry (MS)-based peptidomic description of gluten peptides generated during in vitro gastrointestinal digestion of durum wheat: Implication for celiac disease. *J. Am. Soc. Mass Spectrom.* **2019**, *30*, 1481–1490.

40. Shi, H.; Cooper, B.; Stroshine, R.; Ileleji, K.; Keener, K. Structures of degradation products and degradation pathways of aflatoxin B1 by high-voltage atmospheric cold plasma (HVACP) treatment. *J. Agric. Food Chem.* **2017**, *65*, 6222–6230. [CrossRef]

41. Wieser, H. Chemistry of gluten proteins. *Food Microbiol.* **2007**, *24*, 115–1193. [CrossRef] [PubMed]

42. Shewry, P.; Tatham, A. The prolamin storage proteins of cereal seeds: Structure and evolution. *Biochem. J.* **1990**, *267*, 1–12. [CrossRef] [PubMed]

43. Nagib, A.; Tifenn, D.; Mohammad-Zaman, N.; Setsuko, K. Tissue-specific defense and thermo-adaptive mechanisms of soybean seedlings under heat stress revealed by proteomic approach. *J. Proteome Res.* **2010**, *9*, 4189–4204.

44. Martinez-Esteso, M.J.; NøRgaard, J.; Brohée, M.; Haraszi, R.; Maquet, A.; O'Connor, G. Defining the wheat gluten peptide fingerprint via a discovery and targeted proteomics approach. *J. Proteom.* **2016**, *147*, 156–168. [CrossRef] [PubMed]
45. Li, M.; Wang, Y.; Ma, F.; Zeng, J.; Chang, J.; Chen, M.; Li, K.; Yang, G.; Wang, Y.; He, G. Effect of extra cysteine residue of new mutant 1Ax1 subunit on the functional properties of common wheat. *Sci. Rep.* **2017**, *7*, 7510. [CrossRef]

International Journal of
Molecular Sciences

Review

Cold Atmospheric Pressure Plasma (CAP) as a New Tool for the Management of Vulva Cancer and Vulvar Premalignant Lesions in Gynaecological Oncology

Pavol Zubor [1,2,*], Yun Wang [1], Alena Liskova [3], Marek Samec [3], Lenka Koklesova [3], Zuzana Dankova [4], Anne Dørum [1], Karol Kajo [5], Dana Dvorska [4], Vincent Lucansky [4], Bibiana Malicherova [4], Ivana Kasubova [4], Jan Bujnak [6], Milos Mlyncek [7], Carlos Alberto Dussan [8], Peter Kubatka [3], Dietrich Büsselberg [9] and Olga Golubnitschaja [10]

[1] Department of Gynaecological Oncology, The Norwegian Radium Hospital, Oslo University Hospital, 0379 Oslo, Norway; yunwang@ous-hf.no (Y.W.); anndoe@ous-hf.no (A.D.)

[2] OBGY Health & Care, Ltd., 010 01 Zilina, Slovakia

[3] Department of Medical Biology, Jessenius Faculty of Medicine, Comenius University in Bratislava, 03601 Martin, Slovakia; alenka.liskova@gmail.com (A.L.); marek.samec@gmail.com (M.S.); koklesova5@uniba.sk (L.K.); peter.kubatka@uniba.sk (P.K.)

[4] Biomedical Centre Martin, Jessenius Faculty of Medicine, Comenius University in Bratislava, 03601 Martin, Slovakia; zuzana.dankova@uniba.sk (Z.D.); dana.dvorska@uniba.sk (D.D.); vincent.lucansky@uniba.sk (V.L.); bibiana.malicherova@uniba.sk (B.M.); ivana.kasubova@uniba.sk (I.K.)

[5] Department of Pathology, St. Elizabeth Cancer Institute Hospital, 81250 Bratislava, Slovakia; kkajo@ousa.sk

[6] Department of Obstetrics and Gynaecology, Kukuras Michalovce Hospital, 07101 Michalovce, Slovakia; janbujnak@hotmail.com

[7] Department of Obstetrics and Gynaecology, Faculty Hospital Nitra, Constantine the Philosopher University, 949 01 Nitra, Slovakia; mlyncekmilos@hotmail.com

[8] Department of Surgery, Orthopaedics and Oncology, University Hospital Linköping, 581 85 Linköping, Sweden; cadussanl@gmail.com

[9] Department of Physiology and Biophysics, Weill Cornell Medicine-Qatar, Education City, Qatar Foundation, P.O. Box 24144 Doha, Qatar; dib2015@qatar-med.cornell.edu

[10] Predictive, Preventive Personalised (3P) Medicine, Department of Radiation Oncology, Rheinische Friedrich-Wilhelms-Universität Bonn, 53105 Bonn, Germany; Olga.Golubnitschaja@ukbonn.de

* Correspondence: prof.pavol.zubor@gmail.com or pavzub@ous-hf.no

Received: 9 October 2020; Accepted: 22 October 2020; Published: 27 October 2020

Abstract: Vulvar cancer (VC) is a specific form of malignancy accounting for 5–6% of all gynaecologic malignancies. Although VC occurs most commonly in women after 60 years of age, disease incidence has risen progressively in premenopausal women in recent decades. VC demonstrates particular features requiring well-adapted therapeutic approaches to avoid potential treatment-related complications. Significant improvements in disease-free survival and overall survival rates for patients diagnosed with post-stage I disease have been achieved by implementing a combination therapy consisting of radical surgical resection, systemic chemotherapy and/or radiotherapy. Achieving local control remains challenging. However, mostly due to specific anatomical conditions, the need for comprehensive surgical reconstruction and frequent post-operative healing complications. Novel therapeutic tools better adapted to VC particularities are essential for improving individual outcomes. To this end, cold atmospheric plasma (CAP) treatment is a promising option for VC, and is particularly appropriate for the local treatment of dysplastic lesions, early intraepithelial cancer, and invasive tumours. In addition, CAP also helps reduce inflammatory complications and improve wound healing. The application of CAP may realise either directly or indirectly utilising nanoparticle technologies. CAP has demonstrated remarkable treatment benefits for several malignant conditions, and has created new medical fields, such as "plasma medicine" and "plasma oncology". This article highlights the benefits of CAP for the treatment of VC, VC pre-stages, and postsurgical wound complications. There has not yet been a published report of CAP on vulvar cancer cells, and so this review summarises the progress made in gynaecological oncology and in other cancers, and promotes an important, understudied area for future research. The paradigm shift from reactive to predictive, preventive and personalised medical approaches in overall VC management is also considered.

Keywords: cold atmospheric plasma; gynaecological oncology; vulva cancer; risk factors; plasma tissue interaction; premalignant lesions; cancer development; patient stratification; individualised profiling; predictive preventive personalised medicine (PPPM/3PM); treatment

---

1. Introduction

Cold atmospheric plasma (CAP) is a highly reactive ionised physical state containing a mixture of physical and biologically active agents. The basic components are the variety of reactive oxygen and nitrogen species formed on reaction with molecules (oxygen, nitrogen, and water) present in the ambient air [1]. Plasma-derived reactive species are free radicals, including oxygen forms (ozone O_3, superoxide anion O_2^-), hydroxyl radical (OH), hydrogen peroxide (H_2O_2), nitrogen dioxide radical (NO_2), nitric oxide (NO), peroxynitrite ($ONOO^-$), organic radicals, electrons, energetic ions, and charged particles [2–7].

Study of their interaction with biological cell or tissue components revealed that biological plasma effects are mediated via reactive oxygen (ROS) and nitrogen species (RNS) which affect cellular redox-regulated processes [8,9], initiating many cellular responses with selectively-targeted anti-tumour effects (e.g., inhibition of cell adhesion, selective apoptosis, necrosis or the inhibition of cell proliferation by disrupting the S-phase of cell replication in tumour cells, suppression of metastatic cell migration, induction of membrane permeation or inducing lethal DNA damage) [10].

The mechanisms underlying this selective cancer cells killing are explained as follows: cancer cells are characterised by a more active metabolic status, resulting in higher basal ROS and RNS levels, making these cells more susceptible to the oxidative stress added by CAP, and especially when cancer cells express high DNA replication activity and there is a high percentage of cells in the S-phase [11–13]. This CAP effect on cancer cells can be further augmented by synergic combination with PAM-nanoparticles (plasma activated medium) [14]. The second obvious result is the significant technical progress in tools allowing CAP application in medicine [15]. All these data show that CAP is beginning to be adopted as a new tool in biomedicine.

CAP operates at body temperature, making it feasible for a variety of medical applications, such as chronic wound treatment; skin disinfection [16–18]; tissue regeneration in chronic leg ulcers [19]; dentistry [20]; in dermatology for the treatment of tumours, actinic keratosis, scars, ichthyosis, psoriasis, atopic eczema, as well as for alleviation of pain and itch [21–23]; and in haematology for blood coagulation [24,25]; in ophthalmology (human corneas) [26]. Recently there has been increased interest in clinical applications in anticancer therapy as a novel promising treatment [27], leading to a new field of medicine called "plasma oncology or plasma medicine" [8,28,29].

The evidence from translational and clinical studies of CAP effects on cancer cells or solid tumours has allowed the extensive use of CAP in the clinical management of cancer patients through both intraoperative and postoperative application for local tumour control. CAP applications in oncology have shown remarkable anticancer effects in vitro cell-lines, including, for example, melanoma [30], cutaneous squamous carcinoma [31], pancreatic [32], liver [33], gastric [34], colon [35], prostate or urinary bladder [36,37], breast [38–44], head and neck cancer [45], osteosarcoma [46,47], glioblastoma [48], lymphoma [49], acute myeloid leukaemia [50], multiple myeloma [51], human fibrosarcoma [52], or lung cancer [53], as well as in vivo solid tumour types in animal (mice) models, e.g., colon [54], breast [55,56], prostate cancer [57], cholangiocarcinoma [58], schwannoma [59], glioblastoma [60], or melanoma [61]. A limited number of studies have been published in oncogynaecology, however, mostly restricted to in vitro cell lines, e.g., cervical [12,62–69], endometrial [70–72], or ovarian [11,73–78].

Vulva cancer and vulvar premalignant lesions (VIN) are suitable for the broad clinical application of CAP in an anticancer approach using therapeutic strategies for the following specific reasons:

(a)   Vulva cancer is technically easy to approach using CAP.
(b)   The effect of radioresistance in subtypes of this malignancy is becoming a clinical problem.
(c)   VIN lesions are commonly treated/managed with local drugs or by applying tracer, which may be suitable for large PAM (plasma-activated medium) treatment.
(d)   Recovery from postoperative vulva surgical site wounds is often prolonged, requires special nursing, and is often combined (in 30–75%) with microbial infections [79,80] in need of antibiotics, whereas the antibacterial effect of CAP may facilitate the healing process.
(e)   Anatomical circumstances usually restrict re-excisions after primary surgery, which is often combined with advanced plastic flaps (e.g., in the case of "worrisome" surgical margins).
(f)   The most common type of vulvar cancer is skin squamous carcinoma (70%) [81,82], followed by melanoma (10%) [83,84] and extramammary Paget disease (1–2%) [85,86], for which CAP has already been clinically validated on both cell lines and human tumours.

The current cancer treatment is focused on the complete surgical eradication of cancer cells and minimum non-malignant tissue. It is difficult to obtain satisfactory free surgical margins in vulvar cancer due to its anatomical specificity and in some cases the close location to the urethra and anus. Despite the intentions of radical excision, moreover, there may be a risk of microscopic tumour residue or local spreading beyond surgical margins, and adjuvant treatment with re-excision or radiation/chemoradiation therapy may be required. Importantly, the majority of these patients are elderly, with comorbidities and reduced wound healing. Conversely, patients with vulva cancer precursors are often young, and the various repeated treatments throughout their lives, including skinning surgery or laser treatment, are associated with a risk of developing dyspareunia due to fibrosis, fissures, and loss of normal anatomy. There is thus a need for more specific treatment modalities for vulva cancer.

The PubMed database was searched up until 15th August 2020 to determine the current knowledge of CAP in oncogynaecology, its technological level, and the biology of tissue interactions, using the search terms "cold atmospheric plasma" and "cancer" (in vitro, in vivo, clinical trials, case reports), resulting in 265 matched articles. Relevant papers included in this systematic review were obtained from the English-language literature, mostly dating from 2015–2020. Specific databases related to plasma physics were also reviewed (American Institute of Physics (AIP), IOPscience, IEEE Xplore), including journals focusing on plasma in Scopus, Elsevier, and the Wiley Online Library.

In-depth analysis of the articles showed that plasma studies were mostly conducted in vitro and concerned direct plasma treatments, followed by PAM. In vivo studies were dominantly performed on mice models. Only sporadic clinical studies have been recorded, mostly in dermatology or head and neck malignancies. The data related to gynaecological cancer were scarce. This review thus offers an overview of CAP-related plasma medicine for female malignancies, and especially vulvar cancer.

The review aims to summarise the potential of CAP for the clinical treatment of vulvar cancer and VIN, as it has not been reported previously, apart from sporadic studies on cell lines confirming the anticancer effect of CAP on cellular proliferation, apoptosis, necrosis, or migration. Our study also provides a comprehensive overview of CAP biology, its interaction with the tissues, the origin of biological processes that are crucial steps in carcinogenesis and surgical wound healing, as well as insight into the modern approaches based on CAP for future medicine. The clinical importance of such reviews is now emerging, and plasma medical devices are widely used in current practice, such as the plasma jet kINPen or InvivoPen [87,88]. The benefits of CAP in clinical application are increasing, most recently in immunotherapy [89,90], and in the combination of CAP and nanoparticles [27,40,91]. CAP thus seems to be an auspicious tool for the development of a new cancer treatment strategy in vulva oncology. Non-thermally operated plasma sources could also be a suitable alternative for the treatment of precancerous and cancerous lesions in gynaecological oncology especially, due to small size and high flexibility of the application probes.

2. Epidemiology and the Prevalence of Vulvar Cancer

Vulvar cancer is a rare disease, accounting for some 5–6% of all gynaecological cancers, but is the fifth most common cancer type after uterine corpus, ovarian, cervical, and vaginal cancer, with breast cancer as the most common malignancy in women. Almost 60% of patients are diagnosed at an early stage, without evidence of local lymph node metastasis and infiltration of the surrounding tissue [92,93]. This malignancy often affects older women, between 60–75 years, and around 90% of all vulvar cancers are vulvar squamous cell carcinomas (VSCC) [94–98]. The incidence of vulvar cancers ranges from 0.6–1.0 cases per 100,000 women and has increased profoundly since the 1970s [99,100]. This trend has been observed not only for postmenopausal women, but also in younger women (almost doubled in 30–49 year age group) because of the increase in HPV-mediated disease [101,102], accounting for 34–40% of vulvar cancers [103–105], and immunocompromising conditions in patients, such as renal transplant recipients [106].

A national Norwegian study reported that prevalence has increased in recent decades (>2.5 times), especially among women under 60 (by 150% in the 0–39 year age group, 175% in the 40–49 year age group and 68% in the 50–59 year age group). One factor discussed was altered sexual activity at young ages without the use of condoms. Although the incidence of VSCC has been increasing for decades in most Western countries, there has conversely been a decreasing trend in some southern European states [107].

More precise knowledge of tumour biology and improvements in therapeutic approaches has resulted in less aggressive surgical treatments in clinical practice, with improved survival [94]. The most important prognostic indicator for survival in women with vulvar cancer is inguinofemoral nodal involvement [108] and, deep multivariate analysis of prognostic factors in primary VSCC also indicates newly assessed perineural invasion. This last parameter was determined as the relevant independent prognostic factor for aggressive behaviour and an unfavourable course in VSCC that should be considered in adjuvant treatment planning [109]. The five-year overall survival rate for localised early-stage vulvar cancer (Stage I/II) varies from 86–90%, to 52.6–60% for locally advanced forms or with locoregional groin lymph node metastatic extension (stages III/IVA), decreasing to 20–22.7% for cases with distant metastases (stage IVB) [110,111]. The age-standardised mortality rate for vulvar cancer in Europe is stated as 0.7/100,000 women [99], and worldwide is 0.3/100,000 [100]. The number of women with high-grade VIN tripled during the last decade (five per 100,000 women), mostly in the HPV-related type [112]. In women ≤ 50 years old, the incidence of high-grade VIN increased by four, and of invasive vulvar cancer by 1.6 [113,114].

3. Aetiopathology, Clinical Aspects and Current Treatment of Vulvar Cancer and Its Premalignant Lesions

3.1. Precursors and Classification of the Disease

VSCC initially develops from squamous precursor lesions of the vulva, which are referred to as vulvar intraepithelial neoplasias (VIN), which were initially graded as VIN1, VIN2 and VIN3; the additional VIN3 differentiated type was also introduced recently [115]. VIN1 was removed in recognition of the aetiological and prognostic differences from histopathological, molecular, and clinical studies, due to its negligible risk for cancer progression. A two-tier classification scheme was proposed: (1) uVIN (usual VIN), including lesions previously classified as VIN2 and VIN3, and (2) dVIN (differentiated or simply VIN) [82].

The precursor lesions of VSCC associated with HPV-infection are currently classified as: (1) low-grade squamous intraepithelial lesion (SIL) of the vulva or vulvar LSIL, encompassing flat condyloma or human papillomavirus effect, and (2) high-grade SIL or vulvar HSIL (which was termed uVIN). The vulvar intraepithelial neoplasia differentiated type (dVIN) is the HPV-unrelated precursor lesion of VSCC [116]. Only HSIL/uVIN and dVIN are considered premalignant lesions for vulvar cancer, with a significantly increased incidence in recent decades [114].

These two different pathways with their own precursor lesions are those that have been identified so far in the development of VSCC, based on detailed histological, immunohistochemical, and genetic abnormalities providing genetic evidence for a clonal relationship between VSCC and its precursors. The first pathway is associated with lichen sclerosis (LS) or other chronic vulvar dermatoses [117–120], and dVIN (HPV-independent VIN) [121], correlated with a higher invasive malignancy risk, mutations of p53-p16(INK4a) and the retinoblastoma tumour suppressor gene involved in the process of malignant transformation [122]. The dVIN is the precursor lesion of keratinising SCC, which is the most common subtype of invasive SCC, accounting for 63–86% of all cases of VSCC [118]. The second pathway is caused by a persistent human papillomavirus (HPV) infection (mostly HPV type 16, 33, and 18), with HSIL/uVIN as the associated precursor of warty and basaloid invasive SCC [116], but with better prognosis, longer disease-free survival [123] and better response to radiotherapy [124] than HPV-negative ones, and this is the same for the invasive form of vulvar cancer [103].

Both precursors, HSIL/uVIN and dVIN, show different risks of progression from that of invasive VSCC. The rate of progression from HSIL/uVIN to VSCC has been reported as less than 5%, but dVIN progresses to invasive VSCC in up to 35% of cases [125].

Traditionally, histology and immunohistochemistry (IHC) have been the basis of the diagnosis and classification of VIN. HSIL/uVIN shows conspicuous histological atypia and positivity on p16-IHC, whereas dVIN shows less obvious histological atypia, and overexpression or a null-pattern on p53-IHC. Other diagnostic immunohistochemical markers have also been evaluated for both types of VIN. The molecular characterisation of VIN has been attempted in a few recent studies, and novel genotypic subtypes of HPV-independent VSCC and VIN have been identified [98].

3.2. Current Treatment of the Disease

As the incidence of premalignant vulvar lesions has increased in recent decades, especially in younger women, is it the knowledge of aetiopathology and risk factors that determines its management [114]. The purpose of treatment for vulvar precursor lesions is to relieve symptoms, prevent cancer progression, and preserve anatomy and organ function [126]. The currently preferred treatment modality for HSIL/uVIN or dVIN is surgical excision, or skinning vulvectomy [127]. Recurrence is not uncommon after treatment, however. One study reported a recurrence rate of about 30% and that around 9–18% of patients with high-grade VIN will progress to cancer [128]. The laser vaporisation of small lesions [129] or medical treatment with Imiquimod (Aldara®) are alternative treatments, and the complete response rates after Imiquimod treatment ranged from 5% to 88% [130].

Surgical treatment is a preferred therapeutic approach in the early stages of vulvar cancer. The standard procedure entails radical local excision of the primary tumour and evaluation of groin lymph node status, either by an elective inguinofemoral lymphadenectomy or sentinel node-dissection, depending on tumour size, focality or the presence of suspected metastatic groin lymph nodes [131]. The adequate clearance of groin lymph nodes is important as recurrence occurs early in the groin, and has repeatedly been reported as fatal, with a median OSR of only 6–10 months [132,133]. Recurrent disease confined to the vulva can be treated with surgical resection only, with cure rates of 20–79%. Here, pelvic exenteration is a therapeutic option with acceptable complication rates for patients with large local recurrences, for whom other treatments are not an option [134]. However, the procedure is associated with a high overall mortality rate. The strict selection of patients is necessary to reach satisfactory surgical and oncologic outcomes.

As the surgical treatment of VSCC is associated with significant morbidity and high recurrence rates, which are related to the limited ability to distinguish (pre)malignant from healthy tissue, there is a need for new tools for the real-time detection of occult tumour lesions and the localisation of cancer margins in patients with VSCC. Several tumour-specific imaging techniques have thus been developed to recognise malignant tissue by targeting tumour markers [135], and new technologies such as CAP are considered for the elimination of micrometastases.

An adjuvant radiotherapy should start as soon as possible after surgery when invasive disease extends to the pathological excision margins of the primary tumour, and further surgical excision is not possible, or for cases with more than 1 metastatic lymph node and/or presence of extracapsular lymph node involvement [136,137]. Despite the radical treatment, up to 12–39% of VSCC across all patients (30% local-regional, 18% distant) experience recurrence [81,138,139]. Routine surveillance is recommended following primary treatment. Most recurrences occur within the first two years after treatment: 32.7% of patients with node-positive cancer and 5.1% among women with negative nodes [140]. Patients with nodal metastatic disease recur at the groin at 10.5 months on average [141].

Advanced stage patients should be evaluated in a multidisciplinary setting to determine the optimal choice and order of treatment modalities. Neoadjuvant chemoradiation should be considered in order to avoid exenterative surgery. Definitive chemo-radiation with weekly cisplatin is the treatment of choice in patients with unresectable disease [136,142,143]. The best treatment option for patients with advanced cancer is combined treatment with surgery and radiotherapy ± chemotherapy. Radiotherapy with a dose of ≥54.0 Gy should be considered to achieve better local control if adverse factors are present [144,145]. The GOG 205 trial demonstrated complete clinical response in 78% patients with T3/T4 tumours following chemoradiation [142]. Primary chemoradiation has become the initial treatment choice for locally advanced disease, followed by resection of residual tumour. The management of patients with extrapelvic metastatic disease focuses on palliative care and the improvement of quality of life by chemoradiation and pain-control with supportive care approaches [146].

Outside current practice, the importance of novel therapeutic approaches for local disease control is emerging, as data from the AGO CaRE-1 study, with an exceptionally long follow-up of 80 months, confirmed that the pathologic tumour-free margin distance did not affect the risk of local recurrence (12.6% in patients with margins <8 mm and 10.2% in cases with a margin at least 8 mm). No differences in local recurrences were found between patients who did or did not receive adjuvant radiotherapy [147–149]. Furthermore, any aim to achieve better local margins control can easily result in mutilation, especially when the primary tumour is located close to the clitoris, as it is in up to 25–37% [102]. This aim could be guaranteed by a peritumoural injection of indocyanine green for the intraoperative identification of surgical margins [150] and CAP application for the selective killing of eventual site micrometastases as a novel tool for a surgeon. This data strengthens the recommendation for a more intense, long-term follow-up for VSCC patients with a history of LS or dVIN [133] and supports the proof of concept for starting studies with CAP for better VSCC control.

4. Current Knowledge of In Vitro Cell Lines and Further Potential for Clinical Application of CAP Oncogynaecology

The application of plasma in cancer treatment is currently a highly topical area of research in its many types. New and significant findings have been demonstrated, most of all in the field of skin, head, and neck cancer, as demonstrated in several studies [30,151]. The first clinical study of the local application of CAP was performed by Metelmann et al. (2015) [152] in 12 patients with advanced head and neck cancer and infected ulcerations, followed by palliative treatment. Here, CAP was applied using a plasma jet, kINPen(®) MED (neoplas tools GmbH, Greifswald, Germany; 1 min/cm^2, 3 times/week, 1–9 cycles), with very promising results, showing an increased number of apoptotic cells in tissue areas previously treated with CAP compared to untreated areas. In the CAP group the clinical tumour surface response was expressed as a flat area with vascular stimulation or a contraction of tumour ulceration rims, and no patients showed signs of enhanced or stimulated tumour growth. CAP did reduce the bacterial contamination of cancer ulcerations, and eased local cancer pain felt by patients. Surgeons indicated that CAP application by plasma jet was easy to handle and extremely precise [152,153]. This started further clinical oriented studies. Schuster et al. (2016) [153] applied CAP with 21 patients with advanced squamous cell carcinoma of the head and neck, reporting increased proportions of apoptotic cells in CAP-treated tissue compared to non-treated ones; and Canady (2017) [154] used plasma as a tool for surgery to enable the complete removal of gastrointestinal tumours in Stage IV patients, and minimise the incidence of recurrence.

At the same time, the potential of CAP in the treatment of gynaecological oncologic diseases can be illustrated by the example of current studies evaluating, for example, breast [55] or ovarian cancer [73]. There have not been any large clinical studies on CAP in gynaecological malignancies, however, although VSCC or cervical lesions are suitable for its use at large scale [69]. Its clinical benefits for local solid tumour management are also supported by the ability of different plasma sources to penetrate solid biological tissues both in vivo and in vitro [155,156]. These studies showed penetration of reactive species generated in plasma (e.g., hydrogen peroxide) deep into the tissue, allowing to study plasma effect on dirty, oily, bloody, and morphologically complex surface (e.g., features present in large ulcerated solid malignant tumours) in the future. This is very important for the potential treatment of tumours. The current status of knowledge and results of CAP application on gynaecological malign cell-lines or tissues are summarised in Table 1.

Table 1. Overview on available studies of cold atmospheric plasma (CAP) in gynaecologic cancer cell lines.

Cell Line Origin	Cell Line/s	Main Effects of CAP on Cell Lines Observed in the Studies	Ref.
Cervix	HeLa SiHa HFB	° Reduced viability of cells after plasma treatment in a dose-dependent manner ° Selective inhibition of proliferation in cancer cells compared to HFB ° Higher inhibition effect in the case of SiHa cells in comparison to Hela cells ° Significant increase of cells in subG0 phase cell and vice versa: reduction of populations in S phase and G2/M phase in a cell-type-specific manner ° Identification of caspase-3, -8 and -9 activation as an important mechanism underlying apoptosis in plasma-treated cells	[12]
Cervix	HeLa HFB detroit551	° Induction of HeLa cell apoptosis by facilitating an accumulation of intracellular reactive oxygen and nitrogen species (RONS) in a dose-dependent manner by both dielectric barrier discharge (DBD) plasma and nitric oxide-plasma activated water (NO-PAW) ° Higher selectivity of NO-PAW at given conditions	[62]
Cervix	HeLa	° Inhibited proliferation and induced cell death in an exposure time-dependent manner ° Significant suppression of the migration and invasion ° Reduced activity and expression of the matrix metalloproteinase (MMP)-9 enzyme ° Decreased phosphorylation level of both ERK1/2 and JNK, but not p38 MAPK	[63]
Cervix	CaSki DoTc2-4510 SiHa C-33-A	° Time- and energy-dependent effects of the treatment on cell proliferation ° Higher sensitivity of cervical cancer cells to plasma treatment in comparison to non-cancerous cervical tissue cells ° Decreased metabolic activity in cancer cells lines when compared to NCCT	[64]
Cervix	CaSki	° Distance and flow rate-dependent effect of CAP on tumour cell viability ° Dose-dependent induction of tumour cell death by CAP treatment	[65]

Table 1. *Cont.*

Cell Line Origin	Cell Line/s	Main Effects of CAP on Cell Lines Observed in the Studies	Ref.
Cervix	HeLa	° Augmented number of early apoptotic cells, late apoptotic cells, but rarely necrotic cells by treatment with N2 and air plasma jets ° Induced apoptotic cell death in a dose-dependent manner ° Increased level of ROS and consequently, induction of apoptosis ° Induction of the mitochondria membrane depolarisation, causing increased mitochondrial transmembrane permeability and release of proapoptotic factors ° Blocking of ROS mediated plasma-induced apoptosis by D-mannitol, sodium pyruvate, carboxyl-PTIO or N-acetyl-cysteine ° Generation of different types and compositions of ROS by different plasma sources	[66]
Cervix	HeLa	° After controlled application of plasma with the precision of tens of nanometres observed killing of plasma-treated cells, neighbouring cells were not affected significantly ° Induction of morphological changes as well as indicators of apoptosis in treated cells ° Crucial role of ROS in cancer cell death induction	[67]
Cervix	HeLa	° Induction of cellular lipid membrane collapse by atmospheric-pressure plasma ° Alteration of electrical conductivity of the cells and induction of lipid oxidation by ROS	[68]
Cervix	SiHa + healthy human cervical tissue cells from cervical conus	° Immediate and persisting decrease in CC cell growth and cell viability associated with significant plasma-dependent effects on lipid structures	[69]
Endometrium	AMEC HEC50	° Reduction of cell viability and induction of cell death by PAM ° Increased autophagic cell death ° Inactivation of the mTOR pathway by PAM ° G2/M-phase arrest in all PAM concentrations ° Induction of intracellular ROS accumulation	[70]
Endometrium	HEC-1 HEC-108	° Reduction of cells containing high levels of aldehyde dehydrogenase (ALDH) - a marker of cancer-initiating cells (CICs) ° Synergistic effect of combined treatment with cisplatin, especially at lower doses ° Combination of plasma and cisplatin treatment is effective both in ALDH high and low cells	[71]
Endometrium	HEC-1 GCIY	° Reduction of cell viability ° Reduction of the number of cells with high aldehyde dehydrogenase (ALDH) production	[72]
Ovary	OVCAR-3 SKOV-3 TOV-21G TOV-112D	° Variation of anti-proliferative efficacy of CAP dependent on treatment duration as well as on the OC cell line used ° Decreased motility, invasion, and metastasis potential ° Culture medium treated with plasma before addition mediates the CAP effect on the cells, however, this effect depends on the cell medium composition	[73]
Ovary	SKOV-3 OV-90 HOSE	° Selective anticancer activity of plasma-activated Ringer's Lactate solution (PA-RL) containing reactive oxygen and nitrogen species (RONS)	[74]
Ovary	TOV21G ES-2 SKOV3 NOS2 OHFC HPMC	° Decreased viability of CCC cell line after plasma-activated medium treatment ° Induction of morphological changes in EOC cell lines treated with PAM ° Anti-tumour effects mediated by produced ROS ° Selective anti-proliferative effect on cancer cells without causing adverse reactions in normal cells	[75]

Table 1. *Cont.*

Cell Line Origin	Cell Line/s	Main Effects of CAP on Cell Lines Observed in the Studies	Ref.
Ovary	NOS2 NOS3 NOS2TR NOS2CR NOS3TR NOS3CR	° Decreased viability of ovarian cancer cells treated with PAM in plasma activation time-dependent manner ° Treatment with PAM decreased proliferation rate of paclitaxel and cisplatin-resistant cells derived from parental cell lines ° Addition of ROS scavenger into activated medium decreases anticancer activity, the addition of ROS scavenger inhibitor re-established anticancer activity, thus this point on the crucial role of ROS in an anti-tumour mechanism	[76]
Ovary	K2 K2R100 TOV-21G ES-2	° An anti-tumour effect of PAM on acquired chemo-resistant OC cells ° An anti-tumour effect of aqueous plasma against clear-cell carcinoma, which is natively chemo-refractory OC ° PAM has a selective cytotoxic effect on OC cells	[77]
Ovary	SKOV3 HRA	° Effective killing of ovarian cancer cells lines by the plasma, while plasma-treated fibroblast cells were not damaged ° Plasma treatment induces apoptosis ° The exposure time of treatment affects the proliferation rate	[78]
Ovary	OVCAR-3 NOS2 TOV21G ES-2	° Negative impact of cell density on PAM-induced proliferation inhibition rate ° Selective, cell line dependent sensitivity to PAM ° Dependence of PAM effect on the proportion of ROS and the cell number ° Sensitivity to PAM affected by morphological characteristics of the cells ° TGF-β induced epithelial-mesenchymal morphological transition sensitised cancer cells to PAM	[11]
Ovary	ES2 SKOV3 WI-38 HPMCs	° Inhibition of cell viability of ovarian cancer cells depends on the cell type, cell number, and plasma-activated medium (PAM) dilution ratio ° PAM mediated suppression of cell migration, invasion, and adhesion ° PAM-induced down-regulation of matrix metalloproteinase-9 (MMP-9) prevents cell plantation in co-culture with human peritoneal mesothelial cells ° Inhibition of anti-metastatic effect of PAM by the ROS scavenger	[157]
Breast	MCF-7	° CAP inhibitory effect on the cell proliferation is mediated by miR-19a-3p (miR-19a, oncomiR) ° CAP induces hypermethylation at the promoter CpG sites and subsequent downregulation of miR-19a ° CAP recovers production of ABCA1 and PTEN which are targets of miR-19a	[38]
Breast	MCF-7 MCF-7/TamR	° CAP induces restoration of sensitivity to tamoxifen (Tam) in Tam-resistant cells ° Increase of ROS levels in CAP-treated cells ° Inhibition of the proliferation and promotion of the apoptosis in MCF-7/TamR ° Oppositely altered expression of 20 genes involved in Tam resistance in TamR cells and CAP-treated TamR cells ° *MX1* and *HOXC6* mediated the restoration of sensitivity against Tam	[39]
Breast	MSC MDA-MB-231	° Synergistic inhibition of breast cancer cell growth after treatment with the combination of CAP and drug (5FU) loaded core-shell nanoparticles ° Induction of down-regulation of metastasis-related genes (*VEGF, MTDH, MMP9,* and *MMP2*) ° Facilitation of the uptake of drug-loaded nanoparticles	[40]
Breast	MCF7 MCF10A MTT	° Reduction of the viability of breast cancer cells ° Significantly lower CAP-induced damage on normal cells ° Enhanced reduction of cancer cells viability after addition of 5% oxygen to the helium plasma	[41]
Breast	metastatic BrCa cells MSC	° CAP-induced selective ablation of metastatic BrCa cells in vitro without damaging healthy MSC ° Inhibition of the migration and invasion of BrCa cells after CAP treatment ° Different BrCa cell and MSC responses under varied CAP conditions	[42]
Breast	MCF-7	° Induction of apoptosis in cultured human breast cancer cells ° Significant portion of CAP-treated cells exhibits apoptotic fragmentation, with only limited necrosis	[43]

Table 1. *Cont.*

Cell Line Origin	Cell Line/s	Main Effects of CAP on Cell Lines Observed in the Studies	Ref.
Breast	MDA-MB-231 MCF-7 HMEC	° ROS in a liquid phase is generated via plasma irradiation of gas, producing the reactive species (electrons, ions, and radicals) and these species dissolve into the liquid phase and/or react with water ° Irradiation time, distance to the liquid surface and voltage affects OH radical generation in the extracellular culture medium	[44]
Breast	MDAMB231 MDAMB468 MCF7 MCF10A	° Induction of apoptosis, inhibition of the proliferation and migration of triple-negative breast cancers (TNBC) after PAM treatment ° Significant increase of H_2O_2 concentration in the media after CAP treatment ° PAM selectively inhibits the activity of JNK and NF-κB in TNBC cells	[55]
Breast	4T1	° Inhibition of cell migration after both plasma and doxorubicin treatment, assessed by wound healing assay	[56]
Breast	MCF-7 MCF-7/TxR	° Restoration of sensitivity to paclitaxel in resistant cells ° Identification of altered expression of multiple drug resistance-related genes ° *DAGLA* and *CEACAM1* were essential for the acquisition of resistance and the recovery of sensitivity	[158]

5. Plasma Physical and Chemical Characteristics and Plasma Sources in Medicine

Advancement in medicine was, for decades, characterised by the introduction of innovative technologies from physics to improve the diagnostic and therapeutic management of patients. From X-rays, magnetic resonance, nuclear medicine, PET-CT, and digital mammography to sophisticated radiation therapy (including intraoperative devices), all these technologies revolutionised medicine and brought enormous benefit for patients. In the last decade, a new form of technology is gaining relevance, bringing many opportunities for patient care, called physical plasma. Plasma is commonly known as the fourth state of matter (solid, liquid, gas, and plasma) [159]. Initially used for skin regenerative medicine [160], it is nowadays studied as regards anticancer treatment [27,28,161]. Depending on the plasma force, physical action is based on positive and negative ions, electrons, neutral atoms, photons, and electromagnetic fields, leading to the emission of visible ultraviolet (UV) radiation and thermal effects.

Fundamentally, plasma consists of an ionised gas enriched with biologically and chemically reactive species, including charged electrons and ions, as well as radicals, atoms, and molecules in neutral (e.g., excited) or charged forms, where the electric charge can be positive or negative. In addition to chemical species, plasmas produce electromagnetic radiation, propagating disturbances such as shock waves and heating, among other effects. Medically relevant plasmas (termed CAP) benefit from low intensities of these individual effects, making them a gentle tool that can induce desired biological effects in a controlled manner [20]. CAP is generated under atmospheric pressure at ambient temperatures ranging from 20 °C to 50 °C [162].

Artificial plasma can be classified based on gas pressure (low-pressure vs atmospheric pressure plasma) or based on temperature (thermal/hot vs. nonthermal/cold plasma). Plasmas can be easily generated by applying an electric field to the process gas, typically pure helium or argon, or to a mixture including oxygen. This electric field accelerates electrons and initiates a cascade of chemical reactions that give rise to a diverse range of chemical species. The amount of applied energy and the type and pressure of the processing gas determine both the speed (and thus the temperature) and the chemistry of this cocktail of species. In medicine, low-temperature plasmas that can be generated at atmospheric pressure are desirable, due to the simplicity, versatility, and affordability of such plasma devices.

Clinically, plasma-based electrosurgical devices have long been employed for blood and tissue coagulation, cutting, desiccation, and cauterising during surgery [163,164]. These devices involve heating tissue and their effects are primarily heat mediated. Recently, new sources of CAP with well-controlled temperatures below 40 °C have been designed and clinically applied in plasma medicine. The nature of direct plasma treatment renders it highly suitable for the treatment of primary tumours that arise from skin or mucosal surfaces. This technology may complement surgery as adjuvant therapy or specific therapy in combination with chemotherapeutics or radiation. Of particular clinical interest is the ability of CAP to penetrate tissues and effectively target cancer cells that have infiltrated healthy tissue adjacent to the tumour mass, and to eliminate micrometastases [161].

Sources of Cold Atmospheric Plasma

New CAP sources used in plasma medicine can be classified into three types [162,165,166]:

1. Direct plasma sources: These plasmas use the human body (such as the skin, internal tissues, etc.) as an electrode. Thus, the current produced by plasmas has to pass through the body. The most commonly utilised technology in this category is the dielectric barrier discharge (DBD) plasma source. The major disadvantage of this technique is the application distance (between the electrodes) which must remain within a close range, generally less than three mm^2, thus limiting its use for small areas of the human body [15].

2. Indirect plasma sources: These plasmas are generated between two electrodes. Active species that are created by the plasmas are subsequently transported to target application areas. Several devices are available, ranging from very narrow plasma needles or jets to larger plasma torches such as the kINPen® MED, Atmospheric Pressure MicroPlasma Jet (APMPJ), InvivoPen, and MicroPlaSter® α and β. Plasma jets can be classified according to parameters such as discharge geometry, electrode arrangement, excitation frequency or pattern.

3. Hybrid plasma sources: These plasmas combine the benefits of the two aforementioned plasma source types (e.g., using the plasma production technique of direct plasma sources and the essentially current-free property of indirect plasma sources). This is achieved by introducing a grounded wire mesh electrode, which has significantly smaller electrical resistance than that of the tissue. Thus, in principle, all current can pass through the wire mesh. The MiniFlatPlaSter is an example of a hybrid plasma source.

As a novel technology CAP expanded very quickly to several industrial and medical fields and rapidly increased its applications as a medical device or drug-mediated tool [167]. In biological applications, the most commonly used plasmas are atmospheric pressure plasma jets (APPJs) and dielectric barrier discharges (DBDs) [168].

Various types of APNP-Js with different configurations have been reported, where most of the jets are working with noble gas mixed with a small percentage of reactive gases, such as O_2. Plasma jets operating with noble gases can be classified into four categories, i.e., dielectric-free electrode (DFE) jets, dielectric barrier discharge (DBD) jets, DBD-like jets and single electrode (SE) jets [169].

Several different gases can be used to produce cold atmospheric plasma, such as helium, argon, nitrogen, heliox, and air. Cold atmospheric plasma is created by many methods [170]. Each unique method can be used in different biomedical areas. A variety of different CAP devices have been developed and tested for research and clinical purposes. To date, four plasma devices have been certified for medical purposes. In 2013, the medical device kINPen® MED plasma-pen (INP Greifswald/neoplas tools GmbH, Greifswald, Germany), an APPJ, and PlasmaDerm® VU-2010 (CINOGY Technologies GmbH, Duderstadt, Germany), a DBD source, have been CE-certified in Germany by MEDCERT under the norm ISO 13485, and the InvivoPen system is used for laboratory conditions. The medical device SteriPlas plasma torch (Adtec Ltd., London, United Kingdom) was then certified for use in the treatment of chronic and acute wounds, as well as for reduction of microbial load [87,88,171,172]. Their great advantage, apart from favourable medical use, is their relatively low manufacturing costs [18], allowing for a reduction in the financial burden imposed on health budgets by conventional treatments.

The technology that brought CAP into medicine via experience with clinical applications for local disease control is currently intensively studied as a novel therapeutic agent in oncotherapy. Two methods of applying plasma are described: direct treatment and indirectly using PAM-nanoparticles and PAL (plasma-activated liquids). The first method consists of applying CAP directly to in vitro cells, in vivo animal models, or living human tissue. The second strategy consists of producing PAM and then applying (injecting) it into cell cultures or tumours. These approaches have been studied in recent years, and not only the number of cell lines-type studies, but in vivo studies based on animal models, human tissue medium, or clinically conducted on particular patients, proved its large anticancer potential, with advantages for patients suffering from malignancies [15,162,165,166].

6. Plasma Interaction with Human Tissue

When CAP is applied, it induces both physical effects (production of ultraviolet rays, heat, and electromagnetic fields), and chemical effects (production of ROS/RNS = RONS). Whereas physical effects seem to have a negligible cellular impact, RONS may induce cell membrane alterations, lipid peroxidation, transient poor formation, alterations in protein structure, an increase in intracellular ROS/RNS, DNA double-strand brakes, and subsequently apoptosis (mitochondrial or cellular) [173], without causing thermal damage to the surrounding tissue [174]. Importantly, the source of plasma plays an essential role in cell/plasma interactions. Generally, it is accepted that low-dose plasma is associated with stimulation of processes such as cellular viability, the promotion of cell proliferation and migration. On the other hand, high-dose plasma leads to cellular apoptosis and necrosis, demonstrating apoptosis-independent anti-proliferative cell effects. Furthermore, a dose-dependent increase of cells observed in the G1 phase of the cell cycle indicates the important role of cell cycle regulation for anti-proliferative CAP mechanisms [175].

The first human-based tissue interactions with CAP were observed on fibroblasts and keratinocytes, which are two dominant cell types associated with wound healing, and that can be stimulated via CAP [176]. Ngo et al. (2014) [177] showed that atmospheric N2/Ar micro-plasma stimulated fibroblast proliferation and migration via the release of fibroblast growth factor-7. In another study, the authors used different plasma sources to stimulate keratinocytes. CAP activated molecules are also associated with angiogenesis in skin human epidermal keratinocytes, endothelial cells, and dermal fibroblasts [178]. Wound re-epithelisation after CAP intervention was also detected in a model of full-thickness acute skin wounds in rats [179]. In the same way, the use of N2/Ar plasma therapy to partial thickness skin wounds on murine [180] or mice [181] models resulted in wound healing promotion by altered keratinocyte and fibroblast migration, and changes in adherence junctions and cytoskeletal dynamics as shown by the downregulation of E-cadherin and several integrins, as well as actin reorganisation. The application of CAP on a diabetes model also revealed acceleration in wound healing accompanied by faster re-epithelialisation with the formation of a new epidermis layer, collagen deposition, less inflammation, as well as neovascularisation [182]. In vivo experimental models are now the next appropriate subjects to further analyse the positive impact of CAP on wound healing. There is a great need to address this issue as CAP could become an additional tool in vulva cancer surgery and postoperative management, especially among obese, immobile, or diabetic patients.

7. Plasma Promoted Wound Healing and Its Possibilities in the Surgical Treatment of VSCC

Cutaneous wound healing is a complicated process involving various cells and cytokines. It is divided into an inflammatory, a proliferative, and a remodelling phase. Due to its complexity, it is easily affected by internal and external disturbances, which may lead to chronic or even non-healing wounds, causing serious medical problems [183]. Patients with chronic wounds have a poor health-related quality of life in general, and wound-related costs are substantial. As the prevalence of chronic wounds is greatly increasing [184], the development and implementation of wound management strategies that focus on increasing health-related quality of life and effectively reduce costs for this patient group are urgently needed. Here, CAP exerts its beneficial effects through various mechanisms. CAP may facilitate the transformation of a chronic wound from a stagnating wound to an acute healing wound, by inflammatory and proliferation supporting stimuli [185,186], including neovascularisation [187]. Some studies reported the positive effect of CAP on angiogenesis. ROS and RNS also belong among pro-angiogenic growth factors (e.g., VEGF, EGF, FGF, TGF) and cytokines (e.g., IL-1, 2, 6, 8; TNF).

It seems that ROS/RNS may have an important role in wound vascularisation [188]. This is of enormous importance in patients with postoperative surgical skin flaps or site infection in vulva cancer patients suffering from comorbidities (e.g., obesity, diabetes, and vasculopathies).

Although the trend of surgical treatment in vulvar cancer patients is towards less extended resections, a significant number of cases are still diagnosed with locally advanced diseases, requiring extended resections. The development of early and late postoperative complications following vulvar surgery is thus still a clinically important issue. Bacalbasa et al. (2020) [189] found that the risk of postoperative complications was significantly affected by: (i) the stage of the disease, (ii) the preoperative levels of serum albumin, (iii) the status of the resection margins, (iv) previous history of irradiation, (v) length of hospital stay, and (vi) the association of comorbidities. The most frequent complication was wound dehiscence, necessitating reoperation (21%), followed by urinary tract infection and lower limb lymphedema (both 17.3%). Authors indicate for the precise selection of cases submitted to surgery, which further supports the need for new therapeutic approaches and tools in the management of patients with vulva cancer. Once such complications occur, the first step of treatment is debridement to remove necrotic tissue and exudate, which is conductive to bacterial growth. Simultaneously, systemic or topical antimicrobial agents should be used to eliminate the extensive bacterial burden. The use of antimicrobial agents is often limited by hypersensitivity to antibiotics, however, and the increasing development of drug-resistant bacteria. Novel therapeutic alternatives to improve wound healing, especially on the vulva with problematic healing process are thus greatly needed. In view of all these complications, CAP has enormous potential to achieve a better postoperative outcome for patients.

As mentioned previously, CAP has a broad spectrum of medical applications due to its beneficial properties, including its antimicrobial effect, and the promotion of wound healing. Recent evidence has suggested that CAP intervention enhances the healing process via a reduction of the bioburden, and also via the stimulation of angiogenesis and production of skin cells. An antimicrobial effect was demonstrated in the early 1990s, leading to its application in the clinical sphere. CAP also has great potentials in regenerative medicine as a powerful tool for the treatment of chronic or acute wounds. The promising role of CAP as a medical approach has also been described in dermatology, including the impact of CAP on atopic dermatitis, pruritus, or psoriasis [190]. Several studies demonstrated the positive effect of CAP on the eradication of bacterial infection in chronic wounds associated with the promotion of healing processes [191]. Cold plasma successfully eliminated bacterial colonisation in patients with chronic leg ulcers [19], chronic wounds [192], or chronic venous leg ulcers [193], and resulted in enhanced healing of chronic wounds. Cold atmospheric argon was also observed to have a significant effect in patients with skin graft donor sites on the leg. Data revealed that cohorts of treated patients demonstrated better healing courses than placebo groups the second day after CAP intervention [194]. CAP has demonstrated a positive effect on skin grafts in leg surgery, and it would be interesting to find out whether the same benefit would be observed in vulva surgery, where skin grafts are commonly used after radical tumour resections.

Metelmann et al. (2013) [195] analysed the effect of CAP in volunteers who had received ablative laser skin lesions. Experimental findings showed that the application of CAP promoted the inflammatory reaction necessary for tissue recovery in the early stage of the wound and also prevented posttraumatic skin disorders. There were no side effects of CAP associated with the development of precancerous skin lesions observed in tested individuals [195]. CAP was examined as a medical option for the acceleration of acute wound healing in a comparative study with different treatment groups (control, CAP, local treatment using betamethasone valerate ointment, and the application of basic fibroblast growth factor sprays). The results showed no significance between tested groups in wound healing; however, CAP demonstrated a more rapid recovery accompanied by a reduction in the redness and roughness of the skin. The authors observed no negative side effects from using cold plasma in the CAP group [196]. Recently, CAP was applied for the improvement of wound healing in different types of superficial skin erosion wounds, including patients with pyoderma gangrenosum, trauma wounds, giant genital warts, diabetic foot, and chronic eczema. According to data obtained from the different wound types, CAP accelerates wound healing through the eradication of bacterial colonisation, sterilisation of the wound, changing the local wound environment, and the promotion of tissue restoration [197].

These studies all demonstrated the significant clinical effect of CAP in healing processes with human subjects. The method was successfully used for pathogen eradication from both chronic and acute wounds via its biocidal effect. Evidence also suggests the beneficial role of CAP in the acceleration of healing different wounds without side-effects (i.e., premalignant lesions). As a result, CAP is an appropriate clinical approach for the treatment of wounds after surgical intervention, mostly for minimizing prolonged wound healing, which is associated with a poor prognosis due to delayed adjuvant therapy [198].

8. CAP Specific Abilities Predisposing Its Application in Anticancer Therapy

It is generally accepted that CAP accelerates the healing of wounds with limited side-effects, and also has anticancer properties, and thus it would be very interesting to analyse the potential of its use in the treatment of both premalignant lesions and developed malignancies. The anticancer effects of CAP can be observed at several cellular or molecular levels (Figure 1), and can be briefly described as:

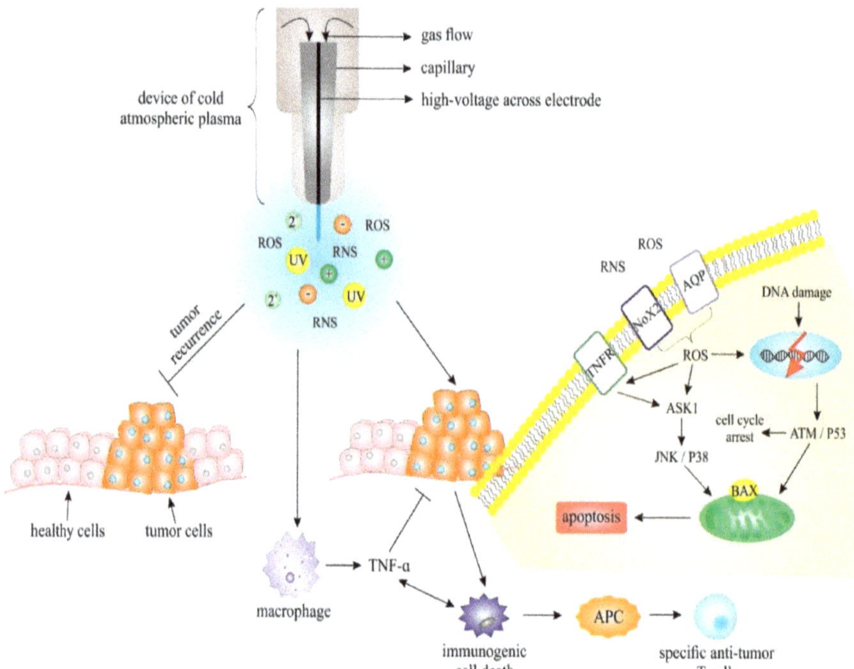

Figure 1. The mechanism of cold atmospheric plasma (CAP) in cancer treatment. Formation of plasma starts in high electric field across the region of gas (pure helium or argon, and/or their mixtures with oxygen) that accelerates electrons. These processes lead to the initiation of a cascade of chemical reactions associated with the generation of various chemical species. CAP is a source of highly reactive species (ROS, RNS, atomic oxygen, hydroxyl radical, superoxide, nitrogen oxides, and singlet delta oxygen), neutral particles (photons and neutrons), electrons, and physical factors (electromagnetic field and UV radiation) [199]. Reactive species produced by CAP have the ability to directly induce DNA damage and cell cycle arrest resulting in the apoptotic signalling of tumour cells. The production of reactive species can activate macrophages leading to higher elevation of TNF-α–mediated NF-κB activation and the expression of proinflammatory genes associated with tumourigenesis. On the other hand, CAP seems to be an effective inhibitor of TNF-α–mediated NF-κB activation with a potential role in anticancer strategies.

CAP can also induce the immunogenic cell death (ICD) of tumour cells that lead to systematic immune response. ICD can also be achieved by the regulation of various cytokines, including TNF, that play a crucial role in the creation of immunogenic microenvironment [200]. Abbreviations: ROS, reactive oxygen species; RNS, reactive nitrogen species; AQP, aquaporin; TNFR, tumour necrosis factor receptor; Nox, NADPH oxidases; ATM, ataxia-telangiectasia mutated kinase; JNK, c-Jun N-terminal kinase; ASK, apoptosis signal-regulating kinase; APC, antigen-presenting cell; TNF-α, tumour necrosis factor alpha; Bax, Bcl-2-associated protein X; UV, ultraviolet radiation; APC, antigen-presenting cell.

8.1. CAP Effect on Cellular and Extracellular Level

The expected basic cellular responses (apoptosis, growth inhibition, selective cancer cell death, cell cycle arrest, DNA and mitochondrial damage, a selective increase of ROS or immunogenic cell death) have been observed after the application of CAP to cell lines and/or tissue [28,201]. Several studies demonstrated the impact of CAP on different cellular processes associated with the suppression of cancer development via modulation of gene expression and other intracellular events [111,177,178,202]. Despite the current focus on CAP as a promising strategy for pathogen eradication contributing to wound healing, the exact mechanisms of the anticancer effect are not known [203]. Additionally, it is important to note that the interaction between plasma and tumour cells is essential, and the impact of plasma on the tumour microenvironment (TME) also plays a significant role in anticancer therapy [204].

Recent evidence revealed the effect of CAP on different compartments of TME (endothelial cells, immune cells, fibroblasts, collagen, fibronectin, elastin, proteoglycan, or glycoproteins). It has been reported that the prolonged application of CAP suppressed the collagen production and cell viability of murine fibroblast cells [205]. Similarly, a reduction of collagen secretion was demonstrated in keloid fibroblasts [206,207] after CAP intervention.

The specific microenvironment of tumour cells causes different responses to increased levels of ROS and RNS, which subsequently leads to apoptosis [208]. Higher levels of cholesterol in plasmatic membrane are also typical for the majority of cancerous cells, and most notably for multidrug resistant cells. This is also accompanied by higher levels of phosphatidyl choline, phosphatidyl ethanolamine and phosphatidyl inositol, which makes the plasmatic membrane of these cells more rigid, and also less permeable for drugs. Conversely, the plasmatic membrane of metastatic cells possesses lower cholesterol, which makes them less rigid, and this facilitates these cells in entering the blood vessels [209,210]. Some studies suppose that lower levels of cholesterol in plasmatic membrane can be tissue specific in some types of breast and prostate cancer, regardless of their metastatic potential [209–211]. Importantly, lower levels of cholesterol in the plasmatic membrane of some tumorous cells also make the membrane more susceptible to peroxidation which results in higher pore formation, enabling the higher diffusion of ROS and RNS into the cell [28,202,212,213]. Lower levels of cholesterol are present only in a smaller portion of tumorous cells, but there are a few other features of the tumour microenvironment that are typical for wider variety of these cells, which means that CAP application results in the induces apoptosis.

Another significant feature of the cancer microenvironment is the generation of superoxide anion O_2^- into the ECM and the presence of protective catalase on the external surface of the cell membrane. The abundance of O_2^- in the vicinity of the cancer cell membrane, which can be achieved by CAP application, triggers specific HOC1 and $ONOO^-$ cell signalling pathways [214,215]. This subsequently leads to the formation of reactive OH radicals, lipid peroxidation and apoptosis. The presence of protective catalases associated with the external membrane can also be disrupted by CAP application [208,216]. A malignant cell microenvironment demonstrates higher activity in the proteasome complex involved in the degradation of intracellular proteins. It affects variable mechanisms in cancer cells, and it is also very significant in the regulation of apoptosis [217,218]. Proteasomes in malignant cells exhibited more sensitivity to the cytotoxic effect of their inhibitor compared to healthy cells, and the medical targeting of proteasomal activities thus became interesting for basic and clinical research [217,219,220].

8.2. CAP and Apoptosis

Apoptosis is the tightly regulated pre-programmed process of cell death essential for physiological homeostasis maintenance. The mechanism of apoptosis is regulated by caspases and occurs through two distinct molecular pathways. The extrinsic pathway is activated by the binding of extracellular death ligands, such as TNF, Fas-L, and TRAIL, to its death receptors. The intrinsic – mitochondrial derived pathway is initiated by intracellular stimuli and involves pro- and anti-apoptotic factors such as Bcl-2 proteins, cytochrome-*c*, and APAF-1 [221–225]. There has been strong interest in the targeted induction of apoptosis in recent years, as it is a very efficient non-invasive treatment [222,224,226].

Cold atmospheric plasma is also a potential targeted cancer treatment tool, as cancer cells are very sensitive to CAP-induced ROS [227]. Several studies have analysed different cell lines in apoptotic content. The loss of cell viability and shrinkage of tumours occurred mainly as a result of apoptotic processes, as evident from the specific morphological changes and higher activity of apoptotic cascade members [28,151,228,229].

A study of SiHa and HeLa cervical cancer cells treated by micro-DBD plasma revealed the different responses of one tissue type to CAP. SiHa cells had a significantly higher caspase-3 activity and thus lower survival rate, and a higher number of aberrantly expressed apoptosis-related genes compared to HeLa cells. CAP treatment also led to the alteration of 166 genes in the control fibroblast lines. The activity of caspases 6, 8, and 9 were similar in SiHa and HeLa cells. It was an interesting observation that CAP-treated cells entered them to the subG0 phase, both of cancer and fibroblast control cell lines [12].

Xia et al. (2019) [151] described the effect of ROS produced by CAP on the extrinsic apoptosis pathway members in A375 and A875 melanoma cell lines. Higher ROS dosage led to the overexpression of antagonistic protein SESTRIN 2, which resulted in the phosphorylation of p38 MAPK and increased expression of iNOS, FAS, and FASL. These changes triggered the activation of caspase 3 dependent apoptosis in the studied cell lines [151]. The increased activation and phosphorylation of JNK and p38 MAPK pathways was also observed after CAP application in HeLa cells [230], head and neck cancer cell lines [45], anaplastic thyroid cancer cell lines [231] and in vivo conditions in tumorous tissues in FaDu mouse xenograft models [45]. The CAP application resulted in the depolymerisation of mitochondrial membrane, accumulation of intracellular ROS and activation of caspase family protein. Similar results were also published by Kaushik et al. [5] in 2015 regarding altered phosphorylated ERK1/2/MAPK protein levels. They analysed various cell lines (MRC-fibroblasts, A549-lung carcinoma, T98G-glioblastoma, and HEK293-human embryonic kidney cells) and observed altered mitochondrial membrane potential and increased activation of caspase apoptotic mechanism. Apoptotic regulators located in the outer membrane of mitochondria, *BAX* and *BAK1* genes were upregulated. A higher expression of *H2AX*, a histone protein, with a phosphorylated form that can be considered a marker of DNA damage, was also observed. On the other side, *BCL-2* was downregulated in solid tumour cells. An increase of *BAX* and decrease of *BCL2* gene expression was also observed in breast cancer cell lines (MCF-7) treated by plasma and a combination of plasma and iron nanoparticles (NPs). The viability of cancer cells was significantly decreased and *BAX/BCL-2* ratio was altered in favour of apoptosis [227]. Yan et al. (2017) [27] also described the activation of apoptosis by ROS-stress response signalisation and regulation by BCL-2 protein family. CAP induced a sub-G(1) arrest in p53 wild-type OSCCs and increased the expression levels of ATM, p53, p21, and cyclin D1, confirming the involvement of DNA damage and triggering sub-G(1) arrest via the ATM/p53 pathway in the apoptosis mechanism [232]. Loss of viability, higher numbers of cell cycle arrests, and the increased activity of caspase 3 connected with a higher apoptosis rate after CAP treatment were observed in several cell lines, including HeLa, squamous carcinoma YD-9 cell lines and melanoma G361 cell lines, however, these changes were more significant in p53 mutated cell lines compared to wild type p53 cells [233].

Whether the application of CAP will initiate apoptotic or other processes depends to a large extent on the duration of exposure, distance, dose and duration of exposure and gas content. Low dose CAP treatment and an exposure less than 60 s leads to increased proliferation and wound healing, but a bigger dose and longer exposure time lead to controlled cell death [12,234–236]. Finally, known data indicates that CAP also seems to have a strong apoptotic effect on cancer cells resistant to current treatments. The mechanisms involved seem to depend to variable extents on p53, p38, NF-KB, JNK or caspase pathways [28].

8.3. CAP and Induced Gene Expressions, Proteomic and Epigenetic Changes

CAP, with its anticancer effects, can induce DNA damage and cell cycle exit into senescence [166,229]. Welz et al. (2015) [237] demonstrated that CAP could decrease cell viability and increase DNA fragmentation leading to cell apoptosis. Furthermore, specific CAP-binding proteins and intracellular ROS can induce the expression of genes involved in cellular apoptosis mediated by TNFα and apoptosis signal-regulating kinase (ASK) [166]. The active genetic expressions with corresponding mRNAs transcriptions were also observed for genes encoding IL-12 (downregulation) and IL-1β, IL-6, IL-8, IL-10, TNFα, VGFR, and interferon-gamma (upregulation) after CAP exposition [217]. In vitro and in vivo studies aimed at wound healing also showed that plasma might induce the expression of IL-6, IL-8, MCP-1, TGF-b1, and TGF-b2 genes, which is crucial for the healing process [186]. The genomic impact of CAP is also demonstrated in the high selectivity for cell death and the removal of tumour cells from the proliferative phase of the cell cycle. Yan et al. (2015) [238] demonstrated that CAP increased the percentage of apoptotic tumour cells by blocking the cell cycle at the G2/M checkpoint, and this effect was mediated by reduced intercellular cyclin B1 and cyclin-depend kinase1, increased p53 and cyclin depending on kinase inhibitor and an increased Bcl-2-like protein4 (BAX)/B cell lymphoma2 (Bcl-2) ratio. Increased amounts of keratinocytes associated with the antiproliferative effects of CAP were also found in the G2/M1 phase [238].

The presence of reactive plasma species can also affect proteins and protein-based structures [239]. Protein modification is mainly initiated by ROS and RNS that can lead to etching, the cross-linking of proteins, oxidative reactions in protein building blocks, and cause the cleavage of proteins into peptides. Some studies report that functional groups such as carboxylic acid or amide bonds can be introduced to the surface of polymers. Tolouie et al. (2018) [239] demonstrated that CAP exposure can selectively alter the protein conformation and function, depending on biological origin, plasma type, and treatment conditions. Interestingly, the effect of CAP on enzymes is inconsistent. In some cases, CAP deactivates enzymes, whereas on the other hand, there are situations where CAP exposure led to increased enzymatic intracellular activity. The inactivation/activation of enzymes after plasma exposition depends on the ability of the cellular defence system to confront stress-induced situations [240].

It is known that CAP-activated media can mediate the anticancer effect on tumour cells. Utsumi et al. (2013) [75] described the effect of CAP-activated media for epithelial ovarian carcinoma cells. The aim of CAP exposure was the inhibition of tumour growth and promotion of apoptosis. CAP exposure can temporarily disrupt the cell membrane and affect intracellular signalling pathways. An interesting study by Schaner et.al (2003) [241] characterised gene expression in epithelial cancers of the ovary. This study showed that the most expressed genes in ovarian carcinomas were PAX8 (paired box gene 8), EFNB1 (ephrin-B1) and mesothelin. The study also revealed that numerous genes have different expression. The authors detected the overexpression of the transcription factor ATF3. The main role of ATF3 is to repress matrix metalloproteinase 2. The expression of ATF3 was higher in the ascites samples. The study also followed the expression of oestrogen receptor 1 and cytochrome P450 4B1. Their production was at relatively low levels in clear cell cancers, compared with other ovarian cancers. It is also interesting that E-cadherin was highly expressed and a member of the discoidin domain receptor family (DDR1) had a lower level of expression in clear cell cancers. It is known that NEAT1 (nuclear paraspeckle Assembly Transcript 1) is overexpressed in many cancers. Knutsen et al. (2020) [242] found that the level of expression of isoform NEAT 1-2 in human is higher upon lactation. This study also reported that the expression of NEAT1-2 correlated with HER2 (human epidermal growth factor receptor 2)-positive breast cancer. The role of NEAT1 is to regulate gene expression at both transcriptional and post-transcriptional levels. Recent studies reported that the loss of Zac1 expression is also associated with the progression of tumours, including cervical cancer, breast cancer and ovarian cancer. Su et al. (2020) [243] found that high Zac1 expression is associated with a poor prognosis of cervical cancer and with epithelial-mesenchymal transition.

Several studies of breast cancer cell lines have reported promising results. Much data has provided evidence that epigenetic changes contribute to breast cancer progression Here, the DNA methylation pattern (induced hypermethylation at the promoter CpG sites) followed CAP application in a breast cancer cell line expressing the oestrogen receptor (MCF-7) [244]. MicroRNA miR-19a-3p (miR-19a) was identified as a mediator of the cell proliferation-inhibitory effect of CAP in the MCF-7 breast cancer cell. *ABCA1* and *PTEN*, which had been suppressed by miR-19a, recovered their expression through CAP treatment. CAP induced damage to DNA in the nucleus by producing a double-strand break (DSB). After exposure to CAP, these cells showed growth retardation, increased DSB, and apoptosis [38]. Many studies identified altered expression in cervical cancer. Another study reported an association between miR-218 expression and various clinicopathological features in cervical cancer. MicroRNA (miR) microarray analysis revealed that miR-218 is downregulated in cervical cancer tissues [245]. According to Su et al. (2020) [243], these results indicate that plasma induces epigenetic and cellular changes in a cell type-specific manner, suggesting that the careful screening of target cells and tissues is necessary for the potential application of plasma as a cancer treatment option.

8.4. CAP Induced DNA Breaks and Modifications

It is known that the biological significance of DNA damage by RONS depends on the extent of damage, where it occurs in the genome, and how fast it can be repaired. As the damage of DNA has importance effects on replication and cell division, the CAP-induced RONS oxidative damage in strand breaks and chemical modification of DNA in the cancer cells leading to sub-lethal or lethal cell reaction is of interest [246]. Here, the advance of CAP is in its specificity to induce DNA strand breaks, surprisingly without any significant rupture of the phospholipid membranes [247]. The interest of studying CAP induced DNA changes is even greater, as cancer cells are more susceptible to the effects of CAP due to a higher percentage of cells in the S-phase of the cell cycle [248], and because CAP has demonstrated the ability to selectively ablate cancer cells while leaving healthy cells mostly unaffected [249].

The significance of damage to DNA by RONS depends on the extent of that damage, where the considerable DNA modifications and breaks usually halt cell replication and cell division. Arjunan et al. (2015) [250] observed that DNA mismatches in nucleobases induced by plasma irradiation can be genotoxic (can hydrolyse the N-glycosidic bond) and lead to cell death. Lackmann et al. (2012) [251] reported the expression of different gene fusions after treating cells with plasma in liquid culture and indicated that CAP emitted particles cause DNA strand breaks, whereas CAP emitted photons provoked cross-link DNA strands. Furthermore, DNA–protein crosslinks [252], DNA chemical modification 8-oxoguanine (8-oxoG), and the up-regulation of the 8-oxoG repair enzyme simultaneously with DNA strand breaks were induced after exposition to CAP [247].

The plasma-treated cells also show an accumulation of gamma-H2A.X, a known marker for DNA double-strand breaks, and higher p53 tumour suppressor gene activity as a response to DNA damage. Interestingly, cytochrome-related changes in mitochondria and its membrane augmented the CAP induced changes on a DNA level [253], and ROS and RNS lead to mitochondria-mediated apoptosis and to further activation of the DNA damage. The plasma effluents, and particularly the plasma-generated particles, also rapidly deactivated proteins in the cellular milieu. In addition to the physical damage to the cellular envelope, therefore modifications to DNA and proteins contribute to the anticancer and anti-bactericidal properties of cold atmospheric-pressure plasma [254].

8.5. CAP and Induced Redox ROS and RNS Effect

As studies have demonstrated that CAP can induce apoptotic cell death in cancer cells, determining the plasma effect on them is a crucial issue. CAP effects on in vivo or in vitro structures, as indicated previously, are mediated by biologically active factors such as the electric field, charged particles (ions and electrons), photons and UV radiations, free radicals, and reactive oxygen and nitrogen species (RONS) [229]. CAP exposure induces redox effects ROS (reactive oxygen species) and RNS (reactive nitrogen species) in cells or tissue, where these reactive species act as antimicrobial molecules produced from nitric oxide and superoxide, causing nitrosative cellular stress. Both ROS and RNS are "double-edged swords", and most atmospheric pressure plasma jet (APPJ) applications focus on the oxidative and/or nitrative stress on bacteria, cells, and tissues [255]. ROS/RNS modulates numerous redox-sensitive biochemical pathways in physiological and pathophysiological cellular processes, affecting cellular integrity. Such induced oxidative modification of biologically essential molecules leads to their functional impairment, such as the loss of biological membranes and structural proteins [256]. At the cellular level, ROS can regulate protein phosphorylation, ion channels activity, and transcription factors involved in critical biosynthetic processes [257]. As the antioxidant mechanism in cancer cells is low, contrary to healthy cells, the RONS-mediated selective effect of CAP mostly affects cancer cell viability. Here the molecular level responses to ROS are related to both redox and phosphorylation signalling with proteins [6].

The biological mechanism of the CAP-induced RONS effects on cells can be explained in two ways. The first involves the insertion of hydrogen peroxide (H_2O_2) to a ROS regulation system. The second involves the changes in mitochondrial transmembrane permeability induced by RNS [258]. The effect of RONS is thus harmful for cells in both its functional and structural being. ROS can damage mitochondrial DNA and cause changes in the permeability of transition pores in mitochondria, which leads to the induction of apoptosis. The most harmful ROS are superoxide anion (O_2^-), hydrogen peroxide (H_2O_2), and hydroxyl radical (OH) [259]. Superoxide and nitric oxide have a role as physiological signalling messengers. Hydrogen peroxide has been suggested as the most crucial signalling messenger in vivo [260]. The generation of ROS begins with the rapid uptake of oxygen, activation of NADPH oxidase, and the production of the superoxide anion radical. The O_2^- is then rapidly converted to H_2O_2. H_2O_2 is further converted to hypochlorous (HOCl) a potent oxidant and antimicrobial agent. Superoxide is removed by superoxide dismutase (SOD), and singlet oxygen is quenched by carotenoids [256]. Under physiological conditions, O_2^- and H_2O_2 appear incapable of directly causing strand breaks or nucleobase modifications in DNA [250].

The regulation-impaired effect of ROS can be explained by the impact on various processes such as proliferation, metabolism, differentiation, and survival, and also by regulating redox-reactive residues on proteins. Most regulators of redox signalling are members of the thioredoxin (Trx)-fold family of proteins. TRX fold proteins, such as thioredoxins (Trxs), glutaredoxins (Grxs), and peroxiredoxins (Prxs), have been characterised as electron donors, guards of the intracellular redox state, and "antioxidants". Today, these redox catalysts are increasingly recognised for their specific role in redox signalling [261].

In today's medicine, RONS has a role in many therapies, including oncology, dermatology, and dentistry. Plasma treatment gives us an opportunity to modulate the healing process and therapeutic response in target cells and tissues.

9. CAP as a Novel Anticancer Treatment Modality, Including Vulvar Pathologies

The use of plasma in the treatment of vulvar pathologies is not unknown. It is not CAP, however, but plasma argon beam coagulation that is used to treat, for example, multifocal VIN III lesions with a favourable clinical outcome. It helped to successfully treat (51.7%) patients with this diagnosis, and no recurrence was demonstrated within the follow-up period of 34.9 months [262]. This experience with plasma medicine in oncogynaecology and positive results from CAP-associated studies in general, is therefore promising for plasma treatment in vulva cancer, which can be used as follows: (a) local induction of immunogenic cell death; (b) induction of cellular immune memory; (c) induction of system response against cancerous cells [263]; (d) surgical removal/reduction of the tumour; (e) elimination of micrometastases through cancer-selective cell killing; and (f) improved chronic wound healing (mainly via antibacterial effects), supporting palliative care.

Consequently, there is increasing interest in oncology-focused research in the application of CAP in anticancer treatment. As shown in Table 2, scientists are now intensively focused on the direct or

9.3. Dual Cancer Therapeutic Approach: Synergy of CAP and Nanotechnology

Cold atmospheric plasma is an emerging biomedical technique that shows great potential for cancer treatment in a novel dual cancer therapeutic method by integrating promising CAP and iron oxide-based magnetic nanoparticles (MNPs) for targeted cancer treatment. Li et al. (2019) [91] showed that the effectiveness of CAP and iron oxide-based MNPs for synergistic application aggressively killed activity against lung cancer cells, and significantly inhibited cell proliferation via a reduction of viability and induction of apoptosis. Importantly, combining CAP with iron oxide-based MNPs induced EGFR downregulation, while CAP inhibited lung cancer cells via depressing pERK and pAKT. The translation of these findings to an in vivo setting demonstrates that CAP combining iron oxide-based MNPs is effective at preventing xenograft tumours. The integration of CAP and iron oxide-based MNPs is thus a promising tool for the development of a new cancer treatment strategy with a significant shift in the current paradigm of cancer therapy [91]. A dual cancer therapeutic method based on the integration of CAP and novel drug-loaded core-shell nanoparticles for the targeted treatment of breast cancer also revealed the synergistic inhibition of cancer cell growth, down-regulation of metastasis-related genes and facilitation of drug-loaded nanoparticle uptake with potential benefits in minimising drug resistance [40].

CAP and silymarin nanoemulsion activated autophagy in human melanoma cells by activating the PI3K/mTOR and EGFR pathways and modulation of the expression of transcriptional factors and specific autophagy-related genes [268]. Co-treatment with PEG-coated gold nanoparticles and CAP also inhibited the proliferation of cancer glioblastoma and lung adenocarcinoma cells through blockading the PI3K/AKT signalling axis, and reversed epithelial-mesenchymal transition (EMT) in solid tumours, thus preventing the growth of tumour cells, which was also observed in vivo [269]. A synergy of CAP and nanoparticles was found to be a promising approach in the therapy of colon cancer as was demonstrated by increased cell death in the presence of gold nanoparticles [270], and helium-based CAP-induced ROS-mediated apoptosis was attenuated by platinum nanoparticles in human lymphoma cells [271]. He et al. (2020) [48] demonstrated the ability of CAP to stimulate clathrin-dependent endocytosis to repair oxidised membrane and promoted the uptake of gold nanomaterial in glioblastoma multiforme cells. Similarly, direct exposure to CAP activates gold nanoparticle-dependent toxicity through an increase in endocytosis and trafficking to lysosomes in the same cell line [272]. A novel dual cancer treatment approach characterised by paclitaxel-loaded core-shell magnetic nanoparticles and CAP has potential as a tool for cancer treatment strategy, as was demonstrated by the growth inhibition of non-small cell lung cancer cells in vitro [273].

Above all, the integration of CAP and nanoparticles is a promising tool for the development of novel cancer treatment strategies. A synergy of nanoparticles, characterised by improved biocompatibility, lower cytotoxicity, and efficacy, and CAP that was explored to selectively target and kill cancer cells represents a new paradigm for a targeted cancer therapeutic approach [14]. Table 2 shows an overview of current trends in the anticancer potential of CAP mediated directly, indirectly, or via synergy with nanoparticles.

9.4. Immunotherapy and CAP

Current trends in the establishment of CAP as a robust approach in anticancer therapy are also associated with the synergic use of plasma and nanotechnology or its application in the modulation of an immune response. The human immune system is significantly associated with cancer, and the ability of the immune system's modulation could overcome the capacity of cancer cells to suppress immune responses through several mechanisms, including cytokines, cell-based therapies, immune checkpoint blockades, and immunogenic cell death [228]. As immunotherapy has become an essential part of anticancer treatment, the CAP effect on the immune system was evaluated in several studies. For example, plasma-activated liquid media (PALM) rich in H_2O_2 reduced proliferation and increased calreticulin exposure and ATP release in pancreatic cancer cells, suggesting its potential to induce immunogenic cell death through activation of the immune system [278]. Van Loenhout et al. (2019) [279] showed that CAP-treated pPBS (plasma-treated phosphate-buffered saline) had the potential to induce immunogenic cell death, and eliminated the immunosuppressive tumour microenvironment in pancreatic cancer cells. pPBS treatment led to the more immunohistomodulatory secretion profiled defined by higher TNF-α and IFN-γ, lower TGF-β in coculture with dendritic cells [279]. These results offer a strong basis for further in vivo evaluation, which is actually partially studied in a clinical setting. Until now, CAP has been applied as an adjunct to immunotherapy in the treatment of glioblastoma multiforme due to its ability to upregulate the immune system by ROS induction [90].

Kaushik et al. (2019) [280] noted another positive effect of plasma on the immune system. They report that plasma treatment stimulated the differentiation of pro-inflammatory (M1) macrophages to a greater extent. This stimulated macrophages to favour anti-tumourigenic immune responses against metastasis acquisition and cancer stem cell maintenance in solid cancers in vitro. The differentiation of monocytes into anticancer macrophages (particularly increasing numbers of mitochondria and lysosomes) could also improve the efficacy of plasma treatment, especially in modifying the pro-tumour inflammatory microenvironment by affecting the highly resistant immunosuppressive tumour cells associated with tumour relapse. [280].

Recently studied plasma-related immunogenic effects include the use of CAP-mediated immune checkpoint blockade (ICB) therapy integrated with microneedles (MN), described for the transdermal delivery of ICB by Chen et al. (2020) [89]. They found that a hollow-structured MN (hMN) patch facilitated the transportation of CAP through the skin, causing tumour cell death. The release of tumour-associated antigens promoted the maturation of dendritic cells in the tumour-draining lymph nodes, subsequently initiating a T cell-mediated immune response. Anti-programmed death-ligand 1 antibody (aPDL1), an immune checkpoint inhibitor, released from the MN patch further augmented antitumour immunity. Their findings indicate that the proposed transdermal combined CAP and ICB therapy can inhibit the tumour growth of both primary tumours and distant tumours, prolonging the survival of tumour-bearing mice [89].

10. Advancements in VC Therapy Based on Better Profiling and Novel Technologies Combining CAP with Existing Treatments

The shift to a modern treatment of vulva cancer is becoming evident, as molecular targeted approaches, evaluated either in monotherapy or as potentiators of chemotherapy, are now widely studied in clinical settings. Wang et al. (2018) [281] found out that the deregulation of CHK1 function often occurs in VSCC and might contribute to tumourigenesis. Targeting CHK1 might thus be a useful antitumour strategy for the subgroup of VSCC harbouring p53 mutations, which is a common finding in VSCC, as well as other genes (NOTCH1, HRAS, and CDKN2A mutations) [282,283], which may activate the PI3K/AKT/mTOR pathway, thus, providing a rationale for new anti-VSCC therapies targeting this signalling actionable pathway. This search for a novel therapy is confirmed by increasing numbers of studies. Brunetti et al. (2017) [284] studied the juxtaposition of two different genes or gene parts due to chromosomal rearrangement, which is a well-known neoplasia-associated pathogenetic mechanism, and found two recurrent fusions with STIP1-CREB3L1 and ZDHHC5-GPR137 present in VSCC. The transcripts were detected only in the tumour samples, not in normal vulvar tissue from healthy controls. They supposed that the detection of such tumourigenic fusions might serve as therapeutic targets for antioncogenic drugs that interact directly with the molecular changes responsible for neoplastic transformation of VSCC. The importance of a molecular approach in VSCC carcinogenesis is also demonstrated by Agostini et al. (2016) [285], who revealed downregulation of the fragile histidine triad (FHIT) and upregulation of the high mobility group AT-hook 2 (HMGA2) gene via miR-30c and let-7a. The results of such translational research involving the molecular landscape of vulva cancer in the past was the basis for today's clinical target-directed therapeutic agents: for example, erlotinib, an inhibitor of the tyrosine kinase activity of the epidermal growth factor receptor (EGFR); bevacizumab, a monoclonal antibody targeting the vascular endothelial growth factor (VGFR); and pembrolizumab, an inhibitor of the programmed death-1 interaction with its ligand, called the PD1-PD-L1 immune checkpoint used in immunotherapy of VSCC [286–289].

As vulva cancer is a very particular malignancy, where most cases gain exceptional inherent radioresistance after standard therapy (surgery, chemo/radiotherapy) [290], the development of other forms of additional treatment would be more advantageous.

Here, the clinical approach is to use electrochemotherapy as a feasible, easy to perform, and reproducible procedure in patients with primary or recurrent vulvar cancer who are unable to undergo surgery. Survival after one year in this population was observed at 50%. Electrochemotherapy may have a role in the management of vulvar cancer, especially as a palliative treatment when other therapies are no longer applicable [291]. However, technical issues mean that this tool is sporadically applied. Conversely, CAP has received much more attention from researchers due to its ability to specifically induce the selective death of cancer cells over normal healthy cells, called antitumour selectivity [291–293], by activating the apoptosis of these cells, decreasing their proliferation and mobility, and by recovering their sensitivity to therapeutic drugs. Previous studies showed encouraging results where CAP returned cisplatin-resistant ovarian cancer cells [75], paclitaxel-resistant breast cancer cells [158], tamoxifen-resistant breast cancer cells [39], and temozolomide-resistant malignant glioma cells [294] to a drug-sensitive state, in addition to inducing apoptosis and growth inhibition. CAP showed enhanced sensitivity to radiation by generating ROS with a simultaneous inhibition of tumour growth [295]. Selectivity in the efficient killing of cancer cells (plasmas self-adaptive toward cancer cells) [296], without adverse toxicity to healthy cells/tissues (as commonly present in radiation therapy), will be one of the most important therapeutic considerations in assessing CAP as a new cancer therapeutic strategy alone or combined with ionising radiation or chemotherapy. These findings may contribute to extending the application of CAP to the treatment protocols in cytotoxic drugs or radiation therapy-resistant cancers. Such effect was previously enhanced by a static magnetic field [297].

CAP was also studied to enhance photodynamic therapy (PDT), which is a non-invasive method for the treatment of superficial malignant cancers. The known limiting challenge of PDT is the hypoxic conditions during treatment, which reduce PDT efficiency. ROS and free radicals in the plasma flame output demonstrated that CAP could improve treatment efficiency of both indocyanine green (ICG) and protoporphyrin IX (PPIX) in breast and colon cancer cell lines [298]. This offers a promising background for conducting future studies in vulva cancer application, for both superficial lesion treatment and ICG guided nodal sampling.

Cold atmospheric plasma has been proposed as a novel therapeutic method due to its anti-cancer potential in combination with hyperthermia (HT) 42 °C or radiation 5 Gy. Synergistic enhancement in cell death with HT and an additive enhancement with radiation were observed following helium-CAP treatment. These findings would be helpful when establishing a therapeutic strategy for CAP in combination with radiation or HT [299], and specifically also used in ovarian cancer patients intraoperatively [300]. CAP application in an in vivo mouse model of intraperitoneal ovarian cancer metastasis via PAM model inhibited the peritoneal dissemination of cancer cells, resulting in prolonged survival [157].

The recurrence of cancer due to the acquisition of chemoresistance to 'classical' cytotoxic chemotherapeutics and molecularly targeted therapies [301] or radioresistance [302] remains a severe problem for the clinical treatment of cancer patients. A relatively high number of cancer patients receiving chemotherapy develop recurrent or metastatic disease due to acquiring drug resistance over time through different mechanisms, including multi-drug resistance, cell death inhibiting (apoptosis suppression), alterations in the drug metabolism, epigenetic and drug targets, enhancing DNA repair and gene amplification [303,304].

The approaches and research activities noted above confirm the importance of significant molecular and technical advancement for further clinical use in the management of patients with vulvar cancer or its premalignant lesions. This "paradigm shift" from surgery with chemo/radiotherapy alone to novel diagnostic, preventive and therapeutic approaches is well supported by the strong preventive and therapeutic potential of HPV vaccines, and is a promising therapeutic response by treatment of VIN [305].

11. A paradigm Shift from Reactive to Predictive, Preventive and Personalised Medicine (3PM)—Prominent Examples in the Context of Vulva Cancer and Premalignant Lesions

As already explained above, vulva cancer is an excellent example of implementing the paradigm shift from reactive medicine (defined as disease care) to cost-effective and patient-friendly 3PM approaches (healthcare) [306]. To this end, it has to be emphasised that the patient stratification based on individualised profiling plays a crucial role in 3PM strategies [307,308]. The levels of patient stratification are classified below, providing prominent examples of3PM implementation in the area.

11.1. The Primary Level of Targeted Prevention

HPV vaccination was mentioned above as one of the prominent approaches in preventing vulvar cancer linked to HPV lesions. In contrast to the human papillomavirus as a possible trigger of the disease, the role of vulvar-vaginal dryness as an essential risk factor is greatly underestimated in currently applied diagnostic and treatment approaches as demonstrated by Olga Golubnitschaja and colleagues [309,310]. At this level of targeted prevention, the main focus is on people in suboptimal health predisposed to the disease development (latent chronic VIN pathologies). The most prominent example is demonstrated for individuals with Flammer syndrome (FS) phenotype, with a higher prevalence in young female populations, and academicians demonstrating signs and symptoms of primary vascular dysregulation, a tendency to perfectionism, strong stress sensitivity, and healing impairments, amongst other things [310]. A topic-dedicated study demonstrated the FS phenotype as potentially characteristic for premenopausal females with vulvar-vaginal dryness [309]. Specifically, in this patient cohort, excessive vasoconstriction, feeling cold, low blood pressure, dizziness, strongly reduced thirst perception, strong smell perception, headache, perfectionistic personality, and tinnitus have been demonstrated as frequently co-exhibited symptoms [309]. Since many of the risk factors linked to FS carry a clearly preventable character, this phenotype is of great clinical utility for screening programs, in order to prevent female genital cancers, which may occur at any age. Contextually, individualised profiling is instrumental for mitigating measures tailored to the person suffering from vulvar-vaginal dryness as part of Sicca syndrome in individuals with the FS phenotype.

11.2. The Secondary Level of Targeted Prevention

The secondary level of prevention deals with complications linked to clinically manifested pathologies such as impaired wound healing, which is one of the prominent examples in this article that is treatable by CAP. If not diagnosed and treated well in time, delayed and/or impaired healing may cause chronic inflammation and cancer development, amongst other problems [311,312]. In addition to non-modifiable risks such as advanced ageing, there are many easily preventable factors involved in impaired healing, such as suboptimal lifestyle and nutritional and vascular deficits [313]. Taking into consideration a highly heterogeneous cohort of patients suffering from impaired healing, individualised profiling as a predictive diagnosis is instrumental for cost-effective targeted secondary prevention, as demonstrated in the multi-professional publication "Wound Healing: Proof-of-Principle Model for the Modern Hospital—Patient Stratification, Prediction, Prevention and Personalisation of Treatment" [314].

11.3. The Tertiary Level of Targeted Prevention

At this level, mitigating measures are applicable, for example, to avoid metastatic disease in vulva cancer patients. As described above, CAP application is of great importance in protecting the patient against local metastatic spread. To estimate the potential for metastatic spread to distanced organs, however, liquid biopsy is instrumental in predictive diagnosis at this stage of cancer progression [315]. To this end, CTC enumeration and the identification of highly specific multiomic patterns in the blood (e.g., miRNAs, CpG-changes, cfDNA) are considered an optimal approach [316,317] followed by personalised chemoprevention [318] and/or targeted therapy [319].

12. Status Quo and Clinically Relevant Perspectives

The application of cold physical plasma in a medical setting is rapidly increasing. It is a well-established therapeutic approach in dermatology, and CAP application in head and neck cancer patients is moving upwards in the pyramid of evidence-based medicine (EBM). The effectiveness of plasma in cancer cell lines, cultivated human tumour cells, human tumour specimens freshly explanted from patients, animal model tumours, and animals with transplanted human tumour stem cells is well documented. There is consensus among experts, referred to as EBM-level IV, regarding the response to plasma in experimental settings and proof of concept by clinical pilot studies, EBM-level III, for plasma treatment of head and neck cancer. It is already a concept in the palliative care of patients with locally advanced head and neck cancer and contaminated ulcerations because of proven effectiveness against microbial pathogens. Patients greatly appreciate that plasma reduces the strong fetid odour and pain, and is not accompanied by severe side effects [320].

The potential for the extensive clinical use of CAP is also significant in other malignancies, including oncogynaecology. An ongoing trial (NCT02658851, Florida, USA) is assessing the effect of CAP on the reduction of lymphocele following pelvic lymph node dissection during robot-assisted radical prostatectomy [321] (which is also a common procedure in cervical, uterine or ovarian cancer surgery), based on the authors' previously noting the high incidence of this pathology [322]. Another ongoing clinical trial (NCT03218436, Tübingen, Germany) is studying the effect of CAP on human cervical neoplasia with histologically confirmed CIN 1-2 lesions [323], as a consequence of proof-of concept studies on cervical tissue from human donors [69]. Both studies have strong potential for oncology. The concept that the anticancer potential of CAP affects the dysplastic cells will shortly be of particular interest, including the treatment of vulvar intraepithelial lesions. Plasma is thus regarded as a potential intraoperative and adjuvant therapy. Its therapeutic efficacy should, therefore, be assessed in combination with current treatment strategies, mostly utilising PAM nanoparticles in maximising therapeutic effect and overcoming radioresistance when applied directly against tumour [295,324]. However, before a specific clinical application of CAP on vulva cancer, animal models are needed to stratify this conclusion in a more biologically relevant system, and several technical parameters need to be solved (e.g., penetration depth, optimal dosage, repetitive applications, type of CAP source device), and medical protocols created in line with safe clinical practice.

Additional perspectives should involve:

(a) Precise drug-directed studies on chemoresistant, hormone resistant or radioresistant cancer cells in single or repetitive CAP applications.

(b) Studies aiming to use CAP as an adjunct tool during intraoperative resections or adjuvant chemotherapy, and as a potential tool against micrometastases outside surgically removed tumours.

(c) Studies using CAP during the palliative care of large inoperable metastatic cancer.

(d) Research aiming to achieve better control for surgical margins sufficiency.

(e) Studies evaluating CAP as an intraoperative tool for local groin lymph nodes silencing in patients undergoing sentinel lymph node biopsy alone, and omitting ipsilateral or contralateral inguinal nodal dissection.

(f) Studies in precision medicine using disease-optimised ROS cocktails via specifically engineered plasmas.

The first clinical case reports should then be conducted in locally advanced stages among patients with palliative care, and positive outcomes should motivate further clinical trials to demonstrate the relevance of CAP in clinical practice for patients with vulvar cancer (basal cell carcinoma, squamous cell carcinoma, malignant melanoma) and its premalignant lesions.

13. Conclusions

Extensive research has been focused on the surgical and adjuvant management of vulvar cancer in the past and huge efforts on deciphering the molecular mechanisms of VSCC carcinogenesis. Technical tools are now providing increasing knowledge, having previously been sporadically applied in oncology. The potential for the extensive clinical use of CAP in oncogynaecology is immense, as CAP has been shown to be successful in various medical applications. Plasma is a potential intraoperative and adjuvant therapy, as intense preclinical studies have demonstrated the unique traits of plasma oncotherapy, such as its multimodal activity, synergic interactions with conventional chemotherapy agents, ability to cause genetic, epigenetic changes affecting processes fundamental to cancer progression and capacity to induce immunogenic cell death. Its therapeutic efficacy should, therefore, be assessed in combination with current treatment strategies (surgery, chemo- or radiotherapy), mostly utilising PAM nanoparticles in maximising therapeutic effect and overcoming radioresistance when applied directly against tumour.

Before the real clinical application of CAP on vulva cancer, however, animal models are needed to stratify this conclusion in a biologic relevant system, and several technical parameters need to be solved (e.g., penetration depth, optimal dosage, repetitive applications, type of CAP source device and administration), and subsequent medical protocols created in line with safe clinical practice. Feasible groups of patients for such clinical trials are those with cancers lacking an effective targeted therapy, tumours that resist radiotherapy, cancers with physical isolation, patients with a relapse of metastases, postoperative patients, cancers that require low penetration depth (e.g., melanoma, skin, vulvar) and cancers with aesthetic requirements (often present with vulvar cancer and skin flaps). This review information may, in the future, serve as a foundation for the design of clinical trials to assess the efficacy and safety of CAP as adjuvant therapy for vulvar skin cancer.

Future strategies in the area could consider highly protective and cost-effective 3PM approaches comprising individualised profiling, predictive diagnosis, innovative screening programmes focused on young populations and individuals in suboptimal health conditions, targeted prevention, and treatments tailored to the individual [309,314,315,325].

Author Contributions: Conceptualization, P.Z., Z.D., A.D., O.G. and Y.W.; methodology, P.Z., Y.W., P.K. and D.B.; formal analysis, P.Z., A.L., I.K. and J.B.; resources, C.A.D., D.D., K.K., B.M. and V.L.; data curation, M.S., D.D., I.K., V.L. and M.M.; writing-original draft preparation, P.Z., P.K., A.L., L.K., Y.W., Z.D., O.G., K.K. and M.S.; writing-review and editing, P.Z., A.D., O.G., Z.D. and P.K.; visualization, M.M., J.B., B.M.; supervision, D.B. and O.G.; All authors have read and agreed to the published version of the manuscript.

Funding: This manuscript was written with the support of the projects VEGA 1/0199/17 and APVV-16-0021, funded by the Scientific Grant Agency and by the Slovak Research and Development Agency, respectively.

Conflicts of Interest: The authors and OBGY Health & Care Ltd. declare no conflict of interest.

References

1.   Graves, D.B. The emerging role of reactive oxygen and nitrogen species in redox biology and some implications for plasma applications to medicine and biology. *J. Phys. D Appl. Phys.* **2012**, *45*, 3001. [CrossRef]

2.   Rehman, M.U.; Jawaid, P.; Uchiyama, H.; Kondo, T. Comparison of free radicals formation induced by cold atmospheric plasma, ultrasound, and ionizing radiation. *Arch. Biochem. Biophys.* **2016**, *605*, 19–25. [CrossRef] [PubMed]

3.   Takenaka, K.; Miyazaki, A.; Uchida, G.; Setsuhara, Y. Atmospheric-pressure plasma interaction with soft materials as fundamental processes in plasma medicine. *J. Nanosci. Nanotechnol.* **2015**, *15*, 2115–2119. [CrossRef] [PubMed]

4.   Yonemori, S.; Ono, R. Effect of discharge polarity on the propagation of atmospheric-pressure helium plasma jets and the densities of OH, NO, and O radicals. *Biointerphases* **2015**, *10*, 029514. [CrossRef] [PubMed]

5.   Kaushik, N.; Uddin, N.; Sim, G.B.; Hong, Y.J.; Baik, K.Y.; Kim, C.H.; Lee, S.J.; Kaushik, N.K.; Choi, E.H. Responses of Solid Tumor Cells in DMEM to Reactive Oxygen Species Generated by Non-Thermal Plasma and Chemically Induced ROS Systems. *Sci. Rep.* **2015**, *5*, srep08587. [CrossRef] [PubMed]

6.   Privat-Maldonado, A.; Schmidt, A.; Lin, A.; Weltmann, K.-D.; Wende, K.; Bogaerts, A.; Bekeschus, S. ROS from physical plasmas: Redox chemistry for biomedical therapy. *Oxidative Med. Cell. Longev.* **2019**, *2019*, 1–29. [CrossRef]

7.   Gorbanev, Y.; O'Connell, D.; Chechik, V. Non-thermal plasma in contact with water: The origin of species. *Chemistry* **2016**, *22*, 3496–3505. [CrossRef]

8. Von Woedtke, T.; Schmidt, A.; Bekeschus, S.; Wende, K.; Weltmann, K.-D. Plasma Medicine: A field of applied redox biology. *In Vivo* **2019**, *33*, 1011–1026. [CrossRef]

9. Ji, W.-O.; Lee, M.-H.; Kim, G.-H.; Kim, E.-H. Quantitation of the ROS production in plasma and radiation treatments of biotargets. *Sci. Rep.* **2019**, *9*, 1–11. [CrossRef]

10. Xu, D.; Luo, X.; Xu, Y.; Cui, Q.; Yang, Y.; Liu, D.; Chen, H.; Kong, M.G. The effects of cold atmospheric plasma on cell adhesion, differentiation, migration, apoptosis and drug sensitivity of multiple myeloma. *Biochem. Biophys. Res. Commun.* **2016**, *473*, 1125–1132. [CrossRef]

11. Utsumi, F.; Kajiyama, H.; Nakamura, K.; Tanaka, H.; Mizuno, M.; Toyokuni, S.; Hori, M.; Kikkawa, F. Variable susceptibility of ovarian cancer cells to non-thermal plasma-activated medium. *Oncol. Rep.* **2016**, *35*, 3169–3177. [CrossRef] [PubMed]

12. Kwon, B.-S.; Choi, E.H.; Chang, B.; Choi, J.-H.; Kim, K.S.; Park, H.-K. Selective cytotoxic effect of non-thermal micro-DBD plasma. *Phys. Biol.* **2016**, *13*, 056001. [CrossRef]

13. Keidar, M.; Shashurin, A.; Volotskova, O.; Stepp, M.A.; Srinivasan, P.; Sandler, A.; Trink, B. Cold atmospheric plasma in cancer therapy. *Phys. Plasmas* **2013**, *20*, 057101. [CrossRef]

14. Aryal, S.; Bisht, G. New paradigm for a targeted cancer therapeutic approach: A short review on potential synergy of gold nanoparticles and cold atmospheric plasma. *Biomedicine* **2017**, *5*, 38. [CrossRef] [PubMed]

15. Isbary, G.; Shimizu, T.; Li, Y.-F.; Stolz, W.; Thomas, H.M.; Morfill, G.E.; Zimmermann, J.L. Cold atmospheric plasma devices for medical issues. *Exp. Rev. Med. Dev.* **2013**, *10*, 367–377. [CrossRef] [PubMed]

16. Nasir, N.M.; Lee, B.K.; Yap, S.S.; Thong, K.L. Cold plasma inactivation of chronic wound bacteria. *Arch. Biochem. Biophys.* **2016**, *605*, 76–85. [CrossRef]

17. Nguyen, L.; Lu, P.; Boehm, D.; Bourke, P.; Gilmore, B.F.; Hickok, N.J.; Freeman, T.A. Cold atmospheric plasma is a viable solution for treating orthopedic infection: A review. *Biol. Chem.* **2018**, *400*, 77–86. [CrossRef] [PubMed]

18. Izadjoo, M.; Zack, S.; Kim, H.; Skiba, J. Medical applications of cold atmospheric plasma: State of the science. *J. Wound Care* **2018**, *27* (Suppl. S9), S4–S10. [CrossRef]

19. Ulrich, C.; Kluschke, F.; Patzelt, A.; Vandersee, S.; Czaika, V.A.; Richter, H.; Bob, A.; Von Hutten, J.; Painsi, C.; Hüge, R.; et al. Clinical use of cold atmospheric pressure argon plasma in chronic leg ulcers: A pilot study. *J. Wound Care* **2015**, *24*, 196–203. [CrossRef]

20. Hoffmann, C.; Berganza, C.; Zhang, J. Cold Atmospheric Plasma: Methods of production and application in dentistry and oncology. *Med. Gas Res.* **2013**, *3*, 21. [CrossRef]

21. Boeckmann, L.; Bernhardt, T.; Schäfer, M.; Semmler, M.L.; Kordt, M.; Waldner, A.; Wendt, F.; Sagwal, S.; Bekeschus, S.; Berner, J.; et al. Aktuelle indikationen der plasmatherapie in der dermatologie. *Hautarzt* **2020**, *71*, 109–113. [CrossRef] [PubMed]

22. Bernhardt, T.; Semmler, M.L.; Schäfer, M.; Bekeschus, S.; Emmert, S.; Boeckmann, L. Plasma medicine: Applications of cold atmospheric pressure plasma in dermatology. *Oxidative Med. Cell. Longev.* **2019**, *2019*, 1–10. [CrossRef]

23. Gareri, C.; Bennardo, L.; De Masi, G. Use of a new cold plasma tool for psoriasis treatment: A case report. *SAGE Open Med. Case Rep.* **2020**, *8*. [CrossRef] [PubMed]

24. Miyamoto, K.; Ikehara, S.; Sakakita, H.; Ikehara, Y. Low temperature plasma equipment applied on surgical hemostasis and wound healings. *J. Clin. Biochem. Nutr.* **2017**, *60*, 25–28. [CrossRef] [PubMed]

25. Filis, K.; Galyfos, G.; Sigala, F.; Zografos, G. Utilization of low-temperature helium plasma (J-Plasma) for dissection and hemostasis during carotid endarterectomy. *J. Vasc. Surg. Cases Innov. Tech.* **2020**, *6*, 152–155. [CrossRef]

26. Rosani, U.; Tarricone, E.; Venier, P.; Brun, P.; Deligianni, V.; Zuin, M.; Martines, E.; Leonardi, A. Atmospheric-pressure cold plasma induces transcriptional changes in ex vivo human corneas. *PLoS ONE* **2015**, *10*, e0133173. [CrossRef]

27. Yan, D.; Sherman, J.H.; Keidar, M. Cold atmospheric plasma, a novel promising anti-cancer treatment modality. *Oncotarget* **2017**, *8*, 15977–15995. [CrossRef]

28. Dubuc, A.; Monsarrat, P.; Virard, F.; Merbahi, N.; Sarrette, J.-P.; Laurencin-Dalicieux, S.; Cousty, S. Use of cold-atmospheric plasma in oncology: A concise systematic review. *Ther. Adv. Med. Oncol.* **2018**, *10*. [CrossRef]

29. Setsuhara, Y. Low-temperature atmospheric-pressure plasma sources for plasma medicine. *Arch. Biochem. Biophys.* **2016**, *605*, 3–10. [CrossRef]

30. Schneider, C.; Gebhardt, L.; Arndt, S.; Karrer, S.; Zimmermann, J.L.; Fischer, M.J.M.; Bosserhoff, A. Acidification is an essential process of cold atmospheric plasma and promotes the anti-cancer effect on malignant melanoma cells. *Cancers* **2019**, *11*, 671. [CrossRef]

31. Wang, L.; Yang, X.; Yang, C.; Gao, J.; Zhao, Y.; Cheng, C.; Zhao, G.; Liu, S. The inhibition effect of cold atmospheric plasma-activated media in cutaneous squamous carcinoma cells. *Futur. Oncol.* **2019**, *15*, 495–505. [CrossRef] [PubMed]

32. Liedtke, K.R.; Bekeschus, S.; Kaeding, A.; Hackbarth, C.; Kuehn, J.-P.; Heidecke, C.-D.; Von Bernstorff, W.; Von Woedtke, T.; Partecke, L.I. Non-thermal plasma-treated solution demonstrates antitumor activity against pancreatic cancer cells in vitro and in vivo. *Sci. Rep.* **2017**, *7*, 1–12. [CrossRef] [PubMed]

33. Gweon, B.; Kim, M.; Kim, D.B.; Kim, D.; Kim, H.; Jung, H.; Shin, J.H.; Choe, W. Differential responses of human liver cancer and normal cells to atmospheric pressure plasma. *Appl. Phys. Lett.* **2011**, *99*, 63701. [CrossRef]

34. Torii, K.; Yamada, S.; Nakamura, K.; Tanaka, H.; Kajiyama, H.; Tanahashi, K.; Iwata, N.; Kanda, M.; Kobayashi, D.; Tanaka, C.; et al. Effectiveness of plasma treatment on gastric cancer cells. *Gastric Cancer* **2014**, *18*, 635–643. [CrossRef]

35. Schneider, C.; Arndt, S.; Zimmermann, J.L.; Li, Y.; Karrer, S.; Bosserhoff, A.-K. Cold atmospheric plasma treatment inhibits growth in colorectal cancer cells. *Biol. Chem.* **2018**, *400*, 111–122. [CrossRef]

36. Weiss, M.; Gümbel, D.; Gelbrich, N.; Brandenburg, L.-O.; Mandelkow, R.; Zimmermann, U.; Ziegler, P.; Burchardt, M.; Stope, M.B. Inhibition of cell growth of the prostate cancer cell model LNCaP by cold atmospheric plasma. *In Vivo* **2015**, *29*, 611.

37. Gelbrich, N.; Stope, M.B.; Burchardt, M. Kaltes atmosphärisches Plasma für die urologische Tumortherapie. *Urol. A* **2018**, *58*, 673–679. [CrossRef]

38. Lee, S.; Lee, H.; Bae, H.; Choi, E.H.; Kim, S.J. Epigenetic silencing of miR-19a-3p by cold atmospheric plasma contributes to proliferation inhibition of the MCF-7 breast cancer cell. *Sci. Rep.* **2016**, *6*, 30005. [CrossRef] [PubMed]

39. Lee, S.; Lee, H.; Jeong, D.; Ham, J.; Park, S.; Choi, E.H.; Kim, S.J. Cold atmospheric plasma restores tamoxifen sensitivity in resistant MCF-7 breast cancer cell. *Free Radic. Biol. Med.* **2017**, *110*, 280–290. [CrossRef] [PubMed]

40. Zhu, W.; Lee, S.-J.; Castro, N.J.; Yan, D.; Keidar, M.; Zhang, L.G. Synergistic effect of cold atmospheric plasma and drug loaded core-shell nanoparticles on inhibiting breast cancer cell growth. *Sci. Rep.* **2016**, *6*, 21974. [CrossRef] [PubMed]

41. Mirpour, S.; Ghomi, H.; Piroozmand, S.; Nikkhah, M.; Tavassoli, S.H.; Azad, S.Z. The selective characterization of nonthermal atmospheric pressure plasma jet on treatment of human breast cancer and normal cells. *IEEE Trans. Plasma Sci.* **2014**, *42*, 315–322. [CrossRef]

42. Wang, M.; Holmes, B.; Cheng, X.; Zhu, W.; Keidar, M.; Zhang, L.G. Cold atmospheric plasma for selectively ablating metastatic breast cancer cells. *PLoS ONE* **2013**, *8*, e73741. [CrossRef] [PubMed]

43. Kim, S.J.; Chung, T.; Bae, S.H.; Leem, S.H. Induction of apoptosis in human breast cancer cells by a pulsed atmospheric pressure plasma jet. *Appl. Phys. Lett.* **2010**, *97*, 23702. [CrossRef]

44. Ninomiya, K.; Ishijima, T.; Imamura, M.; Yamahara, T.; Enomoto, H.; Takahashi, K.; Tanaka, Y.; Uesugi, Y.; Shimizu, N. Evaluation of extra- and intracellular OH radical generation, cancer cell injury, and apoptosis induced by a non-thermal atmospheric-pressure plasma jet. *J. Phys. D Appl. Phys.* **2013**, *46*. [CrossRef]

45. Kang, S.U.; Cho, J.-H.; Chang, J.W.; Shin, Y.S.; Kim, K.I.; Park, J.K.; Yang, S.S.; Lee, J.-S.; Moon, E.; Lee, K.; et al. Nonthermal plasma induces head and neck cancer cell death: The potential involvement of mitogen-activated protein kinase-dependent mitochondrial reactive oxygen species. *Cell Death Dis.* **2014**, *5*, e1056. [CrossRef] [PubMed]

46. Gümbel, D.; Bekeschus, S.; Gelbrich, N.; Napp, M.; Ekkernkamp, A.; Kramer, A.; Stope, M.B. Cold atmospheric plasma in the treatment of osteosarcoma. *Int. J. Mol. Sci.* **2017**, *18*, 2004. [CrossRef]

47. Mateu-Sanz, M.; Tornín, J.; Brulin, B.; Khlyustova, A.; Ginebra, M.-P.; Layrolle, P.; Canal, C. Cold plasma-treated Ringer's saline: A weapon to target osteosarcoma. *Cancers* **2020**, *12*, 227. [CrossRef] [PubMed]

48. He, Z.; Liu, K.; Scally, L.; Manaloto, E.; Gunes, S.; Ng, S.W.; Maher, M.; Tiwari, B.; Byrne, H.J.; Bourke, P.; et al. Cold atmospheric plasma stimulates clathrin-dependent endocytosis to repair oxidised membrane and enhance uptake of nanomaterial in glioblastoma multiforme cells. *Sci. Rep.* **2020**, *10*, 1–12. [CrossRef]

49. Wolff, C.M.; Kolb, J.F.; Weltmann, K.-D.; Von Woedtke, T.; Bekeschus, S. Combination treatment with cold physical plasma and pulsed electric fields augments ROS production and cytotoxicity in lymphoma. *Cancers* **2020**, *12*, 845. [CrossRef]

50. Xu, D.; Ning, N.; Xu, Y.; Wang, B.; Cui, Q.; Liu, Z.; Wang, X.; Liu, D.; Chen, H.; Kong, M.G. Effect of cold atmospheric plasma treatment on the metabolites of human leukemia cells. *Cancer Cell Int.* **2019**, *19*, 135. [CrossRef]

51. Xu, D.; Xu, Y.; Cui, Q.; Liu, D.; Liu, Z.; Wang, X.; Yang, Y.; Feng, M.; Liang, R.; Chen, H.; et al. Cold atmospheric plasma as a potential tool for multiple myeloma treatment. *Oncotarget* **2018**, *9*, 18002–18017. [CrossRef]

52. Chang, C.-H.; Yano, K.-I.; Sato, T. Nanosecond pulsed current under plasma-producing conditions induces morphological alterations and stress fiber formation in human fibrosarcoma HT-1080 cells. *Arch. Biochem. Biophys.* **2020**, *681*, 108252. [CrossRef] [PubMed]

53. Golubitskaya, E.A.; Troitskaya, O.S.; Yelak, E.V.; Gugin, P.P.; Richter, V.A.; Schweigert, I.V.; Zakrevsky, D.E.; Koval, O.A. Cold physical plasma decreases the viability of lung adenocarcinoma cells. *Acta Nat.* **2019**, *11*, 16–19. [CrossRef]

54. Freund, E.; Liedtke, K.R.; Van Der Linde, J.; Metelmann, H.-R.; Heidecke, C.-D.; Partecke, L.-I.; Bekeschus, S. Physical plasma-treated saline promotes an immunogenic phenotype in CT26 colon cancer cells in vitro and in vivo. *Sci. Rep.* **2019**, *9*, 1–18. [CrossRef] [PubMed]

55. Xiang, L.; Xu, X.; Zhang, S.; Cai, D.; Dai, X. Cold atmospheric plasma conveys selectivity on triple negative breast cancer cells both in vitro and in vivo. *Free Radic. Biol. Med.* **2018**, *124*, 205–213. [CrossRef]

56. Mirpour, S.; Piroozmand, S.; Soleimani, N.; Faharani, N.J.; Ghomi, H.; Eskandari, H.F.; Sharifi, A.M.; Mirpour, S.; Eftekhari, M.; Nikkhah, M. Utilizing the micron sized non-thermal atmospheric pressure plasma inside the animal body for the tumor treatment application. *Sci. Rep.* **2016**, *6*, 29048. [CrossRef]

57. Hirst, A.M.; Simms, M.S.; Mann, V.M.; Maitland, N.J.; Oconnell, D.; Frame, F.M. Low-temperature plasma treatment induces DNA damage leading to necrotic cell death in primary prostate epithelial cells. *Br. J. Cancer* **2015**, *112*, 1536–1545. [CrossRef] [PubMed]

58. Vaquero, J.; Judée, F.; Vallette, M.; Decauchy, H.; Arbelaiz, A.; Aoudjehane, L.; Scatton, O.; Gonzalez-Sanchez, E.; Merabtene, F.; Augustin, J.; et al. Cold-atmospheric plasma induces tumor cell death in preclinical in vivo and in vitro models of human cholangiocarcinoma. *Cancers* **2020**, *12*, 1280. [CrossRef] [PubMed]

59. Yoon, Y.J.; Suh, M.J.; Lee, H.Y.; Lee, H.J.; Choi, E.H.; Moon, I.S.; Song, K. Anti-tumor effects of cold atmospheric pressure plasma on vestibular schwannoma demonstrate its feasibility as an intra-operative adjuvant treatment. *Free Radic. Biol. Med.* **2018**, *115*, 43–56. [CrossRef]

60. Chen, Z.; Simonyan, H.; Cheng, X.; Gjika, E.; Lin, L.; Canady, J.; Sherman, J.H.; Young, C.; Keidar, M. A novel micro cold atmospheric plasma device for glioblastoma both in vitro and in vivo. *Cancers* **2017**, *9*, 61. [CrossRef]

61. Saadati, F.; Mahdikia, H.; Abbaszadeh, H.-A.; Abdollahifar, M.-A.; Khoramgah, M.S.; Shokri, B. Comparison of Direct and Indirect cold atmospheric-pressure plasma methods in the B16F10 melanoma cancer cells treatment. *Sci. Rep.* **2018**, *8*, 1–15. [CrossRef] [PubMed]

62. Li, Y.; Kang, M.H.; Uhm, H.S.; Lee, G.J.; Choi, E.H.; Han, I. Effects of atmospheric-pressure non-thermal bio-compatible plasma and plasma activated nitric oxide water on cervical cancer cells. *Sci. Rep.* **2017**, *7*, srep45781. [CrossRef] [PubMed]

63. Lingzhi, B.; Yu, K.N.; Bao, L.; Shen, J.; Cheng, C.; Han, W. Non-thermal plasma inhibits human cervical cancer HeLa cells invasiveness by suppressing the MAPK pathway and decreasing matrix metalloproteinase-9 expression. *Sci. Rep.* **2016**, *6*, 19720. [CrossRef]

64. Feil, L.; Koch, A.; Utz, R.; Ackermann, M.; Barz, J.; Stope, M.B.; Krämer, B.; Wallwiener, D.; Brucker, S.Y.; Weiss, M. Cancer-selective treatment of cancerous and non-cancerous human cervical cell models by a non-thermally operated electrosurgical argon plasma device. *Cancers* **2020**, *12*, 1037. [CrossRef]

65. Ryan, H.A.; Neuber, J.; Song, S.; Beebe, S.J.; Jiang, C. Effects of a non-thermal plasma needle device on HPV-16 positive cervical cancer cell viability in vitro. *Conf. Proc. IEEE Eng. Med. Biol. Soc.* **2016**, *2016*, 537–540. [CrossRef]

66. Ahn, H.J.; Kim, K.I.; Kim, G.; Moon, E.; Yang, S.S.; Lee, J.-S. Atmospheric-pressure plasma jet induces apoptosis involving mitochondria via generation of free radicals. *PLoS ONE* **2011**, *6*, e28154. [CrossRef]

67. Tan, X.; Zhao, S.; Lei, Q.; Lu, X.; He, G.; Ostrikov, K. (Ken) single-cell-precision microplasma-induced cancer cell apoptosis. *PLoS ONE* **2014**, *9*, e101299. [CrossRef]

68. Kim, K.; Ahn, H.J.; Lee, J.-H.; Kim, J.-H.; Yang, S.S.; Lee, J.-S. Cellular membrane collapse by atmospheric-pressure plasma jet. *Appl. Phys. Lett.* **2014**, *104*, 13701. [CrossRef]

69. Wenzel, T.; Berrio, D.A.C.; Reisenauer, C.; Layland, S.; Koch, A.; Wallwiener, D.; Brucker, S.Y.; Schenke-Layland, K.; Brauchle, E.-M.; Weiss, M. Trans-mucosal efficacy of non-thermal plasma treatment on cervical cancer tissue and human cervix uteri by a next generation electrosurgical argon plasma device. *Cancers* **2020**, *12*, 267. [CrossRef]

70. Yoshikawa, N.; Liu, W.; Nakamura, K.; Yoshida, K.; Ikeda, Y.; Tanaka, H.; Mizuno, M.; Toyokuni, S.; Hori, M.; Kikkawa, F.; et al. Plasma-activated medium promotes autophagic cell death along with alteration of the mTOR pathway. *Sci. Rep.* **2020**, *10*, 1–8. [CrossRef]

71. Ikeda, J.-I.; Tsuruta, Y.; Nojima, S.; Sakakita, H.; Hori, M.; Ikehara, Y. Anti-cancer effects of nonequilibrium atmospheric pressure plasma on cancer-initiating cells in human endometrioid adenocarcinoma cells. *Plasma Process. Polym.* **2015**, *12*, 1370–1376. [CrossRef]

72. Ikeda, J.-I. Effect of Nonequilibrium Atmospheric Pressure Plasma on Cancer-Initiating Cells. *Plasma Med.* **2014**, *4*, 49–56. [CrossRef]

73. Koensgen, D.; Besic, I.; Gümbel, D.; Kaul, A.; Weiss, M.; Diesing, K.; Kramer, A.; Bekeschus, S.; Mustea, A.; Stope, M.B. Cold atmospheric plasma (CAP) and CAP-stimulated cell culture media suppress ovarian cancer cell growth–a putative treatment option in ovarian cancer therapy. *Anticancer Res.* **2017**, *37*. [CrossRef]

74. Bisag, A.; Bucci, C.; Coluccelli, S.; Girolimetti, G.; Laurita, R.; De Iaco, P.; Perrone, A.M.; Gherardi, M.; Marchio, L.; Porcelli, A.M.; et al. Plasma-activated Ringer's lactate solution displays a selective cytotoxic effect on ovarian cancer cells. *Cancers* **2020**, *12*, 476. [CrossRef] [PubMed]

75. Utsumi, F.; Kajiyama, H.; Nakamura, K.; Tanaka, H.; Mizuno, M.; Ishikawa, K.; Kondo, H.; Kano, H.; Hori, M.; Kikkawa, F. Effect of indirect nonequilibrium atmospheric pressure plasma on anti-proliferative activity against chronic chemo-resistant ovarian cancer cells in vitro and in vivo. *PLoS ONE* **2013**, *8*, e81576. [CrossRef] [PubMed]

76. Utsumi, F.; Kajiyama, H.; Nakamura, K.; Tanaka, H.; Hori, M.; Kikkawa, F. Selective cytotoxicity of indirect nonequilibrium atmospheric pressure plasma against ovarian clear-cell carcinoma. *SpringerPlus* **2014**, *3*, 1–9. [CrossRef] [PubMed]

77. Kajiyama, H.; Utsumi, F.; Nakamura, K.; Tanaka, H.; Mizuno, M.; Toyokuni, S.; Hori, M.; Kikkawa, F. Possible therapeutic option of aqueous plasma for refractory ovarian cancer. *Clin. Plasma Med.* **2016**, *4*, 14–18. [CrossRef]

78. Iseki, S.; Nakamura, K.; Hayashi, M.; Tanaka, H.; Kondo, H.; Kajiyama, H.; Kano, H.; Kikkawa, F.; Hori, M. Selective killing of ovarian cancer cells through induction of apoptosis by nonequilibrium atmospheric pressure plasma. *Appl. Phys. Lett.* **2012**, *100*, 113702. [CrossRef]

79. Mullen, M.M.; Merfeld, E.C.; Palisoul, M.L.; Massad, L.S.; WoolFolk, C.; Powell, M.A.; Mutch, D.G.; Thaker, P.H.; Hagemann, A.R.; Kuroki, L.M. Wound complication rates after vulvar excisions for premalignant lesions. *Obstet. Gynecol.* **2019**, *133*, 658–665. [CrossRef]

80. Leminen, A.; Forss, M.; Paavonen, J. Wound complications in patients with carcinoma of the vulva. *Eur. J. Obstet. Gynecol. Reprod. Biol.* **2000**, *93*, 193–197. [CrossRef]

81. Weinberg, D.; Gomez-Martinez, R.A. Vulvar cancer. *Obstet. Gynecol. Clin. N. Am.* **2019**, *46*, 125–135. [CrossRef]

82. Crum, C.P.; Herrington, C.S.; McCluggage, W.G.; Regauer, S.; Wilkinson, E.J. Epithelial tumours of vulva. In *WHO Classification of Tumours of Female Reproductive Organs*; Kurman, R.J., Carcangiu, M.L., Herrington, C.S., Young, R.H., Eds.; IARC: Lyon, France, 2014; pp. 232–241. ISBN 978-92-832-2435-8.

83. Boer, F.L.; Eikelder, M.L.T.; Kapiteijn, E.H.; Creutzberg, C.L.; Galaal, K.; Van Poelgeest, M.I. Vulvar malignant melanoma: Pathogenesis, clinical behaviour and management: Review of the literature. *Cancer Treat. Rev.* **2019**, *73*, 91–103. [CrossRef] [PubMed]

84. Moxley, K.M.; Fader, A.N.; Rose, P.G.; Case, A.S.; Mutch, D.G.; Berry, E.; Schink, J.C.; Kim, C.; Chi, D.S.; Moore, K.N. Malignant melanoma of the vulva: An extension of cutaneous melanoma? *Gynecol. Oncol.* **2011**, *122*, 612–617. [CrossRef] [PubMed]

85. Kilts, T.P.; Long, B.; Glasgow, A.E.; Bakkum-Gamez, J.N.; Habermann, E.B.; Cliby, W.A. Invasive vulvar extramammary Paget's disease in the United States. *Gynecol. Oncol.* **2020**, *157*, 649–655. [CrossRef] [PubMed]

86. Van Der Linden, M.; Meeuwis, K.; Bulten, J.; Bosse, T.; Van Poelgeest, M.; De Hullu, J. Paget disease of the vulva. *Crit. Rev. Oncol.* **2016**, *101*, 60–74. [CrossRef] [PubMed]

87. Breathnach, R.M.; McDonnell, K.A.; Chebbi, A.; Callanan, J.J.; Dowling, D.P. Evaluation of the effectiveness of kINPen Med plasma jet and bioactive agent therapy in a rat model of wound healing. *Biointerphases* **2018**, *13*, 051002. [CrossRef] [PubMed]

88. Zhou, X.; Cai, D.; Xiao, S.; Ning, M.; Zhou, R.; Zhang, S.; Chen, X.; Ostrikov, K.; Dai, X. InvivoPen: A novel plasma source for in vivo cancer treatment. *J. Cancer* **2020**, *11*, 2273–2282. [CrossRef] [PubMed]

89. Chen, G.; Chen, Z.; Wen, D.; Wang, Z.; Li, H.; Zeng, Y.; Dotti, G.; Wirz, R.E.; Gu, Z. Transdermal cold atmospheric plasma-mediated immune checkpoint blockade therapy. *Proc. Natl. Acad. Sci. USA* **2020**, *117*, 3687–3692. [CrossRef]

90. Almeida, N.D.; Klein, A.L.; Hogan, E.; Terhaar, S.J.; Kedda, J.; Uppal, P.; Sack, K.; Keidar, M.; Sherman, J.H.; Kedda, N. Cold atmospheric plasma as an adjunct to immunotherapy for glioblastoma multiforme. *World Neurosurg.* **2019**, *130*, 369–376. [CrossRef]

91. Li, W.; Yu, H.; Ding, D.; Chen, Z.; Wang, Y.; Wang, S.; Li, X.; Keidar, M.; Zhang, W. Cold atmospheric plasma and iron oxide-based magnetic nanoparticles for synergetic lung cancer therapy. *Free Radic. Biol. Med.* **2019**, *130*, 71–81. [CrossRef]

92. American Cancer Society: Key Statistics for Vulvar Cancer. Available online: https://www.cancer.org/cancer/vulvar-cancer/about/key-statistics.html (accessed on 20 October 2020).

93. Siegel, R.L.; Mph, K.D.M.; Jemal, A. Cancer statistics, 2020. *CA A Cancer J. Clin.* **2020**, *70*, 7–30. [CrossRef] [PubMed]

94. Meltzer-Gunnes, C.J.; Småstuen, M.C.; Kristensen, G.B.; Trope, C.G.; Lie, A.K.; Vistad, I. Vulvar carcinoma in Norway: A 50-year perspective on trends in incidence, treatment and survival. *Gynecol. Oncol.* **2017**, *145*, 543–548. [CrossRef] [PubMed]

95. Kumar, S.; Shah, J.P.; Malone, J.M. Vulvar cancer in women less than fifty in United States, 1980–2005. *Gynecol. Oncol.* **2008**, *112*, 283–284. [CrossRef] [PubMed]

96. Schuurman, M.S.; Einden, L.V.D.; Massuger, L.; Kiemeney, L.; Van Der Aa, M.; De Hullu, J. Trends in incidence and survival of Dutch women with vulvar squamous cell carcinoma. *Eur. J. Cancer* **2013**, *49*, 3872–3880. [CrossRef]

97. Pleunis, N.; Schuurman, M.; Van Rossum, M.; Bulten, J.; Massuger, L.; De Hullu, J.; Van Der Aa, M. Rare vulvar malignancies; incidence, treatment and survival in the Netherlands. *Gynecol. Oncol.* **2016**, *142*, 440–445. [CrossRef]

98. Dasgupta, S.; Ewing-Graham, P.C.; Swagemakers, S.M.; Van Der Spek, P.; Van Doorn, H.C.; Hegt, V.N.; Koljenović, S.; Van Kemenade, F.J. Precursor lesions of vulvar squamous cell carcinoma–histology and biomarkers: A systematic review. *Crit. Rev. Oncol.* **2020**, *147*, 102866. [CrossRef]

99. Dittmer, C.; Katalinic, A.; Mundhenke, C.; Thill, M.; Fischer, D. Epidemiology of vulvar and vaginal cancer in Germany. *Arch. Gynecol. Obstet.* **2011**, *284*, 169–174. [CrossRef]

100. Bray, F.; Me, J.F.; Soerjomataram, I.; Siegel, R.L.; Torre, L.A.; Jemal, A. Global cancer statistics 2018: GLOBOCAN estimates of incidence and mortality worldwide for 36 cancers in 185 countries. *CA A Cancer J. Clin.* **2018**, *68*, 394–424. [CrossRef]

101. Eva, L.J.; Sadler, L.; Fong, K.L.; Sahota, S.; Jones, R.W.; Bigby, S.M. Trends in HPV-dependent and HPV-independent vulvar cancers: The changing face of vulvar squamous cell carcinoma. *Gynecol. Oncol.* **2020**, *157*, 450–455. [CrossRef]

102. Buttmann-Schweiger, N.; Klug, S.J.; Luyten, A.; Holleczek, B.; Heitz, F.; Du Bois, A.; Kraywinkel, K. Incidence Patterns and Temporal Trends of Invasive Nonmelanotic Vulvar Tumors in Germany 1999–2011. A Population-Based Cancer Registry Analysis. *PLoS ONE* **2015**, *10*, e0128073. [CrossRef]

103. Zhang, J.; Zhang, Y.; Zhang, Z. Prevalence of human papillomavirus and its prognostic value in vulvar cancer: A systematic review and meta-analysis. *PLoS ONE* **2018**, *13*, e0204162. [CrossRef]

104. Holleczek, B.; Sehouli, J.; Barinoff, J. Vulvar cancer in Germany: Increase in incidence and change in tumour biological characteristics from 1974 to 2013. *Acta Oncol.* **2017**, *57*, 324–330. [CrossRef] [PubMed]

105. Forman, D.; De Martel, C.; Lacey, C.J.; Soerjomataram, I.; Lortet-Tieulent, J.; Bruni, L.; Vignat, J.; Ferlay, J.; Bray, F.; Plummer, M.; et al. Global burden of human papillomavirus and related diseases. *Vaccine* **2012**, *30*, F12–F23. [CrossRef]

106. Reinholdt, K.; Thomsen, L.T.; Dehlendorff, C.; Larsen, H.K.; Sørensen, S.S.; Haedersdal, M.; Kjær, S.K. Human papillomavirus-related anogenital premalignancies and cancer in renal transplant recipients: A Danish nationwide, registry-based cohort study. *Int. J. Cancer* **2019**, *146*, 2413–2422. [CrossRef] [PubMed]

107. Mancini, S.; Bucchi, L.; Baldacchini, F.; Giuliani, O.; Ravaioli, A.; Vattiato, R.; Preti, M.; Tumino, R.; Ferretti, S.; Biggeri, A.; et al. Incidence trends of vulvar squamous cell carcinoma in Italy from 1990 to 2015. *Gynecol. Oncol.* **2020**, *157*, 656–663. [CrossRef] [PubMed]

108. Suneja, G.; Viswanathan, A.N. Gynecologic malignancies. *Hematol. Clin. N. Am.* **2020**, *34*, 71–89. [CrossRef] [PubMed]

109. Ferrari, F.; Forte, S.; Ardighieri, L.; Bonetti, E.; Fernando, B.; Sartori, E.; Odicino, F. Multivariate analysis of prognostic factors in primary squamous cell vulvar cancer: The role of perineural invasion in recurrence and survival. *Eur. J. Surg. Oncol.* **2019**, *45*, 2115–2119. [CrossRef] [PubMed]

110. Beller, U.; Quinn, M.; Benedet, J.; Creasman, W.; Ngan, H.Y.S.; Maisonneuve, P.; Pecorelli, S.; Odicino, F.; Heintz, A. Carcinoma of the Vulva. *Int. J. Gynecol. Obstet.* **2006**, *95* (Suppl. S1), S7–S27. [CrossRef]

111. Mantovani, G.; Fragomeni, S.M.; Inzani, F.; Fagotti, A.; Della Corte, L.; Gentileschi, S.; Tagliaferri, L.; Zannoni, G.F.; Scambia, G.; Garganese, G. Molecular pathways in vulvar squamous cell carcinoma: Implications for target therapeutic strategies. *J. Cancer Res. Clin. Oncol.* **2020**, *146*, 1647–1658. [CrossRef]

112. Allbritton, J.I. Vulvar Neoplasms, Benign and Malignant. *Obstet. Gynecol. Clin. N. Am.* **2017**, *44*, 339–352. [CrossRef]

113. Joura, E.A.; Lösch, A.; Haider-Angeler, M.G.; Breitenecker, G.; Leodolter, S. Trends in vulvar neoplasia. Increasing incidence of vulvar intraepithelial neoplasia and squamous cell carcinoma of the vulva in young women. *J. Reprod. Med.* **2000**, *45*, 613–615.

114. Judson, P.L.; Habermann, E.B.; Baxter, N.N.; Durham, S.B.; Virnig, B.A. Trends in the incidence of invasive and in situ vulvar carcinoma. *Obstet. Gynecol.* **2006**, *107*, 1018–1022. [CrossRef] [PubMed]

115. Ridley, C.M.; Frankman, O.; Jones, I.S.; Pincus, S.H.; Wilkinson, E.J.; Fox, P.H.; Friedrich, E.G.; Kaufman, R.H.; Lynch, P.J. New nomenclature for vulvar disease: International society for the study of vulvar disease. *Hum. Pathol.* **1989**, *20*, 495–496. [CrossRef]

116. Bornstein, J.; Bogliatto, F.; Haefner, H.K.; Stockdale, C.K.; Preti, M.; Bohl, T.G.; Reutter, J. The 2015 International society for the study of vulvovaginal disease (ISSVD) terminology of vulvar squamous intraepithelial lesions. *Obstet. Gynecol.* **2016**, *127*, 264–268. [CrossRef] [PubMed]

117. Singh, N.; Ghatage, P. Etiology, Clinical features, and diagnosis of vulvar lichen sclerosus: A scoping review. *Obstet. Gynecol. Int.* **2020**, *2020*, 7480754. [CrossRef] [PubMed]

118. Harmon, M.L. Premalignant and malignant squamous lesions of the vulva. *Diagn. Histopathol.* **2017**, *23*, 19–27. [CrossRef]

119. Bleeker, M.C.G.; Visser, P.J.; Overbeek, L.I.; Van Beurden, M.; Berkhof, J. Lichen Sclerosus: Incidence and risk of vulvar squamous cell carcinoma. *Cancer Epidemiol. Biomark. Prev.* **2016**, *25*, 1224–1230. [CrossRef]

120. Halonen, P.M.; Jakobsson, M.I.; Heikinheimo, O.; Riska, A.E.; Gissler, M.; Pukkala, E.I. Lichen sclerosus and risk of cancer. *Int. J. Cancer* **2017**, *140*, 1998–2002. [CrossRef] [PubMed]

121. Pouwer, A.-F.W.; Einden, L.C.V.D.; Van Der Linden, M.; Hehir-Kwa, J.Y.; Yu, J.; Hendriks, K.M.; Kamping, E.J.; Eijkelenboom, A.; Massuger, L.F.; Bulten, J.; et al. Clonal Relationship Between Lichen Sclerosus, Differentiated Vulvar Intra-epithelial Neoplasia and Non HPV-related Vulvar Squamous Cell Carcinoma. *Cancer Genom. Proteom.* **2020**, *17*, 151–160. [CrossRef]

122. Del Pino, M.; Rodriguez-Carunchio, L.; Ordi, J. Pathways of vulvar intraepithelial neoplasia and squamous cell carcinoma. *Histopathology* **2012**, *62*, 161–175. [CrossRef]

123. Gensthaler, L.; Joura, E.; Alemany, L.; Horvat, R.; De Sanjosé, S.; Pils, S. The impact of p16^{ink4a} positivity in invasive vulvar cancer on disease-free and disease-specific survival, a retrospective study. *Arch. Gynecol. Obstet.* **2020**, *301*, 753–759. [CrossRef] [PubMed]

124. Proctor, L.; Hoang, L.; Moore, J.; Thompson, E.; Leung, S.; Natesan, D.; Chino, J.; Gilks, B.; McAlpine, J.N. Association of human papilloma virus status and response to radiotherapy in vulvar squamous cell carcinoma. *Int. J. Gynecol. Cancer* **2020**, *30*, 100–106. [CrossRef] [PubMed]

125. Hoang, L.N.; Park, K.J.; Soslow, R.A.; Murali, R. Squamous precursor lesions of the vulva: Current classification and diagnostic challenges. *Pathology* **2016**, *48*, 291–302. [CrossRef] [PubMed]

126. Ribeiro, F.; Figueiredo, A.; Paula, T.; Borrego, J. Vulvar intraepithelial neoplasia. *J. Low. Genit. Tract Dis.* **2012**, *16*, 313–317. [CrossRef] [PubMed]

127. Ayhan, A.; Tuncer, Z.S.; Doğan, L.; Yüce, K.; Küçükali, T. Skinning vulvectomy for the treatment of vulvar intraepithelial neoplasia 2-3: A study of 21 cases. *Eur. J. Gynaecol. Oncol.* **1998**, *19*, 508–510.

128. Van Seters, M.; Van Beurden, M.; De Craen, A.J.M. Is the assumed natural history of vulvar intraepithelial neoplasia III based on enough evidence? A systematic review of 3322 published patients. *Gynecol. Oncol.* **2005**, *97*, 645–651. [CrossRef]

129. Penna, C.; Fallani, M.G.; Fambrini, M.; Zipoli, E.; Marchionni, M. CO2 laser surgery for vulvar intraepithelial neoplasia. Excisional, destructive and combined techniques. *J. Reprod. Med.* **2002**, *47*, 915–918.

130. De Witte, C.J.; Van De Sande, A.J.M.; Van Beekhuizen, H.J.; Koeneman, M.M.; Kruse, A.; Gerestein, C.G. Imiquimod in cervical, vaginal and vulvar intraepithelial neoplasia: A review. *Gynecol. Oncol.* **2015**, *139*, 377–384. [CrossRef]

131. Van Der Zee, A.G.J.; Oonk, M.H.; De Hullu, J.A.; Ansink, A.C.; Vergote, I.; Verheijen, R.H.; Maggioni, A.; Gaarenstroom, K.N.; Baldwin, P.J.; Van Dorst, E.B.; et al. sentinel node dissection is safe in the treatment of early-stage vulvar cancer. *J. Clin. Oncol.* **2008**, *26*, 884–889. [CrossRef]

132. Oonkm, M.H.; van Hemel, B.M.; Hollema, H.; de Hullu, J.A.; Ansink, A.C.; Vergote, I.; Verheijen, R.H.; Maggioni, A.; Gaarenstroom, K.N.; Baldwin, P.J.; et al. Size of sentinel-node metastasis and chances of non-sentinel-node involvement and survival in early stage vulvar cancer: Results from GROINSS-V, a multicentre observational study. *Lancet Oncol.* **2010**, *11*, 646–652. [CrossRef]

133. Woelber, L.; Jaeger, A.; Prieske, K. New treatment standards for vulvar cancer 2020. *Curr. Opin. Obstet. Gynecol.* **2020**, *32*, 9–14. [CrossRef] [PubMed]

134. Chiantera, V.; Rossi, M.; De Iaco, P.; Koehler, C.; Marnitz, S.; Fagotti, A.; Fanfani, F.; Parazzini, F.; Schiavina, R.; Scambia, G.; et al. Morbidity after pelvic exenteration for gynecological malignancies: A retrospective multicentric study of 230 patients. *Int. J. Gynecol. Cancer* **2014**, *24*, 156–164. [CrossRef] [PubMed]

135. Huisman, B.; Burggraaf, J.; Vahrmeijer, A.L.; Schoones, J.; Rissmann, R.; Sier, C.F.M.; Van Poelgeest, M. Potential targets for tumor-specific imaging of vulvar squamous cell carcinoma: A systematic review of candidate biomarkers. *Gynecol. Oncol.* **2020**, *156*, 734–743. [CrossRef] [PubMed]

136. Oonk, M.H.M.; Planchamp, F.; Baldwin, P.; Bidzinski, M.; Brännström, M.; Landoni, F.; Mahner, S.; Mahantshetty, U.; Mirza, M.; Petersen, C.; et al. European Society of Gynaecological Oncology Guidelines for the Management of Patients With Vulvar Cancer. *Int. J. Gynecol. Cancer* **2017**, *27*, 832–837. [CrossRef]

137. De Hullu, J.A.; Van Der Zee, A. Surgery and radiotherapy in vulvar cancer. *Crit. Rev. Oncol. Hematol.* **2006**, *60*, 38–58. [CrossRef]

138. Nooij, L.; Brand, F.; Gaarenstroom, K.; Creutzberg, C.L.; De Hullu, J.; Van Poelgeest, M. Risk factors and treatment for recurrent vulvar squamous cell carcinoma. *Crit. Rev. Oncol. Hematol.* **2016**, *106*, 1–13. [CrossRef]

139. Pirot, F.; Chaltiel, D.; Ouldamer, L.; Touboul, C.; Raimond, E.; Carcopino, X.; Daraï, E.; Bendifallah, S. Patterns of first recurrence and outcomes in surgically treated women with vulvar cancer: Results from FRANCOGYN study group. *J. Gynecol. Obstet. Hum. Reprod.* **2020**, *101775*, 101775. [CrossRef]

140. Salani, R.; Khanna, N.; Frimer, M.; Bristow, R.E.; Chen, L.-M. An update on post-treatment surveillance and diagnosis of recurrence in women with gynecologic malignancies: Society of Gynecologic Oncology (SGO) recommendations. *Gynecol. Oncol.* **2017**, *146*, 3–10. [CrossRef]

141. Grootenhuis, N.C.T.; Van Der Zee, A.G.; Van Doorn, H.C.; Van Der Velden, J.; Vergote, I.; Zanagnolo, V.; Baldwin, P.J.; Gaarenstroom, K.N.; Van Dorst, E.B.; Trum, J.W.; et al. Sentinel nodes in vulvar cancer: Long-term follow-up of the groningen international study on sentinel nodes in vulvar cancer (GROINSS-V) I. *Gynecol. Oncol.* **2016**, *140*, 8–14. [CrossRef]

142. Moore, D.H.; Ali, S.; Koh, W.-J.; Michael, H.; Barnes, M.N.; McCourt, C.K.; Homesley, H.D.; Walker, J.L. A phase II trial of radiation therapy and weekly cisplatin chemotherapy for the treatment of locally-advanced squamous cell carcinoma of the vulva: A gynecologic oncology group study. *Gynecol. Oncol.* **2012**, *124*, 529–533. [CrossRef]

143. Gill, B.S.; Bernard, M.E.; Lin, J.F.; Balasubramani, G.K.; Rajagopalan, M.S.; Sukumvanich, P.; Krivak, T.C.; Olawaiye, A.B.; Kelley, J.L.; Beriwal, S. Impact of adjuvant chemotherapy with radiation for node-positive vulvar cancer: A national cancer data base (NCDB) analysis. *Gynecol. Oncol.* **2015**, *137*, 365–372. [CrossRef] [PubMed]

144. Logar, H.B.Z. Long term results of radiotherapy in vulvar cancer patients in Slovenia between 1997–2004. *Radiol. Oncol.* **2017**, *51*, 447–454. [CrossRef]

145. Lupi, G.; Raspagliesi, F.; Zucali, R.; Fontanelli, R.; Paladini, D.; Kenda, R.; di Re, F. Combined preoperative chemoradiotherapy followed by radical surgery in locally advanced vulvar carcinoma. A pilot study. *Cancer* **1996**, *77*, 1472–1478. [CrossRef]

146. Mullen, M.M.; Cripe, J.C.; Thaker, P.H. Palliative care in gynecologic oncology. *Obstet. Gynecol. Clin. N. Am.* **2019**, *46*, 179–197. [CrossRef] [PubMed]

147. Grootenhuis, N.T.; Pouwer, A.; De Bock, G.; Hollema, H.; Bulten, J.; Van Der Zee, A.; De Hullu, J.; Oonk, M.H.M. Margin status revisited in vulvar squamous cell carcinoma. *Gynecol. Oncol.* **2019**, *154*, 266–275. [CrossRef]

148. Grootenhuis, N.C.T.; Pouwer, A.-F.W.; De Bock, G.H.; Hollema, H.; Bulten, J.; Van Der Zee, A.G.J.; De Hullu, J.A.; Oonk, M.H.M. Prognostic factors for local recurrence of squamous cell carcinoma of the vulva: A systematic review. *Gynecol. Oncol.* **2018**, *148*, 622–631. [CrossRef] [PubMed]

149. Woelber, L.; Griebel, L.-F.; Eulenburg, C.; Sehouli, J.; Jueckstock, J.; Hilpert, F.; De Gregorio, N.; Hasenburg, A.; Ignatov, A.; Hillemanns, P.; et al. Role of tumour-free margin distance for loco-regional control in vulvar cancer—A subset analysis of the Arbeitsgemeinschaft Gynäkologische Onkologie CaRE-1 multicenter study. *Eur. J. Cancer* **2016**, *69*, 180–188. [CrossRef]

150. Nguyen-Xuan, H.-T.; Macias, R.M.; Bonsang-Kitzis, H.; Deloménie, M.; Ngô, C.; Koual, M.; Bats, A.-S.; Hivelin, M.; Lécuru, F.; Balaya, V. Use of fluorescence to guide surgical resection in vulvo-vaginal neoplasia: Two case reports. *J. Gynecol. Obstet. Hum. Reprod.* **2020**, *101768*, 101768. [CrossRef]

151. Xia, J.; Zeng, W.; Liu, X.-M.; Wang, B.; Xu, D.; Liu, D.; Kong, M.G.; Dong, Y. Cold atmospheric plasma induces apoptosis of melanoma cells via Sestrin2-mediated nitric oxide synthase signaling. *J. Biophotonics* **2018**, *12*, e201800046. [CrossRef]

152. Metelmann, H.-R.; Nedrelow, D.S.; Seebauer, C.; Schuster, M.; Von Woedtke, T.; Weltmann, K.-D.; Kindler, S.; Metelmann, P.H.; Finkelstein, S.E.; Von Hoff, D.D.; et al. Head and neck cancer treatment and physical plasma. *Clin. Plasma Med.* **2015**, *3*, 17–23. [CrossRef]

153. Schuster, M.; Seebauer, C.; Rutkowski, R.; Hauschild, A.; Podmelle, F.; Metelmann, C.; Metelmann, B.; Von Woedtke, T.; Hasse, S.; Weltmann, K.-D.; et al. Visible tumor surface response to physical plasma and apoptotic cell kill in head and neck cancer. *J. Craniomaxillofac. Surg.* **2016**, *44*, 1445–1452. [CrossRef]

154. Canady, J. Clinical Application of Cold Atmospheric Plasma (CAP) and Hybrid Plasma for the Treatment of Stage IV Gastrointestinal Cancers: Update. IWPCT-2017, Oral Lecture, Session 6. 2017. Available online: https://iwpct2017.sciencesconf.org/resource/page/id/13 (accessed on 20 October 2020).

155. Partecke, L.I.; Evert, K.; Haugk, J.; Doering, F.; Normann, L.; Diedrich, S.; Weiss, F.U.; Evert, M.; Huebner, N.-O.; Guenther, C.; et al. Tissue tolerable plasma (TTP) induces apoptosis in pancreatic cancer cells in vitro and in vivo. *BMC Cancer* **2012**, *12*, 473. [CrossRef]

156. Dobrynin, D.; Fridman, G.; Friedman, G.; Fridman, A.A. Deep penetration into tissues of reactive oxygen species generated in floating-electrode dielectric barrier discharge (FE-DBD): An in vitro agarose gel model mimicking an open wound. *Plasma Med.* **2012**, *2*, 71–83. [CrossRef]

157. Nakamura, K.; Peng, Y.; Utsumi, F.; Tanaka, H.; Mizuno, M.; Toyokuni, S.; Hori, M.; Kikkawa, F.; Kajiyama, H. novel intraperitoneal treatment with non-thermal plasma-activated medium inhibits metastatic potential of ovarian cancer cells. *Sci. Rep.* **2017**, *7*, 1–14. [CrossRef]

158. Park, S.; Kim, H.; Ji, H.W.; Kim, H.W.; Yun, S.H.; Choi, E.H.; Kim, S.J. Cold Atmospheric plasma restores paclitaxel sensitivity to paclitaxel-resistant breast cancer cells by reversing expression of resistance-related genes. *Cancers* **2019**, *11*, 2011. [CrossRef] [PubMed]

159. Burm, K.T.A.L. Plasma: The fourth state of matter. *Plasma Chem. Plasma Process.* **2012**, *32*, 401–407. [CrossRef]

160. Bogle, M.A.; Arndt, K.A.; Dover, J.S. Plasma skin regeneration technology. *J. Drugs Dermatol.* **2007**, *6*, 1110–1112. [PubMed]

161. Dai, X.; Bazaka, K.; Richard, D.J.; Thompson, E.; Rik, W.; Ostrikov, K. The emerging role of gas plasma in oncotherapy. *Trends Biotechnol.* **2018**, *36*, 1183–1198. [CrossRef] [PubMed]

162. Weltmann, K.D.; Kindel, E.; Von Woedtke, T.; Hähnel, M.; Stieber, M.; Brandenburg, R. Atmospheric-pressure plasma sources: Prospective tools for plasma medicine. *Pure Appl. Chem.* **2010**, *82*, 1223–1237. [CrossRef]

163. Manner, H. Argon plasma coagulation therapy. *Curr. Opin. Gastroenterol.* **2008**, *24*, 612–616. [CrossRef]

164. Brunaldi, V.O.; Farias, G.F.A.; de Rezende, D.T.; Cairo-Nunes, G.; Riccioppo, D.; de Moura, D.T.H.; Santo, M.A.; de Moura, E.G.H. Argon plasma coagulation alone versus argon plasma coagulation plus full-thickness

endoscopic suturing to treat weight regain after Roux-en-Y gastric bypass: A prospective randomized trial (with videos). *Gastrointest. Endosc.* **2020**, *92*, 33997–34003. [CrossRef] [PubMed]

165. Isbary, G.; Zimmermann, J.; Shimizu, T.; Li, Y.-F.; Morfill, G.; Thomas, H.; Steffes, B.; Heinlin, J.; Karrer, S.; Stolz, W. Non-thermal plasma—More than five years of clinical experience. *Clin. Plasma Med.* **2013**, *1*, 19–23. [CrossRef]

166. Gay-Mimbrera, J.; García, M.C.; Isla-Tejera, B.; Rodero-Serrano, A.; García-Nieto, A.V.; Ruano, J. Clinical and Biological Principles of Cold Atmospheric Plasma Application in Skin Cancer. *Adv. Ther.* **2016**, *33*, 894–909. [CrossRef] [PubMed]

167. Kramer, A.; Conway, B.; Meissner, K.; Scholz, F.; Rauch, B.; Moroder, A.; Ehlers, A.; Meixner, A.; Heidecke, C.-D.; Partecke, L.; et al. Cold atmospheric pressure plasma for treatment of chronic wounds: Drug or medical device? *J. Wound Care* **2017**, *26*, 470–475. [CrossRef] [PubMed]

168. Šimončicová, J.; Kryštofová, S.; Medvecká, V.; Ďurišová, K.; Kaliňáková, B. Technical applications of plasma treatments: Current state and perspectives. *Appl. Microbiol. Biotechnol.* **2019**, *103*, 5117–5129. [CrossRef] [PubMed]

169. Lu, X.; Laroussi, M.; Puech, V. On atmospheric-pressure non-equilibrium plasma jets and plasma bullets. *Plasma Sour. Sci. Technol.* **2012**, *21*. [CrossRef]

170. Duan, Y.; Huang, C.; Yu, Q. Cold plasma brush generated at atmospheric pressure. *Rev. Sci. Instrum.* **2007**, *78*, 015104. [CrossRef] [PubMed]

171. Winter, J.; Brandenburg, R.; Weltmann, K.-D. Atmospheric pressure plasma jets: An overview of devices and new directions. *Plasma Sources Sci. Technol.* **2015**, *24*, 064001. [CrossRef]

172. Attri, P.; Park, J.H.; Ali, A.; Choi, E.H. How does plasma activated media treatment differ from direct cold plasma treatment? *Anti-Cancer Agents Med. Chem.* **2018**, *18*, 805–814. [CrossRef]

173. Kong, M.G.; Kroesen, G.; Morfill, G.; Nosenko, T.; Shimizu, T.; Van Dijk, J.; Zimmermann, J.L. Plasma medicine: An introductory review. *New J. Phys.* **2009**, *11*, 115012. [CrossRef]

174. Stoffels, E.; Sakiyama, Y.; Graves, D.B. Cold Atmospheric Plasma: Charged Species and Their Interactions With Cells and Tissues. *IEEE Trans. Plasma Sci.* **2008**, *36*, 1441–1457. [CrossRef]

175. Weiss, M.; Barz, J.; Ackermann, M.; Utz, R.; Ghoul, A.; Weltmann, K.-D.; Stope, M.B.; Wallwiener, D.; Schenke-Layland, K.; Oehr, C.; et al. Dose-dependent tissue-level characterization of a medical atmospheric pressure argon plasma jet. *ACS Appl. Mater. Interfaces* **2019**, *11*, 19841–19853. [CrossRef] [PubMed]

176. Wojtowicz, A.M.; Oliveira, S.; Carlson, M.W.; Zawadzka, A.; Rousseau, C.F.; Baksh, D. The importance of both fibroblasts and keratinocytes in a bilayered living cellular construct used in wound healing. *Wound Repair Regen.* **2014**, *22*, 246–255. [CrossRef]

177. Ngo, M.-H.T.; Liao, J.-D.; Shao, P.-L.; Weng, C.-C.; Chang, C.-Y. Increased Fibroblast Cell Proliferation and Migration Using Atmospheric N2/Ar Micro-Plasma for the Stimulated Release of Fibroblast Growth Factor-7. *Plasma Process. Polym.* **2014**, *11*, 80–88. [CrossRef]

178. Haertel, B.; Wende, K.; Von Woedtke, T.; Weltmann, K.-D.; Lindequist, U. Non-thermal atmospheric-pressure plasma can influence cell adhesion molecules on HaCaT-keratinocytes. *Exp. Dermatol.* **2011**, *20*, 282–284. [CrossRef] [PubMed]

179. Kubinova, S.; Zaviskova, K.; Uherkova, L.; Zablotskii, V.; Churpita, O.; Lunov, O.; Dejneka, A. Non-thermal air plasma promotes the healing of acute skin wounds in rats. *Sci. Rep.* **2017**, *7*, srep45183. [CrossRef]

180. Schmidt, A.; Bekeschus, S.; Wende, K.; Vollmar, B.; Von Woedtke, T. A cold plasma jet accelerates wound healing in a murine model of full-thickness skin wounds. *Exp. Dermatol.* **2017**, *26*, 156–162. [CrossRef] [PubMed]

181. Shao, P.-L.; Liao, J.-D.; Wong, T.-W.; Wang, Y.-C.; Leu, S.; Yip, H.-K. Enhancement of wound healing by non-thermal N2/Ar micro-plasma exposure in mice with fractional-CO_2-laser-induced wounds. *PLoS ONE* **2016**, *11*, e0156699. [CrossRef]

182. Cheng, K.-Y.; Lin, Z.-H.; Cheng, Y.-P.; Chiu, H.-Y.; Yeh, N.-L.; Wu, T.-K.; Wu, J.-S. Wound healing in streptozotocin-induced diabetic rats using atmospheric-pressure argon plasma jet. *Sci. Rep.* **2018**, *8*, 1–15. [CrossRef]

183. Olsson, M.; Järbrink, K.; Divakar, U.; Bajpai, R.; Upton, Z.; Schmidtchen, A.; Car, J. The humanistic and economic burden of chronic wounds: A systematic review. *Wound Repair Regen.* **2018**, *27*, 114–125. [CrossRef]

184. Martinengo, L.; Olsson, M.; Bajpai, R.; Soljak, M.; Upton, Z.; Schmidtchen, A.; Car, J.; Järbrink, K. Prevalence of chronic wounds in the general population: Systematic review and meta-analysis of observational studies. *Ann. Epidemiol.* **2019**, *29*, 8–15. [CrossRef] [PubMed]

185. Bender, C.; Partecke, L.-I.; Kindel, E.; Döring, F.; Lademann, J.; Heidecke, C.-D.; Kramer, A.; Hübner, N.-O. The modified HET-CAM as a model for the assessment of the inflammatory response to tissue tolerable plasma. *Toxicol. Vitro* **2011**, *25*, 530–537. [CrossRef] [PubMed]

186. Arndt, S.; Unger, P.; Wacker, E.; Shimizu, T.; Heinlin, J.; Li, Y.-F.; Thomas, H.M.; Morfill, G.E.; Zimmermann, J.L.; Bosserhoff, A.-K.; et al. Cold atmospheric plasma (CAP) changes gene expression of key molecules of the wound healing machinery and improves wound healing in vitro and in vivo. *PLoS ONE* **2013**, *8*, e79325. [CrossRef] [PubMed]

187. Kisch, T.; Helmke, A.; Schleusser, S.; Song, J.; Liodaki, E.; Stang, F.H.; Mailaender, P.; Kraemer, R. Improvement of cutaneous microcirculation by cold atmospheric plasma (CAP): Results of a controlled, prospective cohort study. *Microvasc. Res.* **2016**, *104*, 55–62. [CrossRef] [PubMed]

188. Arndt, S.; Unger, P.; Berneburg, M.; Bosserhoff, A.-K.; Karrer, S. Cold atmospheric plasma (CAP) activates angiogenesis-related molecules in skin keratinocytes, fibroblasts and endothelial cells and improves wound angiogenesis in an autocrine and paracrine mode. *J. Dermatol. Sci.* **2018**, *89*, 181–190. [CrossRef]

189. Bacalbasa, N.; Balescu, I.; Vilcu, M.; Dima, S.; Brezean, I. Risk Factors for Postoperative Complications After Vulvar Surgery. *In Vivo* **2019**, *34*, 447–451. [CrossRef]

190. Gan, L.; Zhang, S.; Poorun, D.; Liu, D.; Lu, X.; He, M.; Duan, X.; Chen, H. Medical applications of nonthermal atmospheric pressure plasma in dermatology. *J. Dtsch. Dermatol. Ges.* **2018**, *16*, 7–13. [CrossRef]

191. Assadian, O.; Ousey, K.J.; Daeschlein, G.; Kramer, A.; Parker, C.; Tanner, J.; Leaper, D.J. Effects and safety of atmospheric low-temperature plasma on bacterial reduction in chronic wounds and wound size reduction: A systematic review and meta-analysis. *Int. Wound J.* **2019**, *16*, 103–111. [CrossRef]

192. Isbary, G.; Morfill, G.; Schmidt, H.; Georgi, M.; Ramrath, K.; Heinlin, J.; Karrer, S.; Landthaler, M.; Shimizu, T.; Steffes, B.; et al. A first prospective randomized controlled trial to decrease bacterial load using cold atmospheric argon plasma on chronic wounds in patients. *Br. J. Dermatol.* **2010**, *163*, 78–82. [CrossRef]

193. Brehmer, F.; Haenssle, H.A.; Daeschlein, G.; Ahmed, R.; Pfeiffer, S.; Görlitz, A.; Simon, D.; Schön, M.; Wandke, D.; Emmert, S. Alleviation of chronic venous leg ulcers with a hand-held dielectric barrier discharge plasma generator (PlasmaDerm®VU-2010): Results of a monocentric, two-armed, open, prospective, randomized and controlled trial (NCT01415622). *J. Eur. Acad. Dermatol. Venerol.* **2014**, *29*, 148–155. [CrossRef]

194. Heinlin, J.; Zimmermann, J.L.; Zeman, F.; Bunk, W.; Isbary, G.; Landthaler, M.; Maisch, T.; Monetti, R.; Morfill, G.; Shimizu, T.; et al. Randomized placebo-controlled human pilot study of cold atmospheric argon plasma on skin graft donor sites. *Wound Repair Regen.* **2013**, *21*, 800–807. [CrossRef]

195. Metelmann, H.-R.; Vu, T.T.; Do, H.T.; Le, T.N.B.; Hoang, T.H.A.; Phi, T.T.T.; Luong, T.M.L.; Doan, V.T.; Nguyen, T.T.H.; Nguyen, T.L.; et al. Scar formation of laser skin lesions after cold atmospheric pressure plasma (CAP) treatment: A clinical long term observation. *Clin. Plasma Med.* **2013**, *1*, 30–35. [CrossRef]

196. Nishijima, A.; Fujimoto, T.; Hirata, T.; Nishijima, J. Effects of Cold Atmospheric Pressure Plasma on Accelerating Acute Wound Healing: A Comparative Study among 4 Different Treatment Groups. *Mod. Plast. Surg.* **2019**, *9*, 18–31. [CrossRef]

197. Gao, J.; Wang, L.; Xia, C.; Yang, X.; Cao, Z.; Zheng, L.; Ko, R.; Shen, C.; Yang, C.; Cheng, C. Cold atmospheric plasma promotes different types of superficial skin erosion wounds healing. *Int. Wound J.* **2019**, *16*, 1103–1111. [CrossRef] [PubMed]

198. Dias-Jr, A.R.; Soares-Jr, J.; De Faria, M.B.S.; Genta, M.L.N.D.; Carvalho, J.P.; Baracat, E.C. Secondary healing strategy for difficult wound closure in invasive vulvar cancer: A pilot case-control study. *Clinics* **2019**, *74*, e1218. [CrossRef] [PubMed]

199. Hirst, A.M.; Frame, F.M.; Maitland, N.J.; O'Connell, D. Low temperature plasma: A novel focal therapy for localized prostate cancer? *BioMed Res. Int.* **2014**, *2014*, 1–15. [CrossRef]

200. Fengyi, W.; Lenardo, M.J. Specification of DNA binding activity of NF-kappaB proteins. *Cold Spring Harb. Perspect. Biol.* **2009**, a000067. [CrossRef]

201. Keidar, M.; Yan, D.; Beilis, I.I.; Trink, B.; Sherman, J.H. Plasmas for Treating Cancer: Opportunities for Adaptive and Self-Adaptive Approaches. *Trends Biotechnol.* **2018**, *36*, 586–593. [CrossRef]

202. Semmler, M.L.; Bekeschus, S.; Schäfer, M.; Bernhardt, T.; Fischer, T.; Witzke, K.; Seebauer, C.; Rebl, H.; Grambow, E.; Vollmar, B.; et al. Molecular Mechanisms of the Efficacy of Cold Atmospheric Pressure Plasma (CAP) in Cancer Treatment. *Cancers* **2020**, *12*, 269. [CrossRef]

203. Arndt, S.; Landthaler, M.; Zimmermann, J.L.; Unger, P.; Wacker, E.; Shimizu, T.; Li, Y.-F.; Morfill, G.E.; Bosserhoff, A.-K.; Karrer, S. Effects of cold atmospheric plasma (CAP) on ß-defensins, inflammatory cytokines, and apoptosis-related molecules in keratinocytes in vitro and in vivo. *PLoS ONE* **2015**, *10*, e0120041. [CrossRef]

204. Roma-Rodrigues, C.; Mendes, R.; Baptista, P.V.; Fernandes, A.R. Targeting tumor microenvironment for cancer therapy. *Int. J. Mol. Sci.* **2019**, *20*, 840. [CrossRef] [PubMed]

205. Xingmin, S.; Jingfen, C.; Guimin, X.; Hongbin, R.; Sile, C.; Zhengshi, C.; Xili, W. Effect of cold plasma on cell viability and collagen synthesis in cultured murine fibroblasts. *Plasma Sci. Technol.* **2016**, *18*, 353–359. [CrossRef]

206. Kang, S.U.; Kim, Y.S.; Kim, Y.E.; Park, J.-K.; Lee, Y.S.; Kang, H.Y.; Jang, J.W.; Ryeo, J.B.; Lee, Y.; Shin, Y.S.; et al. Opposite effects of non-thermal plasma on cell migration and collagen production in keloid and normal fibroblasts. *PLoS ONE* **2017**, *12*, e0187978. [CrossRef] [PubMed]

207. Keyvani, A.; Atyabi, S.M.; Sardari, S.; Norouzian, D.; Madanchi, H. Effects of cold atmospheric plasma jet on collagen structure in different treatment times. *Basic Res. J. Med. Clin. Sci.* **2017**, *6*, 84–90.

208. Privat-Maldonado, A.; Bengtson, C.; Razzokov, J.; Smits, E.; Bogaerts, A. Modifying the tumour microenvironment: Challenges and future perspectives for anticancer plasma treatments. *Cancers* **2019**, *11*, 1920. [CrossRef]

209. Bernardes, N.; Fialho, A.M. Perturbing the dynamics and organization of cell membrane components: A new paradigm for cancer-targeted therapies. *Int. J. Mol. Sci.* **2018**, *19*, 3871. [CrossRef]

210. Zalba, S.; Hagen, T.L.T. Cell membrane modulation as adjuvant in cancer therapy. *Cancer Treat. Rev.* **2017**, *52*, 48–57. [CrossRef]

211. Ding, X.; Zhang, W.; Li, S.; Yang, H. The role of cholesterol metabolism in cancer. *Am. J. Cancer Res.* **2019**, *9*, 219–227.

212. Rivel, T.; Ramseyer, C.; Yesylevskyy, S. The asymmetry of plasma membranes and their cholesterol content influence the uptake of cisplatin. *Sci. Rep.* **2019**, *9*, 1–14. [CrossRef]

213. Van Der Paal, J.; Neyts, E.C.; Verlackt, C.C.W.; Bogaerts, A. Effect of lipid peroxidation on membrane permeability of cancer and normal cells subjected to oxidative stress. *Chem. Sci.* **2016**, *7*, 489–498. [CrossRef]

214. Bauer, G. Tumor cell-protective catalase as a novel target for rational therapeutic approaches based on specific intercellular ROS signaling. *Anticancer Res.* **2012**, *32*, 2599–2624. [PubMed]

215. Bauer, G. Targeting extracellular ROS signaling of tumor cells. *Anticancer Res.* **2014**, *34*, 1467–1482. [PubMed]

216. Bauer, G. Targeting protective catalase of tumor cells with cold atmospheric plasma- activated medium (PAM). *Anti-Cancer Agents Med. Chem.* **2018**, *18*, 784–804. [CrossRef] [PubMed]

217. Dezest, M.; Chavatte, L.; Bourdens, M.; Quinton, D.; Camus, M.; Garrigues, L.; Descargues, P.; Arbault, S.; Burlet-Schiltz, O.; Casteilla, L.; et al. Mechanistic insights into the impact of Cold Atmospheric Pressure Plasma on human epithelial cell lines. *Sci. Rep.* **2017**, *7*, 41163. [CrossRef] [PubMed]

218. Crawford, L.J.; Walker, B.; Irvine, A.E. Proteasome inhibitors in cancer therapy. *J. Cell Commun. Sign.* **2011**, *5*, 101–110. [CrossRef]

219. Zhao, S.; Xiong, Z.; Mao, X.; Meng, D.; Lei, Q.; Li, Y.; Deng, P.; Chen, M.; Tu, M.; Lu, X.; et al. Atmospheric pressure room temperature plasma jets facilitate oxidative and nitrative stress and lead to endoplasmic reticulum stress dependent apoptosis in HepG2 cells. *PLoS ONE* **2013**, *8*, e73665. [CrossRef]

220. Weathington, N.M.; Mallampalli, R.K. Emerging therapies targeting the ubiquitin proteasome system in cancer. *J. Clin. Investig.* **2014**, *124*, 6–12. [CrossRef]

221. Nagata, S. Apoptosis and Clearance of Apoptotic Cells. *Annu. Rev. Immunol.* **2018**, *36*, 489–517. [CrossRef]

222. Jan, R.; Chaudhry, G.-E.-S. Understanding Apoptosis and Apoptotic Pathways Targeted Cancer Therapeutics. *Adv. Pharm. Bull.* **2019**, *9*, 205–218. [CrossRef]

223. Ashkenazi, A.; Fairbrother, A.A.W.J.; Leverson, J.D.; Souers, J.D.L.A.J. From basic apoptosis discoveries to advanced selective BCL-2 family inhibitors. *Nat. Rev. Drug Discov.* **2017**, *16*, 273–284. [CrossRef]

224. Pfeffer, C.M.; Singh, A.T.K. Apoptosis: A target for anticancer therapy. *Int. J. Mol. Sci.* **2018**, *19*, 448. [CrossRef] [PubMed]

225. Brentnall, M.; Rodriguez-Menocal, L.; De Guevara, R.L.; Cepero, E.; Boise, L.H. Caspase-9, caspase-3 and caspase-7 have distinct roles during intrinsic apoptosis. *BMC Cell Biol.* **2013**, *14*, 1–9. [CrossRef]

226. Baig, S.M.; Seevasant, I.; Mohamad, J.A.; Mukheem, A.; Huri, H.Z.; Kamarul, T. Potential of apoptotic pathway-targeted cancer therapeutic research: Where do we stand? *Cell Death Dis.* **2016**, *7*, e2058. [CrossRef] [PubMed]

227. Irani, S.; Mirfakhraie, R.; Jalili, A. Combination of cold atmospheric plasma and iron nanoparticles in breast cancer: Gene expression and apoptosis study. *OncoTargets Ther.* **2016**, *9*, 5911–5917. [CrossRef]

228. Braný, D.; Dvorská, D.; Halašová, E.; Škovierová, H. Cold atmospheric plasma: A powerful tool for modern medicine. *Int. J. Mol. Sci.* **2020**, *21*, 2932. [CrossRef] [PubMed]

229. Turrini, E.; Laurita, R.; Stancampiano, A.; Catanzaro, E.; Calcabrini, C.; Maffei, F.; Gherardi, M.; Colombo, V.; Fimognari, C. Cold atmospheric plasma induces apoptosis and oxidative stress pathway regulation in t-lymphoblastoid leukemia cells. *Oxid. Med. Cell. Longev.* **2017**, *2017*, 1–13. [CrossRef] [PubMed]

230. Ahn, H.J.; Kim, K.I.; Hoan, N.N.; Kim, C.H.; Moon, E.; Choi, K.S.; Yang, S.S.; Lee, J.-S. Targeting cancer cells with reactive oxygen and nitrogen species generated by atmospheric-pressure air plasma. *PLoS ONE* **2014**, *9*, e86173. [CrossRef] [PubMed]

231. Lee, S.Y.; Kang, S.U.; Kim, K.I.; Kang, S.; Shin, Y.S.; Chang, J.W.; Yang, S.S.; Lee, K.; Lee, J.-S.; Moon, E.; et al. Nonthermal plasma induces apoptosis in ATC cells: Involvement of JNK and p38 MAPK-dependent ROS. *Yonsei Med. J.* **2014**, *55*, 1640–1647. [CrossRef]

232. Chang, J.W.; Kang, S.U.; Shin, Y.S.; Kim, K.I.; Seo, S.J.; Yang, S.S.; Lee, J.-S.; Moon, E.; Baek, S.J.; Lee, K.; et al. Non-thermal atmospheric pressure plasma induces apoptosis in oral cavity squamous cell carcinoma: Involvement of DNA-damage-triggering sub-G1 arrest via the ATM/p53 pathway. *Arch. Biochem. Biophys.* **2014**, *545*, 133–140. [CrossRef]

233. Ma, Y.; Ha, C.S.; Hwang, S.W.; Lee, H.J.; Kim, G.C.; Lee, K.-W.; Song, K. Non-thermal atmospheric pressure plasma preferentially induces apoptosis in p53-mutated cancer cells by activating ROS Stress-response pathways. *PLoS ONE* **2014**, *9*, e91947. [CrossRef]

234. Xiong, B.Z. Cold Atmospheric Pressure Plasmas (CAPs) for Skin Wound Healing, Plasma Medicine-Concepts and Clinical Applications. *IntechOpen* **2018**. [CrossRef]

235. Kalghatgi, S.; Friedman, G.; Fridman, A.; Clyne, A.M. Endothelial cell proliferation is enhanced by low dose non-thermal plasma through fibroblast growth factor-2 release. *Ann. Biomed. Eng.* **2010**, *38*, 748–757. [CrossRef] [PubMed]

236. Lendeckel, D.; Eymann, C.; Emicke, P.; Daeschlein, G.; Darm, K.; O'Neil, S.; Beule, A.; Von Woedtke, T.; Völker, U.; Weltmann, K.-D.; et al. Proteomic changes of tissue-tolerable plasma treated airway epithelial cells and their relation to wound healing. *BioMed Res. Int.* **2015**, *2015*, 1–17. [CrossRef] [PubMed]

237. Welz, C.; Emmert, S.; Canis, M.; Becker, S.; Baumeister, P.; Shimizu, T.; Morfill, G.E.; Harréus, U.; Zimmermann, J.L. Cold atmospheric plasma: A promising complementary therapy for squamous head and neck cancer. *PLoS ONE* **2015**, *10*, e0141827. [CrossRef] [PubMed]

238. Yan, D.; Talbot, A.; Nourmohammadi, N.; Cheng, X.; Canady, J.; Sherman, J.; Keidar, M. Principles of using cold atmospheric plasma stimulated media for cancer treatment. *Sci. Rep.* **2015**, *5*, 18339. [CrossRef]

239. Tolouie, H.; Mohammadifar, M.A.; Ghomi, H.; Hashemi, M. Cold atmospheric plasma manipulation of proteins in food systems. *Crit. Rev. Food Sci. Nutr.* **2018**, *58*, 2583–2597. [CrossRef]

240. Puač, N.; Živković, S.; Selaković, N.; Milutinović, M.; Boljević, J.; Malovic, G.; Petrovic, Z.L. Long and short term effects of plasma treatment on meristematic plant cells. *Appl. Phys. Lett.* **2014**, *104*, 214106. [CrossRef]

241. Schaner, M.E.; Ross, D.T.; Ciaravino, G.; Sorlie, T.; Troyanskaya, O.; Diehn, M.; Wang, Y.C.; Duran, G.E.; Sikic, T.L.; Caldeira, S.; et al. Gene expression patterns in ovarian carcinomas. *Mol. Biol. Cell* **2003**, *14*, 4376–4386. [CrossRef]

242. Knutsen, E.; Oslo Breast Cancer Research Consortium (OSBREAC); Lellahi, S.M.; Aure, M.R.; Nord, S.; Fismen, S.; Larsen, K.B.; Gabriel, M.T.; Hedberg, A.; Bjørklund, S.S.; et al. The expression of the long NEAT1_2 isoform is associated with human epidermal growth factor receptor 2-positive breast cancers. *Sci. Rep.* **2020**, *10*, 1–14. [CrossRef]

243. Su, H.; Wu, S.; Yen, L.; Chiao, L.K.; Wang, J.K.; Chiu, Y.L.; Ho, C.L.; Huang, S.M. Gene expression profiling identifies the role of Zac1 in cervical cancer metastasis. *Sci. Rep.* **2020**, *10*, 11837. [CrossRef]

244. Park, S.-B.; Kim, B.; Bae, H.; Lee, H.; Lee, S.; Choi, E.H.; Kim, S.J. Differential epigenetic effects of atmospheric cold plasma on MCF-7 and MDA-MB-231 breast cancer cells. *PLoS ONE* **2015**, *10*, e0129931. [CrossRef] [PubMed]

245. Liu, Z.; Mao, L.; Wang, L.; Zhang, H.; Hu, X. miR-218 functions as a tumor suppressor gene in cervical cancer. *Mol. Med. Rep.* **2020**, *21*, 209–219. [CrossRef] [PubMed]

246. Arndt, S.; Wacker, E.; Li, Y.-F.; Shimizu, T.; Thomas, H.M.; Morfill, G.E.; Karrer, S.; Zimmermann, J.L.; Bosserhoff, A.-K. Cold atmospheric plasma, a new strategy to induce senescence in melanoma cells. *Exp. Dermatol.* **2013**, *22*, 284–289. [CrossRef] [PubMed]

247. Kurita, H.; Haruta, N.; Uchihashi, Y.; Seto, T.; Takashima, K. Strand breaks and chemical modification of intracellular DNA induced by cold atmospheric pressure plasma irradiation. *PLoS ONE* **2020**, *15*, e0232724. [CrossRef] [PubMed]

248. Volotskova, O.; Hawley, T.S.; Stepp, M.A.; Keidar, M. Targeting the cancer cell cycle by cold atmospheric plasma. *Sci. Rep.* **2012**, *2*, 636. [CrossRef]

249. Chung, W.-H. Mechanisms of a novel anticancer therapeutic strategy involving atmospheric pressure plasma-mediated apoptosis and DNA strand break formation. *Arch. Pharmacal Res.* **2015**, *39*, 1–9. [CrossRef] [PubMed]

250. Arjunan, K.P.; Sharma, V.K.; Ptasinska, S. Effects of atmospheric pressure plasmas on isolated and cellular DNA—A review. *Int. J. Mol. Sci.* **2015**, *16*, 2971–3016. [CrossRef]

251. Lackmann, J.W.; Schneider, S.; Narberhaus, F.; Benedikt, J.; Bandow, J.E. Characterization of damage to bacteria and bio-macromolecules caused by (V)UV radiation and particles generated by a microscale atmospheric pressure plasma jet. In *Plasma for Biodecontamination, Medicine and Food Security*; Machala, Z., Hensel, K., Akishev, Y., Eds.; Springer: Berlin/Heidelberg, Germany, 2012; pp. 17–29.

252. Guo, L.; Zhao, Y.; Liu, D.X.; Liu, Z.C.; Chen, C.; Xu, R.; Tian, M.; Wang, X.; Chen, H.; Kong, M.G. Cold atmospheric-pressure plasma induces DNA–protein crosslinks through protein oxidation. *Free Radic. Res.* **2018**, *52*, 783–798. [CrossRef]

253. Kim, G.J.; Kim, W.; Kim, K.T.; Lee, J.K. DNA damage and mitochondria dysfunction in cell apoptosis induced by nonthermal air plasma. *Appl. Phys. Lett.* **2010**, *96*, 021502. [CrossRef]

254. Lackmann, J.-W.; Schneider, S.; Edengeiser, E.; Jarzina, F.; Brinckmann, S.; Steinborn, E.; Havenith, M.; Benedikt, J.; Bandow, J.E. Photons and particles emitted from cold atmospheric-pressure plasma inactivate bacteria and biomolecules independently and synergistically. *J. R. Soc. Interface* **2013**, *10*, 20130591. [CrossRef]

255. Yan, X.; Qiao, Y.; Ouyang, J.; Jia, M.; Li, J.; Yuan, F. Protective effect of atmospheric pressure plasma on oxidative stress-induced neuronal injuries: Anin vitrostudy. *J. Phys. D Appl. Phys.* **2017**, *50*, 095401. [CrossRef]

256. Niki, E. Antioxidants: Basic principles, emerging concepts, and problems. *Biomed. J.* **2014**, *37*. [CrossRef] [PubMed]

257. Abdel-Rahman, E.A.; Mahmoud, A.M.; Khalifa, A.M.; Ali, S.S. Physiological and pathophysiological reactive oxygen species as probed by EPR spectroscopy: The underutilized research window on muscle ageing. *J. Physiol.* **2016**, *594*, 4591–4613. [CrossRef]

258. Murakami, T. Numerical modelling of the effects of cold atmospheric plasma on mitochondrial redox homeostasis and energy metabolism. *Sci. Rep.* **2019**, *9*, 1–12. [CrossRef]

259. Ray, P.D.; Huang, B.-W.; Tsuji, Y. Reactive oxygen species (ROS) homeostasis and redox regulation in cellular signaling. *Cell. Signal.* **2012**, *24*, 981–990. [CrossRef] [PubMed]

260. Forman, H.J.; Maiorino, M.; Ursini, F. Signaling functions of reactive oxygen species. *Biochemistry* **2010**, *49*, 835–842. [CrossRef] [PubMed]

261. Hanschmann, E.-M.; Godoy, J.R.; Berndt, C.; Hudemann, C.; Lillig, C.H. Thioredoxins, glutaredoxins, and peroxiredoxins—Molecular mechanisms and health significance: From cofactors to antioxidants to redox signaling. *Antioxid. Redox Sign.* **2013**, *19*, 1539–1605. [CrossRef]

262. Kushnir, C.L.; Fleury, A.C.; Hill, M.C.; Silver, D.F.; Spirtos, N.M. The use of argon beam coagulation in treating vulvar intraepithelial neoplasia III: A retrospective review. *Gynecol. Oncol.* **2013**, *131*, 386–388. [CrossRef]

263. Miller, V.; Lin, A.; Fridman, A. Why target immune cells for plasma treatment of cancer. *Plasma Chem. Plasma Process.* **2015**, *36*, 259–268. [CrossRef]

264. Schneider, C.; Gebhardt, L.; Arndt, S.; Karrer, S.; Zimmermann, J.L.; Fischer, M.J.M.; Bosserhoff, A.-K. Cold atmospheric plasma causes a calcium influx in melanoma cells triggering CAP-induced senescence. *Sci. Rep.* **2018**, *8*, 10048. [CrossRef]

265. Kim, H.W.; Jeong, D.; Ham, J.; Kim, H.; Ji, H.W.; Choi, E.H.; Kim, S.J. ZNRD1 and its antisense long noncoding RNA ZNRD1-AS1 Are oppositely regulated by cold atmospheric plasma in breast cancer cells. *Oxidative Med. Cell. Longev.* **2020**, *2020*, 1–9. [CrossRef]

266. Mokhtari, H.; Farahmand, L.; Yaserian, K.; Jalili, N.; Majidzadeh-A, K. The antiproliferative effects of cold atmospheric plasma-activated media on different cancer cell lines, the implication of ozone as a possible underlying mechanism. *J. Cell. Physiol.* **2018**, *234*, 6778–6782. [CrossRef] [PubMed]

267. Takeda, S.; Yamada, S.; Hattori, N.; Nakamura, K.; Tanaka, H.; Kajiyama, H.; Kanda, M.; Kobayashi, D.; Tanaka, C.; Fujii, T.; et al. Intraperitoneal administration of plasma-activated medium: Proposal of a novel treatment option for peritoneal metastasis from gastric cancer. *Ann. Surg. Oncol.* **2017**, *24*, 1188–1194. [CrossRef] [PubMed]

268. Adhikari, M.; Adhikari, B.; Ghimire, B.; Baboota, S.; Choi, E.H. Cold atmospheric plasma and silymarin nanoemulsion activate autophagy in human melanoma cells. *Int. J. Mol. Sci.* **2020**, *21*, 1939. [CrossRef] [PubMed]

269. Kaushik, N.K.; Kaushik, N.; Yoo, K.C.; Uddin, N.; Kim, J.S.; Lee, S.J.; Choi, E.H. Low doses of PEG-coated gold nanoparticles sensitize solid tumors to cold plasma by blocking the PI3K/AKT-driven signaling axis to suppress cellular transformation by inhibiting growth and EMT. *Biomaterials* **2016**, *87*, 118–130. [CrossRef]

270. Irani, S.; Shahmirani, Z.; Atyabi, S.M.; Mirpoor, S. Induction of growth arrest in colorectal cancer cells by cold plasma and gold nanoparticles. *Arch. Med. Sci.* **2015**, *6*, 1286–1295. [CrossRef] [PubMed]

271. Jawaid, P.; Rehman, M.U.; Zhao, Q.L.; Takeda, K.; Ishikawa, K.; Hori, M.; Shimizu, T.; Kondo, T. Helium-based cold atmospheric plasma-induced reactive oxygen species-mediated apoptotic pathway attenuated by platinum nanoparticles. *J. Cell. Mol. Med.* **2016**, *20*, 1737–1748. [CrossRef]

272. He, Z.; Liu, K.; Manaloto, E.; Casey, A.; Cribaro, G.P.; Byrne, H.J.; Tian, F.; Barcia, C.; Conway, G.E.; Cullen, P.J.; et al. Cold atmospheric plasma induces ATP-dependent endocytosis of nanoparticles and synergistic U373MG cancer cell death. *Sci. Rep.* **2018**, *8*, 1–11. [CrossRef]

273. Yu, H.; Wang, Y.; Wang, S.; Li, X.; Li, W.; Ding, D.; Gong, X.; Keidar, M.; Zhang, W.-F. Paclitaxel-loaded core–shell magnetic nanoparticles and cold atmospheric plasma inhibit non-small cell lung cancer growth. *ACS Appl. Mater. Interfaces* **2018**, *10*, 43462–43471. [CrossRef]

274. Kletschkus, K.; Haralambiev, L.; Nitsch, A.; Pfister, F.; Klinkmann, G.; Kramer, A.; Bekeschus, S.; Mustea, A.; Stope, M.B. The application of a low-temperature physical plasma device operating under atmospheric pressure leads to the production of toxic NO_2. *Anticancer. Res.* **2020**, *40*, 2591–2599. [CrossRef]

275. Yan, D.; Xu, W.; Yao, X.; Lin, L.; Sherman, J.H.; Keidar, M. The cell activation phenomena in the cold atmospheric plasma cancer treatment. *Sci. Rep.* **2018**, *8*, 1–10. [CrossRef] [PubMed]

276. Yan, D.; Nourmohammadi, N.; Bian, K.; Murad, F.; Sherman, J.H.; Keidar, M. Stabilizing the cold plasma-stimulated medium by regulating medium's composition. *Sci. Rep.* **2016**, *6*, 26016. [CrossRef] [PubMed]

277. Van Boxem, W.; Van Der Paal, J.; Gorbanev, Y.; Vanuytsel, S.; Smits, E.; Dewilde, S.; Bogaerts, A. Anti-cancer capacity of plasma-treated PBS: Effect of chemical composition on cancer cell cytotoxicity. *Sci. Rep.* **2017**, *7*, 16478. [CrossRef] [PubMed]

278. Azzariti, A.; Iacobazzi, R.M.; Di Fonte, R.; Porcelli, L.; Gristina, R.; Favia, P.; Fracassi, F.; Trizio, I.; Silvestris, N.; Guida, G.; et al. Plasma-activated medium triggers cell death and the presentation of immune activating danger signals in melanoma and pancreatic cancer cells. *Sci. Rep.* **2019**, *9*, 1–13. [CrossRef]

279. Van Loenhout, J.; Flieswasser, T.; Boullosa, L.F.; De Waele, J.; Van Audernaerde, J.R.; Marcq, E.; Jacobs, J.; Lin, A.; Lion, E.; Dewitte, H.; et al. Cold Atmospheric Plasma-Treated PBS Eliminates Immunosuppressive Pancreatic Stellate Cells and Induces Immunogenic Cell Death of Pancreatic Cancer Cells. *Cancers* **2019**, *11*, 1597. [CrossRef]

280. Kaushik, N.; Kaushik, N.; Adhikari, M.; Ghimire, B.; Linh, N.N.; Mishra, Y.; Lee, S.-J.; Choi, E.H. Preventing the solid cancer progression via release of anticancer-cytokines in co-culture with cold plasma-stimulated macrophages. *Cancers* **2019**, *11*, 842. [CrossRef]

281. Wang, Z.; Førsund, M.S.; Trope, C.G.; Nesland, J.M.; Holm, R.; Slipicevic, A. Evaluation of CHK1 activation in vulvar squamous cell carcinoma and its potential as a therapeutic target in vitro. *Cancer Med.* **2018**, *7*, 3955–3964. [CrossRef]

282. Nooij, L.S.; Ter Haar, N.T.; Ruano, D.; Rakislova, N.; Van Wezel, T.; Smit, V.T.; Trimbos, B.J.; Ordi, J.; Van Poelgeest, M.I.; Bosse, T. Genomic characterization of vulvar (Pre)cancers identifies distinct molecular subtypes with prognostic significance. *Clin. Cancer Res.* **2017**, *23*, 6781–6789. [CrossRef]

283. Zięba, S.; Kowalik, A.; Zalewski, K.; Rusetska, N.; Goryca, K.; Piascik, A.; Misiek, M.; Bakuła-Zalewska, E.; Kopczyński, J.; Kowalski, K.; et al. Somatic mutation profiling of vulvar cancer: Exploring therapeutic targets. *Gynecol. Oncol.* **2018**, *150*, 552–561. [CrossRef]

284. Brunetti, M.; Agostini, A.; Davidson, B.; Tropé, C.G.; Heim, S.; Panagopoulos, I.; Micci, F. Recurrent fusion transcripts in squamous cell carcinomas of the vulva. *Oncotarget* **2017**, *8*, 16843–16850. [CrossRef]

285. Agostini, A.; Brunetti, M.; Davidson, B.; Trope, C.G.; Heim, S.; Panagopoulos, I.; Micci, F. Expressions of miR-30c and let-7a are inversely correlated with HMGA2 expression in squamous cell carcinoma of the vulva. *Oncotarget* **2016**, *7*, 85058–85062. [CrossRef] [PubMed]

286. Deppe, G.; Mert, I.; Belotte, J.; Winer, I. Chemotherapy of vulvar cancer: A review. *Wien. Klin. Wochenschr.* **2013**, *125*, 119–128. [CrossRef]

287. Mahner, S.; Prieske, K.; Grimm, D.; Trillsch, F.; Prieske, S.; Von Amsberg, G.; Petersen, C.; Mueller, V.; Jaenicke, F.; Woelber, L. Systemic treatment of vulvar cancer. *Exp. Rev. Anticancer. Ther.* **2015**, *15*, 629–637. [CrossRef]

288. Reade, C.J.; Eiriksson, L.R.; Mackay, H. Systemic therapy in squamous cell carcinoma of the vulva: Current status and future directions. *Gynecol. Oncol.* **2014**, *132*, 780–789. [CrossRef] [PubMed]

289. Ott, P.A.; Bang, Y.-J.; Piha-Paul, S.A.; Razak, A.R.A.; Bennouna, J.; Soria, J.-C.; Rugo, H.S.; Cohen, R.B.; O'Neil, B.H.; Mehnert, J.M.; et al. T-cell–inflamed gene-expression profile, programmed death ligand 1 expression, and tumor mutational burden predict efficacy in patients treated with pembrolizumab across 20 cancers: KEYNOTE-028. *J. Clin. Oncol.* **2019**, *37*, 318–327. [CrossRef] [PubMed]

290. Pekkola-Heino, K.; Kulmala, J.; Grenman, S.; Carey, T.E.; Grenman, R. Radiation response of vulvar squamous cell carcinoma (UM-SCV-1A, UM-SCV-1B, UM-SCV-2, and A-431) cells in vitro. *Cancer Res.* **1989**, *49*, 2758419.

291. Corrado, G.; Cutillo, G.; Fragomeni, S.M.; Bruno, V.; Tagliaferri, L.; Mancini, E.; Certelli, C.; Paris, I.; Vizza, E.; Scambia, G.; et al. Palliative electrochemotherapy in primary or recurrent vulvar cancer. *Int. J. Gynecol. Cancer* **2020**, *30*, 927–931. [CrossRef] [PubMed]

292. Biscop, E.; Lin, A.; Van Boxem, W.; Van Loenhout, J.; De Backer, J.; Deben, C.; Dewilde, S.; Smits, E.L.; Bogaerts, A. Influence of cell type and culture medium on determining cancer selectivity of cold atmospheric plasma treatment. *Cancers* **2019**, *11*, 1287. [CrossRef]

293. Tornin, J.; Mateu-Sanz, M.; Rodríguez, A.; Labay, C.; Rodríguez, R.; Canal, C. pyruvate plays a main role in the antitumoral selectivity of cold atmospheric plasma in osteosarcoma. *Sci. Rep.* **2019**, *9*. [CrossRef]

294. Köritzer, J.; Boxhammer, V.; Schäfer, A.; Shimizu, T.; Klämpfl, T.G.; Li, Y.-F.; Welz, C.; Schwenk-Zieger, S.; Morfill, G.E.; Zimmermann, J.L.; et al. Restoration of sensitivity in chemo—Resistant glioma cells by cold atmospheric plasma. *PLoS ONE* **2013**, *8*, e64498. [CrossRef]

295. Lin, L.; Wang, L.; Liu, Y.; Xu, C.; Tu, Y.; Zhou, J. Non-thermal plasma inhibits tumor growth and proliferation and enhances the sensitivity to radiation in vitro and in vivo. *Oncol. Rep.* **2018**, *40*, 3405–3415. [CrossRef]

296. Lin, L.; Yan, D.; Gjika, E.; Sherman, J.H.; Keidar, M. Atmospheric plasma meets cell: Plasma tailoring by living cells. *ACS Appl. Mater. Interfaces* **2019**, *11*, 30621–30630. [CrossRef] [PubMed]

297. Cheng, X.; Rajjoub, K.; Shashurin, A.; Yan, D.; Sherman, J.H.; Bian, K.; Murad, F.; Keidar, M. Enhancing cold atmospheric plasma treatment of cancer cells by static magnetic field. *Bioelectromagnetics* **2017**, *38*, 53–62. [CrossRef] [PubMed]

298. Noghreiyan, A.V.; Imanparast, A.; Ara, E.S.; Soudmand, S.; Noghreiyan, V.V.; Sazgarnia, A. In-vitro investigation of cold atmospheric plasma induced photodynamic effect by Indocyanine green and Protoporphyrin IX. *Photodiagnosis Photodyn. Ther.* **2020**, 101822. [CrossRef] [PubMed]

299. Moniruzzaman, R.; Rehman, M.U.; Zhao, Q.-L.; Jawaid, P.; Takeda, K.; Ishikawa, K.; Hori, M.; Tomihara, K.; Noguchi, K.; Kondo, T.; et al. Cold atmospheric helium plasma causes synergistic enhancement in cell death with hyperthermia and an additive enhancement with radiation. *Sci. Rep.* **2017**, *7*, 1–12. [CrossRef] [PubMed]

300. Van Driel, W.J.; Koole, S.N.; Sikorska, K.; Van Leeuwen, J.H.S.; Schreuder, H.W.; Hermans, R.H.; De Hingh, I.H.; Van Der Velden, J.; Arts, H.J.; Massuger, L.F.; et al. Hyperthermic Intraperitoneal Chemotherapy in Ovarian Cancer. *N. Engl. J. Med.* **2018**, *378*, 230–240. [CrossRef] [PubMed]

301. Holohan, C.; Van Schaeybroeck, S.; Longley, D.B.; Johnston, P.G. Cancer drug resistance: An evolving paradigm. *Nat. Rev. Cancer* **2013**, *13*, 714–726. [CrossRef]

302. Chaiswing, L.; Weiss, H.L.; Jayswal, R.D.; Clair, D.K.S.; Kyprianou, N. Profiles of radioresistance mechanisms in prostate cancer. *Crit. Rev. Oncog.* **2018**, *23*, 39–67. [CrossRef]

303. Mansoori, B.; Mohammadi, A.; Davudian, S.; Shirjang, S.; Baradaran, B. The different mechanisms of cancer drug resistance: A brief review. *Adv. Pharm. Bull.* **2017**, *7*, 339–348. [CrossRef]

304. Longley, D.B.; Johnston, P.G. Molecular mechanisms of drug resistance. *J. Pathol.* **2005**, *205*, 275–292. [CrossRef]

305. Pham, C.T.; Juhasz, M.; Sung, C.T.; Mesinkovska, N.A. The human papillomavirus vaccine as a treatment for human papillomavirus-related dysplastic and neoplastic conditions: A literature review. *J. Am. Acad. Dermatol.* **2020**, *82*, 202–212. [CrossRef] [PubMed]

306. Golubnitschaja, O.; Baban, B.; Boniolo, G.; Wang, W.; Bubnov, R.; Kapalla, M.; Krapfenbauer, K.; Mozaffari, M.S.; Costigliola, V. Medicine in the early twenty-first century: Paradigm and anticipation - EPMA position paper 2016. *EPMA J.* **2016**, *7*, 23. [CrossRef] [PubMed]

307. Golubnitschaja, O.; Flammer, J. Individualised patient profile: Clinical utility of Flammer syndrome phenotype and general lessons for predictive, preventive and personalised medicine. *EPMA J.* **2018**, *9*, 15–20. [CrossRef]

308. Janssens, J.P.; Schuster, K.; Voss, A. Preventive, predictive, and personalized medicine for effective and affordable cancer care. *EPMA J.* **2018**, *9*, 113–123. [CrossRef] [PubMed]

309. Goncharenko, V.; Bubnov, R.; Polivka, J.; Zubor, P.; Biringer, K.; Bielik, T.; Kuhn, W.; Golubnitschaja, O. Vaginal dryness: Individualised patient profiles, risks and mitigating measures. *EPMA J.* **2019**, *10*, 73–79. [CrossRef]

310. Kunin, A.; Polivka, J.; Moiseeva, N.; Golubnitschaja, O. "Dry mouth" and "Flammer" syndromes—Neglected risks in adolescents and new concepts by predictive, preventive and personalised approach. *EPMA J.* **2018**, *9*, 307–317. [CrossRef] [PubMed]

311. Qian, S.; Golubnitschaja, O.; Zhan, X. Chronic inflammation: Key player and biomarker-set to predict and prevent cancer development and progression based on individualized patient profiles. *EPMA J.* **2019**, *10*, 365–381. [CrossRef] [PubMed]

312. Maturo, M.G.; Soligo, M.; Gibson, G.; Manni, L.; Nardini, C. The greater inflammatory pathway—High clinical potential by innovative predictive, preventive, and personalized medical approach. *EPMA J.* **2019**, *11*, 1–16. [CrossRef]

313. Avishai, E.; Yeghiazaryan, K.; Golubnitschaja, O. Impaired wound healing: Facts and hypotheses for multi-professional considerations in predictive, preventive and personalised medicine. *EPMA J.* **2017**, *8*, 23–33. [CrossRef]

314. Golubnitschaja, O.; Stolzenburg-Veeser, L.; Avishai, E.; Costigliola, V. Wound healing: Proof-of-principle model for the modern hospital-patient stratification, prediction, prevention and personalisation of treatment. In *The Modern Hospital: Patients Centered, Disease Based, Research Oriented, Technology Driven*; Latifi, R., Ed.; Springer: Berlin/Heidelberg, Germany, 2018; ISBN 978-3-030-01393-6.

315. Gerner, C.; Costigliola, V.; Golubnitschaja, O. Multiomic patterns in body fluids: Technological Challenge with a great potential to implement the advanced paradigm of 3P medicine. *Mass. Spectrom. Rev.* **2019**. [CrossRef]

316. Lu, M.; Zhan, X. The crucial role of multiomic approach in cancer research and clinically relevant outcomes. *EPMA J.* **2018**, *9*, 77–102. [CrossRef]

317. Golubnitschaja, O.; Polivka, J.; Yeghiazaryan, K.; Berliner, L. Liquid biopsy and multiparametric analysis in management of liver malignancies: New concepts of the patient stratification and prognostic approach. *EPMA J.* **2018**, *9*, 271–285. [CrossRef] [PubMed]

318. Koklesova, L.; Liskova, A.; Samec, M.; Qaradakhi, T.; Zulli, A.; Smejkal, K.; Kajo, K.; Jakubikova, J.; Behzadi, P.; Pec, M.; et al. Genoprotective activities of plant natural substances in cancer and chemopreventive strategies in the context of 3P medicine. *EPMA J.* **2020**, *11*, 261–287. [CrossRef] [PubMed]

319. Zubor, P.; Dankova, Z.; Kolkova, Z.; Holubekova, V.; Brany, D.; Mersakova, S.; Samec, M.; Liskova, A.; Koklesova, L.; Kubatka, P.; et al. Rho GTPases in gynecologic cancers: In-depth analysis toward the paradigm change from reactive to predictive, preventive, and personalized medical approach benefiting the patient and healthcare. *Cancers* **2020**, *12*, 1292. [CrossRef] [PubMed]

320. Metelmann, H.-R.; Seebauer, C.; Rutkowski, R.; Schuster, M.; Bekeschus, S.; Metelmann, P. Treating cancer with cold physical plasma: On the way to evidence-based medicine. *Contrib. Plasma Phys.* **2018**, *58*, 415–419. [CrossRef]

321. Apyx Medical. Available online: https://clinicaltrials.gov/NCT02658851 (accessed on 20 October 2020).

322. Orvieto, M.A.; Coelho, R.F.; Chauhan, S.; Palmer, K.J.; Rocco, B.; Patel, V.R. Incidence of lymphoceles after robot-assisted pelvic lymph node dissection. *BJU Int.* **2011**, *108*, 1185–1189. [CrossRef] [PubMed]

323. CAPCIN. Available online: https://clinicaltrials.gov/NCT03218436 (accessed on 20 October 2020).

324. Marampon, F.; Gravina, G.L.; Popov, V.M.; Scarsella, L.; Festuccia, C.; La Verghetta, M.E.; Parente, S.; Cerasani, M.; Bruera, G.; Ficorella, C.; et al. Close correlation between MEK/ERK and Aurora-B signaling pathways in sustaining tumorigenic potential and radioresistance of gynecological cancer cell lines. *Int. J. Oncol.* **2014**, *44*, 285–294. [CrossRef]

325. Golubnitschaja, O. *Flammer Syndrome—From Phenotype to Associated Pathologies, Prediction, Prevention and Personalisation*; Golubnitschaja, O., Ed.; Springer: Berlin/Heidelberg, Germany, 2019; ISBN 978-3-030-13549-2.

Publisher's Note: MDPI stays neutral with regard to jurisdictional claims in published maps and institutional affiliations.

International Journal of
Molecular Sciences

Review

Insight on Solution Plasma in Aqueous Solution and Their Application in Modification of Chitin and Chitosan

Chayanaphat Chokradjaroen [1], Jiangqi Niu [1], Gasidit Panomsuwan [2] and Nagahiro Saito [1,3,4,5,*]

1   Department of Chemical Systems Engineering, Graduate School of Engineering, Nagoya University, Nagoya 464-8603, Japan; eig@sp.material.nagoya-u.ac.jp (C.C.); niu@sp.material.nagoya-u.ac.jp (J.N.)
2   Department of Materials Engineering, Faculty of Engineering, Kasetsart University, Bangkok 10900, Thailand; gasidit.p@ku.ac.th
3   Conjoint Research Laboratory in Nagoya University, Shinshu University, Nagoya 464-8603, Japan
4   Open Innovation Platform with Enterprises, Research Institute and Academia (OPERA), Japan Science and Technology Corporation (JST), Nagoya 464-8603, Japan
5   Strategic International Collaborative Research Program (SICORP), Japan Science and Technology Corporation (JST), Nagoya 464-8603, Japan
*   Correspondence: hiro@sp.material.nagoya-u.ac.jp

Abstract: Sustainability and environmental concerns have persuaded researchers to explore renewable materials, such as nature-derived polysaccharides, and add value by changing chemical structures with the aim to possess specific properties, like biological properties. Meanwhile, finding methods and strategies that can lower hazardous chemicals, simplify production steps, reduce time consumption, and acquire high-purified products is an important task that requires attention. To break through these issues, electrical discharging in aqueous solutions at atmospheric pressure and room temperature, referred to as the "solution plasma process", has been introduced as a novel process for modification of nature-derived polysaccharides like chitin and chitosan. This review reveals insight into the electrical discharge in aqueous solutions and scientific progress on their application in a modification of chitin and chitosan, including degradation and deacetylation. The influencing parameters in the plasma process are intensively explained in order to provide a guideline for the modification of not only chitin and chitosan but also other nature-derived polysaccharides, aiming to address economic aspects and environmental concerns.

Keywords: solution plasma process; aqueous solutions; chitin; chitosan; degradation; deacetylation

Citation: Chokradjaroen, C.; Niu, J.; Panomsuwan, G.; Saito, N. Insight on Solution Plasma in Aqueous Solution and Their Application in Modification of Chitin and Chitosan. *Int. J. Mol. Sci.* **2021**, *22*, 4308. https://doi.org/10.3390/ijms22094308

Academic Editor: Akikazu Sakudo

Received: 30 March 2021
Accepted: 17 April 2021
Published: 21 April 2021

1. Introduction

In physics and chemistry, plasma is fundamentally defined as one of the four states of matter. Solid, liquid, and gas states are more common on the earth due to the atmospheric condition, whereas the plasma state dominantly exists in the universe (>99%) (e.g., the sun, nebulae, etc.) [1]. Nevertheless, plasma can be artificially formed in the earth by giving sufficient thermal or electric energy supply, which ionizes the neutral gases to a quasi-neutral ionized gas [2]. The ionized gas is composed of negative and positive ions, free electrons, excited molecules, and excited atoms and molecules, as well as the emission of ultraviolet (UV) and high electric field [3]. Artificial plasma is classified as (i) thermal equilibrium plasma and (ii) non-thermal equilibrium plasma [4,5]. Thermal equilibrium plasma can be generated by a strong electrical power and, usually, under gas pressure of more than 5 kPa. Its gas and electron temperatures are nearly equal. On the other hand, in non-thermal equilibrium plasma, the gas temperature is lower than the electron temperature [4]. Non-thermal equilibrium plasma can be induced by giving sufficient energy under a vacuum system; however, it can also be generated under an atmospheric pressure environment by applying transient electrical or electrostatic discharges. Consequently, the non-thermal equilibrium plasma has offered several potential

321

applications. Artificial plasmas in the gas phase are predominantly investigated under a wide range of operating pressures, including both vacuum and atmospheric conditions, and temperatures. A variety of gases (e.g., Ar, N_2, O_2, H_2, and mixtures) is applied to generate the plasmas depending on the purposes [6–8]. Even though plasma is theoretically mentioned as ionized gases, plasma can be in solid form (i.e., the formation of plasmon which is induced by the collective oscillation of free electrons moving around a lattice point) and liquid form (i.e., the generation of plasma in liquid, such as a pulsed electrical discharge in liquids) [4,9,10]. In the same manner, as the plasmas in the gas phase, plasma can be directly generated in the liquid phase by providing a sufficiently high electric field on the electrodes, leading to the electric breakdown of liquids. Over the past several years, the liquid-phase plasmas have been developed, along with the gas-phase plasmas. The liquid-phase plasmas have been mainly focused as a technology for wastewater treatment and water purification, owing to their ability to induce an effective production of reactive species, for example, hydroxyl radical ($^{\bullet}$OH), superoxide anion (O_2^-), and hydrogen peroxide (H_2O_2) [4,11,12]. These reactive species can strongly oxidize and decompose organic pollutions and bacteria in water [12,13]. Recently, as the plasma technology being developed, the liquid-phase plasmas have been expanded and utilized for the synthesis and modification of various materials, such as noble-metal nanoparticles [13,14], metal oxides [15,16], carbon materials [17–20], and natural polysaccharides [21,22]. Articles on the modification of natural polysaccharides, including chitin, chitosan, cellulose, alginate, mushroom polysaccharide, and starch, by liquid-phase plasma have been rapidly published in the past decade (Figure 1), which can imply the growth of the liquid-phase plasma in this field.

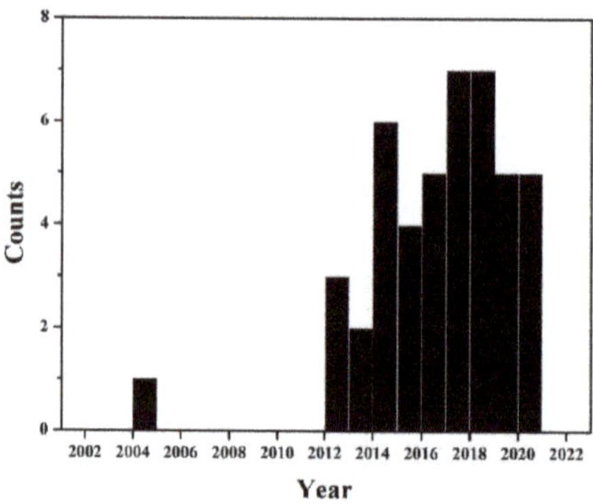

Figure 1. The number of articles relating to the modification of chitin and chitosan by liquid-phase plasma (data are shown in Supplementary Materials).

In this review, the insight into the plasmas in the liquid phase, mainly focused on an aqueous solution, is provided. The plasma technology called "solution plasma process (SPP)" will be dominantly explained. The discussion of recent development on the modification of chitin and chitosan, especially degradation, using the SPP will be given as examples. Modification techniques, reaction mechanisms, and changes in the properties of chitin and chitosan will be explained. In comparison with other existing methods, today, the modification of chitin and chitosan by the SPP is still at an early stage of development. The summary of relevant publications from the recent past to the present will provide benefits and a useful guideline for the researchers in the related fields to achieve the ecologically friendly and efficient method for the modification of chitin and chitosan. In addition, the remaining challenge and future trend of SPP technology in the field of not only chitin and chitosan but also other natural polysaccharides are also discussed to motivate future studies.

2. Solution Plasma Process (SPP): Chemistry and Influencing Parameters

Electrode geometric construction and phase patterns of liquid-phase plasmas can be categorized into four groups: (1) direct electrical discharge between two electrodes [23], (2) contact electrical discharge between electrodes and the surface of the surrounding electrolyte, (3) miscible electrical discharge with external gas injection, and (4) special excitation electrical discharge (i.e., radio frequency, microwave irradiation or laser ablation). In this section, we will discuss on the direct discharge between two electrodes under the liquid-phase, which are presented by various terms. For example, submerged liquid plasma (wire-to-plate configuration using direct current) [24,25], pulsed plasma in liquid (rod-to-rod configuration, using pulsed voltage) [23], and solution plasma (pin-to-pin and wire-to-plate configurations, bipolar pulsed voltage) [26,27]. Apparently, these terms are assigned following their experimental setup, electrical power source, and electrodes configuration. Henceforward, the solution plasma process (SPP) with pin-to-pin electrode configuration will be described. SPP with pin-to-pin electrode configuration was firstly proposed by Takai and Saito's group [13,28]. In the SPP, the plasma is directly discharged between a pair of electrodes submerged under liquids at a short distance (0.2–1 mm), depending on its application, as shown in Figure 2. The power supplies are bipolar-pulsed high voltage supply. The use of pulsed voltage could reduce the current of the discharge. To a certain extent, it reduced the possibility of arc discharge due to thermal ion emission from the electrode and enhanced the number of carriers generated by secondary electron emission, thus enhancing the stability of discharge occurrence. Most studies using the SPP focus on the processing conditions and the properties and performance of the obtained products. The physical and chemical reactions occurring in the SPP are complex and rarely reported. Here, a brief explanation of the electrical discharge under an aqueous solution (e.g., electrical breakdown, formation of reactive species, and influencing factors) in the SPP is provided.

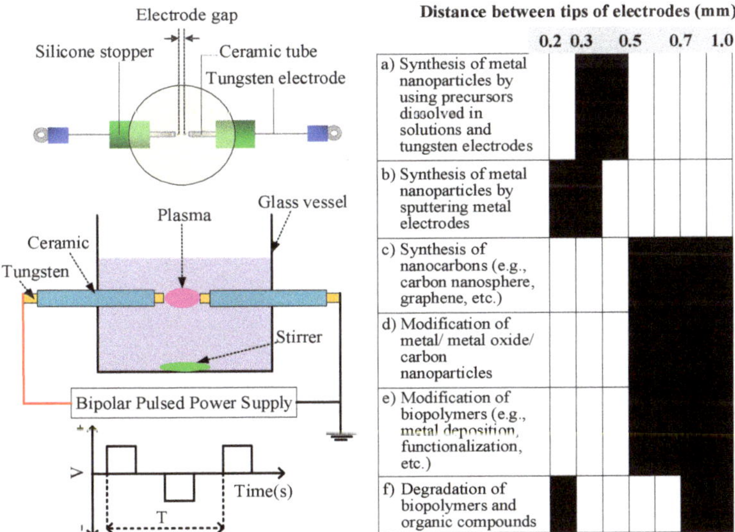

Figure 2. Schematic illustration of the SPP with pin-to-pin electrode configuration and its application with different distances between electrodes: (**a**) [14,29,30], (**b**) [31,32], (**c**) [27,33–35], (**d**) [36–38], (**e**) [15,22], and (**f**) [39–42].

2.1. Electrical Breakdown

In chemical physics, the difference between gas and liquid phases is the molecular density. The difference in molecular density causes different insulation to withstand capability, high collision frequency and energy dissipation rate, and low electron mobility in liquids. Therefore, the plasma chemistry in gas and liquid phases is significantly different. The physical mechanism of gas-phase plasma can be explained based on the electron avalanche [43], which is a process that free electrons in the medium are strongly accelerated by an electric field, resulting in the collision with other atoms or molecules and then ionization. However, the electron avalanche can rarely occur in the liquid mediums because liquids usually have high molecular density, low mobility of charges, and recombination rate. However, as mentioned above, the plasmas in the liquid phase can be carried. The explanation was previously clearly given by Saito et al. in 2008 [44]. The liquid (e.g., aqueous solutions) near the electrode tips turn to gas or the formation of bubbles due to Joule heating, which causes solution vapor and electrolysis, resulting in gases such as H_2 and O_2. When the bubble is formed, the electron avalanche is produced in bubbles. The electrical break down is developed inside the bubbles and then formed as the plasma channel, which is kind of like "unzipping". The electrical breakdown can be controlled by the injection of more electrons into the ionization field in the bubble. Heo et al. revealed the current-voltage waveform by low-pass filter circuits, which could reduce noise signals as compared to that of the conventional circuits when the bipolar pulsed voltage was applied [45]. The stages of the applied voltages, breakdown, and plasma generation in the solution plasma process are also proposed in Figure 3.

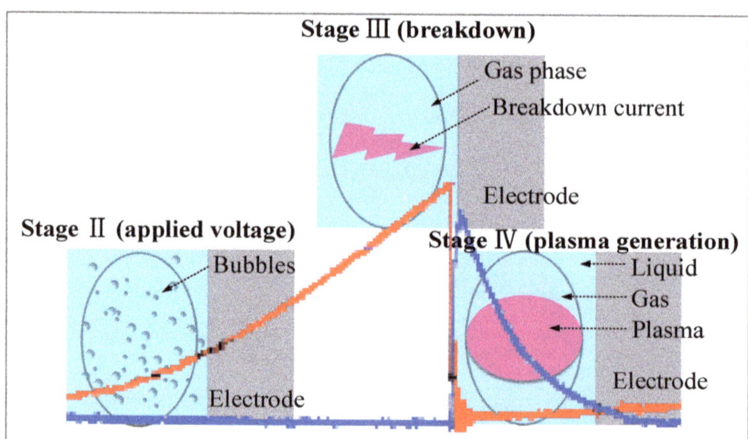

Figure 3. Schematic diagrams of the applied voltage, breakdown, and plasma generation stages in the SPP with the background of the current and voltage waveforms obtained by low-pass filter circuits [45].

Besides, the plasma discharges can be altered in different fluids to induce ionization, accelerate ions, and multiply the initial seed of electrons into the electron avalanche. The breakdown depends on the dipole moment of the fluids, including the dielectric behavior, ionization potential, band gap, and dipole moment [4,44]. For example, in water, the electrical discharge requires a large electric field (67–70 MV/m), while benzene requires a much higher energy field for the breakdown due to its higher dielectric strength (163 MV/m).

2.2. Formation of Reactive Species

Considering how the SPP can be used to modify chitin and chitosan, the plasmas in aqueous solutions or water plays a key role because almost all modifying processes for chitin and chitosan have been reported in aqueous solutions [46–49]. Several reactive species are generated when the electrical discharge is carried in water through molecules collision, mass transfer, vaporization, sputtering, and ultra-violet (UV) [44]. Many previous studies demonstrated that both short-lived reactive species (e.g., hydroxyl radical ($^\bullet$OH) and hydrogen radical ($^\bullet$H), superoxide (O_2^-)) and long-lived reactive species (e.g., hydrogen peroxide (H_2O_2), ozone (O_3)) are continuously formed and further reacted during the activation of water by the electrical discharge without the addition of a catalyst or chemical agents [50,51]. Figure 4 reveals the optical emission spectrum (OES) of reactive species generated in water during the SPP, which was measured in our research group. The short-lived reactive species, like $^\bullet OH$ and $^\bullet H$, are firstly generated from the main reaction via electron impact dissociation and continuous collision of reactive species to surrounding molecules, as shown below in Equation (1) [52]. The activity of $^\bullet OH$ and $^\bullet H$, is found to initiate and prolong modification reactions of chitin and chitosan, such as deacetylation, degradation, and altering the crystal structure, in not only plasma treatment but also in other methods, such as oxidative degradation [53]. Subsequently, hydrogen peroxide (H_2O_2), the most common long-lived reactive species in the plasma-activated water, is formed via the recombination of $^\bullet OH$, as shown in Equation (12). In addition, in the water, there are dissolved oxygen molecules that can also be excited or ionized to reactive oxygen species (ROS), such as exited atomic oxygen ($O(^1D)$ or O_I) and triplet ground-state atomic oxygen ($O(^3P)$), singlet oxygen ($^1O_2$), and triplet oxygen ($^3O_2$). H_2O_2 and these ROS also evidently contribute to the formation of $^\bullet OH$ [51,52].

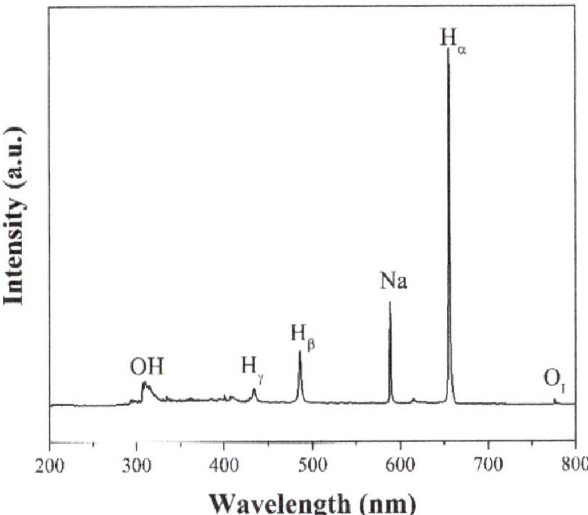

Figure 4. Optical emission spectrum (OES) of reactive species generated in water; its conductivity adjusted by NaCl at a concentration of 0.02 M during the SPP.

The reactions revealing the possible formation of $^\bullet OH$, $^\bullet H$, H_2O_2, and excited O are shown below [51,52]:

$$e^- + H_2O \rightarrow {}^\bullet OH + {}^\bullet H + e^- \text{ (Ionization of water molecule by plasma discharge)} \quad (1)$$

$$e^- + H_2O \rightarrow H_2O^* + e^- \text{ (Excitation of water molecule by plasma discharge)} \quad (2)$$

$$e^- + M \rightarrow M^* + e^- \text{ (Sputtering of metal atom to excited metal atom)} \quad (3)$$

$$M^* + H_2O \rightarrow {}^\bullet OH + {}^\bullet H + M \quad (4)$$

$$e^- + O_2 \rightarrow O(^3P) + O(^1D) + e^- \text{ (Ionization of } O_2 \text{ molecule by plasma discharge)} \quad (5)$$

$$O(^1D) + H_2O \rightarrow {}^\bullet OH + {}^\bullet OH \quad (6)$$

$${}^\bullet H + O_2 \rightarrow {}^\bullet OH + O \quad (7)$$

$$O + O \rightarrow O_2 \quad (8)$$

$$UV + H_2O \rightarrow H_2O^* \text{ (Excitation of } O_2 \text{ molecule by UV)} \quad (9)$$

$$UV + H_2O^* \rightarrow OH^- + H^+ \quad (10)$$

$$e^- + OH^- \rightarrow {}^\bullet OH + e^- \tag{11}$$

$${}^\bullet OH + {}^\bullet OH \rightarrow H_2O_2 \ (\text{Recombination of } {}^\bullet OH \text{ to form } H_2O_2) \tag{12}$$

$${}^\bullet OH + H_2O_2 \rightarrow HO_2 + H_2O \tag{13}$$

$$H_2O_2 \rightarrow {}^\bullet OH + {}^\bullet OH \tag{14}$$

2.3. Influencing Parameters in SPP under Aqueous Solutions

In the SPP, the electrical discharges under a different type of aqueous solutions, and by using different electrode types, and repetition frequency have been investigated and reveals the influence on the electrical breakdown and the formation of reactive species [44]. The number density of ${}^\bullet OH$ was measured in a bipolar electrical discharge between tips of wire-type tungsten electrodes in aqueous solutions containing HCl, KCl, and KOH at the same conductivity (500 µS/cm) by Miron et al. [54]. The breakdown voltage occurred at 920 V for HCl, 850 V for KCl, and 478 V for KOH, which implied that the energy required for the plasma breakdown was higher for the HCl solution than for the solution KCl and KOH. Additionally, the liquid conductivity measurement, which can suggest the ionization and dissociation of water to form hydrogen ions, was found to be relatively higher for the HCl solution. As the conductivity increased (up to approximately 520 µS/cm), the discharge became stronger, accelerating the erosion of the metal surface, due to the process of secondary release. The secondary release results in increasing the possibility of the collision from ions to metal electrode surface in the electron-rich sheath near the tip of electrodes and increases the collision possibility of ions [55,56]. This erosion causes an increase in the distance between the tips of electrodes. The distance between the tips of electrodes in the HCl solution increased from 0.1 mm to 0.17 mm, while 0.12 mm and 0.15 mm were observed in the KCl and KOH solution, respectively. Accordingly, it suggested that acidity has a significant influence on the sputtering of metal electrodes in the SPP, which is considered as an advantage for the synthesis of metal-based nanoparticles without using precursors and reducing agents [32,38,57]. Furthermore, the investigation on the formation of ${}^\bullet OH$ was also conducted via the time-resolved optical emission measurement at various time delays to the positive voltage pulse in the range of 0–38 µs. The density of ${}^\bullet OH$ generated in HCl solution was found to be high (2×10^{17} cm^{-3}), while KOH, which is highly basic and can be an important source of ${}^\bullet OH$, showed the lowest density (5×10^{16} cm^{-3}). This may be caused by the collision of H$^+$, which may easily move in the opposite direction of electron flow owing to low atomic weight, on the surface of metal electrodes, compared to that of K$^+$. The collisions result in the sputtering and the release of electrons or excited metal atoms, which can further react with water molecules and promote the generation of ${}^\bullet OH$ via the reactions showed above in Equation (4). In the nanosecond pulsed plasma discharge system for water-film plasma reactor, the influence of liquid conductivity on the electrical breakdown and the production of H_2O_2 was also studied by Wang et al. [58]. The breakdown voltage was found to be decreased with increasing liquid conductivity (from 100 µS/cm to 36,000 µS/cm). This was explained by the decrease of dielectric relaxation time, which is the time scale of relaxation for moving charge carriers in materials. Increasing the liquid conductivity reduced the resistance of the liquid and facilitated the current to flow through the liquid when the electrical field was applied at the electrodes. This also caused the increment of energy dissipation into the bulk liquid. However, the production rate of H_2O_2, which can be a source for ${}^\bullet OH$, did not change significantly with conductivity. Moreover, the mixture of aqueous solution and organic solvent, like ethanol, was also investigated by Takeuchi et al. [59]. The breakdown was delayed, and the time needed until the steady-state discharge (constant current) was longer for the mixture with 50% ethanol, compared to the aqueous solution alone. This is because the ionization energy of an ethanol molecule (16 eV) is higher than that of a water molecule (13 eV). The total ionization cross-section of ethanol is larger than that of water with electron energy higher than 16 eV [60,61], as shown in Figure 5.

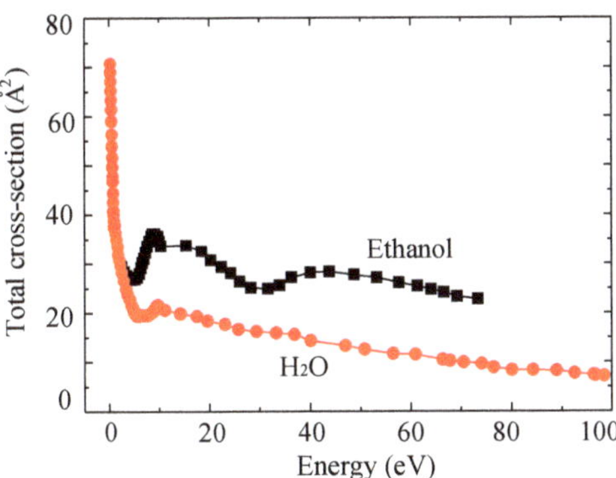

Figure 5. The total ionization cross-section of water and ethanol [60,61].

However, the discharge probability increased with the addition of ethanol to the solution and the power required to produce bubbles and initiated a discharge that decreased with the addition of ethanol. This could be caused by the lower boiling temperature of ethanol, as compared to that of water. Furthermore, the discharge in the solution with ethanol (50%) was found to produce smaller-size bubbles than that obtained in the solution without ethanol, which resulted in easier sustaining discharges and higher discharge probability higher. Besides, the OES spectra from solution without ethanol and with ethanol showed $^\bullet OH$, $^\bullet H$, and $^\bullet O$, while the solution with ethanol showed relatively weak signals of $^\bullet OH$ and $^\bullet O$. However, the emissions of C_2 swan band (440 nm) and $^\bullet CH$ (431 nm) could be observed in the solution with ethanol, due to the dissociation of ethanol vapor. This evidence shows benefit in the synthesis of carbon nanomaterials via the solution plasma process, similar to the systems with other organic solvents [27,62]. According to the evidence previously showed, it pointed out that the choice of solutions, which include conductivity, pH, and type of ions, exhibits a significant effect on the properties of plasmas in liquids.

Types of electrodes also showed a significant effect on the electrical breakdown and generation of $^\bullet OH$ in the solution plasma process. In 2010, Miron et al. studied the current-voltage characteristics and the optical emission spectroscopy in the solution plasma process by using tantalum (Ta) and tungsten (W) electrodes with the circulated and non-circulated water systems [63]. In comparison, the energy required for the electrical breakdown was also found to be higher for the systems with Ta electrodes. The current-voltage characteristics for all systems showed the features of a spark discharge, and the transition to arc plasma was observed in the case of using Ta electrodes. Furthermore, according to the OES spectra, the transition of the $^\bullet OH$ was found to be different for the circulated water and the non-circulated water discharges when different electrodes were used in the discharge. For W electrodes, the band of the $^\bullet OH$ was detected to be the strongest at 307.8 nm, while the band was strongest at 312.6 nm for the Ta electrodes. The reason for this phenomenon was explained relating to the different plasma temperatures in the case of W and Ta electrodes. Moreover, the board emission continuum spectra in the range spectra in the range of 350–940 nm were exhibited only for Ta electrodes, which might be due to the eroded and heated metal particles from the metal surface at the tips of electrodes. Later, in 2011, Miron et al. also studied the influence of electrodes made of W and lanthanum hexaboride (LaB_6), on the electrical breakdown in water [64]. The polished asperities of electrodes showed different morphologies, which caused the difference of electric field required for the electrical breakdown (~190 kV/cm for W electrodes and ~160 kV/cm for LaB_6). The rougher surface of LaB_6 shown in Figure 6 had a higher number of locally concentrated emission sites than that of W electrodes. In addition, the melting points of these two materials (3422 °C and 2220 °C for W and LaB_6, respectively) were believed to be the factor for the erosion of electrodes, which resulted in the formation of different reactive species under corresponding plasma condition. Even though the erosion of metal electrodes could cause contamination to final products in the modification process by the SPP, this phenomenon has also given benefits to the synthesis of metal nanoparticles without using precursors and reducing agents [32]. The influence of electrode types on the generation of reactive species has been continuously investigated. Goto et al. also investigated the electrical discharge in water by using copper (Cu) electrodes. They found the emission line of atomic Cu in the optical emission spectra and proposed that the Cu atoms may act as a catalyst to promote the generation of $^\bullet$OH [51]. Besides, not only the influence of electrode types but also the polishing of electrodes is also an essential step for the SPP. Yui et al. reported that the sharpening could increase the stability of the spatial position of the plasma [5]. Accordingly, the management of electrode tips, such as polishing surface and sharpening, should be done, depending on the application. Sharpening electrodes that can provide the high stability of plasma may be suitably applied in precisely controlled chemical reactions. Meanwhile, reactions like degradation require a large plasma with high production of reactive species, which polishing surface of electrode tips suffices.

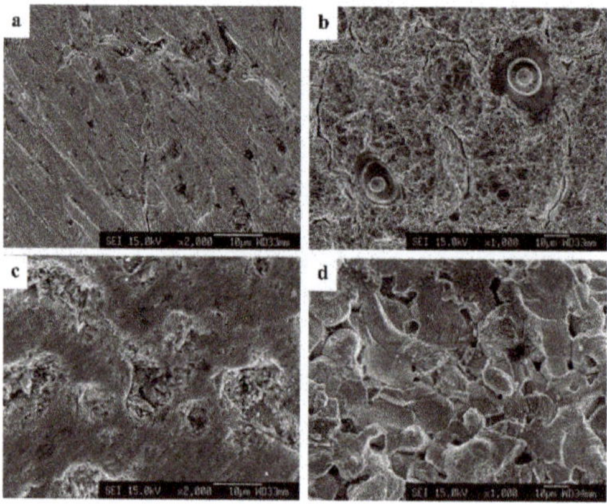

Figure 6. SEM images of the surface of the W electrodes (**a**) before and (**b**) after being exposed to 10 min of the discharge and SEM images of the surface of the LaB$_6$ electrodes (**c**) before and (**d**) after being exposed to 10 min of the discharge [64].

The plasma parameters, such as frequency, were found to be related to the relative amount of injected energy per pulse [20]. Unfortunately, the investigation was conducted by applying the pulse frequency using the bipolar power supply (from 25 to 65 kHz) in only benzene, which led to the formation of carbon materials. The result showed that there were two different operation regimes. The first regime, characterized as glow discharge, occurred when the pulse frequency ranged between 25 and 50 kHz was applied, which referred to lower energy input at a certain interval. In comparison, the second regime, characterized as spark discharge, could be obtained by applying the pulse frequency of 65 kHz, which could result in relatively high energy input in the corresponding period. The plasma/gas temperature was reported to be increased with increasing energy input, which might influence the obtained carbon products from each regime to reveal different morphology. Therefore, it is possible that the electrical discharge in aqueous solutions can also be influenced by applying different frequencies.

Many researchers also attempted to enhance the efficiency of electrical discharge under aqueous solutions by adding bubbles into the system with different configurations, as revealed in Figure 7a,b. The configuration in Figure 7a was proposed by Goto et al. They studied the formation of hydrogen peroxide, which is a powerful oxidant, by applying bipolar pulsed voltage with fine O$_2$ bubbles [51]. The result showed that the high concentration of dissolved O$_2$ insignificantly increases the amount of H$_2$O$_2$. Moreover, Yui et al. developed a direct gas injection system at the tip of the electrode, as shown in Figure 7b [5,65]. The injecting gases were O_2, CO_2, N_2, and Ar. The injection of gases caused the spatial fluctuation of plasma. The electron number density in the plasma was found to be increased by injecting O_2, CO_2, and N_2. It was because the generation of electrons in the plasma was increased by the enhanced collision between the positively charged ions, which can be accelerated by the applied voltage, and other particles, ions, and the cathodic electrode. The kinetic impulses of the collision increased due to the injected gases with larger molecular weights than H_2O. On the other hand, the injection of Ar gas resulted in the reduction of electron number density. This is because Ar is an inert gas that has a larger ionization energy of 1500 kJ/mol, as compared to other gases, such as O_2 (1175 kJ/mol). The contents of the plasma, characterized by the OES, were found to be different, as illustrated in Figure 8.

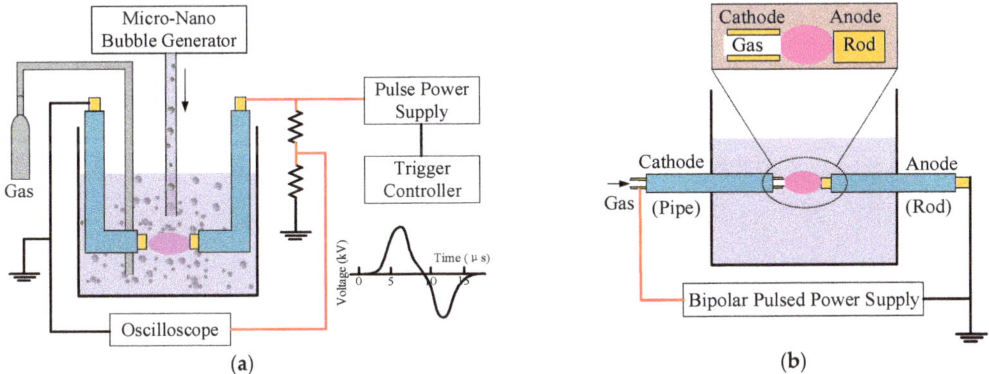

Figure 7. Systems with different configurations of adding bubbles, proposed by (**a**) Goto et al. [51] and (**b**) Yui et al. [65].

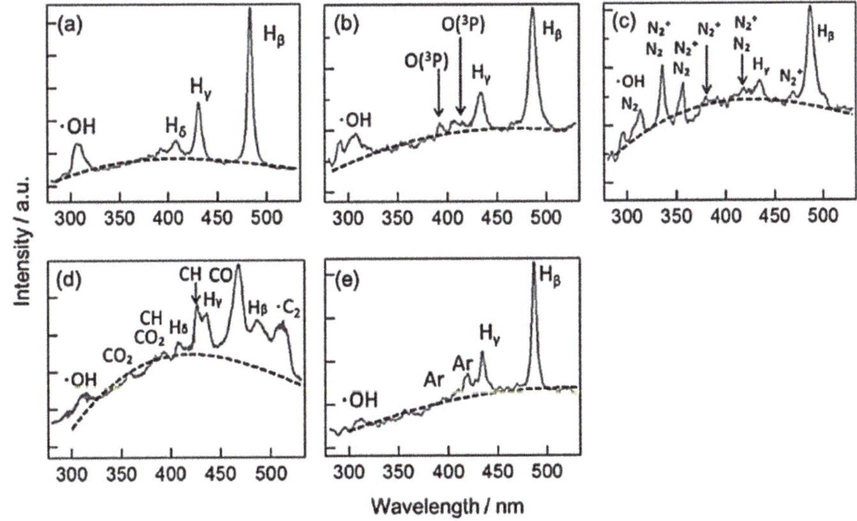

Figure 8. OES spectra from the solution plasma process without the gas injection (**a**), and with the injections of O_2 (**b**), N_2 (**c**), CO_2 (**d**), and Ar (**e**). The spectra were obtained with an integration time of 12.8 s, and 320,000 pulses were integrated during the time span. The black dotted lines represent the best-fitted curves with the blackbody radiation [65].

3. Solution Plasma Process for Modification of Chitin and Chitosan

3.1. Chitin and Chitosan

Chitin was first discovered in the early nineteenth century. It can be extracted from crustacean shell waste (e.g., shrimp and crab shells), insects, and plants [66,67]. Similar to cellulose, it functions as a structural linear polysaccharide. Unlike cellulose, it contains acetamido and amino groups at C–2 position on its pyranose rings [68]. Figure 9 shows the chemical structure of chitin consisting of 2-acetamido-2-deoxy-β-D-glucopyranose as a major repeating unit and glucosamine connected by β (1→4) linkages. Chitosan is one of the most studied chitin derivatives, which is obtained by deacetylation of chitin [69]. Due to the deacetylation reaction, acetyl groups are removed and converted to amino groups, as revealed in Figure 9. The presence of amino groups in the chitosan structure is responsible for its unique functional and biological properties, depending on its molecular weight, as revealed in Figure 10 [70–72]. Moreover, chitin and chitosan also have interesting characteristics, such as biocompatibility, non-toxicity, low allergenicity, and biodegradability. However, their original chemical structures (e.g., high molecular weight and strong hydrogen-band network) cause poor solubility in organic solvents and water, which limit utilization in several fields, especially in biomedical applications [73]. To improve their properties and broaden their applications, chitin and chitosan are intensively studied and modified. For example, degradation of chitin and chitosan to obtain water-soluble degraded products (e.g., chitooligosaccharides), and chemical modification of its functional groups (e.g., deacetylation, carboxymethylation).

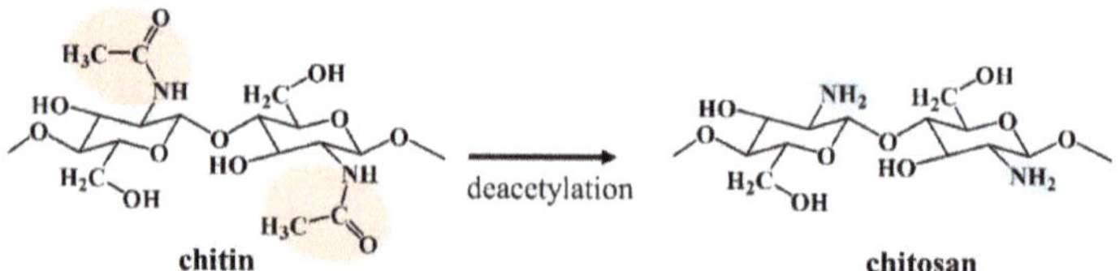

Figure 9. Chemical structures of chitin and chitosan.

Molecular weight (×10³)	1000	500	100	50	30	10	5	<5
Native chitosan	←——→							
Hypocholesterolemic effect		←→						
Anti-microbial: gram-positive			←———→					
Anti-microbial: gram-negative						←→		
Anti-inflammatory							←→	
Anti-tumor/anti-cancer								←→
Anti-allergy							←→	
Calcium absorption								←→
Anti-oxidant								←→

Figure 10. Biological properties of chitosan, depending on its molecular weight. (Note: native chitosan [39,41,67,74], hypocholesterolemic effect [75], anti-microbial: gram-positive [76], anti-microbial: gram-negative [76], anti-inflammatory [77], anti-tumor/anti-cancer [39,49,74,78], anti-allergy [79], calcium absorption [80], anti-oxidant [81]).

Several protocols can be used to modify chitin and chitosan. In general, they can be categorized into three main methods: (i) chemical method [82,83], (ii) enzymatic method [84,85], and (iii) physical method [86–88]. Chemical methods typically give high efficiency or high rate in the modification of chitin and chitosan. However, their drawbacks are concerned with the cost of chemicals, waste management, and severe reactions, resulting in unwanted products (e.g., some over-degraded products). Enzymatic methods can provide a mild reaction and selectively modify chitin and chitosan [89]; for example, it can produce specific oligomers of chitin and chitosan, as shown in Table 1. However, the process is time-consuming, and the cost of handling is relatively high. In recent years, exploring alternative techniques for chitin and chitosan modification, physical methods for the modification of chitin and chitosan, have focused on the utilization of various kinds of energy, including irradiation [90], sonication [91], microwave [92], and plasma. These methods can provide rapid reactions with relatively lower chemical uses and low contamination in the final product, compared to enzymatic and chemical methods, respectively. Among these methods, liquid-phase plasma treatment of chitosan in aqueous solutions is believed to be a novel and effective method, examples of which are given in Tables 1 and 2 (in case of degradation of chitosan). Therefore, it should be clearly understood, aiming at further development.

Table 1. Degradation of chitin and chitosan via chemical and enzymatic methods.

Chemicals	Chemical Concentration	Temperature	Time	MW/ Products	Ref.
Hydrochloric acid	35%	80 °C	1.4 h	$3\text{–}5 \times 10^3$	[93]
Hydrochloric acid	12 M	40 °C	8 h	$1\text{–}3 \times 10^3$	[94]
Nitrous acid	70 mM	0 °C	9 h	$2\text{–}6 \times 10^3$	[95]
Hydrogen fluoride	100%	20 °C	19h	$\sim 2 \times 10^3$	[96]
Hydrogen peroxide	30%	70 °C	2 h	$1.7\text{–}3.81 \times 10^3$	[97]
Chitosanase (Bacillus pumilus BN-262)	0.1 M NaOAc buffer, pH 5.3	37 °C	96 h	2–3 oligomers	[98]
Chitinase (Vibrio furnissii)	Chitin in DMSO/LiCl, pH 7.9	37 °C	24 h	2 oligomers	[99]
Papain	NaOAc–AcOH buffer, pH 4.0	45 °C	24 h	2–50 oligomers	[100]

Table 2. Degradation of chitin and chitosan via physical methods.

Methods	Chemical Concentration	T (°C)	Time	% MW Reduction	Ref.
Microwave 400 W	2% acetic acid	N/A	25 min	79%	[101]
Microwave 100 W	0.1 M acetic acid	N/A	20 min	92.5%	[87]
Ultraviolet 1 kW power	0.1 M acetic acid	25	15 min	98.5%	[88]
Ultrasonication 200 W	0.1 M acetic acid	N/A	120 min	52%	[88]
^{60}Co γ-rays Radiation	2% acetic acid mixed 10 mL hydrogen peroxide	N/A	7 h	95%	[102]
Impinging stream and jet cavitation	acetic acid mixed with sodium acetate trihydrate	40	30 min	88%	[103]
Hydrodynamic cavitation	0.2 acetic acid mixed with 0.1 sodium acetate	40	30 min	95%	[104]
Plasma 350 W	1% acetic acid	N/A	180 min	83 %	[105]
SPP	1 M acetic acid	25–30	300 min	n/a	[46]
SPP	1 M acetic acid	25–30	300 min	96%	[40]
SPP	0.1 M acetic acid 4 M hydrogen peroxide	Room temperature	60 min	85%	[74]
SPP	0.00155 mM carboxylic acids	Room temperature	60 min	86%	[106]
SPP	0.02 M sodium chloride	Room temperature	60 min	96%	[39]

3.2. Reduction of Molecular Weight and Destruction of Crystallinity Via SPP

There are several SPP parameters that have been investigated to understand the degradation process to reduce the molecular weight of chitosan, such as reaction time, electrode configuration, types of electrodes, and frequency. In 2012, the SPP was introduced for the first time to reduce the molecular weight of chitosan by Prasertsung et al. [46]. Chitosan is insoluble in water and organic solvents, but it is soluble in dilute aqueous acidic solution at pH < 6.5. The dissolution resulted in the protonation of the amino group ($R–NH_2$) in glucosamine units into soluble form $R–NH_3^+$. To obtain a homogeneous reaction solution, chitosan was dissolved in acetic acid prior to the SPP. The molecular weight of chitosan sharply dropped in the beginning (0–60 min) and then gradually decreased, approaching a constant. Reactive species, like $^\bullet OH$, was believed to play an important role in the molecular weight reduction of chitosan. As a result, the longer time of the reaction had a lesser effect on the reduction of molecular weight because the number of short-chain chitosan increased while the number of $^\bullet OH$ produced in the system remained the same according to the fixed SPP condition. Moreover, they also reported that the main structure (pyranose ring) was not altered after the solution plasma process. Besides, Prasertsung et al. also applied the SPP on the molecular weight reduction of β-chitosan [40]. β-chitosan has a low hydrogen-bonding network, leading to loss in crystalline structure, compared to normal chitosan, which mostly refers to β-chitosan. Compared to their previous study using α-chitosan, the reaction rate was higher. The molecular weight of β-chitosan (5.5×10^5) was markedly decreased from almost 4 times (1.5×10^5) and 30 times (1.9×10^4) after the solution plasma treatment for 30 min and 300 min, respectively. The water solubility of the obtained products was also found to be improved. Accordingly, the tuning of the SPP treatment time showed potential to obtain chitosan with specific molecular weight, which can be further used in various applications, as shown in Figure 10.

In addition, the effect of SPP on the chitosan derivatives, N,O-carboxymethyl chitosan decorated with gold nanoparticles, water-soluble chitosan with a highly negative charge, was also studied by Chokradjaroen et al. [49]. In their study, chemical reduction of Au^{3+} was firstly conducted in N,O-carboxymethyl chitosan solution, as shown in Figure 11. The aggregation of gold nanoparticles could be observed. However, after the SPP, the distribution of gold nanoparticles was improved and their average size was also reduced from 11 nm to 9 nm, as shown in TEM images (Figure 12). A similar result was also reported and explained that as the SPP prolonged, the pH of the solution decreased, leading to the partial dissolution of metal nanoparticles [28]. Not only the change in the size of gold nanoparticles but also the hydrodynamic size, which referred to the molecular weight of N,O-carboxymethyl chitosan, was also influenced by the $^\bullet OH$ formed in the system. The chain scission of N,O-carboxymethyl chitosan was occurred by the attacking of $^\bullet OH$ to C-1 position of chitosan, resulting in the bond breakage at β-glycosidic linkages, as shown in Figure 11. After the reaction, the result of chemical structure analysis showed that there was no destruction of the main structure and functional groups, including carboxymethyl group and interaction with gold nanoparticles. Nevertheless, the negative charge of N,O-carboxymethyl chitosan decorated with gold nanoparticles was lowered in magnitude, according to the Zeta potential measurement. Consequently, the obtained products were evaluated and showed enhanced cytotoxicity and improved the selectivity toward cancer cells than normal cells.

In addition to the effect of reaction time, the effects of SPP conditions (e.g., electrode types and pulse frequency) also play an important role in the degradation rate and properties of chitosan. Prasertsung et al. used various types of electrodes, including tungsten (W), copper (Cu), and iron (Fe), and varied the applied pulse frequency of the bipolar supply from 15 to 30 kHz [47]. According to the obtained result, the different electrode types differently affected the degradation of chitosan. Within 60 min, the molecular weight of chitosan could be reduced from 1.3×10^5 to 8.7×10^4, 6.2×10^4, and 4.1×10^4 for the system with W, Cu, and Fe at 15 kHz, respectively. The melting points of W, Cu, and Fe are 3422 °C, 1084 °C, and 1204 °C. A stronger promoting effect of the system with Cu and

Fe electrodes could be attributed to the metal atoms or ions from the erosion of the metal surface at the tip of electrodes during plasma treatment. Especially, metal atoms and ions from iron electrodes could be transformed into ferrous ions and effectively participate in the Fenton reaction. The decomposition of H_2O_2 generated in the system led to the increment of $^{\bullet}OH$. Besides, the pulse frequency was found to significantly influence the reduction of the molecular weight of chitosan. After the solution plasma process at the applied pulse frequencies of 15, 22.5, and 30 kHz, the molecular weights of the obtained products were 1.3×10^4, 9.2×10^3, and 6.8×10^3, respectively. This could be described by the raising of the energy input when the pulse frequency increased. However, the molecular weight distributions of the obtained products were still relatively high, which is not suitable for many applications. The polydispersity index (PDI) ranged from 2.5 to 3.5, which was higher than the ideal PDI (1). In biomedical applications, monodisperse low molecular weight chitosan or chitooligosaccharides, COS, (PDI = 1) is desired.

To acquire the specific product of low-molecular-weight chitosan with high purification, several techniques were studied in combination with the SPP. For example, Pornsunthornthawee et al. used the benefit of the chitosan-metal complex to induce the selective chain scission. Metal ions, such as silver ion (Ag^+), zinc ion (Zn^{2+}), copper (II) ion (Cu^{2+}), and ferric ion (Fe^{3+}), were used to form complexes with chitosan, which was dissolved in the acid solution, at a metal-to-chitosan molar ratio of 1:8 [41]. The hydroxyl groups (-OH) and amine ($–NH_2$) groups in the chitosan structure can act as good ligands for coordination with the metal ions. This coordination usually causes the weakening of covalent bonds near the coordinating site, leading to weak points which can promote the chain-scission reaction, as shown in Figure 13. As a result, the complexation of chitosan with Cu^{2+} or Fe^{3+} ions strongly promoted the degradation rate of chitosan, while chitosan–Ag^+ and chitosan–Zn^{2+} complexes exhibited slight change, compared to chitosan alone. After the SPP treatment for 180 min, the only complexation with either Cu^{2+} could produce glucosamine and COS with a molecular weight of 10^3 and PDI of 1.4. However, the reaction time was quite long, and the separation of the metal ions from the glucosamine and COS was required, prior to further use.

Later, the SPP was also combined with an environmental-friendly oxidizing agent (i.e., H_2O_2) and O_2 bubbling in order to enhance the rate of reaction and lower the possibility of contamination to the final products, as well as move toward a greener process. Chokradjaroen et. al. found that the combination of SPP and 4 mM H_2O_2 could effectively promote the chain scission of chitosan, resulting in the significant decrease of molecular weight (from 450×10^3 to 50×10^3) within 60 min [74]. The degradation mechanism of chitosan by applying the SPP in combination with H_2O_2 was also proposed in this work. They explained that excitation and ionization of H_2O molecules should mainly occur during the plasma discharge since the major component in the system was H_2O molecules. Electrons emitted from ionization continuously collided with the surrounding H_2O molecules to produce $^{\bullet}OH$. The addition of H_2O_2 could promote the reaction because H_2O_2 itself can dissociate to form $^{\bullet}OH$. This phenomenon helps to increase the $^{\bullet}OH$ concentration in the system and enhance the degradation of chitosan. Due to the relatively short reaction time and not too severe reaction, the prevention of over-degradation could be done. The obtained COS has an average molecular weight of 1.44×10^3 (8 oligomers), which has been reported to be suitable for anticancer activity. Moreover, Ma et al. reported on the effect of bubbling gas added in the solution plasma process on the molecular weight and physicochemical properties of chitosan [107]. They found that when the bubbling gas (i.e., O_2) was presented in the SPP system, the concentration of $^{\bullet}OH$ increased, which caused not only the enhancement of molecular weight reduction but also influence on physical properties (e.g., destruction of crystallinity). Moreover, the SPP combined with H_2O_2 and O_2 did not show a significant change on the pyranose ring and functional groups ($–OH$, $–NH_2$, etc.) of chitosan, which is a key possessing the biological properties.

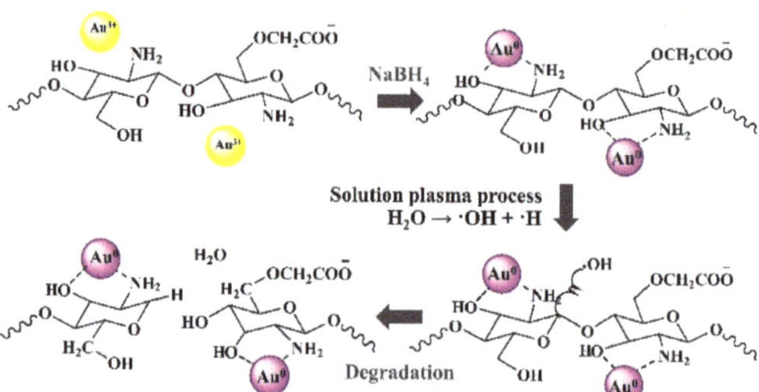

Figure 11. Illustration of a formation of N,O-carboxmethyl chitosan decorated with gold nanoparticles and a possible degradation mechanism by hydroxyl radicals generated by the SPP [49].

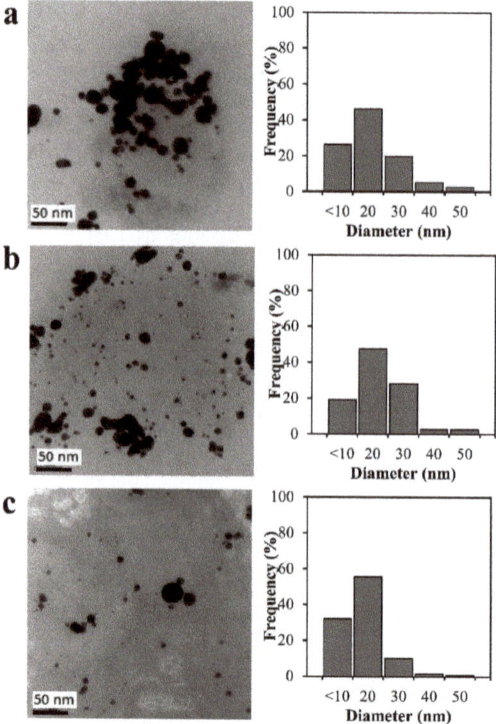

Figure 12. TEM images of the N,O-carboxmethyl chitosan decorated with gold nanoparticles (**a**) and the degraded products after degradation by the SPP for 45 (**b**) and 90 min (**c**) [49].

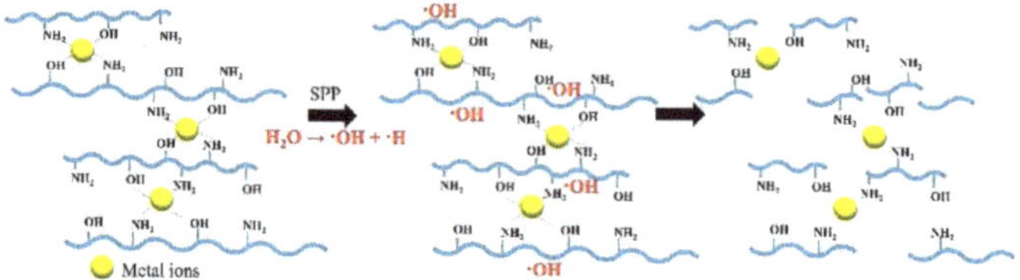

Figure 13. Degradation of chitosan–metal complexes via SPP.

Several methods (e.g., microwave, sonication, and irradiation) have been claimed as green degradation of chitosan. However, in these methods as well as most studies using the SPP, chitosan is mostly dissolved in an acetic acid solution (approximately 1–2 M), to obtain a homogeneous chitosan solution prior to the degradation, as shown in Table 2. After the degradation, separating water-soluble COS from the high-molecular-weight chitosan is usually found to be complicated because they are dissolved together in the solution. Chokradjaroen et al. realized this issue; therefore, they had tried to propose a technique that can effectively produce chitosan with specific molecular weight, reduce the chemical use in the system, and simplify the post-treatment step, including the separation and purification processes. Accordingly, the heterogeneous degradation of chitosan by the SPP was proposed. For example, the heterogeneous degradation of chitosan hydrogel, which could hold a large amount of water in its three-dimensional networks, was conducted by the SPP with the presence of carboxylic acids (i.e., monocarboxylic acid, dicarboxylic acid, and tricarboxylic acid) at an acid-to-chitosan mole ratio of 1-to-8 (~1.55 mM) [106]. The chitosan hydrogel with addition of carboxylic acids was found to have good mobility in the SPP reactor. According to the molecular weight of the obtained products, the number of carboxylic groups in the carboxylic acid exhibited the effect on the rate of reaction, molecular weight, and PDI of the obtained COS. Acetate anions (CH_3COO^-) are small, which should be able to penetrate the three-dimensional networks of chitosan hydrogel and undergo ionic interaction with the protonated amino group ($-NH_3^+$), leading to electrostatic repulsion and expansion between chitosan chains. The expansion was believed to facilitate the accessibility of $^\bullet OH$ to C-1 position of chitosan. Besides, all dicarboxylic acids and tricarboxylic acid can dissociate in the water, based on their pKa. Therefore, ionic interactions between the $-NH_3^+$ groups of chitosan and the $-COO^-$ groups of the carboxylic acids can occur. This can lead to ionic crosslinking or complexation of dicarboxylic and tricarboxylic acids with $-NH_3^+$ groups of some adjacent chitosan chains, which should result in the weakening of the covalent bonds near the complexed sites. The COS obtained from the system with dicarboxylic acid had a similar molecular weight of approximately 2100 (PDI = 1.8), while the tricarboxylic acid system could produce COS with 1500 (PDI = 1.5). As a result, the complexation of dicarboxylic and tricarboxylic acids showed potential for the selective degradation of chitosan. Moreover, due to the use of carboxylic acids with an incredibly low concentration in the reaction, further purification after the centrifugation was unnecessary. The overall process became much simpler, compared to that of the homogenous chitosan solution, like in other studies. In addition, carboxylic acids were reported to be safe for use in food and drug production [108].

Furthermore, the fine power of chitosan was used and provided a good dispersion in the SPP reactor and led to the simpler degradation process of chitosan, compared to other previous techniques, as shown in Figure 14 [39]. A variety of sodium salts (e.g., NaCl, NaI, NaNO$_3$, and Na$_2$SO$_4$) and metal chloride (e.g., CaCl$_2$, MnCl$_2$, and CeCl$_3$) used in this work exhibited the different influence on the rate of degradation on the main structure (i.e., pyranose rings and functional groups) of chitosan. According to the result, the inorganic salts, such as Na$_2$SO$_4$, NaCl, and NaNO$_3$, could promote stable plasma formation as well as the molecular weight reduction of chitosan. After the plasma discharge, water-soluble and water-insoluble products can be easily observed, as revealed in Figure 15a. The morphology and crystallinity of the plasma-treated chitosan were also observed as a function of time (Figure 15b). The result showed evidence that both the degradation and destruction of crystallinity occurred simultaneously, as proposed in Figure 15c. For the presence of NaI, MnCl$_2$, and CeCl$_3$ in the reaction solutions, they could not provide the effective molecular weight reduction of chitosan powder. In general, NaI can be dissociated to iodide ion (I$^-$), which can be oxidized to form iodine molecules. Meanwhile, MnCl$_2$ and CeCl$_3$ are a transition metal and a lanthanide, respectively, which have several oxidation states. Therefore, they can undergo some redox reactions. These reactions could probably compete with the degradation reaction, leading to the lowering of the degradation efficiency. The obtained COS products in their works were analyzed and evaluated for their cytotoxicity and showed that they were highly purified and had potential as a nature-derived anticancer agent.

Recently, the role of reactive species generated in the SPP system was intensively investigated by Ma et al. by using a technique based on radical scavenging (radical $^{\bullet}$OH scavenger, tert-butanol; H_2O_2 scavenger, MnO_2; radical $^{\bullet}$O scavenger, 1,4-benzoquinone; hydrated electrons scavenger, NaH_2PO_4) [109]. The result showed that not only $^{\bullet}$OH but also $^{\bullet}$O and H_2O_2 participate in the degradation of chitosan, while the hydrated electron played a partial role. The addition of $^{\bullet}$O and H_2O_2 scavengers were found to significantly inhibit the degradation of chitosan, compared to that of $^{\bullet}$OH scavenging. This might be because when $^{\bullet}$O and H_2O_2 were scavenged, it might enhance consumption and lower the production of $^{\bullet}$OH, respectively.

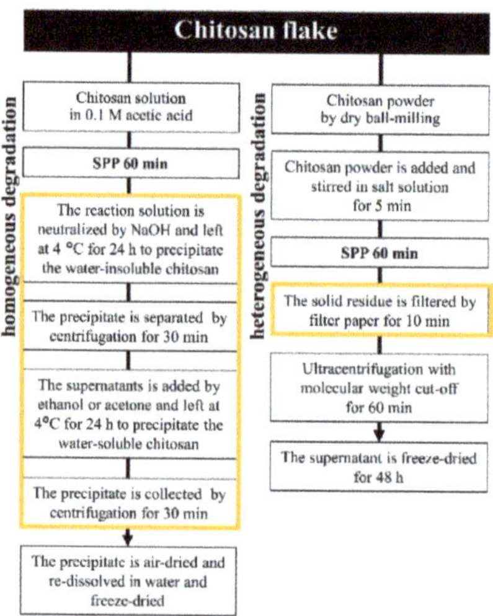

Figure 14. Flow chart of comparison between homogeneous and heterogeneous degradation of chitosan by the SPP [39,74].

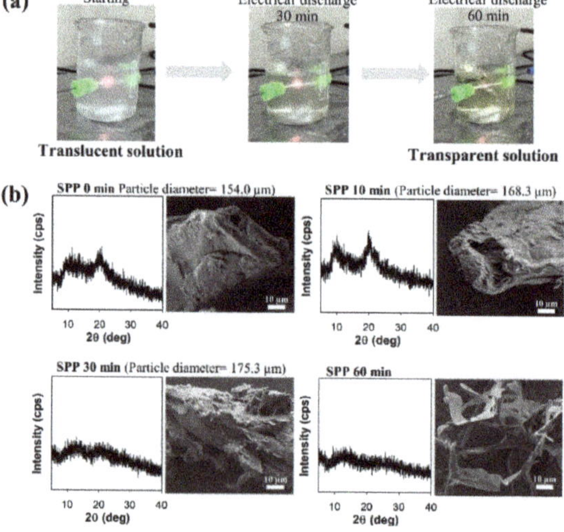

Figure 15. *Cont.*

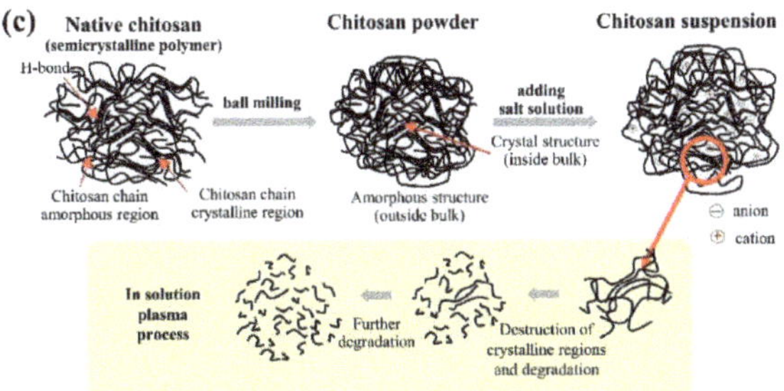

Figure 15. (**a**) Digital images of chitosan powder degradation by the SPP, (**b**) XRD and SEM results of the obtained products from chitosan powder degradation by the solution plasma process, and (**c**) Proposed destruction and degradation mechanism [39].

3.3. Deacetylation of Chitin Via SPP

Deacetylation is considered as a first step to functionalize chitin, which normally cannot dissolve in water and organic solvents, into other various derivates (e.g., chitosan). For several decades, the deacetylation of chitin converting an acetamido group at the C–2 position of a pyranose ring to an amino group has widely been conducted by conventional heat treatment (100–160 °C) using 40–50% NaOH, especially in commercial scale. For example, Kurita et al. used 50% NaOH solution at 130 °C [110], and Galed et al. used the corresponding concentration of NaOH at 110 °C to convert chitin to chitosan [111]. Recently, the SPP was introduced and could provide effective deacetylation of chitin with much lower concentration of chemicals (i.e., 1–12% NaOH) [112]. The key for this green process was the plasma-generated reactive species, including $^{\bullet}$OH and $^{\bullet}$H. The comparison of deacetylation via conventional heat treatment and SPP is revealed in Figure 16. Chitin hydrogel was used as a starting material and dispersed in NaOH/methanol/water solution (i.e., 90% methanol/water solution containing 12% NaOH). During the plasma discharge, $^{\bullet}$OH and $^{\bullet}$H were proposed to be generated via the following reactions:

For the conventional heat treatment,

$$NaOH \rightarrow Na^+ + OH^- \tag{15}$$

For the SPP,

$$H_2O \rightarrow {}^{\bullet}OH + {}^{\bullet}H \tag{16}$$

$$ROH \rightarrow {}^{\bullet}R + {}^{\bullet}OH \tag{17}$$

After the plasma treatment for 5 h, it was found that the degree of deacetylation changed from 35% to 78%, and the molecular weight of chitin decreased from approximately 10^6 to 2×10^5, leading to the improved solubility in diluted acetic acid (2%) and possessing antibacterial activity. Although the degree of deacetylation of the products was still not competitive with that obtained from the conventional heat treatment with a high concentration of NaOH, the finding in this study opened an opportunity for the further development of a more environmentally friendly process for the industrial-scale production of chitosan.

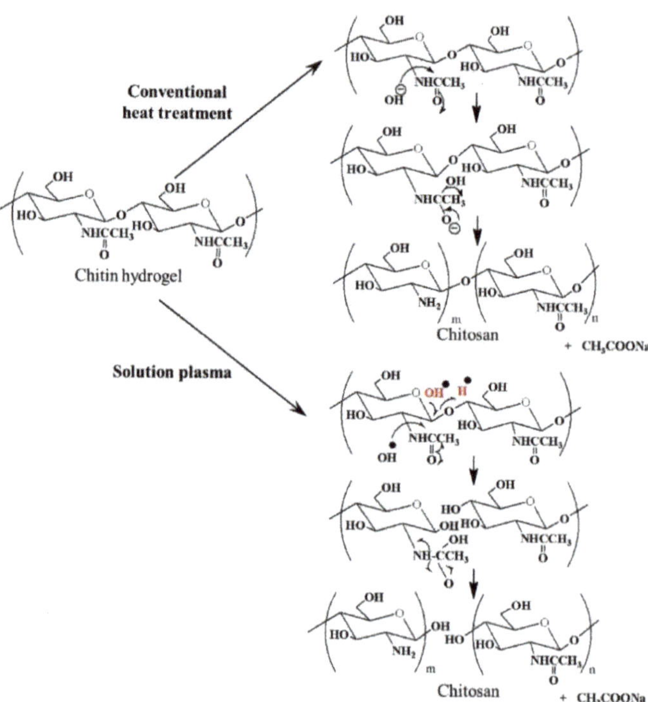

Figure 16. Comparison between deacetylations by conventional heating with NaOH (40–50%) and SPP using NaOH (1–12%) [112].

4. Conclusions and Future Aspects

In this paper, the fundamental electrical discharge in water and aqueous solutions using pin-to-pin electrode configuration, referred to as the solution plasma process (SPP), is discussed. A summary of parameters influencing the electrical breakdown, plasma stability, and reactive species formation is also given. The given fundamental is hoped to be used as a guideline for designing experimental setups and procedures for the SPP, aiming to obtain an effective process not only for modification of chitin and chitosan but also other kinds of application. Moreover, various strategies for the modification of chitin and chitosan, shown in this paper, update the development progress, as well as give ideas for further development to change low-value materials to high-value materials. Even though the production of high-purified chitooligosaccharides via the SPP has been accomplished, selective chain scission of chitosan and chitin to produce chitooligosaccharides with a narrow polydispersity index is challenging. Moreover, the modification of chitin and chitosan via the SPP is still in a beginning stage. It is believed that there is plenty of room to apply the SPP to modify functional groups or pyranose rings of chitin and chitosan, including deacetylation carboxylation, sulfonation, etc.

Supplementary Materials: The following are available online at https://www.mdpi.com/article/10.3390/ijms22094308/s1.

Author Contributions: C.C. wrote the manuscript; J.N. co-wrote the manuscript; G.P. and N.S. offered advice and revised the manuscript. All authors have read and agreed to the published version of the manuscript.

Funding: This research was financially supported by the Open Innovation Platform with Enterprises, Research Institute and Academia (OPERA, Grant JPMJOP1843), and Strategic International Collaborative Research Program (SICORP, Grant JPMJSC18H1) of Japan Science and Technology Agency (JST).

Data Availability Statement: All data will be made available upon request.

Conflicts of Interest: The authors declare no conflict of interest.

References

1. Tiede, R.; Hirschberg, J.; Daeschlein, G.; Von Woedtke, T.; Vioel, W.; Emmert, S. Plasma Applications: A Dermatological View. *Contrib. Plasma Phys.* **2014**, *54*, 118–130. [CrossRef]
2. Rezaei, F.; Vanraes, P.; Nikiforov, A.; Morent, R.; De Geyter, N. Applications of Plasma-Liquid Systems: A Review. *Materials* **2019**, *12*, 2751. [CrossRef] [PubMed]
3. López, M.; Calvo, T.; Prieto, M.; Múgica-Vidal, R.; Muro-Fraguas, I.; Alba-Elías, F.; Alvarez-Ordóñez, A. A Review on Non-thermal Atmospheric Plasma for Food Preservation: Mode of Action, Determinants of Effectiveness, and Applications. *Front. Microbiol.* **2019**, *10*, 622. [CrossRef]
4. Horikoshi, S.; Serpone, N. In-liquid plasma: A novel tool in the fabrication of nanomaterials and in the treatment of wastewaters. *RSC Adv.* **2017**, *7*, 47196–47218. [CrossRef]
5. Yui, H.; Banno, M. Microspectroscopic imaging of solution plasma: How do its physical properties and chemical species evolve in atmospheric-pressure water vapor bubbles? *Jpn. J. Appl. Phys.* **2017**, *57*, 0102A1. [CrossRef]
6. Tendero, C.; Tixier, C.; Tristant, P.; Desmaison, J.; Leprince, P. Atmospheric pressure plasmas: A review. *Spectrochim. Acta Part B Atomic Spectrosc.* **2006**, *61*, 2–30. [CrossRef]
7. De Geyter, N.; Morent, R. Cold plasma surface modification of biodegradable polymer biomaterials. In *Biomaterials for Bone Regeneration*; Dubruel, P., Van Vlierberghe, S., Eds.; Woodhead Publishing: Cambridge, UK, 2014; pp. 202–224.
8. Taghvaei, H.; Kheirollahivash, M.; Ghasemi, M.; Rostami, P.; Rahimpour, M.R. Noncatalytic Upgrading of Anisole in an Atmospheric DBD Plasma Reactor: Effect of Carrier Gas Type, Voltage, and Frequency. *Energy Fuels* **2014**, *28*, 2535–2543. [CrossRef]
9. Šunka, P. Pulse electrical discharges in water and their applications. *Phys. Plasmas* **2001**, *8*, 2587–2594. [CrossRef]
10. Thagard, S.M.; Takashima, K.; Mizuno, A. Plasma Chemistry in Pulsed Electrical Discharge in Liquid. *Trans. Mater. Res. Soc. Jpn.* **2009**, *34*, 257–262. [CrossRef]
11. Sato, M.; Ohgiyama, T.; Clements, J.S. Formation of chemical species and their effects on microorganisms using a pulsed high-voltage discharge in water. *IEEE Trans. Ind. Appl.* **1996**, *32*, 106–112. [CrossRef]
12. Malik, M.A.; Ghaffar, A.; Malik, S.A. Water purification by electrical discharges. *Plasma Sources Sci. Technol.* **2001**, *10*, 82–91. [CrossRef]
13. Takai, O. Solution plasma processing (SPP). *Pure Appl. Chem.* **2008**, *80*, 2003–2011. [CrossRef]
14. Hieda, J.; Saito, N.; Takai, O. Exotic shapes of gold nanoparticles synthesized using plasma in aqueous solution. *J. Vac. Sci. Technol. A* **2008**, *26*, 854. [CrossRef]
15. Janpetch, N.; Saito, N.; Rujiravanit, R. Fabrication of bacterial cellulose-ZnO composite via solution plasma process for antibacterial applications. *Carbohydr. Polym.* **2016**, *148*, 335–344. [CrossRef] [PubMed]
16. Saqib, A.N.S.; Huong, N.T.T.; Kim, S.-W.; Jung, M.-H.; Lee, Y.H. Structural and magnetic properties of highly Fe-doped ZnO nanoparticles synthesized by one-step solution plasma process. *J. Alloy. Compd.* **2021**, *853*, 157153. [CrossRef]
17. Kim, K.; Hashimi, K.; Bratescu, M.A.; Saito, N. The Initial Reactions from Pyridine to Hetero-Carbon Nanomaterials Through Solution Plasma. *Nanosci. Nanotechnol. Lett.* **2018**, *10*, 814–819. [CrossRef]
18. Kang, J.; Li, O.L.; Saito, N. A simple synthesis method for nano-metal catalyst supported on mesoporous carbon: The solution plasma process. *Nanoscale* **2013**, *5*, 6874–6882. [CrossRef]
19. Li, O.L.; Chiba, S.; Wada, Y.; Panomsuwan, G.; Ishizaki, T. Synthesis of graphitic-N and amino-N in nitrogen-doped carbon via a solution plasma process and exploration of their synergic effect for advanced oxygen reduction reaction. *J. Mater. Chem. A* **2017**, *5*, 2073–2082. [CrossRef]
20. Kang, J.; Li, O.L.; Saito, N. Synthesis of structure-controlled carbon nano spheres by solution plasma process. *Carbon* **2013**, *60*, 292–298. [CrossRef]
21. Prasertsung, I.; Chutinate, P.; Watthanaphanit, A.; Saito, N.; Damrongsakkul, S. Conversion of cellulose into reducing sugar by solution plasma process (SPP). *Carbohydr. Polym.* **2017**, *172*, 230–236. [CrossRef]
22. Davoodbasha, M.; Kim, S.-C.; Lee, S.-Y.; Kim, J.-W. The facile synthesis of chitosan-based silver nano-biocomposites via a solution plasma process and their potential antimicrobial efficacy. *Arch. Biochem. Biophys.* **2016**, *605*, 49–58. [CrossRef] [PubMed]
23. Sulaimankulova, S.; Mametova, A.; Abdullaeva, Z. Fusiform gold nanoparticles by pulsed plasma in liquid method. *Sn Appl. Sci.* **2019**, *1*, 1427. [CrossRef]

24. Yoshimura, M.; Senthilnathan, J. Submerged Liquid Plasma for the Formation of Nanostructured Carbon. In *Carbon-related Materials in Recognition of Nobel Lectures by Prof. Akira Suzuki in ICCE*; Kaneko, S., Mele, P., Endo, T., Tsuchiya, T., Tanaka, K., Yoshimura, M., Hui, D., Eds.; Springer International Publishing: Cham, Switzerland, 2017; pp. 61–78.
25. Senthilnathan, J.; Rao, K.S.; Yoshimura, M. Submerged liquid plasma—Low energy synthesis of nitrogen-doped graphene for electrochemical applications. *J. Mater. Chem. A* **2014**, *2*, 3332–3337. [CrossRef]
26. Yu, F.; Wang, C.; Li, Y.; Ma, H.; Wang, R.; Liu, Y.; Suzuki, N.; Terashima, C.; Ohtani, B.; Ochiai, T.; et al. Enhanced Solar Photothermal Catalysis over Solution Plasma Activated TiO2. *Adv. Sci.* **2020**, *7*, 2000204. [CrossRef] [PubMed]
27. Panomsuwan, G.; Saito, N.; Ishizaki, T. Electrocatalytic oxygen reduction on nitrogen-doped carbon nanoparticles derived from cyano-aromatic molecules via a solution plasma approach. *Carbon* **2016**, *98*, 411–420. [CrossRef]
28. Saito, N.; Hieda, J.; Takai, O. Synthesis process of gold nanoparticles in solution plasma. *Thin Solid Film.* **2009**, *518*, 912–917. [CrossRef]
29. Pootawang, P.; Saito, N.; Takai, O. Ag nanoparticle incorporation in mesoporous silica synthesized by solution plasma and their catalysis for oleic acid hydrogenation. *Mater. Lett.* **2011**, *65*, 1037–1040. [CrossRef]
30. Pootawang, P.; Saito, N.; Lee, S.Y. Discharge time dependence of a solution plasma process for colloidal copper nanoparticle synthesis and particle characteristics. *Nanotechnology* **2013**, *24*, 055604. [CrossRef]
31. Hu, X.; Shen, X.; Takai, O.; Saito, N. Facile fabrication of PtAu alloy clusters using solution plasma sputtering and their electrocatalytic activity. *J. Alloy. Compd.* **2013**, *552*, 351–355. [CrossRef]
32. Watthanaphanit, A.; Panomsuwan, G.; Saito, N. A novel one-step synthesis of gold nanoparticles in an alginate gel matrix by solution plasma sputtering. *RSC Adv.* **2014**, *4*, 1622–1629. [CrossRef]
33. Chokradjaroen, C.; Kato, S.; Fujiwara, K.; Watanabe, H.; Ishii, T.; Ishizaki, T. A comparative study of undoped, boron-doped, and boron/fluorine dual-doped carbon nanoparticles obtained via solution plasma as catalysts for the oxygen reduction re-action. *Sustain. Energy Fuels* **2020**, *4*, 4570–4580. [CrossRef]
34. Ishizaki, T.; Chiba, S.; Kaneko, Y.; Panomsuwan, G. Electrocatalytic activity for the oxygen reduction reaction of oxy-gen-containing nanocarbon synthesized by solution plasma. *J. Mater. Chem. A* **2014**, *2*, 10589–10598. [CrossRef]
35. Morishita, T.; Ueno, T.; Panomsuwan, G.; Hieda, J.; Yoshida, A.; Bratescu, M.A.; Saito, N. Fastest Formation Routes of Nanocarbons in Solution Plasma Processes. *Sci. Rep.* **2016**, *6*, 36880. [CrossRef] [PubMed]
36. Panomsuwan, G.; Saito, N.; Ishizaki, T. Fe–N-doped carbon-based composite as an efficient and durable electrocatalyst for the oxygen reduction reaction. *RSC Adv.* **2016**, *6*, 114553–114559. [CrossRef]
37. Panomsuwan, G.; Chantaramethakul, J.; Chokradjaroen, C.; Ishizaki, T. In situ solution plasma synthesis of silver nanopar-ticles supported on nitrogen-doped carbons with enhanced oxygen reduction activity. *Mater. Lett.* **2019**, *251*, 135–139. [CrossRef]
38. Phan, P.Q.; Naraprawatphong, R.; Pornaroontham, P.; Park, J.; Chokradjaroen, C.; Saito, N. N-Doped few-layer graphene encapsulated Pt-based bimetallic nanoparticles via solution plasma as an efficient oxygen catalyst for the oxygen reduction reaction. *Mater. Adv.* **2021**, *2*, 322–335. [CrossRef]
39. Chokradjaroen, C.; Theeramunkong, S.; Yui, H.; Saito, N.; Rujiravanit, R. Cytotoxicity against cancer cells of chitosan oli-gosaccharides prepared from chitosan powder degraded by electrical discharge plasma. *Carbohydr. Polym.* **2018**, *201*, 20–30. [CrossRef]
40. Prasertsung, I.; Damrongsakkul, S.; Saito, N. Degradation of β-chitosan by solution plasma process (SPP). *Polym. Degrad. Stab.* **2013**, *98*, 2089–2093. [CrossRef]
41. Pornsunthorntawee, O.; Katepetch, C.; Vanichvattanadecha, C.; Saito, N.; Rujiravanit, R. Depolymerization of chitosan–metal complexes via a solution plasma technique. *Carbohydr. Polym.* **2014**, *102*, 504–512. [CrossRef]
42. Watthanaphanit, A.; Saito, N. Effect of polymer concentration on the depolymerization of sodium alginate by the solution plasma process. *Polym. Degrad. Stab.* **2013**, *98*, 1072–1080. [CrossRef]
43. Zalikhanov, B.Z. From an electron avalanche to the lightning discharge. *Phys. Part Nucl.* **2016**, *47*, 108–133. [CrossRef]
44. Saito, N.; Bratescu, M.A.; Hashimi, K. Solution plasma: A new reaction field for nanomaterials synthesis. *Jpn. J. Appl. Phys.* **2017**, *57*, 0102A4. [CrossRef]
45. Heo, Y.K.; Lee, S.H.; Bratescu, M.A.; Kim, S.M.; Lee, G.-J.; Saito, N. Generation of non-equilibrium condition in solution plasma discharge using low-pass filter circuit. *Plasma Process. Polym.* **2017**, *14*, 1600163. [CrossRef]
46. Prasertsung, I.; Damrongsakkul, S.; Terashima, C.; Saito, N.; Takai, O. Preparation of low molecular weight chitosan using solution plasma system. *Carbohydr. Polym.* **2012**, *87*, 2745–2749. [CrossRef]
47. Tantiplapol, T.; Singsawat, Y.; Narongsil, N.; Damrongsakkul, S.; Saito, N.; Prasertsung, I. Influences of solution plasma conditions on degradation rate and properties of chitosan. *Innov. Food Sci. Emerg. Technol.* **2015**, *32*, 116–120. [CrossRef]
48. Davoodbasha, M.; Lee, S.-Y.; Kim, J.-W. Solution plasma mediated formation of low molecular weight chitosan and its ap-plication as a biomaterial. *Int. J. Biol. Macromol.* **2018**, *118*, 1511–1517. [CrossRef] [PubMed]
49. Chokradjaroen, C.; Rujiravanit, R.; Theeramunkong, S.; Saito, N. Effect of electrical discharge plasma on cytotoxicity against cancer cells of N,O-carboxymethyl chitosan-stabilized gold nanoparticles. *Carbohydr. Polym.* **2020**, *237*, 116162. [CrossRef]
50. Potocký, Š.; Saito, N.; Takai, O. Needle electrode erosion in water plasma discharge. *Thin Solid Film.* **2009**, *518*, 918–923. [CrossRef]

51. Hayashi, Y.; Takada, N.; Wahyudiono; Kanda, H.; Goto, M. Hydrogen Peroxide Formation by Electric Discharge with Fine Bubbles. *Plasma Chem. Plasma Process.* **2017**, *37*, 125–135. [CrossRef]

52. Mai-Prochnow, A.; Zhou, R.; Zhang, T.; Ostrikov, K.; Mugunthan, S.; Rice, S.A.; Cullen, P.J. Interactions of plasma-activated water with biofilms: Inactivation, dispersal effects and mechanisms of action. *NPJ Biofilms Microbiomes* **2021**, *7*, 1–12. [CrossRef]

53. Tanioka, S.; Matsui, Y.; Irie, T.; Tanigawa, T.; Tanaka, Y.; Shibata, H.; Sawa, Y.; Kono, Y. Oxidative Depolymerization of Chitosan by Hydroxyl Radical. *Biosci. Biotechnol. Biochem.* **1996**, *60*, 2001–2004. [CrossRef]

54. Miron, C.; Bratescu, M.A.; Saito, N.; Takai, O. Optical diagnostic of bipolar electrical discharges in HCl, KCl, and KOH solutions. *J. Appl. Phys.* **2011**, *109*, 123301. [CrossRef]

55. Langendorf, S.; Walker, M. Effect of secondary electron emission on the plasma sheath. *Phys. Plasmas* **2015**, *22*, 033515. [CrossRef]

56. Magnusson, J.M.; Collins, A.L.; Wirz, R.E.; Magnusson, J. Polyatomic Ion-Induced Electron Emission (IIEE) in Electrospray Thrusters. *Aerospace* **2020**, *7*, 153. [CrossRef]

57. Phan, P.Q.; Chae, S.; Pornaroontham, P.; Muta, Y.; Kim, K.; Wang, X.; Saito, N. In situ synthesis of copper nanoparticles encapsulated by nitrogen-doped graphene at room temperature via solution plasma. *RSC Adv.* **2020**, *10*, 36627–36635. [CrossRef]

58. Wang, H.; Wandell, R.J.; Tachibana, K.; Voráč, J.; Locke, B.R. The influence of liquid conductivity on electrical breakdown and hydrogen peroxide production in a nanosecond pulsed plasma discharge generated in a water-film plasma reactor. *J. Phys. D Appl. Phys.* **2018**, *52*, 075201. [CrossRef]

59. Takeuchi, N.; Kawahara, K.; Gamou, F.; Li, O.L.H. Observation of solution plasma in water-ethanol mixed solution for reduction of graphene oxide. *Int. J. Plasma Environ. Sci. Technol.* **2020**, *14*, e01009.

60. Schmieder, F. Neue Wirkungsquerschnittsmessungen an Gasen und Dämpfen. *Z Elektrochem Angew P* **1930**, *36*, 700–704.

61. Szmytkowski, C.; Możejko, P. Electron-scattering total cross sections for triatomic molecules: NO_2 and H_2O. *Opt. Appl.* **2006**, *36*, 543–550.

62. Panomsuwan, G.; Chokradjaroen, C.; Rujiravanit, R.; Ueno, T.; Saito, N. In vitro cytotoxicity of carbon black nanoparticles synthesized from solution plasma on human lung fibroblast cells. *Jpn. J. Appl. Phys.* **2018**, *57*, 0102BG. [CrossRef]

63. Miron, C.; Bratescu, M.A.; Saito, N.; Takai, O.; Brătescu, M.A. Time-resolved Optical Emission Spectroscopy in Water Electrical Discharges. *Plasma Chem. Plasma Process.* **2010**, *30*, 619–631. [CrossRef]

64. Miron, C.; Bratescu, M.; Saito, N.; Takai, O. Effect of the electrode work function on the water plasma breakdown voltage. *Curr. Appl. Phys.* **2011**, *11*, S154–S158. [CrossRef]

65. Banno, M.; Kanno, K.; Yui, H. Development of direct gas injection system for atmospheric-pressure in-solution discharge plasma for plasma degradation and material syntheses. *RSC Adv.* **2016**, *6*, 16030–16036. [CrossRef]

66. Kim, S.-K. *Chitin, Chitosan, Oligosaccharides and Their Derivatives: Biological Activities and Applications*; CRC Press: Boca Raton, FL, USA, 2010.

67. Kumar, M.N.R. A review of chitin and chitosan applications. *React. Funct. Polym.* **2000**, *46*, 1–27. [CrossRef]

68. Rinaudo, M. Chitin and chitosan: Properties and applications. *Prog. Polym. Sci.* **2006**, *31*, 603–632. [CrossRef]

69. Kaczmarek, M.B.; Struszczyk-Swita, K.; Li, X.; Szczęsna-Antczak, M.; Daroch, M. Enzymatic Modifications of Chitin, Chitosan, and Chitooligosaccharides. *Front. Bioeng. Biotechnol.* **2019**, *7*, 243. [CrossRef] [PubMed]

70. Kumar, A.V.; Tharanathan, R. A comparative study on depolymerization of chitosan by proteolytic enzymes. *Carbohydr. Polym.* **2004**, *58*, 275–283.

71. Xia, W.; Liu, P.; Zhang, J.; Chen, J. Biological activities of chitosan and chitooligosaccharides. *Food Hydrocoll.* **2011**, *25*, 170–179. [CrossRef]

72. Merzendorfer, H.; Cohen, E. Chitin/Chitosan: Versatile Ecological, Industrial, and Biomedical Applications. In *Extracellular Sugar-Based Biopolymers Matrices*; Cohen, E., Merzendorfer, H., Eds.; Springer International Publishing: Cham, Switzerland, 2019; pp. 541–624.

73. Pillai, C.; Paul, W.; Sharma, C.P. Chitin and chitosan polymers: Chemistry, solubility and fiber formation. *Prog. Polym. Sci.* **2009**, *34*, 641–678. [CrossRef]

74. Chokradjaroen, C.; Rujiravanit, R.; Watthanaphanit, A.; Theeramunkong, S.; Saito, N.; Yamashita, K.; Arakawa, R. Enhanced degradation of chitosan by applying plasma treatment in combination with oxidizing agents for potential use as an anticancer agent. *Carbohydr. Polym.* **2017**, *167*, 1–11. [CrossRef] [PubMed]

75. Liu, J.; Zhang, J.; Xia, W. Hypocholesterolaemic effects of different chitosan samples in vitro and in vivo. *Food Chem.* **2008**, *107*, 419–425. [CrossRef]

76. No, H.K.; Park, N.Y.; Lee, S.H.; Meyers, S.P. Antibacterial activity of chitosans and chitosan oligomers with different molecular weights. *Int. J. Food Microbiol.* **2002**, *74*, 65–72. [CrossRef]

77. Lee, S.-H.; Senevirathne, M.; Ahn, C.-B.; Kim, S.-K.; Je, J.-Y. Factors affecting anti-inflammatory effect of chitooligosaccharides in lipopolysaccharides-induced RAW264.7 macrophage cells. *Bioorg. Med. Chem. Lett.* **2009**, *19*, 6655–6658. [CrossRef]

78. Prashanth, K.H.; Tharanathan, R. Depolymerized products of chitosan as potent inhibitors of tumor-induced angiogenesis. *Biochim. Biophys. Acta (Bba) Gen. Subj.* **2005**, *1722*, 22–29. [CrossRef] [PubMed]

79. Vo, T.-S.; Kong, C.-S.; Kim, S.-K. Inhibitory effects of chitooligosaccharides on degranulation and cytokine generation in rat basophilic leukemia RBL-2H3 cells. *Carbohydr. Polym.* **2011**, *84*, 649–655. [CrossRef]

80. Jung, W.-K.; Moon, S.-H.; Kim, S.-K. Effect of chitooligosaccharides on calcium bioavailability and bone strength in ovariectomized rats. *Life Sci.* **2006**, *78*, 970–976. [CrossRef] [PubMed]
81. Feng, T.; Du, Y.; Li, J.; Hu, Y.; Kennedy, J.F. Enhancement of antioxidant activity of chitosan by irradiation. *Carbohydr. Polym.* **2008**, *73*, 126–132. [CrossRef]
82. Einbu, A.; Grasdalen, H.; Vårum, K.M. Kinetics of hydrolysis of chitin/chitosan oligomers in concentrated hydrochloric acid. *Carbohydr. Res.* **2007**, *342*, 1055–1062. [CrossRef]
83. Vårum, K.; Ottøy, M.; Smidsrød, O. Acid hydrolysis of chitosans. *Carbohydr. Polym.* **2001**, *46*, 89–98. [CrossRef]
84. Kim, S.; Rajapakse, N. Enzymatic production and biological activities of chitosan oligosaccharides (COS): A review. *Carbohydr. Polym.* **2005**, *62*, 357–368. [CrossRef]
85. Cabrera, J.C.; Van Cutsem, P. Preparation of chitooligosaccharides with degree of polymerization higher than 6 by acid or enzymatic degradation of chitosan. *Biochem. Eng. J.* **2005**, *25*, 165–172. [CrossRef]
86. Shao, J.; Yang, Y.; Zhong, Q. Studies on preparation of oligoglucosamine by oxidative degradation under microwave irra-diation. *Polym. Degrad. Stab.* **2003**, *82*, 395–398. [CrossRef]
87. Wasikiewicz, J.M.; Yeates, S.G. "Green" molecular weight degradation of chitosan using microwave irradiation. *Polym. Degrad. Stab.* **2013**, *98*, 863–867. [CrossRef]
88. Wasikiewicz, J.M.; Yoshii, F.; Nagasawa, N.; Wach, R.A.; Mitomo, H. Degradation of chitosan and sodium alginate by gamma radiation, sonochemical and ultraviolet methods. *Radiat. Phys. Chem.* **2005**, *73*, 287–295. [CrossRef]
89. Poshina, D.N.; Raik, S.V.; Poshin, A.N.; Skorik, Y.A. Accessibility of chitin and chitosan in enzymatic hydrolysis: A review. *Polym. Degrad. Stab.* **2018**, *156*, 269–278. [CrossRef]
90. Choi, W.-S.; Ahn, K.-J.; Lee, D.-W.; Byun, M.-W.; Park, H.-J. Preparation of chitosan oligomers by irradiation. *Polym. Degrad. Stab.* **2002**, *78*, 533–538. [CrossRef]
91. Savitri, E.; Juliastuti, S.R.; Handaratri, A.; Sumarno; Roesyadi, A. Degradation of chitosan by sonication in very-low-concentration acetic acid. *Polym. Degrad. Stab.* **2014**, *110*, 344–352. [CrossRef]
92. Li, K.; Xing, R.; Liu, S.; Qin, Y.; Meng, X.; Li, P. Microwave-assisted degradation of chitosan for a possible use in inhibiting crop pathogenic fungi. *Int. J. Biol. Macromol.* **2012**, *51*, 767–773. [CrossRef]
93. Domard, A.; Cartier, N. Glucosamine oligomers: 1. Preparation and characterization. *Int. J. Biol. Macromol.* **1989**, *11*, 297–302. [CrossRef]
94. Ngo, D.-N.; Lee, S.-H.; Kim, M.-M.; Kim, S.-K. Production of chitin oligosaccharides with different molecular weights and their antioxidant effect in RAW 264.7 cells. *J. Funct. Foods* **2009**, *1*, 188–198. [CrossRef]
95. Furusaki, E.; Ueno, Y.; Sakairi, N.; Nishi, N.; Tokura, S. Facile preparation and inclusion ability of a chitosan derivative bearing carboxymethyl-β-cyclodextrin. *Carbohydr. Polym.* **1996**, *29*, 29–34. [CrossRef]
96. Defaye, J.; Gadelle, A.; Pedersen, C. A convenient access to β-(1 → 4)-linked 2-amino-2-deoxy-d-glucopyranosyl fluoride oligosaccharides and β-(1 → 4)-linked 2-amino-2-deoxy-d-glucopyranosyl oligosaccharides by fluorolysis and fluorohy-drolysis of chitosan. *Carbohydr. Res.* **1994**, *261*, 267–277. [CrossRef]
97. Tian, F.; Liu, Y.; Hu, K.; Zhao, B. Study of the depolymerization behavior of chitosan by hydrogen peroxide. *Carbohydr. Polym.* **2004**, *57*, 31–37. [CrossRef]
98. Fukamizo, T.; Honda, Y.; Goto, S.; Boucher, I.; Brzezinski, R. Reaction mechanism of chitosanase from Streptomyces sp. N174. *Biochem. J.* **1995**, *311*, 377–383. [CrossRef]
99. Yoon, J.H. Enzymatic synthesis of chitooligosaccharides in organic cosolvents. *Enzym. Microb. Technol.* **2005**, *37*, 663–668. [CrossRef]
100. Lin, H.; Wang, H.; Xue, C.; Ye, M. Preparation of chitosan oligomers by immobilized papain. *Enzym. Microb. Technol.* **2002**, *31*, 588–592. [CrossRef]
101. Xing, R.; Liu, S.; Yu, H.; Guo, Z.; Wang, P.; Li, C.; Li, Z.; Li, P. Salt-assisted acid hydrolysis of chitosan to oligomers under microwave irradiation. *Carbohydr. Res.* **2005**, *340*, 2150–2153. [CrossRef]
102. Kang, B.; Dai, Y.-D.; Zhang, H.-Q.; Chen, D. Synergetic degradation of chitosan with gamma radiation and hydrogen peroxide. *Polym. Degrad. Stab.* **2007**, *92*, 359–362. [CrossRef]
103. Huang, Y.; Wang, P.; Yuan, Y.; Ren, X.; Yang, F. Synergistic degradation of chitosan by impinging stream and jet cavitation. *Ultrason. Sonochem.* **2015**, *27*, 592–601. [CrossRef]
104. Yan, J.; Ai, S.; Yang, F.; Zhang, K.; Huang, Y. Study on mechanism of chitosan degradation with hydrodynamic cavitation. *Ultrason. Sonochem.* **2020**, *64*, 105046. [CrossRef] [PubMed]
105. Ma, F.; Wang, Z.; Zhao, H.; Tian, S. Plasma Depolymerization of Chitosan in the Presence of Hydrogen Peroxide. *Int. J. Mol. Sci.* **2012**, *13*, 7788–7797. [CrossRef] [PubMed]
106. Chokradjaroen, C.; Rujiravanit, R.; Theeramunkong, S.; Saito, N. Degradation of chitosan hydrogel dispersed in dilute carboxylic acids by solution plasma and evaluation of anticancer activity of degraded products. *Jpn J. Appl. Phys.* **2018**, *57*, 0102B5. [CrossRef]
107. Ma, F.; Li, P.; Zhang, B.; Zhao, X.; Fu, Q.; Wang, Z.; Gu, C. Effect of solution plasma process with bubbling gas on physico-chemical properties of chitosan. *Int. J. Biol. Macromol.* **2017**, *98*, 201–207. [CrossRef]
108. Skonberg, C.; Olsen, J.; Madsen, K.G.; Hansen, S.H.; Grillo, M.P.; Olsen, J. Metabolic activation of carboxylic acids. *Expert Opin. Drug Metab. Toxicol.* **2008**, *4*, 425–438. [CrossRef] [PubMed]

109. Ma, F.; Zhang, S.; Li, P.; Sun, B.; Xu, Y.; Tao, D.; Zhao, H.; Cui, S.; Zhu, R.; Zhang, B. Investigation on the role of the free radicals and the controlled degradation of chitosan under solution plasma process based on radical scavengers. *Carbohydr. Polym.* **2021**, *257*, 117567. [CrossRef] [PubMed]
110. Kurita, K.; Kamiya, M.; Nishimura, S.-I. Solubilization of a rigid polysaccharide: Controlled partial N-Acetylation of chitosan to develop solubility. *Carbohydr. Polym.* **1991**, *16*, 83–92. [CrossRef]
111. Galed, G.; Miralles, B.; Paños, I.; Santiago, A.; Heras, Á. N-Deacetylation and depolymerization reactions of chitin/chitosan: Influence of the source of chitin. *Carbohydr. Polym.* **2005**, *62*, 316–320. [CrossRef]
112. Rujiravanit, R.; Kantakanun, M.; Chokradjaroen, C.; Vanichvattanadecha, C.; Saito, N. Simultaneous deacetylation and degradation of chitin hydrogel by electrical discharge plasma using low sodium hydroxide concentrations. *Carbohydr. Polym.* **2020**, *228*, 115377. [CrossRef]

International Journal of
Molecular Sciences

Review

Cold Atmospheric Plasma: A Powerful Tool for Modern Medicine

Dušan Braný, Dana Dvorská *, Erika Halašová and Henrieta Škovierová

Biomedical Center Martin, Jessenius Faculty of Medicine in Martin, Comenius University in Bratislava, 036 01 Martin, Slovakia; dusan.brany@uniba.sk (D.B.); erika.halasova@uniba.sk (E.H.); henrieta.skovierova@uniba.sk (H.Š.)
* Correspondence: dana.dvorska@uniba.sk; Tel.: +421-432633654

Received: 2 April 2020; Accepted: 20 April 2020; Published: 22 April 2020

Abstract: Cold atmospheric plasma use in clinical studies is mainly limited to the treatment of chronic wounds, but its application in a wide range of medical fields is now the goal of many analyses. It is therefore likely that its application spectrum will be expanded in the future. Cold atmospheric plasma has been shown to reduce microbial load without any known significant negative effects on healthy tissues, and this should enhance its possible application to any microbial infection site. It has also been shown to have anti-tumour effects. In addition, it acts proliferatively on stem cells and other cultivated cells, and the highly increased nitric oxide levels have a very important effect on this proliferation. Cold atmospheric plasma use may also have a beneficial effect on immunotherapy in cancer patients. Finally, it is possible that the use of plasma devices will not remain limited to surface structures, because current endeavours to develop sufficiently miniature microplasma devices could very likely lead to its application in subcutaneous and internal structures. This study summarises the available literature on cold plasma action mechanisms and analyses of its current in vivo and in vitro use, primarily in the fields of regenerative and dental medicine and oncology.

Keywords: cold atmospheric plasma; wound healing; oncology; regenerative medicine; plasma

---

1. Introduction

William Crookes established the fundamentals of plasma science in 1879 by experimentally ionizing gas in an electrical discharge tube by the application of high voltage through a voltage coil. The ionised gas was named radiant matter. The current plasma term was then initiated almost fifty years later in 1927 by Irvin Langmuir [1]. Plasma has since been applied in many spheres over the past few decades, including medicine [2–4]. Plasma is often defined as an ionised gas produced by disintegration of polyatomic gas molecules or the removal of electrons from monatomic gas shells [5]. However, not every ionised gas that contains charged particles can be considered plasma because of the following strict definition [5–7]: (1) Plasma must have macromolecular neutrality (quasi-neutrality). In the absence of external disturbances in plasma, the net resulting electric charge is zero. Therefore, plasma contains (almost) the same density of positively and negatively charged particles; (2) Plasma must have Debye shielding, where the charged particles in plasma are arranged to effectively shield electrostatic fields within the distance of a Debye length. This is defined as a measure of the distance over which the influence of the electric field of an individual charged particle is felt by other charged particles inside the plasma; (3) Plasma frequency. If plasma loses its equilibrium conditions, the resulting internal space charge-fields promote collective particle motion that tends to restore the original charge neutrality. This motion is characterised by the natural oscillation frequency referred to as plasma frequency. Plasma can therefore be defined as "a quasineutral gas containing many interacting free electrons and ionised atoms and molecules which have collective behavior caused by long-range coulomb forces" [7]. In addition, the charged-particle motion in plasma gives rise to electric fields and generates currents and magnetic fields [6].

Plasma can also be divided into high-temperature, thermal, and non-thermal groups [7,8]. All particles (electrons and heavy particles) have the same temperature in high temperature plasma, and they are therefore in thermal equilibrium. In thermal (quasi-equilibrium plasma) are only areas of thermal equilibrium within the plasma. Finally, non-thermal (non-equilibrium) plasma have particles that are not in thermal equilibrium. This plasma is termed "cold plasma" [7,8].

High-temperature or thermal plasmas have higher electron density and ionization than cold plasma, which has ionization only up to 1% [5–7]. In poorly ionized plasmas, the charge-neutral interactions dominate multiple coulomb interactions [6]. Moreover, although electron–electron collisions achieve thermodynamic equilibrium in cold plasma, and their temperature is then much higher than that of ions and neutrons, they cannot transfer their kinetic energy to larger particles [6].

High temperature plasma can reach 10^8 K, as found in the solar core. Thermal plasma temperature can be approximately 2×10^4 K and this is observed in thunderstorm lightning. Finally, nonthermal plasma temperature can be 300 to 1000 K under artificially created conditions, as in fluorescent tubes [8,9]. Cold plasma discharge can be achieved at both low pressure and atmospheric pressure [2,3,10].

Low pressure cold plasma was first applied in the late 1960s to decontaminate surfaces, and this treatment has in some aspects proven to be more effective than conventional sterilisation [11]. In contrast, the benefits of cold atmospheric plasma (CAP) in reducing microbial load were discovered only in the second half of the 1990s [12,13]. Nevertheless, CAP provides a better alternative for modern medicine than the cold plasma generated at low pressure. CAP application is also an easier process to use than low pressure plasma because it can be generated from a portable device and this enables easier access to affected cells and tissues than the that of the large vacuum generating system required for low pressure cold plasma. In addition, most types of low-pressure plasma generating systems are very expensive [11]. It is also impractical, and almost impossible, to place any body part in this system and leave it to be exposed to plasma discharge. Moreover, all animal and human tissues contain water and its presence is undesirable in low pressure conditions [11]. Finally, although low-pressure plasma discharge can be regulated, its character is quite different and stronger than the discharge generated by atmospheric pressure plasma and is generally not suitable for application to human or animal cells or tissues. Nevertheless, cold plasma generated under low pressure can be utilised in medicine. Examples of this are its use in "flash-sterilisation", implants, and tissue engineering products' surface modifications [11].

Although CAP does not achieve the sterilization and decontamination possibilities of low pressure plasma [11], it is still effective in lowering microbial load [14,15]. Most importantly, CAP's less intense effects enable its direct application to cells and tissues [3,4] and its ability to reduce microbial load makes it a good option to replace antibiotics and to combat bacterial strains that have increasing antibiotic resistance [3]. A further great advantage of CAP-generating devices is their relatively low manufacturing costs [3,4]. Therefore, accessible, efficient, and cheaper CAP devices can very likely reduce the financial burden imposed on the health budget by conventional treatments.

Plasma discharge reduces bacteria viability mainly by formation of UV radiation, induction of reactive oxygen (ROS) and nitrogen (RNS) species, and creation of electric current [4,16–22]. The oxygen-based species generated by plasma discharge include hydroxyl (OH), hydrogen peroxide (H_2O_2), superoxide ($O_2^-\bullet$), hydroxyl radical ($\bullet$OH), singlet oxygen ($^1O_2$), and ozone (O_3) [23,24]. The nitrogen-based species are nitric oxide ($\bullet$NO), nitrogen dioxide ($\bullet NO_2$), dinitrogen tetroxide (N_2O_4), nitrogen trioxide (NO_3), nitrous oxide (N_2O), and peroxynitrite (ONOO-) [23,24]. The reactive species can be formed either by plasma–air interaction or by plasma–liquid interaction. While hydroxyl radicals and nitric oxides are typically formed by plasma–air interactions, nitrites, nitrates, and H_2O_2 with relatively longer lifetimes are formed by plasma–liquid interactions [23,24]. For example, this latter interaction occurs when cells in the cultivation media are exposed to plasma [25]. In addition, CAP generates positively charged ions such as N^{2+} [22] and also electrons [20].

However, prokaryotic bacterial cells differ from eukaryotic human and animal cells, and tumour cells differ from healthy cells. Moreover, all cells can act differently in in vitro conditions and they can respond differently to the cells in living organism tissues. This highlights the possibility of different responses to CAP devices in those cell types and also under in vitro and in vivo conditions [26]. However, it is generally presumed that low dose plasma treatment stimulates cell viability and enhances proliferation, differentiation, and migration, while high doses should induce cell apoptosis [26].

Plasma devices used in clinical practice are designed to avoid danger to healthy cells. They undergo multistep testing and are not currently associated with significant side effects [4,27,28]. However, the appropriate CAP dosage in clinical practice must be closely controlled, and this control depends on treatment type. For example, the distance between the plasma source and treatment object must be carefully adjusted. Here, Nastuda et al. [29] reported that the effect of plasma jet application can vary widely, even with a one centimetre difference between tube nozzle and human skin. The authors found significant differences in plasma spread and current flow through human tissue when the skin–nozzle gap was reduced from 15 mm to 5 mm. Moreover, it cannot be hypothetically excluded that CAP application does not have some minimal negative effects at the molecular level. All these findings and possibilities are subject to ongoing analysis, but current results suggest that the unproven adverse effects are outweighed by CAP's many benefits [3,4,12]. Finally, it is likely that more possibilities and greater benefits for CAP use in clinical practice will be demonstrated in the future, because most CAP studies have just taken place in the last 15 years [3,4,12]. While CAP can be employed in aesthetic procedures [30–32], herein we summarise CAP use in more serious medical conditions. These especially include acute and chronic wounds and oral bacterial infections and tumour therapy. Concurrently, we explore plasma potential in regenerative medicine and biological engineering.

The clinically used and experimentally tested CAP devices are divided into three main categories: 1.) those based on direct-discharge, 2.) those based on indirect-discharge, and 3.) hybrid types [4,33,34].

Direct dielectric barrier discharge (DBD) occurs between a high voltage electrode and a grounded one. The electrodes can both or individually be covered with a dielectric layer or the dielectric material can be placed in the space between the electrodes. DBD devices must also be charged only with alternating or pulsed high voltage to ensure capacitive coupling. Electrons and ions in these devices after initial gas breakdown are "stopped" by the dielectric barrier. Electric field of ions and electrons then shields the electric field from the external source. If this state is maintained, it leads to discharge loss, and the external voltage must be changed to preserve discharge. Breakdown occurs when increasing alternating current voltage is applied, and discharge activity ceases when the maximum voltage is attained. Moreover, two breakdowns occur in the same period when pulse-operated DBDs are employed. The initial breakdown is caused by the high voltage pulse, and the electric field created by the charge carrier is sufficient to cause a further breakdown during voltage decrease. This is in the form of a 'backward discharge' [35–37]. The DBD devices have several variations and designs, and one of the most practical and suitable for medical application is the floating electrode DBD (FE-DBD). Technically, FE-DBD devices do not contain a grounded electrode, and this is replaced by affected cells or tissues. The DBD devices provide a higher intensity and more adaptable and controlled discharge. They can also generate plasma solely in air without the need for carrier gases. The discharge area, however, is relatively limited because it must lie between two electrodes, and a constant distance must be maintained; this demands a smooth, flat surface [4,33,34].

Indirect discharge is generated by devices usually called Plasma jets, Plasma pens, or Plasma torches. These devices are similar to indirect discharge devices in that they comprise two facing electrodes generating a plasma discharge between them, but here, the carrier gas directs the plasma discharge. Therefore, the discharge does not affect the object between the two electrodes but proceeds in the gas flow direction. While this allows the target object to be located outside the device and the affected area is adjustable, lower amounts of ROS and RNS species are produced and the discharge is harder to control than with the DBD devices. Finally, plasma generated by indirect devices is stronger in all UV ranges, but unlike DBD devices, they do not produce electric current.

The hybrid plasma devices combine these principles but are currently applied only at the experimental level. These devices create a discharge by combining micro-discharges on a grounded mesh electrode. There is uniform discharge, no effect on the object between the two electrodes, and the device is relatively easy to control. However, these devices have slightly higher susceptibility to component wear and subsequent deterioration. This is especially noticeable in humid environments and following direct contact with treated cells and tissues [4,33,34].

There are three specific types of plasma devices certified for clinical practice [11]. The very first certified device was the kINPen® MED plasma-pen (INP Greifswald/neoplas tools GmbH, Greifswald, Germany). The second was the PlasmaDerm® VU-2010 DBD device, based on dielectric plasma discharge technology (CINOGY Technologies GmbH, Duderstadt, Germany) and the latest certified device is the SteriPlas plasma torch device (Adtec Ltd., London, UK). However, several more devices with various modifications have been tested under laboratory and experimental conditions and are awaiting possible certification for clinical application.

Finally, plasma application itself can be direct and indirect. In direct plasma application, cell lines are exposed to plasma discharges in vitro, and animal models and human tissues are exposed to the discharges in vivo. In contrast, the medium or solution affected or activated by plasma is used in indirect application. This is then used in cell cultivation or for direct injection into tested subjects, such as mouse xenografts [38]. In addition, long-term species including nitrates, nitrites, and H_2O_2 are usually preserved in plasma-activated medium (PAM) [39].

2. Use of Cold Plasma for Treatment of Chronic and Acute Wounds

The antimicrobial effect of CAP was demonstrated in the 1990s and this led to its use in medicine [12]. The initial 2007 clinical trial used the plasma device in facial rejuvenation procedures [40]. CAP use in regenerative medicine was initially aimed at accelerating acute and chronic wound healing by alleviating bacterial infection, because infection can significantly slow the healing process [4,10,27]. The first randomised pilot trials were performed by Isbary et al. [41]. These authors investigated CAP's effect on microbial infection mitigation in chronic ulcer wounds and confirmed significant infection reduction without side effects. Two years later, they reported that two-minute CAP treatments sufficiently decreased microbial load and improved chronic ulcer healing [42]. Both studies included venous, arterial, diabetic, and traumatic ulcers, and reduced bacterial infection was observed regardless of bacterial type. Chronic leg ulcers are relatively common, especially venous ulcers that affect up to 2% of the global population [43,44]. Ulcer reduction and cure is complex and it requires long-term care. Moreover, 15% of patients have ulcers that never heal and up to 71% have complicated remission [43,44]. Bacterial infection is a major contributor to the complex treatment of ulcers and many strains present in ulcers have increasing resistance to conventional antibacterial treatment [45,46]. Studies, therefore, focused on cold plasma use in ulcer treatment. One randomised study on CAP treatment included 14 patients [47]. Half the patients were treated with conventional procedures and the remainder had concurrent CAP therapy. Treatments were administered tri-weekly for eight weeks to both groups, followed by a four-week observation period. Although reduced arterial ulcers were observed in both groups, the CAP group had more rapid and observable improvement. The effect of CAP on 50 patients with pressure ulcers was then analysed in a similarly conceived study [48]. Subjects were divided into standard treatment and combined CAP-intervention groups, and the treatments were applied weekly for eight weeks. Improvement, including bacterial load decrease, was already significantly greater in the CAP group after the first week.

CAP benefits in pressure ulcer reduction have also been recorded in Wistar rat models. Experimentally created ulcers were treated with CAP for 60 s, 3 times a day for 5 days [49]. There was resultant rapid re-epithelisation, angiogenesis, collagen synthesis, and increased tissue mechanical strength after plasma exposure. Guo and colleagues [50] investigated CAP effects on various chronic wounds, including pyoderma gangrenosum, giant genital warts, and diabetic foot ulcers. The pyoderma gangrenosum patient had previously been unsuccessfully treated with antibiotics. The lesions were irradiated for 5 min once every 2 days, with 60–80 min total irradiation for all lesions. Exudation was significantly lower on the third day, and 6 cycles of exposure enabled the lesions to completely dry and contract. This patient was treated for a further 6 months without remission. A second patient had experienced some improvement of pyoderma gangrenosum after antibiotic treatment and was then treated with this CAP regime. The lesion completely disappeared after 8 CAP treatments. There was again no recurrence noted in the 4-month follow-up period. In addition, a diabetic patient who had an ulcerated foot for two months experienced healed ulcers after four repeated CAP treatments.

Finally, two independent studies reported the possibility of concurrent CAP device use with commonly used octenidine disinfectant for chronic leg ulcer treatment. The authors considered that combined use of cold plasma and this disinfection should achieve better results than separate application [51,52]. CAP applications for chronic wound healing are listed in Table 1.

Table 1. List of cold atmospheric plasma (CAP) applications for chronic wound healing.

Wound Type	Number of Patients/Subjects	Plasma Device Type/Injected Gas	Result	Exposure Time
Chronic leg ulcers [41]	$n = 36$	MicroPlaSter plasma torch/Ar	Faster wound healing, microbial load reduction	5 min/day
Chronic ulcers [42]	$n = 24$	MicroPlaSter alpha and beta plasma torch/Ar	Faster wound healing, microbial load reduction	2 min/day
Chronic venous leg ulcers [47]	$n = 14$	PlasmaDerm(®) VU-2010 DBD device	Strong antimicrobial effect, rapid ulcer size reduction	45 s/cm^2 (max. 11 min)/3× per week/8 weeks
Chronic pressure ulcers [48]	$n = 50$	Plasma Jet/Ar	Better PUSH score, microbial load reduction	1 min/cm^2/1× per week/8 weeks
Pyoderma gangrenosum [50]	$n = 2$	Plasma jet device with variable electrode types	Gradual drying, absorbing, and wound healing	>5 min until all area was not irradiated/every second day/6 and 8 times
Chronic leg ulcers due to diabetes [50]	$n = 1$	Plasma jet device with variable electrode types	Gradual healing of wound	>5 min until all area was not irradiated/every second day/3 times
Pressure ulcers in Wistar rats [49]		Plasma jet/He	Rapid re-epithelialisation, angiogenesis, and collagen synthesis	30 s/3× per day/5 days

Ar: Argon, He: Helium.

The primary intention of CAP use was to provide a modern form of wound disinfection. However, CAP treatment has already proved beneficial in wound healing due to increased cutaneous microcirculation, monocyte stimulation, keratinocyte and fibroblasts proliferation, and cell migration [26,27]. Of these, the keratinocytes and fibroblasts are especially important in the later wound-healing phases [26].

The positive effect of CAP application on keratinocyte and fibroblast proliferation has been demonstrated on cell lines in vitro [53]. The HAcaT keratinocyte cell lines and MRC5 fibroblast cell lines had increased mobility even after brief CAP exposure. This involved decreased gap-junction protein activity at the molecular level, and changes in adherent junctions and cytoskeletal dynamics with concurrent E-cadherin and integrins down-regulation. The in vivo plasma effect in mouse models was also investigated in that study, and the authors demonstrated that the observed wound healing is due to formation and action of NO, UV radiation, ROS, and RNS [53]. Several other studies also confirmed the CAP effect on keratinocyte activity. There was simultaneously increased b1-integrin expression and decreased E-cadherin and EGFR expression in these cells following CAP exposure [54]. Moreover, keratinocytes had significantly increased levels of ROS after CAP application, which resulted in various cell adaptation mechanisms [55]. More than 260 genes, including those coding for cytokines, growth factors, and antioxidant enzymes were differentially expressed. In addition, the HSP-27 cell protective heat-shock protein, which participates in regulating cell development and differentiation, was also highly expressed [55]. Schmidt et al. [56] indicate that the p53 cascade should be a major hub of cold plasma–cell interactions in keratinocytes. The authors also consider that the ATM and ATR redox sensors have higher activity, and that MAP kinase signalling should modulate the p53 signals.

A recent study [57] established interesting 'cross-talk' between CAP-affected fibroblasts and keratinocytes. A co-cultured model of these cells showed that plasma exposure initiated higher HIPPO pathway activity and that the transcriptional coactivator of this pathway YAP was significantly up-regulated. Moreover, the downstream effectors of this pathway and YAP target genes, *CTGF* and *CYR 61* [58], were more highly expressed, but this was only seen in fibroblasts. This increased expression, however, could be reversed by antioxidants application. Most importantly, increased HaCat keratinocyte cell mobility occurred only when they were incubated with CAP-treated fibroblast-conditioned medium. Finally, the authors presumed that CTGF and Cyr61 secretion from fibroblasts activates keratinocytes via paracrine signalling. Evidence for this claim includes that the HaCat lines had greater mobility after exposure to recombinant CTGF and Cyr61, even when CAP treated fibroblasts were not present.

It is interesting that cold plasma in animal mouse models accelerated wound healing in second- and third-degree burns. This was mainly due to increased angiogenesis. At the molecular level, NO formation was accelerated and the expression of the PDGFRβ and CD31 pro-angiogenic markers increased. There was also increased TGFβ1 activity and VEGFA/VEGFR2 signalling stimulation [59–61]. The positive CAP effect was also proven in rat model healing of chemical burns following rat-skin exposure to sulphuric acid. Here, 40 s exposure to CAP each day accelerated healing. The CAP treated wounds almost disappeared after 21 days, while untreated wounds remained clearly visible. The authors also noted that the biochemical profile was altered in CAP treated wounds. This was apparent in the changed levels of the following oxidative stress markers. The malondialdehyde levels were increased, and the reduced glutathione, glutathione peroxidase, and catalase levels were lower in the CAP group than those in unaffected controls, those treated with plasma modified polyurethane wound dressing, and those in the natural wound recovery group. The white blood cell dynamics and the complement component 3 and fibrinogen production differed in these groups during the 21 day treatment. White cell counts were generally highest in the natural wound recovery group and lowest in the CAP group. The fibrinogen and complement component 3 concentrations were similarly highest in the natural wound recovery group and lowest in the CAP treated group [62,63]. Betancourt-Angeles et al. then demonstrated that cold plasma accelerated human burn healing [64]. Pain and itching were relieved after a three-minute CAP application, and a further three-minute repeat application after 16 h produced significantly accelerated healing and new tissue formation. This was a case report study and the authors did not investigate the molecular mechanisms during or after plasma administration. Previous research, however, supports the expectation that angiogenesis and growth factor activation had occurred.

CAP effects on different acute wounds has also been investigated. One study involved patients with different sizes of lower extremity skin transplants. The patients were divided into two groups, with one group receiving a placebo and the other under-going CAP treatment. The final results showed that the CAP group patients had a much better healing course the second day after CAP application [65].

A further interesting study investigated the potential of plasma application to dog bite wounds [66]. Here, the authors analysed the effect of CAP on the following bacterial strains typically found in dog saliva: *Staphylococcus pseudintermedius*, *Staphylococcus aureus*, *Streptococcus canis*, *Pseudomonas aeruginosa*, *Pasteurella multocida*, and *Escherichia coli*. CAP exerted a strong in vitro antiproliferative effect on all these bacteria. However, some differences were noted between the strains, dependent on bacterial growth phase and treatment length. The author's previous similar study [67] revealed that the overall CAP decontamination effect was lower than both polyhexanide-biguanide and saline lavages. These differences, however, were not statistically significant and the employed disinfectant did not affect the overall healing course. Therefore, further analysis would be beneficial to establish definitive conclusions on CAP use in treating dog bite wounds. However, it can be expected that the current potential for CAP application in these acute wounds will increase, especially when these bacterial strains' increasing resistance complicates treatment. Another study involved healthy volunteers who underwent voluntary laser-ablative skin lesions. Their inflicted injuries were then treated with CAP application, and the authors' planned CAP treatments for 30 s intervals over three days was sufficient to improve wound healing [68]. Further work by Nishijima et al. [69] studied cold plasma's ability to accelerate healing in acute surface wounds created by a fractional CO_2 laser. The standard clinical treatment for these wounds is topical ointment application, including steroids, petrolatum, basic fibroblast growth-factor sprays, and gels containing fullerene [69]. The study volunteers were divided into four groups: (1) no treatment, (2) treatment with CAP, (3) applied ointments containing betamethasone valerate, and (4) applied basic fibroblast growth factor sprays. Although no significant differences were observed in the groups' overall wound healing, the CAP group had the most reduced skin redness and mean roughness. Gao et al. [50] investigated the effect of cold plasma on two traumatic wounds. The first patient had used inappropriate self-treatment with glucocorticoids and mupirocin, which resulted in self-inflicted secondary eczema with exudation and crusting. Treatment with CAP for 20 min every second day ceased wound exudation, and visible wound healing was visible after three consecutive treatment rounds. The second patient's wound had been treated with antibiotics without success, and complete wound healing was observed after three consecutive rounds of CAP application. Finally, this study included a patient with a giant genital wart treated with CAP. The patient had this wart for three months during immuno-suppressive drug treatment, and this caused overall reduced immunity. The ablation site was treated with CAP for 40 min after wart removal and this was then repeated after a further three days. Complete healing was accomplished after two repetitions. CAP applications for acute wound healing are listed in Table 2.

Table 2. List of cold atmospheric plasma (CAP) applications for acute wound healing.

Wound Type	Number of Patients/Subjects	Plasma Device Type/Injected Gas	Result	Exposure Time
Burn wound [64]	$n = 1$	Plasma jet/He	Decrease in pain and itching after first application, re-epithelization after second application	3 min/2 times with 16 h between 1st and 2nd applications
Wounds at the donor skin graft sites [65]	$n = 40$	Plasma jet/Ar	Rapid healing of skin graph donor sites in CAP-treated patients	2 min every day/7 days
CO_2 laser skin lesion [68]	$n = 20$	kINPen®MED plasma jet/Ar	Improved scar recovery, no side effects of plasma demonstrated	3–10 s/3 days
Fractional CO_2 laser skin wounds [69]	$n = 12$	kINPen MED® plasma jet/Ar	Wound healing effect similar to standardly used treatment, but with reduced redness and mean roughness of the skin	
Traumatic wound [50]	$n = 2$	Plasma jet device with variable electrode types	Stopping of wound exudation, complete wound treatment after three healing procedures	20 min for whole wound/every two days/three repetitions of healing
Wound after genital wart [50]	$n = 1$	Plasma jet device with variable electrode types	Gradual healing of wound	>5 min until all area was not irradiated/every second day/2 times
Burn wounds in animal models [59–63]	$n = 15$ [60], $n = 12$ [62,63]	Plasma jet/He [59], N_2/Ar [61], Microplasma jet/He [60], Plasma jet/He [62,63]	Anti-inflammatory effect, re-epithelisation, angiogenesis, collagen rearrangement [59–61], increased speed of wound healing, changes in biochemical and haematological profile in plasma treated group [62,63]	1–2 min/8 h interval/5 days [60] 40 s per day/21 days [62,63]
Dog bite wound [66]		KinPEN®VET plasma jet/Ar	Potential antimicrobial effect on bacterial strains typically presented in dog saliva and dog bite wounds	<2 min of exposition under in vitro conditions

Ar: Argon; He: Helium; N_2: Nitrogen.

3. Cold Plasma in Dental Medicine

Standard cleaning procedures and oral cavity disinfection is based on laser device use, mechanical infection removal, or using antimicrobial solutions. The first two procedures, however, may cause thermal or mechanical damage to tissues. This risk can be significantly decreased with CAP devices [70]. A further CAP advantage is that its discharge can be relatively easily applied to uneven surfaces and inaccessible oral cavity sites. The discharge from sufficiently miniature devices can also be applied directly to the dental canal [71]. Finally, in contrast to liquid antimicrobial solutions, CAP can be applied to discrete oral cavity sites, and unpleasant side effects from microbial solution use are not noted after CAP treatment [70,72].

In addition, bacterial strains often present in dental biofilm have increasing resistance. Neglecting dental biofilm removal can lead to serious diseases including aspiration pneumonia, endocarditis, and other systemic disorders [73,74]. Delben et al. [73] demonstrated that CAP application had significant antimicrobial effect on *Candida albicans* and *Staphylococcus aureus*, which are commonly present in dental biofilm. The reduction in microbial load with CAP device was also comparable to that for penicillin G or fluconazole administration. A similarly focused study confirmed the antimicrobial effect of cold plasma on a broad spectrum of bacteria in dental plaques, most importantly *Streptococcus mutans* [75]. Moreover, the resultant CAP antiproliferative in vitro effect on cultured cell lines was greater than that of standard chlorhexidine disinfectant use [75].

Several studies have also investigated the application of CAP devices to reduce microbial load in the dental canal. Here, electron microscope scanning revealed disappearance of the bacterial biofilm to 1 mm depth following 5 min ex vivo CAP application [76]. Armand et al. [77] then simulated *Enterococcus faecalis* infection in a sample of 100 extracted and disinfected teeth. This bacterium is relatively common in inflamed dental canals [78]. He/O_2 plasma was the most effective in reducing the microbial load, as confirmed by electron microscope scanning (SEM), and He plasma had approximately the same efficacy as photodynamic therapy. However, the authors pointed out that the resulting effect was largely influenced by the shape of the dental canals. There was greater effect in the straight canals. Shahmohammadi Beni et al. [79] then provided a beneficial study on the possibilities and limitations of CAP application to the oral cavity. Therein they discussed CAP application related to oral cavity surface structures and they concentrated especially on the dispersal of OH radicals.

In addition to disinfection, CAP has other demonstrated benefits in dental medicine. Dong et al. [80] recorded that plasma has a positive effect on superficial dentine by increasing its binding strength to other dental structures. Moreover, CAP application should have a positive effect on strengthening links between fibre-reinforced posts and composite core material [81]. Cold plasma also has significant effect on titanium structures in the oral cavity. Here, CAP should increase hydrophilicity and also the roughness of titanium surfaces [82]. Both these aspects can improve cell adhesion, proliferation, and differentiation of osteoblasts, and these three factors can provide faster osteointegration [83]. Although the roughness increased by CAP application can induce greater bacteria accumulation [84], CAP's antimicrobial ability can minimise this negative effect. These aspects make CAP devices very suitable for the treatment of peri-implant inflammation. CAP also affects zirconium structures. Yang et al. [85] recorded decreased microbial load and increased hydrophilicity after CAP application, and there was no change in zircon structure topology after CAP application. Finally, a study on the effects of the three types of devices on dental plaques provided interesting results. It was especially noted that there is no currently recognised dental plaque bacterial strain with resistance to CAP treatment [86]. Effects of CAP application to dental and oral cavity structures under in vitro and ex vivo conditions are listed in Table 3.

Table 3. List of cold atmospheric plasma (CAP) application effects to dental and oral cavity structures under in vitro and ex vivo conditions.

Purpose of CAP Application	Device/Injected Gas	Final Effect	Exposure Time
Dental biofilm reduction [73]	Kinpen MED ® plasma jet/Ar	Antimicrobial effect on *Candida albicans* and *Staphylococcus aureus*, proved no side effects on reconstituted oral epithelium	60 s/sample
Dental biofilm reduction on titanium discs [75]	Three different types of CAP devices: (a) kINPen plasma jet/Ar; (b) HDBD device/Ar; (c) VDBD device/Ar	Antimicrobial effect on wide spectrum of saliva bacterial species, especially *Streptococcus mutans*	1–10 min/sample
Dental canal disinfection [76]	Plasma jet device/Ar/O_2	Complete reduction of microbial infection in dental canal ex vivo to a depth of 1 mm	5 min/one extracted tooth
Dental canal disinfection [77]	Plasma jet device/He; He/O_2	Significant reduction of *Enterococcus faecalis* contamination ex vivo, better effectivity when He/O_2 used as a carrying gas	2–8 min/sample
Improvement of dental structures [80–82,85]	Plasma brush/Ar [80]; low-pressure plasma device/O_2, Ar, N_2, and He+N_2 [81]; HDBD device/Ar [82]; plasma jet device [85]	Increased dentin binding [80]; strengthening of links between fibre-reinforced posts and composite core material [81]; increase of hydrophilicity in titanium [82] and zircone [85] structures	30 s/sample [80](62); 10 min [81]; 2–6 min/sample [82]; 30–90 s/sample [85]

Note: Ar: Argon; He: Helium; O_2: Oxygen; N_2: Nitrogen, VDBD: Volume dielectric barrier discharge device; HDBD: Hollow dielectric barrier discharge device.

4. Use of CAP to Activate Proliferation in Stem and Progenitor Cells

It is expected that CAP can also affect stem cell proliferation, and these cells could then be used in regenerative medicine and medical engineering. However, only a few analyses have focused on this possibility. One study by Park et al. showed that CAP devices did actually induce the proliferation of stem cells derived from adipose tissue (ASC) without affecting their vital properties in any way [87]. Those authors reported that stem cell proliferation was 1.57-fold higher as early as 72 h after CAP treatment. The study also reported the influence of NO induced by CAP [87]. This was confirmed by the CAP device's positive increase in stem cell proliferation being dramatically diminished by application of NO scavengers. In addition, when ASC cells that were not exposed to CAP were treated with the DETA-NONOate NO donor, they had a higher rate of proliferation than the control cells, but a significantly lower rate than those affected by CAP. These results indicate that increased NO concentration is a very important contributor to increased proliferation, but it is most likely not the only factor.

No cytotoxic or other negative impacts were observed in osteoprogenitor MC3T3-E1 cells following CAP application. Moreover, CAP induced accrued NO levels. This NO was introduced into the intracellular spaces, and its level in cultured cells could be controlled. There was consequent induction of early osteogenic differentiation in these conditions, and this was achieved even in the absence of pro-osteogenic growth factor [88]. A further study on CAP's effect on osteo-differentiation addressed its impact again on MC3T3-E1 mouse osteoblast precursor lines [89]. There, the CAP stimulatory effect on osteogenic differentiation was comparable to that determined in the osteogenic differentiation medium. The authors also observed significant changes in the molecular cascades following CAP influence. Although PI3K/AKT and MAPK signalling decreased, there was increased expression of the osteogenic genes *RUNX2*, *OCN*, *COL1*, and *ALP*. Furthermore, CAP affected the dephosphorylation of FOXO1, which is important for proliferation and redox balance in osteoblasts and a very important controller of bone formation [90]. Although progenitor cell lines differ from stem cells in some respects, it cannot be discounted that this similar mechanism can also occur in stem cells when CAP treatment initiates osteo-differentiation.

The effect of CAP on human mesenchymal stem cells isolated from periodontal ligaments (hPDL-MSCs) was also analysed. Again, no negative or cytotoxic effect was observed. Furthermore, this CAP treatment inhibited hPDL-MSC migration and induced detachment without loss of overall cell viability. However, although cell viability was not affected, the authors stated that the cells had prolonged population doubling-time after CAP application. Most importantly, CAP promoted osteogenic differentiation in this instance [91].

Use of CAP devices can also initiate proliferation of hematopoietic stem cells and bone marrow-derived stem cells. This proliferation almost doubled in both these cell lines compared to untreated cells [92]. The authors also observed that expression of typical stem cell markers CD44 and CD105 was almost five-times higher in bone marrow cell lines, and that there was increased expression of *OCT4*, *SOX2*, and *NANOG* genes in both cell lines. Finally, the expression of genes that act on G1-S cell-cycle transition also increased and it can be presumed that CAP application can influence this cell cycle phase [92].

Interestingly, the CAP device effects may include initiating stem cell and biomaterial binding, in addition to its beneficial direct application to cells. For example, Alemi et al. [93] assessed whether cold plasma application improved stem cell and biomaterial scaffold adhesion in cartilage tissue engineering. They found that CAP enhanced scaffold surface hydrophilicity, and this significantly reduced the contact angle and helped initial cell binding.

Neurodegenerative diseases and traumatic central nervous system damage are currently difficult to treat. However, neural stem cells (NSC) could improve their treatment, and several studies have investigated NSC proliferation induced by CAP. They found that C17.2-NSC murine NSC proliferated and differentiated significantly faster following CAP treatment [94]. Moreover, almost 75% of neural stem cells differentiated into neuronal cell lines after exposure to CAP, and this percentage is greater than that achieved by specific growth factors. The neurons that differentiated from NSCs after CAP application had high β-tubulin III protein expression levels, and this is considered a typical marker for this cell type [95]. However, recent research indicates that high β-tubulin III expression is also typical in other stem cell types [96]. Furthermore, authors in this study observed only moderate differentiation from NSC to oligodendrocytes, and typical O4 protein marker expression [97] was only slightly higher. These same authors then found that astrocytes exhibited no disparity in differentiation between affected and nonaffected cells, and their typical GFAP protein marker was only minimally expressed [94,98]. Jang et al. [99] also showed that CAP has a positive effect on neural stem cell proliferation, and they recorded the following mechanisms: (1) The excited atomic oxygen in plasma phase initiated ROS and RNS formation, (2) these then interacted with reactive atoms in the extracellular liquid phase to form NO, which induced reversible cytochrome c oxidase inhibition in the mitochondrial complex IV. This resulted in increased mitochondrial O_2^- production and finally (3) cytosolic hydrogen peroxide formed by O_2^- dis-mutation acted as an intracellular messenger to specifically activate the Trk/Ras/ERK signalling pathway. Here, the authors considered that mitochondrial O_2 and cytosolic H_2O_2 must act cooperatively because experimental cytosolic increase in H_2O_2 was not solely sufficient to initiate differentiation. However, some aspects, mainly how the phosphorylation of specific sites in tyrosine kinase receptor signalling occurred, remain unknown. Finally, the CAP induced neural differentiation had advantages over previously used neural differentiation inducers including retinoic acids and resveratrol. This CAP treatment enabled neurites to reach maximum length more rapidly, the differentiation efficiency increased, and almost 70% of differentiated neurons were identified as catecholaminergic DA neurons.

Bourdens et al. [100] provided different results to the previous studies. These authors recorded that, although human dermal fibroblast and adipose-derived stromal cells did not lose viability after 3 min CAP exposure, these cells developed a senescence phenotype associated with a glycolytic switch and increased mitochondria content. Most importantly, the authors noted that the observed proliferation arrest was accompanied by increased p53/p21 damage. Despite this, cell lines maintained some functional properties such as differentiation potential and immunomodulatory effects. Discussion of the role of senescent cells in wound healing has led authors to claim that CAP exposure aimed at provoking a small amount of transient senescence in some cells could be considered to enhance tissue regeneration. The truth of this claim, however, requires further analyses. CAP application effects on stem cells and progenitor cells under in vitro condition are listed in Table 4.

Table 4. List of CAP application effects on stem cells and progenitor cells under in vitro conditions.

Type of Cells	Device/Injected Gas	Result	Exposure Time
Stem cells derived from adipose tissues [87]	DBD device/He	2-fold elevated proliferation of stem cells in vitro after CAP treatment, higher levels of NO, higher activity of Akt, ERK1/2, and NF-κB pathways in these cells	50 s per hour/10 times
Osteoprogenitor (MC3T3-E1) cells [88]	DBD discharge NO-plasma nozzle system	Increase of NO in control media and possibility of its introduction into intracellular space, no cytotoxic effects on cells	30–180 s
Osteoprogenitor (MC3T3-E1) cells [89]	DBD NBP device	Decrease in PIK3/AKT and MAP signalling, increase in p38 signalling, dephosphorylation of FOXO1 transcriptional factor	1–10 min
Human mesenchymal stem cells isolated from periodontal ligaments [91]	Plasma needle/He	Reduced migration of cells, loss of adhesivity, osteo-differentiation of these cells	10–120 s
Hematopoietic stem cells; bone marrow-derived stem cells [92]	DBD Device/He	Increased proliferation of cells, higher expression of surface markers CD44 and CD105, higher expression of *Oct4*, *Sox2*, and *Nanog* genes, effect of CAP on G1-S transition	50 s per hour/0 times
Murine neural stem cells (C.17-2.NSC) [94]	Plasma jet He/O_2	Higher proliferation and differentiation levels of cells, main portion of neural cells differentiated to neurons	60 s
Murine neuroblastoma stem cell (N2a) [99]	DBD device/O_2+N_2	Higher proliferation of cells after CAP exposure, Increase of NO induced inhibition of cytochrome c oxidase, activation of Trk/Ras/ERK pathway	1–10 min
Adipose-derived stromal cells [100]	DBD device/He	Senescence phenotype of cells, proliferation arrest of cells, increase in p53/p21 damage, morphological changer typical for changes in p16 activity	3 min/sample, incubation one hour in PAM

He: Helium; O_2: oxygen; N_2: nitrogen; NO: nitric oxide; NBP: non-thermal biocompatible plasma; DBD: dielectric barrier discharge; PAM: plasma-activated medium.

5. Use of CAP to Regenerate Other than Dermal Tissues

While plasma discharge treatment has focused to a great extent on dermal structures, there is also significant benefit in its application to deeply embedded organs and structures. A study on the use of CAP in the regeneration of nondermal structures assessed nasal mucosa regeneration [101]. Absence of this structure presents a serious problem, and its regeneration is complex and time-consuming. In addition, there is currently no effective way to accelerate its regeneration following severe injury or surgical intervention [101].

It was possible to observe the potential of cold plasma use on nasal mucosa regeneration under in vitro and in vivo conditions. Although the in vitro analyses were performed on BEAS-2B bronchial epithelial cell lines, these are very similar to nasal mucosa cells, and the regeneration complications are similar in severity. The CAP application to the bronchial cells increased cell proliferation and migration [101]. This migration was achieved by increased EGFR receptor activity and EMT signalling. Finally, rapid recovery of nasal mucosa was recorded in mouse models after brief in vivo CAP application [101].

Further effects of CAP devices on S9 bronchial epithelial stem cells have also been noted. CAP use provoked differences in the protein expression profile, and its long- and short-term exposure effects on cells also differed [102]. Further determination of CAP device effects on these cells included (1) induction of the Nrf2-mediated oxidative and endoplasmic reticulum stress response, (2) PPAR-alpha/RXR activation, (3) production of peroxisomes, and (4) prevention of apoptosis in the first hour after CAP administration. It is paradoxical that CAP initially acted as a stress factor on these cells, but subsequently triggered significant cell proliferation and cellular assembly and organisation.

Positive cold plasma effects have also been recognised in the regeneration of several neural cell types isolated from animal models [103]. While CAP stimulated neuronal regeneration and astrocyte growth in vitro, a further increased dosage induced loss of cell culture viability. Therefore, precise determination of the intensity and duration of CAP exposure is required for the treatment of both superficial and deeper structures.

6. Use of CAP for Tumour Treatment

CAP is a promising option for more effective tumour treatment. Its final effect on cancer cells is, however, quite interesting. Cancer cells produce higher amounts of ROS and RNS and although this increases their proliferative activity, even higher levels can cause apoptosis [104]. CAP then increases this to a level lethal to the cancer cell. This effect is caused by changes in their antioxidant system and double-strand breaks, but healthy cells should tolerate this increase [92]. The tumour cells also have more aquaporins [105] and this facilitates ROS and RNS penetration into the cell [106]. The degree of ROS and RNS diffusion into the cell can also be influenced by the membrane's lipid structure [107]. Cancer cells generally have lower levels of cholesterol and this makes cells more susceptible to peroxidation [108,109]. Membrane lipid peroxidation then causes the cell membrane to form more pores so that ROS and RNS can diffuse to a greater extent [109]. A summary of the mechanisms involved indicates that various cellular responses are initiated by CAP application to tumour cells. This increase in reactive species is the main trigger that initiates tumour cell death. The specific responses here are apoptosis, growth inhibition, cell cycle arrest, DNA and mitochondrial damage, and even immunogenic cell death [110]. The actual result, however, is dose-dependent [111–113].

Keidar et al. performed a pilot experiment to test CAP use in tumour treatment. They initially observed that skin melanoma cell lines separated from the culture vessel after CAP application, and this led to their reduced numbers, while the healthy cell lines remained adhered [114]. The authors' study of subcutaneous kidney tumours in mice showed that small tumours were ablated following CAP application and more advanced tumour size was reduced. Kaushik et al. [115] then demonstrated the effects of ROS on tumour cell mortality by assessing the plasma discharge effect on T98G, A549, HEK293, and MRC5 cell lines. The authors showed that the viability of HEK293 and MRC5 non-malignant cells was minimally affected compared to cancer cell lines. The ROS and H_2O_2 generated by plasma discharge altered the mitochondrial membrane potential. This initiated the intrinsic apoptotic pathway and resulted in increased overall pro-apoptotic gene expression and decreased anti-apoptotic gene expression. There was also a change in ERK1/2/MAPK cell signalling activity at the protein level. The resultant effect can be reversed with ROS scavengers.

Cancer cells also differ from healthy cells at the metabolic level. Here, cancer cell metabolic reprogramming assimilates simple carbons into macromolecules, and especially lipid, protein, and nucleic acid macromolecules. This results in intermediate metabolites that tumorous cells can use for growth and proliferation [116–118]. Xu et al. [119] studied CAP effects on altered cancer cell metabolic activity. Their KEGG and GC-TOFMS analyses highlighted that leukaemia cells had different alanine, aspartate, and glutamate metabolism. The authors also recorded reduced glutaminase activity in cancer cells after CAP application. This decreases the conversion of glutamine to glutamic acid. Deficiency of

Glutamic acid, simultaneously with glutamine accumulation can lead to suppression of proliferation and even cell death in leukemia cells [119].

Studies on cell lines and animal models have repeatedly demonstrated the benefits of CAP devices and their anti-tumour activity. One study reported that glioblastoma cell lines lost their viability after plasma treatment [120]. This study also showed that the cell lines previously resistant to a temozolomide alkylating agent regained sensitivity to it [120]. In addition, several authors have reported that CAP exposure caused lost viability and induced apoptosis in other types of brain tumour cell lines [121–123].

Induced apoptosis has been documented in lung cancer TC-1 cell lines after plasma discharge. This has also been recorded in fibroblast lines, but to a significantly lesser extent [124]. However, the main benefit of that last study was the expressed concept of minimising the size of plasma devices to enable easier CAP use in deeper lesions and structures. Herein, apoptosis induction was triggered by plasma discharges generated from devices with 125 to 440 µm diameter. µCAP devices [125] have been used directly on mouse brain tumours and also in vitro on glioblastoma cell lines. In the latter instance, the plasma discharge from a 70 µm device increased ROS and RNS levels and this resulted in marked reduction in U87MG glioblastoma cell viability. Most importantly, this device was also able to apply plasma discharge through an intracranial endoscopic tube to mice brains, which resulted in tumour growth suppression [125]. The authors then optimised the device parameters for the most suitable treatment of brain and breast tumours [126].

The in vivo plasma effect has been demonstrated in animal models subcutaneously injected with 4T breast cell lines. The mouse body tumour growths from these cell lines were exposed to plasma discharge from a 250 µm diameter CAP device. Importantly, plasma administration for 3 min achieved tumour growth reduction comparable with chemotherapy. There was also significant change in the ratio of pro- and anti-apoptotic gene activity at the molecular level [127].

Mashayekh et al. [128] then studied the in vivo effect of CAP on mouse models and the in vitro influence on melanone B16/F10 cell lines. Their work demonstrated lost viability in most cell lines and marked shrinkage of the animal model tumours. The cell-line results were achieved in 48 h after 3 min CAP treatment and in vivo tumour shrinkage was comparable to that achieved by chemotherapy. In summary, analyses of CAP effects on skin tumour treatments were the first performed because plasma is most easily applied directly to dermal structures. For example, G361 melanoma cells lost viability and detached from the surface after CAP application. These cells had lower integrin and FAK expression and altered actin filament structure. This result supports the hypothesis that plasma-facilitated cell death may be related to integrin–ECM interactions [129]. It was also possible to selectively increase the CAP antiproliferative effect on melanoma cell lines using plasma with gold nanoparticles bound to the anti-FAK antibodies [130].

MCF-7 breast cancer cell lines were also subjected to analyses. The initial results indicated reduced cell viability after CAP treatment due to increased apoptosis [22]. Ninomiya et al. [131] then demonstrated that CAP induced injury to 50% of cancerous breast cell lines, regardless of whether they were invasive MB-231 or non-invasive MCF-7 cell lines. Finally, CAP antiproliferative effects were also noted in cell lines derived from human breast cell metastases [131].

Further research was conducted on the HCT 116, SW480, and LoVo colon cancer cell lines. These cells lost viability after CAP application [132], and this effect was accompanied by decreased cell mobility and a higher degree of B-catenin phosphorylation. In addition, although the CAP device was considered to promote apoptosis in up to 70% of B16 and COLO320 colorectal cell lines, there was no effect on the macrophage cell line controls [133]. Ishaq et al. [134] then reported that CAP treatment increased apoptosis in various colorectal cell lines through activation of the Nox2-AKT1 pathway. The authors also demonstrated that although HT29 cell lines are largely resistant to ROS-mediated cancer death, this resistance can be reduced by inhibiting Nrf2/Srx signalling. Finally, CAP exposure has antiproliferative effects on multicellular tumour spheroids, which should mimic the tumour microenvironment in a dose-dependent manner. The loss of spheroid Ki67 and accumulated DNA damage were observed following this CAP exposure [135].

In vitro experimental results also suggest that CAP application should benefit neck and head cancers. While tumours in this area are relatively easier to remove, this procedure is often invasive and CAP should reduce this effect. Guerrero-Preston et al. [136] found that CAP also acted in an antiproliferative manner on these cells. Although these authors considered that this resulted from non-apoptotic processes, contrasting studies suggested that the antiproliferative effect emanated from apoptotic cascade activation [137,138].

Several studies have also assessed CAP application on cervix uteri tumours, primarily on HeLa cells. Apoptotic processes in these cells were observed after CAP treatment due to increased ROS and subsequent changes in the JNK and p38 pathways [139]. A further observed mechanism for this effect was that membrane lipid oxidation caused cell collapse [140]. In addition, Tan et al. [141] employed a microdevice with approximately 1 μm electrode diameter and this selectively induced apoptosis in particular HeLa cells without affecting neighbouring cells.

The possibility of CAP device application to leukaemia cell lines has also been assessed. However, the hypothetical application of CAP in clinical conditions remains unclear, and establishing the best method for its application requires further investigation. Here, the use of CAP-activated liquids could also be considered. However, CAP induced in vitro cell death in THP-1 leukaemia cell lines in a dose dependent manner [142]. CAP application also caused uncontrolled necrotic cell death [143]. The authors then established induced cell apoptosis 45 s after exposure to CAP, and necrosis was observed after treatment for more than 50 s. Finally, two independent studies also demonstrated antiproliferative effects on pancreatic cells [144,145].

CAP treatment has also been directly applied to human tumours in clinical settings [146]. This study involved six patients with advanced neck and head cancer. Two of these patients had initial tumour size reduction after CAP application. Although one patient achieved at least partial remission and improvement, the other relapsed and died. Two patients reported no CAP treatment side effects, but the remaining four had increased dry mouth. Most importantly, four patients reported a significant reduction in pain. These patients, however, were in terminal stages, and five had passed away by the end of the study. Although cold plasma application did not reverse the disease course, it had an overall positive effect without significant harm or side effects.

Plasma discharge has been shown to directly and indirectly affect the cell culture medium [25], and this has encouraged consideration of CAP-activated plasma or other liquid use to reduce tumour cell viability. The greatest advantage of plasma-activated medium (PAM), plasma-activated Ringer's lactate solution (PAL), and other plasma-activated solutions is that they can be stored for a relatively long time. It is also hypothetically believed that activated liquids can be applied to internal structures with relative ease. CAP-activated media generally induce apoptosis due to ROS, RNS, and H_2O_2 content [147]. The resultant effect, however, may be cell-specific and dependent on parameters such as the amount of aquaporins in the cell line [148].

Apoptotic morphological changes have been particularly observed in glioblastoma cell lines following CAP-induced media application. Increased effector caspase 3/7 activity and decreased AKT kinase expression have also been observed at the molecular level [122]. Similar results were obtained when gastric cancer cell lines were affected by PAM [149]. These cells began to show morphological changes associated with apoptosis. Caspase 3/7 activity was increased and the proportion of annexin V positive cells in the affected group was significantly higher than in the control group after two hours PAM exposure. The experimental group also had higher ROS, but this could be eliminated by ROS scavengers. Moreover, cells seeded at 1×10^4 per well showed a degree of resistance to PAM, but this subsided after exposure for more than two minutes. There was more highly-expressed CD44 variant 9 in these resistant cells. This variant is important in the synthesis of reduced glutathione, which is necessary for free radical neutralisation and appropriate responses to stress signalling [150].

The cell viability of ovarian cancer cell lines resistant to paclitalex and cis-platin decreased by almost 30% after CAP treatment [151]. The authors also analysed CAP effects on mouse xenografts. Several chemo-resistant cell lines with serum-free medium and matrigel were initially inoculated into mice, and 200 µL of plasma-activated medium and untreated control RPMI-1640 medium were then inoculated. Dependent on cell lines, the tumours decreased by 66% and 52% in 29 days after the first CAP injection compared to control levels. In addition, PAM application to four types of pancreatic tumour lines also increased ROS levels. This led cells to cell apoptosis, which could be reversed by ROS scavengers [145]. The study also assessed the efficiency of PAM on mouse xenografts, and significant tumour mass shrinkage was noted after 28 days. Moreover, authors Nakamura et al. [152] found that in vivo PAM medium exposure to a different mouse xenograft model inhibited peritoneal dissemination in ES2 cancer cells.

In addition to PAM, PAL can also be used for anti-tumour treatment. PAL induces apoptosis under in vitro conditions in pancreatic cell lines by increasing ROS levels and decreasing cell adhesion. PAL also had anti-tumour effect on peritoneal nodules from these cell lines in in vivo mouse xenograft models [153]. In addition, PAL application exerted strong antiproliferative effects on A549 lung cancer cell lines [154]. In that work, CAP application initiated mitochondrial dysfunction connected with downregulation of the NF-κB-Bcl2 molecular pathway [154]. The authors considered that PAL's final antiproliferative effect on tumour cell lines should be stronger than PAM. This consideration, however, is open to debate and requires further analysis. Regardless of that outcome, PAL's major advantage is that it will be potentially easier to apply in future clinical trials because currently cultivated media cannot be used in medical treatment [154]. Although this provides an appropriate alternative to current methods, it still requires further study. Finally, analyses of PAM effects on tumour cells should also continue, despite possible complications in its use in clinical trials. These analyses may discover additional phenomena and principles beneficial for clinical procedures.

One of the most advanced uses of CAP is its application in cancer immunotherapy. This is possible because the onset of cancer and its course can be significantly influenced by the human immune system. The immune system's modulation ability particularly overcomes the cancer cells' capacity to supress immune responses [155]. These modulations include the use of various cytokines, cell-based therapies, and immune checkpoint blockades [147]. Moreover, some radio- and chemotherapy procedures trigger immunogenic cell death (ICD) [156,157]. Cells damaged or altered by radio- or chemotherapy produce 'damage-associated molecular pattern signals' (DMP). The DMP molecules then lead the immune system to destroy those cells. Most importantly, some studies also suggest that CAP induces ICD, and this leads to subsequent macrophage stimulation [38,158,159].

Miller et al. [160] and Almeida [161] described the detailed principles and possibilities of plasma use in immunotherapy. In particular, the work by Lin et al. [162] demonstrated that CAP should induce ICD in vitro in A549 lung cancer and radio-resistant nasopharyngeal carcinoma cell lines. Therein, extracellular ATP secretion was followed by ICD macrophage stimulation. In addition, CAP increased extracellular ROS levels with a resultant increase in calreticulin production. CAP-based immunotherapy is also a promising option for glioblastoma multiforme treatment. This is a very aggressive cancer with immune system downregulation [161]. Finally, Cheng et al. [163] highlighted the following: 1) 30 s CAP stimulation of macrophages resulted in increased production of IL-6 and IL12 and decreased anti-inflammatory cytokine IL-10 production and 2.) that IL-2 and IFN-g production increased in isolated T-cell lines from mouse spleen following exposure to CAP. The authors attempted to simulate in vivo conditions as they removed mouse model lymph nodes and exposed them to CAP. They then isolated CD4+ T cells from both the control and affected lymph nodes. The cells derived from the affected lymph nodes had increased function, which was manifested in increased production of IL-2 and IFN-g. The authors' work culminated in showing the strong anti-tumour effect in mice who had CAP-affected T cells transferred to their lymph nodes. Effects of CAP application on variable tumour cell lines are listed in Table 5.

Table 5. List of cold atmospheric plasma (CAP) in vitro applications for tumour treatment.

Cell Line Type	Plasma Device/Injected Gas	Exposure Time
Melanoma cell lines [114,128–130]	Plasma jet/He [114]; Plasma jet device/He [128]; Plasma micro jet/He [129]; DBD device [130]	30 s [114]; 3 min [128]; 15 s [129]; 40 s [130]
Breast cancer cell lines [22,115,131]	Plasma jet [115]; Plasma jet/He+O_2 [22]; Plasma jet/He [131]	150 s [115]; 5–30 s [22]; >30 s [131]
Cervical cancer cell lines [139–141]	Plasma jet [139]; Plasma jet [140]; Plasma microjet [141]	5 min [139]; 10–15 s [141]
Brain tumour cell lines [115,120–123,125] [c]	Plasma jet [115]; Surface micro discharge device (hybrid device) [120]; FE-DBD device [121]; NEAPP device [122]; DBD device [123]; Microplasma jet device/He [125]	150 s [115]; 30–120 s [120]; several minutes of medium treatment [122]; 4 min [123]; 5–120 s [125]
Colorectal cancer cell lines [121,132] [b] [133,134]	FE-DBD device [121]; Plasma torch/He+O_2 [132]; Plasma jet/He and Ar [133]; Plasma jet/He [134]	1–4 s [132]; 60–120 s [133]; 5–30 s [134];
Gastric cell lines [149] [b]	NEAPP jet device/Ar	5 min
Lung cancer cell lines [124] [a] [154] [c]	Microplasma jet device/He [124]; Plasma jet/Ar [154]	3 min of solution treatment [154]
Ovarian cancer cell lines [151] [b] [152] [b]	NEAPP jet device/Ar [151]; NEAPP jet device/Ar	30–300 s [151]; 10 min PAM treatment [152]
Head and neck cancer cell lines [136–138]	Plasma jet/He [136]; Plasma jet/He+O_2 [137]; Kinpen® MED [138]	10–45 s [136]; 20–150 s [138]
Leukaemia cell lines [142,143]	Plasma pencil/He [142]	10 s–10 min [142]
Pancreatic cell lines [144,145] [b] [153] [c]	Plasma gun/He [144]; Plasma jet/Ar [145]; Plasma jet/Ar [153]	10–90 s [144]; 30 s–5 min [145]; 3 min of liquid treatment [153]

(a) Animal cell lines; (b) Plasma-activated medium; (c) Plasma-activated solution; He: Helium; Ar: Argon; O_2: oxygen; DBD: dielectric barrier discharge; FE-DBD: floating electrode dielectric barrier discharge; NEAPP: non-equilibrium atmospheric pressure plasma.

7. Conclusions

CAP has significant potential for wide use in modern clinical practice. Analyses already performed indicate that cold plasma application has been beneficial in many areas of medicine, with no significant negative effect on healthy cells. CAP application, however, may produce potentially harmful elements and should therefore be applied under expert guidance and in appropriate dosage. In addition, CAP use should not be limited only to surface structures, because technology advances are likely to enable development of ever-smaller devices capable of applying plasma discharges to internal structures.

These CAP procedures will most likely prove beneficial in the treatment of internal tissues in addition to cancer treatment. However, current analyses of cold plasma effects still focus only on cell lines and animal models. Although these experiments are more accessible and easier to conduct, wider testing of CAP effects in cancer and regenerative medicine studies in human patients is important and should prove beneficial. The potential introduction of CAP in routine clinical practice in regenerative medicine and oncology could significantly reduce the financial burden and costs of these treatments. Most importantly, a CAP treatment regime will be less invasive and stressful for patients.

Authors Contribution

D.D. and D.B. wrote the original draft; H.Š. and E.H. supervised the project and revised and approved the manuscript. All authors have read and agreed to the published version of the manuscript.

Funding: This study was supported by the Slovak Research and Development Agency Grants No. (APVV-15-0217) and by VEGA Grant No. (1/0178/17) of the Scientific Grant Agency of the Ministry of Education of the Slovak Republic and the Slovak Academy of Sciences.

Conflicts of Interest: Authors declare that they have no conflict of interest.

Abbreviations

Ar	Argon
ASC	Adipose tissue stemm cells
CAP	Cold atmospheric plasma
CO_2	Carbon dioxide
DBD	Dielectric barrier discharge
DMP	Damage-associated molecular pattern signals
FE-DBD	Floating electrode dielectric barrier discharge
He	Helium
H_2O_2	Hydrogen peroxide
hPDL–MSC	Stem cells isolated from periodontal ligaments
ICD	Immunogenic cell death
K	Kelvin
NO	Nitric oxide
NO_2	Nitrogrn dioxide
NO_3	Nitrogen trioxide
N_2O	Nitrous oxide
N_2O_4	Dinitrogen tetraoxide
NEAPP	Non-equilibrium atmospheric pressure plasma
NBP	Non-thermal biocompatible plasma
NSC	Neural stem cell
N_2	Nitrogen
OH	Hydroxyl
O_2	Oxygen
O_3	Ozone
PAL	Plasma-activated Ringer's lactate
PAM	Plasma-activated medium
RNS	Reactive nitrogen species
ROS	Reactive oxygen species
SEM	Scanning electron microscope

References

1. Langmuir, I. Oscillations in Ionized Gases. *Proc. Natl. Acad. Sci. USA* **1928**, *14*, 627–637. [CrossRef] [PubMed]
2. Lee, H.W.; Park, G.Y.; Seo, Y.S.; Im, Y.H.; Shim, S.B.; Lee, H.J. Modelling of Atmospheric Pressure Plasmas for Biomedical Applications. *J. Phys. D* **2011**, *44*, 053001. [CrossRef]
3. Izadjoo, M.; Zack, S.; Kim, H.; Skiba, J. Medical Applications of Cold Atmospheric Plasma: State of the Science. *J. Wound Care* **2018**, *27*, S4–S10. [CrossRef] [PubMed]
4. Bernhardt, T.; Semmler, M.L.; Schäfer, M.; Bekeschus, S.; Emmert, S.; Boeckmann, L. Plasma Medicine: Applications of Cold Atmospheric Pressure Plasma in Dermatology. *Oxidative Med. Cell. Longev.* **2019**, *201*, 1–10. [CrossRef] [PubMed]
5. Adhikari, B.R.; Khanal, R. Introduction to the Plasma State of Matter. *Himal. Phys.* **2013**, *4*, 60–64. [CrossRef]
6. Bittencourt, J.A. *Fundamentals of Plasma Physics*; Springer: New York, NY, USA, 2004; pp. 1–28. ISBN 978-1-4419-1930-4.
7. Chaudhary, K.; Imam, A.M.; Rizvi, S.Z.H.; Ali, J. Plasma Kinetic Theory. In *Kinetic Theory*; InTech: Rijeka, Croatia, 2018; pp. 107–127. ISBN 978-953-51-3801-3. [CrossRef]
8. Sakudo, A.; Yagyu, Y.; Onodera, T. Disinfection and Sterilization Using Plasma Technology: Fundamentals and Future Perspectives for Biological Applications. *Int. J. Mol. Sci.* **2019**, *20*, 5216. [CrossRef]
9. Conrads, H.; Schmidt, M. Plasma Generation and Plasma Sources. *Plasma Sources Sci. Technol.* **2000**, *9*, 441. [CrossRef]
10. Fridman, G.; Friedman, G.; Gutsol, A.; Shekhter, A.B.; Vasilets, V.N.; Fridman, A. Applied Plasma Medicine. *Plasma Process. Polym.* **2008**, *5*, 503–533. [CrossRef]

11. Fiebrandt, M.; Lackmann, J.W.; Stapelmann, K. From Patent to Product? 50 Years of Low-Pressure Plasma Sterilization. *Plasma Process. Polym.* **2018**, *15*, 1800139. [CrossRef]

12. Laroussi, M. Plasma Medicine: A Brief Introduction. *Plasma* **2018**, *1*, 47–60. [CrossRef]

13. Napp, J.; Daeschlein, G.; Napp, M.; von Podewils, S.; Gümbel, D.; Spitzmueller, R.; Fornaciari, P.; Hinz, P.; Jünger, M. On the History of Plasma Treatment and Comparison of Microbiostatic Efficacy of a Historical High-Frequency Plasma Device with Two Modern Devices. *GMS Hyg. Infect. Control* **2015**, *10*, Doc08. [CrossRef] [PubMed]

14. Klämpfl, T.G.; Isbary, G.; Shimizu, T.; Li, Y.F.; Zimmermann, J.L.; Stolz, W.; Schlegel, J.; Morfill, G.E.; Schmidt, H.U. Cold Atmospheric Air Plasma Sterilization against Spores and Other Microorganisms of Clinical Interest. *Appl. Environ. Microbiol.* **2012**, *78*, 5077–5082. [CrossRef] [PubMed]

15. Lu, H.; Patil, S.; Keener, K.M.; Cullen, P.J.; Bourke, P. Bacterial Inactivation by High-Voltage Atmospheric Cold Plasma: Influence of Process Parameters and Effects on Cell Leakage and DNA. *J. Appl. Microbiol.* **2014**, *116*, 784–794. [CrossRef] [PubMed]

16. Graves, D.B. Reactive Species from Cold Atmospheric Plasma: Implications for Cancer Therapy. *Plasma Process. Polym.* **2014**, *11*, 1120–1127. [CrossRef]

17. Yan, D.; Sherman, J.H.; Keidar, M. Cold Atmospheric Plasma, a Novel Promising Anti-Cancer Treatment Modality. *Oncotarget* **2017**, *8*, 15977–15995. [CrossRef]

18. Pai, K.; Timmons, C.; Roehm, K.D.; Ngo, A.; Narayanan, S.S.; Ramachandran, A.; Jacob, J.D.; Ma, L.M.; Madihally, S.V. Investigation of the Roles of Plasma Species Generated by Surface Dielectric Barrier Discharge. *Sci. Rep.* **2018**, *8*, 1–13. [CrossRef]

19. Kim, S.J.; Chung, T.H. Cold Atmospheric Plasma Jet-Generated RONS and Their Selective Effects on Normal and Carcinoma Cells. *Sci. Rep.* **2016**, *6*, 20332. [CrossRef]

20. Kalghatgi, S.; Kelly, C.M.; Cerchar, E.; Torabi, B.; Alekseev, O.; Fridman, A.; Friedman, G.; Azizkhan-Clifford, J. Effects of Non-Thermal Plasma on Mammalian Cells. *PLoS ONE* **2011**, *6*, e16270. [CrossRef]

21. Thiyagarajan, M.; Anderson, H.; Gonzales, X.F. Induction of Apoptosis in Human Myeloid Leukemia Cells by Remote Exposure of Resistive Barrier Cold Plasma. *Biotechnol. Bioeng.* **2014**, *111*, 565–574. [CrossRef]

22. Kim, S.J.; Chung, T.H.; Bae, S.H.; Leem, S.H. Induction of Apoptosis in Human Breast Cancer Cells by a Pulsed Atmospheric Pressure Plasma Jet. *Appl. Phys. Lett.* **2010**, *97*, 023702. [CrossRef]

23. Von Woedtke, T.; Schmidt, A.; Bekeschus, S.; Wende, K.; Weltmann, K.D. Plasma Medicine: A Field of Applied Redox Biology. *In Vivo* **2019**, *33*, 1011–1026. [CrossRef] [PubMed]

24. Takamatsu, T.; Uehara, K.; Sasaki, Y.; Miyahara, H.; Matsumura, Y.; Iwasawa, A.; Ito, N.; Azuma, T.; Kohno, M.; Okino, A. Investigation of Reactive Species Using Various Gas Plasmas. *RSC Adv.* **2014**, *4*, 39901–39905. [CrossRef]

25. Tanaka, H.; Nakamura, K.; Mizuno, M.; Ishikawa, K.; Takeda, K.; Kajiyama, H.; Utsumi, F.; Kikkawa, F.; Hori, M. Non-Thermal Atmospheric Pressure Plasma Activates Lactate in Ringer's Solution for Anti-Tumor Effects. *Sci. Rep.* **2016**, *6*, 1–11. [CrossRef] [PubMed]

26. Xiong, Z. Cold Atmospheric Pressure Plasmas (CAPs) for Skin Wound Healing. In *Plasma Medicine-Concepts and Clinical Applications*; Intechopen: London, UK, 2018; Volume 1, pp. 121–133. [CrossRef]

27. Von Woedtke, T.; Reuter, S.; Masur, K.; Weltmann, K.D. Plasmas for Medicine. *Phys. Rep.* **2013**, *530*, 291–320. [CrossRef]

28. Keidar, M.; Shashurin, A.; Volotskova, O.; Stepp, M.A.; Srinivasan, P.; Homepage, J.; Sandler, A.; Trink, B. Cold Atmospheric Plasma in Cancer Therapy Additional Information on Phys. Plasmas Cold Atmospheric Plasma in Cancer Therapy A). *Cit. Phys. Plasmas* **2013**, *20*, 57101. [CrossRef]

29. Nastuta, A.V.; Pohoata, V.; Topala, I. Atmospheric Pressure Plasma Jet-Living Tissue Interface: Electrical, Optical, and Spectral Characterization. *J. Appl. Phys.* **2013**, *113*, 183302. [CrossRef]

30. Gentile, R.D. Cool Atmospheric Plasma (J-Plasma) and New Options for Facial Contouring and Skin Rejuvenation of the Heavy Face and Neck. *Facial Plast. Surg.* **2018**, *34*, 66–74. [CrossRef]

31. Chutsirimongkol, C.; Boonyawan, D.; Polnikorn, N.; Techawatthanawisan, W.; Kundilokchai, T. Non-Thermal Plasma for Acne Treatment and Aesthetic Skin Improvement. *Plasma Med.* **2014**, *4*, 79–88. [CrossRef]

32. Bogle, M.A.; Arndt, K.A.; Dover, J.S. Evaluation of Plasma Skin Regeneration Technology in Low-Energy Full-Facial Rejuvenation. *Arch. Dermatol.* **2007**, *143*, 168–174. [CrossRef]

33. Isbary, G.; Shimizu, T.; Li, Y.F.; Stolz, W.; Thomas, H.M.; Morfill, G.E.; Zimmermann, J.L. Cold Atmospheric Plasma Devices for Medical Issues. *Expert Rev. Med. Devices* **2013**, *10*, 367–377. [CrossRef]

34. Hoffmann, C.; Berganza, C.; Zhang, J. Cold Atmospheric Plasma: Methods of Production and Application in Dentistry and Oncology. *Med. Gas Res.* **2013**, *3*, 21. [CrossRef] [PubMed]

35. Brandenburg, R. Dielectric Barrier Discharges: Progress on Plasma Sources and on the Understanding of Regimes and Single Filaments. *Plasma Sources Sci. Technol.* **2017**, *26*, 053001. [CrossRef]

36. Zhang, S.; Chen, Z.; Zhang, B.; Chen, Y. Numerical Investigation on the Effects of Dielectric Barrier on a Nanosecond Pulsed Surface Dielectric Barrier Discharge. *Molecules* **2019**, *24*, 3933. [CrossRef]

37. Voráč, J.; Synek, P.; Procházka, V.; Hoder, T. State-by-State Emission Spectra Fitting for Non-Equilibrium Plasmas: OH Spectra of Surface Barrier Discharge at Argon/Water Interface. *J. Phys. D Appl. Phys.* **2017**, *50*, 294002. [CrossRef]

38. Azzariti, A.; Iacobazzi, R.M.; Di Fonte, R.; Porcelli, L.; Gristina, R.; Favia, P.; Fracassi, F.; Trizio, I.; Silvestris, N.; Guida, G.; et al. Plasma-Activated Medium Triggers Cell Death and the Presentation of Immune Activating Danger Signals in Melanoma and Pancreatic Cancer Cells. *Sci. Rep.* **2019**, *9*, 1–13. [CrossRef] [PubMed]

39. Bauer, G.; Sersenová, D.; Graves, D.B.; Machala, Z. Cold Atmospheric Plasma and Plasma-Activated Medium Trigger RONS-Based Tumor Cell Apoptosis. *Sci. Rep.* **2019**, *9*, 1–28. [CrossRef]

40. Kilmer, S.; Semchyshyn, N.; Shah, G.; Fitzpatrick, R. A Pilot Study on the Use of a Plasma Skin Regeneration Device (Portrait® PSR3) in Full Facial Rejuvenation Procedures. *Lasers Med. Sci.* **2007**, *22*, 101–109. [CrossRef]

41. Isbary, G.; Morfill, G.; Schmidt, H.U.; Georgi, M.; Ramrath, K.; Heinlin, J.; Karrer, S.; Landthaler, M.; Shimizu, T.; Steffes, B.; et al. A First Prospective Randomized Controlled Trial to Decrease Bacterial Load Using Cold Atmospheric Argon Plasma on Chronic Wounds in Patients. *Br. J. Dermatol.* **2010**, *163*, 78–82. [CrossRef]

42. Isbary, G.; Heinlin, J.; Shimizu, T.; Zimmermann, J.L.; Morfill, G.; Schmidt, H.U.; Monetti, R.; Steffes, B.; Bunk, W.; Li, Y.; et al. Successful and Safe Use of 2 Min Cold Atmospheric Argon Plasma in Chronic Wounds: Results of a Randomized Controlled Trial. *Br. J. Dermatol.* **2012**, *167*, 404–410. [CrossRef]

43. Scotton, M.F.; Miot, H.A.; Abbade, L.P.F. Factors That Influence Healing of Chronic Venous Leg Ulcers: A Retrospective Cohort. *An. Bras. Dermatol.* **2014**, *89*, 414–422. [CrossRef]

44. Bevis, P.; Earnshaw, J. Venous Ulcer RE. *Clin. Cosmet. Investig. Dermatol.* **2011**, *4*, 7–14. [CrossRef] [PubMed]

45. Zmudzińska, M.; Czarnecka-Operacz, M.; Silny, W. Analysis of Antibiotic Susceptibility and Resistance of Leg Ulcer Bacterial Flora in Patients Hospitalized at Dermatology Department, Poznań University Hospital. *Acta Dermatovenerol. Croat.* **2005**, *13*, 173–176. [PubMed]

46. Rit, K.; Sarkar, A.; Maiti, P.; Nag, F. Chronic Venous Leg Ulcer with Multidrug Resistant Bacterial Infection in a Tertiary Care Hospital of Eastern India. *J. Sci. Soc.* **2013**, *40*, 116. [CrossRef]

47. Brehmer, F.; Haenssle, H.A.; Daeschlein, G.; Ahmed, R.; Pfeiffer, S.; Görlitz, A.; Simon, D.; Schön, M.P.; Wandke, D.; Emmert, S. Alleviation of Chronic Venous Leg Ulcers with a Hand-Held Dielectric Barrier Discharge Plasma Generator (PlasmaDerm ® VU-2010): Results of a Monocentric, Two-Armed, Open, Prospective, Randomized and Controlled Trial (NCT01415622). *J. Eur. Acad. Dermatol. Venereol.* **2015**, *29*, 148–155. [CrossRef] [PubMed]

48. Chuangsuwanich, A.; Assadamongkol, T.; Boonyawan, D. The Healing Effect of Low-Temperature Atmospheric-Pressure Plasma in Pressure Ulcer: A Randomized Controlled Trial. *Int. J. Low. Extrem. Wounds* **2016**, *15*, 313–319. [CrossRef] [PubMed]

49. Chatraie, M.; Torkaman, G.; Khani, M.; Salehi, H.; Shokri, B. In Vivo Study of Non-Invasive Effects of Non-Thermal Plasma in Pressure Ulcer Treatment. *Sci. Rep.* **2018**, *8*, 1–11. [CrossRef]

50. Gao, J.; Wang, L.; Xia, C.; Yang, X.; Cao, Z.; Zheng, L.; Ko, R.; Shen, C.; Yang, C.; Cheng, C. Cold Atmospheric Plasma Promotes Different Types of Superficial Skin Erosion Wounds Healing. *Int. Wound J.* **2019**, *16*, 1103–1111. [CrossRef]

51. Klebes, M.; Ulrich, C.; Kluschke, F.; Patzelt, A.; Vandersee, S.; Richter, H.; Bob, A.; von Hutten, J.; Krediet, J.T.; Kramer, A.; et al. Combined Antibacterial Effects of Tissue-Tolerable Plasma and a Modern Conventional Liquid Antiseptic on Chronic Wound Treatment. *J. Biophotonics* **2015**, *8*, 382–391. [CrossRef]

52. Ulrich, C.; Kluschke, F.; Patzelt, A.; Vandersee, S.; Czaika, V.A.; Richter, H.; Bob, A.; Von Hutten, J.; Painsi, C.; Hügel, R.; et al. Clinical Use of Cold Atmospheric Pressure Argon Plasma in Chronic Leg Ulcers: A Pilot Study. *J. Wound Care* **2015**, *24*, 196–203. [CrossRef]

53. Schmidt, A.; Bekeschus, S.; Wende, K.; Vollmar, B.; von Woedtke, T. A Cold Plasma Jet Accelerates Wound Healing in a Murine Model of Full-Thickness Skin Wounds. *Exp. Dermatol.* **2017**, *26*, 156–162. [CrossRef]

54. Haertel, B.; Wende, K.; Von Woedtke, T.; Weltmann, K.D.; Lindequist, U. Non-Thermal Atmospheric-Pressure Plasma Can Influence Cell Adhesion Molecules on HaCaT-Keratinocytes. *Exp. Dermatol.* **2011**, *20*, 282–284. [CrossRef] [PubMed]

55. Schmidt, A.; Von Woedtke, T.; Sander, B. Periodic Exposure of Keratinocytes to Cold Physical Plasma: An In Vitro Model for Redox-Related Diseases of the Skin. *Oxidative Med. Cell. Longev.* **2016**, *2016*, 1–17. [CrossRef] [PubMed]

56. Schmidt, A.; Bekeschus, S.; Jarick, K.; Hasse, S.; Von Woedtke, T.; Wende, K. Cold Physical Plasma Modulates P53 and Mitogen-Activated Protein Kinase Signaling in Keratinocytes. *Oxid. Med. Cell. Longev.* **2019**, *2019*, 1–16. [CrossRef]

57. Shome, D.; von Woedtke, T.; Riedel, K.; Masur, K. The HIPPO Transducer YAP and Its Targets CTGF and Cyr61 Drive a Paracrine Signalling in Cold Atmospheric Plasma-Mediated Wound Healing. *Oxid. Med. Cell. Longev.* **2020**, *2020*, 1–14. [CrossRef] [PubMed]

58. Zhao, B.; Ye, X.; Yu, J.; Li, L.; Li, W.; Li, S.; Yu, J.; Lin, J.D.; Wang, C.Y.; Chinnaiyan, A.M.; et al. TEAD Mediates YAP-Dependent Gene Induction and Growth Control. *Genes Dev.* **2008**, *22*, 1962–1971. [CrossRef] [PubMed]

59. Duchesne, C.; Banzet, S.; Lataillade, J.; Rousseau, A.; Frescaline, N. Cold Atmospheric Plasma Modulates Endothelial Nitric Oxide Synthase Signalling and Enhances Burn Wound Neovascularisation. *J. Pathol.* **2019**, *249*, 368–380. [CrossRef] [PubMed]

60. Lee, O.J.; Ju, H.W.; Khang, G.; Sun, P.P.; Rivera, J.; Cho, J.H.; Park, S.J.; Eden, J.G.; Park, C.H. An Experimental Burn Wound-Healing Study of Non-Thermal Atmospheric Pressure Microplasma Jet Arrays. *J. Tissue Eng. Regen. Med.* **2016**, *10*, 348–357. [CrossRef]

61. Ngo Thi, M.H.; Shao, P.L.; Der Liao, J.; Lin, C.C.K.; Yip, H.K. Enhancement of Angiogenesis and Epithelialization Processes in Mice with Burn Wounds through ROS/RNS Signals Generated by Non-Thermal N2/Ar Micro-Plasma. *Plasma Process. Polym.* **2014**, *11*, 1076–1088. [CrossRef]

62. Nastuta, A.V.; Pohoata, V.; Vasile Nastuta, A.; Topala, I.; Grigoras, C.; Popa, G. Stimulation of Wound Healing by Helium Atmospheric Pressure Plasma Treatment. *Artic. J. Phys. D Appl. Phys.* **2011**, *44*, 105204–105213. [CrossRef]

63. Topala, I.; Nastuta, A. Helium Atmospheric Pressure Plasma Jet: Diagnostics and Application for Burned Wounds Healing. In *NATO Science for Peace and Security Series A: Chemistry and Biology*; Springer: Dordrecht, The Netherlands, 2012; pp. 335–345. [CrossRef]

64. Betancourt-Ángeles, M.; Peña-Eguiluz, R.; López-Callejas, R.; Domínguez-Cadena, N.A.; Mercado-Cabrera, A.; Muñoz-Infante, J.; Rodríguez-Méndez, B.G.; Valencia-Alvarado, R.; Moreno-Tapia, J.A. Treatment in the Healing of Burns with a Cold Plasma Source. *Int. J. Burns Trauma* **2017**, *7*, 142–146.

65. Heinlin, J.; Zimmermann, J.L.; Zeman, F.; Bunk, W.; Isbary, G.; Landthaler, M.; Maisch, T.; Monetti, R.; Morfill, G.; Shimizu, T.; et al. Randomized Placebo-Controlled Human Pilot Study of Cold Atmospheric Argon Plasma on Skin Graft Donor Sites. *Wound Repair Regen.* **2013**, *21*, 800–807. [CrossRef] [PubMed]

66. Winter, S.; Meyer-Lindenberg, A.; Wolf, G.; Reese, S.; Nolff, M.C. In Vitro Evaluation of the Decontamination Effect of Cold Atmospheric Argon Plasma on Selected Bacteria Frequently Encountered in Small Animal Bite Injuries. *J. Microbiol. Methods* **2020**, *169*, 105728. [CrossRef] [PubMed]

67. Winter, S.; Nolff, M.; Reese, S.; Meyer-Lindenberg, A. Vergleich Der Effizienz von Polyhexanid-Biguanid, Argon-Kaltplasma Und Kochsalzlavage Zur Dekontamination von Bisswunden Beim Hund. *Tierärztliche Prax. Ausgabe K Kleintiere / Heimtiere* **2018**, *46*, 73–82. [CrossRef] [PubMed]

68. Metelmann, H.R.; Vu, T.T.; Do, H.T.; Le, T.N.B.; Hoang, T.H.A.; Phi, T.T.T.; Luong, T.M.L.; Doan, V.T.; Nguyen, T.T.H.; Nguyen, T.H.M.; et al. Scar Formation of Laser Skin Lesions after Cold Atmospheric Pressure Plasma (CAP) Treatment: A Clinical Long Term Observation. *Clin. Plasma Med.* **2013**, *1*, 30–35. [CrossRef]

69. Nishijima, A.; Fujimoto, T.; Hirata, T.; Nishijima, J. Effects of Cold Atmospheric Pressure Plasma on Accelerating Acute Wound Healing: A Comparative Study among 4 Different Treatment Groups. *Mod. Plast. Surg.* **2019**, *9*, 18–31. [CrossRef]

70. Ranjan, R.; Krishnamraju, P.V.; Shankar, T.; Gowd, S. Nonthermal Plasma in Dentistry: An Update. *J. Int. Soc. Prev. Community Dent.* **2017**, *7*, 71–75. [CrossRef]

71. Pan, J.; Sun, K.; Liang, Y.; Sun, P.; Yang, X.; Wang, J.; Zhang, J.; Zhu, W.; Fang, J.; Becker, K.H. Cold Plasma Therapy of a Tooth Root Canal Infected with Enterococcus Faecalis Biofilms in Vitro. *J. Endod.* **2013**, *39*, 105–110. [CrossRef]

72. Vandana, B.L. From Distant Stars to Dental Chairs: An Update on Plasma Needle. *Int. J. Dent. Sci. Res.* **2014**, *2*, 19–20. [CrossRef]

73. Aparecida Delben, J.; Evelin Zago, C.; Tyhovych, N.; Duarte, S.; Eduardo Vergani, C. Effect of Atmospheric-Pressure Cold Plasma on Pathogenic Oral Biofilms and in Vitro Reconstituted Oral Epithelium. *PLoS ONE* **2016**, *11*, e0155427.

74. Marsh, P.D.; Zaura, E. Dental Biofilm: Ecological Interactions in Health and Disease. *J. Clin. Periodontol.* **2017**, *44*, S12–S22. [CrossRef]

75. Koban, I.; Holtfreter, B.; Hübner, N.O.; Matthes, R.; Sietmann, R.; Kindel, E.; Weltmann, K.D.; Welk, A.; Kramer, A.; Kocher, T. Antimicrobial Efficacy of Non-Thermal Plasma in Comparison to Chlorhexidine against Dental Biofilms on Titanium Discs in Vitro - Proof of Principle Experiment. *J. Clin. Periodontol.* **2011**, *38*, 956–965. [CrossRef] [PubMed]

76. Jiang, C.; Chen, M.-T.; Gorur, A.; Schaudinn, C.; Jaramillo, D.E.; Costerton, J.W.; Sedghizadeh, P.P.; Vernier, P.T.; Gundersen, M.A. Nanosecond Pulsed Plasma Dental Probe. *Plasma Process. Polym.* **2009**, *6*, 479–483. [CrossRef]

77. Armand, A.; Khani, M.; Asnaashari, M.; AliAhmadi, A.; Shokri, B. Comparison Study of Root Canal Disinfection by Cold Plasma Jet and Photodynamic Therapy. *Photodiagnosis Photodyn. Ther.* **2019**, *26*, 327–333. [CrossRef] [PubMed]

78. Wang, Q.Q.; Zhang, C.F.; Chu, C.H.; Zhu, X.F. Prevalence of Enterococcus Faecalis in Saliva and Filled Root Canals of Teeth Associated with Apical Periodontitis. *Int. J. Oral Sci.* **2012**, *4*, 19–23. [CrossRef] [PubMed]

79. Shahmohammadi Beni, M.; Han, W.; Yu, K.N. Dispersion of OH Radicals in Applications Related to Fear-Free Dentistry Using Cold Plasma. *Appl. Sci.* **2019**, *9*, 2119. [CrossRef]

80. Dong, X.; Chen, M.; Wang, Y.; Yu, Q. A Mechanistic Study of Plasma Treatment Effects on Demineralized Dentin Surfaces for Improved Adhesive/Dentin Interface Bonding. *Clin. Plasma Med.* **2014**, *2*, 11–16. [CrossRef]

81. Yavirach, P.; Chaijareenont, P.; Boonyawan, D.; Pattamapun, K.; Tunma, S.; Takahashi, H.; Arksornnukit, M. Effects of Plasma Treatment on the Shear Bond Strength between Fiberreinforced Composite Posts and Resin Composite for Core Build-Up. *Dent. Mater. J.* **2009**, *28*, 686–692. [CrossRef]

82. Yang, Y.; Guo, J.; Zhou, X.; Liu, Z.; Wang, C.; Wang, K.; Zhang, J.; Wang, Z. A Novel Cold Atmospheric Pressure Air Plasma Jet for Peri-Implantitis Treatment: An in Vitro Study. *Dent. Mater. J.* **2018**, *37*, 157–166. [CrossRef]

83. Monetto, I. The Effects of an Interlayer Debond on the Flexural Behavior of Three-Layer Beams. *Coatings* **2019**, *9*, 258. [CrossRef]

84. Quirynen, M.; Bollen, C.M.L. CA Novel Cold Atmospheric Pressure Air Plasma Jet for Peri-Implantitis Treatment: An in Vitro Study. *J. Clin. Periodontol.* **1995**, *22*, 1–14. [CrossRef]

85. Yang, Y.; Zheng, M.; Yang, Y.; Li, J.; Su, Y.F.; Li, H.P.; Tan, J.G. Inhibition of Bacterial Growth on Zirconia Abutment with a Helium Cold Atmospheric Plasma Jet Treatment. *Clin. Oral Investig.* **2020**, 1–13. [CrossRef] [PubMed]

86. Preissner, S.; Poehlmann, A.C.; Schubert, A.; Lehmann, A.; Arnold, T.; Nell, O.; Rupf, S. Ex Vivo Study Comparing Three Cold Atmospheric Plasma (CAP) Sources for Ebiofllmeremovaleonemicrostructuredetitanium. *Plasma Med.* **2019**, *9*, 1–13. [CrossRef]

87. Park, J.; Lee, H.; Lee, H.J.; Kim, G.C.; Kim, D.Y.; Han, S.; Song, K. Non-Thermal Atmospheric Pressure Plasma Efficiently Promotes the Proliferation of Adipose Tissue-Derived Stem Cells by Activating NO-Response Pathways. *Sci. Rep.* **2016**, *6*, 1–12. [CrossRef] [PubMed]

88. Elsaadany, M.; Subramanian, G.; Ayan, H.; Yildirim-Ayan, E. Exogenous Nitric Oxide (NO) Generated by NO-Plasma Treatment Modulates Osteoprogenitor Cells Early Differentiation. *J. Phys. D Appl. Phys.* **2015**, *48*, 345401. [CrossRef]

89. Han, I.; Choi, E.H. The Role of Non-Thermal Atmospheric Pressure Biocompatible Plasma in the Differentiation of Osteoblastic Precursor Cells, MC3T3-E1. *Oncotarget* **2017**, *8*, 36399. [CrossRef]

90. Klotz, L.O.; Sánchez-Ramos, C.; Prieto-Arroyo, I.; Urbánek, P.; Steinbrenner, H.; Monsalve, M. Redox Regulation of FoxO Transcription Factors. *Redox Biol.* **2015**, *6*, 51–72. [CrossRef]

91. Miletić, M.; Mojsilović, S.; Okićorević, I.; Maletić, D.; Puač, N.; Lazović, S.; Malović, G.; Milenković, P.; Lj Petrović, Z.; Bugarski, D. Effects of Non-Thermal Atmospheric Plasma on Human Periodontal Ligament Mesenchymal Stem Cells. *J. Phys. D Appl. Phys.* **2013**, *46*, 345401–345410. [CrossRef]

92. Park, J.; Lee, H.; Lee, H.J.; Kim, G.C.; Kim, S.S.; Han, S.; Song, K. Non-Thermal Atmospheric Pressure Plasma Is an Excellent Tool to Activate Proliferation in Various Mesoderm-Derived Human Adult Stem Cells. *Free Radic. Biol. Med.* **2019**, *134*, 374–384. [CrossRef]

93. Alemi, P.S.; Atyabi, S.A.; Sharifi, F.; Mohamadali, M.; Irani, S.; Bakhshi, H.; Atyabi, S.M. Synergistic Effect of Pressure Cold Atmospheric Plasma and Carboxymethyl Chitosan to Mesenchymal Stem Cell Differentiation on PCL/CMC Nanofibers for Cartilage Tissue Engineering. *Polym. Adv. Technol.* **2019**, *30*, 1356–1364. [CrossRef]

94. Xiong, Z.; Zhao, S.; Yan, X. Nerve Stem Cell Differentiation by a One-Step Cold Atmospheric Plasma Treatment in Vitro. *J. Vis. Exp.* **2019**, *2019*, e58663. [CrossRef]

95. Jouhilahti, E.M.; Peltonen, S.; Peltonen, J. Class III β-Tubulin Is a Component of the Mitotic Spindle in Multiple Cell Types. *J. Histochem. Cytochem.* **2008**, *56*, 1113–1119. [CrossRef] [PubMed]

96. Foudah, D.; Monfrini, M.; Donzelli, E.; Niada, S.; Brini, A.T.; Orciani, M.; Tredici, G.; Miloso, M. Expression of Neural Markers by Undifferentiated Mesenchymal-like Stem Cells from Different Sources. *J. Immunol. Res.* **2014**, *2014*, 1–16. [CrossRef] [PubMed]

97. Weil, M.T.; Schulz-Éberlin, G.; Mukherjee, C.; Kuo-Elsner, W.P.; Schäfer, I.; Müller, C.; Simons, M. Isolation and Culture of Oligodendrocytes. In *Methods in Molecular Biology*; Humana Press Inc.: Totowa, NJ, USA, 2019; Volume 1936, pp. 79–95. [CrossRef]

98. Hol, E.M.; Pekny, M. Glial Fibrillary Acidic Protein (GFAP) and the Astrocyte Intermediate Filament System in Diseases of the Central Nervous System. *Curr. Opin. Cell Biol.* **2015**, *32*, 121–130. [CrossRef] [PubMed]

99. Jang, J.Y.; Hong, Y.J.; Lim, J.; Choi, J.S.; Choi, E.H.; Kang, S.; Rhim, H. Cold Atmospheric Plasma (CAP), a Novel Physicochemical Source, Induces Neural Differentiation through Cross-Talk between the Specific RONS Cascade and Trk/Ras/ERK Signaling Pathway. *Biomaterials* **2018**, *156*, 258–273. [CrossRef]

100. Bourdens, M.; Jeanson, Y.; Taurand, M.; Juin, N.; Carrière, A.; Clément, F.; Casteilla, L.; Bulteau, A.L.; Planat-Bénard, V. Short Exposure to Cold Atmospheric Plasma Induces Senescence in Human Skin Fibroblasts and Adipose Mesenchymal Stromal Cells. *Sci. Rep.* **2019**, *9*, 1–15. [CrossRef]

101. Won, H.-R.; Kang, S.U.; Kim, H.J.; Jang, J.Y.; Shin, Y.S.; Kim, C.-H. Non-Thermal Plasma Treated Solution with Potential as a Novel Therapeutic Agent for Nasal Mucosa Regeneration. *Sci. Rep.* **2018**, *8*, 13754. [CrossRef]

102. Scharf, C.; Eymann, C.; Emicke, P.; Bernhardt, J.; Wilhelm, M.; Görries, F.; Winter, J.; Von Woedtke, T.; Darm, K.; Daeschlein, G.; et al. Improved Wound Healing of Airway Epithelial Cells Is Mediated by Cold Atmospheric Plasma: A Time Course-Related Proteome Analysis. *Hindawi Oxidative Med. Cell. Longev.* **2019**, *2019*, 1–21. [CrossRef]

103. Katiyar, K.S.; Lin, A.; Fridman, A.; Keating, C.E.; Cullen, D.K.; Miller, V. Non-Thermal Plasma Accelerates Astrocyte Regrowth and Neurite Regeneration Following Physical Trauma in Vitro. *Appl. Sci.* **2019**, *9*, 3747. [CrossRef]

104. Aggarwal, V.; Tuli, H.S.; Varol, A.; Thakral, F.; Yerer, M.B.; Sak, K.; Varol, M.; Jain, A.; Khan, M.A.; Sethi, G. Role of Reactive Oxygen Species in Cancer Progression: Molecular Mechanisms and Recent Advancements. *Biomolecules* **2019**, *9*, 735. [CrossRef]

105. Aikman, B.; De Almeida, A.; Meier-Menches, S.M.; Casini, A. Aquaporins in Cancer Development: Opportunities for Bioinorganic Chemistry to Contribute Novel Chemical Probes and Therapeutic Agents. *Metallomics* **2018**, *10*, 696–712. [CrossRef]

106. Yusupov, M.; Razzokov, J.; Cordeiro, R.M.; Bogaerts, A. Transport of Reactive Oxygen and Nitrogen Species across Aquaporin: A Molecular Level Picture. *Oxid. Med. Cell. Longev.* **2019**, *2019*, 1–11. [CrossRef] [PubMed]

107. Tamma, G.; Valenti, G.; Grossini, E.; Donnini, S.; Marino, A.; Marinelli, R.A.; Calamita, G. Aquaporin Membrane Channels in Oxidative Stress, Cell Signaling, and Aging: Recent Advances and Research Trends. *Oxidative Med. Cell. Longev.* **2018**, *2018*, 1–14. [CrossRef] [PubMed]

108. Rivel, T.; Ramseyer, C.; Yesylevskyy, S. The Asymmetry of Plasma Membranes and Their Cholesterol Content Influence the Uptake of Cisplatin. *Sci. Rep.* **2019**, *9*, 1–14. [CrossRef] [PubMed]

109. Van Der Paal, J.; Neyts, E.C.; Verlackt, C.C.W.; Bogaerts, A. Effect of Lipid Peroxidation on Membrane Permeability of Cancer and Normal Cells Subjected to Oxidative Stress. *Chem. Sci.* **2016**, *7*, 489–498. [CrossRef]

110. Keidar, M.; Yan, D.; Beilis, I.I.; Trink, B.; Sherman, J.H. Plasmas for Treating Cancer: Opportunities for Adaptive and Self-Adaptive Approaches. *Trends Biotechnol.* **2018**, *36*, 586–593. [CrossRef]

111. Yan, D.; Talbot, A.; Nourmohammadi, N.; Cheng, X.; Canady, J.; Sherman, J.; Keidar, M. Principles of Using Cold Atmospheric Plasma Stimulated Media for Cancer Treatment. *Sci. Rep.* **2015**, *5*, 1–17. [CrossRef]
112. Siu, A.; Volotskova, O.; Cheng, X.; Khalsa, S.S.; Bian, K.; Murad, F.; Keidar, M.; Sherman, J.H. Differential Effects of Cold Atmospheric Plasma in the Treatment of Malignant Glioma. *PLoS ONE* **2015**, *10*, e0126313. [CrossRef]
113. Wiegand, C.; Fink, S.; Beier, O.; Horn, K.; Pfuch, A.; Schimanski, A.; Grünler, B.; Hipler, U.-C.; Elsner, P. Dose- and Time-Dependent Cellular Effects of Cold Atmospheric Pressure Plasma Evaluated in 3D Skin Models. *Skin Pharmacol. Physiol.* **2016**, *29*, 257–265. [CrossRef]
114. Keidar, M.; Walk, R.; Shashurin, A.; Srinivasan, P.; Sandler, A.; Dasgupta, S.; Ravi, R.; Guerrero-Preston, R.; Trink, B. Cold Plasma Selectivity and the Possibility of a Paradigm Shift in Cancer Therapy. *Br. J. Cancer* **2011**, *105*, 1295–1301. [CrossRef]
115. Kaushik, N.; Uddin, N.; Sim, G.B.; Hong, Y.J.; Baik, K.Y.; Kim, C.H.; Lee, S.J.; Kaushik, N.K.; Choi, E.H. Responses of Solid Tumor Cells in DMEM to Reactive Oxygen Species Generated by Non-Thermal Plasma and Chemically Induced ROS Systems. *Sci. Rep.* **2015**, *5*, 8587. [CrossRef]
116. DeBerardinis, R.J.; Lum, J.J.; Hatzivassiliou, G.; Thompson, C.B. The Biology of Cancer: Metabolic Reprogramming Fuels Cell Growth and Proliferation. *Cell Metab.* **2008**, *7*, 11–20. [CrossRef]
117. Cha, J.-Y.; Lee, H.-J. Targeting Lipid Metabolic Reprogramming as Anticancer Therapeutics. *J. Cancer Prev.* **2016**, *21*, 209–215. [CrossRef] [PubMed]
118. Ward, P.S.; Thompson, C.B. Metabolic Reprogramming: A Cancer Hallmark Even Warburg Did Not Anticipate. *Cancer Cell* **2012**, *21*, 297–308. [CrossRef]
119. Xu, D.; Ning, N.; Xu, Y.; Wang, B.; Cui, Q.; Liu, Z.; Wang, X.; Liu, D.; Chen, H.; Kong, M.G. Effect of Cold Atmospheric Plasma Treatment on the Metabolites of Human Leukemia Cells. *Cancer Cell Int.* **2019**, *19*, 135. [CrossRef] [PubMed]
120. Köritzer, J.; Boxhammer, V.; Schäfer, A.; Shimizu, T.; Klämpfl, T.G.; Li, Y.-F.; Welz, C.; Schwenk-Zieger, S.; Morfill, G.E.; Zimmermann, J.L.; et al. Restoration of Sensitivity in Chemo—Resistant Glioma Cells by Cold Atmospheric Plasma. *PLoS ONE* **2013**, *8*, e64498. [CrossRef] [PubMed]
121. Vandamme, M.; Robert, E.; Lerondel, S.; Sarron, V.; Ries, D.; Dozias, S.; Sobilo, J.; Gosset, D.; Kieda, C.; Legrain, B.; et al. ROS Implication in a New Antitumor Strategy Based on Non-Thermal Plasma. *Int. J. Cancer* **2012**, *130*, 2185–2194. [CrossRef] [PubMed]
122. Tanaka, H.; Mizuno, M.; Ishikawa, K.; Nakamura, K.; Kajiyama, H.; Kano, H.; Kikkawa, F.; Hori, M. Plasma-Activated Medium Selectively Kills Glioblastoma Brain Tumor Cells by Down-Regulating a Survival Signaling Molecule, AKT Kinase. *Plasma Med.* **2011**, *1*, 265–277. [CrossRef]
123. Kaushik, N.K.; Attri, P.; Kaushik, N.; Choi, E.H. A Preliminary Study of the Effect of DBD Plasma and Osmolytes on T98G Brain Cancer and HEK Non-Malignant Cells. *Molecules* **2013**, *18*, 4917–4928. [CrossRef]
124. Kim, J.Y.; Ballato, J.; Foy, P.; Hawkins, T.; Wei, Y.; Li, J.; Kim, S.O. Apoptosis of Lung Carcinoma Cells Induced by a Flexible Optical Fiber-Based Cold Microplasma. *Biosens. Bioelectron.* **2011**, *28*, 333–338. [CrossRef]
125. Chen, Z.; Simonyan, H.; Cheng, X.; Gjika, E.; Lin, L.; Canady, J.; Sherman, J.H.; Young, C.; Keidar, M. A Novel Micro Cold Atmospheric Plasma Device for Glioblastoma both in Vitro and in Vivo. *Cancers* **2017**, *9*, 61. [CrossRef]
126. Chen, Z.; Lin, L.; Zheng, Q.; Sherman, J.H.; Canady, J.; Trink, B.; Keidar, M. Micro-Sized Cold Atmospheric Plasma Source for Brain and Breast Cancer Treatment. *Plasma Med.* **2018**, *8*, 203–215. [CrossRef]
127. Mirpour, S.; Piroozmand, S.; Soleimani, N.; Jalali Faharani, N.; Ghomi, H.; Fotovat Eskandari, H.; Sharifi, A.M.; Mirpour, S.; Eftekhari, M.; Nikkhah, M. Utilizing the Micron Sized Non-Thermal Atmospheric Pressure Plasma inside the Animal Body for the Tumor Treatment Application. *Sci. Rep.* **2016**, *6*, 29048. [CrossRef] [PubMed]
128. Mashayekh, S.; Rajaee, H.; Akhlaghi, M.; Shokri, B.; Hassan, Z.M. Atmospheric-Pressure Plasma Jet Characterization and Applications on Melanoma Cancer Treatment (B/16-F10). *Phys. Plasmas* **2015**, *22*, 093508. [CrossRef]
129. Lee, H.J.; Shon, C.H.; Kim, Y.S.; Kim, S.; Kim, G.C.; Kong, M.G. Degradation of Adhesion Molecules of G361 Melanoma Cells by a Non-Thermal Atmospheric Pressure Microplasma. *New J. Phys.* **2009**, *11*, 115026. [CrossRef]

130. Kim, G.C.; Kim, G.J.; Park, S.R.; Jeon, S.M.; Seo, H.J.; Iza, F.; Lee, J.K. Air Plasma Coupled with Antibody-Conjugated Nanoparticles: A New Weapon against Cancer. *J. Phys. D Appl. Phys.* **2009**, *42*, 032005. [CrossRef]

131. Ninomiya, K.; Ishijima, T.; Imamura, M.; Yamahara, T.; Enomoto, H.; Takahashi, K.; Tanaka, Y.; Uesugi, Y.; Shimizu, N. Evaluation of Extra- and Intracellular OH Radical Generation, Cancer Cell Injury, and Apoptosis Induced by a Non-Thermal Atmospheric-Pressure Plasma Jet. *J. Phys. D Appl. Phys.* **2013**, *46*, 425401. [CrossRef]

132. Kim, C.H.; Bahn, J.H.; Lee, S.H.; Kim, G.Y.; Jun, S.I.; Lee, K.; Baek, S.J. Induction of Cell Growth Arrest by Atmospheric Non-Thermal Plasma in Colorectal Cancer Cells. *J. Biotechnol.* **2010**, *150*, 530–538. [CrossRef] [PubMed]

133. Georgescu, N.; Lupu, A.R. Tumoral and Normal Cells Treatment with High-Voltage Pulsed Cold Atmospheric Plasma Jets. *IEEE Trans. Plasma Sci.* **2010**, *38*, 1949–1955. [CrossRef]

134. Ishaq, M.; Evans, M.D.M.; Ostrikov, K.K. Atmospheric Pressure Gas Plasma-Induced Colorectal Cancer Cell Death Is Mediated by Nox2-ASK1 Apoptosis Pathways and Oxidative Stress Is Mitigated by Srx-Nrf2 Anti-Oxidant System. *Biochim. Biophys. Acta-Mol. Cell Res.* **2014**, *1843*, 2827–2837. [CrossRef]

135. Plewa, J.M.; Yousfi, M.; Frongia, C.; Eichwald, O.; Ducommun, B.; Merbahi, N.; Lobjois, V. Low-Temperature Plasma-Induced Antiproliferative Effects on Multi-Cellular Tumor Spheroids. *New J. Phys.* **2014**, *16*, 043027. [CrossRef]

136. Guerrero-Preston, R.; Ogawa, T.; Uemura, M.; Shumulinsky, G.; Valle, B.L.; Pirini, F.; Ravi, R.; Sidransky, D.; Keidar, M.; Trink, B. Cold Atmospheric Plasma Treatment Selectively Targets Head and Neck Squamous Cell Carcinoma Cells. *Int. J. Mol. Med.* **2014**, *34*, 941–946. [CrossRef] [PubMed]

137. Kang, S.U.; Cho, J.H.; Chang, J.W.; Shin, Y.S.; Kim, K.I.; Park, J.K.; Yang, S.S.; Lee, J.S.; Moon, E.; Lee, K.; et al. Nonthermal Plasma Induces Head and Neck Cancer Cell Death: The Potential Involvement of Mitogen-Activated Protein Kinase-Dependent Mitochondrial Reactive Oxygen Species. *Cell Death Dis.* **2014**, *5*, e1056. [CrossRef] [PubMed]

138. Hasse, S.; Seebauer, C.; Wende, K.; Schmidt, A.; Metelmann, H.-R.; von Woedtke, T.; Bekeschus, S. Cold Argon Plasma as Adjuvant Tumour Therapy on Progressive Head and Neck Cancer: A Preclinical Study. *Appl. Sci.* **2019**, *9*, 2061. [CrossRef]

139. Ahn, H.J.; Kim, K., II; Hoan, N.N.; Kim, C.H.; Moon, E.; Choi, K.S.; Yang, S.S.; Lee, J.S. Targeting Cancer Cells with Reactive Oxygen and Nitrogen Species Generated by Atmospheric-Pressure Air Plasma. *PLoS ONE* **2014**, *9*, e86173. [CrossRef]

140. Kim, K.; Jun Ahn, H.; Lee, J.H.; Kim, J.H.; Sik Yang, S.; Lee, J.S. Cellular Membrane Collapse by Atmospheric-Pressure Plasma Jet. *Appl. Phys. Lett.* **2014**, *104*, 013701. [CrossRef]

141. Tan, X.; Zhao, S.; Lei, Q.; Lu, X.; He, G.; Ostrikov, K. Single-Cell-Precision Microplasma-Induced Cancer Cell Apoptosis. *PLoS ONE* **2014**, *9*, e101299. [CrossRef]

142. Barekzi, N.; Laroussi, M. Dose-Dependent Killing of Leukemia Cells by Low-Temperature Plasma. *J. Phys. D Appl. Phys.* **2012**, *45*, 422002. [CrossRef]

143. Thiyagarajan, M.; Waldbeser, L.; Whitmill, A. THP-1 Leukemia Cancer Treatment Using a Portable Plasma Device. *Stud. Health Technol. Inform.* **2012**, *173*, 515–517. [CrossRef]

144. Brullé, L.; Vandamme, M.; Riès, D.; Martel, E.; Robert, E.; Lerondel, S.; Trichet, V.; Richard, S.; Pouvesle, J.M.; Le Pape, A. Effects of a Non Thermal Plasma Treatment Alone or in Combination with Gemcitabine in a MIA PaCa2-Luc Orthotopic Pancreatic Carcinoma Model. *PLoS ONE* **2012**, *7*, e52653. [CrossRef]

145. Hattori, N.; Yamada, S.; Tori, K.; Takeda, S.; Nakamura, K.; Tanaka, H.; Kajiyama, H.; Kanda, M.; Fuji, T.; Nakayama, G.; et al. Effectiveness of Plasma Treatment on Pancreatic Cancer Cells. *Int. J. Oncol.* **2015**, *47*, 1655–1662. [CrossRef]

146. Metelmann, H.R.; Seebauer, C.; Miller, V.; Fridman, A.; Bauer, G.; Graves, D.B.; Pouvesle, J.M.; Rutkowski, R.; Schuster, M.; Bekeschus, S.; et al. Clinical Experience with Cold Plasma in the Treatment of Locally Advanced Head and Neck Cancer. *Clin. Plasma Med.* **2018**, *9*, 6–13. [CrossRef]

147. Tanaka, H.; Mizuno, M.; Ishikawa, K.; Toyokuni, S.; Kajiyama, H.; Kikkawa, F.; Hori, M. New Hopes for Plasma-Based Cancer Treatment. *Plasma* **2018**, *1*, 150–155. [CrossRef]

148. Yan, D.; Xiao, H.; Zhu, W.; Nourmohammadi, N.; Zhang, L.G.; Bian, K.; Keidar, M. The Role of Aquaporins in the Anti-Glioblastoma Capacity of the Cold Plasma-Stimulated Medium. *J. Phys. D Appl. Phys.* **2017**, *50*, 055401. [CrossRef]

149. Torii, K.; Yamada, S.; Nakamura, K.; Tanaka, H.; Kajiyama, H.; Tanahashi, K.; Iwata, N.; Kanda, M.; Kobayashi, D.; Tanaka, C.; et al. Effectiveness of Plasma Treatment on Gastric Cancer Cells. *Gastric Cancer* **2015**, *18*, 635–643. [CrossRef] [PubMed]

150. Ishimoto, T.; Nagano, O.; Yae, T.; Tamada, M.; Motohara, T.; Oshima, H.; Oshima, M.; Ikeda, T.; Asaba, R.; Yagi, H.; et al. CD44 Variant Regulates Redox Status in Cancer Cells by Stabilizing the XCT Subunit of System Xc- and Thereby Promotes Tumor Growth. *Cancer Cell* **2011**, *19*, 387–400. [CrossRef]

151. Utsumi, F.; Kajiyama, H.; Nakamura, K.; Tanaka, H.; Mizuno, M.; Ishikawa, K.; Kondo, H.; Kano, H.; Hori, M.; Kikkawa, F. Effect of Indirect Nonequilibrium Atmospheric Pressure Plasma on Anti-Proliferative Activity against Chronic Chemo-Resistant Ovarian Cancer Cells In Vitro and In Vivo. *PLoS ONE* **2013**, *8*, e81576. [CrossRef]

152. Nakamura, K.; Peng, Y.; Utsumi, F.; Tanaka, H.; Mizuno, M.; Toyokuni, S.; Hori, M.; Kikkawa, F.; Kajiyama, H. Novel Intraperitoneal Treatment with Non-Thermal Plasma-Activated Medium Inhibits Metastatic Potential of Ovarian Cancer Cells. *Sci. Rep.* **2017**, *7*, 1–14. [CrossRef]

153. Sato, Y.; Yamada, S.; Takeda, S.; Hattori, N.; Nakamura, K.; Tanaka, H.; Mizuno, M.; Hori, M.; Kodera, Y. Effect of Plasma-Activated Lactated Ringer's Solution on Pancreatic Cancer Cells In Vitro and In Vivo. *Ann. Surg. Oncol.* **2018**, *25*, 299–307. [CrossRef]

154. Matsuzaki, T.; Kano, A.; Kamiya, T.; Hara, H.; Adachi, T. Enhanced Ability of Plasma-Activated Lactated Ringer's Solution to Induce A549 cell Injury. *Arch. Biochem. Biophys.* **2018**, *656*, 19–30. [CrossRef]

155. Chen, D.S.; Mellman, I. Elements of Cancer Immunity and the Cancer-Immune Set Point. *Nature* **2017**, *541*, 321–330. [CrossRef]

156. Radogna, F.; Diederich, M. Stress-Induced Cellular Responses in Immunogenic Cell Death: Implications for Cancer Immunotherapy. *Biochem. Pharmacol.* **2018**, *153*, 12–23. [CrossRef] [PubMed]

157. Hernandez, C.; Huebener, P.; Schwabe, R.F. Damage-Associated Molecular Patterns in Cancer: A Double-Edged Sword. *Oncogene* **2016**, *35*, 5931–5941. [CrossRef] [PubMed]

158. Lin, A.G.; Xiang, B.; Merlino, D.J.; Baybutt, T.R.; Sahu, J.; Fridman, A.; Snook, A.E.; Miller, V. Non-Thermal Plasma Induces Immunogenic Cell Death in Vivo in Murine CT26 Colorectal Tumors. *Oncoimmunology* **2018**, *7*, e1484978. [CrossRef] [PubMed]

159. Khalili, M.; Daniels, L.; Lin, A.; Krebs, F.C.; Snook, A.E.; Bekeschus, S.; Bowne, W.B.; Miller, V. Non-Thermal Plasma-Induced Immunogenic Cell Death in Cancer. *J. Phys. D Appl. Phys.* **2019**, *52*, 423001. [CrossRef] [PubMed]

160. Miller, V.; Lin, A.; Fridman, A. Why Target Immune Cells for Plasma Treatment of Cancer. *Plasma Chem. Plasma Process.* **2016**, *36*, 259–268. [CrossRef]

161. Almeida, N.D.; Klein, A.L.; Hogan, E.A.; Terhaar, S.J.; Kedda, J.; Uppal, P.; Sack, K.; Keidar, M.; Sherman, J.H. Cold Atmospheric Plasma as an Adjunct to Immunotherapy for Glioblastoma Multiforme. *World Neurosurg.* **2019**, *130*, 369–376. [CrossRef]

162. Lin, A.; Truong, B.; Patel, S.; Kaushik, N.; Choi, E.H.; Fridman, G.; Fridman, A.; Miller, V. Nanosecond-Pulsed DBD Plasma-Generated Reactive Oxygen Species Trigger Immunogenic Cell Death in A549 Lung Carcinoma Cells through Intracellular Oxidative Stress. *Int. J. Mol. Sci.* **2017**, *18*, 966. [CrossRef]

163. Cheng, F.; Yan, D.; Chen, J.; Keidar, M.; Sotomayor, E. Cold Plasma with Immunomodulatory Properties Has Significant Anti-Lymphoma Activities in Vitro and In Vivo. *Blood* **2019**, *134*, 5307. [CrossRef]

MDPI

St. Alban-Anlage 66

4052 Basel

Switzerland

Tel. +41 61 683 77 34

Fax +41 61 302 89 18

www.mdpi.com

International Journal of Molecular Sciences Editorial Office

E-mail: ijms@mdpi.com

www.mdpi.com/journal/ijms